The Freshwater Fishes of British Columbia

The Freshwater Fishes of British Columbia

J.D. McPhail

Illustrated by D.L. McPhail

The University of Alberta Press

Published by

The University of Alberta Press
Ring House 2
Edmonton, Alberta, Canada T6G 2E1

Copyright © J.D. McPhail 2007
ISBN 978-0-88864-467-1(bound)

Library and Archives Canada Cataloguing in Publication

McPhail, J. D. (John Donald)
    The freshwater fishes of British Columbia / J. D. McPhail ; foreword,
Joseph S. Nelson ; illustrator, D. L. McPhail.

Includes index.
ISBN 978-0-88864-467-1

    1. Freshwater fishes--British Columbia--Identification. 2. Freshwater
fishes--British Columbia. I. Title.

QL626.5.B7M364 2007   597.176'09711   C2007-903522-1

All rights reserved.
First edition, first printing, 2007.
Printed and bound in Canada by Friesens, Altona, Manitoba.
Copyediting by Paul Payson.

No part of this publication may be produced, stored in a retrieval system, or transmitted in any forms
or by any means, electronic, mechanical, photocopying, recording, or otherwise, without the prior
written consent of the copyright owneror a licence from The Canadian Copyright Licensing Agency
(Access Copyright). For an Access Copyright license, visit www.accesscopyright.ca or call toll free:
1–800–893–5777.

The University of Alberta Press is committed to protecting our natural environment. As part of our
efforts, this book is printed on Enviro Paper: it contains 100% post-consumer recycled fibres and is acid-
and chlorine-free.

The University of Alberta Press gratefully acknowledges the support received for its publishing program
from The Canada Council for the Arts. The University of Alberta Press also gratefully acknowledges the
financial support of the Government of Canada through the Book Publishing Industry Development
Program (BPIDP) and from the Alberta Foundation for the Arts for its publishing activities.

 Canada Council Conseil des Arts   Canadä
for the Arts   du Canada

for C.C. Lindsay—scholar, mentor, and friend

SPONSORS

BC hydro

BC HYDRO—PROTECTING FISH AND THEIR HABITAT

Since water is the source of about 90 per cent of the electricity that BC Hydro generates, we take the health of the province's fresh water ecosystems very seriously. We are committed to keeping the environmental effects of our operations to a minimum, while continuing to provide reliable power, at low cost, for years to come.

OUR COMMITMENT TO SAFEGUARDING FISH

BC Hydro strives to protect fish and their habitat through initiatives that mitigate or compensate for the impact of our operations on reservoirs, rivers and streams. We identify, monitor and record impacts on fish and fish habitat throughout the hydroelectric system, using an environmental management system to manage these effects over time.

At reservoirs—Water levels in reservoirs fluctuate more widely than levels in natural lakes because of annual storage cycles and consumers' daily demand for electricity. When water levels drop, the shoreline is exposed to air and dries out, which hinders invertebrate and plant growth. Plants play an important role in the reservoir food chain and provide habitat for many insects and small fish. In a few situations, low water levels may also affect certain species of fish, such as lake trout, which use shoreline habitat for spawning.

In reservoirs, nutrients can settle on the bottom, where they may become trapped in the sediment and thus removed from the food chain. BC Hydro fertilizes some of its reservoirs to maintain the production of fish food and to help sustain fish populations. Fertilization programs in the Kootenay, Arrow, Jones and Alouette reservoirs have resulted in larger fish and more abundant populations.

At dams and generating stations—In the absence of mitigation, dams can prevent migrating fish from moving upstream and reaching their spawning grounds. For fish moving downstream, dams can entrain or divert fish to water intakes and carry them to spinning turbine blades where they may be disoriented, injured or killed. Fish that survive the passage may be unable to return to the habitat above the dam. Dams can also act as barriers to water, gravel and nutrient flows, all of which affect fish.

BC Hydro uses different methods to help fish bypass our installations safely. Fish ladders allow fish to swim past dams by creating a series of pools where migratory fish can rest during their journey. While this method has proven successful, it is not practical for all dams. For example, it is less feasible at very high dams, where fish ladders would have to be extremely long. Fish screens are also used to allow fish to travel downstream past the dam without entering the water intakes.

Downstream from our facilities—The amount of water diverted through generating stations is linked to the amount of electricity demanded by customers. The volume and changes in river flow and water quality affect fish. If the flow is reduced suddenly, fish can become stranded. In order to accommodate the needs of other resource interests like fish, wildlife and recreation, we have carried out a public operations review process called Water Use Planning, an effort that aims to balance the needs of all reservoir users.

BC Hydro also monitors changes in water quality, such as gas pressure, when dams spill excess water. We try to prevent unnecessary spilling by monitoring weather forecasts and adjusting reservoir levels appropriately. We also undertake downstream enhancement projects, including placing gravel in riverbeds, creating spawning areas and installing large, woody debris, to improve fish habitat and offset impacts.

Along power line corridors—For safety and reliability reasons, BC Hydro must limit the growth of trees and vegetation along power line rights-of-way. Riparian zones—bands of vegetation that border rivers and lakes—provide fish with shade, cover, food, nutrients, bank stability and woody debris.

Fish habitat can be affected when trees in the riparian zone are cut or vegetation is disturbed. When trees are removed, river banks can become unstable and silt can be carried into the stream. Silt can also be introduced to streams when roads are constructed to access power lines. The accumulation of silt can create inferior spawning habitat, as well as reducing resting and hiding areas for juvenile fish.

When vegetation must be cleared along power lines, BC Hydro takes into account climate, steepness and soil stability to prescribe a method that effectively clears vegetation and protects fish. We use a computer mapping system to consider terrain factors at stream crossings across the province. Where tall trees pose a potential hazard, they are frequently pruned rather than removed. In some areas, the planting of more desirable, low-growing plant species follows tree removal.

MITIGATING DAM CONSTRUCTION EFFECTS

BC Hydro is also involved in various programs to mitigate and compensate for the effects of hydroelectric dam construction on fish and wildlife through the Columbia Basin and Peace/Williston Fish and Wildlife Compensation Programs, and the Bridge Coastal Fish and Wildlife Restoration Program. These programs are funded annually to carry out our enhancement projects.

More information is available at *www.bchydro.com*.

PARTNERSHIP PRODUCING RESULTS

The Fish and Wildlife Compensation Program (FWCP)—Columbia Basin—is a joint initiative between BC Hydro, the B.C. Ministry of Environment and Fisheries and Oceans Canada to conserve and enhance fish and wildlife populations affected by the construction of BC Hydro dams in Canada's portion of the Columbia Basin.

Since its formation in 1995, the FWCP has worked diligently to identify, fund and deliver hundreds of fish and wildlife projects, almost one-third of which have focussed on species at risk.

These projects are as diverse as the habitats that have been affected in the southeastern corner of British Columbia. They range from stream restoration projects and inventories of endangered species to the purchase of critical wildlife habitat. The focus of the FWCP is on delivering biologically sound, results-driven work.

One of these projects is the Fertilization Program in Kootenay Lake and upper Arrow Lakes Reservoir, regarded as one of the largest successful lake restoration projects in the world. It has resulted in a dramatic increase in the number of kokanee spawners, benefiting other species such as the large Gerrard rainbow trout, bull trout, bald eagles and bears. In addition, it brings economic benefits to local communities by improving recreation and tourism opportunities.

The FWCP helps deliver scores of other fish conservation and enhancement projects including the operation of two large artificial kokanee spawning channels and the aquaculture component of the recovery initiative for the endangered upper Columbia white sturgeon.

The list of partners the FWCP works with and the projects it delivers is extensive. To learn more about its work visit *www.fwcp.ca*.

SPONSORS

FORTISBC

FortisBC Inc. is a Canadian-owned and operated electrical utility that provides service to customers in south-central British Columbia. At FortisBC, we understand the importance of caring for the communities in which we work by providing safe, reliable power and by working with local organizations to mitigate the impacts of hydro-electric facilities on the Kootenay River.

FortisBC employs best management practices in all of its environmental work in order to ensure the protection of riparian, wetland and other wildlife habitats. FortisBC is investing $500 million in electrical system upgrades and other improvements over the next 5 years to increase system reliability, service, and safety and provide for expected population and business growth in the Okanagan, Boundary, and Kootenay areas. Each project plan is developed to standards consistent with our environmental best management practices.

Externally, the company seeks partnership opportunities that facilitate the achievement of corporate and community environmental priorities. FortisBC is very proud of the accomplishments achieved through its partnership with the Columbia Basin Fish and Wildlife Compensation Program to undertake fish and wildlife projects in the Kootenay region. Successes have been realized through this partnership, including habitat restoration and old growth protection on the Salmo River and Northern Leopard Frog recovery program.

For information on FortisBC and our environmental programs, please visit our website at *www.fortisbc.com*.

HABITAT CONSERVATION TRUST FOUNDATION

ACHIEVING CONSERVATION RESULTS TOGETHER

The Habitat Conservation Trust Foundation (HCTF) is a unique conservation society in British Columbia. Funded largely by the anglers, hunters, guide outfitters and trappers of British Columbia, the society supports projects designed to help fish and wildlife populations and habitats across British Columbia, and to deliver environmental education programs targeted at youth and young adults.

The mission of the HCTF is to invest in projects that maintain and enhance the health and biological diversity of B.C.'s fish, wildlife, and habitats so that people can use, enjoy, and benefit from these resources.

The HCTF came into existence because its major contributors (hunters, anglers, trappers, and guide outfitters) were willing to pay for conservation work above and beyond that required by government for basic management of fish and wildlife resources.

The HCTF funds a variety of conservation work including:
- projects that restore, maintain, or enhance native wild freshwater fish populations and habitats;
- projects that restore, maintain, or enhance native wildlife populations and habitats;
- work required to initiate or to ensure the success of eligible conservation projects, as described above.
- information, education, and stewardship projects that enhance users' enjoyment of fish, wildlife, and habitats or that foster human attitudes and behaviours favourable to management and conservation.
- projects that acquire land or interests in land to secure the value of these areas for conservation purposes.

These investments contribute to healthy and diverse populations of native fish and wildlife by improving knowledge, restoring or managing habitats, and enabling stewardship.

The Habitat Conservation Trust Foundation is proposal-driven. Anyone can submit a proposal for funding from HCTF. Proposed activities must be consistent with the mandate guiding the society. Proposals are subjected to a multi-stage review by experts working in the field of fish and wildlife management and conservation in B.C. The very best proposals that arise out of this technical review are subject to a final detailed review by the Board of Directors, who has final authority over which projects receive funding.

HCTF revenue comes primarily from surcharges on fishing, hunting, guiding and trapping licenses. Additional funds arise from court awards, land management activities, donations and partnerships with individuals, government and non-government organizations, and corporations. Since inception, the HCTF has invested almost $90 million in conservation and education projects and a further $12 million to acquire key habitats throughout B.C. Annual surcharge revenue to the HCTF is approximately $5.5 million.

The Board of the HCTF is pleased to support the publication of this comprehensive book on the systematics of the fishes of British Columbia. It will be an invaluable tool for all people interested in the aquatic heritage of British Columbia.

If you would like more information on how HCTF works, or would like to submit a proposal for funding, visit us on the web at *www.hctf.ca* or call 1-800-387-9853.

Yours in conservation,
The Board of HCTF

Peace/Williston
Fish and Wildlife
Compensation Program

FISH FOR THE FUTURE
The dynamic ecological systems of the Williston and Dinosaur watersheds are home to 23 fish species. Discovering the direction those systems are taking and understanding them in order to conserve and enhance the existing fish resources and their habitats for the future is one of the objectives of the Peace/Williston Fish and Wildlife Compensation Program (PWFWCP).

WATER, WATER
The Williston Reservoir is the largest body of fresh water in British Columbia, and the ninth largest reservoir in the world. It has a surface area of about 1,800 square kilometres with 1,770 kilometres of shoreline. The reservoir was created in north central British Columbia by the construction in 1968 of the WAC Bennett Dam across the Peace River.

In 1980, the much smaller Dinosaur Reservoir was created with the completion of the Peace Canyon Dam about 23 kilometres downstream of the WAC Bennett Dam. Both reservoirs were created to produce hydroelectric power. Together, the two reservoirs control water from a catchment area of approximately 70,000 square kilometres.

THE PWFWCP: A SNAPSHOT
Launched in 1988, the PWFWCP is a joint initiative of BC Hydro, the British Columbia Ministry of Environment and Fisheries, and Oceans Canada. It was designed to compensate for the disruption to fish and wildlife caused by the creation of the Williston and Dinosaur reservoirs.

The program team focuses on improving fish and wildlife populations and their habitats within the watersheds. Inventory studies help determine species distribution and population status. Research on particular species and their habitats provides further insight into their ecological needs. Information collected through these activities then forms the basis for management decisions, including enhancement work done by the PWFWCP within the watersheds. At each stage of the process, evaluations are carried out to determine if program objectives are being met and whether new information can be used to further the success of the program.

TAKING STOCK
Inventory studies of fish and wildlife populations and their habitats identify areas where enhancement projects or conservation efforts might be undertaken. Since the PWFWCP's inception, it has stocked more than 15 lakes. Research projects help identify potential enhancement and conservation opportunities, develop management techniques that reduce negative impacts on populations and on habitat, and provide insight into

alternative means of enhancing populations. They also add to the knowledge of northern ecosystems and species biology in general, which is helpful in land-use planning.

SEEKING ANSWERS

Evaluation of enhancement work includes long-term monitoring of population or habitat changes. For example, the Dinosaur Reservoir has been stocked with rainbow trout for almost 20 years. The PWFWCP recently conducted a survey that was able to estimate the extent of population growth and the contribution these fish provided to local angling.

CREATING NEW OPPORTUNITIES

Habitat protection and enhancement helps provide fish and wildlife with the essentials to maintain or expand their populations and distribution. This often involves increasing the abundance and availability of food sources and improving habitats through natural and artificial means.

INVOLVING THE PUBLIC

One of the PWFWCP's goals is to have individuals and community groups provide local knowledge and active assistance in the implementation of the program's projects when possible. For example, since 1996, students from six area schools have gained hands-on experience raising kokanee from eggs to minnow-sized fry that are then released into streams. Consultation with First Nations, stakeholders and the general public also plays an important role in program direction and development.

To learn more about the PWFWCP, visit our website at *www.bchydro.com/pwcp/*.

PROVINCE OF BRITISH COLUMBIA—FISHERIES PROGRAM

In this work, Dr. McPhail explores the wonderful diversity of fish found in the freshwaters of British Columbia—a resource that is highly valued for recreational, social, cultural, ecological and scientific reasons. This diversity is not just represented by the number of species found here, but also by the way individual species have evolved into so many unique forms to occupy the range of habitat types the province has to offer. As an example, one can see this variety by comparing steelhead from coastal waters, "Kamloops trout" from small lakes in the dry interior and the very large "Gerrard rainbow trout" from Kootenay Lake. All are forms of rainbow trout, but they are quite distinct in appearance, in life history and in the angling opportunity that they provide.

The provincial government is responsible for the management of the province's freshwater fishery resource. Our goals are to ensure the long-term health of our fish stocks, meet our obligations to First Nations and to provide for a diversity of sustainable recreational opportunities. This is accomplished through policy and legislation, resource inventory and assessment, research, restoration programs and enhancements such as stocking programs. We work in partnership with many organizations and rely on the work of stewardship groups. Although finding the balance between these sometimes competing goals is a challenge, the province can boast of providing some of the best fishing opportunities in the world.

Dr. McPhail is widely acknowledged as the scientific authority on the freshwater fishes of British Columbia, and this treatise is an important contribution to our overall understanding of this valuable resource. It will undoubtedly become a standard reference for all British Columbians, and we are very pleased to be able to support Dr. McPhail in this endeavour.

For additional information on provincial fisheries management programs, please visit our website at *www.env.gov.bc.ca/esd*.

CONTENTS

vii   SPONSORS

xxiii   FOREWORD

xxvii   ABOUT THIS BOOK

    Purpose xxvii
    What Fishes Are Included? xxviii
    The Keys xxix
    Abbreviations xxx
    Species Accounts xxx
        THE DISTINCTION BETWEEN EXTINCTION AND EXTIRPATION
    Names of Fishes xxxi
    Conservation Comments xxxii
    Distribution Maps xxxii
    Illustrations xxxiii
    Acknowledgements xxxiv
    Funding xxxv

xxxvii   INTRODUCTION

    Origins of the B.C. Freshwater Fish Fauna xxxvii
        PREGLACIAL
        GLACIATION
        DEGLACIATION
        POSTGLACIAL DISPERSAL

        The Pacific Refugium
        The Great Plains Refugium
        The Bering Refugium

    Present Distribution lii
        NORTH PACIFIC COASTAL ECOREGION
        THE NORTHERN COASTAL SUBREGION
        THE CENTRAL COASTAL SUBREGION
        THE SOUTHERN COASTAL SUBREGION
        GLACIATED COLUMBIA ECOREGION

        The Lower Columbia Subregion
        The Upper Columbia Subregion
        The Upper Kootenay Subregion
        The Okanagan Subregion
        The Flathead Subregion

        THE INTERIOR ECOREGION

        The Middle Fraser Subregion
        The Upper Fraser Subregion
        The Four Rivers Subregion

The Upper Skeena Subregion
The Thompson Subregion
The Upper Nass Subregion

THE MACKENZIE ECOREGION

The Lower Liard Subregion
The Upper Liard Subregion
The Lower Peace Subregion
The Upper Peace Subregion
Hay Subregion

THE YUKON ECOREGION

The Yukon Lakes Subregion
The Upper Taku and Iskut-Stikine Subregion

Conservation of Freshwater Fishes lxi

WHAT TO CONSERVE?

Species
  Parallel evolution
Subspecies
Populations
Evolutionarily Significant Units
Designatable Units

SOME CRITERIA FOR LISTING SPECIES AND POPULATIONS

Declining populations
Restricted geographic ranges
Habitat protection
Scientific value

lxxiv PICTORIAL FAMILY KEY

1 LAMPREYS (PETROMYZONTIDAE)

Lampetra ayresii RIVER LAMPREY 3
Lampetra macrostoma VANCOUVER LAMPREY 8
Lampetra richardsoni WESTERN BROOK LAMPREY 12
Lampetra tridentata PACIFIC LAMPREY 16

23 STURGEONS (ACIPENSERIDAE)

Acipenser medirostris GREEN STURGEON 25
Acipenser transmontanus WHITE STURGEON 30

39 MOONEYES (HIODONTIDAE)

Hiodon alosoides GOLDEYE 40

45 HERRINGS (CLUPEIDAE)

Alosa sapidissima AMERICAN SHAD 46

CONTENTS xix

51 MINNOWS (CYPRINIDAE)

Acrocheilus alutaceus CHISELMOUTH 56
Carassius auratus GOLDFISH 61
Couesius plumbeus LAKE CHUB 65
Cyprinus carpio COMMON CARP 72
Hybognathus hankinsoni BRASSY MINNOW 76
Margariscus margarita nachtriebi NORTHERN PEARL DACE 81
Mylocheilus caurinus PEAMOUTH 86
Notropis atherinoides EMERALD DACE 92
Notropis hudsonius SPOTTAIL SHINER 96
Phoxinus eos NORTHERN REDBELLY DACE 101
Phoxinus neogaeus FINESCALE DACE 106
Pimephales promelas FATHEAD MINNOW 111
Platygobio gracilis FLATHEAD CHUB 116
Ptychocheilus oregonensis NORTHERN PIKEMINNOW 120
Rhinichthys cataractae LONGNOSE DACE 126
Rhinichthys falcatus LEOPARD DACE 133
Rhinichthys osculus SPECKLED DACE 139
Rhinichthys umatilla UMATILLA DACE 144
Richardsonius balteatus REDSIDE SHINER 150
Tinca tinca TENCH 155

159 SUCKERS (CATOSTOMIDAE)

Catostomus catostomus LONGNOSE SUCKER 162
Catostomus columbianus BRIDGELIP SUCKER 169
Catostomus commersonii WHITE SUCKER 175
Catostomus macrocheilus LARGESCALE SUCKER 181
Catostomus platyrhynchus MOUNTAIN SUCKER 187

195 NORTH AMERICAN CATFISHES (ICTALURIDAE)

Ameiurus melas BLACK BULLHEAD 196
Ameiurus natalis YELLOW BULLHEAD 200
Ameiurus nebulosus BROWN BULLHEAD 204

209 PIKES (ESOCIDAE)

Esox lucius NORTHERN PIKE 210

217 SMELTS (OSMERIDAE)

Spirinchus thaleichthys LONGFIN SMELT 219
Thaleichthys pacificus EULACHON 225

231 SALMONIDS (SALMONIDAE)

Salmon, trout, and char (Salmoninae) 233

Oncorhynchus clarkii CUTTHROAT TROUT 240
Oncorhynchus clarkii clarkii COASTAL CUTTHROAT TROUT 241
Oncorhynchus clarkii lewisi WESTSLOPE CUTTHROAT TROUT 249

Oncorhynchus gorbuscha PINK SALMON 256
Oncorhynchus keta CHUM SALMON 262
Oncorhynchus kisutch COHO SALMON 268
Oncorhynchus mykiss RAINBOW TROUT, STEELHEAD 275
Oncorhynchus nerka SOCKEYE SALMON, KOKANEE 287
Oncorhynchus tshawytscha CHINOOK SALMON 296
Salmo salar ATLANTIC SALMON 305
Salmo trutta BROWN TROUT 311
Salvelinus confluentus BULL TROUT 317
Salvelinus fontinalis BROOK TROUT 326
Salvelinus malma DOLLY VARDEN 333
Salvelinus namaycush LAKE TROUT 341

Whitefishes (Coregoninae) 347

Coregonus artedi CISCO 350
Coregonus autumnalis ARCTIC CISCO 355
Coregonus clupeaformis LAKE WHITEFISH 360
Coregonus nasus BROAD WHITEFISH 367
Coregonus sardinella LEAST CISCO 373
Prosopium coulterii PYGMY WHITEFISH 379
Prosopium cylindraceum ROUND WHITEFISH 386
Prosopium williamsoni MOUNTAIN WHITEFISH 392
Stenodus leucichthys nelma INCONNU 398

Graylings (Thymallinae) 405

Thymallus arcticus ARCTIC GRAYLING 406

411 TROUT-PERCH (PERCOPSIDAE)

Percopsis omiscomaycus TROUT-PERCH 412

417 CODS (GADIDAE)

Lota lota BURBOT 418

425 STICKLEBACKS (GASTEROSTEIDAE)

Culaea inconstans BROOK STICKLEBACK 427
Gasterosteus aculeatus THREESPINE STICKLEBACK 432
Pungitius pungitius NINESPINE STICKLEBACK 440

447 SCULPINS (COTTIDAE)

Cottus aleuticus COASTRANGE SCULPIN 452
Cottus asper PRICKLY SCULPIN 457
Cottus cognatus SLIMY SCULPIN 462
Cottus confusus SHORTHEAD SCULPIN 469
Cottus hubbsi COLUMBIA SCULPIN 475
Cottus rhotheus TORRENT SCULPIN 481
Cottus ricei SPOONHEAD SCULPIN 486
Cottus sp. ROCKY MOUNTAIN SCULPIN 490

497 SUNFISHES (CENTRARCHIDAE)

      Lepomis gibbosus PUMPKINSEED 499
      Lepomis macrochirus BLUEGILL 504
      Micropterus dolomieu SMALLMOUTH BASS 509
      Micropterus salmoides LARGEMOUTH BASS 514
      Pomoxis nigromaculatus BLACK CRAPPIE 519

523 PERCHES (PERCIDAE)

      Perca flavescens YELLOW PERCH 524
      Sander vitreus WALLEYE 529

535 RIGHTEYE FLOUNDERS (PLEURONECTIDAE)

537 BIBLIOGRAPHY

603 APPENDIX 1

605 GLOSSARY

615 INDEX TO SCIENTIFIC AND COMMON NAMES

FOREWORD

This book, with its detailed illustrations, is an essential resource for all individuals seeking a comprehensive reference to the fishes in the inland waters of British Columbia. It is for those wishing to know what species are in British Columbia and to learn such things as where they occur, what they feed on, when they reproduce, how big they grow, and how long they live. Information in the book will be welcomed by biologists requiring information on the diversity and biology of B.C. fishes to help make intelligent decisions in conservation, management, and impact studies, as well as by lovers of nature and students of fishes. Professor Emeritus Don McPhail is a long-time leader in ichthyological work in British Columbia, and his refreshing writing style and intimate knowledge of the fishes of the province opens our eyes to the diversity of fishes existing next to Canada's Pacific Ocean. He does a masterful job in presenting what is known and in stimulating further research.

Books on regional ichthyofaunas have a special appeal because they tell us about the biodiversity of fishes in a geographic area of interest. British Columbia has an unusual fish fauna in its lakes and streams: there are marine fishes that invade fresh water, species that occur in northern hot springs, genetically heterogeneous species of Pacific salmon with locally adapted populations, rare and endangered species that occur nowhere else in Canada, examples of parallel evolution, species pairs, and populations that consist of individuals that have recently evolved and are different in appearance from others of the species. Many of these newly evolved species are of immense interest in evolutionary studies. Some forms challenge our concept of the species and certainly challenge our ability to apply it, and there may even be undescribed species. The fauna is of enormous economic value and interest; many species are sport fishes, and others provide good opportunities for fish watching.

For those with an interest in the past, *The Freshwater Fishes of British Columbia* mentions the fossil fish record in British Columbia. Readers are treated to an explanation, resulting in part from the authors research, of where B.C. fishes were during the last ice age and how, historically, they came to be where they are today. For example, some species occur only in Arctic drainage (e.g., goldeye), only in Columbia drainage (e.g., Umatilla dace), or only in Fraser and Columbia drainages (e.g., leopard dace), while others occur in most major drainages throughout the province (e.g., longnose dace).

There are some 65 or 66 native species belonging to 13 families in B.C.'s fresh waters. Another 16 are part of the fauna as a result of human introduction, with four marine species occurring in river mouths and sometimes a short distance upstream. In all then, some 85 species (belonging to 17 families) occur in the inland waters of British Columbia and are covered in this book. Although British Columbia has only a very small proportion of the species known on earth, it has an evolutionarily diverse fauna ranging from primitive fishes that lack jaws, scales, and fins to what ichthyologists consider to be among the most derived of ray-finned fishes. The species represented in this book can be accurately identified with the use of special keys that are supplemented, where appropriate, with diagrams. Species accounts are organized by family and give the official common and scientific names; these are typically followed by headings listing Distinguishing Characters, Sexual Dimorphism, Taxonomy, Distribution (native worldwide and in British Columbia, permitting users to know if they have new distributional records), Life History (with subheadings), Habitat, and Conservation Comments. Many persons naïvely assume that more is known than is actually the case. Don makes it clear what is not known about fishes in British Columbia and what further research is needed (for example, concerning life history), and this should stimulate future research efforts to clear up critical gaps in our knowledge.

There is hardly anywhere in British Columbia where one can go without there being some adjacent or nearby water with fishes. As we learn in the book, every area of British Columbia has a fish fauna with a story to tell. For example, near Princeton, an area where I grew up, there are two lakes, one of which we used to fish for trout in the 1940s, with two perfectly good species in different genera that interbreed under natural conditions. One might think that hybridizing species would look alike. However, these two cyprinids are very different in appearance to us humans and are about as different in feeding habits as one can imagine. One species is the large-mouthed, fish-eating northern pikeminnow, and the other is the chiselmouth, which is adapted to obtain its food by scraping algae and other organisms off rocks. Their hybrid offspring look intermediate and live what seems to be perfectly natural lives, but they are not genetic bridges causing the species to merge. While this example may be surprising to many, it is in fact not all that unusual in British Columbia, because there are many other species in different areas that hybridize without breaking the species integrity.

Don and his students and colleagues have made major research advances in our knowledge of B.C. fishes over the past several decades. This book reflects that progress, and it makes a substantial contribution to our knowledge of the biology of fish species in British Columbia. It presents extensive information that is not readily available elsewhere and also provides a synthesis of the literature. Many biological novelties exist in areas deserving protection as world heritage evolutionary sites because they provide continuing opportunities for the study of fundamental evolutionary problems. For example, sympatric threespine stickleback pairs are expertly described by Don, who was an early driving force in much of the research concerning this phenomenon. British Columbia has key examples of "problem" cases, for example, concerning the species concept and parallel evolution that are discussed. Don justifies controversial cases of not recognizing some forms seemingly behaving as good biological species as named taxonomic species.

I feel lucky to have had Don McPhail as the lab teaching assistant and Cas Lindsey as the professor of my first course in ichthyology in 1958 where, along with world fishes, we learned the fish species in British Columbia. On exams Don sometimes gave us a red herring to keep us alert. As students, we knew that it was difficult to identify some species; however, we relished the challenge of getting it right, and there seemed to us to be little doubt that, in all cases, there was such a thing as getting it right. Now we learn that, with our limited present knowledge, it is sometimes not possible to get it right.

Thank you Don for writing such a fine book and educating us on the legacy of our fish heritage. If we heed the warnings in this book and if fisheries biologists are able to do their work in conserving fish faunas, then future generations will be able to enjoy the riches of the fascinating B.C. fish fauna much as we can at present.

Joseph S. Nelson,

Professor Emeritus,
Department of Biological Sciences,
the University of Alberta,
Edmonton, Alberta.
6 December 2006

ABOUT THIS BOOK

Purpose This book is intended as a reference for fisheries biologists working in British Columbia. It began as a revision of *The Freshwater Fishes of British Columbia* (Carl et al. 1959). That Provincial Museum Handbook was designed for anglers and naturalists. It was small enough to fit in a jacket pocket and contained short species descriptions, identification keys, some information on distribution, and a few snippets on the biology of each species. By the time I began revising the original handbook, it became clear that something more detailed was required. Since the last printing of the handbook in 1977, a major shift had occurred in public and professional attitudes towards native fishes. In the past, professional fisheries biologists concerned themselves with only a few species—the Pacific salmon, the trout and to a lesser extent, the char. The other indigenous fishes were either neglected or actively persecuted. As public concern over deteriorating habitats and the loss of biodiversity grew, this professional bias gradually changed. Now, most fisheries biologists recognize all indigenous species are part of our natural heritage and the Species at Risk Act requires the inclusion of many species in management and conservation plans.

Rational plans require reliable information; however, except for the Pacific salmon and trout, little is known about our native species. An obvious first step towards developing management and conservation plans is to determine what species are present. Unfortunately, most consultants and government biologists cannot identify native fish—a check of species identified by inventory personnel revealed an average error rate of roughly 26% (Haas et al. 2001). For species of conservation concern (those listed as "threatened" or "vulnerable"), the error rate was 42–100%. Clearly, there is a problem with identification!

Most conservation and recovery plans also involve the identification of "critical" habitat. This is a slippery concept in fish. Typically, fish use different habitats at different times in their lives. Thus, determining "critical" habitat requires detailed information on a species' life history and habitat use. Regrettably, there is no reference book that provides recent biological information on native B.C. fishes. This book is an attempt to fill that gap. It includes identification keys and relatively detailed life history and habitat synopses for all species. Where possible, it includes B.C. data; however, except for Pacific salmon and trout, local information is sketchy. Thus, in some ways, the book is a compendium of ignorance rather than a compilation of knowledge. Still, by documenting gaps in our knowledge, I hope to stimulate others to study our native species. To this end, the life-history and habitat-use accounts for indigenous species (other than salmonines) are as detailed as I could make them. In contrast, given the wealth of information available, the same summaries for salmonines are relatively superficial. This is because there are other, excellent books (Behnke 2002; Groot and Margolis 1991; Quinn 2005;

Stolz and Schnell 1991; Trotter 1987) on salmonine life histories and habitat use as well as the detailed U.S. National Marine Fisheries status reviews for the native trout and salmon of the Pacific Northwest (Busby et al. 1996; Gustafason et al. 1997; Hard et al 1996; Johnson et al. 1997; Johnson et al. 1999; Myers et al. 1998; Weitkamp et al. 1995). Similarly, the summaries for introduced species are often cursory. Most of our introduced fishes are of eastern North American or Eurasian origin, and there is little known about their life histories and habitat use in British Columbia. Thus, the best sources of information for introduced fishes are books that describe their biology within their native ranges (e.g., Becker 1983; Scott and Crossman 1973; Trautman 1957; Wheeler 1969).

What Fishes Are Included? In a coastal province like British Columbia, the line between marine and freshwater species can be fuzzy. Many species spend part of their lives in fresh water and part in the sea. Familiar examples are the Pacific salmon. Usually such species are anadromous (i.e., as a normal part of their life history they migrate from rivers to feed in the sea and then return to spawn in fresh water). The reverse migration—from marine or brackish water into rivers to feed and then back to the sea or an estuary to spawn—is catadromy. In British Columbia, anadromy is a common life-history pattern in lampreys, sturgeon, smelts, salmonines, and sticklebacks. In contrast, catadromy is rare; however, some coastal populations of two sculpins (coastrange, *Cottus aleuticus*, and prickly sculpin, *Cottus asper*) breed, or spend a larval stage, in brackish water and spend their adult lives in fresh water.

More troublesome for a book on freshwater fishes are marine species that regularly enter fresh water but in which freshwater residence is not a necessary part of their life history. Examples are the starry flounder (*Platichthys stellatus*) and the staghorn sculpin (*Leptocottus armatus*). Juveniles of these species use estuaries and the lower reaches of rivers along the entire B.C. coast. In the lower Fraser River, juvenile starry flounders are common as far upstream as Fort Langley, and occasional individuals ascend the river almost to Chilliwack (Sumas River). The starry flounder also occur regularly in Pitt Lake. Other marine fishes, such as the tidepool sculpin (*Oligocottus maculosus*) and the sharpnose sculpin (*Clinocottus acuticeps*) are sporadic, but common, visitors to the lower 100 metres or so of many small coastal streams, especially in areas like the Queen Charlotte Islands and the west coast of Vancouver Island where there are relatively few true freshwater fish. Still other marine species (e.g., the shiner perch, *Cymatogaster aggregata*; the Pacific herring, *Clupea pallasii*; and the surf smelt, *Hypomesus pretiosus*) often are abundant in estuaries and the tidal reaches of large rivers like the Fraser River. They are often collected several kilometres up the river alongside freshwater species like minnows and suckers. Still other species, such as the Pacific tomcod (*Microgadus proximus*) and several marine sculpins, occur sporadically in estuaries but are not known to ascend rivers.

Thus, there is a continuum among marine fish that runs from species that habitually enter rivers, especially as juveniles, to those that rarely enter fresh water. In a book on freshwater fishes, this presents the problem of which species to include and which species to omit. Although the decision is arbitrary, I have only included species accounts for fishes that have at least one population that spawns in fresh water. Nonetheless, some marine species like the surf smelt and starry flounder are so common in the lower reaches of large rivers that it is impossible to ignore them. In such cases, I have included these fishes in the identification keys but have not given them species accounts.

Another troublesome group are fishes that breed in fresh water but are only occasionally recorded in British Columbia and are not known to breed in our province. Examples are the green sturgeon (*Acipenser medirostris*), the American shad (*Alosa sapidissima*), the Arctic smelt (*Osmerus dentex*), the emerald shiner (*Notropis atherinoides*), and the ninespine stickleback (*Pungitius pungitius*). For these species, I used available information on their biology to make a decision on whether or not to include them in the book. Basically, I decided that if there were confirmed B.C. records of a species and a viable population nearby, I would include a complete species account. If not, I would leave the species out of the book. Except for the Arctic smelt, I decided to include all of the above species. I decided to leave the Arctic smelt out of the book because there are no confirmed freshwater records from British Columbia (there is a marine record from Barkley Sound), and global warming makes it unlikely that this Arctic species will establish itself in our province.

Finally, there are the aquarium fishes! A surprising diversity of aquarium fishes (everything from guppies to piranhas) are released into the wild every year, especially in urban areas such as the lower Fraser valley, southern Vancouver Island, and the southern Okanagan. Most of these aquarium releases involve tropical fishes and although some make it through the summer, most perish in the winter. Occasionally, however, an aquarium species establishes a reproducing population (e.g., the goldfish, *Carassius auratus*). A full species account is provided in such cases.

The Keys Until recently, the identification of native fish was primarily an academic concern. The focus of management agencies was on salmonids, and technical reports simply lumped suckers, minnows and sculpins together as "other fish" or, worse, made botched attempts at identification. With growing public and governmental interest in biodiversity and conservation, sloppy identifications are no longer acceptable. Resource managers require reliable inventory data, and the first step in acquiring such data is accurate identification.

Traditionally, biologists use dichotomous keys to identify unfamiliar species. The keys in this book differ slightly from traditional keys in that some of the longer keys were modified to reduce repetition. Thus, the keys in this book consist of a series of numbered statements. Each statement contains information (usually counts and body proportions) to which the user answers yes or no. At the end of each statement, there is either a number or the name of a species. You work through a key by starting at the first statement. If this statement fits your specimen, you proceed to the statement indicated by the number at the end (right side) of the statement. If the statement does not fit your specimen, proceed to the statement indicated by the number in parentheses immediately behind the statement number. This statement may end in a number that directs you to another statement. In this case, repeat the procedure until you reach a species name. If the key has been successful, this is the species at hand.

Although simple in concept, the uninitiated find keys difficult to use. A particularly vexing problem with keys is that, once you start into the key, you will come out somewhere—but if you misinterpret a statement, it may not be in the right place! Consequently, understanding the statements is crucial to the use of the key. Unfortunately, keys often contain qualitative statements like "snout short and blunt versus snout long and sharp." Without examples of both states at hand, the user is left to guess how short is short and how round a snout must be to be blunt. Typically, it is this kind of qualitative comparison that leads the user astray. I have attempted to alleviate this problem by illustrating contrasting qualitative characters in the body of the key. Hopefully, this will prevent gross identification errors. Nonetheless, any unknown should be checked against the species' illustration and the distinguishing characters section in the species account. The keys are based on B.C. specimens, and the meristic counts may differ slightly from specimens from other areas.

For people unfamiliar with our freshwater fish fauna, there is a pictorial key to the families at the beginning of the book. If in doubt about the family, run unknown species through the family key. This will reduce the problem of identification to a subset of the available species. Once the family is known, turn to the species key for that family and, hopefully, identify the species.

Abbreviations   Two abbreviations that appear regularly in the conservation comments section are COSEWIC and BCCDC. They stand for the Committee on the Status of Endangered Wildlife in Canada and the British Columbia Conservation Data Centre, respectively. Other abbreviations that occur in the text are FL, fork length; TL, total length; SL, standard length; Ma millions of years; ka, thousands of years, and BP, before present. All abbreviations are spelled out the first time they appear in the text.

Species Accounts   Species accounts are given for all fishes known to breed in the freshwaters of British Columbia. These accounts consist of an illustration, the names (scientific and common) of the fish, its distinguishing characters, taxonomic comments (if there are taxonomic problems), geographic distribution (with a B.C. distribution map), a life-history summary, a habitat-use summary, and conservation comments. Where possible, the life-history and habitat-use summaries contain specific statements about such factors as water velocities and substrates at spawning sites or water velocities, substrates, and depths used by adults, juveniles, and young-of-the-year. I want to stress that such statements are meant as guides and not as absolute values. Most northern temperate freshwater fishes are remarkably adaptable. Consequently, where in a particular stream or lake a species is found is a function, among other things, of the habitats that are available, the other species present, the season, and the time of day. Even relatively sophisticated habitat-use models developed in one area rarely are directly applicable in another area. Nonetheless, where available, I have included specific habitat measurements because, at a minimum, they indicate the characteristics of sites where some other researchers have found that species. Given how little we know about the life histories and habitat uses of most non-game species in British Columbia, it is at least a starting point.

THE DISTINCTION BETWEEN EXTINCTION AND EXTIRPATION
Biologists use two words to distinguish between two phenomena with different conservation consequences—the word extinction is reserved for the global loss of a species, whereas the word extirpation is used for the local loss of part (albeit sometimes an important part) of a species. As an example, no species of Pacific salmon (*Oncorhynchus*) is in immediate danger of extinction; however, many salmon populations in British Columbia are in danger of extirpation. I have tried to maintain this distinction throughout the text, but there are situations—usually involving pairs of recently evolved biological species (*sensu* Mayr 1963)—where it is not clear which of the two words is appropriate.

Names of Fishes Curiously, the names of fishes (both scientific and common) are a never-ending source of confusion and argument among biologists, anglers, naturalists, and ichthyologists. An example is the change—for valid reasons—in the scientific name of rainbow trout from *Salmo gairdneri* to *Oncorhynchus mykiss*. This change produced howls of outrage from anglers and biologists. Apparently, people don't like name changes. Yet changes in scientific names are inevitable. New information is uncovered, or new interpretations are given to old information, and the result is a name change. This is particularly true of generic names. Like other people, ichthyologists follow fashions, and since the boundaries of genera cannot be defined objectively, genera are lumped or split according to the fashion of the day. Thus, in North America, generic names are irritatingly unstable. Specific names are more stable, but their spelling can change with changing interpretations of the International Rules of Nomenclature (e.g., the cutthroat trout's specific name is sometimes spelled *clarki* and, other times, spelled *clarkii*). As of this writing (2005), the trend is to use the original spelling of specific names, with some exceptions. The spellings of the specific names used in this book are those used in the

American Fisheries Society list of the common and scientific names of fishes (Nelson et al. 2004).

In spite of the annoyance and confusion generated by the instability of scientific names, they serve an essential purpose. Each species has a unique scientific name (e.g., *Oncorhynchus nerka* for the sockeye salmon) that is the same in all languages. As a result, a worker in Kamchatka or Japan knows exactly what species scientists in British Columbia mean when they use that name.

Also, there has been an attempt to stabilize the common names of fishes. To this end, the American Fisheries Society publishes a list of "official" common names of North American fishes (Nelson et al. 2004). This list strives to provide a unique common name for each species. Unfortunately, many species have no common name, and the committee had to coin names. By-and-large, they did an admirable job, although some official common names are plain dumb (e.g., Vancouver lamprey, *Lampetra macrostoma,* for a fish known only from Cowichan and Mesachie lakes), whereas others are locally confusing. Catfishes of the genus *Ameiurus* are officially bullheads; however, to anyone raised in British Columbia, they are catfishes, and sculpins (Cottidae) are bullheads. Such local names are useful in communicating with anglers; however, the official common names are the ones used in scientific papers and technical reports.

Another name problem involves collective names for species within families. For example, in its modern sense, the family Salmonidae consists of three subfamilies: Salmoninae, which include salmons, trouts, chars, and other trout-like fishes; Coregoninae, which include the whitefishes; and Thymallinae, which include graylings. Thus, the term "salmonid" should include all members of the family Salmonidae. In the general fisheries literature, however, the term "salmonid" usually means salmon, trout, char, and other trout-like fishes (i.e., the salmonines) and rarely is taken to include whitefishes and graylings. In this book, I use the term "salmonines" for members of the subfamily Salmoninae and the term "salmonids" for all members of the family Salmonidae.

Conservation Comments Most of the species accounts in this book end with a brief section entitled "Conservation comments". This short section details the conservation status of the species. Usually it includes the COSEWIC (Committee On the Status of Endangered Wildlife In Canada) and BCCDC (British Columbia Conservation Data Centre) ranking of the species. These rankings are subject to periodic review and can change. The listings used in this book are the 2005 rankings.

Distribution Maps A spot distribution map is given for each species. For those unfamiliar with the geography of the province, the base map used for distributions—with a key to the rivers mentioned in the text—is given in Figure 1. In most cases, the map dots represent all the records for that species in B.C. For some species (e.g., the rainbow trout), there are so many

Figure 1 Key to the major rivers of British Columbia

1 Lower Fraser
2 Skagit
3 Similkameen
4 Okanagan
5 Kettle
6 Lower Columbia
7 Pend d'Oreille
8 Kootenay
9 Flathead
10 Upper Columbia
11 Thompson
12 Middle Fraser
13 Upper Fraser
14 Homathko
15 Klinaklini
16 Bella Coola
17 Dean
18 Skeena
19 Nass
20 Iskut-Stikine
21 Taku
22 Alsek
23 Yukon
24 Upper Liard
25 Lower Liard
26 Petitot
27 Fort Nelson
28 Finlay
29 Parsnip
30 Upper Peace
31 Lower Peace

records that the spots coalesce into an uninformative blob. In such cases, the records have been edited into abridged distribution maps. On the distribution maps, ● indicates a native species, ■ indicates an introduced species, ◻ indicates a failed introduction, ▲ indicates translocations (introductions of native species outside their natural B.C. range), ○ indicates an unverified (usually out-of-range) record of a native species, ★ indicates a historical record (now extinct), and ✦ indicates the sporadic occurrence of a native species. The spots on the distribution maps were plotted using a database that combined records from government agencies, universities, museums, and consultant reports. Such a large and diverse database inevitably contains errors and ambiguities, and not all locality records were confirmed. However, all the maps were checked for out-of-range spots, and these records were either confirmed or rejected. Thus, the maps are accurate to a river system, but the localities within river systems are sometimes off by several kilometres.

Illustrations Diana McPhail produced all the drawings used in this book. The original illustrations were then scanned and printed from the scanned images. There are advantages and disadvantages associated with the use of scanned images. The main disadvantage is in quality—the illustrations are not as crisp as a good technical drawing. The main advantages are low cost, ease of manipulation, and flexibility in layout. For this book, the advantages far outweighed the disadvantages.

Acknowledgements

When possible, the life-history and ecological information used in this book is local. A remarkable amount of such information lies buried in government files; in unpublished theses; in consultant reports; and in the heads of biologists, fisheries technicians, conservation officers, and other scientists. In compiling the life-history and habitat summaries, I have shamelessly mined these sources, as well as the primary scientific literature. This mountain of information was then condensed into summaries of varying length. In the process, information probably was lost, and mistakes were introduced. The mistakes and errors are mine alone.

No enterprise like this starts without precedents, and I owe a great debt to W.A. Clemens, C.G. Carl, and especially, to C.C. Lindsey—they produced the first handbook on British Columbia freshwater fishes and established the first comprehensive collections of fishes at the Royal British Columbia Museum and the University of British Columbia. These pioneers, along with P.A. Larkin, W.S. Hoar, and T.G. Northcote, established a tradition of excellence in freshwater research at the University of British Columbia that has inspired generations of graduate students. Finally, I want to thank Joe Nelson (University of Alberta) for guiding me through the esoteric rules of zoological nomenclature.

Over the years, a legion of people contributed much of the information used in this book. Indeed, so many people over so many years that a complete list would cover several pages and, inevitably, would omit some names that should have been included. Consequently, instead of a long list of names, I have chosen to give collective thanks to organizations and institutions that helped along the way. Invariably, however, it was individuals in the institutions, and not the institutions, that provided the help. Three people warrant special mention: Clyde Murray, a tower of strength in the field and laboratory; Marvin Rosenau, whose integrity and unyielding advocacy for fish and their environments epitomize what it means to be a fisheries professional; and Arne Langston, who hustled money and handled all the negotiations with publishers. Without his prodding and assistance, this book would never have reached the publication stage.

The B.C. Ministry of Environment (in its present and many previous manifestations) provided permits, information, encouragement, and financial support. BCHydro and the Peace/Williston Fish and Wildlife Compensation Program provided information, field assistance, and great encouragement. Various consulting companies gave me access to unpublished reports, and colleagues at the University of Alberta, the University of Washington, Oregon State University, and the Royal British Columbia Museum provided access to distribution records and specimens. I also want to thank the generations of graduate and undergraduate students for assistance in the field and laboratory. Finally, my deepest thanks to my wife Marjorie, our children, and our grandchildren: they cheerfully put up with the idiosyncrasies of their "nearly normal" husband, father, and papa. Their forbearance and support made this book possible.

Funding A book like this is not a viable commercial enterprise. For me, it is a small repayment to British Columbia for the privilege of spending my adult life doing what I love—studying fish. Still, it costs money to publish a book; although in retirement I can donate my time, many other working people are involved in a book's production. There are illustrators, editors, layout technicians, and printers to pay. These people are professionals, and their time is valuable. Consequently, a book that will never sell enough copies to cover the costs of production requires a subsidy. Some of the beneficiaries of this book—the organizations involved in the management and conservation of our native fishes—responded positively to requests for financial support. Others did not. Curiously, there was an inverse relationship between an organization's budget, its legal obligation to conserve aquatic biodiversity, and its willingness to support the project. BC Hydro, Fortis BC, the Peace/Williston Fish and Wildlife Compensation Program, and the Habitat Conservation Trust Foundation were generous and enthusiastic in their support. The B.C. Ministry of the Environment, savaged by years of deep budget cuts, still managed a small but welcome contribution. The Fish and Wildlife Compensation Program—Columbia Basin also contributed to the project. Together, these organizations made publication possible. I am in their debt.

INTRODUCTION

The native freshwater fish fauna of British Columbia is not large, but a mere count of native species (65 or 66) is deceptive. First, many of our species occur nowhere else in Canada, and a few remarkable fishes occur nowhere else in the world. Thus, the stewardship of this fauna, although a provincial responsibility, also carries national and international obligations. Second, the complex topography and recent glacial history of British Columbia have created an environment conducive to rapid genetic divergence. This has resulted in an explosion of genetic diversity at the level of populations. A familiar example of this phenomenon is the diversity of Pacific salmon stocks found along the B.C. coast. Embedded within single species such as the sockeye salmon (*Oncorhynchus nerka*) are a complex array of genetically distinctive stocks that differ in a myriad of ways (body size, body shape, run timing, egg size, incubation rates, and energy reserves). Although salmon stocks are the most familiar example of this population-level diversity, the same forces that shaped our hundreds of salmon stocks also created divergent stocks in most of our native species. In fact, in some groups (such as sticklebacks) where genetic differences among populations are relatively well studied, there is evidence that such divergences have produced biological species (*sensu* Mayr 1963). These cases are of great scientific interest; however, the processes that produce and maintain diversity among populations are not just an academic concern. The recognition of genetic differences among populations is a fundamental component in strategies used to manage our commercial and recreational fisheries. Thus, population-level diversity has direct economic, as well as conservation, consequences. To successfully sustain and manage this diversity, we must understand both its origins in the complex geological history of the province and the nature of the evolutionary processes that shaped and refined this diversity.

Origins of the B.C. Freshwater Fish Fauna

PREGLACIAL

The roots of our native fish fauna are buried deep in the past with the origins of some groups, such as lampreys, stretching back hundreds of millions of years to the very beginnings of the vertebrates. Although the ancestors of present-day lampreys are ancient, most living species evolved relatively recently—the fossil evidence suggests that most modern families evolved within the last 65 million years (Ma). In western Canada, the first fossils that are clearly related to living B.C. fishes are a pike (*Esox*), a smelt (*Speirsaenigma*), and a trout-perch (*Lateopisciculus*) from Paleocene deposits in Alberta (Murray and Wilson 1996; Wilson 1980; Wilson and Williams 1991). The first B.C. fossils related to our present fauna are from early to mid-Eocene deposits (about 50 Ma). This fauna (Wilson 1977), consists of fishes related to trout (*Eosalmo*), goldeye (*Eohiodon*), suckers (*Amyzon*), and trout-perch (*Libotonius*). Although these Eocene and earlier fossils represent modern families, most belong to different genera than their living counterparts. In addition, the early to mid-Eocene fauna of British Columbia contained extant eastern North American families that are no longer native to western North America: the bowfin family, Amiidae, and the sunfish family, Centrarchidae (Wilson 1977, 1982). This suggests that, sometime during the Tertiary Period, something happened to cause the extinction of these families in the west.

The Tertiary Period started about 65 Ma and ended about 2 Ma. In British Columbia, the Tertiary was characterized by two periods of intense tectonic activity: one in the lower Tertiary (Paleocene and Eocene), and one in the upper Tertiary (late Miocene and Pliocene). This Tertiary mountain building and volcanism profoundly changed both our landscape and our fish fauna. On the Pacific edge of the continent, lower Tertiary uplift produced a parallel set of northwest trending mountains along the axes of the present Coast and Insular Mountains, while the rising Rocky Mountains created a substantial Continental Divide. Apparently, all of the major river systems in British Columbia were initiated at about this time, although the headwaters of the east-flowing Peace and Liard rivers rise west of the Continental Divide and, thus, must pre-date the rise of the northern Rocky Mountains (Holland 1964). This lower Tertiary transformation of western North America into a mountain-dominated environment probably contributed to the extinction in the west of quiet-water fishes like bowfins and sunfish; however, there was a worldwide extinction of many families, genera, and species about 35 to 40 Ma (Cavender 1986).

Following this lower Tertiary mountain building, there was a long episode of erosion during which the Coast and Insular mountains were eroded to a gentle west-sloping plain. The second (upper Tertiary) period of mountain building and volcanism established what is essentially our modern landscape. During the late Miocene, the central interior of British Columbia was covered by immense flows of fluid basalt extruded from vents in the earth. These flows covered an area of over 200,000 km^2 in British Columbia, while contemporaneous flows in Washington, Oregon, and Idaho covered even larger areas. These huge lava flows probably caused the extinction of some of the earlier fauna. Still, in spite of widespread volcanism, the northwestern fish fauna flourished; in the late Miocene, we find the first clear evidence of modern western North American fishes. Late Miocene and Pliocene deposits on the Snake River Plain of Idaho and Oregon (they are associated with various stages of ancient Lake Idaho) contain a remarkable mixture of living and extinct fish genera (Kimmel 1975; Smith 1975, 1981). At this time, the Snake River (now the major Columbia tributary) was not part of the Columbia system but instead flowed southwest to the Pacific Ocean through northern California (Aalto et al. 1998). Many groups familiar to fish biologists in British Columbia make their first appearance in these lacustrine deposits: char (*Salvelinus*), Pacific trout and salmon (*Oncorhynchus*), peamouth (*Mylocheilus*), shiner (*Richardsonius*), pikeminnow

(*Ptychocheilus*), chiselmouth (*Acrocheilus*), sucker (*Catostomus*), and sculpin (*Cottus*). Again, although these late Miocene fossils are recognizable at the generic level, usually they are considered different species than their modern counterparts.

Towards the end of the Tertiary in the Pliocene (about 5 Ma), we find the first fossils of what appear to be modern species—white sturgeon (*Acipenser transmontanus*), largescale sucker (*Catostomus macrocheilus*), cutthroat trout (*Oncorhynchus clarkii*), threespine stickleback (*Gasterosteus aculeatus*), and goldeye (*Hiodon alosoides*) (Smith 1981; Smith et al. 2000). The Pliocene was a period of active mountain building, volcanism, and continued climate cooling. The second uplift of the Coast and Insular Mountains occurred at this time. These mountains now run almost parallel to one another along two northwest-trending axes and are separated by the Coast Trough. Most of our major west-flowing rivers (e.g., the Fraser, Homathko, Klinaklini, Bella Coola, Dean, Skeena, Nass, Iskut–Stikine, Taku, and Alsek) cut across this area of uplift. Since these rivers maintained their westward flow by cutting deep canyons through the Coast Range, they must have occupied their present courses before the Pliocene rise of the Coast Mountains. This does not imply, however, that these rivers are unchanged since the Pliocene. For example, there is geomorphological evidence that suggests that, in preglacial time, the Fraser drainage north of about Williams Lake was tributary to the Peace River (Holland 1964; Read 1993). Additionally, the catchments of the preglacial Stikine and Alsek rivers included large regions in the Yukon Territory that now drain into the Yukon River (Templeman-Kluit 1980), and the Porcupine River (a major Yukon tributary) drained to the east into the Arctic Ocean (Duk-Rodkin and Hughes 1994). To the south, the Pliocene appearance of a cool-water fauna (e.g., *Prosopium* and *Myoxocephalus*) during the Glenns Ferry stage of Lake Idaho suggests a possible connection between the Snake River and some east-flowing (Hudson Bay?) drainage (Smith 1981).

Middle Pliocene (about 3.8 Ma) fossil deposits in eastern Washington and northeastern Oregon suggest that these regions were still connected to the preglacial Snake River (Smith et al. 2000; van Tassell et al. 2001). These deposits contain a mix of extant species (white sturgeon and largescale sucker) and extinct species: a muskellunge (*Esox columbianus*), lake sucker (*Chasmistes*), chiselmouth, peamouth, pikeminnow, bullheads (*Ameiurus*), and sunfish (*Archoplites*) (Smith et al. 2000). Apparently, the Columbia River captured the Snake River sometime in the late Pliocene (Smith et al. 2000).

GLACIATION

As the Tertiary drew to a close about 2 Ma, the climate in northern regions had cooled to the point where permanent ice began to accumulate at high latitudes and on mountain tops. Slowly, these scattered patches of ice grew and coalesced to form ice sheets. In North America, the first major ice sheet formed in the northern Cordillera during the late Pliocene (about 1.9–1.8 Ma). This glaciation covered most of British Columbia and the Alaska Peninsula (Barendregt and Duk-Rodkin 2004). The Quaternary (or Pleistocene) Period started about 1.6 Ma. Classically, the Pleistocene is marked by four major glaciations separated by interglacial periods. Until recently, little was known about the extent and timing of the glaciations that preceded the last (Wisconsinan) glaciation; however, knowledge of the early glaciations and interglacial periods in North America has increased dramatically over the last two decades. Apparently, in the early Pleistocene, there were three Cordilleran glaciations and one to five continental ice sheets (Duk-Rodkin et al. 2004). At least one of these continental glaciations extended as far south as Kansas, but none of them coalesced with the Cordilleran Ice Sheets. Consequently, the northern Great Plains remained ice free. Three more Cordilleran and continental glaciations occurred in the middle Pleistocene (780–130 ka) and, again, the Cordilleran and continental ice sheets did not coalesce (Barendregt and Irving 1998). During the late Pleistocene (about 130 ka), however, a major continental ice sheet formed. This ice sheet (the Laurentide) coalesced with the Cordilleran Ice Sheet and, with the exception of northern and central Alaska, ice covered most of northern North America (Barendregt and Duk-Rodkin 2004). In the west, the ice reached Washington, Idaho, and Montana, and on the Great Plains, it dipped south through South Dakota, Nebraska, Kansas, and Missouri. In most of Canada, this last glaciation reached its maximum about 18,000 BP (before present). In British Columbia, the situation was complicated by the mountainous topography and regional differences in precipitation. Although the direction of ice flow varied in different parts of the province, it generally flowed downhill from the Insular, Coast, and Rocky mountains filling the intervening areas and eventually coalescing into a single ice sheet. At its southern edge, the Cordilleran Ice Sheet did not reach its maximum until about 14,500 BP (Fulton et al. 2004). By this time, the continental ice sheet had begun to withdraw, and parts of the southwestern corner of Alberta were already ice free (Dyke 2004).

Not surprisingly, the Pleistocene expansions and contractions of ice sheets had important impacts on our fish fauna. Thus, as the late Pleistocene ice sheets expanded to cover most of northern North America, the interglacial fish fauna in this area was either destroyed or pushed into unglaciated regions called refugia. Not much is known about the pre-Wisconsinan fish fauna of British Columbia. Fish remains—as yet unidentified—are known from a site near Westwold (Harington 1978) that is overlain by glacial tills and appear to be of Sangamon (the pre-Wisconsinan interglacial period) age (Harington 1996). Fossils found elsewhere in North America indicate that, immediately before the last glaciation, our fish fauna consisted of modern species. For example, threespine stickleback spines are known from a Sangamon deposit near Port Townsend, Washington (Karrow et al. 1995). Additionally, two species now widespread in British Columbia (redside shiner, *Richardsonius balteatus*, and largescale sucker) are known from early Pleistocene deposits in Idaho (Smith 1981), and six other B.C. species (brassy minnow, *Hybognathus hankinsoni*; flathead chub, *Platygobio gracilis*; white sucker, *Catostomus commersonii*; mountain sucker, *Catostomus platyrhynchus*; yellow perch, *Perca flavescens*, and walleye, *Sander vitreus*) are known from early Pleistocene deposits on the Great Plains (Cross et al. 1986). Six common northern species (broad whitefish, *Coregonus nasus*; inconnu, *Stenodus leucichthys*; longnose sucker; *Catostomus catostomus*; burbot, *Lota lota*; Arctic grayling, *Thymallus arcticus*, and slimy sculpin, *Cottus cognatus*) are known from deposits in the Yukon Territory that are about 60 ka old (Cumbaa et al. 1981). Also, *Coregonus, Prosopium, Catostomus,* and *Thymallus* are known from pre-Illinoian deposits in southeastern Indiana (Miller et al. 1993). In British Columbia, late Pleistocene fish fossils are known from Savona near Kamloops (Carlson and Klein 1996). These fossils are Pacific salmon (*Oncorhynchus*) but have not been positively identified to species; however, their small size but strongly developed sexual secondary sexual characteristics—hooked lower jaws and enlarged teeth—suggest kokanee (*Oncorhynchus nerka*) (Carlson and Klein 1996).

Freshwater fish that were present in British Columbia during the last interglacial period probably survived our last glaciation—the Fraser glaciation—in one, or more, of three major ice-free refugia adjacent to the province (Fig. 2). The unglaciated region south of the ice and west of the Continental Divide is sometimes called the Columbia Refugium. Certainly, the unglaciated portions of the Columbia River system were major contributors to B.C.'s freshwater fish fauna; however, coastal areas south of the ice sheet, and south of the Columbia River, were important for euryhaline (saltwater-tolerant) fishes. Thus, I refer to the combined Columbia and southern coastal unglaciated areas as the Pacific Refugium. Similarly, I refer to the unglaciated regions south of the Laurentide Ice Sheet and east of the Rocky Mountains as the Great Plains Refugium. Again, however, this is a combination of two semi-independent refugia: the Missouri Refugium and the Mississippi Refugium. For some species, we can determine whether they entered British Columbia from either the Missouri or the Mississippi refugia, but most of our species of Great Plains origin probably came from both refugia. The third unglaciated area adjacent to British Columbia is the Bering Refugium (the ice-free regions of Alaska and the Yukon). A small number of species found in the northwestern part of our province dispersed into British Columbia from this refugium.

Forty-five of B.C.'s 65 native species—66 if the yellow perch in Swan Lake is native—now occur in only one of these refugia. Thus, they are presumed to have survived the last glaciation in that refugium, and their presence in British Columbia is taken as evidence of postglacial dispersal into the province from that refugium. For British Columbia, the Pacific Refugium was the most important ice-free region. Twenty species (30.7% of our native fishes) colonized British Columbia from the Pacific Refugium (Table 1). The next most important refugial region was the Great Plains. Sixteen species (24.6% of the total) entered British

Figure 2 Approximate extent of Continental glaciation in North America about 18,000 BP (after Dyke 2004). The shaded areas represent the three major ice-free refugia from which freshwater fishes recolonized British Columbia.

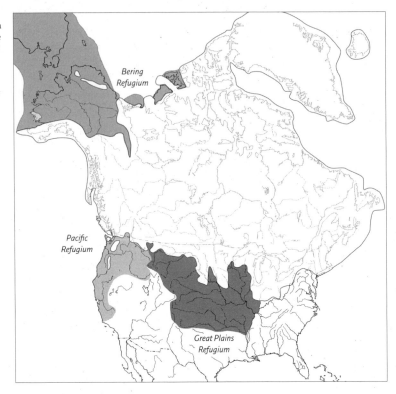

Columbia from this refugium (Table 1). In contrast, the contribution of the Bering Refugium to our fish fauna was relatively small (five species, 7.8% of the total; Table 1).

Each of these major unglaciated regions contained smaller isolated regions that acted as minor refugia. For example, in the Pacific Refugium at the height of the last glaciation, ice from mountains on both sides of the Strait of Georgia coalesced into a lobe that flowed down Puget Sound and stopped just south of Olympia, Washington. South of this ice lobe, but north of the Columbia River system and west of the Cascade Mountains, a small area centred on the Chehalis River acted as a minor refugium (the Chehalis Refugium). This area contributed two unique elements (the Salish sucker and Nooksack dace) to the fish fauna of southwestern British Columbia. There is also evidence of late Pleistocene ice-free pockets along the central coast of British Columbia (Hetherington et al. 2003; Huesser 1971; Warner et al. 1982). For example, both the Brooks Peninsula on the west coast of Vancouver Island and the northeastern tip of Graham Island (Queen Charlotte Archipelago), escaped the last glaciation. These islands now share a number of endemic plant taxa (Ogilvie 1989), but it is not clear if they supported fish during the last glaciation. It is known that northern Vancouver Island and the northeastern tip of Graham Island share a distinctive mitochondrial lineage of threespine stickleback that occurs nowhere else in British Columbia (Thompson et al. 1998). The presence of this lineage on northern Vancouver Island and Graham Island could be interpreted as evidence that sticklebacks survived in the same refugial areas as the endemic plant taxa. Alternatively, since this same stickleback lineage occurs in Alaska and Japan (Orti et al. 1994) and since *Gasterosteus* is a saltwater-tolerant species, this distribution pattern could be a result of postglacial immigration from either Alaska or Asia.

Like the Pacific Refugium, the Great Plains Refugium was not a single refugium. In its broadest sense, the Great Plains Refugium includes the entire Great Plains south of the Laurentide Ice Sheet. This vast area contained at least two semi-isolated refugia: the Missouri and the Mississippi refugia (Cross et al. 1986; Metcalf 1966). Parts of the upper Missouri system and the western Great Plains escaped late Wisconsinan glaciation (White et al. 1979, 1985), and it has been suggested that there was a late Pleistocene ice-free corridor between the Laurentide and Cordilleran ice sheets (Reeves 1973). The presence of such an ice-free corridor throughout the Wisconsinan glaciation is contentious, and recent glaciation maps (see Dyke 2004) indicate that the corridor connecting the Bering and Great Plains refugia did not open until about 12,500 BP. It is clear, however, that the ice retreated from the western plains much earlier than from both the eastern plains and most of British Columbia (Dyke 2004). Consequently, fish had an opportunity to disperse northward from the upper Missouri system well before fish from the upper Mississippi had access to the western plains.

Although 64% of our native species can be assigned to single ice-free regions, 24 species (36% of our fish fauna) now occur in more than one unglaciated area. Consequently, these species (Table 1) could have colonized British Columbia from multiple refugia. Presumably, the advancing ice fragmented the interglacial distributions of these species, and the remnant populations were pushed into two, or more, refugia. A consequence of this range fragmentation was the cessation of gene flow between populations isolated in different refugia. Over time, this geographic isolation, and different selection pressures in the different refugia, led to genetic divergence. Later, as the ice retreated, these divergent forms dispersed back into British Columbia from their separate refugia. As a result, several of our widely distributed species (e.g., longnose dace, *Rhinichthys cataractae*; longnose sucker; and slimy sculpin) colonized the province from more than one refugium. Usually, the forms derived from different refugia are only slightly differentiated and have no taxonomic status; however, sometimes they differ in life history or habitat use and, thus, are of some management concern. Indeed, much of the within-species diversity found in B.C. freshwater fish is a product of this process of range fragmentation and genetic divergence followed by recolonization from different refugia. Within the province, the exact boundaries and zones of contact between such forms are not well known, but the advent of sensitive techniques for detecting genetic divergence now make it possible to reconstruct cases of postglacial dispersal from multiple refugia (e.g., McPhail and Taylor 1999; Seeb et al. 1987; Stamford and Taylor 2004).

TABLE 1 The native freshwater fishes of British Columbia, the glacial refugia used by B.C. populations, and the Aquatic Ecoregions in which the B.C. populations occur. Abbreviations for the Aquatic Ecoregions are NPC=North Pacific Coastal, GC=Glaciated Columbia, INT=Interior, MAC=Mackenzie, YUK=Yukon.

Common name	Scientific name	Glacial refugia	Ecoregion
River lamprey	Lampetra ayresii	Pacific	NPC
Vancouver lamprey	Lampetra macrostoma	Postglacial (in situ origin)	NPC
Western brook lamprey	Lampetra richardsoni[1]	Pacific	NPC
Pacific lamprey	Lampetra tridentata	Pacific	NPC
Green sturgeon	Acipenser medirostris	Pacific	NPC
White sturgeon	Acipenser transmontanus	Pacific	NPC
Goldeye	Hiodon alosoides	Great Plains	MAC
Chiselmouth	Acrocheilus alutaceus	Pacific	GC, INT
Lake chub	Couesius plumbeus	Pacific, Great Plains, Nahanni(?)	GC, INT, MAC, YUK
Brassy minnow	Hybognathus hankinsoni[2]	Great Plains	NPC, INT
Northern pearl dace	Margariscus margarita[2]	Great Plains	MAC
Peamouth	Mylocheilus caurinus	Pacific, Chehalis	NPC, GC, MAC
Emerald shiner	Notropis atherinoides	Great Plains	MAC
Spottail shiner	Notropis hudsonius	Great Plains	MAC
Northern redbelly dace	Phoxinus eos[2]	Great Plains	MAC
Finescale dace	Phoxinus neogaeus[2]	Great Plains	MAC
Northern pikeminnow	Ptychocheilus oregonensis	Pacific	NPC, GC, INT, MAC
Longnose dace	Rhinichthys cataractae	Pacific, Chehalis, Great Plains	NPC, GC, INT, MAC,
Leopard dace	Rhinichthys falcatus	Pacific	NPC, GC, INT
Speckled dace	Rhinichthys osculus	Pacific	GC
Umatilla dace	Rhinichthys umatilla	Pacific	GC
Redside shiner	Richardsonius balteatus	Pacific	NPC, GC, INT, MAC
Longnose sucker	Catostomus catostomus	Bering, Great Plains, Pacific, Chehalis	NPC, GC, INT, MAC, YUK
Bridgelip sucker	Catostomus columbianus	Pacific	NPC, GC, INT
White sucker	Catostomus commersonii	Great Plains	INT, MAC
Largescale sucker	Catostomus macrocheilus	Pacific, Chehalis	NPC, GC, INT, MAC
Mountain sucker	Catostomus platyrhynchus	Pacific	NPC, GC, INT
Northern pike	Esox lucius	Bering, Great Plains	MAC, YUK
Longfin smelt	Spirinchus thaleichthys	Pacific	NPC
Eulachon	Thaleichthys pacificus	Pacific	NPC
Cutthroat trout	Oncorhynchus clarkii	Pacific	NPC, GC, INT
Pink salmon	Oncorhynchus gorbuscha	Pacific, Bering	NPC, INT
Chum salmon	Oncorhynchus keta	Pacific, Bering	NPC, INT, MAC, YUK
Coho salmon	Oncorhynchus kisutch	Pacific, Bering	NPC, INT
Rainbow trout	Oncorhynchus mykiss	Pacific, Bering	NPC, GC, INT, MAC
Sockeye salmon	Oncorhynchus nerka	Pacific, Bering	NPC, GC, INT, MAC, YUK
Chinook salmon	Oncorhynchus tshawytscha	Pacific, Bering	NPC, INT, MAC
Bull trout	Salvelinus confluentus	Pacific	NPC, GC, INT, MAC, YUK
Dolly Varden	Salvelinus malma	Chehalis, Bering	NPC, INT, MAC, YUK
Lake trout	Salvelinus namaycush[4]	Bering, Great Plains	INT, MAC, YUK
Cisco	Coregonus artedi	Great Plains	MAC
Arctic cisco	Coregonus autumnalis[3]	Bering	MAC
Lake whitefish	Coregonus clupeaformis	Bering, Nahanni, Great Plains	INT, MAC, YUK
Broad whitefish	Coregonus nasus	Bering	YUK
Least cisco	Coregonus sardinella[3]	Bering	YUK
Pygmy whitefish	Prosopium coulterii	Pacific, Great Plains, Bering	GC, INT, MAC, YUK
Round whitefish	Prosopium cylindraceum	Bering	MAC, YUK
Mountain whitefish	Prosopium williamsoni	Pacific, Chehalis	NPC, GC, INT, MAC
Inconnu	Stenodus leucichthys[3]	Bering	MAC, YUK
Arctic grayling	Thymallus arcticus	Bering, Nahanni	MAC, YUK

TABLE 1 *(continued)*

Common name	Scientific name	Glacial refugia	Ecoregion
Trout-perch	*Percopsis omiscomaycus*	Great Plains	MAC
Burbot	*Lota lota*	Pacific, Bering, Great Plains	GC, INT, MAC, YUK
Brook stickleback	*Culaea inconstans*	Great Plains	MAC
Threespine stickleback	*Gasterosteus aculeatus*	Pacific, Chehalis, Bering	NPC, YUK
Ninespine stickleback	*Pungitius pungitius*	Great Plains	MAC
Coastrange sculpin	*Cottus aleuticus*	Pacific, Chehalis(?), Bering(?)	NPC, INT, YUK
Prickly sculpin	*Cottus asper*	Pacific, Chehalis	NPC, GC, INT, MAC
Slimy sculpin	*Cottus cognatus*	Pacific(?), Bering, Great Plains	NPC, GC, INT, MAC, YUK
Shorthead sculpin	*Cottus confusus*	Pacific	GC
Columbia sculpin	*Cottus hubbsi*	Pacific	GC
Torrent sculpin	*Cottus rhotheus*	Pacific	GC, INT
Spoonhead sculpin	*Cottus ricei*	Great Plains	MAC
Rocky Mountain sculpin	*Cottus* sp.	Great Plains	GC
Yellow perch	*Perca flavescens*[5]	Great Plains	MAC
Walleye	*Sander vitreus*	Great Plains	MAC

1 Probably multiple in situ postglacial origins.
2 Mississippi species that reached British Columbia by way of the Clearwater spillway.
3 These fishes also may have survived the last glaciation in ice-free areas associated with the lower Mackenzie region and the Arctic coast of Siberia.
4 Most native lake trout in British Columbia appear to have originated from the Nahanni Refugium.
5 Swan Lake (near Tupper) population may be native. Other British Columbia populations are introduced.

DEGLACIATION

Ice retreat in British Columbia was not simply the ice advance played backwards (Clague 1981). Although ice retreat generally began first in the south and later in the north, local topography and climate complicated this pattern. Thus, ice retreat from coastal areas was rapid, and in the interior, ice stagnation in valleys and lateral withdrawals back into highland regions and mountain ranges were common (Clague 1981; Tipper 1971). At any latitude, however, events were not necessarily synchronous, and minor readvances of ice tongues down valleys periodically dammed rivers and produced glacial lakes that spilled over into other river systems. These ephemeral drainage connections provided important dispersal routes for fish recolonizing the province. Also, huge blocks of ice were sometimes separated from the main ice sheets and slowly decayed where they stood (Tipper 1971). Another consequence of deglaciation was isostatic rebound. At the height of glaciation, the weight of the ice depressed the underlying land, and the land rebounded when this weight was removed during deglaciation. On the coast, this rebound lagged behind rising sea levels (caused by melting ice), and this resulted in marine submergences. These submergences were most pronounced in the Strait of Georgia region but occurred to a lesser extent along the entire coast (Clague et al. 1982). As the land rebounded, some coastal fjords were cut off from the sea and became lakes. Thus, the bottom waters of some of these fjord lakes (e.g., Sakinaw, Powell, and Nitinat lakes) still retain stagnant seawater (Northcote 1964; Northcote et al. 1964b; Williams et al. 1961).

This period of marine submergence and isostatic rebound added yet another dimension to the B.C. freshwater fish fauna. On the coast, rising land isolated populations of anadromous species (e.g., Pacific lamprey, *Lampetra tridentata*; threespine stickleback; and longfin smelt, *Spirinchus thaleichthys*) in fresh water. Cut off from gene exchange with other populations and landlocked in new environments, they rapidly diverged. In some cases (e.g., the Vancouver lamprey, *Lampetra macrostoma*, and the limnetic and benthic pairs of threespine stickleback), this process produced "biological species" that are unique to British Columbia (Beamish 1982; McPhail 1994). Likewise, in the interior of the province, differential rebound altered drainage systems and created impassable barriers on many rivers. Typically, fish isolated upstream of these barriers are cut off from gene flow with populations below the barriers, and once again, isolation, founder effects, and differential selection produced genetic divergence (Northcote and Hartman 1988).

POSTGLACIAL DISPERSAL
The Pacific Refugium—Species that survived the last glaciation in the Pacific Refugium recolonized British Columbia via two major dispersal routes: one through the sea and the other through temporary drainage connections. Euryhaline fish (e.g., lampreys, sturgeons, smelts, trouts, salmon, and sticklebacks) could use the marine route, whereas exclusively freshwater fish (e.g., minnows and suckers) were unable to disperse through the sea and needed drainage connections to disperse northward. For this inland route, ecological conditions at the time of a drainage connection were critical, and not all species used every connection (McPhail and Lindsey 1986). Presumably, conditions early in deglaciation were harsh, and the drainage connections were cold, turbid, and turbulent. Fish that could tolerate these conditions (whitefish, trout, char, and perhaps some suckers and minnows) were able to leapfrog from drainage system to drainage system and disperse far to the north. Thus, species of unquestioned Columbia origin (peamouth, *Mylocheilus caurinus*; northern pikeminnow, *Ptychocheilus oregonensis*; and redside shiner) reached the Skeena, Nass, and upper Peace river systems. In contrast, other Columbia species (e.g., leopard dace, *Rhinichthys falcatus*, and bridgelip sucker, *Catostomus columbianus*) apparently required more benign conditions and only colonized the Fraser system after the northern dispersal routes to the Skeena, Nass, and Peace systems were closed. The northward dispersal of other Columbia species (e.g., Umatilla dace, *Rhinichthys umatilla*; speckled dace, *Rhinichthys osculus*; shorthead sculpin, *Cottus confusus*; and Columbia sculpin, *Cottus hubbsi*) was even later and occurred after the connections between the Columbia and Fraser drainages were severed. Consequently, these species never made it to the Fraser River system.

Northward dispersal from the Pacific Refugium started about 13 ka. On the coast, the rapid retreat of ice from lowland areas allowed euryhaline species to quickly spread northward. Thus, lampreys, trout, salmon, stickleback, and some sculpins rapidly dispersed up the entire coast and are now found on all islands with permanent fresh water.

A number of euryhaline species (e.g., Pacific salmon) survived the last glaciation in both the Pacific and Bering refugia. Presumably, at some point along the B.C. coast, individuals dispersing north from the Pacific Refugium met individuals dispersing south from the Bering Refugium. In a few cases (e.g., Chinook, *Oncorhychus tshawytscha*, and coho salmon, *Oncorhychus kisutch*), the geographic distributions of genotypes originating from different refugia are known and suggest a relatively broad contact zone along the central coast (Wilson et al. 1987; Gharrett et al. 2001).

About the time northward dispersal began along the coast, the Puget Ice Lobe started to retreat northward up Puget Sound. Because the Puget Lowlands slope to the north, large proglacial lakes formed in the wake of the retreating ice (Bretz 1913; Thorson 1980). These lakes, and their associated meltwater channels, provided saltwater-intolerant fish in the Chehalis Refugium with a dispersal route into the lower Fraser valley at a time when the Fraser Canyon was still blocked with ice (Armstrong 1981). Two Chehalis forms, the Salish sucker and the Nooksack dace, reached southwestern British Columbia through this route. Although these Chehalis forms are closely related to widespread Columbia species, they have diverged from their Columbia counterparts. Thus, the Salish sucker (a distinctive form of the longnose sucker) displays diagnostic mitochondrial haplotypes (McPhail and Taylor 1999) that suggest isolation from the Columbia system since the last interglacial period. The Salish sucker has no formal taxonomic status but is an evolutionary significant unit (see section on conservation for a definition and discussion of the ESU concept) and is of conservation concern. The Nooksack dace has a geographic distribution similar to the Salish sucker, but its mitochondrial genome is about 2% divergent from the widespread Columbia–Fraser form of longnose dace (J.D. McPhail and E.B. Taylor, in preparation). This suggests that the Nooksack dace has been isolated from its Columbia counterpart (the longnose dace) since the early Pleistocene and may warrant formal taxonomic status.

Another Columbia species, the peamouth probably dispersed north into British Columbia from both the Chehalis and Columbia systems. The peamouth is a minnow that tolerates moderate salinities and is commonly found in both the Fraser and Columbia estuaries. It is also the only minnow native to Vancouver Island, and under exceptional flood conditions, peamouth are physiologically capable of crossing the Strait of Georgia on the freshwater plume associated with the Fraser River (Clark and McInerney 1974). Appropriately, peamouth populations on eastern Vancouver Island are clustered on the central Nanaimo Lowlands almost directly across from the mouth of the Fraser River; however, there are also peamouth populations on the west coast of Vancouver Island. Not only are these populations isolated by a mountain range from those on the east coast, but also they are scattered over a wider latitudinal range than the peamouth on the east coast of Vancouver Island (see peamouth species account for a distribution map). The simplest explanation for the presence of peamouth on the west coast of Vancouver Island is dispersal north from the Olympic Peninsula. Peamouth occur in the Chehalis system and in rivers on the west side of the Olympic Peninsula as far north as Lake Ozette but are absent from rivers on the eastern and northern shores of the peninsula. Presumably, during deglaciation, meltwater lowered inshore salinities, and peamouth were carried on the prevailing currents north from the mouth of the Ozette River to the west coast of Vancouver Island.

The southern edge of the Cordilleran Ice Sheet ended just south of the U.S. border, and the unglaciated portions of the Columbia system were the main sources of postglacial immigrants into the interior of British Columbia. The details of ice retreat in the Canadian portion of the Columbia system are complex and were strongly influenced by local topography and climate. Thus, the ice rapidly withdrew back up into the mountains, and this often isolated large blocks of ice that stagnated in the valleys. This blocked rivers and created large glacial lakes. Consequently, the chronology of ice retreat in the Interior is uncertain and, in places, contradictory. Nonetheless, most radiocarbon dates indicate that, about 14,000 BP, the ice was retreating northward along the entire southern edge of the Cordilleran Ice Sheet (Carra et al. 1996) and that most of the valleys in the southern interior of the province were ice free by about 11,000 BP (Dyke 2004; Fulton and Archard 1985; Kershaw 1978). There are, however, some anomalous dates. For example, *Oncorhynchus* fossils from near Savona have been dated at approximately 18,000 BP (Carlson and Klein 1996). This puts Pacific salmon in the Kamloops region about 3,000–4,000 years before the Cordilleran Ice Sheet reached its maximum (Dyke 2004). Since the southern margin of the ice sheet was never static, this apparently anomalous date may reflect some local episode of contraction and re-expansion. Recent glacial maps show that, at about 18,000 BP, the southern margin of the Cordilleran Ice Sheet was farther north than it was at the glacial maximum. Additionally, a portion of southeastern British Columbia appears to be ice free at this time (Dyke 2004). Perhaps, at this time, there was a connection between the Columbia River system and a lake in the Kamloops area that later was covered by ice.

At its maximum, the Cordilleran Ice Sheet covered the entire Columbia and Kootenay systems in British Columbia. Apparently, the Elk River valley (upper Kootenay system) was the first area to become ice free about 16,000 BP (Ferguson and Osborn 1981). At this time, the southern ice margin was still active, and at times, local readvances blocked the mainstem Columbia. Huge lakes were ponded upstream whenever the river was blocked, but eventually, these lakes lifted the obstructing ice and released cataclysmic floods. On the Columbia River, at least 40 of these floods occurred between 15,000 and 12,800 BP (Allen et al. 1986). One of these ice-dammed lakes (Lake Missoula) inundated over 8,000 km^2 in the Kootenay and Pend d'Oreille systems in the United States. It is unclear if Missoula Lake ever reached into British Columbia; however, since the Elk River valley was ice free before Glacial Lake Missoula formed, this river probably was tributary to the lake. Thus, for fish associated with Missoula Lake, there was an early dispersal route into the upper Kootenay system and, later, from there into the upper Columbia via Canal Flats. This connection probably explains the puzzling presence of the torrent sculpin (*Cottus rhotheus*) in the upper Columbia and in the Kootenay drainage above Bonnington Falls. Further evidence for an early dispersal route into the upper Kootenay system, and then into the upper Columbia, comes from a recent genetic analysis of bull trout (*Salvelinus confluentus*) in the Arrow Lakes (Latham 2002). Latham (2002) presents evidence that bull trout entered the Arrow Lakes from two sources: an early colonization of the upper Columbia via the Kootenay system followed by downstream dispersal into the upper Arrow Lakes, and a later upstream colonization of the Arrow Lakes via the mainstem Columbia.

As the Cordilleran Ice Sheet retreated northwards, the Columbia mainstem became a dispersal corridor into British Columbia for freshwater fish that survived glaciation in the ice-free portions of the Columbia system. In south-central British Columbia, a complex series of large glacial lakes characterized this ice retreat (Fulton 1969; Mathews 1944). These large lakes (e.g., glacial lakes Kamloops, Thompson, and Shuswap) flowed first into the Columbia system and later became part of the Fraser system. Thus, they provided direct connections between the present Columbia and Fraser drainages. Presumably, it was through these connections that species like the peamouth, longnose dace, northern pikeminnow, and redside shiner were able to disperse far to the north. Other Columbia species (e.g., leopard dace and bridgelip sucker) also used the connections between the Columbia and Fraser but apparently required more benign conditions, and by the time the climate ameliorated, the northern connections between the Fraser and other drainages had closed. Thus, they reached the Fraser system but were unable to disperse farther north. Apparently, Bonnington Falls on the lower Kootenay had formed by this time and their distributions are now limited to the lower Columbia and lower Kootenay systems. Still other Columbia species (e.g., Umatilla dace, shorthead sculpin, Columbia sculpin, and speckled dace) apparently entered British Columbia after the Columbia–Fraser connection closed since they never reached the Fraser system and, in British Columbia, only occur in the lower Columbia.

The early connections between the Fraser and the Columbia may have been functional in both directions. Two, and perhaps more, cold-adapted species (lake chub, *Couesius plumbeus*, and longnose sucker) may have dispersed into the Columbia system from the Fraser system. The southern limits of the northwestern ranges of these species now coincide nicely with the southern margin of the Cordilleran Ice Sheet in the coldwater parts of the Kootenay and Columbia systems.

The distribution of species, and the pattern of genetic variation within species, suggests colonization from the ice-free parts of the Columbia through two routes. An eastern route went from Glacial Lake Missoula or its successors, glacial lakes Clark, Kootenay, and Invermere into the upper Kootenay drainage system and from there, with the reversal of the outflow of Glacial Lake Invermere (Sawicki and Smith 1992), a route into the headwaters of the Columbia River. A more direct route was up the mainstem Columbia. The route through Glacial Lake Missoula (and its successors) opened first and allowed cold-adapted species to colonize British Columbia through the upper Kootenay system. Later, but before the closure of the Columbia–Fraser connection, fish adapted to less harsh conditions moved up the Columbia River. By the time these species reached the lower Kootenay River, there was a barrier at Bonnington Falls. This prevented fish dispersing up the mainstem Columbia from reaching Kootenay Lake and the upper Kootenay River. After the closure of the Columbia–Fraser connection, other species (e.g., the shorthead and Columbia sculpins and the Umatilla dace) dispersed into British Columbia. These latecomers reached the Castlegar–Trail area (including the Slocan system) but did not reach the Fraser River system or the Kootenay River system above Bonnington Falls. It is not clear why they did not disperse farther up the mainstem Columbia River. They are riverine species and, perhaps, the Arrow Lakes formed an ecological barrier.

The Great Plains Refugium—Parts of the upper Missouri River system were ice free during the last glaciation, and by about 13,000 BP, a corridor between the Cordilleran and Laurentide ice sheets extended as far north as the lower Peace River area in British Columbia (Dyke 2004). At this time, a large proglacial lake covered much of what is now the Peace River drainage basin. This lake, referred to as Glacial Lake Peace, existed as a number of stages of different sizes and with different drainage connections (Lemmen et al. 1994; Mathews 1980; Teller and Clayton 1983). The earliest stage of Lake Peace (about 13,000 BP) drained southeast into the Missouri system. Later, about 11,500 BP, the drainage was north into the Mackenzie system. By about 11,000 BP, Lake Peace drained north via Lake Hay (Hay River system) into the Fort Nelson River and, from there, into the Liard River. This complex series of drainage connections allowed species that survived glaciation in the upper Missouri system (e.g., goldeye and flathead chub) to colonize northeastern British Columbia.

During an early stage of Glacial Lake Peace, the Peace River Canyon was inundated (Mathews 1980), but later in deglaciation, the canyon became a barrier to the upstream dispersal of the Great Plains fauna. Consequently, species that reached Glacial Lake Peace before the canyon was exposed could disperse westward across the low Continental Divide into Glacial Lake Prince George. Species that arrived later had no access to Pacific drainages. Apparently, only three Great Plains species (white sucker; lake whitefish, *Coregonus clupeaformis*; and brassy minnow) were able to reach the upper Fraser system before this route closed.

Although most of the Great Plains species that reached northeastern British Columbia probably dispersed north from the unglaciated portion of the upper Missouri system, there was, for a brief time, a dispersal route into British Columbia from the upper Mississippi system by way of Lake Agassiz. The Clearwater spillway carried water from Lake Agassiz to Lake McConnell and then north to the Beaufort Sea. This dispersal route became available about 10,300 BP and only operated for a short time (Dyke 2004); however, several species (e.g., northern redbelly dace, *Phoxinus eos*; finescale dace, *Phoxinus neogaeus*, and northern pearl dace, *Margariscus margarita nachtriebi*) evidently reached northeastern British Columbia through this route (Remple and Smith 1998). Apparently, by the time these species reached the Peace River, the Peace Canyon was a barrier, and none of these species made it into the upper Peace system (Lindsey and McPhail 1986).

The Grand Canyon on the Liard River also provided a barrier to the upstream dispersal of the Great Plains fauna. The only Great Plains species found above this barrier are lake chub, longnose dace, and white sucker. It is possible, however, that the lake chub entered the upper Liard from an ice-free region in the Nahanni system (Foote et al. 1992b; Stamford 2001b; Stamford and Taylor 2004; Wilson and Hebert 1998). From the upper Liard, lake chub have crossed into the upper Yukon, Taku, and Stikine rivers.

There is one other as yet unnamed species, the Rocky Mountain sculpin (*Cottus* sp.) that apparently is a recent immigrant into British Columbia from the east side of the Rocky Mountains. This sculpin is indigenous to the upper South Saskatchewan and upper Missouri drainages. In postglacial time, it managed to cross the Continental Divide into the lower Flathead River (a Pacific drainage). The contention that this crossing of the Continental Divide is relatively recent is based on the morphological and genetic similarity of populations in the Flathead, upper South Saskatchewan, and upper Missouri rivers (J.D. McPhail and E.B. Taylor, in preparation). A likely route into the Flathead system from the upper Missouri system was Summit Lake in Marias Pass. At one time, this lake had two outlets: Bear Creek that flowed west into the Flathead River system, and Summit Creek that flowed east into the upper Missouri River system (Schultz 1941).

The Bering Refugium—Although the climates of central and northern Alaska and the Yukon Territory were bitterly cold during the last glaciation, precipitation was low. Consequently, there was not enough snow to build massive ice sheets, and much of the lower and central Yukon system and the Arctic Slope of Alaska remained free of ice. A distinctive set of species survived glaciation in this refugium but, because the northwestern parts of British Columbia were among the last regions in North America to be deglaciated, this fauna has had relatively little time to disperse southwards. In British Columbia, two Bering species (least cisco, *Coregonus sardinella*, and broad whitefish) are restricted to Yukon tributaries. Nevertheless, there have been some opportunities for freshwater dispersal south from the Bering Refugium. For example, the Nakina River (a Taku tributary) postglacially captured a number of upper Yukon tributaries (Holland 1964). Apparently, these connections allowed round whitefish (*Prosopium cylindraceum*), and pike (*Esox lucius*), to enter the upper Taku system. In the Liard system, round whitefish occurs only above, or immediately below, the Liard Canyon (see the round whitefish distribution map). This suggests round whitefish reached the Liard from upper Yukon tributaries. Also, Arctic grayling (*Thymallus signifer*) probably entered the Taku, upper Stikine, and upper Liard rivers from the same source. In addition, there is some evidence from genetic studies on lake whitefish, Arctic grayling, and lake trout (*Salvelinus namaycush*) for a minor refugium in the upper Nahanni system and subsequent dispersal south into the lower Liard River (Foote et al. 1992b; Stamford 2001b; Stamford and Taylor 2004) and, in the case of lake trout, into most of the species' native B.C. range (Wilson and Hebert 1998).

For euryhaline species, the sea provided convenient dispersal routes to the south and east from the Bering Refugium. Thus, Pacific salmon of Bering origin were able to disperse south along the Pacific coast until they encountered salmon dispersing north from the Pacific Refugium. Also, although never collected in B.C. fresh water, another Bering species (Arctic smelt, *Osmerus dentex*) is apparently present in southeastern Alaska (Morrow 1980). The inconnu occurs in the Mackenzie, Yukon, and several Siberian drainage systems. It probably survived the Wisconsinan glaciation in river systems on both coasts of the Bering Landbridge. In contrast, the Arctic cisco (*Coregonus autumnalis*) does not occur in Bering Sea drainages and, thus, may have survived only in unglaciated areas along the Siberian and North American Arctic coasts. In British Columbia, the Arctic cisco occurs only in the lower Liard River.

Present Distribution British Columbia is the third largest province in Canada. North to south, it extends over 11 degrees of latitude and has a land area of almost 1,000,000 km^2. Summarizing the distribution of plants and animals in this vast area is a major task. For terrestrial organisms, British Columbia uses a system of biogeographic regions based on topography, climate, and plant communities. Unfortunately, such terrestrial biogeographic systems are inadequate for freshwater fishes (Abell et al. 2000; Hughes et al 1987). This is because of the limited dispersal abilities of freshwater fishes. With the exception of euryhaline species, freshwater fishes need drainage connections to disperse, and such connections are rare (except during deglaciation). Consequently, in a recently glaciated region like British Columbia, the most important factor governing what species reached which rivers is the historical sequence of drainage connections.

The most convenient way of summarizing freshwater fish distribution is by river systems; however, there are hundreds of separate drainage systems in British Columbia, and rather than separately listing the fish faunas of each river, I have grouped our rivers into five aquatic ecoregions: the North Pacific Coastal Ecoregion, the Columbia Glaciated Ecoregion, the Interior Ecoregion, the Mackenzie Ecoregion, and the Yukon Ecoregion (Fig. 3).

Figure 3 The Aquatic Ecoregions of British Columbia: A) North Pacific Coastal, B) Yukon, C) Mackenzie, D) Interior, E) Glaciated Columbia

These aquatic ecoregions are relatively large geographic areas that usually encompass the catchment of a major drainage system (e.g., the Mackenzie, Yukon, and Columbia drainage systems). Typically, they contain distinctive assemblages of fish species. Most of the B.C. ecoregions are parts of, and have the same names as, 5 of the 76 North American freshwater ecoregions (Abell et al. 2000).

In British Columbia, the ecoregions are defined by the primary postglacial source of their fauna and the presence of a diagnostic fish community (i.e., usually they contain species not found in the other provincial ecoregions). There are two exceptions: the Interior Ecoregion and the North Pacific Coastal Ecoregion. The Interior Ecoregion is specific to British Columbia and consists of the middle and upper Fraser River system. This ecoregion contains a distinctive mix of Columbia and Great Plains species. In contrast, the North Pacific Coastal Ecoregion is a physiographic rather than a biogeographic unit. It consists of a large number of small, independent rivers that flow directly into the North Pacific Ocean and a few larger rivers that transect the Coast Mountains. It coincides (with a few minor differences) to the Coast Mountains. Holland (1964) divided the Coast Mountains into three natural grouping of individual mountain ranges: the Boundary Ranges, the Kitimat Ranges, and the Pacific Ranges. I have modified the North Pacific Coastal Ecoregion (as defined in Abell et al. 2000) to reflect faunal differences between these physiographic subunits. Thus, I recognize three subregions in the North Pacific Coastal Ecoregion—northern, central, and southern coastal subregions. The northern coastal subregion extends from the

Yukon border down to, but not including, the Nass River; the central coastal subregion includes the Queen Charlotte Archipelago, the Nass River and the rivers between the Nass and Bella Coola rivers; and the southern coastal subregion includes the rivers between the Bella Coola River and the U.S. border, as well as Vancouver Island.

Most fishes in the North Pacific Coastal Ecoregion are euryhaline species. Thus, the region is characterized by a dearth of true freshwater fishes. Nonetheless, the upper reaches of the larger rivers that transect the coastal mountain ranges all contain some true freshwater species that are absent in the lower reaches of these same rivers. Additionally, the freshwater species in the different subregions are derived from different sources. To accommodate these differences, I have divided rivers that transect the Coast Mountains into upper and lower reaches, with the dividing line in the canyons and velocity barriers that occur where the rivers cut through the mountains. The upper reaches of such rivers are included in the ecoregions that provided their freshwater faunas. For example, based on their fish faunas, the upper reaches of the Alsek, Taku, and Iskut–Stikine rivers belong in the Yukon Ecoregion rather than in the North Pacific Coastal Ecoregion. Similarly, the upper reaches of the Nass, Skeena, Dean, Bella Coola, Klinaklini, and Homathko rivers belong in the Interior Ecoregion rather than in the North Pacific Coastal Ecoregion.

The two largest rivers—the Skeena and the Fraser—that breach the Coast Mountains present some problems. The lower reaches of these rivers, especially the Fraser, contain more true freshwater fishes than other coastal rivers. Nonetheless, there are clear faunal differences between the lower and upper sections of these rivers. Consequently, even though they contain a freshwater fauna, I have included the lower Skeena and lower Fraser in the North Pacific Ecoregion.

THE NORTH PACIFIC COASTAL ECOREGION

This ecoregion includes the short drainage systems along the B.C. coast as well as the coastal islands and the lower reaches of larger rivers that transect the Coast Mountains. It covers about a third of the province and, in British Columbia, is defined by the Coast Mountains. It is divided into three subregions—northern, central, and southern subregions.

The Northern Coastal Subregion—This subregion contains the lower reaches of the Alsek, Taku, and Iskut–Stikine rivers. Except for Pacific salmon, this subregion is poorly known but it probably contains the same suite of euryhaline species that occur along the entire coast: anadromous lampreys, smelts, Pacific salmon, trout, Dolly Varden (*Salvelinus malma*), threespine stickleback, coastrange sculpin (*Cottus aleuticus*) and prickly sculpin (*Cottus asper*). In *Fishes of Alaska*, Morrow (1980) lists the Arctic smelt (under the name *Osmerus mordax*, the rainbow smelt) from the lower Alsek and Taku rivers. In addition, the round whitefish occurs in the upper parts of both these rivers, and it may also occur in the lower rivers. These species suggest a Beringian component in the fish fauna of these rivers that is absent from the central and southern coastal subregions.

A minnow, the lake chub, occurs at least as far downstream as Tulsequah in the Taku River, and the slimy sculpin extends as far downstream as the Alaskan border. There are also bull trout in the upper Taku system that may reach the lower river.

The fish fauna of the lower Iskut–Stikine system is not well documented; however, judging from the upstream fauna, this river system forms a transition between rivers with some Bering influences and those that were colonized mainly from the Pacific Refugium. Again, the maps in Morrow (1980) indicate the presence of the Arctic smelt in the estuary, and there are some Beringian components in the upper river (e.g., lake trout and Arctic grayling); however, the round whitefish does not occur in this system. Here, a similar Pacific species, the mountain whitefish (*Prosopium williamsoni*), is present and may reach the lower river.

The Central Coastal Subregion—This subregion contains numerous small coastal streams, the Queen Charlotte Islands, and the lower reaches of four major rivers (the Nass, Skeena, Dean, and Bella Coola rivers). Together these major rivers drain about 100,000 km^2 of British Columbia. Except for Pacific salmon and the eulachon (*Thaleichthys pacificus*), the fish fauna of the lower Nass River is not well known. The upper Nass River, however, contains primary freshwater fishes of Columbia origin (e.g., peamouth, northern pikeminnow, longnose dace, redside shiner, largescale sucker, and longnose sucker). Apparently, at least two of these species (the redside shiner and the longnose sucker) reach the lower river. The rest of the fauna consists of euryhaline species.

The lower Skeena contains the most distinctive fish fauna on the northern and central coasts. Three minnows of Columbia origin (peamouth, northern pikeminnow, and redside shiner) and one sucker (the largescale sucker) extend downstream in the Skeena system to at least Lakelse Lake. The rest of fauna is made up of euryhaline species (lampreys, smelts, salmonines, sticklebacks, and sculpins) and is similar to the fauna on the Queen Charlotte Islands and other central coast rivers.

There are no true freshwater fish on the Queen Charlotte Archipelago (Haida Gwaii); however, it is included in the North Pacific Coastal Ecoregion on the basis of five Pacific Refugium species (Pacific lamprey; western brook lamprey, *Lampetra richardsoni*; coastal cutthroat trout, *Oncorhynchus clarkii clarkii*; rainbow trout, *Oncorhynchus mykiss*; and prickly sculpin). Although most of these fishes probably are recent (postglacial) immigrants, the islands are home to a remarkable diversity of threespine sticklebacks, some of which are shared with northern Vancouver Island.

The Dean and Bella Coola rivers rise on the Interior Plateau and flow through the Coast Mountains. There are slight differences in the fish faunas of the upper reaches of these rivers but no recorded differences in the lower reaches. Although the usual suite of euryhaline species—lampreys, Pacific salmon, trout, sticklebacks, and sculpins—occur in the lower rivers, one species (the mountain whitefish) and, perhaps, another (the bull trout) occur downstream of the canyons that separate the upper and lower rivers. The absence of minnows and suckers distinguish the lower reaches of these rivers from the lower Skeena River.

The Southern Coastal Subregion—This subregion contains the southern coastal rivers, Vancouver Island, and the lower Fraser River. The southern coastal rivers consist of a series of small independent rivers that rise in the Coast Mountains as well as the lower reaches of two larger rivers (the Homathko and Klinaklini rivers) that flow through the Coast Mountains. All of these rivers drain directly into Georgia, Johnstone, or Queen Charlotte straits. Their freshwater fish fauna is made up mainly of euryhaline species; however, there is one true freshwater fish (the peamouth) on the Sechelt Peninsula and Nelson Island. As well, the bull trout occurs in the lower reaches of some of the rivers (e.g., Squamish, Southbridge, lower Homathko, and lower Klinaklini rivers).

Vancouver Island includes Vancouver Island and associated islands in Georgia Strait. Again, the freshwater fishes are almost entirely euryhaline species. Nonetheless, the peamouth also occurs on both the east and west coasts of Vancouver Island. The peamouth is a Columbia endemic, and its presence (along with two euryhaline Pacific species: western brook lamprey and coastal cutthroat trout) place the island in the North Pacific Coastal Ecoregion. There is also an endemic lamprey (the Vancouver lamprey) and endemic pairs of sympatric sticklebacks on Vancouver, Texada, and Lasqueti islands. These endemics are all of postglacial origin and are found nowhere else in the world.

The Fraser Canyon separates the lower Fraser River from the middle and upper Fraser subregions of the Interior Ecoregion. The lower river extends from the river's delta upstream to about Boston Bar. Although the lower Fraser contains the usual suite of euryhaline species (e.g., lampreys, sturgeon, smelts, salmon and trout, sticklebacks, and sculpins), it differs from other coastal areas by the presence of a substantial number of true freshwater fishes. There are six minnows (brassy minnow, peamouth, northern pikeminnow, longnose dace, leopard dace, and an unnamed dace [the Nooksack dace]); four suckers (bridgelip sucker, largescale sucker, mountain sucker, and a genetically distinctive form of the longnose sucker, the Salish sucker). Most of these freshwater species survived glaciation in the unglaciated portion of the Columbia River system; however, the Nooksack dace and Salish sucker survived in the Chehalis Refugium. These two Chehalis forms also occur in two Puget Sound drainages (the Skagit and Nooksack rivers) that rise in British Columbia. Consequently, I have included the B.C. portions of these drainages in the lower Fraser system.

Like Vancouver Island, the lower Fraser system also contains forms that evolved after the last glaciation. The taxonomic status of these forms is uncertain, but they are a distinctive part of the fish fauna of the lower Fraser system. For example, two life-history forms of coastrange sculpin coexist in Cultus Lake: a "normal" form that is common in tributary streams and less common in the lake and a "pygmy" form that is restricted to the lake. This latter form appears to be neotenous and migrates from deep water towards the surface at night (Ricker 1960). Also, there are non-migratory forms of the longfin smelt in both Harrison and Pitt lakes. These smelts also appear to be neotenous: they mature at a small size (relative to their anadromous counterparts) and retain some larval characteristics (e.g., a translucent body) into adult life. Those in Pitt Lake are of special interest because "normal" anadromous smelts migrate into the lake in the fall. Thus, the "normal" and "pygmy" forms seasonally coexist in Pitt Lake.

GLACIATED COLUMBIA ECOREGION

The Columbia is the master river of Cascadia and contains one of the most distinctive fish faunas in North America (Miller 1965). Although only about 15% of the drainage lies within British Columbia, it is still the third-largest river system in the province. Together, the Columbia River and its major B.C. tributaries (the Pend d'Oreille, Kootenay, Kettle, Okanagan, and Similkameen rivers) drain slightly over 100,000 km^2 of southeastern British Columbia. The native fish fauna of this ecoregion is dominated by true freshwater species. Historically, however, anadromous species reached the B.C. portion of this system—Chinook salmon ascended the river as far upstream as Windemere Lake, a remnant population of anadromous sockeye salmon (*Oncorhynchus nerka*) still reaches the lower Okanagan River, and lamprey once reached the Slocan River system (Parkham 1937). In British Columbia, natural barriers (both physical and environmental) divide the region into five subregions: upper Columbia, lower Columbia, upper Kootenay, Okanagan, and Flathead.

The Lower Columbia Subregion—The lower Columbia subregion consists of the Columbia mainstem between the Arrow Lakes and the U.S. border and of the Kootenay River (and tributaries) below Bonnington Falls. Several indigenous species (e.g., Umatilla dace, shorthead sculpin, and Columbia sculpin) occur in this subregion but are absent from the upper Columbia and upper Kootenay subregions.

The Upper Columbia Subregion—This subregion extends from the Arrow Lakes to Columbia Lake. No indigenous species are restricted to this subregion, and it is defined by the absence of species that occur in the lower Columbia subregion. Why these species are absent from the upper Columbia is unclear. There is no obvious barrier between the Arrow Lakes and the lower Columbia subregion; yet, the geographic ranges of both shorthead and Columbia sculpins end just before lower Arrow Lake. Perhaps, the lakes themselves are barriers to these fluvial species.

The Upper Kootenay Subregion—Like the previous subregion, the upper Kootenay subregion is defined by the absence of several species that occur in the lower Columbia subregion. In this case, however, the barrier is obvious. The lower boundary of this subregion is Bonnington Falls.

Apparently, these falls prevented species such as Umatilla dace and shorthead sculpin from dispersing up the Kootenay River.

The Okanagan Subregion—This subregion includes the Kettle, Okanagan, and Similkameen rivers. It is the warmest subregion in British Columbia, and the abundance of warm-water species and absence of bull trout distinguish it from other Columbia subregions. Also, the Kettle River is the only river in Canada that contains speckled dace, and both shorthead and Columbia sculpins occur in the short section of the river between Cascade Falls and the U.S. border. The Similkameen River contains mountain sucker, Umatilla dace, and Columbia sculpin but not shorthead sculpin. All of these species are absent from the Okanagan system.

The Flathead Subregion—The Flathead subregion consists of a small section of the upper Pend d'Oreille drainage system in the extreme southeastern part of British Columbia. It is the only place in the province where an unnamed species of sculpin, the Rocky Mountain sculpin, occurs.

THE INTERIOR ECOREGION

Most of the Interior Ecoregion lies within the Fraser Basin. The Fraser River has the second-largest drainage basin in British Columbia. It drains about 25% (238,000 km^2) of the province and is the largest drainage basin contained almost entirely within British Columbia (a few small streams tributary to the Sumas and Chilliwack rivers rise in Washington State). The primary freshwater fish fauna of the Interior Ecoregion is dominated by species that clearly originated from the Columbia system (e.g., chiselmouth, peamouth, northern pikeminnow, leopard dace, and redside shiner). The downstream boundary of the Interior Ecoregion is in the Fraser Canyon near Boston Bar. This ecoregion also includes the upper Homathko, upper Klinaklini, upper Bella Coola, upper Dean, upper Skeena, and upper Nass rivers. These rivers all transect the Coast Mountains and contain redside shiner and assorted other species of Columbia origin in their upper reaches. The Interior Ecoregion contains six subregions: the middle Fraser, the Upper Fraser, the Four Rivers, the upper Skeena, the Thompson, and upper Nass subregions.

The Middle Fraser Subregion—This subregion extends from about Boston Bar in the Fraser Canyon upstream to the confluence of the Bowron and Fraser rivers. The presence of chiselmouth, lake chub, and white sucker, as well as the absence of euryhaline species like river lamprey (*Lampetra ayresii*), longfin smelt, eulachon, and threespine sticklebacks differentiate this subregion from the lower Fraser subregion in the North Pacific Coastal Ecoregion.

The Upper Fraser Subregion—Although there is no obvious barrier, the upper Fraser subregion is defined by the absence of species that are present in the middle Fraser subregion. For example, the upstream distributions of six species (Pacific lamprey, brassy minnow, leopard dace, bridgelip sucker, white sucker, and prickly sculpin) all end somewhere between Prince George and the confluence of the Fraser and Bowron rivers. It is not clear why this happens; however, relative to the Nechako River at Prince George, summer water temperatures decrease and gradients increase in the upper Fraser subregion. This change in the physical environment probably influences the distribution of some species. For example, the bridgelip sucker is abundant in the middle Fraser subregion, and the primary food source of adults is periphyton. Perhaps, the cold waters of the upper river reduce periphyton production to levels that can no longer sustain bridgelip suckers.

The Four Rivers Subregion—This subregion contains the headwaters of four coastal rivers (the Dean, Bella Coola, Klinaklini, and Homathko rivers). These rivers each drain small parts of the Interior Plateau before they cut through the Coast Mountains. Presumably, at some time during deglaciation, the canyons that divides these rivers into its upper and lower reaches were blocked with ice. At this time, the rivers were tributary to the middle Fraser system. Apparently the connections were brief, and relative to the middle Fraser, these rivers contain only a small freshwater fauna. Bull trout, redside shiner, and longnose sucker are present in all the river; however, mountain whitefish, peamouth, lake chub, longnose dace, and largescale sucker are present in some of these rivers but not in others.

The Upper Skeena Subregion—Postglacial connections between the upper Skeena and the middle Fraser subregions are reflected in the freshwater fish fauna. The upper Skeena contains most of the freshwater species that occur in the middle Fraser system; however, five species (bridgelip sucker, chiselmouth, brassy minnow, leopard dace, and slimy sculpin) that are present in the middle Fraser are absent from the upper Skeena subregion.

The Thompson Subregion—This subregion contains two Columbia species that are found nowhere else in the Fraser system. The South Thompson River contains native populations of the westslope cutthroat trout (*Oncorhynchus clarkii lewisi*), and the North Thompson River contains torrent sculpins.

The Upper Nass Subregion—This subregion appears to have received its freshwater fauna from the Skeena system sometime before many of the species that are now shared between the Fraser and Skeena systems (e.g., lake trout, lake whitefish, lake chub, and white sucker) colonized the Skeena River. These species are absent from the upper Nass River; however, for a north coast river the upper Nass contains a substantial freshwater fauna: peamouth, northern pikeminnow, longnose dace, redside shiner, longnose sucker, and largescale sucker.

THE MACKENZIE ECOREGION

The Mackenzie Ecoregion contains a fish fauna derived primarily from the Great Plains Refugium. The Mackenzie is the second largest river in North America. It drains a huge catchment (about 1,800,000 km^2); although only a small fraction of the drainage basin lies within British Columbia, it is still the largest drainage system in the province (almost 280,000 km^2). This ecoregion is divisible into five subregions: the Upper and Lower Liard subregions, the Upper and Lower Peace subregions, and the Hay subregion.

The Lower Liard Subregion—The fauna of the Lower Liard Subregion is dominated by Great Plains species (e.g., goldeye, flathead chub, northern pearl dace, northern redbelly dace, and finescale dace). In addition to Great Plains species, the lower Liard also contains two migratory Bering species (Arctic cisco and inconnu) that are absent from the upper river. Also, a third species, the chum salmon (*Oncorhynchus keta*), occasionally appears in the B.C. portion of the Liard system. It is not known if chum salmon maintain self-sustaining populations in Liard River in British Columbia.

The Upper Liard Subregion—With the exceptions of the white sucker, lake chub, and longnose dace, there are no clearly Great Plains species in the upper Liard subregion. This subregion, however, contains two widespread Pacific species (rainbow trout and Dolly Varden) that are not present in the lower Liard, as well as a Bering species (round whitefish). The latter species is moderately common in the upper Liard but only occurs for a short distance downstream of the Liard Canyon. Persistent rumours of unsanctioned trout introductions into Liard headwaters (e.g., the Turnagain and Eagle rivers) suggest that rainbow trout may not be indigenous to the system. The Dolly Varden, however, appears to be native.

The Lower Peace Subregion—Like the Lower Liard Subregion, the Lower Peace Subregion is dominated by Great Plains species but lacks the Beringian elements found in the lower Liard. Also, several Great Plains species (e.g., cisco, *Coregonus artedi*; emerald shiner, *Notropis atherinoides*; spottail shiner, *Notropis hudsonius*; and ninespine stickleback, *Pungitius pungitius*) that are present in the Lower Liard Subregion are absent from the lower Peace. Additionally, several Pacific species (e.g., largescale sucker, northern pikeminnow, redside shiner, and prickly sculpin) are present in the Lower Peace Subregion but absent from the lower Liard.

The Upper Peace Subregion—Fishes of Pacific origin (e.g., sockeye salmon, Dolly Varden, peamouth, northern pikeminnow, redside shiner, and prickly sculpin) dominate parts of the upper Peace subregion. Nonetheless, some Great Plains species (e.g., white sucker and brassy minnow) occur in the upper Peace, as well as one Bering species (Arctic grayling).

Hay Subregion—Most of the Hay River system lies in northern Alberta and the Northwest Territories, but about 7500 km^2 of the catchment are in northeastern British Columbia. Until recently, there was no road access to this system, and even now, roads only reach the headwaters of the system. Consequently, the Hay subregion is poorly known. Nonetheless, its geographic location and the presence of the finescale dace, brook stickleback, northern pike, and walleye indicate a Great Plains fauna. The few collections available from this system suggest a depauperate version of the lower Peace fauna.

THE YUKON ECOREGION

This is the smallest of the four freshwater ecoregions in British Columbia. It consists of the Yukon, Alsek, Taku, and Iskut–Stikine rivers. These rivers drain the extreme northwestern corner of the province and contain a mixed fauna. Their upper reaches are characterized by the absence of saltwater-intolerant Pacific species and the presence of species, or forms, derived from the Bering Refugium. The region is divisible into two subregions: the Yukon Lakes Subregion and the upper Taku and Iskut–Stikine Subregion.

The Yukon Lakes Subregion—With a total catchment area of 829,000 km^2, the Yukon River is the third-largest river system in North America; however, only about 3% (25,000 km^2) of the basin lies within British Columbia. The Yukon Lakes Subregion contains two species (broad whitefish and least cisco) found nowhere else in British Columbia.

The Upper Taku and Iskut–Stikine Subregion—This subregion consists of three independent drainage systems: the Alsek, Taku, and Iskut–Stikine rivers. Together these rivers drain about 74,000 km^2 in the northwestern corner of the province. Their lower reaches are in the North Pacific Coastal Ecoregion and contain euryhaline species (e.g., Pacific salmon, coastal cutthroat trout, coastrange sculpin, and prickly sculpin); however, in their upper reaches, there is evidence of inland connections. There are Beringian species (e.g., Arctic grayling and northern pike) in their headwaters, and Columbia (bull trout) and Great Plains (lake chub) species in the headwaters of the Taku and Iskut–Stikine systems. Both the Alsek and Taku systems also contain round whitefish; however, this species is absent from the Stikine drainage.

Conservation of Freshwater Fishes Among vertebrates, freshwater fishes exhibit exceptionally high rates of extinction (Leidy and Moyle 1997; Ricciardi and Rasmussen 1999). Although other factors may be involved, habitat destruction is implicated in the majority of these extinctions. Our province is no longer an unspoiled wilderness—over 40% of the provincial forest is cut; about 80% of the lower Fraser floodplain is gone; many rivers are dammed, diverted, or polluted; and our cities and suburbs continue their untrammelled growth. All of this "progress" is bought at the price of a dwindling native flora and fauna. Consequently, the preservation and conservation of our biotic diversity have become important environmental issues. So far, only two, or perhaps four, native fish species have gone extinct in British Columbia: the benthic–limnetic stickleback species pair on Lasqueti Island and what may have been a lake whitefish species pair in Dragon Lake near Quesnel. Still, we have lost hundreds of local populations, and many other populations are in imminent danger of extirpation (Slaney et al. 1996). Not surprisingly, the threat to local populations is greatest in areas undergoing rapid urbanization (e.g., the lower Fraser valley, southern Vancouver Island, and the southern Okanagan). Development in these and other regions of the province is unlikely to stop. It will wax and wane with the economy; however, in the long term, the human population will continue to grow, and our native flora and fauna will continue to dwindle. Inevitably, the incremental loss of populations will become the loss of species, and one of the major challenges facing the next generation of resource managers will be the preservation of what is left. For fish, this will not be an easy task. Rational conservation strategies are built on a foundation of knowledge, and we are genuinely ignorant about the biology and ecology of most of our native fishes.

At first, this lack of knowledge seems odd. After all, we have been managing aquatic resources in this province for over a century. Management, however, implies a goal, and the original goal of those managing living resources such as forests and fisheries was to maximize yield to industry. This simple strategy continued until well into the second half of the 20th century and, sadly, its legacy—the notion that the primary role of resource managers is to serve industry—still pervades the government agencies (federal and provincial) responsible for the stewardship of our living resources. Under this view of conservation, fishes used for food, recreation, or profit are "valuable," whereas fishes like suckers and sculpins have no "value." Given this utilitarian view, it is not surprising that most of our native fish were of no interest to either management agencies or lobby groups like the B.C. Wildlife Federation that, ostensibly, are strongly committed to conservation.

Consequently, we know little about these "worthless" species. Nonetheless, Canada is signatory to the Convention on Biological Diversity, and this Convention obliges the signatory nations to develop domestic legislation to protect biodiversity. In Canada, this led to federal legislation (the Species at Risk Act) designed to protect and preserve biodiversity in Canada. However, Canada is a confederation of provinces, and constitutional ambiguity over jurisdiction complicates conservation. In spite of some jurisdictional complications, a federal–provincial Accord for the Protection of Species at Risk was signed in 1996. This accord obligates the provinces and territories either to enact provincial species at risk legislation or use existing legislation for this purpose. British Columbia opted to use an existing piece of legislation, the Wildlife Act. The effectiveness of the Wildlife Act as protection for species at risk has been reviewed (see Wood and Flahr 2004 for details) and, so far, B.C.'s performance has been abysmal. This is not surprising in a province that includes facilitating economic development in the mandate of the provincial agency charged with protecting the environment. Sadly, preserving biodiversity is not a high priority in British Columbia and, for those fishes perceived as worthless only the ones listed by COSEWIC are protected under the Species at Risk Act.

WHAT TO CONSERVE?

Most conservation biologists agree that the major goals of conservation are the preservation of biodiversity and the prevention of human-induced extinctions. In this regard, freshwater fish present conservation biologists with special problems. Freshwater fish are poor dispersers. This is because "fish gotta swim;" they can't walk, and they can't fly. Lakes are aquatic islands surrounded by land, and the only way dispersal between lakes occurs is through connecting streams. Dispersal between river systems is even more difficult, since salt water is lethal to most freshwater fishes, and natural connections between drainage basins are rare. Even anadromous species like Pacific salmon home to specific breeding areas, and this restricts gene flow among their populations. Consequently, freshwater fish readily form demes (genetically divergent populations). The depths of these divergences depend on founder effects, population size, the intensity of local selection, the amount of gene exchange among adjacent populations, and time. The divergence can be slow or remarkably rapid and ranges from slight differences amongst adjacent populations to major differences among species.

Thus, in freshwater fish, there is a continuum in the degree of evolutionary divergence that spans the spectrum from populations to species; however, with the dubious exception of subspecies, evolutionary divergences below the species level have no taxonomic status. This does not mean that diversity below the species level is unimportant; however, in setting conservation priorities, taxonomic status matters, and threats to species usually are viewed as more serious than threats to populations. This is because the extinction of species results in the irretrievable loss of entire genomes, whereas the extirpation of populations usually means the loss of only parts (albeit unique parts) of a genome.

Species—Unfortunately, even evolutionary biologists cannot agree on what is meant by a species (Hey et al. 2003). Thus, it is not always clear what is, and what is not, a species; in the context of the B.C. freshwater fish fauna, no single species definition fits all situations. Nonetheless, two species concepts are particularly useful in the context of our freshwater fishes—the evolutionary species concept (Simpson 1944, 1953) and the biological species concept (Mayr 1942, 1963).

The evolutionary species concept emphasizes the temporal nature of what we call species (i.e., species have a history). Through fossils, we can trace the ancestry of some B.C. fishes back to the Miocene.

Thus, we know that the pikeminnows (*Ptychocheilus*) and chiselmouths (*Acrocheilus*) have been separate evolutionary lineages for millions of years. Also, they share derived morphological features that suggest a common ancestry somewhere in the even more distant past. Given what we know about their evolutionary history and the striking differences in the morphology of the living members of these lineages, most biologists accept that not only are they different species but also that they belong in different genera. For most fishes, however, no clear fossil record is available, and indirect methods must be used to make inferences about their evolutionary history. Typically, phylogenetic analyses based on morphological or molecular data are used to infer evolutionary histories. The theoretical foundations and methodologies of phylogenetic analyses are beyond the scope of this book; however, Mayden and Wiley (1992) provide an excellent introduction to the subject.

In British Columbia, most of our native species are living members of lineages that have existed for a million or more years, and thus, their evolutionary histories can be inferred through a phylogenetic analysis. However, there has been a remarkable amount of rapid postglacial divergence in British Columbia and other recently glaciated areas. Most of this evolution involves divergences among populations, but occasionally, we find two morphological and ecological forms of the same taxonomic species in a single lake. Typically, these sympatric pairs are reproductively isolated and maintain themselves as separate ecological and genetic entities. These situations are recent (usually less than 12 ka) and not amenable to phylogenetic analyses. Consequently, the evolutionary species concept is inapplicable in these cases; however, Mayr's biological species concept is applicable.

Originally, the biological species concept was proposed as a general species definition (Mayr 1942, 1963), but it has more to do with the contemporary maintenance of the integrity of species than with their evolutionary history. Nonetheless, it emphasizes an important characteristic of species—reproductive isolation. One of its limitation as a species definition is that it can only be applied where species are in contact (i.e., in sympatric or parapatric situations), but its great strength is that it is directly testable. Under the biological species concept, species are genetic and ecological entities that remain reproductively isolated where they coexist. Despite its limitations, the biological species concept has stimulated much of the modern research on the process of speciation.

The biological species concept's emphasis on reproductive isolation has led some biologists to argue that "good" species do not hybridize. This is not so. Interspecific hybrids are common in nature and especially common in freshwater fishes (Hubbs 1955). Hybridization, however, does not necessarily lead to introgression (fusion of genomes). For example, although Dolly Varden and bull trout have a long history as independent lineages, they hybridize (and backcross) in most places where they come in contact. Nonetheless, they continue to maintain themselves as distinct genetic and ecological entities (Hagen and Taylor 2001; Redenbach and

Taylor 2002). This is not an isolated case. In British Columbia, rainbow and cutthroat trout, as well as many species of suckers, minnows, whitefish, and sculpins regularly hybridize (and backcross) but still persist as distinct entities that we recognize as "good" species.

In the context of conservation priorities for B.C. fishes, the biological species concept is important. It emphasizes the biodiversity significance of pairs of biological species (sympatric, reproductively isolated forms) even if they belong to the same taxonomic species. Typically, such species pairs occur in recently glaciated regions. They are not unique to British Columbia; however, we have some, and scientifically, they are arguably the most important component of our fish fauna. Still, some conservation managers question the importance of these pairs of biological species. They ask why, if these pairs are 'real' species, have they not been given scientific names? The short answer is parallel evolution.

Parallel evolution—Parallel evolution is the independent evolution, from the same ancestral form, of similar phenotypes (morphological, behavioural, and ecological) in similar environments. For conservation managers, parallel evolution presents difficult problems in prioritizing conservation concerns, especially when it involves parallel divergences of sympatric or parapatric forms of the same taxonomic species. In such cases, where there is usually some degree of reproductive isolation and habitat partitioning, parallel evolution becomes a messy extension of the problem of defining species.

For the B.C. fish fauna, the most common examples of parallel evolution involve the repeated colonization of fresh water by anadromous species. As ice from the last glaciation retreated inland from our coast, anadromous fishes were able to colonize new freshwater environments. Consequently, as rivers and streams became available, anadromous fishes ascended these new fresh waters and founded new resident populations. For a number of reasons, this divergence between freshwater residents and their anadromous progenitors was rapid. First, the initial freshwater residents probably were not a random subset of the genetic variation in the founding anadromous populations. Second, relative to recently deglaciated fresh waters, the sea was more productive and provided a greater forage base. An important consequence of this difference in productivity between the sea and fresh water is a shift in body size—inevitably freshwater residents are smaller than their anadromous counterparts, and this change in body size can occur in a single generation. Additionally, fish do not produce carotenoids but can only acquire them from prey, and carotenes are important in the formation of red spawning colours. Thus, adult freshwater residents typically are smaller and less colourful than their anadromous siblings. Since size-assortative mating is common in anadromous species (Foote 1988; Foote and Larkin 1988; Hanson and Smith 1967) and differences in nuptial colours influence mate preferences (Craig and Foote 2001; Foote et al. 2004; McLennan and McPhail 1990), these changes in body size and nuptial colour inevitably lead to some reproductive isolation. In turn, this reduces gene flow between resident and anadromous populations, and differential selection for life in the two environments leads to genetic divergence. In such cases, parallel evolution occurs when the phenotypes (morphological, physiological, or behavioural) of independently derived freshwater populations diverge from anadromous populations in similar (but not necessarily identical) ways.

The classic example of this process is the evolution of kokanee from sockeye salmon. Kokanee are non-migratory sockeye that forage in lakes rather than in the sea. They are clearly derived from anadromous sockeye, but the kokanee life history is not unique to a single population or even a geographically cohesive group of populations. Kokanee are scattered over a wide geographic area and occur in many different river systems. Morphological, behavioural, and genetic studies indicate that they have evolved repeatedly, and independently, from anadromous sockeye (Foote et al. 1992a; Taylor et al. 1996; Wood and Foote 1996). In some cases, kokanee are no longer in contact with their anadromous ancestors, but in other populations, spawning kokanee and sockeye are sympatric. Although there is usually some hybridization associated with these sympatric spawning populations, the two life-history forms persist and maintain themselves as distinct genetic entities (Foote and Larkin 1988; Foote et al. 1989).

This phenomenon of the repeated divergence of freshwater residents from anadromous species transcends phyletic lines, and in British Columbia, parallel evolution is common in lampreys, smelts, trout, Pacific salmon, and sticklebacks (McCusker et al. 2000; McPhail 1994; Taylor et al. 1996). For conservation managers, sympatric pairs of biological species pose a problem. Since reproductive isolation between the pairs, and selection in the two alternative environments, are usually sufficient to maintain them as discrete entities, an argument can be made that they are good biological species (*sensu* Mayr 1963). Taxonomically, however, life-history dichotomies like sockeye–kokanee and anadromous–resident sticklebacks usually are treated as single taxonomic species. There are several reasons for treating such biological species as taxonomically conspecific. One reason is practical—taxonomic chaos would result from recognizing dozens of independently evolved freshwater–resident forms as separate species. For example, in freshwater–anadromous pairs of sticklebacks (*Gasterosteus*), there are usually clear morphological and genetic differences between parapatric freshwater and anadromous forms but not between geographically separated freshwater populations. Consequently, the only sure way of distinguishing among freshwater residents is their geographic location.

Another reason is more theoretical. Because each freshwater-resident population is founded from the same ancestral species (although often from different, slightly divergent, ancestral populations), and the resulting freshwater phenotypes (morphological, behavioural, and physiological) are remarkably similar, the same genes probably are involved in each divergence. Indeed, recent genetic studies have confirmed that the same major genes are involved in the parallel evolution of several morphological traits in sticklebacks (Colosimo et al. 2004, 2005; Shapiro et al. 2004). Not only are the same major genes involved in stickleback parallel evolution but also modifiers that influence the expression of the major genes (Colosimo et al. 2004). Since parallel evolution is a pervasive element not only in postglacial divergences in sticklebacks but also in lampreys, salmonines, and smelts, it seems likely that a similar mechanism is involved in all these groups. Thus, major genes and modifiers probably were involved in the divergence of most freshwater populations from anadromous populations, and the speed of these divergences suggests that the major genes and modifiers were already present in the ancestral (anadromous) founding populations. A possible scenario involves shifts in allele frequencies in response to new selection regimes in fresh water. If shifts in allele frequencies were involved, we would expect genetic distances between such species pairs to be small and fixed genetic differences to be rare. This is the case in sticklebacks (McPhail 1994; Taylor and McPhail 2000). Consequently, many postglacial biological species pairs may be transient. Unlike taxonomic species with long histories of evolution in allopatry, there are usually no fixed genetic differences between members of these species pairs. Thus, recombination in hybrid generations is not as great a barrier to introgression as it is when older species with fixed genetic differences hybridize. Typically, reproductive isolation in recent species pairs is a product of disruptive selection for life in two alternative environments. If these environments are disturbed, reproductive isolation can break down. This makes recent species pairs especially vulnerable to environmental disturbances, and in British Columbia, we appear to be witnessing the collapse of such a pair in Enos Lake on Vancouver Island (Kraak et al. 2001).

Given that most postglacial species pairs don't have scientific names and that there are often multiple independent origins of the pairs, they present a practical problem in setting conservation priorities. For example, along the coast of British Columbia, there are dozens, perhaps hundreds, of parapatric or sympatric pairs of anadromous and freshwater-resident lampreys, salmon, trout, and sticklebacks. These pairs have considerable scientific value in that they provide opportunities to study fundamental evolutionary questions. Should conservation managers attempt to protect all, or some subset, of these pairs? There is no simple answer. Some biological species pairs are not only common in British Columbia but also have evolved elsewhere, even on other continents (e.g., anadromous and stream-resident sticklebacks or sockeye and kokanee). This argues that the conditions necessary for these divergences are widespread and common, and although multiple replicates of evolutionarily important processes (e.g., reproductive isolation and resource partitioning) are a major scientific resource, it is probably sufficient to protect a subset of the examples. In contrast, other cases of parallel evolution are geographically restricted (e.g., the limnetic–benthic stickleback pairs). This argues that the conditions necessary for these divergences are rare and local—the limnetic–benthic species pairs occur only in British Columbia and, here, only in the Strait of Georgia region. Consequently, we in British Columbia are the sole stewards of these biological treasures and have an international obligation to protect them. Thus, in setting conservation priorities for these biological species, the number of examples of a particular pair and their geographic distribution are important.

Subspecies—Subspecies are infraspecific taxonomic units that have been given formal scientific names. Because of difficulties in defining subspecies, modern ichthyologists rarely use this taxon. Nonetheless, from a conservation perspective, subspecific names are useful in calling attention to situations where there is an evolutionary split within a species. Characteristically, the geographic ranges of subspecies are separated except for a relatively narrow zone of contact and, typically, populations in the contact zone are intergrades between the two subspecies. Among B.C. freshwater fishes, the best example of subspecies are the two geographically separated forms of the cutthroat trout—the coastal cutthroat (*Oncorhynchus clarkii clarkii*) and the westslope cutthroat (*O. c. lewisi*). These subspecies differ morphologically, behaviourally, and genetically. They even have different chromosome numbers. Both subspecies are of conservation concern, and I have given them separate write-ups. I have given one other subspecies, the northern pearl dace (*Margariscus margarita nachtriebi*), a full species account. In this case, I use the subspecific name because the American Fisheries Society recognizes this subspecies (Nelson et al. 2004), and some recent authors (e.g., Bailey et al. 2004) have argued that northern and southern pearl dace are different species.

Populations—Most of the freshwater fish diversity in the province occurs at the level of populations, and the protection of populations is an important conservation concern. Pacific salmon stocks are familiar examples of this interpopulation diversity. Embedded within a single species like the sockeye salmon are a complex array of genetically distinctive stocks that are "fine tuned" in a myriad of ways to local environments. For salmonines, our society regards the erosion of this population-level diversity as a serious conservation problem. This concern is reflected in the considerable effort and resources that government fisheries agencies expend to preserve and manage wild salmonine stocks. Although salmonines are the most familiar examples of interpopulation diversity, the conditions and processes that created this diversity also affected most of our native species, and they are every bit as subtly adapted to their local environment as are our Pacific salmon.

Evolutionarily Significant Units—Potentially, with modern statistical and molecular techniques, it is possible to demonstrate that most freshwater fish populations are to some degree different. For managers, this raises the question of just how different do populations have to be to warrant special protection?

In the United States, the listing of the first salmon population under the Threatened and Endangered Species Act released a flood of petitions for the listing of local populations, and some criteria for making conservation decisions about populations became a necessity. Consequently, the concept of an evolutionary significance unit (ESU) was developed. To qualify as an ESU, a population that is in serious decline must show "substantial" reproductive isolation from other conspecific populations and form a "significant" part of the evolutionary history of a species (Waples 1995). An ESU can be a single population or a group of populations.

The ESU concept has two major strengths. First, it captures the continuous nature of genetic divergences—from differences among populations to differences among groups of populations. Secondly, the concept provides criteria for determining whether a population (or group of populations) is, or is not, an ESU. The major problem with this approach is that the criteria for recognizing an ESU are somewhat subjective—how much reproductive isolation is "substantial," and what constitutes a "significant" part of the evolutionary history of a species? In practice, these problems are not insurmountable. If a population (or group of populations) is unlikely to be rescued from extirpation by immigration from other populations, then it is substantially isolated. The barrier to gene flow can be physical (natural or man made) or it can be biological (e.g., spawning time, spawning site, or behaviour). For conservation purposes, the nature of the barrier is irrelevant. If a population (or group of populations) displays some unique characteristics, it is probably a significant part of the evolutionary history of the species.

The important feature of an ESU is that something unique would be lost if an ESU is extirpated. The key word is unique. The Oxford Dictionary gives two meanings to this word. In the strict sense, it is an absolute concept (there can be only one instance of occurrence); in a looser sense, it is used for something unusual or rare, but there can be more than one instance of occurrence. For example, the Salish sucker (a distinctive form of the longnose sucker) is an ESU. Its geographic distribution is consistent with the Pleistocene history of Puget Sound and the southern Strait of Georgia, and all Salish sucker populations share molecular markers that are not present in other longnose sucker populations (McPhail and Taylor 1999). Thus, if the Salish sucker went extinct, a unique part of the evolutionary history of the longnose sucker would be lost. In contrast, a summer-run steelhead population does not necessarily qualify as an ESU. Seasonal runs (e.g., summer and winter runs) have evolved many times in the two major clades of *O. mykiss* in British Columbia (McCusker et al. 2000) and are geographically scattered along the Pacific coast. Consequently, this run-timing attribute in itself is not a unique part of the evolutionary history of *O. mykiss*. Thus, although severely depleted summer steelhead runs may be a serious management concern, they do not necessarily qualify as ESUs.

The ESU concept works well with freshwater fishes, but there is a problem with genetically based ESUs in deciding how much genetic differentiation is enough to qualify as an ESU. If we apply a strict definition of unique (only one), relatively few populations or groups of populations would qualify. This problem has led to a number of variant definitions of an ESU, but so far, no modification of the concept has met with universal approval.

Designatable Units—In Canada, COSEWIC has adopted a pragmatic approach to assessing the conservation status of wildlife (Green 2005). Designatable units (DUs) are infraspecific units that are distinguishable from, and have different extinction probabilities than, the species as a whole. DUs are an attempt to cut through the philosophical fog surrounding academic debates about infraspecific subdivisions (subspecies, metapopulations, and populations) and get on with the job of conservation. Basically, a taxonomic species is assessed to see if a uniform status rank captures the probability of extinction for that species. If there are subdivisions within the species such that a single status designation does not adequately portray the probability of extinction within the species, DUs within the species may be given different conservation status. The approach uses all available information (taxonomic, phylogenic, genetic, range disjunctions, and biogeographic data).

Designatible units are a good idea and probably will prove to be a useful conservation tool; however, for fishes with no utilitarian value, they may be a mixed blessing—DUs have the potential to overwhelm the assessment system. For example, in British Columbia, there are hundreds of salmon and trout populations that would probably qualify as designatible units, and there is no question that many of these populations warrant protection. However, considerable resources (human and economic) will be required to assess all these populations. It would be tragic if limited conservation resources were diverted from "worthless" species to assess populations of species that are under the stewardship of management agencies that already have multimillion dollar budgets.

SOME CRITERIA FOR LISTING SPECIES AND POPULATIONS

Under the Species at Risk Act, any interested party can request that a plant or animal be federally listed for protection; however, it is COSEWIC that makes the final recommendations for listing, and the federal minister that makes the final decision. In addition to federal protection, most provinces also provide protection for species at risk within their boundaries. Since species may be at risk in one province but not in other provinces, there often are differences between federal and provincial lists. Regardless of the jurisdiction, however, the criteria for listing species (or populations) are similar and usually involve an assessment of population status (dangerously low or steadily declining), the extent and nature of the species' geographic distribution, and its scientific value (i.e., does it have unique evolutionary, ecological, or life-history characteristics).

Declining populations—Evidence of a steady downward trend in numbers usually indicates a species or population is in difficulty. In British Columbia, the bull trout and Arctic grayling are good examples. At one time, the bull trout ranged as far south as northern California; however, the California populations are now gone, and the southern Oregon, southern Idaho, and northern Nevada populations are in decline. As a result of these extirpations and declines in the United States, the bull trout was listed as an endangered species in 1997. The U.S. listing prompted an evaluation of bull trout populations in British Columbia, and although the species is in no imminent danger of extinction, some southern populations were found to be in decline. Similarly, grayling at the southern margin of their B.C. distribution (the upper Peace watershed) have declined precipitously in the last few decades. Consequently, these species are ranked by the BCCDC as S3 (may have lost peripheral populations) for bull trout and S1 (critically imperiled) for grayling in the Peace/Williston watershed, and special regulations are provided for their protection.

Likewise, some populations of Chinook, coho, and sockeye salmon are known to be in imminent danger of extinction, and some of these are now candidates for federal listing. Also, there is general concern about the status of wild salmonines throughout the province, and no-kill regulations now apply to most fluvial populations of trout and char. How many non-salmonid species or populations are also in decline or in danger of extirpation is unknown.

Restricted geographic ranges—The argument for protecting fishes with restricted ranges is based on the assumption that species with restricted distributions are more vulnerable to extinction than widely distributed species. Empirical evidence supports this argument: the majority of fish extinctions in North America have involved species with restricted geographic ranges (Williams et al. 1989). Although a restricted distribution is a valid argument for special protection, many restricted distributions are artifacts of man-made boundaries. For example, the only Canadian populations of the shorthead sculpin are found in the Columbia draingage system in south-central British Columbia; however, this species also occurs immediately to the south in Washington State, and there are no obvious barriers between the Canadian and American populations. The shorthead sculpin's restricted Canadian distribution is an artifact of a political boundary that means nothing to the fish. This is not to say that the Canadian populations should not be protected, but as long as suitable habitat is maintained in Canada, immigration from the U.S. populations will probably prevent extirpation. In such cases, the pressing problem is ignorance about the species' habitat requirements—we can't protect critical habitat if little, or nothing, is known about the species' habitat requirements. In fish, this is not a minor problem. Most fish grow throughout life and change their habitat requirements as they grow. Ignorance about the habitat requirements of indigenous fish species at different stages in their life cycle is the single biggest impediment to designing rational management or recovery plans.

In British Columbia, there are also situations where a population (or group of populations) of a widely distributed species is isolated from the species' main geographic range. In this case, the restricted distribution is important. An example is the Canadian distribution of the speckled dace. In Canada, this species is found in a single river system (the Kettle River); however, the speckled dace is common throughout the lower and middle Columbia River system in the United States, and its total geographic range extends south into northern Mexico. At first glance, this species' restricted Canadian distribution appears to be another artifact of a political boundary, but this is not so. There is an impassable barrier (Cascade Falls) on the Kettle River about 5 km upstream from the U.S. border. Consequently, most of the Canadian population of speckled dace is completely closed to immigration from elsewhere in the species' range. If, for some reason, the population above Cascade Falls were extirpated (an unlikely event since the species is widely distributed above the falls), there is no chance that the speckled dace would recolonize the upper Kettle system without human assistance.

Many of the differences between federal and provincial conservation listings also result from political boundaries. For example, the cisco reaches its western distributional limit in the extreme northeastern corner of our province. Here, it is restricted to a single lake; however, the species is widely distributed across central Canada. For this reason, the cisco is not listed by COSEWIC; however, since there is no contact between the B.C. population and populations to the east, the species is vulnerable in British Columbia and is ranked provincially as critically imperiled (S1). Thus, decisions about the protection of species with restricted distributions hinge on the likelihood that immigration from outside the local jurisdiction will rescue the population from extirpation.

Habitat protection—Habitat protection probably is the single most important component in conservation strategies directed at freshwater fishes. Although the protection of entire ecosystems (multiple habitats and the biotic and physical processes that maintain them) is a central theme in terrestrial conservation, habitat protection in freshwater fish tends to target individual species or groups of related species (e.g., salmonines) and ignores other species using the same waters. Additionally, protective measures typically are taken only after a problem becomes apparent and often are aimed only at a perceived "critical" habitat component (e.g., spawning habitat). This approach ignores the fact that most fish species grow throughout their lives and that their habitat requirements change as they grow. Moreover, it is becoming increasingly clear that the biotic environment strongly influences habitat use, especially for early life-history stages. Given our ignorance of the habitat requirements of most of our indigenous fishes and the huge research effort required to gain this information, a reasonable approach is to attempt to preserve the integrity of entire aquatic ecosystems. This would involve setting aside drainages within the different aquatic ecoregions and subregions in the province and protecting the ecological integrity of these drainages.

Scientific value—Another reason for protecting species, or populations, is that they possess attributes that are of unusual scientific value. Obvious B.C. examples are the pairs of benthic and limnetic sticklebacks. These pairs occur nowhere else in the world. In addition, they are a textbook example of rapid speciation and the evolution of reproductive isolation and resource partioning. As a scientific resource, they are as precious as the Galapagos finches. Similarly, populations of large piscivorous rainbow trout are rare but have evolved independently several times in British Columbia (Keeley et al. 2005). They are a remarkable example of the parallel evolution of a suite of morphological and behavioural traits that adapt them to an unusual (for this species) ecology. Although not confined to British Columbia, the Umatilla dace and the all-female gynogenic lineages of *Phoxinus neogaeus* – *Phoxinus eos* are valuable examples of the evolutionary significance of hybridization in vertebrates (Goddard et al. 1998; Haas 2001) and, thus, deserve protection within our province.

PICTORIAL FAMILY KEY

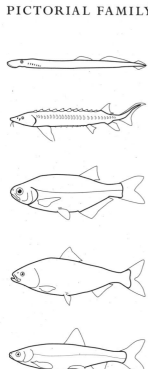

LAMPREYS
PETROMYZONTIDAE

STURGEONS
ACIPENSERIDAE

GOLDEYE
HIODONTIDAE

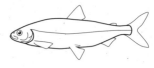

WHITEFISH
SALMONIDAE;
Subfamily Coregoninae

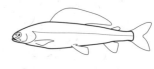

GRAYLINGS
SALMONIDAE;
Subfamily Thymallinae

TROUT-PERCHES
PERCOPSIDAE

HERRINGS, SHAD
CLUPEIDAE

CODS
GADIDAE

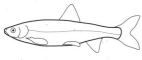

MINNOWS
CYPRINIDAE

STICKLEBACKS
GASTEROSTEIDAE

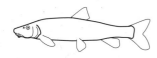

SUCKERS
CATOSTOMIDAE

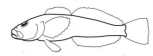

SCULPINS
COTTIDAE

BULLHEADS
ICTALURIDAE

SUNFISH, BASS
CENTRARCHIDAE

PIKES
ESOCIDAE

PERCHES
PERCIDAE

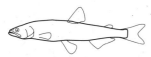

SMELTS
OSMERIDAE

FLOUNDERS
PLEURONECTIDAE

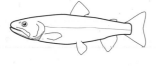

SALMON, TROUT,
CHARS
SALMONIDAE;
Subfamily Salmoninae

FAMILY PETROMYZONTIDAE — LAMPREYS

The lampreys are an ancient group of jawless fishes whose ancestry stretches directly back to the earliest known vertebrates. There are about 40 living species: 34 in the Northern Hemisphere and 4 in the Southern Hemisphere. All lampreys breed in fresh water, but some are anadromous (they migrate to sea and return to fresh water to spawn). Anatomically, lampreys differ from most other fishes in that they lack bone. The skull and the unsegmented rod that functions as a backbone are made entirely of cartilage. They also have a single nostril on the top of the head and seven small, circular gill pores on each side of the head. Lampreys lack normal vertebrate jaws and teeth. In adults, the mouth is a circular sucking disk containing a complex array of "teeth" made of keratin.

The life history of lampreys is also unusual. They have two morphologically and ecologically distinct life histories: a burrowing larval phase (the ammocoetes, called ditch eels in the Fraser valley) and a free-living adult phase. The worm-like ammocoetes lack eyes and a proper mouth. They live in deposition areas where they burrow into the soft sediments and feed by filtering small organisms and organic debris through a sieve-like structure. Water and food particles (usually diatoms and detritus) are pumped through the sieve by muscular contractions and the elastic recoil of a cartilaginous branchial apparatus. The food is trapped in a mucus sheet and slowly drawn back into the intestine. The length of time spent in this ammocoete phase varies among species, but eventually, all ammocoetes transform into small lampreys. This transformation is energetically expensive, and newly transformed lampreys usually are smaller and weigh less than the largest untransformed ammocoetes. During transformation, lampreys develop eyes and the typical circular sucking mouth.

Adult lampreys display two strikingly different ecologies: some species are parasitic and feed on other fish, whereas other species never feed as adults. In these non-parasitic species, the adults are simply an ammocoete's way of making more ammocoetes—they spawn shortly after transformation and then die. In contrast, the adults of parasitic species can live and feed for several years. They feed by attaching to their prey with the sucker-like mouth. This sucker is muscular and contains an array of "teeth" and a cartilage ring. When the muscles spread the sucker and it is placed on the prey, the elastic recoil of the cartilage sets the teeth and holds it in much the same way as a rubber sucker holds to a surface. The lamprey then rasps through the prey's skin with "teeth" located on the tongue and feeds on the tissue and blood that issues from the wound. A glandular secretion prevents clotting and also contains enzymes that partially digest tissue. Although large lampreys can kill prey, the attacks are not necessarily fatal, and up to 67% of the Adams River sockeye salmon (*Oncorhynchus nerka*) run show evidence (scars or wounds) of having survived attacks by lampreys (Williams and Gilhousen 1968).

Ostensibly, there are four species of lampreys in British Columbia—the river lamprey (*Lampetra ayresii*), the western brook lamprey (*L. richardsoni*), the Pacific lamprey (*L. tridentata*), and the Vancouver lamprey (*L. macrostoma*); however, the taxonomy of some of the species is in disarray. This is because two of the species (river and Pacific lampreys) are anadromous, and like other anadromous fishes, they sometimes give rise to freshwater-resident populations. In British Columbia, anadromous river lampreys appear to have repeatedly founded the non-parasitic freshwater resident populations we call western brook lamprey. At the mitochondrial level, river and western brook lampreys are identical (Docker et al. 1999), and the two "species" are probably analogous to the life-history dichotomies seen in sockeye salmon and threespine stickleback (*Gasterosteus aculeatus*). In both the sockeye–kokanee and anadromous–resident stickleback dichotomies, the two life-history forms differ in their behaviour and morphology and are at least partially reproductively isolated. Nonetheless, these different forms are treated as single species (the reasons are detailed in the parallel evolution section in the introduction). If the river and western brook lamprey represent a similar life-history dichotomy, they probably should not be treated as separate species. Similarly, the status of Vancouver lamprey as a species distinct from the Pacific lamprey is questionable.

KEY TO ADULT LAMPREYS

1(2) Four pairs of lateral tooth plates, the centre pairs with three cusps; supraoral tooth bar with three sharp teeth (Fig. 4A)	PACIFIC LAMPREY, *Lampetra tridentata*, or VANCOUVER LAMPREY, *Lampetra macrostoma*
2(1) Two or three pairs of lateral tooth plates, the centre pair with two or three cusps; supraoral tooth bar with two sharp teeth *(see 3)*	
3(4) Teeth sharp; a large median tooth on tongue; usually seven (rarely six) sharp cusps on infraoral tooth bar (Fig. 4B)	RIVER LAMPREY, *Lampetra ayresii*
4(3) Teeth blunt; no distinct median tooth on tongue; seven (rarely six or eight) blunt cusps on infraoral tooth bar (Fig. 4C)	WESTERN BROOK LAMPREY, *Lampetra richardsoni*

Figure 4 Tooth patterns in adults of the B.C. species of lampreys: (A) Pacific and Vancouver lampreys, (B) river lamprey, and (C) western brook lamprey.

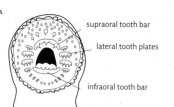
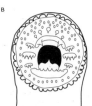

Lampetra ayresii (GÜNTHER)
RIVER LAMPREY

1 cm

Distinguishing Characters This mid-sized lamprey (adults vary in length from 16 to 30 cm) has 2 sharp supraoral teeth, 7 (5–8) sharp infraoral teeth, 3 multicuspid inner lateral teeth (Fig. 4A), and 62–70 myomeres between the last gill opening and the cloacal slit. The river lamprey differs from the Pacific lamprey (*Lampetra tridentata*) in adult size (anadromous Pacific lampreys often exceed 60 cm in length) and dentition: two sharp supraoral teeth in river lampreys vs. three sharp supraoral teeth in Pacific lamprey, and three sets of inner lateral teeth vs. four in Pacific lamprey. River lamprey also are distinguished from the western brook lamprey (*Lampetra richardsoni*) by size (adult brook lamprey rarely exceed 16 cm in length) and dentition (sharp teeth in the river lamprey vs. blunt teeth in the western brook lamprey). The ammocoetes are difficult to identify, but apparently, reduced pigmentation in the head region is useful in separating river lamprey ammocoetes from the other B.C. species (Richards et al. 1982).

Taxonomy Lampreys often occur as pairs of closely related parasitic and non-parasitic species (Vladykov and Kott 1979). In British Columbia, the parasitic river lamprey is anadromous and apparently has given rise, repeatedly, to a non-parasitic freshwater-resident form that we call the western brook lamprey. These two "species" seem to be analogous to the life-history dichotomies seen in other anadromous species (see the introduction for a general discussion of the problem of parallel evolution).

The geographic distribution of the western brook lamprey supports a hypothesis of recent, multiple divergences of non-parasitic brook lampreys from the anadromous river lamprey. Western brook lampreys are supposedly intolerant of saltwater (Beamish and Withler 1986), yet they are widely distributed on islands along the entire B.C. coast. Much of this distribution must have been achieved through postglacial marine dispersal. Additionally, although most river lamprey die if prevented from migrating to the sea, some individuals can survive to maturity in fresh water (Beamish and Youson 1987). Thus, the capacity to complete their life cycle in fresh water is present in the river lamprey. Furthermore, the mitochondrial DNA sequences of these nominal species are identical (Docker et al. 1999). Also, analysis of allozyme data revealed no unique alleles in the two forms, and the average genetic distance (0.017) between them is typical of interpopulation divergences rather than species divergences (Beamish and Withler 1986). Finally, there are populations in Morrison Creek on Vancouver Island (Beamish 1987) and, perhaps, in tributaries to Harrison Lake that contain intermediate individuals.

For now, although the available evidence points to a single genome containing two life-history forms, I have retained river and western brook lamprey as separate species; however, the problem clearly requires further study.

Sexual Dimorphism   Outside the spawning season, the sexes are difficult to distinguish. Spawning males possess a conspicuous genital papilla, lack a pseudoanal fin (a prominent swelling behind the vent), and have a higher dorsal fin than females. In contrast, reproductive females lack the conspicuous genital papilla, have a relatively lower dorsal fin, and possess a pseudoanal fin.

Distribution   NATIVE DISTRIBUTION
The river lamprey is native to western North America. Here, it occurs from the Sacramento – San Joaquin river system in central California north to about Juneau, Alaska.

BRITISH COLUMBIA DISTRIBUTION
River lamprey rarely penetrate far inland, but in northern California, they are known to ascend the Eel River for 250 km (Moyle 2002). In the Fraser River system, ammocoetes identified as this species have been collected as far upstream as Yale, and newly transformed adults are common near the mouth of the river. Most B.C. records are from the sea, but there are scattered freshwater records from rivers north of the Fraser River (Map 1), and a river lamprey was observed feeding in the lower Skeena River (Withler 1955). Anadromous river lamprey probably occur in most large coastal drainages as far north as the Taku River. In addition, freshwater populations (landlocked or non-migratory) are reported from Powell (Carl et al. 1959) and Harrison (Dryfoss 1965) lakes.

Life History   The major B.C. sources for river lamprey life-history information are Beamish (1980) and Beamish and Youson (1987). The following account is based on these sources supplemented with a few personal observations and published data from other areas along the Pacific Coast (e.g., Moyle 2002; Vladykov and Follett 1958).

REPRODUCTION
River lamprey spawn in the spring (April–June) and die shortly (a few hours to weeks) after spawning (Beamish 1980). In the laboratory, they spawned at 12 °C and constructed nests that were approximately 15 cm

Map 1 The B.C. distribution of the river lamprey, *Lampetra ayresii*

• native

in diameter (Beamish 1980). Courtship and spawning behaviour was similar to that of other lampreys (see description of spawning behaviour in the western brook lamprey species account). In June, river lampreys are reported to spawn in small streams associated with the lower Homathko River. Curiously, although the river lamprey is very abundant in the lower Fraser system (Beamish and Youson 1987), where they spawn is a mystery. Perhaps they use the gravel deposition section in the main river (approximately Hope to Chilliwack) at a time when high water and turbidity make them difficult to observe.

Presumably, fecundity varies with female body size, but so far, only two estimates of egg number (11,398 and 37,288) are available (Moyle 2002). The incubation period is unknown but is probably similar to that reported for the western brook lamprey—about 30 days at 10 °C and about 12 days at 18 °C (Meeuwig et al. 2005). Lamprey ammocoetes collected in the Fraser River ranged from 8 to 93 mm (Beamish and Youson 1987). Presumably, the smallest larvae were newly hatched and, like the western brook lamprey, emerged from the gravel 2–3 weeks after hatching. The small ammocoetes are then swept downstream into quiet water where they burrow into the mud.

AGE, GROWTH, AND MATURITY

The length of the ammocoete life-phase is unknown in river lampreys; however, laboratory-reared ammocoetes reached lengths ranging from 27 to 46 mm TL in their first year (Beamish and Youson 1987), and in the

Fraser River, the largest ammocoetes identified as this species were about 120 mm long (Richards et al. 1982). Given a similar growth rate as western brook lamprey ammocoetes, river lampreys probably spend about 5 years as ammocoetes (Meeuwig and Bayer 2005). Also, growth probably varies seasonally: faster in summer and slower in winter. Metamorphosis begins in early June but is delayed over the winter and completed the next spring. In the Fraser system, the earliest recorded out-migrant was taken in April. Thus, the river lamprey takes about 9 months to complete the transformation from ammocoete to young adult and then migrates to the sea (Beamish and Youson 1987). The out-migrants range in size from 80 to 120 mm and are counter coloured—silver below and dark blue above. In the Fraser system, lampreys in the process of transformation appear to aggregate just above the upper limit of saltwater intrusion. Metamorphosis is complete about the time of maximum river discharge (early June). The young adults enter the sea and feed voraciously for about 10 weeks and then return to the river from September to February. Over their parasitic period, they increase in length by about 11–14 cm.

FOOD HABITS

Like other lampreys, river lamprey ammocoetes are filter feeders and use the movement of cilia to pass a current through the oral hood and out the gill pores. Presumably, their diet is similar to the ammocoetes of western brook lamprey: mostly diatoms and desmids during the summer and particulate organic detritus in the fall and winter. The largest ammocoetes probably stop feeding in the late summer or fall and begin their transformation into adults. Adult river lamprey are parasitic and feed on a variety of fishes, including salmon; however, Pacific herring (*Clupea pallasii*) appears to be a preferred prey (Beamish 1980).

Habitat Not much is known about habitat use in either the ammocoete or adult stages of the river lamprey; however, there is some data on the marine phase of their life cycle (Beamish 1980).

ADULTS

Upon entering the sea, the Fraser River population concentrates around the mouth of the river, in nearby Howe Sound, along the west side of the Sechelt Peninsula, and among the Gulf Islands (Beamish 1980). Since these inshore areas are influenced by the Fraser River discharge, river lampreys may prefer regions with reduced salinities. They appear to prefer surface waters and seldom occur at depths below 50 m.

Adult river lampreys are occasionally collected in large lakes (e.g., Harrison, Nimpkish, Kitsumkalum, and Lakelse lakes). Nothing is known about the biology of lacustrine river lampreys and they may be either freshwater-residents or returning sea-run adults. Most lacustrine river lampreys are caught incidentally at night in trawls fished near the surface.

JUVENILES

Juvenile (>60 mm) ammocoetes are rare in the habitats used by young-of-the-year ammocoetes. In the Salmon River, the largest western brook lamprey ammocoetes were associated with the deepest pools (Pletcher 1963) and, presumably, the largest juvenile river lamprey ammocoetes also shift downstream into deeper water. In the mainstem Fraser River downstream of the gravel deposition region, the river deepens, and the substrate gradually shifts to fine sand and silt. If large ammocoetes aggregate in this area, they would be difficult to collect.

YOUNG-OF-THE-YEAR

In the area between Chilliwack and Hope, small ammocoetes that fit the description (Richards et al. 1982) of river lamprey are especially common in mud, silt, and leaf litter associated with sites where small streams enter side channels (e.g., Herrling Channel) off the mainstem Fraser River.

Conservation Comments Although the river lamprey is abundant near the mouth of the Fraser River, it does not appear to be common elsewhere along the B.C. coast. This may be a sampling artifact; however, until we know more about its abundance and its relationship to the western brook lamprey, this should be a species of special concern. Regardless of their taxonomic status, populations with intermediate forms (e.g., the Morrison Creek population) should be protected for their scientific value. The river lamprey is not listed by COSEWIC or the BCCDC, but the Morrison Creek population is listed as threatened by COSEWIC.

Lampetra macrostoma BEAMISH
VANCOUVER LAMPREY

1 cm

Distinguishing Characters   The Vancouver lamprey is a recent (postglacial) derivative of the Pacific lamprey (*Lampetra tridentata*), and mature Vancouver lamprey have the same morphology (most body proportions, myomere counts, and dentition) as the Pacific lamprey (Beamish 1982). The oral disc, however, is longer on average (6.5–11.7% of total length) in the Vancouver lamprey than in the Pacific lamprey (6.4–8.2%). The ammocoetes of Pacific and Vancouver lampreys are indistinguishable (Richards et al. 1982).

Taxonomy   The taxonomic status of the Vancouver lamprey is unclear. It is obviously a recent (postglacial) derivative of the Pacific lamprey, and Beamish (1982, 1985) argues that the Vancouver lamprey is a biological species (*sensu* Mayr 1963) because the two forms are sympatric and reproductively isolated. Although there are no mitochondrial differences between the species (Docker et al. 1999), there are major ecological and physiological differences between Vancouver and Pacific lampreys (Beamish 1982). For example, the Vancouver lamprey completes its entire life cycle in fresh water; in contrast, if retained in fresh water, immature (but transformed) Pacific lampreys die before reaching maturity (Beamish 1982; Clarke and Beamish 1988). These physiological differences, combined with sympatry and some average (but overlapping) differences in body proportions, are the most compelling arguments for the status of Vancouver lamprey as a biological species. Similar arguments have been made for sympatric populations of sockeye salmon and kokanee (*Oncorhynchus nerka*), and marine and freshwater-resident threespine sticklebacks (*Gasterosteus aculeatus*). Yet, no recent authors have argued that these biological species should be given taxonomic names. The reasons for treating such cases as single species are detailed in the introductory section on parallel evolution, but a major component involves the multiple origins of kokanee and freshwater-resident sticklebacks, and the taxonomic chaos that would result from giving each independently evolved pair taxonomic names.

If the Vancouver lamprey is the only postglacial freshwater derivative of the Pacific lamprey, then the taxonomic chaos argument would not hold. Consequently, it is important to establish whether the Vancouver lamprey is, or is not, unique. As of 2005, the answer to this question is unknown. It is known, however, that Pacific lampreys that are landlocked through human activities rarely, if ever, establish permanent, self-sustaining populations in fresh water (Beamish and Northcote 1989). Nonetheless, there are naturally occurring Pacific lamprey populations in British Columbia that may be permanent freshwater residents (e.g., Sakinaw and Ruby lakes on the Sechelt Peninsula, Village Bay Lake on Quadra Island, and West Lake on Nelson Island). It is not clear that these populations are permanent freshwater residents, but there is strong evidence that, like the Vancouver lamprey, they breed at a small body size and feed in fresh water.

Again, although the status of the Vancouver lamprey requires further study, it is retained as a separate taxon until more information becomes available.

Sexual Dimorphism Outside the breeding season the sexes are difficult to distinguish. As spawning approaches, however, the sexes develop different morphologies—males develop a conspicuous, slender urogenital papilla that protrudes up to 6 mm from the body. In females, the urogenital papilla is inconspicuous, but they develop a pseudoanal fin (a prominent swelling behind the vent) that is absent in males. There also are slight sex differences in the height of the second dorsal fin and some body proportions (Beamish 1982).

Distribution The Vancouver lamprey appears to be restricted to two connected lakes (Cowichan and Mesachie) in the Cowichan River system on southern Vancouver Island (Map 2).

Life History The only published information on the life history of the Vancouver lamprey is in the original description (Beamish 1982). Presumably, with the exception of its totally freshwater life cycle, its life history is similar to that of the Pacific lamprey.

Reproduction Apparently, the spawning period is protracted, and some males and females were in spawning condition from May to August (Beamish 1982). An unusual aspect of their spawning behaviour is that they spawn on shallow gravel bars in the lakes rather than in flowing water. However, they may also spawn in flowing water since ammocoetes were collected in inlet streams, albeit usually within 100 m of the lake. Adults die after spawning.

Map 2 The B.C. distribution of Vancouver lamprey, *Lampetra macrostoma*

• native

AGE, GROWTH, AND MATURITY

There is no published information on the duration of ammocoete life; however, the ammocoetes reach a large size (10–17 cm). This suggests ammocoetes have lifespan similar to Pacific lamprey and, thus, may live for up to 7 years (Beamish and Levings 1991). The transformation to immature adults occurs in the fall (mid-September to mid-November), and they are thought to begin feeding the following spring. The immature adults range in length from 11.8 to 27.3 cm and mature adults range from 17.9 to 25.6 cm. The minimum estimated lifespan from metamorphosis to spawning is 2 years (Beamish 1982).

FOOD HABITS

Adult Vancouver lampreys are parasitic and a serious predator on resident fish in Cowichan and Mesachie lakes (Beamish 1982). Lamprey-killed juvenile coho salmon (*Oncorhynchus kisutch*) are occasionally picked from the lake bottom and along the shores (Beamish 1982) and 8 out of 10 fish examined from Lake Cowichan bore lamprey scars (Carl 1953). Presumably, the diet of ammocoetes in the lakes and inlet streams are similar to other ammocoetes (i.e., particulate organic detritus, diatoms, and desmids).

Habitat The only published information available on habitat use by the Vancouver lamprey is in Beamish (1982).

ADULTS
Since they feed on a variety of fish, including cutthroat trout (*Oncorhynchus clarkii*) and juvenile coho salmon, the adults probably forage in most areas of the lakes.

JUVENILES
Large ammocoetes (10–17 cm) were found in silt deposition areas along the shores of Lake Cowichan and Mesachie Lake.

YOUNG-OF-THE-YEAR
Ammocoetes as small as 16 mm were also found in silt deposition areas along the shores of the lakes.

Conservation Comments Although the status of the Vancouver lamprey as a species distinct from the Pacific lamprey is contentious, the Vancouver lamprey is of scientific interest. The selective pressures that cause anadromous species to give rise to freshwater-resident populations are not well understood. This is a general problem that cuts across diverse phyletic lineages. COSEWIC lists the Vancouver lamprey as threatened, and the BCCDC lists it as S1 (critically imperiled).

Lampetra richardsoni VLADYKOV & FOLLETT
WESTERN BROOK LAMPREY

1 cm

Distinguishing Characters This small (usually less than 16 cm) non-parasitic lamprey has 2 blunt supraoral teeth, 7 (6–9) blunt infraoral teeth, 3 multicuspid inner lateral teeth (Fig. 4B), and 60–68 myomeres. It differs from the Pacific lamprey (*Lampetra tridentata*) in maximum adult size (16 cm vs. 60 cm in length) and dentition—two blunt supraoral teeth (vs. three sharp supraoral teeth in Pacific lamprey), and three (vs. four in the Pacific lamprey) sets of inner lateral teeth. Western brook lamprey also differ from the river lamprey (*Lampetra ayresii*) in size (adult river lamprey reach lengths of 30 cm) and dentition (blunt teeth vs. sharp teeth in the river lamprey). The ammocoetes of the western brook lamprey and the river lamprey are difficult to distinguish; however, the pigmentation in the head region is well developed in western brook lamprey ammocoetes relative to river lamprey ammocoetes (Richards et al. 1982).

Taxonomy A fascinating aspect of speciation in lampreys is the repeated evolution of closely related pairs of parasitic and a non-parasitic species (Vladykov and Kott 1979). The western brook lamprey and the river lamprey appear to be such a species pair. The available evidence argues for the multiple, recent (postglacial) evolution of western brook lamprey populations from the anadromous river lamprey (see the river lamprey taxonomy section for a discussion of this specific case, and the introductory section of this book for a discussion of the general problem of the parallel evolution of freshwater-resident populations from anadromous species).

For now, although the available evidence points to a single genome with two life-history forms, I have retained river and western brook lampreys as separate species. The problem clearly requires further study.

Sexual Dimorphism Outside the breeding season, the sexes are difficult to distinguish. As spawning approaches, however, the sexes develop different morphologies (Pletcher 1963). Thus, males develop a conspicuous, slender urogenital papilla that protrudes up to 6 mm from the body. Females have an inconspicuous urogenital papilla but develop a pseudoanal fin (a prominent swelling behind the vent) that is absent in males. There also are sex differences in dorsal fin height and body proportions (Pletcher 1963).

Distribution NATIVE DISTRIBUTION
The western brook lamprey is endemic to western North America. Here, it ranges from the Coquille River on the south-central Oregon coast north to at least the Skeena River system and Queen Charlotte Islands.

BRITISH COLUMBIA DISTRIBUTION
Western brook lamprey are common in small streams throughout the Fraser valley from Hope downstream to Musqueam Creek. They also occur in streams on both coasts of Vancouver Island, on King and Princess Royal islands, and are present in low numbers in many streams on the Queen Charlotte Islands (Northcote et al. 1989). They are abundant in the lower Skeena system. Their northern limit is uncertain, but they are rumoured to occur in both the lower Nass and lower Iskut–Stikine systems (Map 3).

Life History In British Columbia, the primary source for life-history data on the western brook lamprey is an unpublished M.Sc. thesis (Pletcher 1963). The following account is derived from this source supplemented with personal observations and published data from other areas along the Pacific Coast.

REPRODUCTION
Western brook lamprey ammocoetes over 90 mm in length begin to transform into adults in the late summer and spend the following winter burrowed in the substrate. They emerge in early spring and, in some rivers, make a short migration to suitable spawning sites (gravel riffles). Actual spawning begins when water temperatures rise above 10 °C. McIntyre (1969) and Pletcher (1963) provide descriptions of nest construction and spawning behaviour. The male initiates nest construction, but the female assists once the nest is started. Typically, the nest is located in a riffle or at the tail-out of a pool. The surface substrate usually is gravel less than 2 cm in diameter. In the Salmon River near Langley, water velocities at nest sites averaged 0.3–0.4 m/s (Pletcher 1963).

Nest construction involves the removal of surface gravel. A lamprey attaches to a stone, lifts it, and then drops it (usually at the downstream lip of the nest). Larger stones are rolled out of the nest site. When most of the surface gravel is removed, the male attaches to a stone at the upstream lip of the nest, turns on his side and performs rapid tail beats. This behaviour moves small pebbles out of the nest and disperses sand over the bottom of the nest. When the nest is complete, the female attaches to a stone at the upstream end of the nest and, gently undulating her body, she rests on the bottom. The male approaches from downstream and glides his buccal disk along her body. After several bouts of this "courting" behaviour, the male attaches to the female's head, arches

Map 3 The B.C. distribution of western brook lamprey, *Lampetra richardsoni*

the forward part of his body and coils his tail around the female. Both lampreys vibrate vigorously, and their bodies press against the substrate. The actual release of eggs and sperm often occurs with the lower parts of the pair's bodies buried in the sand. About 10 eggs are released at each spawning (McIntyre 1969), and multiple spawning bouts are separated by periods of egg covering and nest enlargement. Although spawnings usually involve a single male and female, group spawnings consisting of multiple males and females are relatively common (McIntyre 1969; Pletcher 1963). Different pairs or groups of lampreys may sequentially use the same nest site.

Fecundity varies with female body size and ranges from 1,000 to 3,700 eggs. The light-green eggs are slightly oblong (about 1 mm wide and a bit longer than 1 mm), denser than water, and adhesive. They take about 30 days to hatch at 10 °C and about 12 days at 22 °C; however, above 18 °C, mortalities and abnormalities increase sharply (Meeuwig et al. 2005). In the laboratory, newly hatched larvae (7–10 mm) remained on the surface for approximately 2 weeks before burrowing into the mud substrate; however, in nature, the larvae did not emerge from the gravel until 2–3 weeks after they hatched (Pletcher 1963). They emerged at night and were swept downstream into quiet water areas where they burrowed into the mud.

AGE, GROWTH, AND MATURITY

Western brook lamprey spend about 4 years as ammocoetes. Growth varies seasonally, with the fastest growth during the summer and, apparently, a slowdown during the winter (Pletcher 1963; Schultz 1930). In the John Day River, Oregon, young-of-the-year ammocoetes averaged about 18 mm (14–21) by September and 108 mm (85–131) after five summers (Meeuwig et al. 2005). In the Salmon River near Langley, 0^+ ammocoetes reached about 20 mm by October and 115–120 mm before beginning to transform into adults. Apparently, both ammocoete and adult lengths vary among populations (Pletcher 1963), but ammocoetes rarely exceed 150 mm TL. Transformation occurs from August to November, and the newly transformed adults remain in the gravel until spring. The adults die shortly after spawning.

FOOD HABITS

Adult western brook lamprey do not feed. Ammocoetes are filter feeders and use the movement of cilia to pass a current through the oral hood and out the gill pores. Diatoms and desmids are the main food; however, in the fall and winter, particulate organic detritus is an important energy source. Typically, the largest ammocoetes stop feeding in the late summer or fall and begin their transformation into adults.

Habitat Again, the main source of habitat data for B.C. populations of western brook lamprey is an unpublished M.Sc. thesis (Pletcher 1963).

ADULTS

Upon emergence from the substrate, newly transformed adults move onto riffles and spawn. The spawning period can be protracted, and depending on location, adults are found associated with riffles from mid-April to early July.

JUVENILES

In the Salmon River, monthly collections from different substrates found ammocoetes aggregated in deposition areas (back-eddies) in pools (Pletcher 1963). The largest ammocoetes were associated with sand and leaf substrates in the deepest quiet water areas.

YOUNG-OF-THE-YEAR

In contrast to large ammocoetes, young-of-the-year and yearling ammocoetes aggregate in fine substrate (mud and silt) in shallow water along pool edges (Pletcher 1963).

Conservation Comments Most brook lamprey populations along the B.C. coast probably are relatively healthy, and this species is not listed by either COSEWIC or the BCCDC. Nonetheless, in the lower Fraser valley and on southern Vancouver Island, urbanization has destroyed many of the small streams favoured by western brook lamprey. Although no one keeps score, many brook lamprey populations in Greater Vancouver, Victoria, and Nanaimo probably have been extirpated.

Lampetra tridentata (GAIRDNER)
PACIFIC LAMPREY

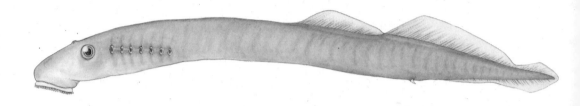

1 cm

Distinguishing Characters   This large lamprey can reach lengths of about 70 cm; however, adult body size varies among populations (Beamish 1980; Farlinger and Beamish 1984). Typically, there are 3 sharp supraoral teeth, 5 sharp infraoral teeth, 4 multicuspid inner lateral teeth (Figs. 4A, B), and 64–74 muscle grooves between the last gill opening and the cloacal slit. The Pacific lamprey differs from the river lamprey (*Lampetra ayresii*) in adult size (Pacific lamprey often exceed 60 cm, whereas river lamprey rarely exceed 30 cm) and dentition: three sharp supraoral teeth in Pacific lamprey vs. two supraoral teeth in the river lamprey and four sets of inner lateral teeth in Pacific lamprey vs. three in river lamprey). Pacific lamprey also differs from the western brook lamprey (*Lampetra richardsoni*) in adult size (western brook lamprey rarely exceed 16 cm in length) and dentition (sharp teeth in the Pacific lamprey vs. blunt teeth in the western brook lamprey). Lamprey ammocoetes are difficult to identify, although Pacific lamprey ammocoetes are said to have a unusual lightly pigmented area in the tail region (Richards et al. 1982).

Taxonomy   There are two taxonomic problems associated with this species. One is its generic placement, and the other is the relationship between anadromous and freshwater-resident populations. The latter problem is discussed in the taxonomy section under Vancouver lamprey. The generic status problem centres on whether the Pacific lamprey belongs in *Lampetra* or in a separate genus, *Entosphenus*. In the past, the generic name was *Entosphenus*, but the Pacific lamprey was placed in *Lampetra* in the 1980s (Bailey 1980; Potter 1980). The recent American Fisheries Society list of the names of fishes (Nelson et al. 2004) still retains this species in *Lampetra*, despite strong evidence (Docker et al. 1999; Gill et al. 2003) that the Pacific lamprey does not belong in this genus.

Lampetra tridentata PACIFIC LAMPREY

Sexual Dimorphism Outside the breeding season, the sexes are difficult to distinguish, but their morphologies diverge as they approach spawning (Pletcher 1963). Males develop a slender, short (relative to other B.C. lampreys) urogenital papilla. The female's urogenital papilla is inconspicuous, but females develop a pseudoanal fin (a prominent swelling behind the vent) that is absent in males. There also are sex differences in dorsal fin height and body proportions (Pletcher 1963).

Distribution

NATIVE DISTRIBUTION

Anadromous Pacific lamprey occur on both sides of the North Pacific Ocean. They range from northern Japan (Hokkaido) north and east through the Aleutian Islands to Alaska and south along the North American coast to Baja California (Rio Santo Domingo).

BRITISH COLUMBIA DISTRIBUTION

The Pacific lamprey is common in rivers (large and small) along the entire coast (Map 4). Although most Pacific lamprey are anadromous, there are freshwater-resident populations in the Strait of Georgia region (Sakinaw and Ruby lakes on the Sechelt Peninsula, Village Lake on Quadra Island, and West Lake on Nelson Island) and on northern Vancouver Island. Adults of these populations are much smaller than anadromous Pacific lamprey.

In the Fraser, Skeena, and (in the past) Columbia river systems, anadromous Pacific lamprey penetrated well into the interior of the province—at least to the Nechako River and Shuswap Lake in the Fraser system; to Babine Lake, the upper Bulkely River, and the upper Morice River in the Skeena system; and, at one time, to Vaseux Lake and the Slocan River in the Columbia system (Parkham 1937). In the Fraser River system, ammocoetes are common in the Prince George area; however, only one adult has been collected, although others have been observed. This single specimen was small (145 mm TL) but in reproductive condition (McPhail and Lindsey 1970). Usually, Pacific lampreys that make long upstream migrations are large (Farlinger and Beamish 1984; Pletcher 1963). Consequently, the presence of small adults in the upper Fraser population hints that some individuals in the population may be non-migratory.

Life History In British Columbia, the primary sources for life-history information on Pacific lampreys are Beamish (1980) and Pletcher (1963). The following account is based on these sources supplemented with personal observations and published data from other areas around the North Pacific Rim.

REPRODUCTION

Pacific lamprey spawn in the spring (April–June) and usually die shortly after spawning. Sometimes in coastal streams, however, substantial numbers of adults survive spawning and migrate back downstream towards the ocean (Michael 1980, 1984). Some of these downstream migrants may return to spawn a second time. Typically, spawning

Map 4 The B.C. distribution of Pacific lamprey, *Lampetra tridentata*

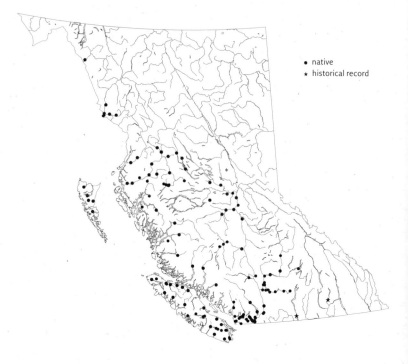

commences when water temperatures rise to about 10 °C (Pletcher 1963). Nest construction behaviour is similar to that described for other lampreys, with males moving most of the gravel but with some participation by females. Normally, the nest is located at the tail-out of a pool, and the gravel substrate often exceeds 4 cm in diameter. In the Salmon River near Langley, water velocities of 0.37–0.46 m/s were recorded over nest sites (Pletcher 1963). The nest is larger (50–60 cm in diameter) and deeper (4–5 cm) than the nests of sympatric western brook lampreys (Pletcher 1963). In the upper and lower Babine rivers and Nilkitkwa Lake, nests were found on gravel shoals at water depths ranging from 30 cm to 4 m (Farlinger and Beamish 1984). The deepest nests were in cutbank areas. Although most lampreys spawn in flowing water, Pacific lamprey were observed spawning in both the North Arm of Babine Lake (at a depth of about 50 cm) and the lower end of Nilkitkwa Lake (at depths of 2–3 m). No unidirectional flow was detectable at either of these sites (Russell et al. 1987).

The spawning behaviour of Pacific lampreys is similar to that described for other lampreys (see description of spawning behaviour in the western brook lamprey account): courtship "gliding" is pronounced, there are multiple egg releases in a spawning bout, and occasionally, communal spawning (Fukutomi et al. 2001; Pletcher 1963). Fecundity increases with female body size and ranges from 34,000 to 238,400 eggs (Wydoski and Whitney 2003). In the lower Fraser system, fecundities ranged from 10,000 to 106,000 eggs in females 262–406 mm in length (Pletcher 1963).

The eggs are slightly elliptical (about 1.0–1.1 mm wide and 1.1–1.2 mm long). Salmon River Pacific lamprey eggs began hatching after 15 days at 15 °C (Pletcher 1963), whereas in Japan, eggs incubated at 15 °C began hatching at 13–17 days (Fukutomi et al. 2001). Eggs obtained from lampreys at Bonneville Dam on the Columbia River were incubated at 10 °C, 14 °C, 18 °C, and 22 °C (Meeuwig et al. 2005). Hatching occurred in 29 days at 10 °C and 9 days at 22 °C. There was a sharp increase in abnormalities and deaths at the highest incubation temperature (Meeuwig et al. 2005). Apparently, the larvae exhaust their yolk reserves about 20–30 days after hatching and begin to burrow (Pletcher 1963). Presumably, it is at this stage that they first emerge from the gravel. Emergence occurs at night, and the ammocoetes are swept downstream into quiet water areas where they burrow into soft substrates. In the Thompson and Nicola rivers, the smallest ammocoetes collected were 7–8 mm in length. A similar minimum size (about 8 mm) is reported from the Babine River and Nilkitkwa Lake (Farlinger and Beamish 1984). Presumably, these small ammocoetes had only recently emerged.

AGE, GROWTH, AND MATURITY

Length–frequency data from the Nicola River (Pletcher 1963) indicate that young-of-the-year reach about 10–25 mm by mid-August, and the largest ammocoetes are about 120 mm; however, ammocoetes as large as 176 mm occur in the upper Skeena system (Farlinger and Beamish 1984). Most age and growth studies involving Pacific lampreys use length–frequency data, but a recent study using statoliths suggests that the overlap between year classes increases with ammocoete age (Meeuwig and Bayer 2005). This makes the length–frequency method unreliable for aging older ammocoetes. Still, the ammocoete stage can last for up to 7 years (Beamish and Levings 1991).

Ammocoete growth probably varies seasonally, with the fastest growth during the summer and a slowdown during the winter. Metamorphosis begins in July, and entry into salt water occurs from December until June (Beamish 1980). This variation in the timing of downstream migrations appears to be related to the distance of the migration, the size of the watershed, and flow conditions (Beamish and Levings 1991). Thus, in the Fraser system, young adults from interior streams (e.g., the Nicola River) begin to migrate as early as late September and continue migrating into late June or early July of the following year. In contrast, young adults did not appear until late March in the lower Fraser River (Beamish and Levings 1991). The size of out-migrants varies among river systems but ranges from about 100 to 160 mm (Beamish 1980). On average, young adults are about 130 mm when they first enter salt water. At this stage, they are strikingly counter coloured—silver below and dark blue above.

Adults spend up to 3.5 years in the sea and reach lengths of up to 720 mm before returning to fresh water; however, there is considerable variation in adult length. Beamish (1980) gives a size range for maturing adults of 160–720 mm. On average, the smallest adults are associated with short coastal rivers, and the largest adults, with long migrations into the interior of the province. Although there is little direct evidence of homing in Pacific lampreys, consistent differences in migration timing and life histories among populations in different rivers imply local adaptation and homing.

Pacific lamprey return to fresh water from May to September, but most populations enter fresh water from April to June (Beamish 1980) and spend the winter in streams before spawning in the spring or early summer. In the upper Skeena, spawning was observed in mid-July (Farlinger and Beamish 1984). This is about a month before the arrival of next year's spawners. Thus, the adults spend almost a year in fresh water before spawning.

FOOD HABITS

Like other lampreys, the ammocoetes of Pacific lamprey are filter feeders and use the movement of cilia to pass a current through the oral hood and out the gill pores. Their diet is similar to that of other lamprey ammocoetes: mostly diatoms and desmids during the summer and particulate organic detritus in the fall and winter (Pletcher 1963). Typically, the largest ammocoetes stop feeding in the late summer or in autumn and begin their transformation into adults.

Some newly metamorphosed Pacific lamprey begin feeding in fresh water. In the sea they attack a wide variety of prey: salmon, rockfish, halibut and other flatfish, as well as several species of cods (Beamish 1980).

Habitat There are some data on the marine phase of the Pacific lamprey's life cycle (Beamish 1980); however, this aspect of their biology is not well known. In contrast, we know quite a bit about the habitat use of ammocoetes (Pletcher 1963; Stone and Barndt 2005; Toregson and Close 2004).

ADULTS

Upon entering the sea, Pacific lamprey move offshore into relatively deep water (>70 m). Young adults have been caught at depths of 100–250 m in the Strait of Georgia and off the west coast of Vancouver Island (Beamish 1980) and over 100 km off the Oregon coast. How far, and to what depth, the larger adults roam is unknown, but there is evidence of their presence (e.g., scars and wounds on other fish) all along the B.C. coast.

JUVENILES

In Cedar Creek, western Washington, ammocoetes were associated with slow (0.00–0.10 m/s), deep (60–80 cm) pools and eddy habitats (Stone and Barndt 2005). Only the largest ammocoetes were found in faster water with large gravel substrates. Additionally, ammocoetes in the headwaters of the creek were larger than those in lower reaches. A similar pattern of habitat use was reported from the Middle Fork John Day River in Oregon (Toregson and Close 2004). The presence of large ammocoetes in fast-water reaches was attributed to the downstream displacement of small ammocoetes during freshets. In the B.C. interior, however, the largest ammocoetes leave the Nicola River during spring high water and settle downstream in the Thompson or Fraser rivers before migrating to the lower river (Beamish and Levings 1991).

YOUNG-OF-THE-YEAR
In the Salmon River, and in the Nicola and Thompson rivers, the smallest ammocoetes (10–25 mm) are found in shallow, quiet water sites with mud, silt, and leaf litter substrates located close to the river edges (Pletcher 1963).

Conservation Comments In Canada, the Pacific lamprey is not listed federally or provincially but is considered endangered in the U.S. portion of the Columbia River system. Since the mid-1960s, the numbers of Pacific lamprey passing through dams in Washington and Oregon have declined dramatically (Close et al. 2002). Apparently, loss of rearing habitat and impediments to migration are major causes of the decline. As far as is known, the Pacific lamprey is extirpated in the B.C. portion of the Columbia system. Their status in Fraser River drainages has not been assessed, but in the urbanized streams of the lower Fraser valley (e.g., the Salmon River), spawning adults are much less common than they were in the 1960s.

FAMILY ACIPENSERIDAE — STURGEON

The sturgeons are living remnants of an ancient lineage of ray-finned fishes—the Chondrostei. The internal skeletons of modern chondrosteans are made up mostly of cartilage. Once, the absence of a bony internal skeleton was viewed as a primitive trait, but we now know that the Mesozoic ancestors of sturgeons had bony internal skeletons. Thus, in sturgeons, the cartilaginous skeleton has secondarily replaced the bony skeleton. Still, many morphological features of sturgeons are primitive: a shark-like heterocercal tail (the upper lobe is much larger than the lower lobe); a spiracle (the small hole above and behind the eye that is thought to be a remnant of a gill-opening present in some ancient fishes); a spiral valve in the intestine (a feature shared with sharks); and, instead of scales, rows of stud-like plates (scutes) located on the back, the midlateral sides, the lower sides (between the pectoral base and the pelvic fins), and midventrally between the pelvic and caudal fins. These scutes are abraded in adults but sharp and with a spine-like protrusion in juveniles. The ventral, toothless mouth is protrusible. It can be shot out quickly and, if a big sturgeon gets close enough, it can inhale even agile prey like adult salmon. On the underside of the snout, in front of the mouth, there are four conspicuous barbels. These barbels are covered with chemoreceptors, and sturgeon can detect prey in the dark or in turbid water. Additionally, the underside of the snout is dotted with mucous-filled pores. Their function is uncertain, but they are remarkably similar to the electro-receptors on the snouts of paddlefish and may aid in prey detection.

Newly hatched sturgeon larvae are hardly recognizable as fish. They have a large, orange-yellow yolk sac, continuous dorsal and anal fin folds (broken only at the cloaca), and small pectoral fin buds. They look more like tiny translucent tadpoles than sturgeon. As the larvae grow, the barbels appear; the fin folds shrink; and the dorsal, anal, and caudal fins begin to form. Later, the barbels and snout lengthen, sharp scutes begin to form on the back and sides, and the caudal fin takes on its characteristic heterocercal shape. By about 60–80 mm TL, metamorphosis is complete, and the young fish are clearly recognizable as small sturgeon.

Another curious feature of sturgeons is the role that polyploidy (the doubling of chromosome numbers) has played in their evolution. A related chondrostean, the paddlefish (*Polyodon spathula*) has a diploid number of 120 chromosomes. This is thought to be the original chromosome number in sturgeons. Now, most modern sturgeons have about 240 chromosomes, and at least one sturgeon has about 500 chromosomes. Although polyploidy is a common mode of speciation in the higher plants, it is relatively rare in vertebrates.

There are about 23 living sturgeons, most of them in the genus *Acipenser*. They are renowned for their large size and long life. The beluga—this name applies to a sturgeon as well as a whale—is the largest freshwater fish in the world. Beluga are said to attain lengths of over 8 m, weights of

1300 kg, and ages of almost 120 years. In British Columbia, we have two sturgeons: the green sturgeon (*A. medirostris*), and the white sturgeon (*A. transmontanus*). The white sturgeon is the largest fish in North America, and there are records from the Fraser River of white sturgeon 3.8 m long and surpassing 600 kg in weight. In 1992, a large (about 4 m) white sturgeon washed up on the banks of the lower Fraser River. This behemoth was starting to decompose and couldn't be accurately weighed or measured; however, it was aged at 138 years. Much less is known about the green sturgeon, but it is reputed to reach lengths of over 1.5 m and weights exceeding 100 kg. All sturgeons spawn in fresh-water, but some species (e.g., the green sturgeon) spend most of their lives in marine or brackish waters. Others species, like the white sturgeon, either spend their entire lives in fresh water or make sporadic forays into the sea.

KEY TO ADULT STURGEONS

1(2) Barbels about equidistant between tip of snout and mouth (Fig. 5A); 20–30 lateral scutes; a single row of 1–4 scutes is present on the ventral surface between the pelvic and anal fins	GREEN STURGEON, *Acipenser medirostris*
2(1) Barbels closer to tip of snout than to mouth (Fig. 5B); 38–40 lateral scutes; two rows of 4–8 scutes are present on the ventral surface between the pelvic and anal fins	WHITE STURGEON, *Acipenser transmontanus*

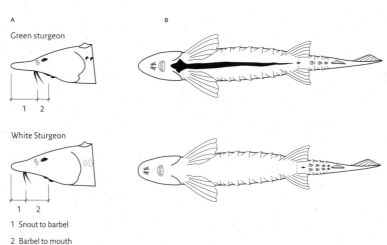

Figure 5 (A) Placement of barbels relative to the mouth, and (B) the ventral surfaces of the two B.C. sturgeon species

A Green sturgeon

B

White Sturgeon

1 Snout to barbel
2 Barbel to mouth

Acipenser medirostris AYRES
GREEN STURGEON

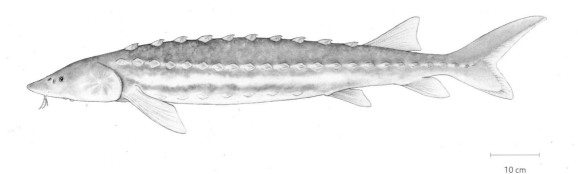

10 cm

Distinguishing characters — This sturgeon has a long narrow snout and barbels that are about equidistant between the tip of the snout and the mouth (Fig. 5A). There are about 20–30 midlateral scutes and a single midventral row of 1–4 scutes between the pelvic fins and the origin of the anal fin. As the common name implies, the dorsal surface and upper flanks usually are dark green. A midventral dark, arrow-like stripe on the belly is a useful field character that separates most adult green sturgeons from white sturgeons (*Acipenser transmontanus*): the stripe is present on green sturgeon and not on white sturgeon.

Taxonomy — The taxonomic status of the green sturgeon is still unclear: some authors (e.g., Zhang et al. 2001) argue that Asian and North American green sturgeon are the same species, but others (Birstein and DeSalle 1998; Ludwig et al. 2001; North et al. 2002) argue that they are different species. Regardless of the outcome of this debate, the scientific name of the North American green sturgeon will remain *Acipenser medirostris*.

Within North America, the U.S. National Marine Fisheries Service recognizes two distinct population segments: a southern segment that extends from the Eel River, northern California, south to northern Baja California and a northern segment that extends from the Klamath River north to southeastern Alaska. So far, all of the sonically tagged green sturgeon detected in B.C. waters originated from the northern population segment. Nonetheless, a recent genetic study (Israel et al. 2004) suggests that the division into southern and northern groups may be an oversimplification. This study found significant differences between green sturgeon collected in the Rogue and Klamath Rivers and those collected in San Pablo Bay, California, and in the Columbia River estuary.

Sexual Dimorphism There are no foolproof external indicators of sex in green sturgeon; however, live adults can be differentiated by examining the urogenital opening: in males it is Y-shaped and in females it is shaped like a doughnut. Apparently, this method is over 85% accurate on live fish but does not work on dead fish (Vescei et al. 2003).

Distribution NATIVE DISTRIBUTION

The green sturgeon occurs on both sides of the North Pacific Ocean: its North American marine range extends from northern Baja California to southeastern Alaska, and its Asian range, from northern Japan to Kamchatka. Although it is uncommon in British Columbia, the species is seasonally abundant in the lower Columbia River and, until recently, supported a small fishery in Grays Harbor on the outer Washington Coast. There is no evidence that green sturgeon spawn in the rivers associated with these northern estuaries.

BRITISH COLUMBIA DISTRIBUTION

Although the green sturgeon is rare in fresh water (Map 5), there are records from the lower Fraser River (the most recent in 2005 from near Fort Langley). The majority of recent B.C. records are either from the sea off the west coast of Vancouver Island or from northern estuaries (e.g., the Skeena, Nass, and Taku estuaries). There are no known spawning sites in British Columbia.

Life History Little is known about the life history of green sturgeon in British Columbia. Therefore, the following account is based mainly on data from green sturgeon studies in Washington, Oregon, and northern California. The principal source of this information is a National Marine Fisheries Service status report (Adams et al. 2002). This report is available on the Southwest Fisheries Science Centre web site and summarizes most of what is known about the life history and ecology of North American green sturgeon. Much of the information in the report is unpublished and difficult to access, and I have not seen most of the letters, agency reports, consultant reports, and governmental memoranda synthesized in the green sturgeon status report (Adams et al. 2002). Consequently, unless another source is indicated, all of the following information is from the green sturgeon status report.

REPRODUCTION

There is no evidence that green sturgeon spawn in British Columbia, and the nearest known spawning sites are in northern California (Sacramento and Klamath rivers) and southern Oregon (the Rogue River). Historically, however, they spawned in other northern California rivers (e.g., the San Joaquin and Eel rivers) and, perhaps, in the Umpqua River, Oregon. Apparently, they spawn in deep, fast mainstem channels (Moyle 2002). Spawning occurs at water temperatures ranging from

Map 5 The B.C. distribution of green sturgeon, *Acipenser medirostris*

+ native, sporadic

8 °C to 14 °C. In the Sacramento River, spawning peaks in the spring (April to June); however, some fish spawn as early as March, and others, as late as July (Moyle 2002). Spawning behaviour has not been observed, but surface behaviours like breaching and rolling commonly occur at known spawning sites.

Fertilized and water-hardened eggs are large (about 4.5 mm), demersal, and slightly adhesive (Deng et al. 2002). Fecundities for two Klamath River females were 52,000 and 82,000 eggs (Van Eenennaam et al. 2001), but apparently, fecundities as high as 224,000 eggs have been reported from the same river. Mature females spawn every 3–5 years, but males spawn more frequently. There is no spawning-site preparation, although females may seek out locations with appropriate substrate (Kynard et al. 2005). Presumably, the eggs are fertilized as they are released. The eggs are dense, sink rapidly (Kynard et al. 2005), and drift downstream a short distance before lodging in crannies among rocks and cobbles on the bottom. Sperm remain motile for at least 5 minutes, and this relatively long period of sperm motility may be an adaptation to spawning in fast water. At 15.7 °C, the eggs hatch in 7–9 days. Apparently, temperatures above 20 °C are lethal to developing embryos. Newly hatched larvae are about 12.5–14.5 mm long, and at about 10 days after hatching (23–25 mm), they begin exogenous feeding (Deng et al. 2002; Van Eenennaam et al. 2001). By about 45 days after hatching (62–95 mm TL), metamorphosis is complete (Deng et al. 2002).

GROWTH, AGE, AND MATURITY

Growth is rapid, and green sturgeon in the Klamath River reach 30 cm in their first year of life and 60 cm after 2 or 3 years. Males become sexually mature at about 15 years, and females typically mature 2 or more years later. Green sturgeon are not as long lived as white sturgeons but their maximum lifespan is unknown. One estimate suggests a maximum age of 42 years, but another estimate suggests a lifespan of about 70 years.

FOOD HABITS

There is little information on the diet of green sturgeon in fresh water. In the freshwater portions of the Sacramento Delta, juvenile green sturgeon feed primarily on mysids and amphipods (Radtke 1966). Apparently, adults do not feed in fresh water, but their marine diet includes crabs, shrimps, isopods, amphipods, and small fish.

Habitat Most of the habitat information available for North American green sturgeon is summarized in a National Marine Fisheries Service status report (Adams et al. 2002). Little is known about their marine habitat requirements.

ADULTS

Except for spawning events, most of the adult life is spent at sea. Apparently, adults of the northern population segment leave their spawning streams when water temperature drops below 10 °C. Once in the sea, they migrate north. This northward migration is rapid and, since trawlers and long-line gear make most marine catches, green sturgeon appear to remain close to the bottom. Marine catches in Canadian waters also suggest that adult green sturgeon school: the catches are rare, sporadic and, typically, one-quarter to one-half of the yearly green sturgeon catch comes from a single net-haul (Slack and Stace-Smith 1996).

JUVENILES

Little is known about juvenile freshwater habitat use; however, they probably use sites similar to the overwintering habitats described in the section on young-of-the-year. In freshwater, juveniles probably are only active at night. After 1–4 years in fresh water or estuaries, juvenile green sturgeon enter the sea and migrate north. A peculiar aspect of their subadult marine phase is the tendency to aggregate in the estuaries of major rivers in the late summer and early fall. These are not pre-spawning concentrations, because most estuaries where they have been reported (e.g., Yaquina Bay, the Columbia Estuary, Willipa Bay, Grays Harbor, and the Fraser Estuary) do not support spawning runs. Presumably, the subadult marine phase of the green sturgeon's life history lasts for 15–20 years.

YOUNG-OF-THE-YEAR

Newly hatched green sturgeon are weak swimmers. They clump together on the bottom (at least in the laboratory) and appear to preferentially seek cover (Kynard et al. 2005). About 6–12 days after hatching, they become active at night and swim up into the water column and migrate downstream. In California, larval green sturgeon are sometimes caught in salmon out-migrant traps (Moyle 2002). This suggests that they either passively drift, or actively migrate, downstream until they metamorphose. At metamorphosis, they become diurnal and forage actively. Peak activity is at night, and when foraging in the laboratory, they tend to move upstream; however, at about 3.5–6 months after hatching, they migrate downstream to overwintering habitats. Laboratory observations suggest that overwintering habitat consists of deep pools with low light and some rock structures (Kynard et al. 2005).

Conservation Comments Although COSEWIC lists the green sturgeon as a species of special concern, the BCCDC does not list this species. Since green sturgeons rarely enter fresh water in British Columbia and don't breed here, there is little we can do to assure its continued survival. We can, however, ensure that those incidentally caught in our fisheries are returned alive to the water. In British Columbia, it is illegal for anglers to keep green sturgeon.

Acipenser transmontanus RICHARDSON
WHITE STURGEON

10 cm

Distinguishing characters Relative to green sturgeons (*Acipenser medirostris*), adult white sturgeon have short, rounded snouts (but see taxonomy and sexual dimorphism sections for comments on snout length). A transverse row of four barbels is located between the tip of the snout and the mouth. These barbels are positioned closer to the tip of the snout than to the mouth. There are about 38–48 midlateral scutes and two rows of 4–8 midventral scutes between the pelvic fins and the origin of the anal fin. Rarely, there is an extra row of scutes between the dorsal and midlateral rows. Usually, the back is dark grey and the upper flanks shade from dark grey to white on the underparts. A useful field character for separating adult white and green sturgeons is ventral pigmentation: the ventral surface of white sturgeon is immaculate, whereas in green sturgeon there is a dark, arrow-like stripe on the ventral surface.

Taxonomy Although there are no taxonomic problems associated with this species in British Columbia, there are striking morphological differences between white sturgeon living above and below the Fraser Canyon. Adult sturgeon below the Fraser Canyon typically have short, blunt snouts and rounded fins and tails, while those from above the canyon have longer, pointed snouts and sickle-shaped fins and tails. The reasons for these differences are unknown, but they may have some hydrodynamic function. There are six genetically identifiable white sturgeon populations in British Columbia—lower, middle, and upper Fraser River populations; a Nechako population; a Columbia population; and a Kootenay (above Bonnington Falls) population (Bartley et al. 1985; Brown et al. 1992, 1993; Smith et al. 2002). These populations have different life-history characteristics and are managed as separate units.

Sexual Dimorphism There are no foolproof external indicators of sex in white sturgeon; however, the urogenital opening can differentiate live adults: in males this structure is Y-shaped, and in females, it is shaped like a doughnut (Vescei et al. 2003). Apparently, this method is over 85% accurate on live fish but does not work on dead fish. Mature fish also can be sexed by blood plasma analyses (Webb et al. 2002).

Snout length and width in a small sample from the Columbia River were associated with sex—of the 10 individuals examined, 5 males had long narrow snouts and 4 females had short, rounded snouts (Crass and Gray 1982). A similar snout dimorphism occurs in juvenile sturgeon in the lower Fraser River (e.g., Nicomen Slough), but it is not known if the dimorphism is related to sex.

Distribution NATIVE DISTRIBUTION

White sturgeon are indigenous to the Pacific Coast of North America. Here, they occur sporadically in the sea from Monterey, California, to the Gulf of Alaska (Cook Inlet). The latitudinal extent of their fresh water range is much narrower than this marine distribution, and self-sustaining populations probably only occur in the Sacramento – San Joaquin, Columbia, and Fraser river systems. There are records from smaller drainages in northern California, Oregon, and Washington, but it is not clear that white sturgeon reproduce in these smaller rivers. Similarly, there are sporadic records in British Columbia, usually in estuaries from coastal drainages north of the Fraser River; however, it is possible that there are small, self-sustaining populations in the Skeena and Nass rivers.

BRITISH COLUMBIA DISTRIBUTION

White sturgeon are known from most of the large west-flowing rivers in the province (Map 6). In the Fraser River, white sturgeon occur in the mainstem from the estuary to well upstream of Prince George (at least as far up as Rearguard Falls) and in the Nechako River from Prince George upstream to well above Vanderhoof. They are also found in many of the large lakes associated with the Fraser and Nechako rivers (e.g., Pitt, Harrison, Fraser, Stuart, Takla, and Trembleur lakes). In the Thompson River, there are unconfirmed reports from as far upstream as Kamloops.

White sturgeon often occur in estuaries and sometimes spend protracted periods in the marine environment (Veinott et al. 1999). Based on tagging and recapture observations, individuals occasionally move between river systems, and fish tagged in the Columbia River sometimes are recaptured in the lower Fraser River. White sturgeon also are recorded from the lower Skeena and lower Nass rivers. These records are rare and suggest that they probably do not reproduce in these rivers. Additionally, white sturgeon occasionally turn up in rivers on both the west and east coasts of Vancouver Island (e.g., the lower Somass and Cowichan rivers) and in Howe Sound (Mamquam River). In spite of

Map 6 The B.C. distribution of white sturgeon, *Acipenser transmontanus*

• native
+ native, sporadic

records from other coastal rivers, there is no evidence for permanent residents in any B.C. river systems except the Fraser and Columbia.

In the Columbia River system, white sturgeon originally occurred from the estuary upstream to Kinbasket Lake (almost 1300 km from the sea). In the Pend d'Oreille River, they are found as far upstream as Flathead Lake in Montana and, in the Kootenay River, they occur above and below Bonnington Falls. As in the Fraser system, white sturgeon occur (or have occurred) in many of the large lakes and reservoirs associated with the Columbia River (e.g., Arrow, Kootenay, and Duncan lakes).

Life History Until recently, we knew relatively little about the life history of white sturgeon in British Columbia; however, since they were listed as a protected species by both the federal and provincial governments, information has accumulated rapidly. There is an excellent review of their life history in the Fraser system (Echols 1995), and there are a number of consultants' reports (Golder Associates Ltd, 2004; Perrin et al. 1999, 2003; R. L. & L. Environmental Services Ltd.1995b, 1998, 2001) on their reproductive biology and movements. These sources, and the primary literature, form the basis of the following life-history account.

REPRODUCTION

In British Columbia, white sturgeon usually begin spawning shortly after the peak spring discharge. Thus, over the species' geographic range, spawning can begin anytime between late February and late June. In the lower Columbia River, spawning occurs at water temperatures ranging from 8 °C to 18 °C (McCabe and Tracy 1994). In the upper Columbia River, however, different sturgeon populations spawn at different times and at different temperatures. The Arrow Lakes population spawns at the lower end of the temperature range (8.5–11.1 °C) and late in the summer—July to mid-August (Golder Associates Ltd. 2004). Sturgeon in the Kootenai River, Idaho, spawn from May to June but at water temperatures (8–12 °C) similar to the Arrow Lakes population. In contrast, the population in the Columbia River between Keenleyside Dam and the U.S. border spawns from mid-June to late July at water temperatures ranging from 16.0 °C to 21.5 °C. In the Fraser River, between the confluence of the Fraser and Coquihalla rivers and Chilliwack, sturgeon eggs and larvae were collected from mid-June to late July in water ranging from 14.8 °C to 18.4 °C (Perrin et al. 2003).

Actual spawning has not been observed, but surface activities (breaching and rolling) are commonly observed at known spawning sites. In the Fraser River, eggs and larvae were collected mainly (five out of six sites) in side-channels even though main-channel areas were also sampled (Perrin et al. 2003). Eggs and larvae were captured at similar depths (3.0–4.5 m) but at different near-bed velocities (larvae at 0.5–1.5 m/s and eggs at velocities averaging 1.7 m/s). The flow in the side-channels was not turbulent, and the substrate was usually a mix of gravel and sand. The Fraser is an unregulated river, and the characteristics of these side-channel spawning sites are strikingly different from spawning sites in the impounded sections of the Columbia River.

In the impounded sections of the Columbia River, there is a strong association between dams and spawning sites: most known spawning sites are located in turbulent tailraces below dams (Parsley and Kappenman 2000). The same is true in the B.C. portion of the Columbia system: two of the confirmed spawning sites are in the tailraces of Waneta Dam (R. L. & L. Environmental Services Ltd. 1995b, 1998, 2001) and the Revelstoke Dam. Fertilized eggs were collected below the Revelstoke Dam from mid-June to mid-July. At this time, water temperature varied from 14 to 17°C, which is consistent with laboratory studies on incubation temperature and egg survival (Wang et al. 1985), and discharge from the dam varied from about 900 to 1,200 m^3/s. The substrate in the tailrace of the Waneta Dam consisted of smooth bedrock grading downstream into small boulders and cobbles (R. L. & L. Environmental Services Ltd.1995b). At this site, spawning occurs as a series of discrete events (usually five) spread over about a month. The last spawning event at this site often occurs at temperatures of about 21°C. This is well above the optimal incubation temperatures (14–17°C) for this species (Wang et al. 1985).

There are exceptions to this general description of spawning sites. The Kootenai River spawning-reach in Idaho differs physically and hydraulically from typical Columbia mainstem spawning sites (Paragamian et al. 2001). It is located in the mainstem Kootenai River near Bonner's Ferry about 100 km downstream of Libby Dam. In autumn and early spring,

Kootenay Lake sturgeon migrate up the Kootenay River and hold in pre-spawning reaches in Idaho (Paragamian and Kruse 2001). Later, they migrate farther upstream to the spawning-reach. Rising water levels and water temperatures appear to be the main migration triggers (Paragamian and Kruse 2001): males migrated at 5.5–12.1°C (about 2 weeks before spawning), and females migrated about a week after the males and at slightly warmer temperatures. During spawning, water velocities varied from 0.2 to 1.0 m/s, and water temperature ranged from 8.5 to 12.0°C. At the Kootenai site, the main difference from "normal" white sturgeon spawning sites in the Columbia and Fraser rivers was the substrate. In 2 out of 8 years, 15 eggs were collected over gravel–cobble substrates, but in the other 6 years, much larger numbers of eggs were collected on sand bottoms (Paragamian et al. 2001).

Fecundity in white sturgeon increases with female size. In the upper Columbia system, sturgeon over 2.0 m are rare (R. L. & L. Environmental Services Ltd. 1995b) and breeding females probably have a similar fecundity to those in the lower Columbia system where about 47,000 eggs were reported for a 1.0 m female and 210,000 eggs in a 1.5 m fish (Beamesderfer et al. 1989). In the lower Fraser population, there are some females over 2.0 m in length. These large females have fecundities commensurate with their large size, and the estimated fecundity for a 2.39 m female was 700,000 eggs (Scott and Crossman 1973).

Fecundity estimates for lower Columbia white sturgeon range from 39,400 to 713,000 eggs (Wydoski and Whitney 2003). White sturgeon do not spawn ever year. The interval between spawnings is about 4 years for young females and 9–11 years for older females (Semakula and Larkin 1968). Mature white sturgeon eggs are about 3.4–3.6 mm in diameter (Deng 2002), a brownish colour and, initially, adhesive. The incubation period varies with water temperature: at a constant 11°C in the laboratory, hatching begins 230 hours after fertilization and extends over 3.5 days, whereas at 17°C, hatching starts after 131 hours and is complete after 1 day (Wang et al. 1985). Newly hatched larvae are 10–11 mm TL, negatively phototactic, and seek cover during the day (Loew and Sillman 1998). At 16.5–30°C, 2-day-old larvae leave the substrate and actively swim. At lower temperatures, they are less active and return to the substrate, but by 18 days, they become day-active (Brannon et al. 1985). The yolk sac is absorbed and exogenous feeding begins 10 days after hatching at about 17–19 mm TL (Deng 2002).

AGE, GROWTH, AND MATURITY
Sturgeon are renowned for their great age, and in the lower Fraser system, the oldest recorded sturgeon was 138 years old. In B.C. populations, however, most adult sturgeons are less than 35 years old, and the modal age group is usually 10–15 years. Exceptions are populations with a history of prolonged recruitment failure. Thus, in the Nechako population, the modal age is between 36 and 39 years, and only about 2% of the population is in the 6- to 10-year age group. Similarly, in the Columbia

River between Keenleyside Dam and the border, the population consists entirely of adult fish that range in age from 23 to 48 years and in size from 102 to 271 cm (R. L. & L. Environmental Services Ltd. 1995b).

Young white sturgeon grow rapidly and, in the lower Columbia, reach a length of almost 30 cm in their second growing season and exceed 1 m by age 12 (Beamesderfer et al. 1989). Also in the Columbia system, males and females reach sexual maturity at about 12 and 15–20 years, respectively (Galbreath 1985). In the lower Fraser population, maturity was estimated to be in the early teens to early 20s (at about 90 cm TL) for males and in the middle to late 20s (at about 180 cm) for females (Semakula and Larkin 1968). The largest white sturgeon recorded from B.C. was reputed to be over 3.5 m long and weighed almost 650 kg.

FOOD HABITS

The gut morphology of sturgeons suggests that they are primarily carnivorous (Buddington and Christofferson 1985). Once they begin exogenous feeding, sturgeon larvae forage on chironomid larvae, cyclopoid copepods, and amphipods (Muir et al. 2000). As they grow, the size and importance of amphipods increases in their diet. Juveniles eat larger prey than young-of-the-year and consume amphipods, chironomids, freshwater clams, aquatic insect nymphs, and small fish (McCabe et al. 1989). In the lower Fraser River, adults forage primarily on fish (eulachon, sculpins, sticklebacks, and lampreys), large crustaceans (e.g., crayfish), and adult salmon. In 1994, there was a die-off of large sturgeon in the lower Fraser River. Not all the fish autopsied at this time had food in their stomachs, but those that did contained sockeye salmon (M.L. Rosenau, personal communication). Also, in Kootenay and Columbia populations, spawning kokanee are important seasonal components in the diet.

Habitat White sturgeon occur in the three largest rivers systems along the Pacific coast of North America—the Sacramento – San Joachim, Columbia, and Fraser river systems. Within these drainage systems, they are found in the river mainstems, large tributaries, reservoirs, and large lakes. Although they occur in all these environments, the specific habitats they use are not well known. A habitat summary for Fraser River white sturgeon is provided by Echols (1995). For the Columbia River between Keenleyside Dam and the border, there is a discussion of the habitat characteristics of areas where sturgeon concentrate (R. L. & L. Environmental Services Ltd. 1995b).

ADULTS

In the B.C. portion of the mainstem Columbia below Keenleyside Dam, there are four high-use areas for adult white sturgeon: from the dam down to the mouth of Norns Creek, the confluence of the Kootenay and Columbia rivers, the Fort Shepherd Eddy, and the Waneta Eddy (R. L. & L. Environmental Services Ltd. 1995b). These high-use areas have lower than average water velocities and depths (>15 m) for the mainstem Columbia (R. L. & L. Environmental Services Ltd. 1995b). Sturgeon occur in other areas but in low numbers. Most white sturgeon tagged within these high-use areas exhibited only local movements (<5 km), and these occurred mostly in the summer. There were, however, local movements out of deep water into shallow areas in spring and summer (R. L. & L. Environmental Services Ltd. 1995b). Similar shifts from deep to shallower water in the spring and summer occur in the Kootenai River (Apperson and Anders 1991) and Roosevelt Reservoir (Brannon and Setter 1992). Apparently, in both the Columbia and Fraser systems, white sturgeon overwinter in deep water (Haynes et al. 1978; McCabe and Tracy 1994), and at this time, movements are small and localized (often <0.2 km).

In the lower Fraser River, adults are abundant in the main river channel for most of the year (Nelson et al. 2004). In the main channel downstream of Mission, the water is turbid and up to 10–20 m deep; the substrate is sandy silt. Upstream of Mission, the main channel becomes shallower (often <5.0 m), and the substrate shifts to gravel. Again, movements of tagged individual usually are local, but there are longer seasonal movements (upstream spawning migrations and downstream movements in response to changes in the seasonal abundance of food). From October to March, activity is reduced and adults remain in deep low-velocity sites over winter.

In addition to their freshwater habitats, fluctuations in the concentration of strontium in their pectoral fin rays indicate that adult white sturgeon use the brackish Fraser estuary and occasionally spend extended periods of time in the sea (Veinott et al. 1999). Thus, at irregular intervals, white sturgeons appear in the lower reaches of some Vancouver Island rivers, and sometimes, tagged sturgeon move between the Columbia and Fraser rivers (Nelson et al. 2004). There is even a record of a white sturgeon tagged in the Sacramento – San Joachim drainage system, California, and recovered in the Columbia River (Chadwick 1959).

JUVENILES

In British Columbia, the upper Columbia, Kootenay, and Nechako rivers' white sturgeon populations have experienced a prolonged period of recruitment failure. Consequently, data on juvenile habitat use for these populations is meagre. In the upper Fraser, juveniles are most common in the lower reaches of large tributaries or at their confluence with the main river (R. L. & L. Environmental Services Ltd. 2001). In the lower Fraser River, juveniles are common, and there are some data on juvenile habitat use (Lane and Rosenau 1995). Here, during the summer, juveniles use both the main river channel and areas peripheral to the main channel (e.g., sloughs and backwaters). They are most abundant in warm, turbid, low-velocity areas, and at depths greater than 5 m. They appear to move out of these areas when water temperatures drop below 13–15°C. Presumably, they migrate to the main river. In the main channel, salmon test fishing also provides evidence of seasonal sturgeon movements.

It appears that there is a spring and fall pulse in abundance near Albion. Some juveniles may rear in brackish tidal channels in the river's upper estuary. The dominant substrates in the lower Fraser sites where juveniles are abundant are sand and silt.

In the lower Columbia River below McNary Dam, juveniles (<1.0 m) are found at depths ranging from 2 to 58 m in slow water areas with bottom velocities of 0.1–0.8 m/s (Parsley et al. 1993).

YOUNG-OF-THE-YEAR
Laboratory studies of the behaviour and habitat use of newly hatched and larval white sturgeon are consistent with the sparse field data on their early life history (Kynard and Parker 2005). Newly hatched larvae display a brief (about 2 day) period of downstream dispersal. This is achieved by alternate bouts of swimming about 50 cm up into the water column and then drifting back to the bottom. They do not move far and are negatively phototactic. During the day, they shelter among gravel or in other bottom cover. The use of bottom cover gradually decreases, and by about 3 weeks, the larvae are foraging in the open about 1 m above the substrate. At this time, they aggregate (at least in the laboratory) and do not disperse downstream. They are most active at night.

About 45–50 days after hatching the larvae have transformed (i.e., developed scutes, an elongated snout, and a heterocercal tail). A second downstream dispersal occurs at this time. It is not clear how long this second dispersal lasts or how far the young sturgeon move, but in rivers like the lower Fraser, they probably move many kilometres. In British Columbia, recently metamorphosed sturgeon are rarely caught. Where they go and what habitat they use are still a bit of a mystery.

Conservation Comments   COSEWIC lists the white sturgeon as an endangered species, and the BCCDC lists it as S2 (imperiled because of rarity). There are six genetically identifiable white sturgeon populations in British Columbia. Three of these stocks (the lower, middle, and upper Fraser River populations) appear to be breeding successfully, and at least one of the stocks—the lower Fraser River population—is growing with about 60,000 individuals in the 40 to 220 cm length range (Nelson et al. 2004). Two decades ago, this population was in decline; however, a series of management decisions (a ban on commercial exploitation, limited First Nations harvest, and a catch-and-release recreational fishery) have reversed the decline, and the population rebounded. In spite of this apparent success, there is some evidence from the last 2 years that the lower Fraser population may be experiencing another decline. It is not known whether this is a natural fluctuation or something more serious.

The fate of the three remaining stocks (the Nechako, Columbia, and Kootenay populations) is still in doubt. Most years, there is some reproduction in these populations but no survival of young-of-the-year. The reasons for this persistent recruitment failure are not clear and may differ among the stocks. For the Nechako population, there is evidence of a massive sediment discharge into the river in the early 1960s. This produced a sediment wave that slowly shifted downstream and filled the interstices in the gravel of the major spawning site upstream of Vanderhoof (McAdam et al. 2005). Since white sturgeon larvae use gravel substrate for shelter for parts of their early life history, this sediment wave may be involved in the failure of the Nechako population to produce young. In the Columbia and Kootenay populations, dams have subdivided the populations and changed the historic hydrograph. These changes could affect recruitment success. In British Columbia, the three threatened stocks are under intensive study, and large-scale efforts are underway to save them from local extirpation.

FAMILY HIODONTIDAE — MOONEYES

The mooneyes are part of an ancient group of basal teleost fishes—the Osteoglossomorpha. They are distinguished from other basal teleosts by a number of osteological features including the presence of teeth on a median bone (the parasphenoid) on the roof of the mouth. These parasphenoid teeth, and the teeth on the tongue, provide a bite that is unique among living bony fishes. Most living osteoglossomorphs occur in the tropical freshwaters of Africa, Asia, Australia, and South America. The mooneyes (Hiodontidae) are an exception to this generally tropical distribution. This family is restricted to North America east of the Continental Divide. In the past, however, mooneyes also occurred in East Asia and in North America west of the Continental Divide (Li et al. 1997). In British Columbia, an extinct genus (*Eohiodon*) is known from mid-Eocene deposits near Princeton, Kamloops, and in the Horsefly region (Wilson 1977). There are two living species in the family—the goldeye (*Hiodon alosoides*), and the mooneye (*H. tergisus*). Only the goldeye occurs in British Columbia.

Hiodon alosoides (RAFINESQUE)
GOLDEYE

1 cm

Distinguishing Characters   The goldeye has a deep, laterally compressed body and sides that taper downward into a smooth ventral keel; the dorsal fin is placed well behind the middle of the body and originates directly over, or behind, the origin of the anal fin; the length of the anal fin base is over twice the length of the dorsal fin base; and, in adults, the iris of the eye is a golden yellow. Although no other freshwater fish in British Columbia resembles this species, it is occasionally mistaken for the lake whitefish; however, the goldeye lacks the adipose fin that is present in all whitefishes.

Taxonomy   The goldeye is the only member of the mooneye family in British Columbia. Thus, identification is not a problem in this province. In Alberta, however, there are two species of *Hiodon*—the goldeye and its close relative the mooneye (*H. tergisus*). The two species are easily confused, but *The Fishes of Alberta* provides a detailed list of the characters that separate these species (Nelson and Paetz 1992).

Sexual Dimorphism   Mature female goldeye are, on average, larger than mature males. The shape of the anal fin in adult males is distinctive—the anterior fin-rays are thickened and over twice as long as the posterior fin-rays, and the transition between the two parts of the fin is abrupt. This gives the anterior portion of the anal fin a distinctive convex lobe (the accompanying goldeye illustration is a male). This sexual dimorphism is ancient, and fossil *Eohiodon* from Eocene lake deposits show the same distinctive male anal fin as found in living goldeye (Wilson 1996). In adult females and immature fish, the anterior rays of the anal fin also are long, but their length tapers smoothly back toward the tail.

Distribution NATIVE DISTRIBUTION
With the exception of an apparently isolated set of populations in northern Ontario and Quebec, the goldeye is a species of the large turbid rivers of the North American Plains. Here, goldeye range from just above the Mackenzie River Delta (Arctic Red River) to Alabama.

BRITISH COLUMBIA DISTRIBUTION
All B.C. records of goldeye are from turbid rivers associated with the lower Peace and lower Liard drainage systems (Map 7). The goldeye in the B.C. portion of the Peace system probably represents a segment of the upper Peace population in Alberta. Apparently, this population is biologically different from the goldeye that migrate into the Peace–Athabasca Delta: they are smaller at any given age, and there is evidence that they do not spawn every year (Donald and Kooyman 1977a). Similarly, the Liard population is probably part of a larger Mackenzie River population. Although goldeye are known to spawn in rivers, there are no records of goldeye spawning, and no fry have been collected, in British Columbia.

Life History Virtually nothing is known about the biology of goldeye in British Columbia; however, their biology probably is similar to populations in northern Alberta and the western Northwest Territories. Thus, information on these populations (Bond 1980; Donald and Kooyman 1977a, 1977b; Kennedy and Sprules 1967; Tallman et al. 1996b; Tripp et al. 1981), augmented with a few B.C. observations, form the basis of the following life-history account.

REPRODUCTION
Apparently, goldeye (adults and juveniles) overwinter in large turbid rivers like the lower Peace and lower Liard. Shortly after ice-out (usually sometime in late March or early May), the adults migrate to spawning sites. Some goldeye that spawn in the Peace–Athabasca Delta perform a return migration of over 700 km (Donald and Kooyman 1977b), and a tagged fish was recovered 320 km from its original tagging site (Bond 1980). The spring migration of juveniles from the lower Peace River into the Peace–Athabasca Delta occurs about 2 weeks after the adult migration (Donald and Kooyman 1977b). Presumably, this juvenile migration is a shift to summer feeding sites. Spawning in goldeye has not been observed, but data on ripe adults and newly hatched larvae suggest that spawning occurs in pools and backwaters in large turbid rivers as well as in lakes (Battle and Sprules 1960; Kennedy and Sprules 1967). In winter, many of these putative spawning sites freeze to the bottom, and the migratory behaviour of both adults and juveniles may be a response to severe winter conditions.

Spawning in the Peace–Athabasca Delta region begins about mid-May and lasts for 4 or 5 days, although a few fish are still spawning in mid-June (Donald and Kooyman 1977b). Water temperatures at this time range from 10 °C to 12 °C, and spawning is thought to occur at night (Kennedy

Map 7 The B.C. distribution of goldeye, *Hiodon alosoides*

and Sprules 1967). Although there are relatively little data on fecundity, it is clear that egg number is a function of female body size. For eight adult goldeye from the Slave River Delta, egg counts ranged from 2,000 to 16,345 (Tripp et al. 1981). Fertilized eggs are translucent with a steel-grey cast and are relatively large (roughly 4 mm in diameter); however, about half of this diameter is perivitelline space. For a freshwater fish, the eggs are unusual in that they are semi-buoyant and, apparently, at some depths they are free-floating (Battle and Sprules 1960). Presumably, in rivers, these semi-buoyant eggs would drift downstream. The eggs take roughly 15 days to hatch, and the newly hatched larvae are about 7.5 mm TL. The newly hatched larvae have a large yolk sac and have been observed (Battle and Sprules 1960) floating head-up at the surface. The yolk is absorbed at 12–13 mm, and by this size, the larvae can swim normally and begin exogenous feeding.

AGE, GROWTH, AND MATURITY

Most age data on goldeye are based on scales; however, the operculum gives more reliable age estimates, especially for older fish (Donald et al. 1992). In relatively productive areas like the shallow lakes of the Peace–Athabasca Delta region, growth in the first summer is rapid, and by September, young-of-the-year can reach almost 100 mm FL (Donald and Kooyman 1977a). However, entirely fluvial populations grow more slowly at all ages (Donald and Kooyman 1977b), and young-of-the-year in the Athabasca River reached about 55 mm by the end of their first growing season (Bond 1980). In the Fort Nelson River in British Columbia, the operculae of goldeye 100–120 mm in length indicated that they were in their second growing season.

Growth slows as goldeye approach sexual maturity; because males, on average, mature 1 or 2 years earlier than females, the growth trajectories of the sexes diverge. In Athabasca River goldeye, there is a divergence in the growth rates of males and females at 5 or 6 years (Bond 1980); in the Slave River Delta, about half the males are mature by age 3, but only a small percentage of females are mature by age 5 (Tripp et al. 1981). In the lower Liard River in British Columbia, an immature female was aged at 6 years (McLeod et al. 1979). Goldeye normally reach lengths of about 350 to 400 mm and live about 10–12 years. The maximum age recorded for this species is 23 years (Tripp et al. 1981).

FOOD HABITS

Adult goldeye eat a wide variety of organisms, although aquatic insects (especially corixids and stonefly nymphs) are the dominant items in their diet. Large individuals also prey on small fishes. At times, adult goldeye forage extensively on aerial insects and occasionally on frogs and mice (Bond 1980; Donald 1997; Donald and Kooyman 1977b; Tallman et al. 1996b). Newly hatched larvae forage primarily on copepods and cladocerans and, as they grow, select the larger organisms from the plankton community. Eventually, the fry add insects to their diet. Apparently, June and July are the main growing season for yearlings and juveniles (Donald and Kooyman 1977b).

Habitat ADULTS

In North Dakota, adult goldeye are abundant in backwaters of the Missouri River (Moon et al. 1998). In Alberta and the Northwest Territories, adult goldeye occur in both rivers and lakes. Apparently, the use of lakes in the Peace–Athabasca Delta region is seasonal. Mature adults overwinter in the Peace River and migrate into the delta lakes in May (just after ice-out). They migrate out of the lakes and back to the river from mid-July to mid-August. This migration from the river to the lakes and back to overwintering sites in the river can cover a distance of 700–800 km (Donald and Kooyman 1977a).

There is evidence for two goldeye populations in the Peace River (Donald and Kooyman 1977a)—a population in the lower Peace River discussed above and an upper Peace population that presumably spawns in the upper river. The goldeye in the B.C. portion of the Peace River probably are part of this upper Peace population. Generally, in British Columbia, adult goldeye are confined to turbid rivers. In the summer, they appear to aggregate in deep, quiet water areas near the mouths of tributary streams or in the lee of islands.

JUVENILES

In the Missouri River, juvenile goldeye usually are found in backwaters (Moon et al. 1998). In British Columbia, juveniles are sometimes collected in shallow water (<1 m) along relatively slow moving stretches of river. The substrate is usually sand and silt. It is rare to catch more than one juvenile in a single seine haul, and this suggests that they may be solitary. This is consistent with aquarium observations (Fernet and Smith 1976) of a well-developed agonistic repertoire in young goldeye that may function as a spacing mechanism.

YOUNG-OF-THE-YEAR

In the shallow Peace–Athabasca Delta lakes, fry are most abundant within 100 m of shore and clearly are not randomly distributed around the lakes (Donald and Kooyman 1977a). In the Missouri River, they occur in backwaters (Moon et al. 1998). No goldeye fry have been collected in British Columbia, and there are no local data on their habitat use.

Conservation Comments   Although goldeye are modestly common in the mainstems of the lower Peace and lower Liard rivers (including large tributaries like the Fort Nelson River), it is not clear that there is a breeding population in British Columbia. Neither the BCCDC nor COSEWIC list this species.

FAMILY CLUPEIDAE — HERRINGS

The herrings (Clupeidae) belong to a large and economically important group of basal teleost fishes, the Clupeomorpha, that are thought to be the sister group of the Ostariophysi (Lecointre and Nelson 1996). Within the Clupeomorpha, the herrings belong in the order Clupeiformes. This order contains about 330 species. Most of these species are marine; however, some species are anadromous, and a few are restricted to fresh water. Two herrings sporadically enter the fresh waters of British Columbia—the native Pacific herring (*Clupea pallasii*) and an introduced anadromous species, the American shad (*Alosa sapidissima*). The Pacific herring is common in estuaries and often enters tidal freshwater in large rivers like the Fraser and Skeena. In the Fraser, herring occasionally occur upstream to about the Port Mann Bridge (about 20 km upstream). Because the Pacific herring does not breed in fresh water, I include it in the species key but do not give a synopsis of its distribution, life history, or habitat use.

The anadromous American shad is native to the east coast of North America where it ranges from southern Newfoundland to Florida. It was introduced into the North Pacific (California) in 1871 and reached the coast of British Columbia by 1876 (Hart 1973). A second west coast introduction into the lower Columbia River was made in 1885–1886 (Wydoski and Whitney 2003). The shad in the North Pacific Ocean range from San Pedro, California, to southeastern Alaska and, perhaps, Kamchatka. Their breeding range, however, is much narrower than this marine distribution—the Sacramento River system, California, to the Chehalis River, Grays Harbor, Washington. By the late 1950s, shad appeared to be increasing in the lower Fraser River (Carl et al. 1959); however, they have never become established, although occasional individuals are still caught in salmon nets in the estuary and on the banks just off the river delta. Still, with global warming, it is possible that this species may yet establish a self-sustaining run in the Fraser River.

KEY TO ADULT HERRINGS

1 (2) Body deep (maximum depth about 3–4 times into standard length); well-developed striae on the operculum (Fig. 6A); enlarged scales on the base of the caudal fin just above and below the midline of the fin	AMERICAN SHAD, *Alosa sapidissima*
2 (1) Body moderately deep (depth about 4.5 times into standard length); no striae on the operculum (Fig. 6B); no enlarged scales on the base of the caudal fin	PACIFIC HERRING, *Clupea pallasii*

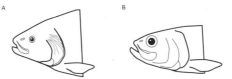

Figure 6 The heads of an American shad (A) and a Pacific herring (B), illustrating the presence or absence of striations on the gill covers.

Alosa sapidissima (WILSON)
AMERICAN SHAD

1 cm

Distinguishing Characters   This large (up to 60 cm in the Columbia River) herring-like fish has thin scales that are easily shed, a deep compressed body with strong saw-like serrations (scutes) on the midline of the belly, and a large number (59–73) of long thin gill rakers on the lower limb of the first gill arch.

Most of the B.C. records of shad in fresh water are from the lower Fraser River. Pacific herring (*Clupea pallasii*) also occur in the Fraser River below the Port Mann bridge. Consequently, it is possible, although unlikely, that both species could be encountered downstream of the Port Mann bridge. They are easily distinguished by body shape (noticeably deeper in shad), saw-like scales on the ventral midline (strong in shad and weak in herring), and by the strong, downward slanting striae on the operculum of shad (absent in herring).

Taxonomy   Like many anadromous fish, shad home to their natal streams and consequently form demes (local interbreeding units). Thus, there are often small, but consistent, genetic differences even among introduced populations on the west coast. Within their native range—the Atlantic Coast of North America—the genetic differences among shad populations are greater than on the west coast; however, this variation tends to be clinal (Leggett and Carscadden 1978). Thus, although populations at the extreme northern and southern ends of the natural geographic range are strikingly different, intermediate populations tend to grade into one another. For this reason, no subspecies are recognized.

Sexual Dimorphism   There are no obvious morphological differences between the sexes.

Distribution NATIVE DISTRIBUTION
American shad are indigenous to the Atlantic coast of North America. Here, their geographic range extends from Labrador and Newfoundland south to the St. Johns River, Florida.

BRITISH COLUMBIA DISTRIBUTION
Occasional individuals turn up in our coastal waters (Map 8), and small numbers appear sporadically in the lower Fraser River (Carl et al. 1959). So far, the maximum upstream penetration is to about Hope, and presumably, the Fraser Canyon is a barrier to their upstream movement. Although there is no evidence of reproduction in British Columbia, shad may eventually breed in lower Fraser tributaries if water temperatures in the Fraser River continue to rise. At present, however, the closest breeding population is in the Chehalis River, Grays Harbor, Washington (Wydoski and Whitney 2003). The largest adjacent breeding population (estimated to be about 30 million) is in the Columbia River.

Life History Little is known about shad in British Columbia. Consequently, the following account is based on data from Washington (Wydoski and Whitney 2003) and California (Moyle 2002).

REPRODUCTION
There is a well-established run of shad in the Columbia River. They enter the Columbia River in April and migrate upstream from mid-April to August. In California, they do not enter freshwater until the water temperature rises above 14 °C, and in the Maritime Provinces, the runs taper off when the water temperature exceeds 20 °C (Leim 1924). There is some evidence that shad are adapting to conditions in the Pacific Northwest. For example, the entry time of shad runs into the Columbia River has changed over 50 years (Quinn and Adams 1996)—50% of the shad run entered the river by late July (water temperature about 18 °C) in 1938, whereas 50% of the run entered the river by June 15 (water temperature about 16 °C) in 1993. In the Columbia River, actual spawning occurs at 15.5–18.3 °C and, in California, at 17–24 °C.

Spawning occurs in the main channels of rivers. The substrate does not appear to be critical, although most spawning occurs over sand or gravel at depths of 1–10 m and water velocities of 0.30–0.90 m/s (Moyle 2002). Shad are group spawners, and spawning activity appears to be triggered by the onset of darkness. Males initiate the spawning act when one or more males press against a female. The fish then swim side-by-side in a circle, releasing gametes as they go. The dorsal and upper caudal fins often break the surface and, when many groups are spawning at the same time, they produce a distinctive splashing sound. Not all the gametes are released in a single spawning, and over a period of days, most fish spawn several times. Fecundity is high, and shad in California produce from about 100,000 to over 400,000 eggs. The fertilized eggs

Map 8 The B.C. distribution of American shad, *Alosa sapidissima*

are about 2.5–3.5 mm in diameter and semi-buoyant (Leim 1924). They drift downstream and hatch in about 8–12 days at temperatures of 11–15 °C. Newly hatched shad are about 6–10 mm TL, planktonic, and begin exogenous feeding at 9–12 mm. They metamorphose into actively swimming fry at about 25 mm (Moyle 2002).

AGE, GROWTH, AND MATURITY

In California, freshwater growth rates appear to be related to temperature and food availability (Moyle 2002). In the Columbia system, most fry migrate to sea in the fall of their first year. At this time, they are about 65–115 mm FL (Wydoski and Whitney 2003). Once in the sea, they grow rapidly; males typically mature at about 36–43 cm (ages 3 or 4 years), whereas females mature at age 4 or 5 years (usually at about 40–50 cm). The maximum size in the Columbia system is about 61 cm, and the maximum age in California is 7 years (Moyle 2002).

FOOD HABITS

At sea, shad feed primarily on zooplankton (especially copepods and mysids), but occasionally, they take small fishes (Leim 1924). In fresh water, fry forage on small zooplankton; however, as they grow, they add insect larvae and pupae (especially chironomids) to their diet. They are visual predators and hunt during the daylight hours.

Habitat ADULTS
Adult shad spend most of their lives in the marine pelagic zone and only enter fresh water to spawn. There are well-established runs in the lower Columbia River, and some individuals migrate as far upstream as the lower Snake River (about 700 km from the sea). At present, there is no evidence of a self-sustaining run in the Fraser River. Shad do not necessarily die after spawning (Leim 1924), and some individuals return to the sea and eventually spawn again; however, in California, most adults are first-time spawners (Moyle 2002).

JUVENILES
In California, some juveniles remain in shallow (1–1.5 m) estuarine waters for 1–2 years (Moyle 2002).

YOUNG-OF-THE-YEAR
The fry spend their first summer in fresh water. Apparently, they prefer relatively warm water (17–23 °C) and concentrate in the sloughs and embayments in the lower reaches of their natal streams. In the Columbia system, the fry spend their first summer in freshwater and migrate to sea in the fall, but in California, most fry also move directly out to sea (Moyle 2002).

Conservation Comments Although shad are not necessarily a warm-water species, they do best at warmer temperatures than salmonines. Consequently, if shad do establish themselves in the Fraser River, it may signal that something is seriously wrong with the river. Their history in the Columbia system suggests that, once they become established, they can adapt to new conditions and become very numerous over a relatively short time (about 50 years). What millions of new planktivores would do to the ecological balance in the Strait of Georgia is hard to predict, but probably, it would not be good.

FAMILY CYPRINIDAE — MINNOWS

The minnows and carps (Cyprinidae) comprise the largest known family of freshwater fishes with over 2,400 living species. Although, cyprinids have the largest number of species of any family in the fresh waters of the Northern Hemisphere, they reach their greatest diversity in Southeast Asia. Together with a number of smaller families (e.g., the suckers, Catostomidae; the loaches, Cobitdae; and river loaches, Balitoridae), the minnows make up the order Cypriniformes. In this order and two large related orders (the Characiformes and Siluriformes), the first four or five vertebrae and associated ligaments are modified to provide a direct link between the swimbladder and the inner ear. Apparently, this structure (the Weberian apparatus) enhances sound detection in water. They also possess specialized cells in the skin that release a "fright substance" if the skin is torn. Although conspecifics have the strongest reaction to this substance, related species also react (albeit less strongly), and even fish in other families may learn to respond to this substance.

In North America, the earliest known minnow fossils (mid-Oligocene) are from the Pacific Northwest (Cavender 1992). There are now almost 300 indigenous species of minnows in North America, but only 12 species are native to British Columbia. In addition to our native species, we have four introduced species—three of these (the common carp, *Cyprinus carpio*; goldfish, *Carassius auratus*; and tench, *Tinca tinca*) are Eurasian species and one (the fathead minnow, *Pimephales promelas*) is native to North America but not to British Columbia. Although we have only 12 native minnows, five of them (chiselmouth, *Acrocheilus alutaceus*; peamouth, *Mylocheilus caurinus*; leopard dace, *Rhinichthys falcatus*; speckled dace, *Rhinichthys osculus*; and Umatilla dace, *Rhinichthys umatilla*) occur nowhere else in Canada.

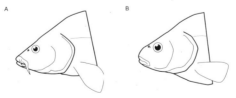

Figure 7 The heads of the common carp (A) and goldfish (B) illustrating the presence and absence of barbels on their upper jaws

Figure 8 Body profiles of the redside shiner (A) and northern pearl dace (B) illustrating differences in the origin of the dorsal fins relative to the origins of the pelvic fins

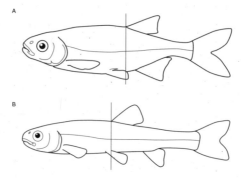

Figure 9 Body profiles of peamouth (A) and chiselmouth (B), illustrating the differences in the origin of the dorsal fins relative to the origins of the pelvic fins.

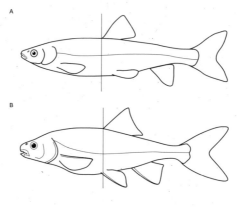

Figure 10 Head profiles of northern pikeminnow (A) and peamouth (B), illustrating the difference in jaw length relative to the eye.

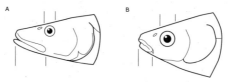

Figure 11 Head profiles of longnose (A) and speckled (B) dace, illustrating the difference in the attachment of the upper jaw to the snout.

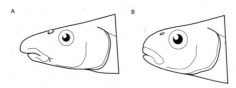

Figure 12 Ventral view of a chiselmouth head showing the keratinized lower jaw.

FAMILY CYPRINIDAE—MINNOWS   53

Figure 13 Body profiles of fathead minnow (A) and northern pearl dace (B) illustrating differences in the origins of the dorsal fins relative to the pelvic fins.

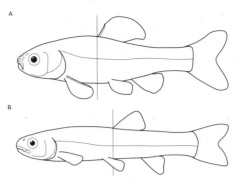

Figure 14 Head profiles of leopard (A) and speckled (B) dace, illustrating the differences in snout overhang.

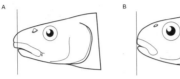

Figure 15 Diagrams illustrating dorsal fin shapes in leopard (A) and speckled (B) dace.

Figure 16 Ventral views of mouth size relative to the eyes in finescale (A) and northern redbelly (B) dace.

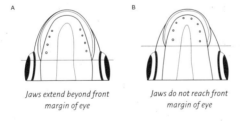

Jaws extend beyond front margin of eye   *Jaws do not reach front margin of eye*

Figure 17 Diagram illustrating the dfferences in complexity of intestinal loops in finescale (A) and northern redbelly (B) dace.

KEY TO ADULT MINNOWS

1 (4) Dorsal fin base much longer than the head *(see 2)*

    2 (3) Two pairs of barbels on the sides of upper jaw (Fig. 7A) — CARP, *Cyprinus carpio*

    3 (2) No barbels on the sides of upper jaw (Fig. 7B) — GOLDFISH, *Carassius auratus*

4 (1) Dorsal fin base shorter than head *(see 5)*

    5 (6) Caudal peduncle deep, its depth more than half the head length; all the fins are dark; tail has no fork; iris of the eye red — TENCH, *Tinca tinca*

    6 (5) Caudal peduncle narrow, its depth less than half the head length; not all the fins are dark; tail is forked; iris of the eye is not red *(see 7)*

        7 (8) Dorsal fin originates far back on body; viewed from the side, its origin is almost directly above the tips of the pelvic fins (Fig. 8A) — REDSIDE SHINER, *Richardsonius balteatus*

        8 (7) Dorsal fin originates at mid-body; viewed from the side, the tips of the pelvic fins extend well past origin of dorsal fin (Fig. 8B) *(see 9)*

            9 (16) Viewed from the side, a perpendicular line projected downward from the insertion of the dorsal fin does not overlap the pelvic fins (Fig. 9A) *(see 10)*

                10 (13) Mouth large, upper jaw reaches to, or extends beyond, the front margin of the eye (Fig. 10A) *(see 11)*

                    11 (12) Outer pectoral rays long, equal to head length and nearly reaching pelvic fins; top of head flat; snout overhangs mouth; prominent barbels at corners of mouth — FLATHEAD CHUB, *Platygobio gracilis*

                    12 (11) Pectoral fins shorter than head; top of head convex; snout does not overhang mouth; no barbels at corners of mouth — NORTHERN PIKEMINNOW, *Ptychocheilus oregonensis*

                13 (10) Mouth small, upper jaw does not reach front margin of the eye (Fig. 10B) *(see 14)*

                    14 (15) Small barbels at the corners of the mouth; more than 60 scales in the lateral line — PEAMOUTH, *Mylocheilus caurinus*

                    15 (14) No barbels at the corners of the mouth; fewer than 45 scales in lateral line — BRASSY MINNOW, *Hybognathus hankinsoni*

            16 (9) Viewed from the side, a perpendicular line projected downward from the insertion of the dorsal fin clearly overlaps the pelvic fins (Fig. 9B) *(see 17)*

                17 (18) Snout attached to upper lip, no groove separating snout from upper lip; upper jaw not protrusible (Fig. 11A) — LONGNOSE DACE, *Rhinichthys cataractae*

                18 (17) Snout not attached to upper lip, a groove separates snout from upper lip; upper jaw protrusible (Fig. 11B) *(see 19)*

                    19 (20) Chisel-like lower jaw nearly straight in adults (Fig. 12); 9 or 10 anal rays — CHISELMOUTH, *Acrocheilus alutaceus*

                    20 (19) Lower jaw normal, not chisel-like; 7 or 8 anal rays *(see 21)*

                        21 (22) Viewed from the side, the dorsal fin originates directly above, or slightly in front of, the origin of the pelvic fins (Fig. 13A); predorsal scales crowded and noticeably smaller than the scales along the sides — FATHEAD MINNOW, *Pimephales promelas*

                        22 (21) Viewed from the side, the dorsal fin originates behind the origin of the pelvic fins (Fig. 13B); predorsal scales not crowded and not noticeably smaller than the scales along the sides *(see 23)*

                            23 (26) Mouth subterminal, snout overhangs upper jaw (Fig. 14A); trailing edge of dorsal fin concave (Fig. 15A) *(see 24)*

                                24 (25) Barbels conspicuous and clearly protrude beyond corners of the mouth; well-developed fleshy membranes (stays) connect inner rays of pelvic fins to body; base of pelvic fin longer than free portion of the last ray — LEOPARD DACE, *Rhinichthys falcatus*

KEY TO ADULT MINNOWS *(continued)*

25 (24) Barbels inconspicuous and do not protrude beyond corners of the mouth; fleshy membranes (stays) that connect inner rays of pelvic fins to body not well developed; base of pelvic fin shorter than free portion of the last ray	UMATILLA DACE, *Rhinichthys umatilla*
26 (23) Mouth terminal, snout does not overhang upper jaw (Fig. 14B); trailing edge of dorsal fin straight or convex (Fig. 15B) *(see 27)*	
27 (28) Prominent black spot at base of caudal fin	SPOTTAIL SHINER, *Notropis hudsonius*
28 (27) No prominent black spot at base of caudal fin *(see 29)*	
29 (30) In life, a silvery midlateral band (dark in preserved specimens); 9–13 (usually 11) anal rays	EMERALD SHINER, *Notropis atherinoides*
30 (29) No midlateral silver band; 6–9 (usually 7 or 8) anal rays *(see 31)*	
31 (34) Peritoneum silvery; barbels present (but inconspicuous in northern pearl dace) *(see 32)*	
32 (33) Conical barbels visible at posterior corners of the mouth; snout moderately pointed; fewer than 70 lateral line scales	LAKE CHUB, *Couesius plumbeus*
33 (32) Flap-like barbels (rarely visible without magnification) are located slightly in front of the corners of the mouth; snout blunt; more than 70 lateral line scales	NORTHERN PEARL DACE, *Margariscus margarita nachtriebi*
34 (31) Peritoneum brown, black, or darkly speckled; no barbels at the corners of the mouth *(see 35)*	
35 (36) Adults without dark lateral stripes; anal rays 6 or 7 (rarely 8); in British Columbia, found only in the Kettle River system	SPECKLED DACE, *Rhinichthys osculus*
36 (35) Adults with one or two dark lateral stripes; anal rays usually 8 (rarely 7 or 9); in British Columbia, found only in the Mackenzie River system *(see 37)*	
37 (38) Viewed from below, jaw extends beyond front margin of eye (Fig. 16A); single dark lateral stripe; intestine with a single loop (Fig. 17A)	FINESCALE DACE, *Phoxinus neogaeus*
38 (37) Viewed from below, jaw does not reach the front margin of eye (Fig. 16B); two dark lateral stripes; intestine with multiple loops (Fig. 17B)	NORTHERN REDBELLY DACE, *Phoxinus eos*

Acrocheilus alutaceus AGASSIZ & PICKERING
CHISELMOUTH

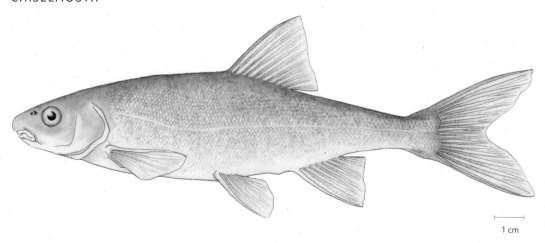

1 cm

Distinguishing Characters — This large (adults often exceed 200 mm TL), fine-scaled minnow has an unusually narrow caudal peduncle and a large caudal fin (in juveniles, the tail is asymmetrical with the upper lobe noticeably longer than the lower lobe). Viewed from the side, the strongly subterminal mouth gives the impression of a fish with a severe overbite. In adults, the lower jaw is almost a straight cutting-edge sheathed in a keratinized "chisel." This distinctive chisel is used to scrape algae from hard surfaces. Like most fish that consume large amounts of plant material, the peritoneum is jet black. Although adults are morphologically distinctive, the fry are not. Their "chisel" is not obvious, and their tail is not asymmetrical. Consequently, they do not stand out when collected in mixed schools of northern pikeminnow (*Ptychocheilus oregonensis*), redside shiner (*Richardsonius balteatus*), and peamouth (*Mylocheilus caurinus*) fry.

The relative positions of the dorsal and anal fins separates chiselmouth fry from young redside shiner fry—the dorsal fin inserts well behind the origin of the anal fin in redside shiners, whereas the dorsal fin inserts in front of the anal fin in chiselmouth fry. Mouth position and pigment pattern distinguish chiselmouth fry from pikeminnow fry: the snout overhangs in chiselmouth fry (it is terminal in pikeminnow fry), and chiselmouth fry lack the dark spot at the base of the caudal fin that is conspicuous in pikeminnow fry. Chiselmouth and peamouth fry are similar in general appearance; however, the chiselmouth's dorsal fin originates directly above (or slightly behind) the origin of the pelvic fins, whereas the peamouth's dorsal fin originates in front of the pelvic fins.

Sexual Dimorphism — Although sexual differences in size are not obvious in chiselmouth, adult females are, on average, slightly larger than adult males (Lassuy 1990). The most apparent external difference between the sexes is pectoral fin length. In adult males, the pectoral fins are longer than the head and extend back almost to the origin of the pelvic fins. In adult females, the pectorals are shorter than the head and stop well in front of the pelvic fins. Breeding fish of both sexes develop tubercles on the pectoral and pelvic fins, the head, the operculum, and the back; however, these tubercles are less obvious in females than in males. Also, in breeding males, there are distinctive patches of roughened, comb-like scales on the breast immediately in front of each pectoral fin, and the dorsal surfaces of the pectoral fins are more heavily pigmented than those in females.

Hybridization   In British Columbia, hybrids between chiselmouth and the northern pikeminnow are common in Missezula Lake and Wolfe Creek (both near Princeton). Ecologically and behaviourally, this is an unusual hybrid combination: one parent is an algae scraper, and the other a piscivore. Still, first-generation hybrids somehow survive, and there is evidence of backcrossing into the parental genomes (Stewart 1966). In the same area, there is evidence of hybridization between chiselmouth and the redside shiner, and between chiselmouth and peamouth (Stewart 1966).

Distribution   **NATIVE DISTRIBUTION**
The chiselmouth is a Columbia endemic: it occurs only in the Columbia River system and adjacent drainages that received their fauna from the Columbia River. It is a species of arid interior regions and occurs in Idaho (below Shoshone Falls), Nevada, Oregon, Washington, and British Columbia.

BRITISH COLUMBIA DISTRIBUTION
Chiselmouths are an interior species (Map 9). They are found in the Okanagan, Kettle, and Similkameen rivers, and in the Kootenay River below Bonnington Falls. They also occur at scattered sites in the Fraser system from just above the Fraser Canyon (Big Bar area) to the Prince George region. Their distribution within the Fraser system is curious: there is a northern cluster of populations that includes sites in the Chilcotin, Nazko, and Euchiniko rivers and single records from the mainstem Fraser near Woodpecker (between Quesnel and Prince George); the Salmon River near Prince George; and a southern cluster in the Thompson River system that includes populations in Nicola, Vidette, and Mara lakes.

Historically, chiselmouth were abundant in the Similkameen and Okanagan systems (especially in small lowland lakes), but in the 1950s and 1960s, most of the small lowland lakes in the south Okanagan were "rehabilitated" (poisoned). Consequently, except for large lakes (e.g., Osoyoos, Vaseaux, Skaha, and Okanagan) and rivers, chiselmouth are now less common in the south Okanagan. They are still relatively abundant in the Kettle River (both above and below Cascade Falls), but except for a dubious record from Windemere Lake (Carl et al. 1959), they appear to be absent from the upper Columbia River system. The Windemere Lake record is suspect for two reasons: (*i*) no other chiselmouth

Map 9 The B.C. distribution of chiselmouth, *Acrocheilus alutaceus*

has been recorded anywhere in the upper Columbia system in over 50 years of fisheries work, and (*ii*) recent intensive collecting, specifically for chiselmouth, failed to find this species in either Windemere Lake or adjacent waters.

Life History Two studies provide information on the life history of chiselmouth in British Columbia: one involves a lacustrine population in Wolfe Lake near Princeton (Moodie 1966; Moodie and Lindsey 1972), and the other involves a fluvial population in the Blackwater River (Porter and Rosenfeld 1999; Rosenfeld et al. 2001). The following account is based on these studies, supplemented with a few incidental field observations.

REPRODUCTION

Chiselmouth spawn in the early summer, apparently when water temperature rises above 17 °C (Stewart 1966). Consequently, the exact time of spawning varies among localities: gravid females first appeared in the inlet of Wolfe Lake in mid-June, whereas spawning started in Missezula Lake outlet in early June (Stewart 1966). In the much larger and deeper Nicola Lake, spawning females were not collected until early August. Spawning occurs in both streams and lakes, and the eggs are deposited at night over gravel substrates. The eggs are demersal and adhesive, and although there is no mention of egg size in the literature, gravid museum specimens contain eggs about 1.8 mm in diameter. Presumably, ripe unpreserved eggs are somewhat larger. The only egg count available for chiselmouth is the mean count (6,200) for six females from Wolfe Lake. Development rate is temperature sensitive, and eggs reared at 12 °C hatched in 16 days while those reared at 18 °C hatched in 6 days. Newly hatched larvae are about 8 mm TL.

AGE, GROWTH, AND MATURITY

In the Kettle River, by the end of their first growing season (October) young-of-the-year chiselmouth ranged from 35 to 62 mm FL. In the same river, a second size class ranged from 77 to 115 mm. These sizes-at-age are similar to those recorded in Washington (Moodie and Lindsey 1972) and are slightly larger than those found in the Wolfe Lake population. Males reach sexual maturity in their third year, and most females mature a year later. The maximum age recorded for this species is 22 years (Lassuy 1990).

FOOD HABITS

Trophically, the chiselmouth is unusual for a B.C. freshwater fish. Its morphology, behaviour, and physiology are adapted for herbivory (Lassuy 1990)—a relatively common trophic role in tropical fresh waters but rare in northern freshwater fish. Like many fish, chiselmouth diet changes with season, age, and habitat. In Oregon, adults in a fluvial population began feeding on algae in the early spring (before spawning), whereas diatoms became important in their diet later in the summer and early autumn (Lassuy 1990). In winter, adult chiselmouth became omnivorous and included non-algal plants and animal matter in their diet (Lassuy 1990). In a fluvial population, young-of-the-year and juveniles (<100 mm) consumed mostly insects in summer (Moodie 1966). Most of these were taken from the surface. In contrast, the diet of adults (>100 mm) in the same area consisted primarily of diatoms and algae. A similar diet shift from insects to algae and diatoms was observed in the Yakima system in Washington (Titus 1996); however, it occurred at a smaller size (about 40 mm). A comparison of summer diets in lacustrine and fluvial chiselmouth (Moodie 1966) revealed that the diets of young chiselmouths were similar in both habitats but that the diet of fluvial adults was more restricted than that of lacustrine adults. The latter consumed a wider range of animal material (mostly chironomid larvae and pupae) than fluvial adults.

Habitat In British Columbia, information on habitat use by chiselmouth is limited to two studies: one in Wolfe Lake near Princeton (Moodie 1966; Moodie and Lindsey 1972) and the other in the rivers of the Blackwater system near Prince George (Porter and Rosenfeld 1999; Rosenfeld et al. 1998). The following account is based on these studies supplemented with incidental field observations.

ADULTS

In British Columbia, chiselmouth occur in both lakes and rivers, but they are primarily a fluvial species (Moodie 1966). Certainly, their streamlined shape, narrow caudal peduncle, and enlarged caudal fin suggest a strong swimmer. An analysis of habitat use by chiselmouth (Rosenfeld et al. 1998) suggests that water temperature is the primary factor influencing the presence of chiselmouth in large drainage basins within British Columbia; however, habitat variables (e.g., stream widths, velocities, depths, and substrates) govern distribution within drainage basins. Thus, in the Blackwater River system, chiselmouth were confined to the mainstems of rivers with a bankfull width of over 10 m (Porter and Rosenfeld 1999). Within these rivers, adult chiselmouth were collected in a range of habitats: runs, riffles, glides, backwaters, and pools; however, they were most common in the deeper (>1.0 m), higher velocity (0.4–0.8 m/s) runs and glides with a high proportion of cobble and boulder substrate.

In Washington, adults were spaced out over the bottom suggesting that they may be territorial (Moodie 1966). In the Kettle River, adults occupied similar habitats to those described in the Blackwater system: water over 1 m deep in areas with a boulder bottom and moderate to strong currents. In Wolfe Lake during the spring, summer, and early fall, adult chiselmouths occur in the littoral zone; however, in October, they shift into deeper water and may remain quiescent over the winter (Moodie 1966). A similar seasonal habitat shift from rivers into small lakes occurs in the late fall in the Blackwater River system (Rosenfeld et al. 1998).

JUVENILES

In the Blackwater system, juvenile chiselmouth were most abundant in habitats associated with low water velocities and dense vegetation. These habitats usually were marginal to the main river channels (e.g., side-channels and shallow, vegetated areas). In Wolfe Lake in the summer, juveniles (<100 mm) occurred in mixed schools of minnows in shallow water close to shore.

YOUNG-OF-THE-YEAR

In rivers, the fry are found in quiet (<4 cm/s), shallow backwaters and pools, usually over cobble or large gravel substrates. They often occur in mixed schools with other cyprinid fry.

Conservation Comments The fragmented distribution of chiselmouth in the southern half of British Columbia suggests that they were once more widely distributed within the province. This warm-water species probably expanded its northern range during the postglacial warm period about 6,000 years ago and later retracted into isolated pockets of suitable habitat. Because there is little or no movement between the population groups, each fragment is vulnerable and has little chance of being rescued through immigration. Although the chiselmouth is not of immediate conservation concern, the populations should be monitored but the chiselmouth could be one species that will benefit from global warming. It is considered not to be at risk by COSEWIC but is listed as S_3 (rare or uncommon) by the BCCDC.

Carassius auratus (LINNAEUS)
GOLDFISH

1 cm

Distinguishing Characters   A deep-bodied minnow with single spines positioned at the origins of both the dorsal and anal fins. The dorsal spine is serrated on its trailing edge. Most feral populations are a dull, olive colour on the back, white on the belly, and resemble young common carp (*Cyprinus carpio*); however, goldfish lack barbels at the corners of the upper jaw. Barbels are well developed in the common carp.

Sexual Dimorphism   Like most minnows, mature females usually are larger than males. Breeding males develop nuptial tubercles on their opercula, backs, and pectoral fins. Tubercles are absent, or poorly developed, in females, but breeding females develop bloated abdomens.

Distribution   NATIVE DISTRIBUTION
The goldfish is indigenous to eastern Asia and eastern Europe. It has a long history of domestication as an ornamental fish in China and first appeared in western Europe in the late 17th or early 18th century. In the late 19th century, goldfish were brought to North America (Scott and Crossman 1973), and feral populations are now well established in parts of central and eastern North America. Although not widely distributed in western North America, scattered self-sustaining populations occur in most of the western states and in British Columbia.

62 FAMILY CYPRINIDAE — MINNOWS

Map 10 The B.C. distribution of goldfish, *Carassius auratus*

■ introduction

BRITISH COLUMBIA DISTRIBUTION

Every year, goldfish are released into waters all over the province, and occasional individuals regularly appear in ponds, lakes, and streams near human population centres. In spite of these persistent "aquarium" introductions, self-sustaining populations of goldfish are rare in British Columbia and usually confined to man-made environments (e.g., park ponds, golf courses, and private housing developments) in the southern half of the province. There are self-sustaining feral populations in the lower mainland (e.g., Wonnock and Mill lakes), the southern interior, and southern Vancouver Island (Map 10). Feral goldfish are easily distinguished from recent aquarium releases by their colour: in the wild, goldfish revert to a drab greenish brown colour in a few generations.

Life History There is no study of the life history of feral goldfish in British Columbia, but there are summaries of the biology of feral populations in other jurisdictions (e.g., Becker 1983; Muus and Dahlstrom 1971; Scott and Crossman 1973; Wydoski and Whitney 2003).

REPRODUCTION

Although most goldfish populations contain males and females, some European populations contain only females (Muus and Dahlstrom 1971). These females are sexual parasites on other cyprinids (i.e., they reproduce by soliciting fertilizations from males of other species). The sperm activates the eggs, but the normal fusion of male and female nuclei never occurs. Consequently, only the female genome participates in further development. This mode of reproduction is called gynogenesis. For normal goldfish, spawning begins in the spring, usually when water temperature rises above 15 °C, and continues throughout the summer. Vegetation is essential for spawning; consequently, goldfish usually spawn in shallow littoral areas. Typically, two or three males pursue a female through dense vegetation and, as she goes, she scatters eggs onto plants and other submerged structures. The eggs are small (about 1.5 mm) and adhesive. They usually stick as single eggs (occasionally as bunches of two or three eggs) and are immediately fertilized by the pursuing males (Jones et al. 1978a). Fecundity is a function of female size and ranges from about 300 to 20,000 eggs. The eggs develop rapidly: 8–10 days at 15 °C, 5 days at 20 °C, and 3 or 4 days at 30 °C (Battle 1940; Jones et al. 1978a). The newly hatched fry are about 7 mm TL and remain on plants or on the bottom for 1 or 2 days before they swim up.

AGE, GROWTH, AND MATURITY
Growth rates are variable, and stunting is common in feral populations. Males usually mature in their second year, and females typically mature a year later. They probably live only 4–6 years in the wild, but in aquaria, they can survive for up to 30 years. In the Wonnock Lake population, adults rarely exceed 20 cm TL; however, a pond in Salmon Arm is said to produce much larger goldfish.

FOOD HABITS
Goldfish have a long gut and a large number (about 40) of closely spaced gill rakers. This morphology suggests that they feed on microplankton. Presumably, they use a filtering system to concentrate microorganisms in the pharynx, and palatal protrusions to retain large particles, similar to those described for the common carp (Callan and Sanderson 2003). Although young-of-the-year goldfish feed on microscopic plants and animals, adults are omnivorous. In mid-water, they graze on phytoplankton but also include diatoms, zooplankton (especially cladocerans), and even terrestrial insects in their diet. On the bottom, they take macrophytes, algae, detritus, molluscs, and small insect larvae (mostly chironomids).

Habitat Habitat use by feral goldfish is unstudied in British Columbia, and the following account is derived from the meagre habitat-use information available on feral populations in North America (Becker 1983; Scott and Crossman 1973; Wydoski and Whitney 2003).

ADULTS
Usually, feral goldfish occur in small lakes, ponds, or sluggish sloughs. They often are associated with dense vegetation and waters with a strong diel oxygen pulse. Goldfish tolerate a wide range of temperatures and are unaffected by ice cover; however, they require water of 15–25 °C

to breed. Consequently, they are unlikely to establish self-sustaining populations in central or northern British Columbia.

JUVENILES

Basically, juveniles use the same habitat as adults.

YOUNG-OF-THE-YEAR

Goldfish fry are strongly associated with aquatic vegetation and typically occur near the surface in dense weed beds.

Conservation Comments The major conservation concern with goldfish is that people persist in releasing them into the wild. Most of these aquarium releases fail, and viable feral populations are rare in British Columbia. Thus, the impact of goldfish on our native species is minimal; however, they can be difficult to remove once they become established (Wydoski and Whitney 2003).

Couesius plumbeus (AGASSIZ)
LAKE CHUB

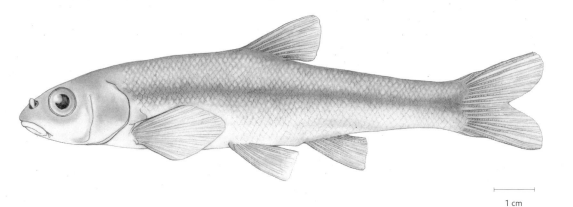

1 cm

Distinguishing Characters This moderate-sized (adults typically less than 120 mm FL) minnow, usually has a dark midlateral band on the back half of the body and a terminal mouth that extends back to the anterior margin of the eye. There are 54–65 lateral line scales, and when viewed from below, maxillary barbels are clearly visible just forward of the corners of the mouth.

The lake chub probably is the most consistently misidentified minnow in British Columbia: in southern parts of the province, it is confused with juvenile peamouth (*Mylocheilus caurinus*); in the Prince George area, it is confused with the brassy minnow (*Hybognathus hankinsoni*); and in the northeast, it is confused with the northern pearl dace (*Margariscus margarita nachtriebi*). Additionally, sometimes it is confused with juvenile flathead chub (*Platygobio gracilis*). This identification problem stems from the absence of any unambiguous characters that identify this species.

The placement of the dorsal fin relative to the pelvic fins and the mouth size distinguish adult lake chub from juvenile peamouth: the origin of the dorsal fin usually is slightly behind the origin of the pelvic fins in lake chub and slightly forward of the pelvic fins in peamouth. Additionally, the upper jaw in lake chub extends back to the anterior margin of the eye while the upper jaw in peamouth falls far short of the anterior margin of the eye.

At first glance, the lake chub and brassy minnow look similar; however, the resemblance is superficial, and the two species are easily distinguished by peritoneum colour (black in brassy minnow and silvery in lake chub), mouth position (the snout overhangs slightly in brassy minnows but is terminal in lake chub), and the presence of maxillary barbels in lake chub and their absence in the brassy minnow. Similarly, there is a strong resemblance between lake chub and northern pearl dace. They have similar body shapes, and except during breeding season, both are nondescript dark brown or grey minnows with dull silver undersides. To make matters worse, they often hybridize. This makes identification difficult in the lower Peace and lower Liard systems (the only areas in British Columbia where these species co-occur). Usually, the position of the dorsal fin relative to the pelvic fins separates the two species: in lake chub the dorsal fin originates directly above the origin of the pelvic fins, whereas in northern pearl dace, the dorsal fin originates about an eye diameter behind the origin of the pelvic fins. The convex space between the eyes (this area is flat in flathead chub) and the rounded pectoral fins (the pectorals are sickle shaped in flathead chub) distinguishes adult lake chub from juvenile flathead chub.

Taxonomy Some authors (e.g., Wells 1978) recognize three subspecies of lake chub: *C. p. greeni* west of the Continental Divide, *C. p. dissimilis* on the western Great Plains, and *C. p. plumbeus* in the east. The morphological differences among these putative subspecies are subtle, and the level of genetic divergence among them is not clear. Nonetheless, there is evidence of significant (3–4%) mitochondrial sequence divergence between lake chub west of the Continental Divide and those in eastern Canada (E.B. Taylor, personal communication). If the subspecies are real, then two subspecies probably occur in British Columbia: *C. p. greeni* in the Columbia, Fraser, Skeena and associated drainage systems and *C. p. dissimilis* in the lower Peace, Hay and lower Liard river systems. In addition, the populations in the upper Peace system (above the Peace Canyon) appear to be a mosaic of intermediate populations. It is possible that the upper Liard populations (including those associated with hot springs) and those in the upper Yukon survived the last glaciation in the Nahanni Refugium.

Sexual Dimorphism Typically, adult females are up to 30% larger than adult males. Spawning males and females both develop small tubercles on the head; however, the tubercles are better developed in males than in females. Apparently, there is a difference in the distribution and development of tubercles in lake chub on the western and eastern sides of the Continental Divide. In Pacific drainages, the first six pectoral rays in males are thickened and heavily pigmented. There are also tubercles on both the dorsal and ventral surface of the pectoral fin and distinctive patches of roughened comb-like scales on the breast immediately in front of each pectoral fin. In females, the pectoral rays are neither thickened nor heavily pigmented, but they have tubercles on their dorsal and ventral surfaces. The distinctive patches of rough scales on the breast are absent in females. On the east side of the Continental Divide (at least on the Great Plains), the sexual dimorphism in tubercles is similar; however, the first pectoral ray in males is usually kinked, and the rough patches of breast scales are larger (they extend onto the gill margins and operculum) and more exaggerated than in Pacific drainage males.

Hybridization   In British Columbia, lake chub commonly hybridize with pearl dace at disturbed sites in the lower Peace and lower Liard river systems and, less commonly, with redside shiners (*Richardsonius balteatus*) and longnose dace (*Rhinichthys cataractae*) in Pacific drainages. Usually, but not always, these hybrids are morphologically intermediate between the parental species.

Distribution   NATIVE DISTRIBUTION

Lake chub are found only in North America. In eastern North America, lake chub range from northern Quebec to the upper Delaware River. On the Great Plains, their range extends from the Mackenzie Delta to Colorado. West of the Continental Divide, lake chub occur from the southern margin of glaciation in the upper Columbia drainage system to the middle Yukon River. The record of *Couesius* from Twin Lakes, Stillaguamish River system, western Washington (Wells 1978; Wydoski and Whitney 2003) is an error. Of all the North American minnows, the lake chub appears to the most highly adapted to northern climates. Apparently, low winter temperatures (5–12 °C) are necessary for the normal development of sperm in this species (Ahsan 1966).

BRITISH COLUMBIA DISTRIBUTION

The lake chub is an inland species and is rarely found within 100 km of the coast (Map 11). It occurs in all major drainages—Columbia, Fraser, Skeena, Peace, Liard, Stikine, Taku, and Yukon—but with the exception of the Dean River, it is absent from most rivers that rise on the Interior Plateau and flow westward through the Coast Mountains. At the southern end of its B.C. range (the Columbia River system), the lake chub occurs as scattered, isolated populations, but on the Interior Plateau and in the northeastern portion of the province, lake chub are ubiquitous.

Life History   Relatively little is known about the life history of lake chub in British Columbia. Consequently, most of the following information comes from studies in Saskatchewan (Brown et al. 1970) and the eastern United States (Fuiman and Baker 1981); however, an unpublished B.A. thesis (Geen 1955) contains some data on lake chub reproduction and feeding in a small lake near Clinton.

REPRODUCTION

In northern areas, lake chub spawn in the early spring just after ice-out, whereas in more southern areas, spawning peaks in July and continues into late August. Spawning appears to be temperature dependent, and the peak can vary by 2–3 weeks in sites separated by only a few kilometres (Brown et al. 1970). In the Lac la Ronge area, Saskatchewan, spawning aggregations start to form as early as the first week of May (water temperature 4 °C), but actual spawning does not occur until late May when the water temperature rises above 10 °C (Brown et al. 1970). In early July, a small lake in the Clinton area contained both newly emerged fry (about 5.5 mm TL) and young-of-the-year as large as 15 mm (Geen 1955)

68 FAMILY CYPRINIDAE — MINNOWS

Map 11 The B.C. distribution of lake chub, *Couesius plumbeus*

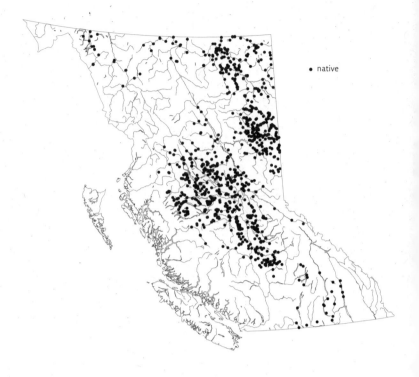

• native

This suggests spawning starts about mid-June or, perhaps, 1–2 weeks earlier. In the lower Liard system, fry appeared in early August (McLeod et al. 1979), and in a tributary to the Prophet River 110 km south of Fort Nelson, a spawning aggregation of lake chub was observed (Stewart et al. 1982) in mid-June (water temperature about 10 °C).

Spawning occurs in both flowing and standing water, and the type of substrate appears to be unimportant. Usually, several males are associated with a single spawning female. The males chase the female and quiver against her until she releases eggs. Lake chub are fractional spawners (females do not release all their eggs at one time). In a New Brunswick stream, the lake chub spawning migration occurred at night (Reebs et al. 1995). Fecundity is dependent on female size and ranges from 500 to over 10,000 eggs (Becker 1983; Brown et al. 1970; Bruce and Parsons 1976). Ripe lake chub eggs range from 1.8 to 2.4 mm in diameter (Becker 1983; Fuiman and Baker 1981). They are demersal and non-adhesive. Depending on water temperature, the eggs hatch in 5–10 days, and the newly emerged larvae are about 6 mm TL (Fuiman and Baker 1981).

AGE, GROWTH, AND MATURITY

By late July, young-of-the-year in the Clinton area average about 20 mm (16–24 mm FL), and by the end of the growing season (early November), the cohort has an average fork length of about 30 mm (range 28–34 mm). In late July, this population contained four clear length modes that suggest at least four age classes. Breeding fish were restricted to the two largest

(90 and 115 mm) size classes (Geen 1955). Thus, sexual maturity probably is achieved at either the end of the second, or the beginning of the third, summer of life. This is consistent with other studies (e.g., Bond 1980; Bruce and Parsons 1976) that indicate males mature in their third summer, and females in their fourth summer. Data from the Hay and Petitot rivers also show four clear size classes; however, the mean sizes of the two largest size groups are about 10 mm smaller than the two largest size groups from the Clinton area.

FOOD HABITS

The diet of lake chub consists primarily of benthic organisms (amphipods, chironomid larvae, oligochaetes, and some plant material), but they also take terrestrial insects and occasionally the fry of suckers and minnows. Although they are primarily carnivores, lake chub in the thermal areas of the upper Liard system forage almost exclusively on *Chara*. It is not known if they actually digest the cellular contents of this siliceous plant or obtain nutrition from the bacteria and protozoans associated with the plant.

In New Brunswick, lake chub were most active at dawn and dusk (Reebs et al. 1995). In the upper Liard drainage system, however, they forage throughout the day and at a wide range of temperatures. In late October, adjacent populations (one in a thermal area, Liard Hotsprings, and one in a cold stream, Kledo Creek) were observed actively foraging at water temperatures of 25 and 1.2 °C, respectively.

Habitat Although lake chub are an abundant and widespread species in British Columbia, almost no quantitative data is available on their habitat use. Consequently, the following account draws heavily on information from outside our area.

ADULTS

At the southern edge of its B.C. distribution, the lake chub is primarily a lacustrine species; however, over most of the rest of the province, it occurs in everything from the mainstems of large turbid rivers to tiny rivulets, and in standing waters ranging from small eutrophic ponds to large oligotrophic lakes. On the Interior Plateau, the lake chub is often the only fish in alkaline lakes and ponds, and in the northeastern part of the province, it occurs in both clear-water and lightly stained streams. Over this wide range of environments, the lake chub not only flourishes but also often is the numerically dominant species. In the Petitot drainage system, however, lake chub are not abundant and are noticeably absent from the heavily stained waters dominated by brook sticklebacks (*Culaea inconstans*) and finescale dace (*Phoxinus neogaeus*). In contrast, in the adjacent Hay River and its B.C. tributaries, the lake chub is found everywhere (including deeply stained waters).

In the spring, lacustrine adults occur close to the bottom throughout the littoral zone (depths from about 0.5 to 10 m). By late June, they move closer to shore, and during the day, small schools are common in shallow water. In lakes with predatory fishes, large numbers of lake chub are often observed sheltering in the submerged tangled branches of fallen trees during the day. At night, they move away from shelter sites, and individual adults often occur at the surface 50 m or more offshore. In lakes containing predatory fishes but without submerged cover, adults form dense schools and move slowly through the littoral zone.

There is some evidence that, in the northern parts of its range, lake chub migrate into tributary streams in the spring and migrate out to larger rivers (presumably to overwinter) in the fall (Stewart et al. 1982). There are several thermal areas in the upper Liard and upper Yukon systems where the only fish associated with these hotsprings are lake chub. Usually, they occur in water <27 °C but make occasional forays into slightly warmer water.

JUVENILES
Basically, juvenile (1+) lake chub use the same habitats as adults. In lakes with predatory fish, they are strongly associated with cover during the day and often aggregate among the branches of trees that have fallen into the water. At night, however, individual fish spread out into exposed littoral areas.

YOUNG-OF-THE-YEAR
In the late spring, lake chub fry concentrate within 1 m of shore in water less than 1 m deep. They occur in midwater over a variety of substrates (sand, organic litter, and fine gravel) but appear to avoid exposed areas with coarse gravel and cobble bottoms. Cover appears to be important, and most individuals are associated with weeds or other cover. At this time, under-yearlings are solitary or loosely associated with two or three other individuals. As summer progresses, small schools of fry form and venture 2–3 m from shore. By October, they join the juvenile population. Lake chub are abundant in the central and northern parts of British Columbia. Originally, there were scattered populations in the southern half of the province, but many of these were eradicated with rotenone in the 1950s and 1960s. Although the western populations of lake chub were named as a separate subspecies (*C. p. greeni*), the validity and distribution of this subspecies have never been critically examined. Nonetheless, recent molecular data (E. B. Taylor, personal communication) suggest that there are major sequence differences between lake chub populations in British Columbia and those in Ontario. Consequently, the remaining populations in south-central British Columbia may deserve some protection.

Conservation Comments In the northern parts of the province, some populations of lake chub are associated with hot or warm springs. Given the severe winters in the region, these populations probably are subject to unusual local selection regimes. So far, there is no strong evidence that they are morphologically differentiated from adjacent cold-water populations (McPhail 2001).

If, however, they are specialized for life in these unusual environments, there may be detectable differences in their life histories, physiology, and enzyme kinetics. Their physiology and biochemistry is currently under study; however, regardless of the level of their differentiation from adjacent coldwater populations, their unusual habitats and, by extension these fish, should be protected.

Cyprinus carpio LINNAEUS
COMMON CARP

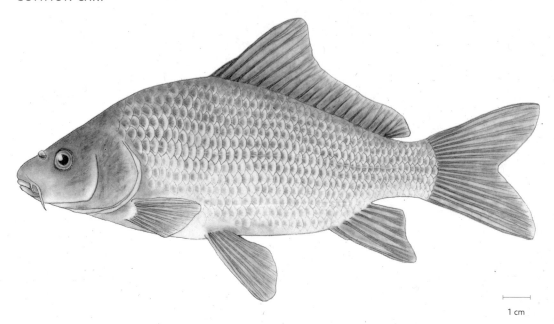

1 cm

Distinguishing Characters   This large minnow (adults reach lengths of over 40 cm) has large scales (35–39 along the lateral line), a long dorsal fin (18–20 soft rays), and a strong spine (serrated along the trailing edge) at the origins of both the dorsal and anal fins. Juvenile common carp can be confused with feral goldfish (*Carassius auratus*) (they are usually the same colour); however, carp have well-developed barbels at the corners of the upper jaw. These barbels are absent in goldfish.

Taxonomy   The taxonomy of the common carp and, especially, the origins of domesticated carp are unclear. Some authors argue for a single glacial refuge in the Caspian Sea area with postglacial dispersal into central Europe and, perhaps, Asia. Other authors argue for survival in Asia as well as in the Caspian region. Recent molecular data (Zhou et al. 2003) support the contention that the carp native to the Caspian region and parts of China are different subspecies: *C. c. carpio* in the Caspian region and *C. c. haematopterus* in China. There is, however, disagreement about the origins of the European domesticated carp (Balon 1995). Some authors argue that the common carp of Europe originated from domesticated Asian carp, whereas others argue that it was independently domesticated in eastern Europe. Genetic data (Zhou et al. 2003) suggest both views are correct. Apparently, different strains of the domesticated European carp were derived from different ancestral wild carp, and both the Asian and eastern European subspecies were involved in the origins of the domesticated European carp.

Sexual Dimorphism As in many fishes, female common carp are, on average, larger than males. Breeding fish have small white tubercles on the head and scales. These tubercles are more developed in males than in females.

Distribution **NATIVE DISTRIBUTION**
Although centuries of fish culture activities have obscured the original distribution of the common carp in both eastern Europe and Asia, the species apparently was native to both regions. From Europe, common carp have been spread to temperate waters around the globe. They were introduced into eastern North America in the mid-19th century and into the Columbia River system during the 1880s. In North America, except for the far northern rivers, they are now established in most major drainage systems from coast to coast and from southern Canada to Mexico.

BRITISH COLUMBIA DISTRIBUTION
Initially, common carp entered British Columbia from Washington State. Apparently, they first appeared in Okanagan Lake in 1917 and reached the Fraser system (southern Shuswap drainages) by 1928 (Clemens et al. 1939). By the 1940s, they reached the lower Fraser valley. In the Kootenays, they occur from the Arrow Lakes to the U.S. border and in the Kootenay River below Brilliant Dam. In the Fraser system, they occur from the North and South Thompson rivers to the Fraser Delta (Map 12). At one time, they also occurred in Glen Lake on Vancouver Island; however, this population appears to have died out.

Life History The life history of the common carp in our area is unstudied; however, the species' biology is well known. The following account is derived primarily from Clemens et al. (1939), MacCrimmon (1968), Scott and Crossman (1973), and Wydoski and Whitney (2003).

REPRODUCTION
Carp spawn in the spring or early summer when water temperatures rise above 15 °C. At this time, they aggregate in areas close to shore and are conspicuous because of their habit of frequently breaking the surface in slow, lazy rolls. Eventually, small groups of one or two females and a few males move into shallow, weedy areas, especially temporarily flooded lake margins. Here they spawn with much splashing about. Carp are very fecund—egg numbers ranging from 36,000 to over 2,000,000 (Swee and MacCrimmon 1966). Females are fractional spawners and release only a few hundred eggs per spawning bout. Thus, they spawn many times in a season. The eggs are small (about 1.0 mm in diameter), a pale green colour, and adhesive. They stick to weeds, grasses, and other debris in the spawning area. The fertilized eggs take 3–6 days to hatch and produce larvae about 5.0 mm TL. The newly hatched larvae stay on the bottom for a few days absorbing yolk and then begin exogenous feeding.

Map 12 The B.C. distribution of the common carp, *Cyprinus carpio*

■ introduced
□ failed introduction

AGE, GROWTH, AND MATURITY

Carp growth varies with water temperature, food availability, and length of growing season. In British Columbia, they commonly reach lengths of 60–70 mm in their first summer and double their length in their second growing season. Growth slows at sexual maturity, which is usually 4 years for males and 5 for females in British Columbia. In the wild, the common carp lives 10–15 years, but in captivity, they are reputed to reach 50 years. Apparently, in Europe, they sometimes achieve weights exceeding 30 kg, but a fish of 5 kg would be noteworthy in British Columbia.

FOOD HABITS

Adult carp have a reputation as bottom feeders, and when rooting about in the substrate, they ingest a mix of organic and inorganic particles. They use a filtration system to concentrate small food particles in the pharynx and expel inorganic particles out through their gill slits. Large food particles are retained by specialized palatal protrusions, and large inorganic particles are spit out through the mouth (Callan and Sanderson 2003). They are omnivores (Powles et al. 1983) but consume mostly invertebrates and very little plant material. They often feed heavily on insect larvae (especially chironomids), molluscs, worms, amphipods, and even zooplankton. Like many fishes, their diet changes with age. Thus, young-of-the-year carp feed primarily on plankton and algae, whereas juveniles gradually shift to the adult diet.

Habitat   Habitat use by common carp in British Columbia is unstudied; however, elsewhere, the species' biology is well known. The following account is derived primarily from Clemens et al. (1939), MacCrimmon (1968), Scott and Crossman (1973), and Wydoski and Whitney (2003).

ADULTS
Typically, adult carp are found in quiet, warm, eutrophic waters (i.e., the shallows of lakes and sluggish, soft-bottomed streams and sloughs). Typically, they are viewed as bottom dwellers; however, they are versatile, and common carp in Lake Washington have been caught in the limnetic zone over water almost 60 m deep (Wydoski and Whitney 2003). They are hardy fish and can tolerate a wide range of temperatures (0–32 °C), oxygen conditions (<2.0 mg/L), and turbidity. In lakes, they usually occupy the littoral zone in summer and favour weedy areas, but they move offshore into deeper water in winter. Similarly, in rivers, they usually are associated with weedy areas in the summer and move into deeper water in the winter.

JUVENILES
Juvenile common carp occupy habitats similar to those used by adults but tend to occur in shallower water than adults.

YOUNG-OF-THE-YEAR
Common carp fry are strongly associated with shallow water and dense beds of aquatic vegetation. They remain in the shelter of vegetation until the end of their first growing season.

Conservation Comments   There is no official desire to conserve common carp in British Columbia, and many biologists and anglers consider this alien species a major nuisance. Nonetheless, near large cities like Vancouver, carp provide a recreational fishery for a small but devoted band of anglers. Also, it is said to be an excellent food fish. Regardless of its good or bad points, the common carp is firmly established in the lower Mainland and in the Okanagan and Shuswap drainage systems. If our climate continues to warm, the common carp may expand its range within the Fraser drainage system. If that occurs, care should be taken to assure that they do not reach the Skeena and upper Peace rivers.

Hybognathus hankinsoni HUBBS
BRASSY MINNOW

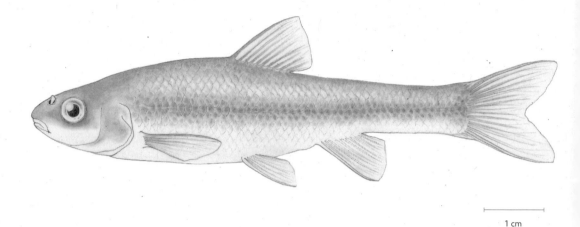

1 cm

Distinguishing Characters   This small (usually less than 70 mm FL), plain (no conspicuous markings) minnow has a snout that slightly overhangs the upper jaw. The dorsal fin originates in front of the pelvic fins; the mouth is small (it does not extend back as far as the front margin of the eye), and the scales are large (37–40 along lateral line). It has a dark peritoneum and an intestine with multiple coils. In British Columbia, most erroneous records of the brassy minnow (and there are many) are misidentified lake chub (*Couesius plumbeus*). The resemblance is superficial, and the two species are easily distinguished by peritoneum colour (black in brassy minnow and silvery in lake chub), mouth position (the snout overhangs slightly in the brassy minnow but is terminal in lake chub), and the absence of maxillary barbels in the brassy minnow.

Taxonomy   Based on a multivariate morphometric analysis (Wells 1978), there may be two forms of brassy minnows in Alberta: one that survived the last glaciation in the upper Mississippi system and another that survived in the Missouri system. The differences are subtle, but it appears that the northern Alberta and British Columbia populations are derived from the Mississippi form (Wells 1978). At this writing, a molecular study of the brassy minnow is in progress.

Sexual Dimorphism   Sexually mature males have slightly longer paired fins than females, and on average, females are larger than males of the same age. In breeding males, the first pectoral ray has a kink about two-thirds of the way towards the tip (the kink is usually less developed, but still apparent, in the second and third rays) and, when viewed from above, the first four or five pectoral rays are heavily pigmented. The pectoral fins in females lack these features.

In the stained bog waters of the upper Fraser and upper Peace drainage systems, there is a striking colour difference between breeding males and females: the flanks of males are a bright brassy colour, whereas those of females are silver. Curiously, this colour difference is absent in fish from the less stained waters of the lower Fraser River system. Here, both sexes remain silver during the breeding season.

Distribution NATIVE DISTRIBUTION
The brassy minnow is a North American species with an unusual geographic distribution. It occurs in a belt across the entire central section of the continent—from the Saint Lawrence system in the east to the Fraser system in the west—and from Fort McMurray, Alberta (Nelson and Paetz 1992), to eastern Colorado (Scheurer et al. 2003). Apparently, it is abundant in the south-central part of this distribution; however, although locally abundant in western Canada (Alberta and British Columbia), populations are widely scattered. Because most of its distribution lies east of the Continental Divide and because all the other species in the genus are also from east of the Rocky Mountains, it was assumed that the brassy minnow was introduced into British Columbia (Bailey 1954) when it was first discovered in the province (Keenleyside 1954). We now know, however, that it is widely, albeit sporadically, distributed in the Fraser and Peace systems. This suggests that the brassy minnow probably crossed the Continental Divide from the Peace River drainage into the Fraser system sometime during deglaciation.

BRITISH COLUMBIA DISTRIBUTION
The brassy minnow has a scattered distribution in the Fraser and upper Peace systems (Map 13). In the Fraser system, it is abundant in lakes and sluggish streams near Vanderhoof and Prince George, and it is abundant in the lower Fraser valley from Chilliwack to the Fraser River delta (including areas that are brackish at high tide). In the intervening 500 km of the Fraser River, there is only one confirmed record: a beaver pond in the Horsefly drainage system near the village of Horsefly, central British Columbia. In the upper Peace system, the only reliable records are from the headwater lakes, streams, and bogs associated with the Crooked River.

Life History Although some aspects of the ecology of the brassy minnow have been described (Copes 1975), its life history is not well known. The only B.C. reference is an unpublished M.Sc. thesis (Ableson 1973) on a lacustrine population. Thus, the brassy minnow's life history in flowing water remains a bit of a mystery. For example, at some sites (e.g., ditches in Richmond) they are abundant seasonally every year, whereas at other adjacent (and often connected) sites, they only sporadically appear in large numbers and then disappear for several years. Even at sites where they are abundant every year, they abruptly vanish in July and reappear in September or October. This sporadic appearance of large numbers of brassy minnow at different sites, and their regular seasonal appearance and

Map 13 The B.C. distribution of brassy minnow, *Hybognathus hankinsoni*

• native

disappearance at other sites, suggests that they school and that migrations may be a common feature of their life history in fluvial environments.

The following account is based on Ableson (1973) and Copes (1975), supplemented with observations made in a large aquarium and field data on the brassy minnow in the lower Fraser valley.

REPRODUCTION

In the lower Fraser valley, spawning occurs from mid-May (when water temperatures reach about 14 °C) to early June. In Wyoming, spawning occurred at water temperatures between 16 °C and 27 °C (Copes 1975), but there was a relatively short spawning period (7–10 days). In the Prince George region, spawning starts in early June and continues into early August (Ableson 1973). In Wyoming during the spawning season, females move in schools, whereas males lurk individually in adjacent vegetation (there is no territorial behaviour). When a school of females passes by, 1–15 males approach a female on the periphery of the school and swim above, below, or beside her. The female quickly swims (in either a spiral or a straight path) into a patch of vegetation. Males pursue her, and one or more of them press against her and rapidly vibrate. Presumably, eggs are released at this time.

Aquarium observations on brassy minnows from the lower Fraser valley also indicate that several males are involved in the spawning of a single female. In this case, spawning occurred slightly after noon, and only a few eggs were released at each spawning. Fertilization occurred near the surface or in mid-water and over vegetation. The eggs were adhesive and denser than water. They sunk into the vegetation or settled on the substrate. Adults that were not spawning accompanied the spawning fish and excitedly dashed about eating the sinking eggs.

Depending on body size, females produce about 100 to 1,000 eggs. Not all the eggs are released in a single spawning, and females probably spawn several batches of eggs over a period of about a week. The ripe eggs are about 1 mm in diameter, but once fertilized, they almost double in diameter. Embryo development is rapid (hatching occurs within 70 hours at 18 °C). The newly hatched larvae are small (about 5 mm TL), transparent, and lack eye pigment. Melanophores develop over the next 4 days, and by the sixth day, the larvae have filled their swim bladders. By day 8, they are about 6 mm long and begin to feed. In some years in the lower Fraser valley, there is a second spawning in the fall that produces fish about 15 mm long by mid-November.

AGE, GROWTH, AND MATURITY

Growth in the first months of life is rapid, and in a lacustrine population, fry averaged 27.9 mm FL by early August (Ableson 1973). In the lower Fraser valley, populations that disappear in the summer tend to reappear in the fall (early October). At this time, the young-of-the-year average about 42 mm FL. In these stream populations, the entire population (adults and young-of-the-year) overwinters at breeding sites and spawns again early in the next summer. Both males and females reach sexual maturity after one winter (except for young produced in fall spawnings). By the time they reach sexual maturity, females usually are slightly larger than males. The maximum age recorded in British Columbia is 4 years (3+), and so far, all individuals of this age were females.

FOOD HABITS

The long intestine and black peritoneum suggest that plant material is an important part of the brassy minnow's diet. In the lower Fraser valley, the intestine is usually filled with algae, organic detritus, chironomids, cladocerans, and copepods. A lacustrine population near Prince George had a similar diet: organic debris, algae, and small insects (Ableson 1973).

Habitat In British Columbia, brassy minnows typically occur in small lakes, small slow-moving streams, beaver ponds, and drainage ditches. In the upper Peace and upper Fraser systems, they usually occur in stained waters, but in the lower Fraser valley, they are found in both clear and turbid water.

ADULTS

In lakes, adults often are associated with soft, mud bottoms and dense vegetation. In the summer, they are rarely observed in water more than 1.5 m deep. In streams, adult brassy minnow remain close to vegetation and appear to avoid waters with surface velocities >50 cm/s.

JUVENILES AND YOUNG-OF-THE-YEAR
There are no obvious habitat differences between juveniles and adults; however, fry tend to occur in shallower and quieter water than adults.

Conservation Comments The brassy minnow is an enigmatic little fish. Its B.C. distribution is highly fragmented, but ecologically, it appears adaptable: it occurs in both clear and stained lakes, swampy streams and bogs, turbid rivers, polluted and unpolluted ditches, and brackish estuaries. This fragmented distribution makes individual populations vulnerable to habitat alterations. The largest known B.C. populations are in the lower Fraser valley (especially the ditches in Richmond), and many of these populations are threatened by urbanization. At one time, this species was abundant in the esker lakes between Prince George and Vanderhoof; however, when the lakes became a provincial park, brook trout (*Salvelinus fontinalis*) were introduced, and the brassy minnow disappeared. Presumably, they were casualties of the double standard B.C. Parks applies to fishes relative to the rest of the fauna and flora in parks. COSEWIC does not list this species, but the BCCDC lists it as S3 (rare or uncommon).

Margariscus margarita nachtriebi (COX)
NORTHERN PEARL DACE

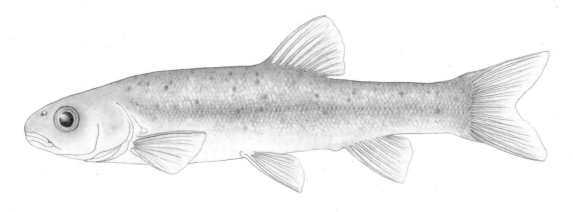

Distinguishing Characters This moderate-sized minnow (usually less than 160 mm FL) has small scales (62–75 along lateral line), and the origin of the pelvic fins is well forward of the dorsal origin. In northeastern British Columbia, most of the erroneous records of pearl dace are misidentified lake chub (*Couesius plumbeus*). The general appearance (and size) of the two species is similar, but the position of the dorsal fin relative to the pelvic fins and the development of the maxillary barbels distinguish these species. In pearl dace, the dorsal fin originates about an eye diameter behind the origin of the pelvic fins, whereas the dorsal fin in lake chub originates almost directly above, or slightly behind, the dorsal origin. Additionally, when viewed from below, barbels are clearly visible near the corners of the mouth in lake chub but not in pearl dace. In pearl dace, the barbels are hidden and usually are not visible without magnification. In addition, breeding male pearl dace have an orange or red stripe along the lower flanks, whereas breeding male lake chub only have reddish coloured patches in the axillae of the pectoral and pelvic fins.

Taxonomy Traditionally, two subspecies of pearl dace are recognized: a southern subspecies (*Margariscus m. margarita*) and a northern subspecies (*M. m. nachtriebi*). The two supposed subspecies barely overlap in lateral line scale counts (Bailey and Allum 1962). The American Fisheries Society Common and Scientific Names Committee (Nelson et al. 2004) retains the two forms as subspecies but in a footnote indicates that some authors (e.g., Bailey et al. 2004) treat them as separate species. I have followed the Common Names Committee and retained *M. m. nachtriebi* as the scientific name for pearl dace in British Columbia. If, however, the two subspecies are elevated to the rank of species, the proper scientific name for pearl dace in British Columbia becomes *Margariscus nachtriebi*.

Sexual Dimorphism In British Columbia, males rarely exceed 100 mm FL, but females can reach lengths of over 150 mm. Spawning males develop a bright-coloured band (orange to red) on their lower flanks, the first six pectoral rays are thickened, and there are distinctive patches of roughened comb-like scales on the breast immediately in front of each pectoral fin. Also, there are minute tubercles on the branchiostegal rays and on the ventral sides of the head. In breeding females, the lateral band usually is reduced to a pale rose wash over the lower flanks, and the pectoral fins lack the thickened rays and tubercles.

Hybridization In British Columbia, pearl dace occasionally hybridizes with both lake chub and finescale dace (*Phoxinus neogaeus*). These hybrids are common enough to make field identifications difficult. Consequently, most unverified records of pearl dace in British Columbia are suspect, and voucher specimens should be deposited in a permanent collection.

Distribution

NATIVE DISTRIBUTION

The northern pearl dace is an eastern North American subspecies distributed from Nova Scotia to northeastern British Columbia and from just south of Great Slave Lake to the Sand Hills in Nebraska.

BRITISH COLUMBIA DISTRIBUTION

Northern pearl dace are confined to the Mackenzie River system. Within this drainage system, they are distributed sporadically in the lower Peace and lower Liard river systems. Although there are relatively few verified records from the province, pearl dace are commonly reported by consultants working in northeastern British Columbia (Map 14). Some of these records are undoubtedly real, but others probably are misidentifications.

Life History No life-history study is available for pearl dace in British Columbia. Consequently, most of the following information comes from central Canada, and the United States but, wherever possible, this information is supplemented with information gleaned from B.C. populations.

REPRODUCTION

Spawning occurs in the early spring when water temperatures reach about 12–13 °C; however, ripe males and a few ripe females (and many spent females) were collected in northeastern British Columbia as late as June 21 (water temperature 17 °C). In northwestern Ontario, spawning occurs in both streams and shallow, vegetated areas in lakes (Tallman et al. 1984). The spawning behaviour of the northern pearl dace was observed in two streams in lower Michigan (Langlois 1929). The streams were about 5 m wide and 50 cm deep. Apparently, the current at individual spawning sites varied from strong to almost still. In these populations, the larger males were territorial and defended a small (about 20 cm) site on a sand or gravel substrates. In other populations, however, spawning occurs over mud or silt bottoms (Bendell and McNichol 1987). Although

Map 14 The B.C. distribution of northern pearl dace, *Margariscus margarita nachtriebi*

the spawning sites are cleared, there is no evidence of nest building. Small, but mature, males do not hold territories and frequently intrude on the territories of larger males. This suggests that small males may use sneaking as a reproductive strategy.

Apparently, the spawning behaviour is quite complex (Langlois 1929) and involves direct contact and contorted positions in both sexes. Females release only a few eggs at each spawning and, presumably, spawn with several males during the course of the breeding season. In Wisconsin at the start of the breeding season, female pearl dace about 120 mm in length contained 4,000–4,200 ripe eggs (Becker 1983). The unfertilized eggs were about 1.3–1.4 mm in diameter. After fertilization, the eggs probably swell to a larger size. Like lake chub, the eggs in female pearl dace almost reach maturity in the autumn and are held in that state over the winter (Becker 1983).

AGE, GROWTH AND MATURITY

By mid-August in northeastern British Columbia, young-of-the-year range from 30 to 37 mm FL, and in Wisconsin, they reach a range of 34–48 mm (TL) by mid-September (Becker 1983). Apparently, females grow faster than males, and both sexes mature in their second year (Fava and Tsai 1974; Stasiak 1978a, 1978b). In Wisconsin (Becker 1983) and southern Ontario (Chadwick 1976), there appears to be a high post-spawning mortality in males, and very few males live into their third growing season. Despite this male mortality, most populations contain at least three size classes by midsummer. This suggests that females may live for about 4 years. Curiously, pearl dace in some northeastern B.C. populations grow to an exceptionally large size, and the two largest recorded individuals (158 and 165 mm FL) are from the Peace River District. No age data are available for these unusually large females, but they probably were older than 4 years.

FOOD HABITS

During the summer in northwestern Ontario, there are distinct evening and early morning peaks in foraging activity (Tallman et al. 1984). In streams, adult pearl dace eat mainly benthic invertebrates: amphipods, pea clams, mayfly nymphs, chironomid larvae, and occasional terrestrial insects (Stasiak 1978a). In lakes, the diet is similar, except copepods, cladocerans, and detritus are important items (Tallman et al. 1984). Some of the unusually large pearl dace collected in northeastern B.C. contained longnose sucker (*Catostomus catostomus*) fry.

Habitat Little is known about pearl dace habitat use in British Columbia; however, its scattered distribution and propensity to hybridize with other species in northeastern British Columbia suggests that it may be close to its physiological or habitat limits in this region. Some habitat data on this species are available from Manitoba (Tallman and Gee 1982) and northwestern Ontario (Tallman et al. 1984). These reports are the source of most of the habitat use information given below, supplemented, when appropriate, with local observations.

ADULTS

Most pearl dace in British Columbia are collected in sluggish streams or small lakes. They rarely occur in large numbers; however, an exception is a small, un-named eutrophic stream near the Alberta border in the Red Willow drainage system. Here, adult pearl dace occur in remarkable densities and grow to an exceptional size. At most B.C. sites where pearl dace occur, the water is stained and slow moving. They usually are found in pools 1–2 m deep; although the streambed may contain coarse gravel and cobbles, the pearl dace are usually found over substrates of fine gravel, sand, or silt. In lakes during the summer, they typically occur in water less than 2 m deep, over silt or sand bottoms, and close to vegetation. In Tsinhia Lake, pearl dace are rare relative to other sympatric minnows, and during the day, they occur in dense, mixed schools of northern redbelly dace (*Phoxinus eos*), finescale dace, and hybrids of various combinations of these species.

JUVENILES

In both lakes and streams, juvenile pearl dace occur closer to shore and nearer to vegetation than adults.

YOUNG-OF-THE-YEAR

Typically, young-of-the-year pearl dace are found in shallow water along the margins of streams and lakes. Early in their first summer, they often occur in vegetation.

Conservation Comments In British Columbia, the pearl dace exists as scattered, isolated populations in the northeastern corner of the province. There is probably little, or no, gene flow among many of these populations. Consequently, the populations are vulnerable to local disturbances (especially those that affect water quality). At least two B.C. populations once contained unusually large individuals. The reasons for this large body size are unknown, and one of these unusual populations (in Charlie Lake) is now extirpated—pike (*Esox lucius*) and walleye (*Sander vitreus*) were introduced into the lake. Neither COSEWIC or the BCCDC list this species.

Mylocheilus caurinus (RICHARDSON)
PEAMOUTH

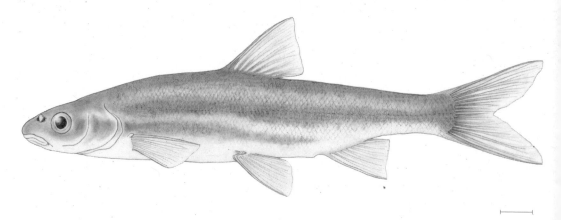

1 cm

Distinguishing Characters   This large minnow (adults reach lengths of over 250 mm) has a relatively small mouth, and small barbels at the corners of the upper jaws. Usually, the dorsal fin originates in front of the pelvic fins, and in juveniles, the caudal fin is asymmetrical (the lower lobe is longer than the upper lobe) and deeply forked (the middle rays are half, or less than half, the length of the longest rays). Adults have two dark bands on the flanks. The upper band extends from the operculum to the base of the caudal fin, and the lower band extends from the eye to about the end of the pelvic fins.

The peamouth is a member of a western North American lineage of minnows that includes the northern pikeminnow (*Ptychocheilus oregonensis*), redside shiner (*Richardsonius balteatus*), and chiselmouth (*Acrocheilus alutaceus*). Although adult peamouth are distinctive, the fry are not, and young-of-the-year peamouth often occur in mixed schools of northern pikeminnow, redside shiner, and chiselmouth fry. The relative position of the dorsal fin separates peamouth fry from the young of these other species: in peamouth, the dorsal fin originates in front of the pelvic fins, whereas in the other species, the dorsal fin insertion is directly over, or behind, the origin of the pelvic fins. Additionally, peamouth fry lack the small dark spot at the base of the caudal fin that is characteristic of pikeminnow fry. Also, the mouth in peamouth is almost terminal, whereas the mouth in chiselmouth fry is strongly subterminal, and peamouth fry lack the distinctive band of dark pigment along the base of the anal fin that is characteristic of redside shiner fry.

Sexual Dimorphism   As in most minnows, mature females are larger than males. This size dimorphism results from a more rapid growth rate, and a longer lifespan in females than in males (Hill 1962). Breeding adults have red lips and red pigment in the pectoral and pelvic axillae. Spawners of both sexes develop small, white tubercles on the head, operculum, and back; however, the tubercles on males are noticeably larger and more densely packed than those on females. Apparently, the sexes also can be distinguished by colour—spawning males have dark green backs and females have brown backs (Schultz 1935).

Hybridization Occasional hybrids between peamouth and redside shiners are known from several localities in the Columbia, Fraser, and Skeena drainage systems. In Stave Lake (a reservoir in the lower Fraser valley), this hybrid combination is common (Aspinwall et al. 1993). The hybrids are not sterile and regularly backcross into the parental gene pools; however, in spite of this persistent hybridization, the parental species maintain themselves as separate ecological entities (Aspinwall and McPhail 1995). In British Columbia, peamouth also hybridize with northern pikeminnow and chiselmouth (Stewart 1966).

Distribution NATIVE DISTRIBUTION
The peamouth is another Columbia endemic: its natural range is restricted to the Columbia River system and adjacent drainages that received their fauna from the Columbia River. In the Columbia system, peamouth range from the river mouth upstream to, but not above, Shoshone Falls on the Snake River in Idaho.

BRITISH COLUMBIA DISTRIBUTION
Peamouth range from the upper Columbia River system to the Nass River (Map 15). In the B.C. interior, peamouth reach the headwaters of both the Columbia and Fraser drainage systems. From the middle Fraser River, peamouth have colonized the Skeena River system and, from the Skeena system, the Nass River. They also have crossed the Continental Divide into the upper Peace system but never reached the Liard, Yukon, Iskut–Stikine, or Taku river systems. Although absent from most rivers along the central B.C. coast, they are present in the upper portions of the Klinaklini and Dean rivers.

At first glance, the peamouth's distribution is typical of a minnow that colonized the province from the unglaciated portion of the Columbia River system—it is mainly an interior species and only reaches the coast in large rivers like the Fraser and Skeena. For a primary freshwater fish, however, its southern coastal distribution is unique: it occurs on Vancouver Island, Nelson Island, and the Sechelt Peninsula. To reach these areas, peamouth had to cross substantial marine barriers. How did a fish that is ostensibly saltwater intolerant manage to cross the Strait of Georgia? One unusual aspect of the peamouth's physiology is that it can temporarily tolerate moderate salinities, and peamouth probably reached Vancouver Island, Nelson Island, and the Sechelt Peninsula by crossing marine barriers on a dilute seawater bridge (Clark and McInerney 1974). Although surface salinities in the Strait of Georgia usually are higher than peamouth can tolerate, surface salinities in the Strait are reduced when the Fraser River is in flood. The observed distribution of peamouth in the

Map 15 The B.C. distribution of peamouth, *Mylocheilus caurinus*

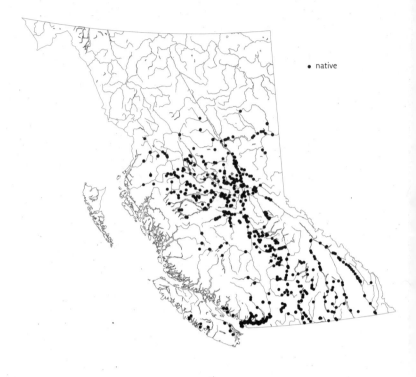

• native

Strait of Georgia region neatly fits this hypothesis; however, it does not explain the presence of peamouth in Kennedy and Cecilia lakes on the west coast of Vancouver Island. These populations are isolated from the populations on the east coast of Vancouver Island by the Island Mountains. Originally, it was argued that the Cecilia Lake record was probably an error and that the Kennedy Lake population was "almost certainly an introduction" (Clark and McInerney 1974). We now know that the Cecilia Lake record is not an error and that peamouth are present at other west coast sites (e.g., Easter Lake and a small lake on Meares Island). It is unlikely that all these scattered peamouth populations on the west coast of Vancouver Island are introductions, and they certainly did not reach the west coast on a low-salinity plume originating from the Fraser River. Where, then, did they come from? A possible source is the Olympic Peninsula. During the last glaciation, Ozette Lake on the northwestern tip of the Olympic Peninsula was not covered in ice. This lake contains a diverse primary freshwater fish fauna (including peamouth). Perhaps, during deglaciation, melting ice lowered the surface salinity of Juan de Fuca Strait enough to allow peamouth to ride north-trending surface currents to the west coast of Vancouver Island. Thus, peamouth may have colonized British Columbia from two sources—the Columbia River and the unglaciated southern and western portions of the Olympic Peninsula.

Life History REPRODUCTION

Peamouth spawn in the spring. Breeding is triggered by increasing day-length and rising water temperatures, and the threshold water temperature for spawning is about 9 °C. Spawning usually occurs in flowing water over clean gravel substrates. Although some populations spawn in lakes over gravel beaches, most lacustrine populations spawn in inlet or outlet streams. At these sites, peamouth rarely ascend, or descend, the stream more than a few hundred metres, and spawning usually occurs on the first or second riffle above or below the lake. Some populations spawn at night (Aspinwall and McPhail 1995), but most populations spawn during the day. Night spawning in the lower Fraser valley and Vancouver Island sometimes involves large numbers of fish spawnings en masse.

Day spawnings in lakes and streams also involves aggregations of fish, but they do not spawn en masse. In streams, ripe fish mill slowly about in a pool or low-velocity area below a riffle until a female swims rapidly upstream into the riffle where she is joined immediately by an entourage of four or five males. They crowd against the female and vibrate rapidly, which causes her to release a few eggs. Over the course of the spawning period, females spawn many times with many different males. Spawning in lakes has been described (Schultz 1935) and is similar to spawning in streams.

The eggs are demersal, adhesive, and about 2 mm in diameter when fertilized. As in most fish, fecundity is a function of body size, and egg number is estimated to range from about 10,000 to 20,000 eggs (Hill 1962). The eggs take about 6 days to hatch at 18 °C, and the newly hatched fry are about 7 mm TL and strongly nocturnal (Gadomski and Barfoot 1998).

AGE, GROWTH, AND MATURITY

Newly hatched peamouth begin feeding at about 9 mm and grow rapidly in their first summer. Depending on latitude and food availability, they reach 35–60 mm by the end of their first growing season. Typically, males reach sexual maturity in their third summer (2+), and most females mature a year later. Males rarely live longer than 8 years, but females can live to 19 years (the maximum lifespan recorded in British Columbia).

FOOD HABITS

Young-of-the-year peamouth consume a variety of prey including organisms taken from both the substrate and the surface. In lakes, they feed heavily on planktonic crustaceans and chironomid pupae but also take some benthic organisms (Chisholm et al. 1989; Clemens et al. 1939; Miura 1962). In rivers, peamouth fry forage on benthic organisms (amphipods, benthic copepods, chironomid larvae, and oligochaetes) but will feed on zooplankters if they are present (e.g., in sloughs and backwaters). As they grow, peamouth take larger prey, and in lakes, adults forage in both the littoral and limnetic zones. Here, they are primarily water column foragers (Aspinwall et al. 1993), but they also take prey from the bottom (larvae and nymphs of aquatic insects) and the surface (winged insects). Although in Lake Koocanusa peamouth appear to prefer *Daphnia* as prey, the largest size classes occasionally eat fish (Chisholm et al. 1989). In turbid rivers, adult peamouth feed primarily on benthic prey but also take winged insects from the surface.

Habitat   ADULTS

Peamouth occur in lakes, large rivers, and in the spring, small streams. In lakes, habitat use by adults changes seasonally. During the winter in Nicola Lake, adults are closely associated with the bottom, often at depths of over 20 m (MacLeod 1960). In the spring, they move inshore to spawn, and they begin a daily migration in the summer that brings them towards the surface and inshore in the evening with a reverse migration in the morning. In late fall, they cease this diel migration and move into their winter habitat.

In large rivers like the Columbia and Fraser, seasonal habitat shifts are not as well documented as they are in lakes. In the Fraser River near New Westminster, a distinct peak in the catch of peamouth occurred in August and then tapered off in the fall and early winter to no catch in January, February, and March (Whitehouse and Levings 1989). This was followed by slowly increasing catches in spring and early summer and then another late-summer peak. A similar seasonal pattern of catch occurs in the Columbia River below Keenleyside dam (Hildebrand 1991).

JUVENILES

In lakes, juvenile peamouth school in littoral areas but usually in deeper water than the young-of-the-year. In rivers, juvenile peamouth show a strong preference for slow (<0.1 m/s), shallow (<0.5 m) water over fine substrates (Porter and Rosenfeld 1999). In large rivers, they congregate near the mouths of tributary streams and, as summer progresses, often penetrate considerable distances up low-gradient tributaries. In the fall, however, they return to the main river, where they spend the winter.

YOUNG-OF-THE-YEAR

In lakes, during the summer, schools of young-of-the-year peamouth occur in shallow littoral areas. These schools often contain mixtures of peamouth, redside shiner, and northern pikeminnow. Like adults, under-yearling peamouth often exhibit a diel migration; however, their daily migration is the reverse of adults—the young school in shallow, littoral regions during the day and disperse into deeper water at night. In the spring, in the Columbia River near Trail, peamouth fry congregate in shallow, quiet water areas, especially near the mouths of tributary streams.

Conservation Comments   As the only native primary freshwater fish on Vancouver Island, peamouth are a unique part of the island's biotic heritage. Unfortunately, the populations on the east coast of Vancouver Island (Holden, Michael, and Quennell lakes near Nanaimo and Somenos Lake near Duncan) are in trouble. They are either extirpated or have declined dramatically in numbers, apparently casualties of urbanization and the introduction of exotic species. In contrast, the populations on the west coast of Vancouver Island, and on the Sechelt Peninsula and Nelson Island, are still strong. Neither COSEWIC nor the BCCDC list the peamouth.

Notropis atherinoides RAFINESQUE
EMERALD SHINER

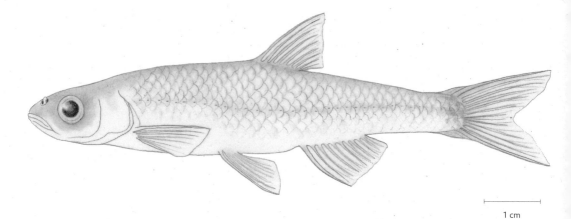

1 cm

Distinguishing Characters   This slim, elegant little minnow has large scales (35–41 along the lateral line), no barbels, and a long anal fin (9–13 rays). In life, there is an iridescent silver stripe along the midline, the back is iridescent green, and the lower flanks are almost translucent. The only other *Notropis* in British Columbia is the spottail shiner (*N. hudsonius*). The two species are easily separated by colour pattern (no caudal spot in the emerald shiner), anal fin ray count (7 or 8 rays in spottail shiner and (9) 10–13 rays in emerald shiner), and dorsal fin position (the dorsal fin originates well behind the origin of the pelvic fins in the emerald shiner and immediately above (or slightly before) the origin of the pelvic fins in the spottail shiner).

Sexual Dimorphism   Like many minnows, female emerald shiner grow faster than males and, on average, are larger than males of the same age (Fuchs 1967). Also, breeding males have small tubercles on the dorsal surface of the pectoral fins (Becker 1983) that are absent, or less well developed, in females (Nelson and Paetz 1992).

Distribution   **NATIVE DISTRIBUTION**
The emerald shiner is another of the widely distributed eastern North American minnows. It ranges from Quebec to northeastern British Columbia and from the Mackenzie River system just downstream of Fort Simpson south to Texas.

BRITISH COLUMBIA DISTRIBUTION
There is one verified B.C. record of the emerald shiner (Map 16). It was collected in 1960 at the mouth a small tributary just downstream from the confluence of the Fort Nelson and Muskwa rivers. The same site has been collected several times since 1960, but no additional emerald shiners have been collected. Nonetheless, they may occur sporadically in British

Columbia. The emerald shiner is known from two adjacent localities: the lower Liard River roughly 80 km below the British Columbia – Northwest Territories border and Bitscho Lake, northwestern Alberta (Nelson and Paetz 1992). There is no physical barrier between the lower Liard River in British Columbia and the Northwest Territories, and the outlet of Bitscho Lake (the Petitot River) flows through the extreme northeastern corner of our province. A recent survey of the Petitot River failed to find the emerald shiner, but occasional individuals probably disperse downstream into British Columbia from Alberta.

Life History There is no local life-history information available for this species. Consequently, the following account is based on data from outside the province. An unpublished thesis on the life history of the emerald shiner in Lake Erie (Flittner 1964) was not available, and the following references to this work are based on excerpts published in Becker (1983).

REPRODUCTION

Emerald shiner spawning appears to be triggered by temperature, and the threshold spawning temperature is about 22 °C (Flittner 1964). In southern Ontario, females were gravid at temperatures ranging from 20 °C to slightly over 23 °C (Campbell and MacCrimmon 1970). In Manitoba (Stewart and Watkinson 2004), the emerald shiner spawns from late June to early August at temperatures of 21–24 °C, and in northeastern Alberta, they spawn in late June and July (Bond 1980). Apparently, spawning occurs at night (Flittner 1964) at 30–60 cm below the surface (Stewart and Watkinson 2004). Sandy shoals 2–6 m deep are typical spawning sites, and the substrate can vary from mud to hard sand, gravel, cobbles, and coarse rubble.

Mature females sometimes contain two developmental stages of eggs (Fuchs 1967). Because the spawning season appears to last about 2 months, individual fish probably spawn more than once in a season. Mature, but unfertilized, eggs are about 0.8–0.9 mm in diameter. Fecundity in a small sample of females from northeastern Alberta ranged from roughly 1,000 to 2,000 eggs (Bond 1980), whereas in Wisconsin, fecundities reach almost 3,000 eggs (Becker 1983). The eggs are not adhesive and sink to the bottom. They develop rapidly (24–32 hours at typical spawning temperatures). Newly hatched young are small (about 4 mm) but reach a size of about 9 mm in 11 days (Flittner 1964).

AGE, GROWTH, AND MATURITY

The emerald shiner grows quickly, and in Ontario, they reach a length of about 50 mm by the end of the summer. In northern Alberta, however, they are roughly half that size at the end of their first growing season (Bond 1980). Sexual maturity is reached in the second summer, and relatively few individuals survive to the autumn of their third year (Campbell and MacCrimmon 1970).

Map 16 The B.C. distribution of emerald shiner, *Notropis atherinoides*

- native
- native, unverified record

FOOD HABITS

Apparently, this species is a facultative surface feeder (Hebert and Haffner 1991). In lakes, adults feed mainly on plankton (especially *Daphnia*), but in rivers, they forage on chironomid larvae and the nymphs of aquatic insects. They also take terrestrial insects at the surface. The first food of newly hatched emerald shiner usually is rotifers, but they also consume copepod nauplii, protozoans, and algae (Siefert 1972). The amount of algae in their diet decreases as the proportion of plankton increases.

Habitat Obviously, with only a single emerald shiner known from British Columbia, there are no data on its habitat use in this province; however, faunal works in other regions (Becker 1983; Nelson and Paetz 1992; Scott and Crossman 1973; Stewart and Watkinson 2004) indicate that emerald shiners are an open-water species typically found in large rivers, lakes, and reservoirs. Atypically, the only B.C. record is from a small stream tributary to the Fort Nelson River. At the time of the collection, the stream was clear but stained, had a moderate current, and a silt substrate. However, the collection was made within about 10 m of the mainstem Fort Nelson River, and the river is large (about 0.25 km wide), fast flowing, turbid, and deep. Thus, the single specimen in the tributary stream may have been a stray, and it is possible that there is an emerald shiner population in the main river.

ADULTS

In lakes in the summer, schools of adult emerald shiner follow the diel movements of plankton: ascending to near the surface at dusk and descending into deeper water (but not below the thermocline) at dawn. There is also a seasonal movement inshore in the fall and then offshore into deep water in the winter (Becker 1983). Their ecology in large rivers is not well known.

JUVENILES

Presumably, juveniles occupy the same habitat as adults.

YOUNG-OF-THE-YEAR

In lakes in the summer, young-of-the-year emerald shiner form large schools that move about near the surface. Such schools often occur far offshore (Becker 1983). The schools move inshore in the fall and, with the onset of winter, offshore into deeper water. Again, the biology of emerald shiner fry in rivers is poorly known.

Conservation Comments There is no direct evidence of a self-sustaining population of emerald shiner in British Columbia but it probably occasionally strays into the province from adjacent populations in Alberta or the Northwest Territories. Nonetheless, it is possible that there is an as yet undiscovered population in northeastern British Columbia. If so, it would be of special interest and deserve protection. The emerald shiner is listed by the BCCDC (S1, extremely rare) but is not listed by COSEWIC.

Notropis hudsonius (CLINTON)
SPOTTAIL SHINER

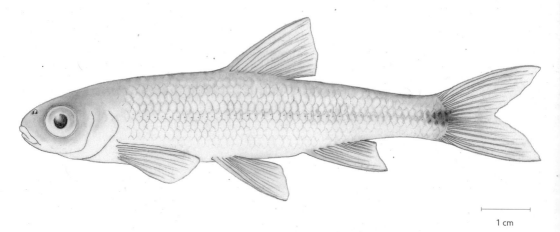

1 cm

Distinguishing Characters   This attractive little minnow has large scales (36–41 along the lateral line), no barbels, and a short anal fin (usually 8 rays). In life, the sides are silvery with a yellow cast, the back is dull green, and there is a conspicuous black spot at the base of the caudal fin. In preserved adults, there is a conspicuous silver midlateral stripe that, in smaller fish, shows as a dark lateral band. The only other *Notropis* in British Columbia is the emerald shiner (*N. atherinoides*). The two species are easily separated by colour pattern (a conspicuous caudal spot in the spottail shiner), anal fin ray count (7 or 8 rays in spottail shiner and (9) 10–13 rays in emerald shiner), and dorsal fin position (the dorsal fin originates immediately above (or slightly before) the origin of the pelvic fins in the spottail shiner and well behind the origin of the pelvic fins in the emerald shiner).

    Other B.C. fish with caudal spots are juvenile longnose dace (*Rhinichthys cataractae*), juvenile northern pikeminnow (*Ptychocheilus oregonensis*), finescale dace (*Phoxinus neogaeus*), and northern redbelly dace (*Phoxinus eos*), but it is unlikely that they could be confused with juvenile spottail shiner. Longnose dace have maxillary barbels (absent in the spottail shiner), and the upper jaw is attached to the snout (the upper jaw is free from the snout in spottail shiner). The northern pikeminnow has a larger mouth (it extends beyond the anterior margin of the eye) and smaller scales (65–77 along the lateral line) than the spottail shiner. The finescale dace and the northern redbelly dace have exceptionally small scales (over 70 in the lateral line) and a conspicuous, dark midlateral stripe (single in the finescale dace and double in the northern redbelly dace).

Sexual Dimorphism   Like many minnows, female spottail shiner grow faster than males and, on average, are larger than males of the same age (Smith and Kramer 1964). Also, breeding males have small tubercles on the dorsal surface of the head and upper surface of the pectoral rays, and the pectoral fins are longer in males than in females (McPhail and Lindsey 1970).

Distribution **NATIVE DISTRIBUTION**

The spottail shiner is another widely distributed eastern North American minnow. From east to west, it ranges from Quebec to British Columbia and from north to south from the area around Arctic Red River (lower Mackenzie River) south to Tennessee and Georgia on the Atlantic Coastal Plain.

BRITISH COLUMBIA DISTRIBUTION

The spottail shiner barely makes it into British Columbia (Map 17). The only known indigenous population is in Maxhamish Lake (Petitot River drainage) in the lower Liard River system. There are no other records from the Liard system in British Columbia; however, they do occur in the Liard River upstream from its confluence with the Mackenzie River to about 80 km below the British Columbia – Northwest Territories border. Spottail shiner from Alberta were introduced into Charlie Lake near Fort St. John as a forage fish. They have now spread into other lower Peace River tributaries (Map 17).

Life History The only site in British Columbia where the spottail shiner is abundant (aside from the introduced population in Charlie Lake) is Maxhamish Lake. Not surprisingly, little is known about their life history in our province, and most of the following information was gleaned from populations in Alberta and elsewhere (Bond 1980; Nelson and Paetz 1992; Scott and Crossman 1973; Stewart and Watkinson 2004).

Reproduction Normally, spottail shiners begin breeding in the spring when water temperatures reach or exceed 11 °C. Depending on latitude, the breeding season can extend from spring to late summer (May to August). Spawning usually occurs at dusk over sandy shoals along lakeshores or on gravel riffles near stream mouths. In Wisconsin in early June, thousands of individuals were milling about in what appeared to be a spawning aggregation (Becker 1983). In Maxhamish Lake, fry are about 12–15 mm FL in early July. This suggests that spawning in British Columbia starts sometime in June. In northeastern Alberta, spawning occurs from late June to early July (Bond 1980).

Little is known about the actual spawning behaviour of this species; however, dense aggregations of adults have been observed over gravel shoals and at creek mouths, and in some areas, there appear to be spawning runs into small streams (Becker 1983). In Lake Michigan, spottail shiner were observed spawning at a depth of about 5 m above a water intake (Wells and House 1974). The eggs were deposited on clumps of vegetation. Ripe, but unfertilized eggs, are about 1.0 mm in diameter. Presumably, egg diameter increases after fertilization. Fecundity varies with female size and egg number ranges from about 700 to 9,000.

Map 17 The B.C. distribution of spottail shiner, *Notropis hudsonius*

No fecundity data are available from British Columbia, but egg counts from northeastern Alberta ranged from 746 to 1,400 (Bond 1980). Development is rapid at temperatures above 20 °C, and the newly hatched fry are 4–5 mm long.

AGE AND GROWTH

The fry grow rapidly, and in Maxhamish Lake, some reach lengths of about 65 mm by the end of August (DeGisi 1999). Apparently, females grow more rapidly than males, and there is a clear size difference between the sexes after the third growing season (Wells and House 1974). The minimum size of mature females is about 55 mm, and this suggests that sexual maturity is reached in their second summer. In Lake Michigan, about half the males and 40% of the females reach sexual maturity in their second summer, and all individuals were mature by their third summer. A few females survive into their fifth growing season and can reach lengths of almost 140 mm (Becker 1983). In Maxhamish Lake in late July, young-of-the-year ranged from 12 to 25 mm FL, and by late August (in a different year), they ranged from 13 to 65 mm. This variability in length late in the summer suggests a protracted spawning season (mid-June to late July). Four distinct size classes (including young-of-the-year) were present in Maxhamish Lake in early July. This suggests that, in British Columbia, the spottail shiner lives at least 4 years.

FOOD HABITS

Spottail shiner school and are facultative planktivores (Hebert and Haffner 1991). In Ontario (Lake Nipigon), they feed extensively on plankton and also forage on aquatic insect larvae (Dymond 1926). Apparently, diet changes with fish size (Smith and Kramer 1964), and below 10 mm, the fry forage on rotifers and algae; however, as they grow, cladocerans, copepods, and ostracods become important foods. Above 70 mm, insects (especially chironomids) become major food items. In Maxhamish Lake in July, adults collected during the day contained mostly chironomid larvae and some *Daphnia*, whereas juveniles collected at the same time contained mostly plankton and some filamentous algae.

Habitat Little is known about the habitat use of the spottail shiner in British Columbia. Consequently, the following synopsis is derived from sources outside the province (Becker 1983; Nelson and Paetz 1992; Scott and Crossman 1973; Stewart and Watkinson 2004).

ADULTS

Adult spottail shiner usually are considered an open-water species, typically found in large rivers and lakes. In lakes during the day, they usually occur out beyond any emergent vegetation, but at night, they move in closer to shore (Becker 1983). In flowing water, they occupy river margins in areas with little current (Stewart and Watkinson 2004). In Maxhamish Lake, adults (>70 mm FL) occurred farther offshore and in deeper water (2–3 m) than juveniles and were most common over sand substrates. In Lake Michigan, this species moves inshore in the spring and summer and offshore in October. Consequently, they occur at different depths during different seasons—in relatively shallow water in the summer but down to about 30 m in the winter.

JUVENILES

In Maxhamish Lake, juvenile spottail shiner (35–50 mm in length) are abundant in the summer in shallow water (<2 m deep) over sandy substrates and around weed beds.

YOUNG-OF-THE-YEAR

In Maxhamish Lake in the summer, small schools of fry typically occur within 1 m of shore and at depths of less than 0.25 m.

Conservation Comments At times, introduced species can cloud conservation issues. The spottail shiner is such a case. As far as is known, only the Maxhamish Lake population is native to British Columbia. There is, however, a thriving population of spottail shiner in Charlie Lake near Fort St. John. These introduced shiners have spread to other sites in the lower Peace system, and many people, including some biologists, think that the spottail shiner is native to the B.C. portion of the Peace River system. It is not.

The Maxhamish Lake population is located outside the continuous distribution of this species and, as a peripheral isolate, our native spottail shiner is of some scientific interest (see Scudder 1989 for a discussion of the biodiversity value of peripheral isolates). This species is listed as S1 by the BCCDC but is not listed by COSEWIC. Maxhamish Lake Provincial Park provides this population with some protection.

Phoxinus eos (COPE)
NORTHERN REDBELLY DACE

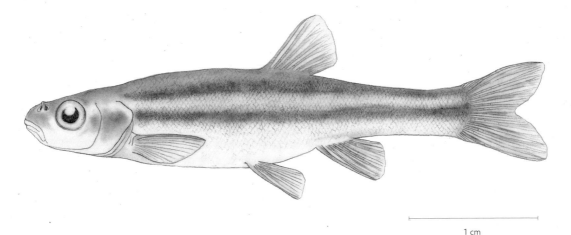

1 cm

Distinguishing Characters   This small (usually less than 70 mm FL) minnow has very fine scales (usually more than 70 along the midline), two dark lateral stripes, a long coiled gut, and a short oblique mouth. In British Columbia, this species is most likely to be confused with the finescale dace (*Phoxinus neogaeus*). Usually, they can be distinguished by mouth size and the amount of gut coiling. Viewed from below, the mouth of the northern redbelly dace almost reaches the anterior margin of the eyes, whereas the finescale dace's mouth extends past the anterior margin of the eyes. The gut in the finescale dace has a single loop, whereas in the northern redbelly dace the gut has a main loop and at least two coils. Beware, however, of hybrids. Hybridization between northern redbelly and finescale dace (see the hybridization section below) can yield a wide array of morphological phenotypes.

Taxonomy   Interbreeding between northern redbelly and finescale dace can produce hybrids that are clones of their mother (for details, see the hybridization section below). Since these clones reproduce by gynogenesis, they are reproductively isolated from the sympatric parental species, and some authors (e.g., Goddard and Dawley 1990; Goddard et al. 1998) treat them as a distinct species: *Phoxinus eos-neogaeus*. Although there are sites in British Columbia where northern redbelly and finescale dace hybridize, research on the hybrids has just begun as of this writing. At present, it is not known if the any of the gynogenetic hybrid species occurs in our province.

Hybrids aside, there are sometimes genetic differences in body shape among adjacent populations of northern redbelly dace (Toline and Baker 1994). These shape differences appear to result from natural selection fine-tuning populations to local foraging opportunities (Toline and Baker

1993). Again, it is not known if such body shape differences occur in British Columbia.

Sexual Dimorphism   Breeding males have a red band below the midline. Usually, this band is absent in breeding females, but occasionally there is a faint wash of red in the same area. The first pectoral ray in reproductive males is thickened, and when viewed from above, the first three or four pectoral rays are heavily pigmented and bear a series of sharp, conical, nuptial tubercles. Also, there are a series of modified comb-like scales in front of the pectoral fin base. Usually, the gill membranes cover these modified scales. In addition, tiny tubercles are visible on the chin and snout of some males. Females lack these features.

Hybridization   Where they coexist, the northern redbelly dace and the finescale dace often hybridize, and at some sites, these hybrids form a complex of different chromosomal forms: diploids, triploids, and diploid–triploid mosaics. All the hybrids are females, and some are exact copies of their mother (clones). However, others are heterozygous for loci fixed at different alleles in the parental species and, thus, express the paternal genome (Goddard and Dawley 1990; Goddard et al. 1998; Schlosser et al. 1998). At gametogenesis, females that express the paternal genome produce both haploid and diploid ova. If congeneric sperm penetrates either a haploid or diploid egg, the egg begins to develop, but the fusion of the maternal and paternal nuclei (syngamy) does not necessarily occur. The resulting diploid offspring are exact genetic copies of their mother. However, depending on the chromosome number in the egg, the occurrence of syngamy can produce offspring that are diploid, triploid, or triploid–diploid mosaics that express both the paternal and maternal genome (Goddard et al. 1998).

At one site in northeastern British Columbia (Tsinhia Lake), northern redbelly, finescale, and northern pearl dace (*Margariscus margarita nachtriebi*) coexist and morphological evidence suggests the presence of three-way hybrid combinations. Apparently, this same tri-hybrid combination occurs in Quebec (Legendre 1970).

Distribution   NATIVE DISTRIBUTION

The northern redbelly dace is a North American species found mostly in the eastern and central parts of the continent. Its longitudinal distribution extends from Nova Scotia to northeastern British Columbia, and its latitudinal distribution is from the lower Mackenzie River near the Arctic Red River south to Nebraska.

BRITISH COLUMBIA DISTRIBUTION

The northern redbelly dace is uncommon and restricted to lowland areas east of the Continental Divide (Map 18). Here, it is confined to the Mackenzie River system, where it is sporadically distributed in the lower Peace and lower Liard drainages.

Phoxinus eos NORTHERN REDBELLY DACE 103

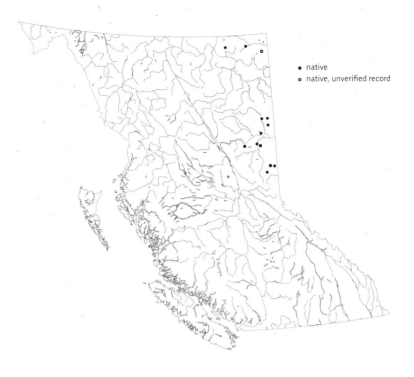

Map 18 The B.C. distribution of northern redbelly dace, *Phoxinus eos*

• native
○ native, unverified record

Life History Because of its rarity in British Columbia, little is known about the local biology of the northern redbelly dace. Consequently, most of the following information is from populations in Alberta and northeastern North America.

REPRODUCTION

In Alberta, the northern redbelly dace begins spawning in mid-June when water temperature rise above 13 °C and continues spawning until late July (Das and Nelson 1990). In Ontario, the spawning season extends from mid-June to mid-August (Powles et al. 1992). Near Chetwynd, B.C., sympatric northern redbelly dace, finescale dace, and their hybrids were in spawning condition in early July (water temperature 18 °C). Spawning behaviour has not been observed in British Columbia. Elsewhere, however, more than one male is involved in a spawning, and the eggs are released over aquatic plants or mats of algae (Hubbs and Cooper 1936; Nelson and Paetz 1992). Females are fractional spawners (Powles et al. 1992) and sometimes have more than one spawning period in a season (Hubbs and Cooper 1936). Apparently, the eggs are not adhesive. Fecundity increases with female size and ranges from about 100 to 1,000 eggs (Das and Nelson 1990). Before fertilization, the ripe eggs are about 0.9–1.25 mm in diameter. At 16 °C, the eggs take about 5 or 6 days to hatch, and the newly hatched fry are about 6 mm TL.

AGE, GROWTH, AND MATURITY

In Tsinhia Lake, in the extreme northeastern corner of British Columbia, northern redbelly dace fry reach about 25 mm FL by mid-July. Including the young-of-the-year, there are at least four size classes in this population. Among mature fish, the smallest size class (40–50 mm) is made up mostly of males, whereas the larger size classes are predominately females. This suggests that most males reach sexual maturity in their second summer, and most females, in their third summer. Cursory examination of otoliths suggests that some females may live up to 4 years.

FOOD HABITS

The long, coiled gut and black peritoneum of the northern redbelly dace suggests a diet dominated by plants or algae; however, stomach contents indicate that northern redbelly dace forage on both benthic and water-column prey (Litvak and Hansell 1990). Nonetheless, where they coexist, northern redbelly dace consumes relatively more plant matter (mostly green algae) than finescale dace (Cochran et al. 1988). In northern redbelly dace, the relative availability of benthic and water-column prey types may select for different body shapes (Toline and Baker 1993).

The few B.C. specimens of northern redbelly dace examined for stomach contents contained plant material, detritus, algae, chironomids, and zooplankton (mostly *Daphnia*). The stomachs of fry (about 25 mm) contained mostly plankton.

Habitat There has been no formal study of habitat use in B.C. northern redbelly dace; however, the few sites known in this province correspond closely to habitat descriptions in Alberta and eastern North America (Becker 1983; Nelson and Paetz 1992; Scott and Crossman 1973; Stewart and Watkinson 2004). All B.C. collections of northern redbelly dace are from slow, boggy streams and shallow, boggy lakes. Usually, the water in such habitats is the colour of dark tea. All of the recent B.C. collections of the northern redbelly dace are from sites that also contain finescale dace. This suggests that the species have similar habitat requirements. There is, however, one old record (Carl 1945) of northern redbelly dace from Stoddard Creek, the outlet of Charlie Lake. Finescale dace have never been collected at this site.

ADULTS

Adult northern redbelly dace are associated with stained waters, usually bogs, beaver ponds, and sluggish streams. Relative to their gynogenetic hybrids, the frequency of both northern redbelly and finescale dace is highest in well-oxygenated environments and decreases in poorly oxygenated waters (Schlosser et al. 1998). In British Columbia, northern redbelly dace also occur in shallow, stained lakes. In such lakes, adults are usually found along the lake margins during the day. They occur in mixed schools of northern redbelly, finescale, and less often, northern pearl dace. Typically, they remain close to cover (usually vegetation) in water less than 2 m deep and over soft silt bottoms (these lakes contain northern pike). In some Quebec lakes, northern redbelly dace leaves the littoral zone at sunset and migrates into open water (Naud and Magnan 1988). At night, the schools break up, and the fish feed individually on plankton. At sunrise, the schools re-form, and the fish move back into the littoral zone. It is not known if similar diel habitat shifts occur in British Columbia.

JUVENILES

The habitat of juvenile northern redbelly dace is similar to that of the adults. During the day, they occur in mixed schools of juvenile cyprinids. Typically, they are mixed in with adults but stay closer to vegetation and in shallower water than adults. It is not known if the juveniles migrate offshore at night.

YOUNG-OF-THE-YEAR

During the day, northern redbelly dace fry remain in shallow water (usually <1.0 m) and close to, or in, vegetation.

Conservation Comments The fragmented distribution, and the rarity of "pure" populations, suggest that the northern redbelly dace is near its physiological or habitat limits in northeastern British Columbia. Consequently, it may be especially vulnerable to habitat disturbances. Moreover, recent collections in Stoddard Creek have failed to find northern redbelly dace. Thus, one of the few known B.C. populations appears to be gone. Given the scientific interest in the all-female clones associated with the hybridization of finescale and northern redbelly dace, sites where northern redbelly and finescale dace coexist should be protected. This species is not listed by COSEWIC, but the finescale and northern redbelly dace hybrids are listed as S1 by the BCCDC (imperiled because of extreme rarity). As of this writing, research on sympatric populations of northern redbelly and finescale dace has begun in British Columbia, and we should know soon if any gynogenetic hybrids occur in our province.

Phoxinus neogaeus COPE
FINESCALE DACE

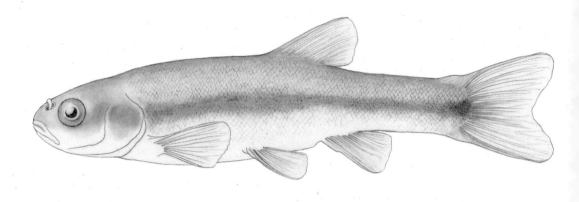

1 cm

Distinguishing Characters This small (usually less than 100 mm FL) minnow has very fine scales (in British Columbia, usually more than 75 along the midline), a single dark midlateral stripe, a relatively short gut with a single main loop, and a longer jaw and less oblique mouth than the northern redbelly dace. In British Columbia, this species is most likely to be confused with the northern redbelly dace (*Phoxinus eos*). Usually, these species can be distinguished by mouth size and the amount of gut coiling. Viewed from below, the mouth of the finescale dace reaches past the anterior margin of the eyes, whereas the mouth in the northern redbelly dace does not quite reach the anterior margin of the eyes (Fig. 16A and B). The gut in the finescale dace has a single loop, and the gut in the northern redbelly dace has a main loop and at least two subsidiary coils. Beware, however, of hybrids. Hybridization between northern redbelly and finescale dace (see the hybridization section in the northern redbelly dace species account) can yield a wide array of morphological phenotypes.

Taxonomy Interbreeding between finescale and northern redbelly dace can produce hybrids that are clones of their mother. Since these clones reproduce by gynogenesis (see hybridization section in the northern redbelly dace species account), they are reproductively isolated from the sympatric parental species and some authors (e.g., Goddard and Dawley 1990; Goddard et al. 1998) treat them as a distinct species: *Phoxinus eos-neogaeus*. There are sites in British Columbia where northern redbelly and finescale dace hybridize, and research on these hybrids is in progress as of this writing.

Sexual Dimorphism The nuptial colour in males is unusually variable. In most B.C. populations, the lower flanks of breeding males are bright yellow; however, in some populations the flanks are bright red (the red pigment usually forms a band just below the midline), and in other populations, males have a mix of yellow and red colour on the flanks. Typically, the coloured band on the lower flanks is absent in breeding females, but occasionally, there is a faint wash of colour in the same area. When viewed from above, the first pectoral ray in reproductive males is thickened and, in some populations, has a distinct swelling at its anterior base. In addition, the first pectoral ray has a posteriorly directed kink near its tip. The dorsal surfaces of the first three to six rays are heavily pigmented and bear sharp, conical tubercles. Also, there are a series of comb-like scales in front of the pectoral fin base. Typically, the gill membranes cover these modified scales. There are also tubercles on the ventral side of the caudal peduncle that extend forward to the origin of the anal fin base. Females lack these features.

Hybridization Where they coexist, the finescale dace and the northern redbelly dace often hybridize. For a discussion of hybridization in these species, see the northern redbelly dace species account. At some B.C. sites, finescale dace co-occur with the northern pearl dace (*Margariscus margarita nachtriebi*), and at these sites, the presence of morphologically intermediate individuals also suggests hybridization. Hybridization can make identification difficult, and specimens of finescale dace from sites that contain either northern redbelly or northern pearl dace (or both) should be deposited in a permanent collection.

Distribution NATIVE DISTRIBUTION
The finescale dace is a North American species that occurs primarily in the eastern and central portion of the continent. Longitudinally, it is distributed from New Brunswick to northeastern British Columbia and, latitudinally, from the middle Mackenzie River system south to Wyoming.

BRITISH COLUMBIA DISTRIBUTION
The finescale dace is restricted to lowland areas east of the Continental Divide. Here, it is confined to the lower Peace and lower Liard rivers (Mackenzie River system). It has a wider distribution in these drainages (Map 19) than the northern redbelly dace.

Life History Relatively little is known about the life history of the finescale dace in British Columbia. Consequently, most of the following information is from eastern North America and may not be entirely applicable to our populations. Except where otherwise noted, the primary source of life-history information is a Minnesota population (Stasiak 1978a, 1978b).

REPRODUCTION
Finescale dace spawning appears to be triggered by the sudden rise in water temperature that follows ice-out (Stasiak 1978b). In west-central Alberta, spawning begins in early June when water temperature rises above 12–13 °C (Das and Nelson 1990), whereas in Minnesota, spawning

Map 19 The B.C. distribution of finescale dace, *Phoxinus neogaeus*

begins in late April and peaks in May when the water temperature rises above 15 °C (Stasiak 1978b). In British Columbia, spawning fish have been collected in mid-June (18 °C), and by mid-July, most females are spent. In Minnesota, Wisconsin, and Alberta, finescale dace apparently begin spawning earlier than northern redbelly dace, although there is broad overlap in their spawning seasons.

Actual spawning has not been observed in the field; however, Stasiak (1978b) observed a spawning aggregation and describes behaviours associated with spawning. Apparently, spawning involves a single female and one to three males. Females appear to initiate spawning bouts by swimming with a zigzag motion towards cover. Males respond by chasing the female, and the spawning group disappears into a depression under some form of cover (usually a log or submerged tree branches). Presumably, gamete release occurs in these depressions. The finescale dace is a fractional spawner and spawns 20–30 eggs several times in a season (Stasiak 1978b).

Fecundity in the finescale dace increases with female size, and egg number ranges from about 400 to about 3,000 eggs. Before fertilization, the ripe eggs are about 1.4 mm in diameter (Das and Nelson 1990). Development is rapid, and at 20 °C, the eggs hatch in about 6 days (Stasiak 1978b). The newly hatched larvae are small (about 4–5 mm TL) and do not start feeding until 7 days after hatching.

AGE, GROWTH, AND MATURITY

In northeastern British Columbia (Tsinhia Lake) in mid-July, finescale dace fry ranged from 16 to 35 mm FL. This range in size suggests a protracted spawning season. In contrast, in early September, finescale dace fry in small stream populations in the Petitot River system ranged from 20 to 30 mm in length. By September, most B.C. populations contain at least five size classes. Both sexes mature are mature by the beginning of their third summer, and the largest size classes are predominately females. Cursory examination of otoliths suggests that some individuals probably live for 5 or 6 years.

FOOD HABITS

The black peritoneum and long gut suggest some energetic dependence on plant material; however, relative to northern redbelly dace, finescale dace appear to consume only modest amounts of algae. Benthic invertebrates (often chironomid larvae) are their primary prey, but they also take occasional terrestrial insects. Juveniles less than 35 mm FL appear to have more detritus in their diet than adults. Samples taken at midday in Tsinhia Lake had empty stomachs but partially digested food in the intestine. This hints at early morning foraging.

Habitat

There has been no study of finescale dace habitat use in British Columbia; however, the sites where they occur in this province correspond closely to habitat descriptions from Alberta and eastern North America (Becker 1983; Das and Nelson 1990; Stasiak 1978a).

ADULTS

Most finescale dace in British Columbia are collected in slow-moving streams or small lakes. In the Petitot and Hay river systems in northeastern British Columbia, finescale dace were strongly associated with small, sluggish, heavily stained headwater streams. Usually, the only other fish present at these sites is the brook stickleback (*Culaea inconstans*). Occasionally, however, adults are collected in large, turbid rivers like the Fort Nelson River where they are usually found in close proximity to tributary streams. They also occur in lakes (e.g., One Island Lake near Tupper). In lakes, at least during the day, they remain inshore in water <2 m deep, where they often cruise around the lake margins in dense schools.

JUVENILES

In lakes, juveniles school with adults, and in sluggish streams, aggregations of juveniles often are associated with vegetated areas.

YOUNG-OF-THE-YEAR

Fry usually are found in shallower water than either adults or juveniles and remain close to vegetation.

Conservation Comments   The finescale dace has a wide distribution in northeastern British Columbia, and most populations appear to be healthy. This suggests that the finescale dace is in no immediate risk in the province. Nonetheless, the sites where this species coexists with northern redbelly dace are of scientific interest and should be protected. Neither the BCCDC nor COSEWIC list this species; however, the BCCDC lists the northern redbelly dace and finescale dace hybrids as S1 (imperiled because of extreme rarity). See the conservation comments in the northern redbelly dace account for more details.

Pimephales promelas RAFINESQUE
FATHEAD MINNOW

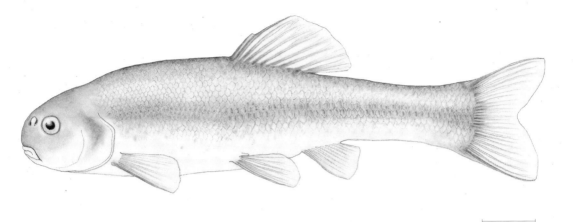

1 cm

Distinguishing Characters This chunky little fish has a blunt head, a small oblique mouth, no maxillary barbels, crowded predorsal scales, and a unique (at least among B.C. minnows) short, thickened first dorsal ray that is not closely bound to the second ray. The secondary sex characteristics of male fathead minnow are so distinctive that they are unlikely to be misidentified; however, females and juveniles are sometimes mistaken for the brassy minnow (*Hybognathus hankinsoni*), lake chub (*Couesius plumbeus*), and finescale dace (*Phoxinus neogaeus*). They are separated from the brassy minnow by scale size (40–50 in fathead minnow and 35–40 in brassy minnow) and anal ray number (usually seven in fathead minnow and usually eight in brassy minnow). Lake chub have larger mouths than fathead minnow (the mouth reaches the front margin of the eye in lake chub and falls short of the eye in fathead minnow) and a maxillary barbel at the corner of the mouth (absent in the fathead minnow). The fathead minnow is separated from finescale dace by scale size (40–50 in fathead minnow and over 70 in finescale dace) and anal ray number (usually seven in fathead minnow and usually eight in finescale dace).

Sexual Dimorphism Like many cyprinids, the sexes in the fathead minnow differ in body size. In most minnows, females are larger than males; however, in fathead minnows, this size dimorphism is reversed, and males are larger than females. Large size is probably an advantage in male–male competition for mates and spawning sites. In addition, breeding males develop large tubercles on the snout and lower jaw as well as fine tubercles on the top of the head and the dorsal surface of the pectoral fins. Breeding males also have a dark spot at the anterior base of the dorsal fin, and an elongate spongy pad that extends from the back of the head to the dorsal fin. Breeding females develop an enlarged ovipositor-like structure. Fathead minnows attach eggs to the undersides of rocks and sticks, and males rub their dorsal pad on egg deposition sites and on eggs after they are deposited. This rubbing behaviour may prepare the site for egg deposition, clean the eggs, and assist in the hatching of the eggs (Smith and Murphy 1974). The distinctive male secondary sex characteristics appear about a month before actual breeding begins and disappear shortly after the breeding season (Markus 1934).

Distribution NATIVE DISTRIBUTION

The fathead minnow is native to North America and is one of the most widely distributed North America minnows. Its natural range includes most of the Great Plains and the Appalachian region. Here, the fathead minnow extends from Quebec to Alberta and from near Fort Smith, northern Alberta, to Chihuahua, Mexico. They have been introduced into most western states as forage for introduced bass.

BRITISH COLUMBIA DISTRIBUTION

The fathead minnow was first recorded from One Island Lake near Tupper (Smith and Lamb 1976); however, it is not native to British Columbia. The fathead minnow was absent from early collections from One Island Lake, and no native populations of fathead minnow are known from elsewhere in the Peace River system in British Columbia. Additionally, native fathead minnow populations appear to be absent from adjacent Peace River drainages in Alberta (Nelson and Paetz 1992). Presumably, the fathead minnow reached One Island Lake in a bait bucket. In the 1980s, this species began appearing in scattered B.C. drainages west of the Continental Divide: the lower Fraser valley, the Little Campbell River near Whiterock, and the Sunshine Coast near Powell River (Map 20). Because the fathead minnow is tolerant of high temperatures and low oxygen conditions (Robb and Abrahams 2003), they are easily transported; however, in a province where using live fish for bait is illegal, it is not clear why people move them around.

Life History Because they are a relatively recent addition to the B.C. fish fauna, little is known about their life history in our province. Presumably, however, it is similar to that of fathead minnows elsewhere in the northern parts of their range (e.g., Alberta, Saskatchewan, Ontario, and the north-central United States).

REPRODUCTION

Within their native range, the fathead minnow breeds in the spring and continues spawning throughout most of the summer. In the lower Fraser valley, spawning begins when the water temperature rises above 15 °C (usually about mid-May). The males are territorial and defend spawning sites under rocks, logs, or sticks. Males rub the spongy dorsal pad (mentioned in the sexual dimorphism section) on egg deposition sites.

Map 20 The B.C. distribution of fathead minnow, *Pimephales promelas*

■ introduction

This rubbing behaviour deposits mucous on the deposition sites and may act both to mark the sites and clean them in preparation for egg deposition (Smith and Murphy 1974). Males are very aggressive and defend their nest site from other males. The courtship behaviour of the fathead minnow is relatively well known (McMillan 1972). After courtship, the female deposits some eggs in the nest and then is driven away by the male. The male guards and fans the eggs and keeps them clear of silt and algae. During incubation, he rubs (cleans?) the eggs with his dorsal pad. Newly fertilized eggs are sticky and about 1.2 mm in diameter. Fecundity is reported to range from about 250 to 2,500 eggs; however, the females are fractional spawners and only release a portion of their eggs (80–370) at each spawning (Becker 1983). Typically, individual females spawn with several males and deposit eggs in several different nests. Hence, most nests contain egg batches obtained from different females, and consequently, the egg mass often contains eggs at different stages of development. Development rate is rapid (4–7 days, depending on water temperature), and newly hatched young are about 5 mm in length.

AGE, GROWTH, AND MATURITY
Fathead minnow fry grow rapidly and reach a length of about 50 mm FL in their first summer (Held and Peterka 1974). In the Fraser valley, males appear to grow faster than females, and both sexes reach maturity early in their second summer. Elsewhere, however, some individuals reach maturity and spawn before the end of their first summer (Markus 1934).

In both North Dakota (Held and Peterka 1974) and western Ontario (Tallman et al. 1984), relatively few fathead minnows survive beyond 2 years, and most mature adults die 1–2 months after spawning (Markus 1934).

FOOD HABITS

Fathead minnows have the long, coiled intestines and black peritoneum typical of fish that consume plant material or detritus. Although they eat algae and detritus, they are opportunistic feeders and take a wide variety of seasonally abundant animal prey (Held and Peterka 1974; Price et al. 1991; Tallman et al. 1984). Young-of-the-year begin feeding on small prey (e.g., rotifers and nauplii) and, as they grow, add increasing numbers of cladocerans, copepods, and chironomids to their diets. Adult males are larger than females and juveniles and include larger prey (amphipods) in their diet (Price et al. 1991).

Habitat Again, because of their relatively recent introduction, nothing is known about habitat use by fathead minnow in British Columbia. Presumably, it is similar to that in their natural range.

ADULTS

Where they are native, the fathead minnow occur in sloughs, sluggish creeks, lakes, and ponds. They are strongly associated with heavy vegetation and low oxygen conditions. On the Great Plains, they often occur in highly saline waters. In the lower Fraser valley, introduced fathead minnow are especially abundant in slow moving, highly eutrophic sloughs and ditches around Chilliwack and less abundant in the flowing portions of streams in the same area.

In a small Alberta pond that contained only fathead minnow, there were differences in the seasonal and diel habitat use among juvenile and adult fish (Price et al. 1991): adults of both sexes were more active at night than during the day. In contrast, there was no evidence of a daily peak in adult activity in Ontario (Tallman et al. 1984).

JUVENILES

In lakes early in the summer, more juveniles occur inshore during the day than at night, but later in the season, juveniles also became night-active (Price et al. 1991). In Ontario, however, juveniles appear to be less active at night and in the morning than they are in the afternoon (Tallman et al. 1984).

YOUNG-OF-THE-YEAR

Given their rapid growth rate and their propensity to mature (and even spawn) in their first summer of life, there is no clear distinction between fry and juveniles in this species. Nonetheless, in the summer in the lower Fraser valley, fry (15–25 mm FL) remain in shallower water and closer to the margins of streams and sloughs than larger (30–40 mm) juveniles.

Conservation Comments   This introduced species has no conservation status in British Columbia. Although it probably is a relatively benign introduction, males are aggressive and defend breeding territories. Consequently, it is possible that they have a negative impact on native species spawning at the same time and in the same areas.

Platygobio gracilis (RICHARDSON)
FLATHEAD CHUB

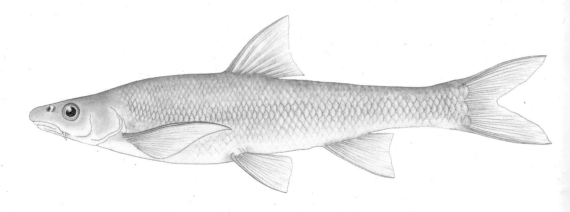

1 cm

Distinguishing Characters   The flathead chub is a large minnow that can reach a length of 37 cm (Kristensen 1980). It has a distinctive flattened head, large barbels at the corners of the mouth, and sickle-shaped fins (especially the pectoral fins). In life, the lower flanks are silvery, and the back is usually a dull brown. In preservative, young individuals have a muted midlateral band that is most pronounced on the posterior half of the body. Occasionally, small flathead chub are mistaken for lake chub (*Couesius plumbeus*), but even at a small size, the two species are easily distinguished by head shape (broad and flattened in flathead chub and slightly rounded in lake chub) and fin shape (from above, the trailing edge of the pectoral fin is sickle-shaped in flathead chub and rounded in lake chub).

Sexual Dimorphism   The sexes are difficult to distinguish, although mature females are larger on average than males. Both sexes develop breeding tubercles, and there are subtle differences between the sexes in the development and distribution of tubercles. Both sexes have minute tubercles on the head, back, and dorsal sides of the pectoral fins, but males also develop tubercles on the anterior body scales and on most dorsal fin rays. In females, tubercles usually are absent from body scales and restricted to the dorsal side of the first ray of the pectoral fin.

Distribution   NATIVE DISTRIBUTION
The flathead chub is indigenous to North America. Here, it is found in the large, turbid rivers of the Great Plains and is distributed in a longitudinally narrow but latitudinally broad band that extends from the Mackenzie River Delta to the upper Rio Grande and lower Mississippi rivers.

BRITISH COLUMBIA DISTRIBUTION

Flathead chub is restricted to the lower portions of the large, turbid rivers that drain the northeastern portion of the province. In both the Peace and Liard systems, it is found below, but not above, the velocity barriers that separate the upper and lower sections of these river systems (Map 21).

Life History Little is known about the life history of the flathead chub in British Columbia, and the following account is based on information from elsewhere on the Great Plains (Bishop 1975; Bond 1980; Peters and Holland 1994; Olund and Cross 1961; Young et al. 1997). These sources are supplemented with scattered field observations on B.C. populations.

REPRODUCTION

There is no information on flathead chub spawning sites and even the time of spawning is in doubt. In Alberta, spawning is thought to occur in July and early August (Bishop 1975; Nelson and Paetz 1992); however, exceptionally large catches of flathead chub from middle to late May on the lower Slave River suggest a spring spawning aggregation (Tallman et al. 1996b). Based on sampling young-of-the-year, spawning in British Columbia appears to begin about mid-June and may last as long as mid-August. In the Muskwa River, fry (14–24 mm FL) were collected in late July. Additionally, in a collection made during the first week of June at the same site (but in a different year), the smallest size class ranged from 43 to 60 mm. These early June fish probably hatched in the previous year; thus, these samples suggest that spawning occurs sometime in June. In the mainstem Liard River, two-thirds of adult flathead chub sampled from July 20 to August 10 were recently spent fish (McLeod et al. 1979). The remaining adults either had spawned earlier or were not going to spawn that year. Some of this confusion over the spawning time of flathead chub may reflect a protracted spawning period.

In the Athabasca River, females ranging in size from 235 to 297 mm FL had fecundities of 7,000–15,000 eggs (Bond 1980). There is no information on egg size or incubation time.

AGE, GROWTH, AND MATURITY

In the lower Peace River, 0+ fry had a mean fork length of 61 mm, and 1+ fry averaged 91 mm (Bishop 1975). Sexual maturity occurred at about 180 mm FL and was reached at age 4 (Bishop 1975). Farther north in the lower Liard River, some individuals in their sixth and seventh years (averaging 189 and 201 mm FL, respectively) had not achieved sexual maturity (McLeod et al. 1979). Farther south in the Missouri River system, growth is faster, and sexual maturity is reached at a smaller size: a few males mature at 75 mm, and a few females, at 85 mm (Fisher et al. 2002; Olund and Cross 1961). All males are mature at 110 mm, and all females, by 170 mm. In British Columbia, the oldest recorded individual was in its eleventh year (McLeod et al. 1979), but fish as old as 14 years were recorded from the Slave River Delta (Tripp et al. 1981).

118 FAMILY CYPRINIDAE—MINNOWS

Map 21 The B.C. distribution of flathead chub, *Platygobio gracilis*

• native

FOOD HABITS

The flathead chub's large barbels and well-developed nostrils suggest a fish adapted to feed in turbid environments. In the Fort Nelson area, native people catch flathead chub by exploiting the fish's chemosensory capability (see McPhail and Lindsey 1970). In northeastern British Columbia, nymphs of aquatic insects dominate the summer diet of adult flathead chub, but they also contained terrestrial insects and molluscs. In the lower Slave River, gastropods and corixids were the commonest items in the diet of flathead chub, but they also took chironomids, ants, and beetles (Tallman et al. 1996b). In the Muskwa River, adults often contain sucker and cyprinid fry, and in northern Alberta, a shrew was recorded from one stomach (Bond 1980). Fish less than 150 mm appear to forage heavily on ostracods.

Habitat ADULTS

In British Columbia, adult flathead chub occur in the mainstems of the lower Peace and Liard rivers and their large, turbid tributaries (e.g., the Fort Nelson, Muskwa, Prophet, Beatton, Pine, Halfway, and Moberly rivers). Although they are found over a wide variety of substrates (silt, sand, and gravel), they appear to be most common over sand bottoms. Their streamlined bodies and falcate fins suggest a fish adapted to fast water; however, most B.C. collections are from relatively slow (<0.5 m/s) shallow (<1.0 m) waters. These collections are dominated by juveniles (<200 mm) but also contained adults. The B.C. observations are consistent with the habitat use data given for flathead chub in the Missouri, lower Platte, and lower Yellowstone rivers (Fisher et al. 2002; Peters and Holland 1994 Young et al. 1997). These authors usually found flathead chub in turbid water over sand substrates at depths of 0.10–1.0 m and water velocities of <0.40 m/s.

JUVENILES

Juveniles occupy shallower (<50 cm) and slower water (<0.3 m/s) than adults. They occur in deeper and somewhat faster water than the young-of-the-year.

YOUNG-OF-THE-YEAR

In the summer, young-of-the-year occur in quiet (<0.1 m/s), shallow (<20 cm) water along the edges of embayments and side-channels of large rivers. The substrate usually is sand or gravel overlain by silt. Typically, the only cover in such areas is the turbidity of the water.

Conservation Comments   At present, the flathead chub is not a conservation concern in British Columbia. This could change if the proposed Site C hydroelectric development proceeds. Large turbid rivers are important, perhaps critical, habitat for flathead chub, and there is evidence that reservoirs in such rivers adversely affect downstream flathead chub abundance (Fisher et al. 2002). Neither COSEWIC nor the BCCDC lists this species.

Ptychocheilus oregonensis (RICHARDSON)
NORTHERN PIKEMINNOW

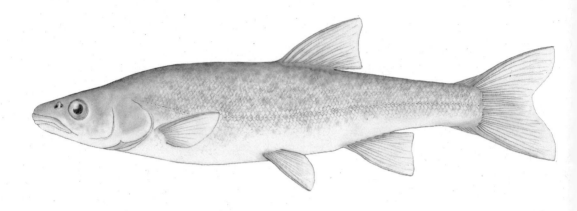

1 cm

Distinguishing Characters   The northern pikeminnow is the largest native minnow in British Columbia: adults often exceed 300 mm FL, and some individuals reach lengths of over 400 mm. There is an old record from Shuswap Lake (Carl et al. 1959) of a 29 pound (13 kg) northern pikeminnow—some minnow! Its size and large mouth—the upper jaw extends back beyond the front margin of the eye—easily distinguishes adult pikeminnows from other B.C. minnows.

Although adult and juvenile pikeminnows are distinctive, the fry often occur in mixed schools with redside shiner (*Richardsonius balteatus*) and peamouth (*Mylocheilus caurinus*) fry. These small fish are easily confused. For quick field identification, the small, dark spot at the base of the caudal fin separates young northern pikeminnow from most other minnow species. An exception is the longnose dace (*Rhinichthys cataractae*). The fry of this species also have a black caudal spot but are distinguished from pikeminnow fry by a band of dark pigment on the snout in front of the eye. This snout pigment is absent in northern pikeminnow fry.

Taxonomy   The northern pikeminnow is the Columbia representative of the genus *Ptychocheilus*. Other pikeminnow species occur in the Umpqua, Sacramento – San Joaquin, and Colorado river systems. There are no taxonomic issues with this distinctive minnow. There are, however, populations with exceptionally small adult size. As far as is known, these "dwarf" pikeminnows are confined to a few small lakes on the Bonaparte Plateau. Their ecology and life history are under investigation, but the absence of other cyprinids in these lakes suggest that these are introduced populations.

Sexual Dimorphism As in most minnows, mature females are larger than mature males. This size dimorphism results from a more rapid growth rate and a longer lifespan for females than for males. Breeding males have small tubercles on the head and a double row of tooth-like tubercles on the dorsal surfaces of the anterior pectoral and pelvic fin rays. Most females show no tubercle development on the head, but they have tubercles on the dorsal surfaces of the pectoral and pelvic rays. These tubercles, however, are small and not as sharp as those in males. Outside the breeding season, the sexes are difficult to separate, and the only clear difference is in pelvic fin length. The pelvic fins in both males and females fall short of the vent; however, the pelvic fins in mature males come within less than an eye diameter of the vent, while the pelvic fins in mature females fall short of the vent by more than an eye diameter.

Hybridization In British Columbia, occasional hybrids are found between northern pikeminnow, peamouth, redside shiner, and chiselmouth (*Acrocheilus alutaceus*). Usually, but not always, these hybrids are morphologically intermediate between the parental species. In Missezula and Wolfe lakes (near Princeton), hybrids between chiselmouth and northern pikeminnow are relatively common. Ecologically and behaviourally, this is an unusual hybrid combination: one parent is a periphyton scraper, and the other is a piscivore. Still, the hybrids somehow survive, and a few even backcross into the parental genomes (Stewart 1966).

Distribution NATIVE DISTRIBUTION
The northern pikeminnow is a Columbia endemic and is restricted to the Columbia River system and adjacent drainages that received their fish fauna from the Columbia River. Like many Columbia endemics, it is absent from the Snake River above Shoshone Falls.

BRITISH COLUMBIA DISTRIBUTION
Northern pikeminnow are widely distributed throughout the interior of the province (Map 22). They occur throughout the Columbia and Fraser drainage systems, and from the Fraser, they have colonized the Skeena, Nass, and upper Peace systems. The northern pikeminnow is absent from the Liard, Yukon, Iskut–Stikine, and Taku drainage systems. Along the central B.C. coast, they occur in the upper portions of the Klinaklini and Dean rivers. Although primarily an interior species, northern pikeminnow reaches the coast in large rivers like the Fraser and Skeena, but they are intolerant of high salinities and have not reached Vancouver Island or any of the other coastal islands.

Life History Curiously, for such an abundant species, no detailed life-history study is available for the northern pikeminnow. Consequently, the following synopsis is derived mostly from journal articles and theses (Beamesderfer 1992; Cartwright 1956; Jeppson and Platts 1959) and from unpublished observations made in our area.

Map 22 The B.C. distribution of northern pikeminnow, *Ptychocheilus oregonensis*

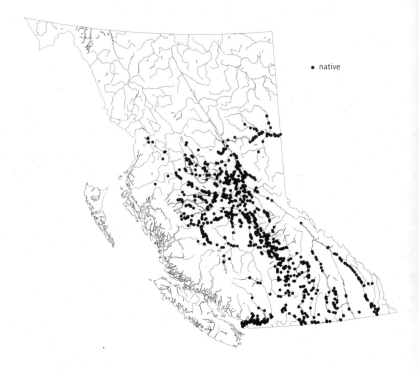

• native

REPRODUCTION

Northern pikeminnow spawn in the spring. Apparently, rising water temperatures trigger the spawning migrations. The threshold spawning temperature of about 12 °C is reached at different times in different parts of the province. Thus, spawning may begin as early as late April in the Lower Mainland and as late as July in the upper Peace and Nass systems. Although most known northern pikeminnow spawning sites in British Columbia are in inlet streams, spawning can occur in both flowing water and in lakes. It is rare for northern pikeminnow to ascend spawning streams for more than a few hundred metres, and spawning usually occurs on either the first or second riffle above the lake. A lake-spawning population (Missezula Lake, near Princeton) started spawning at dusk (Stewart 1966). First, males aggregated in shallow water, while females cruised offshore in deeper water. Eventually, single females entered the aggregation of males, and spawning occurred in water so shallow that the fish closest to shore were mostly out of the water. Several males were involved in the spawning of each female. Males in a fluvial population in the St. Joe River (a Columbia tributary in northern Idaho) also formed a dense aggregation, while females cruised slowly around the edges of the aggregation (Beamesderfer 1992). Females that were ready to spawn entered the aggregation and swam more rapidly than the cruising females. Up to eight males followed the fast-moving females down to within a few centimeters of the substrate, where eggs and sperm were released. Females repeatedly released eggs at 4- to 6-second intervals.

Suitable spawning sites appear to require water velocities of <0.4 m/s and a sand-free substrate of gravel or cobbles (Beamesderfer 1992).

The eggs are demersal, adhesive, and about 2.5 mm in diameter when fertilized. Fecundity in females is a function of body size and ranges from about 5,000 to 95,000 eggs (Cartwright 1956; Olney 1975; Parker et al. 1995). The eggs take about 6 days to hatch at 18 °C, and the fry are about 8 mm TL at hatch.

AGE, GROWTH, AND MATURITY

Northern pikeminnow grow rapidly in their first summer and, in British Columbia, usually reach between 35 and 65 mm (average is about 50 mm FL) by the end of their first growing season. In exceptional circumstances (e.g., a new reservoir), they may reach almost 100 mm in their first summer (Chisholm et al. 1989). In males, sexual maturity usually is reached in 3 or 4 years, whereas females typically mature a year later (i.e., at 5 or 6 years). In British Columbia, most northern pikeminnow over 500 mm FL are females about 15–20 years old. On average, females live 3 or 4 years longer than males (Parker et al. 1995).

FOOD HABITS

Young-of-the-year northern pikeminnow consume a wide variety of prey including organisms taken from both the bottom and the water surface (cladocerans, copepods, ostracods, and chironomid larvae and pupae). As they grow, they take larger and larger prey and, at about 100–125 mm, begin to include fish in their diet (Olney 1975; Steigenberger and Larkin 1974). Above a length of 300 mm, they are primarily piscivores, although they will eat almost any prey of suitable size (e.g., crayfish, frogs, toads, and even small rodents). Adult feeding occurs mainly from dusk to dawn (Chisholm 1975; Steigenberger and Larkin 1974).

Habitat ADULTS

In British Columbia, the northern pikeminnow is found primarily in lakes and large, slow-moving rivers. Tests of their sustained swimming ability (Mesa and Olson 1993) suggest that water velocities >1.0 m/s may exclude northern pikeminnow. In lakes in the summer, adults often cruise the littoral zone about 1 m above the bottom and on the offshore side of weed beds. In winter, northern pikeminnow move into deeper water and are not as bottom oriented. In some lakes, a few adults occupy the limnetic zone in both winter and summer (Beauchamp et al. 1995). In rivers, there is a similar pattern of habitat use (Beamesderfer 1992), with adults occupying deeper water than juveniles. Incidental tag recoveries from a few fish tagged in the Columbia River between Keenlyside Dam and the U.S. border suggest that adult northern pikeminnow do not move far (up to 3 km) from the sites where they were tagged (Hildebrand et al. 1995).

JUVENILES

In the summer in lakes, juveniles typically occur in shallower water and are more surface oriented than adults. Loose schools often occur at the outer edges of weed beds. They move offshore into deeper water in the fall. In rivers, juveniles also are associated with shallow (<1.0 m), quiet water.

YOUNG-OF-THE-YEAR

In the summer, northern pikeminnow fry are found along lake margins in shallow water (<0.30 m) close to cover (usually weeds) and in mixed schools with other cyprinids (Miura 1962). In rivers, fry also occur in mixed cyprinid schools in shallow (<0.25 m) water along the river edges (Beamesderfer 1992).

Conservation Comments In British Columbia, the northern pikeminnow is not a conservation concern; however, it is a persecuted species. Its crime is that it is a superbly adapted piscivore—the adults will consume any fish they can swallow. Consequently, they are unpopular with anglers and many fisheries professionals. Like most predators, northern pikeminnow have a role in the natural balance of northwestern aquatic ecosystems. Unfortunately, humans have changed this balance, especially in the Columbia River system. Here, dams drop water from one reservoir into the next in an almost unbroken series from Bonneville upstream to the Canadian border (and beyond). It is not the pikeminnows' fault that engineers have provided them with convenient feeding stations below the dams, and that fish culturists annually release a generous supply of inexperienced prey. Under these circumstances, northern pikeminnow do what comes naturally—eat and make more northern pikeminnows.

Unfortunately for the northern pikeminnow, this brings them in conflict with attempts to restore the Columbia River salmon runs to their pre-dam levels (without, of course, restoring the natural environment). So, the northern pikeminnow has become a scapegoat for human mismanagement of the aquatic environment and prompted various attempts, including a bounty system, to reduce their numbers. Although it is estimated that control programs (Beamesderfer et al. 1996) could reduce the number of juvenile salmon consumed by northern pikeminnow by up to about 50% at the highest pikeminnow exploitation rate (20%), it would take about 15 years of sustained control efforts to determine if the control program actually worked. Then, even if the estimated benefits are achieved—and there are a lot of assumptions in the calculations—the program would have to be continued in perpetuity. Given their reproductive potential, the northern pikeminnow will probably bounce right back to their pre-control numbers if there is any relaxation of the control program.

Unfortunately, the publicity given to control programs, and especially the payment of a bounty, reinforces angler prejudice against the northern pikeminnow. As a consequence, while northern pikeminnow and their prey mostly are still in their natural balance in British Columbia, decent anglers treat these fish with shameful cruelty. Although rarely used as food, they are seldom returned to the water unharmed. At best, they are killed quickly, but most often they are mutilated and left to flop their lives away on land. The rationale for this barbarity is that the northern pikeminnow is "bad" and, therefore, killing them is "good." Nonetheless, in spite of this persecution, the northern pikeminnow will be there when the Columbia River again flows freely to the sea.

Rhinichthys cataractae (VALENCIENNES)
LONGNOSE DACE

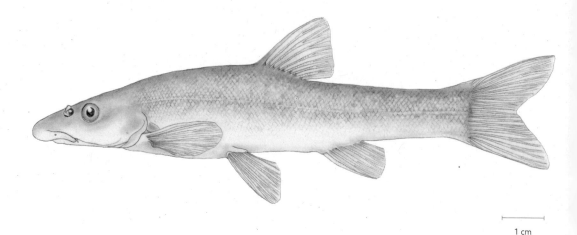

1 cm

Distinguishing Characters   This ventrally flattened minnow has a long snout (2–2.7 times into head length) that clearly overhangs the mouth. The upper lip is firmly attached to the snout by a frenum (a fleshy connection between the upper lip and the snout), and the horizontal mouth has well-developed barbels at the corners of the jaws. In most of British Columbia, the caudal peduncle is relatively deep (2.5–2.8 times into head length), and the caudal fork is shallow. In specimens less than 100 mm FL, there is usually a distinct dark midlateral stripe that is constricted at the caudal flexure and then expands again on the tail as a dark spot. Fry (down to 12 mm) have a distinctive dark stripe on the snout.

    Three other species of *Rhinichthys* occur in British Columbia (leopard dace, *R. falcatus*; speckled dace, *R. osculus*, and Umatilla dace, *R. umatilla*). The speckled dace is restricted to the Kettle River and its tributaries; the Umatilla dace is found in the Columbia River and its tributaries below Arrow Lakes and below Bonnington Falls on the Kootenay River. The Umatilla dace also occurs in the Similkameen drainage system. The leopard dace occurs in both the Columbia and Fraser systems. These other dace lack the frenum (the upper jaw is free from the snout) characteristic of the longnose dace. There is of a fourth, unnamed, dace (the Nooksack dace) in British Columbia. Although its taxonomy is uncertain—it may be a distinct species or a subspecies—it is clearly related to the longnose dace. This taxonomic ambiguity is discussed below; however, it has fewer lateral line scales (54–61 vs. 59–70) and a narrower caudal peduncle than typical western longnose dace. Additionally, the Nooksack dace's cytochrome *b* sequence differs from that of typical Columbia or Fraser longnose dace by about 2.5%.

Taxonomy Given its extensive geographic range, it is not surprising that the longnose dace has been divided into subspecies. Two of these nominal subspecies occur in British Columbia—*R. c. cataractae* in the northeast and *R. c. dulcis* in the rest of the province. However, there are unresolved taxonomic problems with the longnose dace. The supposed western subspecies (*R. c. dulcis*) was named from a Missouri River tributary, the Sweetwater River in Wyoming. Thus, it is unlikely that it occurs west of the Continental Divide. Additionally, mitochondrial data indicate about a 4% sequence difference in cytochrome *b* between longnose dace on the eastern and western sides of the Continental Divide. These data argue that the two forms diverged sometime before the beginning of the Pleistocene. Thus, *dulcis*, as either a subspecific or a specific name, is inappropriate for northwestern longnose dace.

Furthermore, behavioural observations (Bartnik 1972) indicate major differences in spawning behaviour and nuptial colouration between eastern and northwestern longnose dace. The main differences are in male nuptial colour (breeding males on the Great Plains develop a vivid red lateral stripe, whereas northwestern males lack, or have only faint, nuptial colours), time of spawning (day in the Great Plains form and night in the northwestern form), and female territoriality (absent in spawning Great Plains females and present in northwestern females). Given the extent of the mitochondrial divergence and the reproductive differences between the two forms, northwestern longnose dace may warrant specific recognition; however, the two forms probably come in contact in the Peace River system downstream of the Peace Canyon (present site of Bennett Dam). Thus, any attempt to determine the taxonomic status of western longnose dace will require investigation of this potential contact zone.

An added complication to the taxonomy of northwestern longnose dace is the presence of a third form, the Nooksack dace, in the southwestern portion of the lower Fraser valley. The Nooksack dace's taxonomic status is uncertain, but it is clearly related to other northwestern longnose dace. It is endemic to western Washington and southwestern British Columbia. In Washington, it occurs in rivers on the west side of the Olympic Peninsula, in the Chehalis River system, and in rivers entering the east side of Puget Sound. In British Columbia, the Nooksack dace is restricted to tributaries of the Nooksack River in the Abbotsford–Aldergrove region of the lower Fraser valley (e.g., Fishtrap, Bertrand, and Pepin creeks) and one river on the north side of the Fraser River (the Brunette River). Morphologically, the Nooksack dace is distinguished from the longnose dace by differences in body proportions and scale counts (Bisson and Reimers 1977; McPhail 1967). At the molecular level (cytochrome *b*), Nooksack dace have diverged (about 2.5%) from other northwestern longnose dace. This is a greater divergence than that found between some other northwestern species of *Rhinichthys* (e.g., leopard and Umatilla dace).

The unresolved aspect of the taxonomic status of the Nooksack dace involves its hybridization with longnose dace. The evidence for hybridization comes from mitochondrial DNA sequences of dace from two rivers immediately upstream of the Brunette River: the Coquitlam and Alouette rivers. Both these rivers contain some dace with mitochondrial sequences that are characteristic of Nooksack dace and other dace with the mitochondrial sequence typical of northwestern longnose dace. About 40% and 28% of the dace have Nooksack dace mitochondrial sequences in the Coquitlam and Alouette rivers, respectively. Because the mitochondrial genome is maternally inherited and passed unchanged to the mother's offspring, the presence of Nooksack dace mitochondrial sequences in these rivers requires explanation. It could be the footprint of past (postglacial) hybridization; alternatively, it could reflect present hybridization. If there is current hybridization, it implies both forms are present (sympatric) in these rivers. Unfortunately, the original samples from these rivers were discarded without careful morphological analysis. Thus, it is not clear if the Nooksack dace mitochondrial signal in these rivers is associated with a Nooksack dace morphological signal. Research to clarify this situation is underway as of this writing; however, given the present uncertainty, I have not given the Nooksack dace a separate species account.

Sexual Dimorphism As in many minnows, mature female longnose dace are larger than males. This size dimorphism results from both a more rapid growth rate and a longer lifespan in females than in males. The sexes also differ in the development of nuptial tubercles; however, the differences are subtle. Both sexes develop tubercles on the dorsal aspects of the pectoral and pelvic fins, but these tubercles are noticeably larger in males. There are also fine tubercles on the scales, especially the scales anterior to the dorsal fin; again, these tubercles are better developed and more widespread in males than in females. Additionally, both the pectoral and pelvic fins are slightly longer, and the pectoral fin is more heavily pigmented, in males than in females.

Hybridization In British Columbia, hybrids between longnose dace and other species are rare; however, hybrids with redside shiner (*Richardsonius balteatus*) and lake chub (*Couesius plumbeus*) have been recorded in the province (Carl et al. 1959). In the Peace River region, hybrids between longnose dace and lake chub are usually associated with disturbed sites, especially sites where beavers have dammed streams and flooded riffles.

Distribution **NATIVE DISTRIBUTION**
The longnose dace is endemic to North America. Here, it has the widest geographic distribution of any indigenous minnow: it occurs in rivers from the Atlantic to the Pacific coast and from the Arctic (just upstream of the Mackenzie River Delta) to northern Mexico.

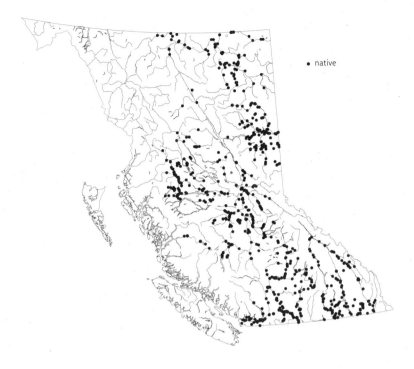

Map 23 The B.C. distribution of longnose dace, *Rhinichthys cataractae*

• native

BRITISH COLUMBIA DISTRIBUTION

Longnose dace are found in suitable habitats throughout most of the province (Map 23). They occur from the Columbia and Fraser river systems in the south to the lower Liard River in the north and are present in one of the coastal drainages that arise on the Interior Plateau and flow west through the Coast Mountains—the Klinaklini River. Longnose dace are absent from coastal islands, most of the short rivers along the B.C. coast, and most of the large north-coast rivers (e.g., the Iskut–Stikine, Taku, Alsek, and Yukon river systems); however, they are present in the Skeena and Nass river systems.

Life History Some life-history observations are available for fluvial populations in British Columbia; however, most of the following information is based on unpublished theses, field observations, and examination of museum specimens. There is no published information on lacustrine populations in our area, but data from Lake Winnipeg (Gee and Machniak 1972) and Lake Michigan (Brazo et al. 1978) suggest that the life history of lacustrine longnose dace is similar to that of fluvial populations.

REPRODUCTION

In British Columbia, longnose dace begin spawning in the spring (May or June) when water temperatures rise above 10 °C. Spawning occurs on riffles (surface velocities of about 0.40–1.0 m/s) over coarse gravel substrates. Males are territorial and defend a small area (about 10 cm in diameter) of cleaned gravel. Male territories are clustered and appear to be closer together than required by the available spawning area. Females of the eastern form are not aggressive and remain segregated from males until they are ready to spawn. In contrast, females of the northwestern form are aggressive and vigorously defend territories against intrusions by both females and males (Bartnik 1972). Also, females of the northwestern form often court males. Courtship is complex and involves a number of distinct behaviours (e.g., substrate probing, lateral and snout nudging, fin nibbling, quivering, and trembling). During courtship, males quiver (position themselves parallel to females and pass muscular contractions down the length of the body), but females only quiver during actual spawning. Only males tremble (high-frequency vibrations in which the whole body trembles). Actual spawning takes place in a depression cleaned by repeated substrate probing (pushing the snout into the substrate and quivering). Both fish then assume a side-by-side position in the depression. They quiver, and eggs and sperm are released. Females do not deposit all of their eggs in a single spawning and usually spawn with more than one male.

Fecundity varies with female size. The smallest mature females are about 55 mm FL and produce around 150 eggs, whereas the largest females (over 120 mm) produce in excess of 3,000 eggs. Like most cyprinids, the fertilized eggs develop a large space between the chorion and the yolk. After fertilization, the eggs average about 2.4 (range 2.3–2.6) mm in diameter, but the actual yolk is about 1.5 mm in diameter. The eggs are demersal and, initially, adhesive. They are deposited in clumps amongst the gravel within the male's territory.

Depending on temperature, the eggs start to hatch in about a week; however, hatching is protracted and takes at least 2 days at 18 °C. Newly hatched fry are about 6.0 mm TL and reach about 8 mm before the yolk is absorbed and exogenous feeding begins. At some sites in the Columbia system, spawning may occur in two pulses—there are two temporally separated peaks in the frequency of gravid females, and by early July, there is strong bimodality in the size distribution of the young-of-the-year. More than one spawning in a season also is known in North Carolina (Roberts and Grossman 2001).

AGE, GROWTH, AND MATURITY

At most sites in British Columbia, young-of-the-year longnose dace reach 20–35 mm FL by the end of their first growing season. Typically, longnose dace reach sexual maturity at the end of their second summer and spawn for the first time the next spring (2+). Most spawning aggregations consist of fish in their third (2+) or fourth (3+) summers. Within spawning aggregations, males are usually smaller and younger than females. This suggests that, on average, males mature a year earlier than females. Typically, most of the largest adults are females; some of these are in their fifth summer (4+), and a few are in their sixth (5+).

FOOD HABITS

Adult longnose dace forage primarily at night (Culp 1989). Experimental examination of the influence of light on their foraging activity (Beers and Culp 1990) indicates maximum efficiency under twilight conditions; however, under starlight, longnose dace change their searching behaviour and increase the frequency of benthic rooting. This rooting may be associated with locating prey by smell (Beers and Culp 1990). Certainly, of the species of dace in our area, the longnose dace has the longest maxillary barbels, and Schmidt (1983) suggests these barbels function in food location.

The diet of fluvial adults consists primarily of the larvae of aquatic insects, especially those associated with riffles. In the fall, however, some terrestrial insects appear in stomach contents. Adults occasionally prey on larval fish. For example, P. Dill (personal communication) found newly emerged kokanee in longnose dace from Mission Creek (Kelowna), and in the spring, adults in lower Fraser tributaries often are gorged with largescale sucker larvae. In lacustrine populations, the adult diet is similar to that found in flowing water (i.e., primarily the larvae of aquatic insects) but also includes snails, oligochaetes, and pea clams.

Unlike adults and juveniles, young-of-the-year longnose dace forage actively during the day. They also forage in midwater as well as on the substrate. Their diet consists mostly of chironomid larvae but also includes significant quantities of periphyton (algae and diatoms) and occasionally plankton.

Habitat Some observations on habitat use by fluvial longnose dace are available for British Columbia (e.g., Gee and Northcote 1963; Glozier et al. 1997; Porter and Rosenfeld 1999; R. L. & L. Environmental Services Ltd. 1995a). The following account uses these sources plus published information derived from eastern North American and Great Plains populations.

ADULTS

The body and fin shapes of longnose dace suggest a fish adapted to fast water, and in British Columbia, they have been taken in riffles with surface velocities of up to 1.8 m/s (Glozier et al. 1997). Typical adult habitats have substrates of loose, fist-sized (or larger) rocks. Although associated with fast water, the actual microhabitats used by adults may have much lower water velocities (R. L. & L. Environmental Services Ltd. 1995a). This may explain why this ostensibly torrent-adapted species is common in lakes with sufficient fetch to provide wave-swept cobble beaches (Brazo et al. 1978; Gee and Machniak 1972). Nonetheless, an experimental study of size-specific habitat segregation in longnose dace (Mullen and Burton 1998) found that adults preferred the fastest available velocity (in this case, 0.4–0.5 m/s) and largest substrate.

West of the Continental Divide, there is no evidence of habitat segregation between the sexes, but during the pre-spawning and spawning periods in Manitoba, there is a partial segregation by substrate between males and females (Gibbons and Gee 1972): females are associated with larger

rocks than males. In the B.C. interior, adult longnose dace shift from riffles to slower, deeper water in the winter. In the lower Fraser valley, however, adults often remain in riffles throughout the year. In the B.C. interior, there is some evidence (casual observations at smolt counting fences) of major seasonal movements (migrations?) by adults.
In the mainstem Columbia River between Keenleyside Dam and the U.S. border, longnose dace were most abundant in daytime catches in the winter, spring, and fall but in the summer they were most abundant at night (R. L. & L. Environmental Services Ltd. 1995a).

JUVENILES

As young longnose dace grow, swimbladder growth does not keep pace with the rest of the body. This results in a gradual increase in density. Consequently, during their first winter, the juveniles usually become bottom-dwellers and move into riffles, where they occupy habitats similar to, but with less overhead turbulence than, those used by adults. An experimental study of size-specific habitat segregation (Mullen and Burton 1998) indicates that juvenile longnose dace avoid low-velocity areas (0.0–0.1 m/s), and when adults are removed from a riffle, juveniles expand their velocity niche to include faster water. Field observations, however, suggest that, during freshets, yearlings move back into quiet water and seek shelter along river edges.

YOUNG-OF-THE-YEAR

In small streams, longnose dace fry are found in shallow pools, backwaters, and other low-velocity areas. In the Nazko River, juvenile longnose dace (this study's juvenile category includes both young-of-the-year and 1+ fish) showed a strong preference for average current velocities ranging from 0 to 0.1 m/s (Porter and Rosenfeld 1999). Such areas usually have silt or sand substrates, and the neutrally buoyant fry swim above the bottom. In large rivers, fry often aggregate above the bottom in areas with no measurable current (Porter and Rosenfeld 1999). In lakes, young-of-the-year longnose dace also are found in quiet water, usually close to shore and in areas where there is cover (Brazo et al. 1978; Gee and Machniak 1972).

Conservation Comments Over most of their B.C. distribution, longnose dace populations are healthy; however, in some lower Fraser valley rivers (e.g., the Coquitlam River), gravel washing has degraded riffle habitats and probably reduced population from their original levels. One of the three B.C. forms of longnose dace, the Nooksack dace, is listed as endangered by COSEWIC and as S1 (critically imperiled) by the BCCDC. In British Columbia, the Great Plains form of longnose dace is restricted to the northeastern portion of the province. Its provincial status needs to be evaluated.

Rhinichthys falcatus (EIGENMANN & EIGENMANN)
LEOPARD DACE

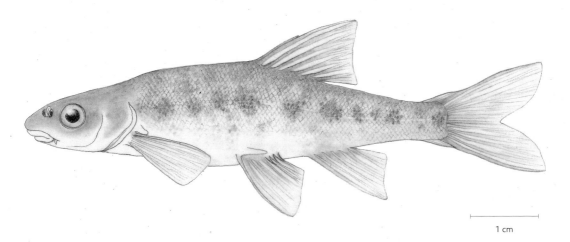

1 cm

Distinguishing Characters This slim-bodied dace has the snout free from the upper lip. The mouth is horizontal with well-developed barbels at the corners of the jaw and a narrow caudal peduncle (usually more than 3 times into head length). The dorsal and anal fins are falcate, and the upper lobe of the caudal fin is more pointed than in longnose dace (*Rhinichthys cataractae*). There usually are 9 or 10 dorsal rays, 51–63 scales along the lateral line, and dark, irregular spots covering the back and sides.

The snout is also free from the upper jaw in two other B.C. species (speckled dace, *Rhinichthys osculus*, and Umatilla dace, *Rhinichthys umatilla*). Where they coexist, these species can be confused with the leopard dace; however, leopard and speckled dace do not occur in the same drainage systems in British Columbia—the speckled dace is restricted to the Kettle River system. In contrast, leopard and Umatilla dace coexist in several areas: the Columbia River below Arrow Lakes, the Kootenay River below Bonnington Falls, and the lower Similkameen River. Morphologically, these species are similar, but the leopard dace has a noticeably narrower caudal peduncle (the caudal peduncle depth is less than the snout length in leopard dace, and longer than snout length in Umatilla dace). Also, the relatively long maxillary barbels are exposed in leopard dace and contained within the maxillary groove in Umatilla dace.

Taxonomy Compared with longnose and speckled dace, the leopard dace exhibits relatively little geographically patterned morphological variation. Thus, most of the taxonomic problems associated with this species are connected to its relationship with the Umatilla dace (see the Umatilla dace taxonomy section for a discussion of this problem) and its position within the genus *Rhinichthys*.

Some authors (Hubbs et al. 1974) recognize two subgenera in the genus *Rhinichthys*: *Apocope* containing speckled dace, leopard dace, and Umatilla dace and a subgenus containing the blacknose daces (*Rhinichthys obtusus* and *Rhinichthys atratulus*), longnose daces, and Umpqua dace (*Rhinichthys evermanni*). Other authors (Miller 1984) recognize three species groups within *Rhinichthys*: the *R. cataractae* group (containing longnose dace and *R. evermanni*), the *R. atratulus* group containing the two species of blacknose dace, and the speckled dace group (containing the many geographic forms of speckled dace; the Las Vegas dace, *Rhinichthys deaconi*; and leopard dace). More recently, Woodman (1992) argued that *Rhinichthys* should not be divided into subgenera or species groups. Aside from the placement of the longfin dace (*Agosia chrysogaster*) and the loach minnow (*Tiaroga cobitis*) in *Rhinichthys*, the major difference between Woodman's arrangement of relationships within the genus and that of earlier workers (e.g., Hubbs et al. 1974; Miller 1984) is the placement of the leopard dace.

The early workers viewed speckled dace and leopard dace as sister species, whereas Woodman (1992) places the leopard dace close to blacknose dace and states unequivocally that leopard dace and speckled dace are not sister species. Although the argument over relationships within *Rhinichthys* remains unresolved, unpublished molecular data support the early view that the leopard dace and speckled dace are close relatives.

Sexual Dimorphism As in most western minnows, mature females are on average larger than mature males. This size dimorphism probably results from a more rapid growth rate and a longer lifespan in females than in males. Both sexes develop breeding tubercles; however, the tubercles in females are fewer and smaller than those in males. Spawning males have a single white tubercle at the posterior margin of most of the body scales. These tubercles are most prominent on the dorsal surface anterior to the dorsal fin but also occur on the upper flanks. They also have conspicuous, tooth-like tubercles on the dorsal surface of the pectoral rays, and small tubercles on the membranes between the pelvic rays. Additionally, breeding males develop red lips and red pigment in the pectoral and pelvic fin axillae. Outside the breeding season, the sexes differ in the length of the pelvic fins: the pelvic fins extend back beyond the origin of the anal fin in males, whereas those of females are short and do not reach the origin of the anal fin.

Hybridization Although there is strong evidence that the Umatilla dace evolved through a relatively old hybridization event involving speckled and leopard dace (Haas 2001), there is no evidence of contemporary hybridization between the leopard dace and any other species. Given their morphological similarity, however, hybrids between Umatilla and leopard dace would be difficult to detect by morphology alone.

Distribution **NATIVE DISTRIBUTION**

The leopard dace is a Columbia endemic that also occurs in the adjacent Fraser River. In the Columbia drainage system, leopard dace range from the Cowlitz and Willamette rivers (lower Columbia tributaries) upstream to the Bruneau River. Like many Columbia endemics, they are absent above Shoshone Falls on the Snake River (Simpson and Wallace 1978).

BRITISH COLUMBIA DISTRIBUTION

The leopard dace is strongly associated with the mainstem Columbia and Fraser rivers and their larger tributaries (Map 24). In the Fraser system, the leopard dace is common in appropriate habitats between Chilliwack and Prince George but occurs only sporadically downstream of the confluence of the Sumas and Fraser rivers. In the Prince George area, it is common in the mainstem Fraser upstream to at least the confluence of the Fraser and Willow rivers. It is also locally abundant in major Fraser tributaries like the Nechako, Blackwater, Thompson (North and South), and Nicola rivers but does not occur in the headwaters of these rivers. In small tributaries, leopard dace are rarely found more than 1–2 km above a confluence with a larger river.

Compared with the Fraser system, the leopard dace is relatively uncommon in the B.C. portions of the Columbia drainage system. Here, it is most abundant in the Similkameen and Okanagan systems but occurs sporadically in the mainstem Columbia River from lower Arrow Lake to the U.S. border and in the Kootenay River system below Bonnington Falls. It is absent from the Kettle River above Cascade Falls.

Life History Life-history data on B.C. leopard dace are sparse; however, Gee and Northcote (1963) and Peden and Orchard (1993), present some B.C. data. Their accounts, supplemented with unpublished observations, form the basis for the following summary.

REPRODUCTION

Leopard dace spawning has not been observed in the wild; however, an early July spawning period is suspected in the lower Fraser system (Gee and Northcote 1963), and "running-ripe" adults were collected in the Nechako River in mid-July. Leopard dace have been spawned in the laboratory (Haas 2001), and the newly fertilized eggs are about 2.0–2.5 mm in diameter, but like many cyprinids, the actual yolk is only about 1.5 mm in diameter. The eggs are demersal and, initially, adhesive. Apparently, they are released above the substrate and drop into spaces between rocks or are swept under rocks by the current. At 18 °C, they hatch in 6 days, and the larvae are about 7 mm TL at hatch. The larvae probably remain in the gravel for about a week before emerging. Newly emerged fry are about 9.5 mm long and, in the wild, begin feeding by the first week in August. Fecundity in leopard dace is unknown but, presumably, is similar to other dace species.

Map 24 The B.C. distribution of leopard dace, *Rhinichthys falcatus*

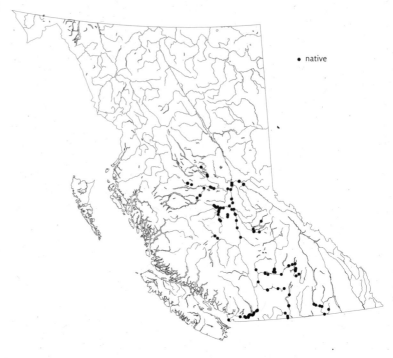

AGE, GROWTH AND MATURITY

In our area, leopard dace reach a size of about 20 mm by the end of their first growing season. Most males attain sexual maturity at the end of their second summer (1+) and spawn for the first time during their third summer. Typically, females mature a year later than males. The maximum size recorded in our area is 120 mm FL, and most individuals over 80 mm are females. So far, in British Columbia, the oldest individual aged by otoliths was a female in her fifth summer (4+).

FOOD HABITS

During the summer, adult leopard dace feed primarily on the larvae of aquatic insects but include some periphyton (algae and diatoms) and a few terrestrial insects in their diet. The dominant diet items in fish collected in the fall (September and October) are terrestrial insects. Adults collected during flood periods often contain earthworms. The stomach contents of longnose dace and leopard dace were examined from nine sites (eight in the Fraser system and one in the Similkameen River) where the two species co-occurred (Johannes 1958). There were no consistent differences in diet between the species; however, at one site, leopard dace contained plecopteran (stonefly) larvae, whereas longnose dace contained other insect larvae but no plecopterans.

The diets of young-of-the-year and juvenile leopard dace are similar to those of adults (larvae of aquatic insects). The main difference is the

Habitat The main sources of habitat data for the leopard dace are Gee and Northcote (1963), Peden (1991), Porter and Rosenfeld (1999), and Simpson and Wallace (1978). These references form the basis for the following account, supplemented with a few personal observations on habitat use.

ADULTS

Although leopard dace occur in both rivers and lakes (e.g., Okanagan and lower Arrow lakes), this species is primarily a river dweller. In rivers, adult leopard dace are associated with slower velocities (<40 cm/s) and bottom velocities (about 2 cm/s) than the longnose dace and, consequently, often occur over finer substrates than the latter species (Peden 1991). In the Nazko River, adult leopard dace also preferred low-velocity habitats with relatively fine substrates (Porter and Rosenfeld 1999). In the Fraser River, adults occurred at depths between 0.3 and 1.0 m (Gee and Northcote 1963; Haas 2001). Also, in the Fraser system, adult leopard dace are associated with gravel deposition areas and braided channels. They are common over fine gravel and silted, cobble substrates on midchannel bars. In turbid rivers, adults are more common in shallow water (<30 cm) at night and deeper water (>30 cm) during the day (Gee and Northcote 1963).

The leopard dace may have been always rare (relative to longnose and Umatilla dace) in the mainstem Columbia River and the Kootenay River below Bonnington Falls; however, it is possible that their present rarity is a consequence of habitat and flow alterations caused by dams (Peden 1991).

JUVENILES

In rivers, juveniles occupy similar habitats to adults but at sites with lower water velocities and shallower water than those occupied by adults (Porter and Rosenfeld 1999). In the late summer, juveniles often are more common on the slower side-channel sides of bars than are adults. At such sites, the substrate is usually coarse gravel or cobbles. During freshets, juveniles move into quiet water and often seek shelter in flooded vegetation along river edges.

YOUNG-OF-THE-YEAR

In rivers, young-of-the-year leopard dace occur in shallow water (often <10 cm deep) and are associated with low-velocity areas like shallow pools and backwaters. In large rivers, such areas usually have silt or sand substrates, but at low water, fry often are abundant in the lee of cobble bars at sites where the cobbles are covered in silt. In lakes, young-of-the-year leopard dace usually are found close to shore and in areas where wave action is dampened by reed beds or large rocks (Peden 1991).

Conservation Comments   The leopard dace is abundant in the gravel deposition reaches of the Fraser River and its major tributaries, but with some exceptions (e.g., the Similkameen River below Keremeos), leopard dace are not abundant in the B.C. portion of the Columbia system. The contrast in the abundance of leopard dace between the mainstem Fraser and the mainstem Columbia is puzzling. It is not clear if the scarcity of leopard dace in the mainstem Columbia is a recent (post-dam) phenomenon or if the species was never abundant in this area. In any event, leopard dace are sufficiently rare in the Canadian portion of the Columbia system that they should be monitored and regarded as a species of regional concern. Leopard dace were assessed by COSEWIC and listed as not at risk.

Rhinichthys osculus (GIRARD)
SPECKLED DACE

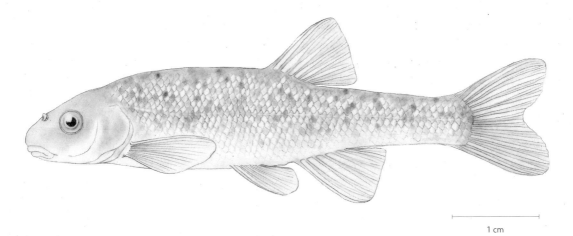

1 cm

Distinguishing Characters — This blunt-nosed dace has a snout that scarcely overhangs the upper lip. The mouth is oblique, and the upper lip is free from the snout. In the Kettle River above Cascade Falls, they lack barbels at the corners of the mouth. The caudal peduncle is moderately narrow (2.3–2.7 times into head length); the dorsal and anal fins and the lobes of the caudal fin are rounded. The fork of the caudal fin is shallow; its depth equals about 6% of total length. There are 8 or 9 dorsal fin rays and 59–69 scales along the lateral line. The back and sides of juveniles and females are without irregular, dark spots, but mature males develop markings that approach the lateral spots seen in the Umatilla dace (*Rhinichthys umatilla*).

Taxonomy — In the arid regions of western North America, this "species" exists as hundreds of isolated populations. Many of these populations are morphologically distinctive, and numerous subspecies have been named (e.g., Hubbs et al. 1974). As a result, the taxonomy of speckled dace is in disarray over much of its geographic range; however, a recent molecular analysis (Oakey et al. 2004) suggests that the species is divisible into eight major clades, and one of these major clades is centred in the Columbia River basin. Within the Columbia system, morphological variation in speckled dace falls into three broad categories (Gilbert and Evermann 1895). In the lower Columbia and its tributaries, adult speckled dace typically possess a distinctive dark lateral band (the "*nubalis*" form). In the middle and upper Columbia (including the Snake River below Shoshone Falls), adults are lightly speckled and usually lack the dark lateral band. In the Spokane region, some individuals lack the typical barbels. Above Shoshone Falls and in the Wood River system, adult dace usually have light-green sides with large black spots (the "*carringtoni*" form).

Both geography and morphology align B.C. speckled dace with Gilbert and Evermann's "Spokane" group of speckled dace: they occur only in the Kettle River (a middle Columbia tributary), and they lack barbels. Except for the short reach of the Kettle River between Cascade Falls and the U.S. border, most of B.C.'s speckled dace are isolated above Cascade Falls. Mitochondrial sequence data indicate that Kettle River speckled dace above Cascade Falls have diverged from lower Columbia and Snake River populations; however, it is not known if they differ from other mid-Columbia populations.

Sexual Dimorphism Like most western minnows, mature females are on average larger than mature males. This size dimorphism results from both a more rapid growth rate and a longer lifespan in females than in males. In addition, breeding males develop tooth-like tubercles on the upper surface of the pectoral fins that are absent in females. Outside the breeding season, mature males are separable from mature females by the length of the pectoral fins (they reach the origins of the pelvic fins in males and fall short by about an eye diameter in females). Additionally, breeding males develop red lips and red pigment in the pectoral and pelvic fin axillae.

Distribution **NATIVE DISTRIBUTION**
The speckled dace has the widest distribution of any western North American minnow. In one form or another, it ranges from the Columbia River system south to the upper Colorado River in Arizona, New Mexico, and Sonora, Mexico. Over this extensive geographic range, there are hundreds of divergent populations isolated in the arid regions of Oregon, California, Nevada, Utah, and Arizona as well as in coastal drainages between western Washington and southern California.

BRITISH COLUMBIA DISTRIBUTION
In Canada, the speckled dace is restricted to a single river system: the Kettle River. Here, it occurs from below Cascade Falls upstream in both the west and east forks of the Kettle River and in the Granby River (Map 25).

Life History The main sources of life-history information for the B.C. populations of speckled dace are Peden (1994) and Peden and Hughes (1981, 1984). These references form the basis of the following account, supplemented with a few personal observations.

REPRODUCTION
Apparently, a number of environmental cues can trigger spawning. Increasing photoperiod and increasing water temperature released spawning behaviour in the laboratory (Kaya 1991), and in Arizona (John 1963), speckled dace spawning peaks were associated with seasonal rains and flooding. In the Kettle River, spawning probably begins in mid-July (individuals collected in early July were almost in spawning condition). At this time, water levels are high but receding, and water temperature

Map 25 The B.C. distribution of speckled dace, *Rhinichthys osculus*

is rising. Apparently, speckled dace spawning can be protracted (Kaya 1991) and may include more than one spawning peak (John 1963; Kaya 1991). Interestingly, in mid-July, Kettle River females that were close to spawning contained relatively few (usually <100) large eggs. Given a protracted spawning period, this observation suggests speckled dace may be fractional spawners.

Although there is no published information on either their spawning sites or spawning behaviour in British Columbia, speckled dace spawning has been described in Arizona (John 1963) and New Mexico (Mueller 1984). Except for a few minor details, these descriptions are similar and probably reflect most basic aspects of speckled dace spawning sites and spawning behaviour. In both cases, spawning occurred over clean gravel (about 1.0–5.5 cm in diameter) in shallow (2.5–10 cm deep) water. The Arizona study (John 1963) suggests some site preparation by males (described as a nest) and, at the beginning of spawning activity, evidence of a dominance hierarchy among males. As activity increased, any defense of the nest broke down, and up to 60 males occupied a single nest site (about 30 cm in diameter). In contrast, in New Mexico, there was no evidence of site preparation (Mueller 1984), but this author describes several spawning clusters (each containing over 25 fish) in an area of about 1 m^2. Both authors indicate that females aggregate in deeper quiet water near the spawning sites and that, when a female enters the spawning site, she is "swarmed" by males. Apparently, once ovulated, females enter spawning sites repeatedly and deposit a few eggs on each visit.

Speckled dace have been spawned in the laboratory (Haas 2001; Kaya 1991). Newly fertilized eggs are about 1.8 mm in diameter, demersal, and adhesive. In aquaria, they are deposited in corners, on filters, and at the base of any available stones (Kaya 1991). In the laboratory and in nature, egg cannibalism is intense, and only those eggs that are swept under rocks or fall into crannies between rocks are likely to survive.

Like most fish, fecundity in the speckled dace is a function of female size. In the Kettle River, population fecundity varies from about 500 to over 2,000 eggs (Peden 1994); however, in the lower Columbia River and elsewhere in western Washington, female speckled dace often exceed 120 mm FL and have fecundities in excess of 4,000 eggs.

Development from fertilization to hatching is rapid: 4–5 days at 24 °C and 6–7 days at 18 °C. Newly hatched larvae average about 6 mm TL and, depending on temperature, become free swimming about 5–7 days after hatching. At about 8 mm, they emerge from the substrate and begin exogenous feeding.

AGE, GROWTH, AND MATURITY

In British Columbia, newly emerged fry (about 9 mm long) appear in the Kettle River in early August. By the end of the growing season (late October), the fry usually are 20–30 mm FL. Most males attain sexual maturity at the end of their second summer (1+) and spawn for the first time during their third summer. Typically, females mature a year later than males. No detailed data exists on the age structure of the B.C. population; however, individuals in their second or third summer (<60 mm FL) dominate the adult population. Occasional fish (all females) reach lengths of over 90 mm and are in their fourth summer (3+).

FOOD HABITS

Adults feed mainly on the larvae of aquatic insects but also include significant amounts of filamentous algae in their diets. A few specimens collected in the early fall (September) contained winged insects. The few juveniles and young-of-the-year examined for stomach contents suggest a diet similar to that of adults but with a higher proportion of periphyton (algae and diatoms) and chironomids in the diet.

Habitat

The main sources for habitat use by B.C. speckled dace are Peden (1994), Peden and Hughes (1981, 1984), and Haas (2001). These references form the basis of the following account, supplemented in places with some personal observations.

ADULTS

Because of the strong spring–summer (May–July) peak in discharge in the Kettle River (and its major tributary, the Granby River), adult speckled dace are difficult to collect during the early summer. In the early spring (March), they are found in deep (>1.0 m) runs often in the lee of large rocks, logs, and bridge abutments. The near-ripe fish collected in mid-July were taken in baited minnow traps placed behind a bridge abutment in water about 1 m deep. Later in the year (late July to October), adults are usually found in relatively shallow (0.1–0.65 m) water in areas with slow surface currents (<0.25 m/s) and bottom velocities of about 2 cm/s. In swimming-tube experiments, speckled dace had lower critical holding velocities than either leopard (*Rhinichthys falcatus*) or Umatilla dace (Haas 2001). Adults appear to shelter in pools and back-eddies during the day. In contrast, at night, minnow trap sets in areas with overhead cover (e.g., large woody debris or undercut banks) often make large catches. Males are rare in most collections, and there may be some microhabitat differences between males and females: males may occupy deeper or swifter water than females.

JUVENILES

Juvenile speckled dace are commonly caught in shallow water (<20 cm) in areas of slow (about 10 cm/s) current over substrates of coarse gravel or small stones. During freshet, they are often found sheltering in seasonally flooded vegetation.

YOUNG-OF-THE-YEAR

In early August, fry (about 10 mm in length) were dip-netted along the river margins in shallow (<2 cm), still water over silt or sand substrates.

Conservation Comments   In Canada, the distribution of the speckled dace is restricted to the Kettle River. Although the species is abundant in much of the Columbia river system, most of the Kettle River population is isolated from the rest of the Columbia drainage system by a velocity barrier at Cascade Falls. Although there are about 5 km of river between this barrier and the U.S. border, there is no possibility of gene flow into the Kettle River above the barrier from downstream speckled dace populations. Consequently, if a catastrophe occurred, most of the Canadian portion of the Kettle River could not be repopulated by natural immigration from downstream. Since a catastrophe that would eliminate speckled dace from the entire upper river (i.e., the west and east forks of the Kettle River and its major tributary, the Granby River) is extremely unlikely—it would have to simultaneously affect all three watersheds—the speckled dace is in no immediate danger of extirpation in Canada. Nonetheless, it is a species of concern, and some attempt should be made to systematically document its distribution, life history, and habitat use in the Canadian portion of the Kettle River. The speckled dace is listed as endangered by COSEWIC and as S_1S_2 (imperiled because of rarity) by the BCCDC.

Rhinichthys umatilla (GILBERT & EVERMANN)
UMATILLA DACE

1 cm

Distinguishing Characters   This small minnow (usually <100 mm FL) has an upper lip that is free from the snout and a slightly subterminal mouth. The mouth is horizontal, and in British Columbia, the maxillary barbels are small and often hidden in the maxillary groove. The caudal peduncle is moderately narrow (2.3–2.7 times into head length); the dorsal and anal fins are strongly falcate; and the lobes of the caudal fin are pointed. The depth of the caudal fork equals about 10% of total length. There are 9 or 10 dorsal rays and 56–72 scales along the lateral line. Adults usually are heavily marked with irregular dark blotches on the sides and back.

The Umatilla dace is easily confused with the leopard dace (*Rhinichthys falcatus*), and the two species coexist in the Columbia River below Arrow Lakes, in the Kootenay River below Bonnington Falls, and in the Similkameen River system. Superficially, the species are similar; however, the Umatilla dace has a noticeably deeper caudal peduncle than the leopard dace (the caudal peduncle depth is greater than the snout length in the Umatilla dace and less than the snout length in the leopard dace). Also, the maxillary barbels are exposed in the leopard dace and contained within the maxillary groove in the Umatilla dace.

In British Columbia, there is also limited sympatry between the speckled (*Rhinichthys osculus*) and Umatilla dace. They coexist in a short stretch of the Kettle River (about 5 km) between Cascade Falls and the U.S. border. Here, the two species are easily separated by mouth position and barbel development: the mouth in Umatilla dace is subterminal (it is almost terminal in speckled dace), and barbels are visible at the corners of the mouth, whereas B.C. speckled dace lack barbels.

Taxonomy   The Umatilla dace has a checkered taxonomic history. Apparently, even the original describers (Gilbert and Evermann 1895) had doubts about the specific validity of the species and alluded to its unusual morphology (intermediate between the speckled and leopard dace) and the possibility that it might be a hybrid. Since then, the Umatilla dace has been treated as a distinct species (Schultz 1936), a subspecies of speckled dace (Bond 1973), or simply ignored (Wydoski and Whitney 1979). The 1991 edition of the American Fisheries Society (AFS) list of common and scientific names (Robins et al. 1991) mentioned Umatilla dace in a footnote but did not list it as a species; however, the new AFS list of common and scientific names (Nelson et al. 2004) and the new edition of the *Inland Fishes of Washington* (Wydoski and Whitney 2003) both treat the Umatilla dace as a distinct species.

Here, the Umatilla dace also is treated as a distinct species for three reasons: (*i*) it is morphologically separable (Haas 2001; Peden and Hughes 1988) from other Columbia species of *Rhinichthys*; (*ii*) at the mitochondrial molecular level, it differs from speckled dace at 3.8%, and from leopard dace at 1.5%, of the 1140 cytochrome *b* sites examined; and (*iii*) although its morphology and ecology tend to be intermediate between leopard and speckled dace, at many localities it occurs with only one, or none, of the putative parents (Haas 2001). Thus, although there is compelling evidence that the Umatilla dace originated from a past hybridization event (Haas 2001), it is now capable of maintaining self-perpetuating populations. Similar situations occur in other western North American minnows (DeMarais et al. 1992; Dowling and DeMarais 1993; Dowling and Secor 1997), and speciation through hybridization is now recognized as a significant evolutionary mechanism in the lower vertebrates (Arnold 1997; Stone 2000). The relatively low level of molecular differentiation between leopard and Umatilla dace suggests a quite recent (mid-Pleistocene?) origin of the species.

The taxonomic problems associated with Umatilla dace are not confined to the question of whether or not it is a valid species. In British Columbia, there are morphological differences between Umatilla dace in the Similkameen River system and those in the mainstem Columbia River (Peden and Hughes 1988). If molecular data support this divergence, and the sketchy cytochrome *b* sequence data available for the Similkameen population appear to (Haas 2001), this raises the possibility of more than one form of the Umatilla dace—a biologically interesting but taxonomically difficult situation.

Sexual Dimorphism As in most species of *Rhinichthys*, females are, on average, larger than males. This size dimorphism is probably a result of both a faster growth rate and a longer lifespan in females. Most species of *Rhinichthys* develop spawning tubercles that differ between the sexes in degree of development and distribution. So far, however, spawning tubercles have not been observed on Umatilla dace. Nonetheless, mature males are separable from mature females by the length of the pelvic fins: the pelvic fins reach back to the origin of the anal fin in males, whereas the pelvic fins in females fall short of the anal origin.

Hybridization So far, there are no reports of hybrids involving Umatilla dace in British Columbia. Nonetheless, given the propensity of dace species to hybridize, and this species' close relationship to the leopard dace, occasional hybrids probably occur, especially in the Similkameen River where Umatilla and leopard dace are sympatric.

Distribution NATIVE DISTRIBUTION
The Umatilla dace is endemic to the middle and upper Columbia River and its large tributaries. They are absent from the Snake River (the major Columbia tributary) above Shoshone Falls. Unlike many other Columbia endemics, the Umatilla dace did not spread postglacially into the Fraser River and other adjacent drainage systems.

BRITISH COLUMBIA DISTRIBUTION
The Umatilla dace occurs in the lower Similkameen River and the Columbia River below Keenleyside Dam, the Kootenay River below Bonnington Falls, the Slocan River, the lower Pend d'Oreille River, and the Kettle River below Cascade Falls (Map 26).

Life History Little is known about the life history of Umatilla dace. The following account is derived from a report to B.C. Hydro (R. L. & L. Environmental Services Ltd. 1995a) and a Habitat Conservation Trust Fund report on the Similkameen River population (Peden and Orchard 1993) supplemented with personal observations made in the Slocan and Columbia rivers.

REPRODUCTION
Like the closely related leopard and speckled dace, Umatilla dace probably spawn in mid-summer (near-ripe individuals were collected in early July). Fecundity in five females (80–115 mm FL) ranged from 300 to 2,000 eggs. These eggs were not fully ripe and averaged 1.6 mm in diameter. Although there is no published information on their spawning sites or spawning behaviour, the Umatilla dace has been spawned in the laboratory (Haas 2001). Newly fertilized eggs are about 2.0 mm in diameter and adhesive. At 18 °C, they hatch in 6 days, and fry at hatch are about 7 mm TL. The fry probably spend about a week in the gravel before emerging. Dace fry about 10 mm long were collected along the margins of the Slocan River in early August. They were not longnose dace (*Rhinichthys cataractae*; the only other dace recorded from the area) and, thus, were probably Umatilla dace. These fry had begun exogenous feeding.

AGE, GROWTH, AND MATURITY
In nature, Umatilla dace rarely reach 30 mm by the end of their first growing season. Most males reach maturity by the end of their second summer (1+) but do not spawn until mid-summer of the next year. Most females mature a year later than the males. Adults rarely achieve a fork length of more than 120 mm, and most of the largest fish (>70 mm FL)

Map 26 The B.C. distribution of Umatilla dace, *Rhinichthys umatilla*

are females. In our area, the oldest individual aged by otoliths was a 118 mm female in her sixth summer (5+).

FOOD HABITS

Adults feed mainly on the larvae of aquatic insects (especially chironomids). They also consume, at least in winter, significant amounts of periphyton and detritus (R. L. & L. Environmental Services Ltd. 1995a). The few juveniles and young-of-the-year examined for stomach contents suggest a diet similar to the adults.

Habitat Only sparse information is available on the ecology of the Umatilla dace. The following account is derived from a report to B.C. Hydro (R. L. & L. Environmental Services Ltd. 1995a) and a Habitat Conservation Trust Fund report on the Similkameen River population (Peden and Orchard 1993) supplemented with personal observations made in the Slocan and Similkameen rivers.

ADULTS

In the mainstem Columbia River, adult Umatilla dace are most abundant in areas with water velocities >0.50 m/s, a bank slope of 6–15%, and a cobble to boulder substrate (R. L. & L. Environmental Services Ltd. 1995a). Within these glides, however, Umatilla dace shelter on the bottom between rocks where the water velocity usually is <20 cm/s. In both of the above reports, backpack electro-shockers were used to collect adults, and both studies comment on the occurrence of adults out to, and beyond, a safe electro-fishing depth (about 1 m). In addition, the size of fish appears to increase with depth (Peden and Orchard 1993). This suggests that adult Umatilla dace may be more abundant than museum collections indicate. In the Similkameen River, snorkel surveys observed what appear to be adult Umatilla dace in areas about 1 m deep with surface velocities of about 0.50 m/s. This is consistent with depth ranges (0.2–0.9 m) recorded in the field (Haas 2001). In the lower Similkameen River, large Umatilla dace were associated with large (>0.8 m) boulders (Peden and Orchard 1993). Also in the lower Similkameen River, the sex ratio was highly skewed at some sites (Peden and Orchard 1993). This suggests the possibility of habitat differences between adult males and females.

In the mainstem Columbia River between Keenleyside Dam and the U.S. border, there is some evidence for seasonal and diurnal habitat shifts in adult Umatilla dace (R. L. & L. Environmental Services Ltd. 1995a). For example, they were present at all sites in the middle of this section of the river in the winter but were absent from the same sites in the spring. Also, in the summer and fall, they were recorded in shallow, nearshore waters but were not recorded in nearshore environments in the winter and spring. Additionally, during the summer, there were no differences in catch rates between day and night samples; however, in the spring, fall, and winter catch rates were higher at night.

JUVENILES

In the mainstem Columbia River between Keenleyside Dam and the U.S. border, juvenile (1+) Umatilla dace were abundant in shallow nearshore environments during the summer, but they had shifted into the adult habitat by fall (i.e., deeper, faster water). During freshet in the much smaller Slocan River, juvenile Umatilla dace (up to 40 mm FL) sheltered in shallow, quiet areas and were often associated with flooded vegetation. Under less severe water conditions, juveniles shift into habitats similar to those used by adults but closer to shore and in shallower, slower water. Apparently, juveniles are not as bottom oriented as the adults; in the mainstem Columbia in the summer, they often aggregate in large groups in midwater behind structures that break the current (R. L. & L. Environmental Services Ltd. 1995a).

YOUNG-OF-THE-YEAR

In early August, Umatilla dace fry (about 10 mm TL) were dip-netted in quiet water along the edges of the Slocan River. At these sites, the water was shallow (<10 cm deep), the current was not measurable, and the substrate was sand or silt. During the day, the fry were foraging in midwater. Scattered debris and some vegetation provided cover. By late August, fry were still found in shallow water (<10 cm) but had shifted to sites dominated by cobbles and smaller rounded-rock substrates. Casual observation suggested that they were still foraging in midwater.

In the mainstem Columbia River, young-of-the-year were recorded in shallow nearshore areas throughout the year (R. L. & L. Environmental Services Ltd. 1995a).

Conservation Comments B.C. is the only province in Canada where this species occurs, and there are two forms of Umatilla dace here—a Similkameen River form, and a Columbia–Kootenay–Kettle river form. These forms differ morphologically (Peden and Hughes 1988; Peden and Orchard 1993), and there may be some differences in their mitochondrial genome (Haas 2001). The form in the Columbia, lower Kootenay, and Slocan rivers is locally abundant; however, the Similkameen form appears to be in trouble. Recent sampling in the Similkameen River has either failed to find this species or has found only a single specimen. Additionally, the Otter Creek population appears to have been extirpated. In Otter Creek (a Similkameen tributary by way of the Tulameen River), Umatilla dace went from modest abundance in the 1950s to extirpation by the mid-1990s. Although Otter Creek has changed over this time, the cause of this local extinction is not clear. However, regardless of the reasons for this extirpation, the Umatilla dace's spotty distribution, relatively low abundance, and range fragmentation (by power dams) make some populations vulnerable. The Umatilla dace is listed by COSEWIC as a species of special concern and by the BCCDC as S2 (a species imperiled because of rarity).

Richardsonius balteatus (RICHARDSON)
REDSIDE SHINER

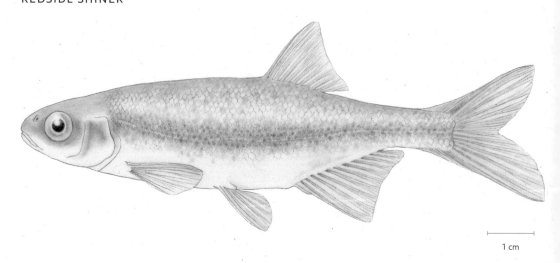

1 cm

Distinguishing Characters   This small (typically <200 mm FL) minnow has a deeply forked caudal fin, an unusually high number of anal rays (10–22), and a deep, laterally compressed body. In addition, the origin of the dorsal fin lies well behind the origin of the pelvic fins. In British Columbia, adult redside shiner are unlikely to be confused with other minnow; however, the fry often occur in mixed schools and sometimes it is difficult to separate them from peamouth (*Mylocheilus caurinus*) and northern pikeminnow (*Ptychocheilus oregonensis*) fry. The relative position of the dorsal and pelvic fins distinguishes peamouth fry from redside shiner fry: the dorsal fin in peamouth originates directly above, or in front of, the origin of the pelvic fins. Pikeminnow fry have a dark spot at the base of the caudal fin. This spot is absent in young shiners. There is a distinctive band of dark pigment along the base of the anal fin in redside shiner fry that is absent in our other minnow species.

Taxonomy   Although, morphologically, the redside shiner is one of the most variable species in northwestern North America, there are no taxonomic problems that affect the B.C. populations. However, there are recognized subspecies elsewhere. For example, redside shiner occurs above and below Shoshone Falls on the Snake River, and the form found above the falls is said to be a different subspecies, *R. b. hydrophlox*, than the Columbia River form. *Richardsonius* is also native to a cluster of rivers (the Siuslaw, Umpqua, and Coos rivers) on the central Oregon coast. This Oregon coastal form was originally named as a separate species, *R. siuslawi*, but has not been recognized as such for many years. Nonetheless, there are morphological and mitochondrial difference between the Oregon coastal form and Columbia redside shiners that suggest that the coastal form may a valid species.

Sexual Dimorphism   As in most minnows, mature females are usually larger than males. This size dimorphism results from a more rapid growth rate and a longer lifespan in females than in males. Spawning males often are brilliantly coloured with a gold patch under the eye, a bright yellow-gold stripe above the midlateral line, and crimson flanks. The colour in females is more subdued with the lower flanks usually a pale rose rather than crimson. Both sexes develop breeding tubercles; however, the distribution and development of the tubercles differs between the sexes, and the tubercles on males are noticeably larger than those on females. Also, males have tubercles on the lateral aspects of the dorsal and anal rays, and these tubercles are usually absent in females. Spawning individuals of both sexes have tubercles on the head and scales, but again, these are more developed and widespread on males than on females. A striking difference between the sexes is the distinctive patches of roughened comb-like scales on the male's breast immediately in front of each pectoral fin. These patches are absent in females.

Outside the breeding season, the length of the pectoral and pelvic fins distinguishes the sexes: in females, the paired fins are short, whereas in males the pectoral fins extend back to the origin of the pelvic fins, and the pelvic fins reach back to the origin of the anal fin.

Hybridization   In British Columbia, occasional hybrids occur between redside shiner and longnose dace (*Rhinichthys cataractae*), peamouth, northern pikeminnow, and perhaps, chiselmouth (*Acrocheilus alutaceus*) (Carl et al. 1959). Usually, but not always, these hybrids are morphologically intermediate between the putative parental species.

Distribution

NATIVE DISTRIBUTION

The redside shiner is another Columbia endemic. Originally, it ranged from the Nass River system to the Great Basin (Utah and Nevada); however, bait-bucket introductions have spread redside shiner into the upper Missouri and Colorado drainage systems.

BRITISH COLUMBIA DISTRIBUTION

Although mainly an interior species, redside shiner reaches the coast in large rivers like the Fraser and Skeena; however, they have a low tolerance for seawater and have not reached any of our coastal islands (Map 27). From the Fraser system, redside shiner has colonized the Skeena system (and, from the Skeena, the Nass River) and the upper Peace River, but they have never reached the Liard, Yukon, Taku, or Iskut–Stikine river systems. Among central coast rivers, redside shiners are present only in those that rise on the Interior Plateau adjacent to Fraser tributaries (e.g., the Homathko, Klinaklini, and Dean rivers).

Map 27 The B.C. distribution of redside shiner, *Richardsonius balteatus*

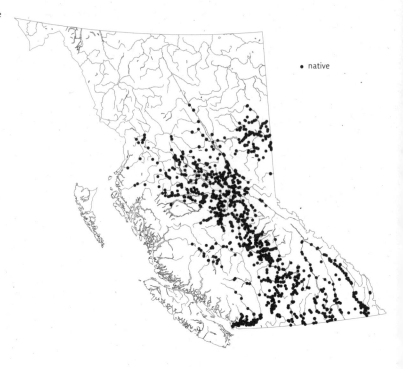

• native

There is one detailed life-history account of a B.C. population of redside shiner (Lindsey and Northcote 1963). It is the main source of the following life-history synopsis.

Life History REPRODUCTION

Redside shiner spawn in the spring. Apparently, increasing day length and rising water temperatures trigger spawning migrations. The threshold temperature for breeding is about 10 °C, and this temperature is reached at different times in different parts of the province. Thus, spawning may begin in April in the Lower Mainland and as late as early June in the interior of the province. Typically, spawning occurs in flowing water over clean gravel substrates, but some populations spawn in lakes. Most lacustrine populations, however, spawn in inlet or outlet streams associated with lakes. At these sites, redside shiners rarely ascend, or descend, the stream for more than a few hundred metres, and spawning usually occurs on the first or second riffle above or below the lake. Some populations spawn at night, others spawn during the day (Weisel and Newman 1951), and still others spawn in both daylight and darkness (Lindsey and Northcote 1963).

Typically, ripe fish slowly mill about in a pool or low-velocity area below a riffle until a female swims rapidly upstream into the riffle. An entourage of males (often four or five) immediately joins the female and crowd against her. A few eggs are released. The female may release several batches of eggs over a period of 1–2 minutes and then return to the holding area. Over the course of the spawning season, each female spawns many times with many different males. The eggs are demersal, adhesive, and when fertilized, about 2 mm in diameter. As in most fish, fecundity is a function of body size, and egg numbers range from about 700 to 4,000 eggs. At 18 °C, the eggs take about 5 days to hatch, and the newly hatched fry are about 6 mm long. After about 10 days in streams associated with lakes, the fry migrate to the lake at night.

AGE, GROWTH, AND MATURITY

Young-of-the-year begin feeding at about 8 mm TL and grow rapidly in their first summer. Depending on altitude and food availability in British Columbia, they are 25–40 mm long by the end of their first growing season. Males reach sexual maturity in their third summer, and most females mature a year later. The maximum lifespan recorded in the province is 7 years (6+).

FOOD HABITS

Young-of-the-year redside shiners consume a variety of prey including organisms taken from both the bottom and the water surface (diatoms, cladocerans, copepods, ostracods, and chironomid larvae and pupae). As they grow, shiners take larger prey. Adults forage in the littoral zone, where they take prey from the bottom, from midwater, and from the surface. Nymphs and pupae of aquatic insects, and adult terrestrial insects, dominate their diet, but they also take cladocerans, copepods, molluscs, and in season, the eggs and fry of fish (including their own species).

Habitat There is a detailed description of habitat use in a lacustrine population of redside shiner (Crossman 1959) and other sources (Beauchamp et al. 1995; Narver 1967; Scarsbrook and McDonald 1973, 1975) give some information on habitat use in lakes. Also, habitat use has been described (Porter and Rosenfeld 1999) in a B.C. fluvial population. These articles, plus unpublished observations, form the basis for the following synopsis.

ADULTS

In British Columbia, redside shiners are ubiquitous in streams, rivers, ponds, lakes, and reservoirs. In lakes, in the summer, adults cruise the littoral zone during the day, usually in water less than 4 m deep (Crossman 1959). They forage in small, alert groups, constantly moving, aggressively pushing into weed beds, checking out items on the bottom, and darting to the surface. At night, they move offshore, and scattered individuals often are caught in the limnetic zone of large oligotrophic lakes (Scarsbrook and McDonald 1973; 1975). In rivers, adult redside shiners occur in relatively deep (1–2 m), slow (<0.20 m/s) water over fine substrates (Porter and Rosenfeld 1999).

JUVENILES

In lakes during the summer, juvenile redside shiners occur in loose schools around lake margins. They usually are associated with the outer margins of weed beds and stay closer to shore than adults.

YOUNG-OF-THE-YEAR

In the summer, young-of-the-year redside shiners inhabit shallow water (usually less than 1 m) along lake and stream margins. They often occur in aquatic vegetation and typically associate with mixed schools of underyearling peamouth and northern pikeminnow (Miura 1962). In streams and rivers, young redside shiners prefer quiet water (<0.1 m/s), fine-grained substrates, and water less than 0.5 m deep (Porter and Rosenfeld 1999). They are abundant in shallow backwaters and weedy bays and, again, usually are associated with mixed schools of other minnows.

Conservation Comments The redside shiner is probably the most abundant minnow in the B.C. interior and is not a conservation concern in those regions. In the lower Fraser valley, however, it has disappeared from many of the sloughs and shallow lakes where it was once abundant. The causes of this decline are not clear but may be related to the spread of largemouth bass (*Micropterus salmoides*) in these habitats. Neither COSEWIC nor the BCCDC list this species.

Tinca tinca (LINNAEUS)
TENCH

1 cm

Distinguishing Characters	The tench is unlikely to be confused with any other B.C. minnow. It has a deep body, exceedingly small, embedded scales, and is remarkably slimy. The mouth is small, oblique, and with a single barbel at the corner of the jaw; the fins are dark; the tail has no fork; and the iris of the eye is red.
Sexual Dimorphism	In adult males, the first ray in the pelvic fins is thickened. In females and juveniles of both sexes, this ray is normal.
Distribution	**NATIVE DISTRIBUTION**

The tench is indigenous to Europe (south of Scandinavia) and western Siberia, where it occurs as far east as the Ob and Yenisei rivers (Berg 1948). Tench were imported into North America in the late 19th century. Apparently, tench were introduced into lakes in the Columbia system (eastern Washington) in 1895–1896 (Wydoski and Whitney 2003). Presumably, our tench are derived from this source.

BRITISH COLUMBIA DISTRIBUTION
Tench occur in reservoirs in the Pend d'Orielle River system and in Osoyoos, Vaseaux, Skaha, and Okanagan lakes (Map 28).

Life History — The biology of tench is unstudied in British Columbia. Consequently, the following account is derived mainly from sources outside North America (Maxwell 1904; Weatherley 1959; Wheeler 1969).

Map 28 The B.C. distribution of tench, *Tinca tinca*

■ introduction

REPRODUCTION

Tench spawn in the summer when the water temperature reaches 18 °C. At this time, they aggregate in areas of dense vegetation. In early June, an overnight fyke net set in a warm slough off Lake Washington caught about 100 tench (Wydoski and Whitney 2003). This may have been a spawning aggregation. Males (often more than one) are so active and persistent in their pursuit of gravid females that, at times, they drive the females ashore (Maxwell 1904). Tench are exceedingly fecund. Females produce about 500,000 eggs per kilogram of body weight (Varely 1967).

The fertilized eggs are greenish, adhesive, and about 1.0 mm in diameter. They stick to aquatic vegetation and take about a week to hatch. The newly hatched larvae are about 3 mm in length.

AGE, GROWTH, AND MATURITY

Although growth is slow, tench reach a large size for a minnow: about 45 cm in Washington (Wydoski and Whitney 2003) and, in northern England, tench are said to reach weights of almost 5 kg (Maxwell 1904). Males mature in their third or fourth year, and females usually mature a year later. Apparently, tench live 20–30 years (Scott and Crossman 1973). In British Columbia, the largest recorded tench (about 40 cm TL) was caught in the mainstem Columbia River below Keeenleyside Dam.

FOOD HABITS

Tench are omnivorous. Adults feed primarily on aquatic insects (especially chironomids) but include amphipods, molluscs, and some plant material in their diet. They also prey on plankton. In an experimental study using a size range of zooplankton (0.2–4.0 mm), tench preferentially foraged on the largest prey (Ranta and Nuutinen 1984). Juveniles typically forage on cladocerans and copepods associated with aquatic plants. Newly hatched fry take smaller prey (rotifers, diatoms, and copepodites).

Habitat Except in the most general sense, habitat use by tench in British Columbia is unstudied. Thus, the following account is based on the same sources as the life-history account.

ADULTS

Except when breeding, adult tench are solitary and creatures of quiet water. They favour heavily vegetated ponds and sluggish streams with mud bottoms and do best in warm water (over 15 °C), but they can tolerate cool water and ice cover. Given their preference for heavily vegetated warm water, they probably are also resistant to low oxygen conditions. Although the reservoirs and lakes tench occupy in British Columbia are probably not ideal habitat, they contain sufficient warm, shallow areas to sustain tench. Apparently, tench become quiescent in winter and "buries itself in the mud, and lies low till the return of warmth in the spring" (Maxwell 1904).

JUVENILES

Small schools of juveniles are strongly associated with cover (usually dense vegetation but, in reservoirs, often with root wads and woody debris).

YOUNG-OF-THE-YEAR

Tench fry form loose schools and, in the summer, occupy weedy littoral areas.

Conservation Comments Although an alien species, the tench is relatively benign and has no known negative impacts on native species. Consequently, it is not a conservation concern in British Columbia.

FAMILY CATOSTOMIDAE — SUCKERS

The suckers are a family of medium-sized fishes related to the minnows (Cyprinidae). Like minnows, their mouths lack teeth and their anterior vertebrae and associated ossicles and ligaments are modified to provide a link between the swimbladder and the inner ear (the Weberian apparatus). There are about 60 living species in North America, and two species (the Chinese sucker, *Myxocyprinus asiaticus*, and longnose sucker, *Catostomus catostomus*) in Asia (Smith 1992). The longnose sucker also occurs in North America. Its restricted distribution in eastern Siberia and wide distribution in northern North America suggest the longnose sucker is a recent (Pleistocene) emigrant from North America (Walters 1955). Paradoxically, however, longnose suckers are absent from rivers on the Siberian side of the Bering Landbridge (the Anadyr and Amguema rivers) but do occur farther west in the Kolyma River system. A possible explanation for this curious Siberian distribution is that longnose suckers dispersed into Siberia sometime before the Illinoian glaciation and, subsequently, were extirpated by a later glaciation from areas east of the Kolyma River (Lindsey and McPhail 1986). The other Asian species, the Chinese sucker, is restricted to eastern China. It appears to be a morphologically primitive sucker that once was placed in the subfamily Cycleptinae (Smith 1992; Suzuki 1992) but is now considered sufficiently different to warrant in its own subfamily, the Myxocyprininae (Harris and Mayden 2001). The oldest known fossil suckers (*Amyzon*) are from Eocene deposits in British Columbia (Horsefly River). They appear to be related to another primitive subfamily, the buffalo suckers (Ictiobinae).

All five living species of suckers in British Columbia are members of the genus *Catostomus*. Three species (longnose sucker, *C. catostomus*; white sucker, *C. commersonii*; and largescale sucker, *C. macrocheilus*) are ecologically generalized suckers and, as adults, inhabit a wide range of freshwater environments. Their diet consists mainly of benthic insect larvae and pupae, supplemented with detritus and periphyton. In contrast, the mountain sucker (*C. platyrhynchus*) is an ecologically specialized periphyton scraper and, in British Columbia, is associated with relatively large, swift rivers (the Fraser, Similkameen, and North Thompson rivers). Ecologically, the fifth species—the bridgelip sucker (*C. columbianus*)—is somewhat intermediate between the other suckers. It forages primarily on periphyton and detritus but also takes some benthic insects. Although most bridgelip sucker populations are associated with flowing water, there are lacustrine populations. In addition to these five species, a distinctive form of longnose sucker, the Salish sucker, occurs in the lower Fraser valley, in Puget Sound drainages, and on the Olympic Peninsula (McPhail and Taylor 1999; Pearson and Healey 2003).

Figure 18 Diagram illustrating the differences in the depth of the lip clefts and shape of the lower lips in mountain (A) and longnose (B) suckers

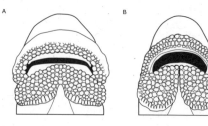

Figure 19 Diagrams illustrating differences in the depth of the notches at the corners of the mouths in mountain (A) and bridgelip (B) suckers

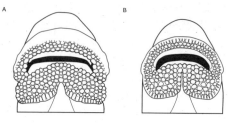

Figure 20 Ventral views of the heads of longnose (A) and white (B) suckers illustrating the difference in snout length

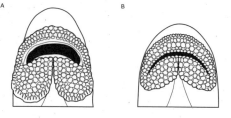

Figure 21 Differences between largescale (A) and white (B) suckers in the relationship between dorsal fin base and caudal peduncle depth

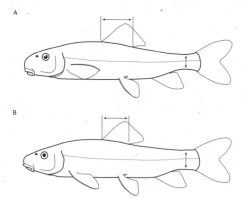

FAMILY CATOSTOMIDAE — SUCKERS 161

Figure 22 Pelvic fin of a mountain sucker showing membranous stay and the pelvic axillary process

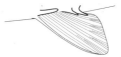

Figure 23 Ventral view illustrating the difference in the shape of the lip clefts in white (A) and largescale (B) suckers

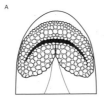 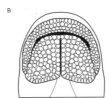

KEY TO ADULT SUCKERS

1 (4) Cleft between lower lips shallow; lower jaw with a slightly curved cartilaginous sheath that is clearly visible in the opened mouth (Fig. 18A) *(see 2)*

    2 (3) Posterior margins of lower lips almost straight; conspicuous notches at corners of mouth (Fig. 19A); outer face of upper lip usually without papillae; pelvic axillary process usually well developed (Fig. 22) — MOUNTAIN SUCKER, *Catostomus platyrhynchus*

    3 (2) Posterior margins of lower lips rounded; notches at corners of mouth inconspicuous or absent (Fig. 19B); outer face of upper lip usually with papillae; pelvic axillary process usually absent — BRIDGELIP SUCKER, *Catostomus columbianus*

4 (1) Cleft between lower lips deep; lower jaw with a strongly curved cartilaginous sheath that is scarcely visible in opened mouth (Fig. 18B) *(see 5)*

    5 (6) Scales small, 90 or more along lateral line; snout clearly overhangs mouth (Fig. 20A) — LONGNOSE SUCKER, *Catostomus catostomus*

    6 (5) Scales large, 75 or fewer along lateral line; snout scarcely overhangs mouth (Fig. 20B) *(see 7)*

        7 (8) Caudal peduncle depth less than half the length of dorsal fin base (Fig. 21A); 13 or more dorsal rays; cleft between the lower lips forms a broad inverted "V" (Fig. 23A) — LARGESCALE SUCKER, *Catostomus macrocheilus*

        8 (7) Caudal peduncle depth more than half the length of dorsal fin base (Fig. 21B); 12 or fewer dorsal rays; cleft between the lower lips forms a narrow inverted "V" (Fig. 23B) — WHITE SUCKER, *Catostomus commersonii*

Catostomus catostomus (FORSTER)
LONGNOSE SUCKER

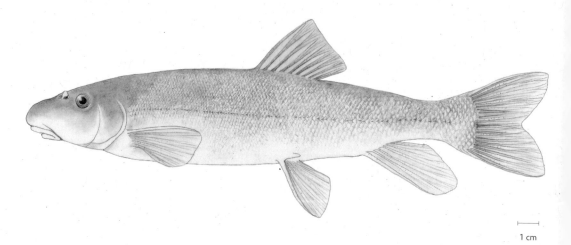

1 cm

Distinguishing Characters This fine-scaled sucker (there are more than 90 scales along the lateral line) has 9–11 dorsal rays and a deep cleft between the lower lips. In profile, the body tapers gradually back from the dorsal fin to a relatively deep caudal peduncle (the peduncle depth is well over half the width of the dorsal fin base). The snout is long and, when viewed from below, projects beyond the upper lip. The lower lips are large, and in profile, their hind margin extends back to about the middle of the nostril.

Two other species of fine-scaled suckers occur in British Columbia—the bridgelip sucker (*Catostomus columbianus*) and the mountain sucker (*Catostomus platyrhynchus*). Both species have a shallow cleft between the lower lips (deep in longnose sucker), and both possess membranous stays that connect the pelvic fins to the body (absent in longnose sucker).

Taxonomy In British Columbia, there are three kinds of longnose sucker: a widely distributed "normal" form that breeds at a relatively large size (>300 mm FL), a set of geographically scattered "dwarf" populations that breed at a relatively small size (150–200 mm FL), and the genetically distinctive Salish sucker (McPhail and Taylor 1999). Whether "dwarf" longnose sucker populations represent a genetically programmed life-history type or are simply stunted populations is unknown. However, in Edwards Lake in the Kootenays, "dwarf" and "normal" longnose suckers once coexisted (Geen 1958). This suggests that, in some cases, there may be a genetic difference between "dwarf" and "normal" longnose suckers. Unfortunately, Edwards Lake was poisoned before these sympatric forms could be studied.

There are mitochondrial DNA and morphological differences between Salish suckers and other northwestern longnose suckers; however, this divergence, although consistent, is not great. Consequently, although clearly a unique evolutionary lineage, the Salish sucker is not sufficiently divergent to warrant a separate taxonomic name. Nevertheless, it is protected in British Columbia because of its precarious conservation status. The Salish sucker's distribution and biology are summarized in Pearson and Healey (2003).

Sexual Dimorphism On average, mature females are larger than males. Breeding male longnose sucker have a distinct red or pink lateral band that is sometimes sandwiched between an upper yellow-gold band and a lower black band. Breeding colours are more subdued in females. The sexes also differ in the extent and distribution of nuptial tubercles: breeding males have large tubercles on their anal fin and on the lower lobe of the caudal fin, whereas females usually lack tubercles on these fins or have only small tubercles. Males also possess weak tubercles on the dorsal and ventral surfaces of the pectoral fins, fine tubercles on the dorsal and ventral surfaces of the head, fine tubercles on scales along the back and upper flanks, and tubercles on most of the scales around the caudal peduncle. Again, these tubercles are absent or weakly developed in females. The only patch of tubercles that is more strongly developed in females than in males is on the dorsal surface of the caudal peduncle immediately in front the caudal fin. Adults differ in the size and shape of some fins: the anal fin and the lower lobe of the caudal fin are noticeably longer in adult males than in females. Also, the posterior margins of the pelvic fins are squared off in males (all rays are about equal in length) and bluntly pointed in females. Apparently, this difference in pelvic fin shape is a reliable indicator of sex in both immature individuals and adults outside the breeding season (Stanley 1988).

Hybridization Although hybridization is relatively common among sucker species, hybrids involving the longnose sucker are rare in British Columbia. In upper Blueberry Creek (a Columbia tributary near Castlegar), there is a hybrid swarm involving longnose and bridgelip suckers. The suckers in this stream are morphologically variable, but most individuals are intermediate between the two supposed parents. In addition, occasional individuals intermediate between longnose and white suckers (*Catostomus commersonii*) are encountered in the Liard River system near Fort Nelson. They may be hybrids.

Distribution NATIVE DISTRIBUTION
The longnose sucker occurs in both North America and Asia. In North America, it is distributed in cool waters from Labrador to the Pacific Coast, and on the Great Plains from the Arctic Coast to Colorado. In the east, it occurs as far south as Maryland. In contrast, its Asian distribution is restricted to a few Arctic coastal drainages in eastern Siberia.

BRITISH COLUMBIA DISTRIBUTION

Longnose suckers are found in cool waters throughout the province (Map 29)—from the Columbia River system in the south to the Liard and Yukon rivers in the north. Although primarily an interior species, longnose sucker approaches the coast in the lower Fraser River system (downstream to about Hope). It also occurs in the upper parts of some of the west-flowing rivers that rise to the east of the Coast Mountains (Dean, Klinaklini, and Homathko rivers). In British Columbia, the Salish sucker is restricted to streams in the lower Fraser valley: streams near Chilliwack and Agassiz, Nooksack River tributaries, Salmon River, and Little Campbell River (the latter population is gone).

Life History Although there is only one local study of longnose sucker life history (Geen et al. 1966), there is a rich literature on this species' biology elsewhere in North America. Thus, the following account is based primarily on the local study (Geen et al. 1966) but supplemented, where appropriate, with information from central Canada as well as incidental observations from British Columbia.

REPRODUCTION

Typically, longnose sucker spawning migrations precede those of other suckers (Geen 1958). The primary migration trigger appears to be temperature, and spawning runs often begin shortly after ice-out or when water temperatures reach about 5 °C (Geen et al. 1966). There is also evidence that increased spring flows may initiate spawning migrations (Barton 1980). Curiously, although most longnose sucker populations spawn in the early spring, some populations in the Peace and Cariboo regions delay spawning until mid-June and breed at water temperatures of about 15–16 °C. The spawning period in longnose sucker is usually short (about 2 weeks), but in the Salish sucker, it is protracted (6–8 weeks) and lasts from March to early July (Pearson and Healey 2003).

Longnose suckers usually spawn in streams over gravel substrates (0.5–10.0 cm in diameter) in moderate currents (0.30–0.45 m/s). Nevertheless, some populations spawn in shallow (often <20 cm) water along lakeshores. Spawning usually occurs during the day, and except during the actual spawning act, the sexes are partially segregated. In flowing water, males normally hold position just above the substrate over spawning sites, while females aggregate along the banks or in quiet water areas downstream of spawning sites. Multiple males (a minimum of two and sometimes three or four) simultaneously spawn with a single female. There is no site preparation, but spawning activity often cleans the substrate. The eggs are demersal and adhesive. Some eggs stick to the gravel surface and others fall into crevices among the rocks. In minutes, most exposed eggs are eaten either by other suckers or other fishes.

Egg number varies with female size. In small-sized populations, females of about 150 mm FL contain approximately 3,000 eggs. In normal populations, large females (some over 400 mm) produce over 44,000 eggs. The

Map 29 The B.C. distribution of longnose sucker, *Catostomus catostomus*

unfertilized eggs are yellow and about 2.5 mm in diameter. On fertilization, the eggs swell to about 3 mm. Development rate is temperature dependent, and hatching occurs in about 11 days at 10 °C and at about a week at 16 °C. The newly hatched larvae are unpigmented and about 10 mm TL. Their yolk sac is tubular and about 7 mm long. The fry remain in the gravel until they are about 12 mm long (1–2 weeks after hatching). On emergence, the larvae are still dependent on their yolk reserves but exogenous feeding begins at about 14 mm. The mouth becomes subterminal at approximately 16–20 mm.

AGE, GROWTH, AND MATURITY

Initial growth is slow and varies with density, food availability, and latitude. In southern British Columbia, fry reach lengths of approximately 50–60 mm FL by the end of their first summer. Farther north, young-of-the-year are approximately 30–40 mm long by the end of their first growing season, and at the extreme northern border of the province (the lower Liard system), 15 mm fry are still present in early August (McLeod et al. 1979). Growth accelerates in the second and third years and then slows as the animals approach sexual maturity.

Most males reach maturity about a year before females, but the age at first maturity varies among populations. In central British Columbia, males mature at 5 or 6 years, and females, at 6 or 7 years (Geen et al. 1966), whereas in the extreme north of the province, males and females mature at about 7 and 9 years, respectively (McLeod et al. 1979). After maturity, growth slows in both sexes, but the decrease in growth rate is more marked in males than in females. Consequently, the largest fish in most populations are females. In British Columbia, relatively few longnose sucker exceed 500 mm FL, and the oldest recorded age (from otoliths) is 19 years; however, suckers are notoriously difficult to age, and individuals of up to 28 years have been reported in the Slave River, NWT (Tallman et al. 1996b).

The only information on the age, growth, and maturity in B.C. populations of "dwarf" longnose suckers comes from a study of the Salish sucker (Pearson and Healey 2003). In the lower Fraser valley, male Salish suckers mature at about 2 years, and females mature at least a year later. So far, the smallest recorded mature Salish sucker male was 96 mm FL, and the largest was 206 mm (Pearson and Healey 2003). The largest mature female was 287 mm, but only 10% of the females in the study population exceeded 200 mm.

FOOD HABITS

As adults, longnose sucker are benthivores (Welker and Scarnecchia 2003), but like many fish, they begin life foraging in shallow water on a variety of prey. Early in life, the mouth is terminal but slowly shifts to a ventral position. In some populations, the mouth is clearly subterminal by about 18–20 mm FL, but in other populations, the mouth is still terminal in fish 25–30 mm long. In lacustrine populations, plankton (*Daphnia*, *Cyclops*, and *Bosmina*) are common in the diet of fry before the mouth becomes subterminal. In contrast, chironomid larvae are the most common prey in fluvial populations. As the mouth changes position, fry become more substrate oriented, and their diet shifts towards benthic prey, especially chironomid larvae and ostracods. Juvenile and adult longnose suckers are predominately insectivores and consume large numbers of chironomid larvae and pupae, and trichopteran and plecopteran larvae. Periphyton appears to be relatively unimportant in the diet of B.C. longnose suckers, but elsewhere, they consume significant amounts of periphyton (Sayigh and Morin 1986). In situations where zooplankton is abundant, adult suckers will forage on zooplankton (Barton 1980).

Habitat Although the longnose sucker is usually viewed as a coldwater species, they occur in a variety of habitats ranging from small headwaters streams to large rivers, large and small lakes, oligotrophic and eutrophic lakes, beaver ponds, lowland sloughs, and swamps in British Columbia. There has been no attempt to gather quantitative data on their habitat use in the province. Thus, the following account is based primarily on the local field observations supplemented, where appropriate, with information from elsewhere in Canada.

ADULTS

In British Columbia, the longnose sucker appears to be a habitat generalist, and they are found throughout the length and breadth of the province. In the southern interior with the exception of spawning aggregations, longnose sucker are more common in lakes than in streams; however, "dwarf" populations often occur in headwater streams in the east Kootenay region, and there are stream-resident populations in coolwater streams as far south as the U.S. border. In the lower Fraser valley, the Salish sucker is primarily a species of small, often sluggish streams. There are, however, lacustrine populations of Salish suckers in adjacent Washington state. In the central and northern interior, the longnose sucker is common in both flowing and standing water, and in large northern rivers (e.g., the lower Liard River), they are often the most abundant fish species (McLeod et al. 1979). In summer, lake-dwelling adults forage inshore during the night but usually remain below the thermocline during the day. Little is known about the habitat use of adult longnose sucker in rivers. In northern British Columbia, there is some evidence for complex migrations between spring spawning sites, summer foraging sites, and overwintering areas (McLeod et al. 1978). In the mainstem Columbia River below the Arrow Lakes, tagged adults were recaptured after 3 years close to where they were originally caught (R. L. & L. Environmental Services Ltd. 1993). They probably moved in the intervening years (perhaps to spawn), but their recapture near the original tagging site suggests some fidelity to specific areas. In a swamp in the lower Fraser valley, the home range of the Salish sucker averaged 170 linear metres (Pearson and Healey 2003). In the same area, in the summer, the species is most active in the early morning and late evening.

JUVENILES

In lakes and rivers, juvenile longnose sucker use habitats similar to those used by adults; however, juveniles usually occur in shallower, and in streams, in quieter water than adults. In streams, they often are abundant in beaver ponds and other quiet water areas, and in large rivers, they occur in shallow side-channels and embayments. In lakes in the early summer, juveniles stay closer to shore than adults but move into deeper water if the lake stratifies.

YOUNG-OF-THE-YEAR

In rivers, newly emerged fry aggregate in quiet, shallow water (often less than 10 cm deep). They are associated with soft substrates and often found in seasonally flooded vegetation. As they grow, the fry move into deeper water but still aggregate in quiet areas (side-channels and backwaters behind deposition bars). In lakes, newly emerged fry stay close to the shore and remain near cover (typically vegetation or woody debris). As they grow, the fry move through shallow littoral areas in loose schools. Usually, these schools are encountered in shallow water over sand or mud substrates. As summer proceeds, the young increase in size, and by fall, the largest fry become solitary and move to deeper water.

Conservation Comments So far, the Salish sucker is the only evolutionarily significant unit of the longnose sucker that is specifically protected in Canada. It is listed as endangered by COSEWIC and as S1 (critically imperiled) by the BCCDC. A recovery strategy is being crafted and habitat improvement work is in progress. Some of the other "dwarf" longnose populations found in British Columbia may merit designatible unit status.

Catostomus columbianus (EIGENMANN & EIGENMANN)
BRIDGELIP SUCKER

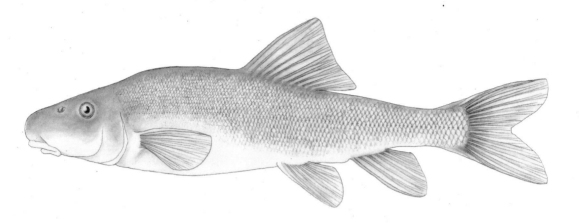

1 cm

Distinguishing Characters   This fine-scaled (usually more than 90 lateral line scales) sucker has 11–14 dorsal rays, a shallow cleft between the lower lips, and membranous stays that connect the pelvic fins to the body. Two other species of fine-scaled suckers occur in British Columbia—the longnose sucker (*Catostomus catostomus*) and the mountain sucker (*Catostomus platyrhynchus*). Dorsal fin ray numbers and body shape separate adult and juvenile bridgelip suckers from longnose suckers. Usually, bridgelip suckers have 11 or more dorsal rays, whereas longnose suckers have 10 dorsal fin rays. The body shape difference is clear in profile: in bridgelip sucker, the upper body tapers sharply downwards from the origin of the dorsal fin to the caudal peduncle, whereas the taper is much more gradual in longnose sucker.

Lip shape separates adult bridgelip and mountain suckers: in the bridgelip sucker the lower lips are about as deep as they are wide and only slightly, if at all, notched at the corners of the mouth, whereas the lower lips in the mountain sucker are wider than they are deep and are conspicuously notched at the corners of the mouth. Also, adult mountain sucker usually has a well-developed pelvic axillary process, which the bridgelip sucker usually lacks. On dead specimens, peeling back the skin on the head and checking the top of the skull for a fontanelle (hole) provides a positive identification—the fontanelle is well developed in the bridgelip sucker and absent or reduced to a narrow slit in the mountain suckers.

In bridgelip sucker fry (<20 mm TL), the conspicuous black peritoneum clearly shows through the body wall and distinguishes them from other sympatric sucker fry.

Taxonomy The taxonomy of bridgelip sucker is not an issue in British Columbia; however, there are geographically restricted morphological forms (subspecies?) in the Columbia Basin. The bridgelip suckers in the Wood River system, Idaho, are considered a distinct subspecies (*C. c. hubbsi*), and the isolated populations above Palouse Falls, Washington, were described as *C. c. palouseanus* (Schultz and Thompson 1936). Additionally, the bridgelip suckers in the upper Crooked River, Oregon, may warrant subspecific status (Smith 1966).

Like other suckers, some B.C. populations of bridgelip suckers mature at unusually small body sizes (<135 mm FL), but it is not known if this growth pattern has a genetic basis or is simply a reflection of limited food availability.

Many of the morphological characteristics of the bridgelip sucker are intermediate between the same traits in largescale (*Catostomus macrocheilus*) and mountain suckers. Hubbs et al. (1943) suggested that the bridgelip sucker might have originated through a hybridization event involving these two species. It is now clear that hybridization has played a significant role in the evolution of other western North American fishes (Dowling and Secor 1997; Haas 2001; Hubbs et al. 1943), and the notion that the bridgelip sucker had a hybrid origin deserves a closer look.

Sexual Dimorphism On average, female bridgelip sucker are larger than adult males, and breeding males develop a reddish-orange lateral stripe that is absent, or faint, in females. Additionally, the sexes differ in the extent and distribution of nuptial tubercles. Breeding males develop large, sharp tubercles on their anal and caudal fins (those on the lower caudal lobe are noticeably larger than those on the upper lobe), whereas females usually lack tubercles on these fins or have less well developed tubercles. The one patch of tubercles that is more strongly developed in females than in males is on the dorsal surface of the caudal peduncle immediately anterior to the caudal fin. Like other suckers, adults differ in the size and shape of some fins. The anal fin and the lower lobe of the caudal fin are noticeably longer in mature males than in females. Additionally, the posterior margins of the pelvic fins are squared off in males (all rays about equal in length) and bluntly pointed in females. This difference in pelvic fin shape appears to be a reliable indicator of gender in immature individuals and non-breeding adults.

Hybridization In British Columbia, there is a hybrid swarm involving longnose and bridgelip suckers in upper Blueberry Creek near Castlegar. In the mainstem Columbia River from the Arrow Lakes to the U.S. border, hybrids between largescale and bridgelip suckers are relatively common. Information on the identification of this hybrid combination is provided in Dauble and Buschbom (1981).

Distribution **NATIVE DISTRIBUTION**

The bridgelip sucker is a Columbia endemic that occurs only in the Columbia (below Shoshone Falls) and Fraser drainages systems. It is an interior species that ranges from Columbia tributaries in Nevada and Idaho to the middle Fraser River. It also occurs in the Harney Basin, a disjunct former tributary of the Snake and Columbia rivers in Oregon (Bisson and Bond 1971).

BRITISH COLUMBIA DISTRIBUTION

The bridgelip sucker is common in the lower Columbia, Similkameen, lower Kootenay, and Pend d'Oreille rivers (Map 30). It is absent from the upper Kootenay River above Bonnington Falls and from the Kettle River above Cascade Falls. It does not appear to ascend the mainstem Columbia River above the Arrow lakes. In the Fraser system, it is abundant in the Prince George region, especially in the Nechako system but appears to reach its northern distribution limit about 50 km upstream of Prince George. The bridgelip sucker also occurs in the Fraser River below the Fraser Canyon, where it is common in the gravel deposition region between Hope and the mouth of the Sumas River. Occasional individuals occur as far downstream in the mainstem Fraser River as the Fraser Delta.

Life History Although the bridgelip sucker is abundant in the central and southern interior of British Columbia, their life history has never been studied in the province. The only detailed life-history investigations of this species are from central Washington (Dauble 1980; Wydoski and Whitney 2003). The following account is based on these sources supplemented, where appropriate, with incidental observations from British Columbia.

REPRODUCTION

The bridgelip sucker spawns in the spring as water levels start to rise and water temperature reaches about 6 °C. In the middle Fraser, spawning starts in mid-April, peaks in May, and is over by mid-June. In a tributary to the Yakima River, Washington, bridgelip sucker were observed spawning on a riffle about 15 cm deep with a water velocity of 0.40 m/s and a substrate of pebbles, gravel, and cobbles (Wydoski and Whitney 2003). Ripe fish were collected at a physically similar site in early June in the Salmon River near Prince George. Here, they were aggregated on a swift, shallow riffle over a coarse gravel and cobble substrate. The water temperature was 16 °C, and there was a pool a few metres below the riffle. Unfortunately, the river was in freshet and turbid, so no actual spawning was observed. Observations in the Yakima River suggest that there is some site preparation by the female (if true, this is unusual for *Catostomus*) and that males array themselves in a wedge-shaped formation downstream of the female (Wydoski and Whitney 2003). Although the details of spawning behaviour in this species are undescribed, they probably are similar to those described for other *Catostomus* species.

Map 30 The B.C. distribution of bridgelip sucker, *Catostomus columbianus*

• native

Bridgelip sucker eggs are adhesive and demersal. Once fertilized, some eggs stick to the gravel surface, and others fall into crevices among the rocks. Fecundity ranges from about 10,000 to 20,000 eggs (Dauble 1980). Mature unfertilized eggs are about 2 mm in diameter, and newly fertilized eggs are 10–15% larger (2.5–3.0 mm). Incubation time is uncertain, although in Washington fry emerge approximately 25 days after spawning (Wydoski and Whitney 2003). The newly emerged fry are 12–13 mm TL. In both the Nechako River and the Columbia River below Arrow Lakes, yolk-sac fry are common in early July. They have a conspicuous black peritoneum that shows through the body wall and range in total length from 13 to 15 mm. Exogenous feeding starts at about 15 mm, and by about 18–20 mm, the gut is highly coiled, and the mouth is subterminal.

AGE, GROWTH, AND MATURITY

Initial growth is slow, and in the middle Fraser system, fry range from 20 to 45 mm FL by late September. In the Columbia system (especially in the Okanagan and Similkameen rivers), fry grow more rapidly and reach 60–80 mm FL by the end of their first growing season. Growth is more rapid in the second, third, and fourth years but slows as the animals approach sexual maturity: 5 or 6 years for males and 6 or 7 years for females (Dauble 1980). Maximum adult body size varies widely among populations: in the middle Fraser, the maximum body size in some populations is about 350 mm FL; however, in other populations, it is around 400 mm. Unusually large individuals (>500 mm) are reported in the mainstem Columbia River near Castlegar (R. L. & L. Environmental Services Ltd. 1993). These individuals might be hybrids between large-scale and bridgelip suckers. Adult suckers are notoriously difficult to age, but one 409 mm female from the Columbia River near Castlegar gave an otolith age of 11 years (R. L. & L. Environmental Services Ltd. 1993). Again, this fish might have been a hybrid. The oldest bridgelip sucker recorded from British Columbia was a 272 mm (FL) female from Birkenhead Lake. It was 31+ years.

FOOD HABITS

The diet of the bridgelip sucker varies with size, place, and season. In Washington, small individuals (46–149 mm FL) ate mostly periphyton but included chironomid larvae and pupae as well as zooplankton and miscellaneous aquatic invertebrates in their diet (Dauble 1980). The proportion of periphyton in the diet increases with size until adults are almost exclusively periphyton grazers. In the winter, small amounts of filamentous algae are present in some stomachs. Cursory examination of stomachs in the B.C. portion of the Columbia suggests a similar diet; however, in the turbid waters of the middle Fraser system, bridgelip suckers appear to be highly dependent on detritus. The guts of all size classes were packed with silt, and only the smallest fish examined (17–20 mm TL) contained traces of chironomid larvae. Presumably, the energy source in this silt is organic detritus. No comparable data are available for bridgelip suckers, but the proportion of detritus in the diet of white suckers (*Catostomus commersonii*) increases when invertebrate prey are scarce (Ahlgren 1990).

Habitat In British Columbia, there is little published on habitat use by the bridgelip sucker. Consequently, the following account is based on a Washington study (Dauble 1980) supplemented, where appropriate, with local field observations.

ADULTS

In British Columbia, bridgelip sucker usually are found in flowing water, but they also occur in lakes. They are common in the mainstem Columbia and middle Fraser rivers and abundant in low- to moderate-gradient portions of tributary streams and rivers. During the day in the mainstem Columbia River, adult bridgelip sucker are most abundant in deep water with strong currents but move inshore at night (Dauble 1980). A similar diel habitat shift occurs in middle Fraser tributaries: adults occur in deeper water (0.5–1.0 m) with faster surface velocities (0.4–0.9 m/s) during the day than at night. In the Similkameen River during the day, adults were observed in deep runs (up to 2 m) and pools (>2 m deep) with cobble–boulder substrates and relatively strong surface currents. In the turbid Fraser River between Hope and Chilliwack, adults are moderately abundant in the shallow water (<1.0 m) along the edges of cobble bars.

Although primarily a riverine species in British Columbia, bridgelip sucker also occur in lakes. Often the lakes are small and shallow. For example, bridgelip sucker are abundant in the small lakes upstream of Otter Lake near Tulameen. They also occur in low-altitude lakes in the Okanagan system. They are present, but apparently not abundant, in the large lakes (Okanagan, Skaha, and Osoyoos lakes) in this system. Their habitat use in lakes is unstudied.

JUVENILES

During the day in middle Fraser River tributaries near Prince George, juvenile bridgelip sucker occurred in backwaters up to 0.5 m deep with relatively slow (0.1–0.2 m/s) surface velocities and fine substrates.

YOUNG-OF-THE-YEAR

In mid-June, newly emerged fry are found in shallow (about 10 cm) water along the current edge and within 1 m of the shore. In the Columbia River below Trail, they are especially common in seasonally flooded vegetation (typically near the mouths of tributary streams). By July or August, fry shift into shallow quiet water (<0.2 m/s) in back channels and embayments with mud or sand substrates. In the turbid rivers of the middle Fraser system, newly emerged fry concentrate in shallow quiet water areas and are often associated with the fry of minnows and other suckers.

Conservation Comments The bridgelip sucker is common in the Columbia and Fraser systems and, at present, is not at risk. Neither COSEWIC nor BCCDC list this species; however, systematic studies of its habitat requirements and reproductive behaviour are needed.

Catostomus commersonii (LACEPÈDE)
WHITE SUCKER

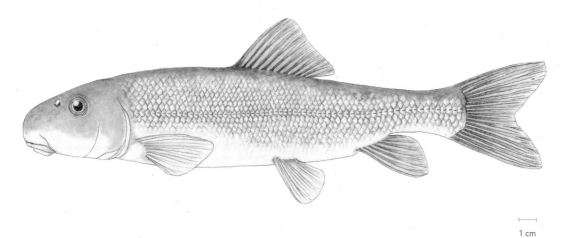

1 cm

Distinguishing Characters This large-scaled sucker (usually fewer than 70 scales along the lateral line) has 10–13 (usually 11 or 12) dorsal rays. In profile, the body gradually tapers back from the dorsal fin to the relatively deep caudal peduncle, and the caudal peduncle depth is usually more than half the width of the dorsal fin base. The snout is short, blunt, and when viewed from below, scarcely projects beyond the upper lip. In profile, the hind margin of the lower lips barely reaches the front half of the nostril.

In British Columbia, the only other sucker with big scales is the largescale sucker (*Catostomus macrocheilus*). There is a superficial resemblance between the two species, but they usually can be separated by dorsal fin ray counts (typically 11 or 12 in white suckers and 13–16 in largescale suckers) and the shape of the inverted "V" where the lobes of the lower lips come together—narrower in white suckers and wider in largescale suckers. In British Columbia, spawning white sucker lack a conspicuous lateral band, whereas breeding largescale sucker have a black lateral band (more conspicuous in males than in females).

Taxonomy Aside from problems identifying interspecific hybrids, there are no taxonomic issues involving white suckers in British Columbia. However, their presence in Barney Lake in the upper Liard system (see Map 31) is curious. The white sucker, lake chub (*Couesius plumbeus*), and longnose dace (*Rhinichthys cataractae*) appear to be the only Great Plains species to have reached the upper Liard system.

Sexual Dimorphism Mature females are, on average, larger than mature males. The sexes also differ in the extent and distribution of nuptial tubercles. Breeding males develop large, sharp tubercles on their anal and caudal fins (those on the lower caudal lobe are noticeably larger than those on the upper lobe), whereas females usually lack tubercles on these fins or have only small tubercles. Males also possess weak tubercles on the dorsal and ventral surfaces of the pectoral fins, fine tubercles on the dorsal and ventral surfaces of the head, fine tubercles on scales along the back and upper flanks, and tubercles on most of the scales around the caudal peduncle. Again, these tubercles are absent or weakly developed in females. The one patch of tubercles that is more strongly developed in females than in males is on the dorsal surface of the caudal peduncle immediately anterior to the caudal fin.

Like other suckers, adults differ in the size and shape of some fins. The anal fin and the lower lobe of the caudal fin are noticeably longer in mature males than in females. Additionally, the posterior margins of the pelvic fins are squared off in males (all rays about equal in length) and bluntly pointed in females. This difference in pelvic fin shape is a reliable indicator of gender in immature individuals and non-breeding adults (Stanley 1988).

Hybridization Where the geographic ranges of white and largescale suckers overlap in the upper Peace and middle Fraser rivers, hybrids are relatively common (about 6% in some populations), and Nelson (1968, 1974) provides useful information on the identification of hybrids in these drainage systems. Additionally, occasional hybrids between white and longnose suckers (*Catostomus catostomus*) are countered in the lower Liard system near Fort Nelson.

Distribution **NATIVE DISTRIBUTION**
The white sucker is widespread in eastern North American and appears to have crossed the Continental Divide into central British Columbia sometime during the deglaciation of the upper Peace drainage system. On the Great Plains, white suckers range from the Ramparts River, lower Mackenzie system, to New Mexico and Oklahoma and, in eastern North America, from Labrador and Nova Scotia to Georgia.

BRITISH COLUMBIA DISTRIBUTION
White suckers are widespread and common in the northeastern portion of the province (e.g., upper and lower Peace system and lower Liard system). White suckers also occur in the upper Liard (above the Grand Canyon) but are rare except in Barney Lake near the British Columbia – Yukon border. In this lake, they are moderately abundant (Craig and Bruce 1982). White suckers also are abundant in the Fraser River system in the Cariboo region but are not found in the Fraser system upstream of the Bowron River or downstream of 70 Mile House (Map 31). White suckers also dispersed from the Fraser system into the upper Skeena River system.

Catostomus commersonii WHITE SUCKER 177

Map 31 The B.C. distribution of white sucker, *Catostomus commersonii*

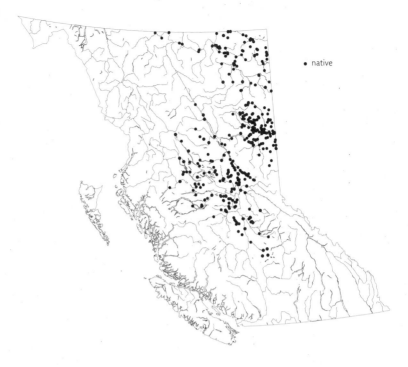

• native

Life History   Two life-history studies are available on B.C. white suckers (Geen et al. 1966; Nelson 1968). On the northern Great Plains and elsewhere in central Canada, their life history is well known. Thus, the following account is based primarily on these references, supplemented with information from Alberta and central Canada.

REPRODUCTION

In British Columbia, the white sucker spawns in the spring, usually after water temperatures reach about 10–12 °C. Although temperature is the primary migration cue, increasing spring flows may also initiate spawning migrations (Barton 1980). Apparently, the white sucker returns repeatedly to specific spawning streams (Olsen and Scidmore 1963). This suggests homing behaviour, and white sucker probably use odour to locate spawning streams (Werner 1979). In British Columbia, spawning begins about mid-May and continues to at least mid-June. Most lacustrine populations spawn in streams (usually inlet streams), but lake spawning is also common (Geen et al. 1966).

In streams and rivers, spawning usually occurs on shallow riffles (<1 m) adjacent to deeper areas. Apparently, substrate size is important and typically consists of coarse gravel rather than sand (Corbett and Powles 1983; Geen et al. 1966; Nelson 1968). In lakes, spawning also occurs over gravel on submerged stream deltas (Stewart 1926) and shallow gravel shoals (Nelson 1968). White sucker spawn mainly at night, but some activity usually is visible during the day. Breeding fish of both sexes aggregate in the vicinity of spawning sites, but except during the actually spawning act, the sexes are partially segregated—in rivers and streams, males aggregate in deeper and faster water than females, whereas in lakes, males aggregate close to the spawning site and in shallower water than females.

Spawning occurs when a female leaves the female aggregation and moves onto the spawning site. She is immediately joined by a minimum of two males (sometimes three to five). The males press against the female on both sides and vibrate rapidly. Eggs and sperm are released just above the bottom and, although there is no site preparation, the substrate is often cleaned by the thrashing of spawning fish. Egg number is a function of female size. In British Columbia, white sucker fecundity varies roughly between 10,000 and 50,000 eggs; however, a 565 mm female from northeastern Alberta contained over 85,000 eggs (Bond 1980). Unfertilized eggs are pale yellow and about 2 mm in diameter. On fertilization, the eggs swell and a wide space forms between the yolk and the outer membrane. The diameter of fertilized eggs is roughly 3.0 mm. The eggs are demersal and adhesive. Some stick to the gravel surface, but others are carried by currents into crevices and crannies among and beneath the rocks. Development rate is temperature dependent and hatching occurs in about a week at 16 °C and at about 21 days at 10 °C (Stewart 1926). The newly hatched larvae are approximately 9 mm TL and remain in the gravel for 11–13 days (Corbett and Powles 1983). On emergence, the larvae are around 11 mm TL and still dependent on their yolk reserves. Exogenous feeding starts at 12–13 mm, and the mouth becomes ventral at roughly 20 mm.

AGE, GROWTH, AND MATURITY

Growth rates in the white sucker are variable and probably constrained by both population density and food supply (Chen and Harvey 1994a). In British Columbia, the largest young-of-the-year reach a length of slightly over 60 mm FL by the end of their first summer. Growth slows as the animals approach sexual maturity. Apparently, age at maturity is dependent on early growth rate, and faster growing young produce early maturing adults (Chen and Harvey 1994b). In some northern populations, there is a delay of 1 or 2 years between sexual maturity and the first spawning (Barton 1980). Minimum ages at sexual maturity in British Columbia are 3+ for males and 4+ for females.

Once mature, white sucker do not necessarily spawn every year (Stewart 1926). After maturity, growth slows in both sexes, but the decreased growth rate is more marked in males than in females. Consequently, the largest fish are mainly females. The use of otoliths for aging white suckers has been validated (Thompson and Beckman 1995) and gives a maximum age (in Missouri) of 18 years.

FOOD HABITS

As adults, white suckers are benthivores (Welker and Scarnecchia 2003), but they begin life as planktivores. Early in life the mouth is terminal and slowly shifts to a ventral position. It is clearly subterminal by about 20 mm TL. There often is a growth surge associated with this change in mouth position that may result from an increase in the diversity of prey available to benthic feeders (Corbett and Powles 1983). When the mouth is terminal, the larvae feed primarily on water column prey (especially cladocerans and rotifers); however, as the mouth shifts to a ventral position, their diet diversifies with the addition of algae, ostracods, and the larvae and pupae of aquatic insects, especially chironomids (Corbett and Powles 1983).

Adults feed on a wide variety of aquatic insects, but chironomids usually remain their primary food item (Tallman et al. 1996b). Although primarily benthivores, adult white sucker will forage heavily on zooplankton when zooplankters are abundant (Barton 1980). Interestingly, at some times and in some populations, the diet of juvenile white sucker consists mainly of detritus. Apparently, the protein content of detritus is too low to sustain growth, but white sucker intentionally ingest detritus when invertebrates are scarce (Ahlgren 1990).

Habitat In British Columbia, white sucker occur in both rivers and lakes. Nonetheless, local information on their habitat use is scarce. West of the Rocky Mountains, they are relatively uncommon in large, swift rivers like the Fraser and Skeena and occur mainly in lakes and streams associated with lakes. In the upper Peace system, they also occur in lakes but are common in rivers. In the northeastern portion of the province, they also occur in large rivers (e.g., Muskwa, Fort Nelson, and Liard rivers) but are still primarily associated with lakes.

ADULTS

Information on habitat use by adult white sucker in lakes and rivers is scarce in British Columbia. In thermally stratified lakes in Ontario, adult white sucker were abundant in different temperature strata within the littoral zone (Logan et al. 1991), but no white sucker were taken below a depth of 20 m. In B.C. rivers during the summer, adults are found in low-gradient sections of mainstem rivers. They are usually encountered at depths of 1–2 m over soft substrates and are often found in slow side-channels. In the Parsnip River (upper Peace system) in late September or early October, adult white sucker migrate downstream into the lower reaches of the river near Williston Reservoir (McLeod et al. 1978).

JUVENILES

During summer in thermally stratified lakes, juvenile white sucker (<200 mm) are most abundant in shallow water in the warmest temperature zone (Logan et al. 1991). In upper Peace lakes (Crooked River system), juvenile white sucker concentrate during summer within a few metres of the shore and appear to favour shallow water with weedy areas.

In rivers during the summer, they are found in quiet water (e.g., side-channels and embayments) over silt–sand substrates. They are often associated with juveniles of other sucker species and with juvenile minnows.

YOUNG-OF-THE-YEAR

In the Prince George region during the summer, white sucker fry (35–45 mm TL) are abundant in the littoral zones of many of the warmer lakes. Here, they are associated with shallow weedy regions and soft substrates. Relative to other sucker species, white sucker fry are not common in rivers like the Fraser and Nechako, but where they occur, they are also associated with shallow weedy areas and soft substrates. In the Parsnip River system, white sucker fry concentrate in the lower reaches of tributary (spawning?) streams and in marshy areas associated with small lakes (McLeod et al. 1978). By late September or early October, the fry move out of these rearing areas and into larger rivers.

Conservation Comments Although the white sucker is widely distributed east of the Continental Divide, the only native populations west of the Rocky Mountains are in the upper Peace, middle Fraser, and upper Skeena river systems. At present, these populations are healthy and of no conservation concern. The white suckers in the upper Liard River system are a different matter. Their presence above the Grand Canyon of the Liard River is a bit of an anomaly. This canyon prevented most Great Plains species from reaching the upper Liard drainage system, and it is not clear how the white sucker reached the upper river or if they differ genetically different from the Fraser and Skeena populations.

Catostomus macrocheilus GIRARD
LARGESCALE SUCKER

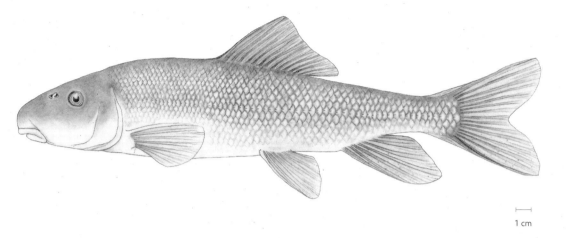

1 cm

Distinguishing Characters This large-scaled sucker (it usually has fewer than 75 scales along the lateral line) has 12–17 (usually 13–16) dorsal rays. In profile, the body tapers steeply back from the dorsal fin to the relatively narrow caudal peduncle (its depth usually less than half the width of the dorsal fin base). The snout is short and, when viewed from below, scarcely projects beyond the upper lip. In profile, the hind margin of the lower lips barely reaches the front half of the nostril.

The only other large-scaled sucker in British Columbia is the white sucker (*Catostomus commersonii*). Superficially, the two species resemble one another; however, they usually can be distinguished by dorsal fin ray counts (typically 13–16 in largescale suckers and 11 or 12 in white suckers) and the shape of the inverted "V" where the lobes of the lower lips come together—wide in largescale suckers and narrow in white suckers. Spawning largescale sucker (both sexes) display a black lateral band (more conspicuous in males than in females) set below a pale greenish band. This band is absent in spawning white suckers.

Taxonomy There are no taxonomic issues with this species in British Columbia; however, the relationship between Columbia–Fraser largescale suckers and those in rivers along the central Oregon coast is an unresolved problem.

Sexual Dimorphism Mature largescale sucker females are, on average, larger than mature males. The sexes also differ in the extent and distribution of nuptial tubercles. Breeding males develop large tubercles on their anal and caudal fins (those on the lower caudal lobe are noticeably larger than those on the upper lobe), whereas females usually either lack tubercles on these fins or have only small tubercles. Males also possess weak tubercles on the dorsal and ventral surfaces of the pectoral fins and small tubercles on some scales (especially on the posterior half of the body). Again, these tubercles are either absent or weakly developed in females. The one patch of tubercles that is more strongly developed in females than in males is on the dorsal surface of the caudal peduncle immediately anterior to the caudal fin.

Like other suckers, adults differ in the size and shape of some fins. The anal fin and the lower lobe of the caudal fin are noticeably longer in mature males than in females. Additionally, the posterior margins of the pelvic fins are squared off in males (all rays about equal in length) and bluntly pointed in females. Apparently, this difference in pelvic fin shape is a reliable indicator of sex in immature individuals and non-breeding adults (Stanley 1988).

Hybridization   Where the geographic ranges of largescale and white suckers overlap in the upper Peace and middle Fraser rivers, hybrids are relatively common (about 6% in some populations). Nelson (1968, 1974) provides useful information on the identification of hybrids in these drainage systems. In addition, hybrids between largescale and bridgelip suckers (*Catostomus columbianus*) are encountered in the mainstem Columbia River from the Arrow Lakes to the U.S. border and in the middle Fraser system. Dauble and Buschbom (1981) provide information on the identification of this hybrid combination.

Distribution   NATIVE DISTRIBUTION

The largescale sucker is endemic to western North America. Here, it ranges from the Nass and Peace rivers in the north to the Columbia River in the south. There is a cluster of rivers (Siuslaw, Umpqua, Coos, Coquille, Flora, and Sixes rivers) on the central Oregon coast that also contain largescale suckers; however, molecular evidence suggests that these Oregon coastal suckers are not largescale sucker. Like most Columbia endemics, the largescale sucker is absent above Shoshone Falls on the Snake River.

BRITISH COLUMBIA DISTRIBUTION

The largescale sucker is abundant throughout the Columbia, Fraser, Skeena, Nass, and upper Peace drainages (Map 32). From the Fraser, they have colonized the upper portions of at least three central coast drainages—the Homathko, Klinaklini, and Dean rivers. Apparently, they never reached the northwestern and northeastern portions of the province and are absent from the Liard, Yukon, Iskut–Stikine, and Taku rivers. Although the largescale sucker is an interior species, they reach tidewater in large rivers like the Fraser and Skeena, and in these rivers, they are regular inhabitants of tidal sloughs. On most high tides, these sloughs are brackish. This suggests that the largescale sucker has some salinity tolerance but apparently not enough to colonize coastal islands or coastal rivers adjacent to a source like the Fraser River.

Map 32 The B.C. distribution of largescale sucker, *Catostomus macrocheilus*

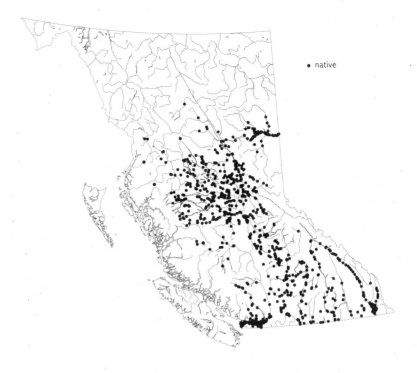

Life History   Curiously, no detailed study of the life history of the largescale sucker is available for British Columbia; however, McCart and Aspinwall (1970) and McLeod et al. (1978) describe spawning behaviour and spawning sites in British Columbia, and Miura (1962) describes the early life history of a lacustrine population. The only comprehensive life-history description for this species is a study in the Columbia River in Washington (Dauble 1986). The following account is based on these sources supplemented, where appropriate, with incidental field observations.

REPRODUCTION

Typically, the largescale sucker spawns in the spring. Spawning usually begins when water temperatures reach about 8 °C. Thus, in the lower Fraser valley, spawning can start as early as the beginning of April and is often over by mid-June. In contrast, interior populations start spawning in late May and peak in June, usually at water temperatures approaching 15 °C. In the upper Peace system, largescale suckers were observed spawning in the outlet of Summit Lake (Crooked River) during the first week of June (water temperature 15 °C), and in the Kettle River (Columbia system), they have been observed spawning in early July at temperatures of about 16 °C. Spawning occurs in both flowing water and lakes. In rivers and streams, riffles adjacent to areas of slower water are typical spawning sites. In lakes, spawning usually occurs in shallow water (<2 m deep) over areas of coarse gravel; however, substrate size is not critical and apparently can range from fine gravel to cobbles.

Breeding fish of both sexes aggregate in the vicinity of spawning sites, but the sexes are partially segregated except during the actually spawning act. In rivers and streams, males occur in deeper, faster water than females, whereas in lakes, males aggregate closer to spawning sites and in shallower water than females. Apparently, in areas with little cover or in areas subject to disturbance (human or animal), both sexes remain in deep water and move onto riffles at night (McCart and Aspinwall 1970). In the Kettle River during the day, small groups of males (usually two or three) were spaced out on cobble and coarse gravel riffles. These riffles were about 1 m deep and had surface velocities of 0.8–1.1 m/s. In contrast, females were in slower water close to the river edges. In Norrish (Suicide) Creek near Dewdney, both sexes congregate in deep (>1.5 m) pools immediately below swift cobble riffles. Presumably, spawning occurs on these riffles at night.

Spawning occurs when a female moves onto the spawning site. She is immediately joined by a minimum of two (sometimes three to five) males. The males press against both sides of the female and vibrate rapidly. Eggs and sperm are released just above the bottom, and although there is no site preparation, the substrate is often cleaned and a shallow depression created by the thrashing of spawning fish. The eggs are demersal and adhesive. Some eggs stick to the gravel surface, but the current carries others into cracks and crannies among and beneath the rocks. Within minutes of spawning, most of the exposed eggs are eaten by other suckers.

Egg number is a function of female size, and fecundity in largescale suckers varies between about 9,000 and 30,000 eggs. Unfertilized eggs are pale yellow and about 3 mm in diameter. Once fertilized, the eggs swell, and a wide space forms between the yolk and the outer membrane. The diameter of fertilized eggs is about 3.5–3.7 mm. Development rate is temperature dependent, and hatching occurs in about 20 days at 10 °C and in about a week at 16 °C. The newly hatched larvae are about 10 mm TL and remain in the gravel until they are about 12 mm long. On emergence, the larvae are still dependent on their yolk reserves. Exogenous feeding starts at about 15 mm.

AGE, GROWTH, AND MATURITY

Initial growth is moderate, and the largest fry can reach a length of about 65 mm FL by the end of their first summer; however, the size variation in late summer young-of-the-year is wide, and yolk sac fry in the upper Fraser system are still emerging from the gravel in early August. Apparently, growth slows as the animals approach sexual maturity: 5 or 6 years for males and 6–9 years for females (Dauble 1986). After maturity, growth is slow, but adult survival is high. The largest recorded B.C. specimen was a female 740 mm FL (R. L. & L. Environmental Services Ltd. 1993), but relatively few adults reach 600 mm. Although adult suckers are notoriously difficult to age, one 620 mm female gave an unverified otolith age of 18 years. This is probably an underestimate. In Washington, ages of 27 and 28 years are reported for largescale suckers less than 500 mm long (Wydowski and Whitney 2003).

FOOD HABITS

As adults, largescale sucker are benthivores, but they begin life as planktivores. Initially, the mouth is terminal but eventually shifts to a ventral position; it is clearly subterminal by about 20 mm FL. When the mouth is terminal, about 80% of their food is plankton (e.g., *Daphnia*, *Cyclops*, *Bosmina*, and various nauplii), but as the mouth shifts position, the diet shifts towards benthic prey (*Chydorus*, chironomid larvae, and periphyton). Periphyton remains the dominant food throughout the juvenile and adult life-stages, but aquatic insects, especially tricopteran larvae, are a major item in the diet of adults. Adults exploit seasonally abundant food sources (e.g., fish eggs and larvae in the spring) and, in winter, eat filamentous algae (Dauble 1986). Although primarily bottom feeders, largescale suckers are versatile, and they have been observed foraging at the surface on plankton and emerging aquatic insects in lakes in both British Columbia and Washington.

Habitat No quantitative study of habitat use by largescale sucker is available for British Columbia; however, they appear to be especially common in large lakes and rivers.

ADULTS

In British Columbia, largescale sucker are abundant in low- to moderate-gradient rivers and low-altitude lakes throughout the southern two-thirds of the province. In upper Peace drainages, they are common in the Crooked and Parsnip river systems but less common in the much colder Finlay system. This suggests that they may prefer warmer water. In a brief survey of the Similkameen River, sites with and without largescale suckers were compared (Rosenfeld 1996), and sites with largescale suckers were significantly warmer than sites where this species was absent. Also, in the same drainage system, largescale suckers were associated mainly with lakes and lake tributaries or outlets.

In British Columbia, little is known about the seasonal and diurnal movements of adults in lacustrine environments, but in summer, they are found both above and below the thermocline. Video footage taken at depths of over 100 m in Okanagan Lake shows large numbers of this sucker foraging over the mud–silt bottom. Adults are associated with relatively slow currents (<1.0 m/s); however, in many rivers, they sometimes aggregate in deep pools. In large rivers like the Columbia and Fraser, they often occur at mid-channel sites where currents exceed 1.0 m/s. In the Columbia River, adult largescale sucker moved inshore at night and offshore during the day (Dauble 1986). Except for spawning migrations, largescale sucker are relatively sedentary. In Washington, most tagged fish were recaptured within 500 m of the original tagging site, although a few were recaptured 60 km downstream (Dauble 1986). The pattern of movement is similar in British Columbia—in the Columbia River, most tagged fish were recaptured within 5 km of their original tagging site (some after almost 3 years); however, a few individuals moved up to 75 km (R. L. & L. Environmental Services Ltd. 1993). A curious phenomenon in Okanagan Lake is the fall aggregation of adults at some sites. These aggregations consist of hundreds of closely packed fish swimming in relatively tight circles. It is not known why this happens.

JUVENILES

In the Nazko River, juvenile sucker preferred relatively shallow (0.25–0.5 m), slow-water areas (0–0.1 m/s) with sand or silt bottoms (Porter and Rosenfeld 1999). Although this study did not identify juvenile suckers to species, the largescale sucker is the most common species in that area. During the summer in lakes, juvenile suckers forage in similar, but deeper, areas than fry.

YOUNG-OF-THE-YEAR

Some data are available on the diel and seasonal habitat use of young-of-the-year in Nicola Lake (Miura 1962). Here, fry were most abundant in shallow water over rock and gravel substrates in July but shifted to open sand areas in August. They appeared to avoid heavily vegetated areas. The fry made diurnal onshore–offshore movements: onshore at dawn and offshore at dusk. There also was a seasonal shift in their vertical distribution that corresponded to the diet shift from plankton to benthos. In early summer, small schools of fry cruised close to the surface in shallow water, whereas the schools were closer to the bottom in late summer. Less is known about fry habitat use in flowing water; however, yolk-sac fry often concentrate in seasonally flooded vegetation along the edges of streams. Apparently, such areas are important foraging sites when fry first start feeding. Initially, the fry forage off the bottom, but they become bottom oriented as they grow. When water levels drop in the summer and autumn, fry concentrate in shallow side-channels, pools, and embayments. These areas usually have soft bottoms or silt-covered gravel or cobble substrates and no vegetation. In the Parsnip River, there are seasonal fry migrations between tributaries and the main river (McLeod et al. 1978).

Conservation Comments The largescale sucker is abundant in British Columbia (especially west of the Rocky Mountains). Most populations appear healthy, and at present, they are not a conservation concern.

Catostomus platyrhynchus (COPE)
MOUNTAIN SUCKER

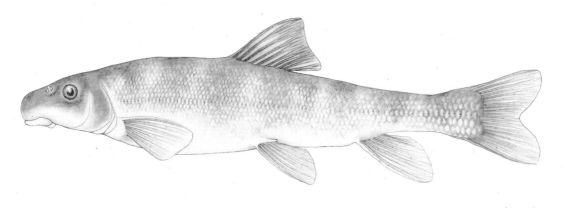

1 cm

Distinguishing Characters This small (usually <250 mm FL) sucker has fine scales (more than 70 lateral line scales) and 10–12 dorsal rays (usually 10 or 11 in British Columbia). The mouth is distinctive; it has a shallow cleft between the lower lips, a deep notch at each corner of the mouth, and a slightly curved cartilaginous sheath on the lower jaw (Fig. 19A). Two other species of fine-scaled suckers occur in B.C.—the longnose sucker (*Catostomus catostomus*) and the bridgelip sucker (*Catostomus columbianus*). Adult and juvenile mountain and longnose suckers differ in lip shape: the cleft between the lower lips in mountain suckers is shallow, and the lips are conspicuously notched at the corners of the mouth; in longnose suckers, the cleft between the lower lips is deep (usually complete), and there are no notches at the corners of the mouth.

Mountain and bridgelip suckers are more difficult to separate: in both species the cleft between the lower lips is shallow, and sometimes, bridgelip sucker have small notches at the corners of the mouth. Although these notches are not as well developed as in mountain suckers, they can lead to misidentifications. Both species have a chisel-like lower jaw, but this trait is more prominent in the mountain sucker than in the bridgelip suckers. Also, each of the lower lips in mountain suckers is wider than it is deep, whereas in the bridgelip sucker, each lower lip is about as deep as it is wide. On dead specimens, peeling back the skin on the head and checking the top of the skull for a fontanelle (an opening in the top of the skull) provides a positive identification: a fontanelle is absent, or reduced to a narrow slit, in the mountain suckers and well developed in the bridgelip sucker.

In British Columbia, a useful field character for juveniles, and many adults, is the presence of three dark bars across the dorsal surface. No other sucker in the province has such bars on their dorsal surface.

Taxonomy In the past (e.g., Carl et al. 1959), the mountain sucker was called *Pantosteus jordani*, a species originally named from an upper Missouri tributary (the Red Rock River) in Montana (Evermann 1893). Earlier, however, a mountain sucker was described (Cope 1874) as *Miniomus platyrhynchus* from Provo, Utah (presumably from the Provo River). Later, *jordani* was synonymized with *platyrhynchus* and *Pantosteus* was reduced to a subgenus of *Catostomus* (Smith 1966). Thus, the species in British Columbia became *Catostomus platyrhynchus*.

In a monograph on mountain suckers and their relatives, Smith (1966) recognized that mountain suckers from above Shoshone Falls on the Snake River together with those in the Great Basin formed a group of populations distinguishable from the group of populations in the Columbia, Missouri, and Saskatchewan river systems. However, because the morphological variation within groups was as great as the variation among groups, they were all assigned to *C. platyrhynchus* (Smith 1966). Interestingly, within the Columbia–Missouri complex, the Fraser River sample appeared as a morphological outlier. Recent mitochondrial DNA data (the entire cytochrome *b* gene) suggest *C. platyrhynchus* is a complex of species. Mountain suckers on the east and west sides of the Continental Divide have diverged by about 6%. Furthermore, the mountain suckers from the upper Snake River (above Shoshone Falls) are about 5% divergent from mountain suckers both east of the Continental Divide and in the Columbia–Fraser watersheds. The depths of these divergences suggest that what we now call a single species is actually several species. Obviously, their taxonomy needs re-examination.

Sexual Dimorphism On average, adult females are larger than mature males. Also, breeding males develop a rosy midlateral stripe that is absent, or faint, in females. This stripe is not obvious in males taken from the turbid waters of the Fraser River. In addition to body size, the sexes differ in the extent and distribution of nuptial tubercles. Breeding males develop tubercles on their anal and caudal fins (those on the lower caudal lobe are noticeably larger than those on the upper lobe). In females, these tubercles are small or absent. Males also possess weak tubercles on the dorsal and ventral surfaces of the pectoral fins and have small tubercles scattered over most of the body. Again, these tubercles are either absent or only weakly developed in females.

Like other suckers, adults differ in the size and shape of some fins. The anal fin and the lower lobe of the caudal fin are noticeably longer in mature males than in females. Additionally, the posterior margins of the pelvic fins are squared off in males (all rays about equal in length) and bluntly pointed in females. Apparently, this difference in pelvic fin shape is a reliable indicator of sex in immature individuals and non-breeding adults (Stanley 1988).

Hybridization   In the lower Fraser River between Hope and Chilliwack, mountain and bridgelip suckers coexist. Occasional individuals from this area do not fit neatly into either species, and it is possible that some are hybrids. Hybrids between mountain suckers and both longnose and white suckers (*Catostomus commersonii*) are known from Wyoming and South Dakota (Hubbs et al. 1943). In British Columbia, the mountain suckers does not coexist with either of these species.

Distribution   NATIVE DISTRIBUTION

The mountain sucker is widely, but sporadically, distributed in mountainous regions throughout western North America. East of the Continental Divide, it occurs from the Saskatchewan River in Alberta and southwestern Saskatchewan, southwards along the eastern slope of the Rocky Mountains, out onto the Great Plains in Montana and Wyoming, and east into the Black Hills of South Dakota. West of the Continental Divide, the mountain sucker occurs in the middle and lower Fraser system; in the Columbia system in Washington, Oregon, Idaho (both above and below Shoshone Falls); and in the Great Basin in Nevada, California, and Utah (including headwaters of the Green River, a Colorado River tributary).

BRITISH COLUMBIA DISTRIBUTION

The mountain sucker has a scattered distribution in southern British Columbia. There are three local areas where they are modestly abundant: the gravel deposition region in the lower Fraser River (from near Hope downstream to about the mouth of the Sumas River), the North Thompson River in the region of Heffley, and the Similkameen River system (including the Tulameen River) from above Princeton to the U.S. border (Map 33). In British Columbia, there are no confirmed records from the mainstem Columbia or its major tributaries (except the Similkameen River); however, there is an unconfirmed record from near the confluence of the Salmo and Pend d'Oreille rivers (Baxter et al. 2003).

Life History   Little is known about the life history of the mountain sucker in British Columbia. Elsewhere, there are two life-history studies available for this species: one from the upper Missouri system in southwestern Montana (Hauser 1969) and the other from the Great Basin in Utah (Wydoski and Wydoski 2002). Both of these studies are based on populations that are molecularly different from the mountain sucker in the Columbia–Fraser systems (and from each other). Consequently, some life histories details may differ in the B.C. populations. Nonetheless, the general aspects of their life history probably are similar and, thus, form the basis of the following account.

Map 33 The B.C. distribution of mountain sucker, *Catostomus platyrhynchus*

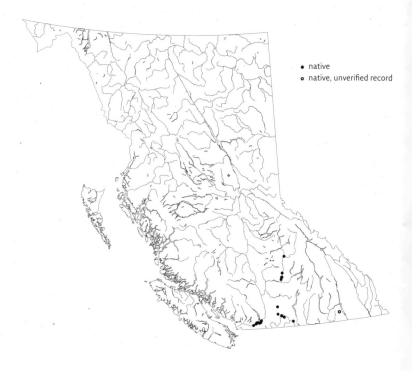

- native
- native, unverified record

REPRODUCTION

Mountain sucker spawn in the spring when water temperatures reach about 10 °C (Hauser 1969). In Lost Creek, Utah, mountain sucker were observed spawning from late May to late June at stream temperatures ranging from 9 °C to 11 °C (Wydoski and Wydoski 2002). The spawning peak lasted about 1 week in early June, and the spawning site was a shallow (about 11–30 cm deep) riffle. The substrate was gravel, the water surface was broken, and water velocities ranged from 0.06 to 0.20 m/s. In British Columbia, the only pre-spawning mountain sucker examined was a female (222 mm FL) captured in the Fraser River near Chilliwack in late April. This fish contained maturing eggs (about 1.5 mm in diameter) and probably would have spawned within a few weeks. This suggests that, if the lower Fraser mountain suckers spawn in the main river, they probably spawn in June just before the peak of the hydrograph. Where they spawn is unknown. There are several side-channels in the unconfined floodplain area between Chilliwack and Laidlaw that might have the right characteristics. Not enough is known about the distribution of mountain sucker in either the North Thompson or Similkameen rivers to speculate on where they might spawn.

Fecundity in mountain sucker increases with female size, and egg numbers from the mountain sucker in eastern Montana ranged from 990 to 3,700 (Hauser 1969). For the Lost Creek mountain sucker, females 131–182 mm FL had fecundities ranging from 1,239 to 2,863 eggs (Wydoski and Wydoski 2002). A study of the early development of three species of suckers (including the mountain sucker) indicates that the fertilized eggs ranged in diameter from 2.5 to 3.0 mm, and mountain sucker eggs were at the low end of this range (Snyder 1983). The mean egg diameter of ripe but unfertilized eggs in Lost Creek females was 1.77 mm (range 1.61–1.93 mm). In Lost Creek females, there was a positive but highly variable relationship between female size and egg size. Presumably, mountain sucker eggs are adhesive and demersal. They probably are released close to the bottom and lodge in cracks and crannies in the gravel substrate. At 15 °C, the eggs take 7 days to begin hatching (Snyder 1983). The newly hatched larvae are about 8.5 mm TL and probably remain in the gravel until they are about 10 mm in length. Exogenous feeding begins at about 13 mm.

AGE, GROWTH, AND MATURITY

There are no age and growth data available for the North Thompson and Similkameen populations. In the lower Fraser, however, growth appears to be slower than that reported for the Lost Creek population (Wydoski and Wydoski 2002) but similar to upper Missouri populations (Hauser 1969). In the lower Fraser, young-of-the-year average 35 mm FL by mid-September and 65 mm by the end of their second growing season. Most males in the lower Fraser are mature by their fourth year, and most females mature by their fifth year. These maturity estimates are based on small samples (<30 fish of each sex) and unverified otolith readings. They should be considered approximations. The oldest mountain sucker known from B.C. was a female 222 mm FL that had an estimated otolith age of 9+.

FOOD HABITS

Mountain sucker have exceptionally long, highly coiled guts and a jet-black peritoneum. In fishes, this combination typically implies herbivory, and the specialized mouth and lower jaw suggests mountain suckers feed by scraping periphyton off rocks. In aquaria, Fraser River mountain sucker were regularly observed scraping algae off of rocks. Nonetheless, the entire guts of 5 adults and 20 juveniles collected in the Fraser River in August were packed with silt (including sand grains and flakes of mica), some *Closterium*, a few strands of filamentous algae, and small numbers of larval insects (mostly chironomids). Except for the apparent absence of diatoms and the relatively low abundance of filamentous algae, this diet is similar to that reported in eastern Montana (Hauser 1969). The stomach contents of the Fraser River sample suggest that, although mountain suckers use their specialized lower jaw to scrape periphyton in clear rivers, they also ingest material directly off the substrates in turbid waters. In the Fraser sample, fry contained mainly diatoms and *Closterium*.

Habitat In British Columbia, no systematic attempt has been made to characterize the habitats used by mountain sucker. Thus, the following remarks are based on casual field observations. A quantitative survey of mountain sucker habitats is needed in British Columbia, especially since the

Columbia–Fraser mountain suckers are molecularly divergent from both the mountain suckers in the upper Snake River and northern Great Basin, and those east of the Continental Divide.

ADULTS

All the B.C. records of mountain sucker are from flowing water; however, in the southern parts of its range, the species does occur in lakes and reservoirs (Smith 1966; Snyder 1983; Wydoski and Wydoski 2002). Mountain sucker are described as inhabiting clear, cold mountain streams usually less than 12 m wide and with sand, gravel, cobble, or boulder substrates (Brown 1971; Hauser 1969; Nelson and Paetz 1992; Scott and Crossman 1973; Simpson and Wallace 1978). This description does not adequately portray mountain sucker habitats in British Columbia. Here, they occur in three rivers that differ in width, annual discharge, water clarity, temperature, timing of the peak hydrograph, and primary productivity. Only the upper reaches of the Tulameen River and the Similkameen River below Princeton approach the kind of habitat typically described for mountain sucker. Curiously, the species does not appear to be abundant in these rivers. For example, a recent survey of the Similkameen River (Rosenfeld 1996) failed to find any mountain suckers and concluded that the species had either declined in numbers or has an extremely patchy distribution within the drainage system. The Fraser River between Hope and Chilliwack is a large (in places up to 1 km wide) and turbid river. At low water, the river fragments into a complex network of channels, islands, gravel bars, and side-channels. Most (but not all) of the side-channels go dry in the late summer and fall. The flows in the main channels are swift, and the substrate can vary from cobbles to fine gravels and silts depending on water velocity. During late summer and fall, adult mountain sucker occur in channels among the gravel bars. They appear to be most abundant on the lee sides of bars in areas of moderate depth (up to 1.5 m) and current (<0.7 m/s). Presumably, light penetration and primary production is low in the turbid water.

During the summer, adult mountain sucker in the North Thompson and Similkameen rivers are most commonly observed in deep (>1.5 m) glides and pools. Often there are two or three adults in the same pool.

JUVENILES

Similar to the adults in the Fraser River in the summer, juveniles are also associated with mid-river gravel bars; however, they usually occur in shallower (<1 m) and slower (<0.5 m/s) water than adults. In the North Thompson and Similkameen rivers, juveniles usually are found where small tributary streams enter the main rivers.

YOUNG-OF-THE-YEAR

In the Fraser River, young-of-the-year mountain sucker are common during summer and fall in shallow (<20 cm deep) embayments and blind side-channels associated with mid-river gravel bars. They are relatively rare in similar habitats along the river's edges. In the North Thompson and Similkameen rivers, fry aggregate out of the main current in warm (15–20 °C) shallow embayments and near the mouths of tributary streams.

Conservation Comments   The status of the mountain sucker in Canada needs to be re-examined. COSEWIC assessed the status of mountain suckers and decided they were not at risk; however, their evaluation was based on the assumption that all mountain sucker populations in Canada are the same species (*C. platyrhynchus*). Molecular data suggest that this is not the case. The BCCDC ranks the mountain sucker as S3 (rare or uncommon), and at least one of the three B.C. populations may be in decline (Rosenfeld 1996).

FAMILY ICTALURIDAE — NORTH AMERICAN CATFISHES

The catfishes (Order Siluriformes) are primarily tropical freshwater fishes; however, there are some temperate freshwater species and two marine families. Like the minnows (Cyprinidae), they posses a Weberian apparatus that aids in the detection of sounds transmitted in water, and specialized cells in the skin that release a "fright substance" when the fish is injured. The North American catfishes (about 46 living species) belong to the family Ictaluridae. The earliest fossil record of ictalurids dates back to the Paleocene (roughly 30 Ma). Ictalurid catfishes lack scales and have an adipose fin, stout spines at the leading edges of the dorsal and pectoral fins, and four pairs of barbels on the head. One pair of barbels is attached to the snout near the posterior nostrils, another pair of long barbels is attached to the upper lip, and two pairs of shorter barbels are found under the chin.

North of Mexico, all of the living species in this family originally were restricted to waters east of the Continental Divide; however, they have been widely introduced into western North America. Nonetheless, there once were native ictalurids in western North America. Fossils from late Pliocene deposits in central Washington State (Smith et al. 2000) and eastern Oregon (Van Tassell et al. 2001) establish that catfishes (*Ameiurus*) were present in the Columbia system until the late Pliocene (about 3 Ma).

British Columbia has three introduced species: the black bullhead (*Ameiurus melas*), the yellow bullhead (*Ameiurus natalis*), and the brown bullhead (*Ameiurus nebulosus*). The official common name for *Ameiurus* catfishes are bullheads (Nelson et al. 2004). This is a good example of the potential for confusion associated with common names—in British Columbia, bullheads are called catfishes, and sculpins are called "bullheads."

KEY TO THE ADULT BULLHEADS

1 (2) Barbels on chin light (pale yellow or white); 24–28 anal rays (usually 25–28)	YELLOW BULLHEAD, *Ameiurus natalis*
2 (1) Barbels on chin dark (dark brown or black); 16–23 anal rays (see 3)	
3 (4) Fin membranes in adults noticeably darker than adjacent fin rays; 16–22 anal fin rays (usually 17–21); no (or weak) serrations on the trailing edge of the pectoral spine (Fig. 24A)	BLACK BULLHEAD, *Ameiurus melas*
4 (3) Fin membranes in adults not noticeably darker than adjacent fin rays; 21–24 anal fin rays (usually 22 or 23); strong (especially in young fish) serrations on the trailing edge of the pectoral spine (Fig. 24B)	BROWN BULLHEAD, *Ameiurus nebulosus*

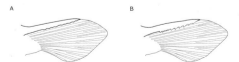

Figure 24 Diagram illustrating the difference in the development of serrations on the trailing edges of the pectoral spines in (A) black and (B) brown bullheads

Ameiurus melas (RAFINESQUE)
BLACK BULLHEAD

1 cm

Distinguishing Characters   The black bullhead has darkly pigmented barbels, weak serrations along the posterior edge of the pectoral spine, and 15–19 anal rays. In our area, adults are black above and white or lemon yellow below. Juveniles are also black above and white below. The black bullhead is easily confused with the brown bullhead (*Ameirus nebulosus*) and the yellow bullhead (*Ameirus natalis*); however, adults can be separated by the development of serrations on the trailing edge of the pectoral spines. These serrations are weak or absent in black bullheads and strong in brown and yellow bullheads. Also, the pigment on the membranes between the dorsal and anal rays is noticeably darker than the adjacent rays in black bullhead and about the same colour as the rays in the other species. In addition, there usually is a pale vertical bar at the base of the caudal fin in black bullhead that is absent in both brown and yellow bullheads. The young of all three species are uniformly dark black.

Sexual Dimorphism   In our area, breeding males usually are lemon yellow below, whereas the undersides of females are white or pale yellow. Additionally, adult males have a single urogenital opening behind the vent, and females have two openings posterior to the vent: a urinary and a genital opening (Moen 1959).

Distribution   NATIVE DISTRIBUTION
The black bullhead is indigenous to central North America. Its natural distribution centres on the Mississippi River, where it ranges from the river's northernmost tributaries downstream to the Gulf of Mexico. The native range also extends east along the southern margins of the Great

Lakes as far as northwestern New York and to the west as far as southeastern Saskatchewan and southwestern Manitoba.

Black bullhead have been widely introduced into western North America, and there are now self-sustaining populations in Arizona, British Columbia, California, Nevada, Oregon, Utah, and Washington. Although black bullhead is said to be rare in Washington and found mainly in the southeastern quarter of the state (Wydoski and Whitney 2003), the B.C. populations probably are derived from populations in northeastern Washington.

BRITISH COLUMBIA DISTRIBUTION

Black bullhead occur in the wetlands where the Kootenay River enters Kootenay Lake (Creston Valley Wildlife Management Area), and in Osoyoos, Vaseaux, Skaha, and Okanagan lakes. They also occur in quiet side-channels of the Okanagan River between Osoyoos and Skaha lakes (Map 34).

Life History In British Columbia, the only published report on the biology of black bullhead is Forbes and Flook (1985). It deals with the population in the Creston Valley Wildlife Management Area. Consequently, most of the following review is based on eastern Canadian and western U.S. sources.

REPRODUCTION

In British Columbia, the black bullhead spawns in the early summer when water temperatures rise above 20 °C (Forbes and Flook 1985). Females excavate a shallow, circular depression in the bottom. This nest is usually in water <1 m deep and near clumps of vegetation or other cover. Wallace (1967) gives a detailed description of black bullhead courtship and spawning behaviour. Courtship includes butting and both sexes slide their barbels over their partner's body. Egg deposition is preceded by a behaviour where the male curls his caudal fin around the female's head. Spawning occurs intermittently with small batches of eggs released into the nest where they are fertilized by the male. Between spawnings, the female fans the spawned eggs. Fecundity is related to female size and varies between 1,000 and 7,000 eggs (Dennison and Bulkely 1972). The eggs stick together in clumps and are enclosed in a jelly-like covering. The parents aerate the eggs by fanning with their pelvic fins. Fertilized eggs are about 3.0 mm in diameter and hatch in about 5–10 days. The newly hatched fry are about 4 mm in length. They are maintained in a tight ball by their parents for about a week and then herded in schools for another 1–2 weeks. The parents then leave the fry, but the young remain in schools until winter.

AGE, GROWTH, AND MATURITY

Black bullhead growth is variable but generally quite slow, and under crowded conditions, they are prone to stunting. In Washington, they reach a length of 70–90 mm at the end of their first year and about 170 mm

Map 34 The B.C. distribution of black bullhead, *Ameiurus melas*

■ introduction

after 2 years (Wydoski and Whitney 2003). Sexual maturity is reached at 2–4 years, and males usually mature a year before females. In the wild, black bullhead rarely live more than 6 or 7 years, and the maximum age appears to be about 10 years.

FOOD HABITS

Black bullhead are omnivorous and forage mostly at night. They have chemoreceptors on their barbels and head and use chemical cues to detect prey at night (Bardach et al. 1967). Apparently, they are unable to visually focus on near objects (Sivak 1973). Their diet includes aquatic insects, earthworms, crustaceans, and molluscs. Given their benthic foraging behaviour, it is not surprising that their stomachs usually contain some detritus and plant material. Young-of-the-year and juveniles forage mostly on chironomid larvae. Apparently, they display two distinct foraging periods: one just before dawn and another shortly after the onset of darkness (Darnell and Meieretto 1965).

Habitat Again, the only B.C. information on habitat use for this species is Forbes and Flook (1985). Thus, much of the following information is derived from sources outside the province.

ADULTS

In Osoyoos Lake, black bullhead are common in shallow, weedy ponds and lagoons connected to the lake, and less common in the main lake. Also, they are relatively rare in the artificially constrained Okanagan River but abundant in the murky, sluggish original river channel. In the Creston Valley Wildlife Management Area, they occupy small, shallow, eutrophic lakes and sluggish streams. Thus, although most of the B.C. sites where black bullhead are common are altered environments, the habitat appears to be typical of natural black bullhead habitat. In 23 natural lakes in South Dakota, black bullhead abundance increased with decreasing lake size, increasing nutrients, and decreasing water transparency (Brown et al. 1999). Adults are nocturnal and, during the day, lie quiescent in weed beds or under cover. At this time, they often lie in groups and are in physical contact with other adults. At night, they become active and schools of black bullhead forage in shallow water over mud bottoms.

JUVENILES

Unlike adults, juvenile black bullhead are day active. They are most abundant in shallow water (typically <1 m deep) in areas with patches of vegetation rather than dense weed beds (Weaver et al. 1997).

YOUNG-OF-THE-YEAR

Young-of-the-year black bullhead also are day active. They move in dense schools through the same shallow water habitats as those used by juveniles.

Conservation Comments   In British Columbia, black bullhead occur in altered environments that contain a high proportion of non-native species (e.g., the south Okanagan). It is not clear what impacts, if any, the black bullhead has had on our indigenous fishes.

Ameiurus natalis (LESUEUR)
YELLOW BULLHEAD

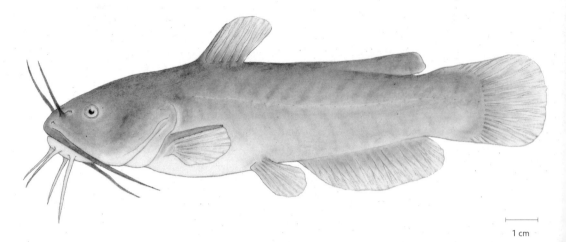

1 cm

Distinguishing Characters — The yellow bullhead has pale chin barbels, a rounded caudal fin (the caudal fin is square or slightly forked in the other species), and usually 22–25 anal rays. In British Columbia, this species is most likely to be confused with either the brown (*Ameirus nebulosus*) or black (*Ameirus melas*) bullheads. The yellow bullhead differs from the other two B.C. species in the colour of the chin barbels (pale vs. dark) and the length of the anal fin: when depressed, the anal fin overlaps the anterior rays of the caudal fin in the yellow bullhead but does not reach the anterior rays of the caudal fin in black or brown bullheads.

Taxonomy — The only taxonomic problem associated with the yellow bullhead is its phyletic relationship to the black and brown bullheads. On the basis of morphology, two clades of living bullhead catfishes were recognized by Lundberg (1992): a *natalis* clade consisting of black, brown, and yellow bullheads, and a *catus* clade consisting of the other four species of *Ameiurus*. A recent alternative molecular phylogeny rejects both of these clades and places the yellow bullhead as the living basal member of the genus and black and brown bullheads as the most recently derived members (Hardman and Page 2003).

Sexual Dimorphism — Adult males have a single urogenital opening behind the vent, and females have two openings posterior to the vent: a urinary and a genital opening (Moen 1959).

Distribution — **NATIVE DISTRIBUTION**
Originally, the yellow bullhead was confined to the eastern United States: Vermont to Florida, and Minnesota to western Texas. The only native Canadian populations occur in southeastern Ontario. Apparently,

the yellow bullhead was introduced into the Columbia River system in 1905 (Lampman 1946) and is now locally common in appropriate habitats in eastern Washington and Oregon. The yellow bullhead also occurs in the lower Colorado River system in southern California (Moyle 2002). The two populations closest to British Columbia are in western Washington (Wydoski and Whitney 2003).

BRITISH COLUMBIA DISTRIBUTION

The yellow bullhead was discovered recently in Silvermere Lake near Ruskin in the lower Fraser valley (Map 35) by Dr. G.F. Hanke of the Royal British Columbia Museum. Silvermere Lake is man made, and yellow catfish probably were introduced into the lake along with illegal introductions of largemouth bass. One of the two populations of yellow bullhead in adjacent Whatcom County, western Washington, is a likely source of the B.C. population. The species appears to be spreading: in the summer of 2006, it was collected in a slough on Westham Island in the Fraser Delta.

Life History The life history of the yellow bullhead has not been investigated in British Columbia. Useful sources for life-history information on this species in western North America are Moyle (2002) and Wydoski and Whitney (2003). These sources, supplemented with literature from eastern North America, form the basis of the following summary.

REPRODUCTION

Even within their native range, there is little known about the biology of the yellow bullhead. They spawn in the spring and early summer (April through July). Apparently, both sexes are involved in nest construction. The nest can be anything from a circular depression in the substrate (sand, gravel, or rarely, mud) to an actual burrow. Usually, the nest is in shallow water (<1.5 m deep), and in lakes, it is often associated with cover (e.g., clumps of vegetation, logs, or tree roots). Spawning behaviour in aquaria has been described (Wallace 1972). Spawning occurs with small batches of eggs intermittently released into the nest and fertilized by the male. Fecundity varies between 1,650 and 7,000 eggs (Jones et al. 1978b). Since most nests contain 300–700 eggs (Becker 1983), females probably spawn more than once in a season. The eggs stick together in clumps and are enclosed in a jelly-like covering. The male aerates the eggs by fanning with the pelvic fins. Fertilized eggs are about 3.0 mm in diameter and hatch in about 5–10 days (Jones et al. 1978b). The male guards the eggs and young and contains the newly hatched fry in a tight mass in the nest until the yolk sac is absorbed. The male continues to guard the fry until they reach a length of about 50 mm. The fry remain in compact groups until fall when they seek shelter for the winter.

Map 35 The B.C. distribution of yellow bullhead, *Ameiurus natalis*

■ introduction

AGE, GROWTH, AND MATURITY

Yellow bullhead growth is variable. In Wisconsin, they reach an average length of about 50 mm at the end of their first year and about 130 mm after 2 years (Becker 1983). Sexual maturity is reached at a length of about 140 mm. Males usually mature earlier than females. In the wild, yellow bullhead rarely live more than 6 or 7 years.

FOOD HABITS

Like other bullheads, adult yellow bullhead forage at night. Visually, they probably are unable to focus on close objects (Sivak 1973); however, their barbels are covered with chemoreceptors, and they can locate food in the dark (Bardach et al. 1967). Yellow bullhead are omnivorous, and adults eat a variety of benthic insects, crustaceans, molluscs, fish, and plant material. If crayfish are present, they appear to be the preferred prey (Keast 1985). Apparently, juveniles are crepuscular (Reynolds and Casterlin 1978) and exploit a wider array of prey than the adults (Keast 1985). The young-of-the-year forage during the day, mostly on chironomid larvae; however, in quiet water, they school close to the surface (Harlan and Speaker 1956) and consume some zooplankton.

Habitat Habitat use by the yellow bullhead has not been studied in British Columbia. Consequently the following summary is based mainly on information from eastern North America.

ADULTS

Yellow bullhead tolerate a wide range of temperatures and oxygen levels; however, they generally appear to prefer clearer waters than either black or brown bullheads. Typically, they do best in small clear lakes or sluggish streams with soft bottoms and lots of vegetation; however, they also occur—but are not common—in large lakes and reservoirs. In Wisconsin, they are most abundant in clear water at depths of 0.6–1.5 m over sand or mud substrates (Becker 1983). Typically, adults are nocturnal and, during the day, lie quiescent in weed beds or under cover. At night, adults engage in complex (sometimes aggressive) social behaviours that are mediated through chemical communication (McLarney et al. 1974; Todd et al. 1967).

JUVENILES

Unlike adults, juvenile yellow bullhead are most active at dawn and dusk. They are often abundant in shallow water (typically <1 m deep) in areas with patches of vegetation or other cover.

YOUNG-OF-THE-YEAR

Young-of-the-year yellow bullhead are day active and use the same shallow water habitat as the juveniles. At first, they move in compact schools among vegetation and near the surface. These schools break up in the fall.

Conservation Comments — The yellow bullhead was first recorded from Silvermere Lake near Ruskin. This lake is man made and mostly contains introduced species. Consequently, the yellow bullhead in Silvermere Lake probably was not a serious threat to native fishes; however, it has spread downstream into the Fraser River delta and, presumably, into the intervening waters. Since this species prefers clearer water than either of the other two introduced bullheads, eventually it will probably spread throughout the lower Fraser Valley. It is not clear what effects the yellow bullhead will have on our native species, but they are unlikely to be beneficial.

Ameiurus nebulosus (LESUEUR)
BROWN BULLHEAD

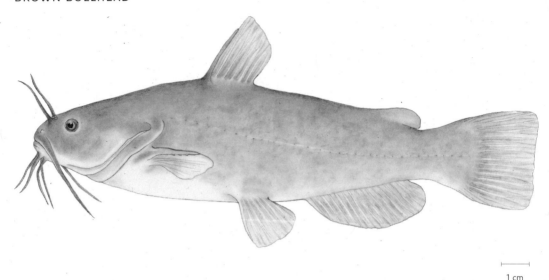

1 cm

Distinguishing Characters   The brown bullhead has darkly pigmented barbels, strong serrations on the posterior edge of the pectoral spine, and usually 19–21 anal rays. Adults are yellow-brown above and dirty white below. Juveniles are dark above and white below. The brown bullhead is easily confused with the black bullhead (*Ameiurus melas*); however, adults can be separated by the development of serrations on the trailing edge of the pectoral spines—these serrations are strong in the brown bullhead and weak in the black bullhead. There is also a difference in colour pattern. In brown bullhead, the pigment on the membranes between the rays of the dorsal and anal fins is about the same colour as the rays, whereas in black bullhead, the membranes between the rays on these fins are noticeably darker than the rays. In addition, there usually is a pale vertical bar at the base of the caudal fin in black bullhead that is absent in brown bullhead. The young of both species are uniformly dark black and difficult to distinguish.

Sexual Dimorphism   In British Columbia, breeding males usually are larger than females. Additionally, adult males have a single urogenital opening behind the vent, and females have two openings posterior to the vent: a urinary and a genital opening (Moen 1959).

Distribution   NATIVE DISTRIBUTION
The original distribution of the brown bullhead was similar to that of the black bullhead but included the Atlantic Coastal Plain from Nova Scotia south to Florida. Also, it did not extend as far to the west as the black bullhead. Again, the northernmost natural populations are in southeastern

Saskatchewan. The brown bullhead has been widely introduced in western North America, and there are now populations in Arizona, British Columbia, California, Nevada, Oregon, Utah, and Washington.

BRITISH COLUMBIA DISTRIBUTION

Brown bullhead occur in the lower Fraser valley and in numerous lakes on the southern third of Vancouver Island (Map 36). There are also reports of this species in the Kootenays and in the Okanagan River system. So far, however, all the bullheads received for verification from the Canadian portion of the Columbia system have been black bullheads. Although there are no verified records of brown bullhead from south-central British Columbia, this species is well established in the Pend d'Oreille system in Washington and Idaho, and eventually, they probably will disperse downstream into British Columbia.

Life History   The life history of the brown bullhead has not been investigated in our area; however, there is a life-history summary available for a population in Washington (Imamura 1975). This source, along with life-history reviews from California (Moyle 2001) and Washington (Wydoski and Whitney 2003) as well as recent eastern North American literature form the basis of the following summary.

REPRODUCTION

Brown bullhead spawns in the spring or early summer when water temperatures reach about 20 °C—May in western Washington (Imamura 1975) and May or June in British Columbia. Bulmer (1985a) gives some details of their reproductive cycle. The female excavates a circular depression in the substrate (sand, gravel, or rarely, mud). Usually, the nest is in shallow water (<1 m deep) and near clumps of vegetation or other cover. Spawning occurs with small batches of eggs intermittently released into the nest and fertilized by the male. Between spawnings, the female fans the spawned eggs. Fecundity is related to female size and varies between 1,000 and 10,000 eggs (Jones et al. 1978c). The eggs stick together in clumps and are enclosed in a jelly-like covering. The parents aerate the eggs by fanning with their pelvic fins. Fertilized eggs are about 3.0 mm in diameter and hatch in about 4 days (Jones et al. 1978c). Nest predation is intense; although both sexes provide parental care, relatively few nests survive to the end of the normal parental care cycle (Bulmer 1985b, 1986). The parents maintain the newly hatched fry in a tight mass in the nest until the yolk sac is absorbed (7–10 days). At swim-up, the fry are about 8 mm long. For another 1–2 weeks, the fry are herded into loose schools by the parents. The parents then leave the fry; however, the young bullheads remain in schools until winter.

AGE, GROWTH, AND MATURITY

Brown bullhead growth is variable but generally quite slow. In western Washington, they reach a length of about 70 mm at the end of the first

Map 36 The B.C. distribution of brown bullhead, *Ameiurus nebulosus*

■ introduction

year and about 130 mm after 2 years (Imamura 1975). Sexual maturity is reached at about 3 years, and males usually mature a year before females. In the wild, brown bullhead rarely live more than 6 or 7 years.

FOOD HABITS

Brown bullhead are omnivorous (Keast 1985; Kline and Wood 1996) and often forage at night (Keast and Welsh 1968). Their barbels are covered with chemoreceptors, and they are able to locate food sources in the dark (Bardach et al. 1967). Typically, the diet of brown bullhead includes aquatic insects, amphipods, crustaceans, and molluscs, but they also consume fish eggs. Given their benthic foraging behaviour, it is not surprising that their stomachs often contain some detritus and plant material. Young-of-the-year and juveniles mostly forage on chironomid larvae but also eat zooplankton.

Habitat Habitat use by the brown bullhead has not been studied in British Columbia. Consequently the following summary is based mainly on information from central North America.

ADULTS

Brown bullhead are remarkably adaptable and tolerate a wide range of temperatures and oxygen levels; generally, however, they appear to prefer clearer, cooler waters than black bullhead. Typically, they do best in small lakes or sluggish streams with soft bottoms and lots of vegetation;

however, they also occur in large reservoirs, and on Vancouver Island, they occasionally occur in oligotrophic lakes. Typically, adults are nocturnal and, during the day, lie quiescent in weed beds or under cover; however, in small ponds, they are often day active. Adults are social and forage at night in loose schools. Generally, they forage in somewhat deeper water than the black bullhead. As winter approaches, brown bullheads move into shallow water and bury themselves in mud and detritus (Loeb 1954).

JUVENILES

Unlike adults, juvenile brown bullhead are day active and, at low densities, may be territorial (Carr et al. 1987). They are most abundant in shallow water (typically <1 m deep) in areas with patches of vegetation rather than dense weed beds (Weaver et al. 1997).

YOUNG-OF-THE-YEAR

Young-of-the-year brown bullheads are also day active and use the same shallow water habitat as the juveniles. At first, they school, but the schools break down when the parents stop guarding.

Conservation Comments For some native species, the introduction of brown bullhead into small lakes has been a disaster. For example, on Vancouver Island, Lasqueti Island, and in the lower Fraser valley, native populations of threespine stickleback (*Gasterosteus aculeatus*)—including one of the unique benthic–limnetic species pairs—have disappeared following the introduction of the brown bullhead. Threespine sticklebacks build their nests on the substrate, and these extirpations probably result from nocturnal predation on their eggs.

FAMILY ESOCIDAE — PIKES

The pikes (Order Esociformes) are another problematic order of relatively primitive teleost fishes. Some workers (Johnson and Patterson 1996) consider them to be more advanced than the Salmoniformes but other workers (Ishiguro et al. 2003; López et al. 2004) argue that they are the sister order of the Salmoniformes. This problem is still unresolved.

The family Esocidae has a Holarctic distribution with most of the living species clustered in eastern North America. There is one extant genus (*Esox*), but at least two extinct genera, *Estesesox* and *Oldmanesox*, occur in late Cretaceous deposits in southern Alberta and Montana (Wilson et al. 1992). Morphologically, the pikes have changed little in the last 60 million years, and a fossil *Esox* from early Tertiary (Paleocene) deposits in Alberta (Wilson 1980) is remarkably similar to the five living species of *Esox*. Although no fossil pike are known from British Columbia, an extinct muskellunge (*Esox columbianus*) existed in Washington State until the late Pliocene (Smith et al. 2000). The native distributions of four of the living species are relatively restricted: three occur only in eastern North America, and one is confined to northeastern Asia (the Amur River). In contrast, the fifth species (the northern pike, *Esox lucius*) has one of the widest longitudinal distributions of any freshwater fish and is found in cool waters throughout the Northern Hemisphere. This is the species native to northeastern British Columbia.

All pikes are voracious ambush predators, anatomically specialized for rapid acceleration. The dorsal and anal fins are roughly similar in area and positioned close to the caudal fin. This arrangement allows for rapid, and hydrodynamically efficient, acceleration (Firth and Blake 1991, 1995). The snout resembles a duck's bill, and the large mouth is full of teeth. The posterior teeth on the dentaries, and the teeth on the head of the vomer and inner edges of the palatines are canine; however, most of the hundreds of small needle-like teeth on the premaxillaries, tongue, vomer, and palatine bones slant backward towards the gullet and are depressible. Presumably, this arrangement facilitates the retention of active prey, although it can be the pike's undoing (see Hughes 1960).

Esox lucius LINNAEUS
NORTHERN PIKE

1 cm

Distinguishing Characters   The long, flattened duck-like snout and the posterior positioning of the dorsal and anal fins are unique among B.C. fishes.

Taxonomy   Although there are no taxonomic problems involving northern pike in British Columbia, this species probably colonized British Columbia from at least two glacial refugia. Fossil evidence indicates that they were present in both the Bering and Mississippi refugia before the Wisconsinan glaciation (Cavender 1986). Additionally, there are biochemical differences between pike on the Great Plains and elsewhere in North America (Seeb et al. 1987). Thus, in B.C., the pike in the Taku, Yukon, and upper Liard systems probably originated from the Bering Refugium, whereas those in the lower Liard and lower Peace systems may have originate on the Great Plains.

Sexual Dimorphism   Adult females are, on average, larger than adult males (Casselman 1996), and mature northern pike can be sexed (even outside the breeding season) by examining the urogenital region (Casselman 1974). The urogenital pore is located just behind the anus; in females, there is an area of convoluted, plicate (longitudinally folded), protuberant tissue behind the anus and continuous with the urogenital pore. In males, this protuberance is absent, and there is a slit-like, concave depression associated with the urogenital opening (Casselman 1974).

Distribution   **NATIVE DISTRIBUTION**
The northern pike occurs in cool waters throughout the Northern Hemisphere. It ranges from the British Isles eastward across Europe and northern Asia (except for the Amur River system) to the Bering Strait. In North America, it ranges from Alaska (the Seward Peninsula) east in a broad band across the northern Great Plains to Ontario, Quebec,

and Labrador and from the Arctic coast south to eastern Montana, Nebraska, and Missouri. It is absent from the Maritime Provinces and is absent from the Atlantic Coastal Plain in the United States, except for the Hudson and Connecticut river drainages.

BRITISH COLUMBIA DISTRIBUTION

Northern pike are abundant in the muskeg lowlands associated with the lower Peace, lower Liard, and Hay rivers. It also occurs in the upper Liard (above the Grand Canyon) and various Yukon River tributaries that rise in British Columbia. The presence of pike in the upper Liard and the headwaters of two Pacific drainages—the Taku and Alsek rivers—suggest that this species entered British Columbia from both the Bering and western Great Plains refugia (Map 37). Although pike have been introduced into the U.S. portion of the Columbia system (McMahon and Bennett 1996), they have not yet reached south-central British Columbia. Apparently, an illegal introduction of pike into the upper Peace River system (Crooked River) failed.

Life History Although little is known about the biology of northern pike in British Columbia, there is a rich literature on this species in central Canada and the Prairie Provinces. In addition, there is a review (Craig 1996) of life-history information on both North American and European pike populations. These are the primary sources used for the following life-history account.

REPRODUCTION

Northern pike spawn in the spring: in northern British Columbia, this is usually about mid-May. In some large northern rivers (e.g., the Liard and lower Peace) immediately after ice-out, there is a pre-spawning movement of pike into tributaries and areas that are subject to seasonal flooding. Spawning occurs in shallow water (usually <0.6 m deep), and evidently, flooded vegetation is essential for successful reproduction (Bry 1996; McCarraher and Thomas 1972). Pike spawn during the day at water temperatures ranging from 4 °C to 18 °C (Clark 1950; Franklin and Smith 1963; Frost and Kipling 1967). Typically, spawning activity peaks in the afternoon when daily water temperatures are at their highest (Franklin and Smith 1963; Lucas 1992).

Clark (1950) gives a brief description of spawning behaviour in the wild, and there is a detailed analysis of pike spawning behaviour in large aquaria (Fabricius and Gustafason 1958). Individual females cruise slowly through vegetation in shallow water (usually accompanied by one to three males). The males swim alongside, or slightly behind, the female. At intervals, the female reduces her forward speed, and the males position themselves parallel to the female with their urogenital openings close to that of the female. Both sexes then increase the intensity and amplitude of their swimming movements and increase their forward speed. The males suddenly deliver powerful tail slaps, and sperm and eggs are

Map 37 The B.C. distribution of northern pike, *Esox lucius*

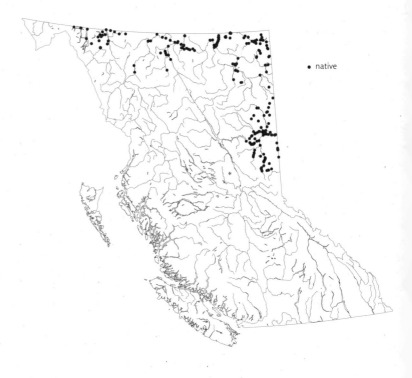

• native

released. These tail thrusts move the fish forward and scatter the eggs. About 5–60 eggs are released during each spawning (Svärdson 1949), and the spawning acts are clustered into series that last for 1–11 minutes (Fabricius and Gustafason 1958). Apparently, some females spawn several times on consecutive days (Clark 1950).

Fecundity is a function of female size and ranges from about 15,000 to well over 100,000 eggs. However, even in similar-sized females, there are major fecundity differences among localities (Billard 1996). The eggs are amber in colour and, when water hardened, are 2.3–3.0 mm in diameter (Clark 1950; Frost and Kipling 1967). They are denser than water, adhesive, and stick to vegetation and debris in the spawning area. Development rate is a function of temperature, and hatching takes 30.9 days at 5.8 °C, 7 to 10 days at 8.4 °C, and 5 days at 20 °C (Billard 1996; Clark 1950; Swift 1965). There are also population differences in egg size, and apparently, these translate into differences in the lengths of newly hatched larvae. The length range for newly hatched larvae in Europe is 8.5–9.0 mm (Billard 1996), whereas the length range for eastern North America is 6.5–8.0 mm (Franklin and Smith 1963). Newly hatched larvae lack gills and gas exchange is through the skin (Braum et al. 1996). The larvae possess paired cement glands on the head immediately in front of the eyes and use the sticky secretion from the cement glands to attach themselves to vegetation (usually they hang vertically). They remain attached until the yolk sac is absorbed (about 10 days to 2 weeks after hatching). They are 11–14 mm long at this stage and about 12–15 mm when they fill their swimbladder (Billard 1996).

AGE, GROWTH, AND MATURITY

Growth rate is temperature and food dependent but, generally, is rapid and, by late summer, the young-of-the-year range in size from 30 to 150 mm (Casselman 1996). In North America, growth was examined over a wide latitudinal range, and there was a significant negative correlation between latitude and body length at age 1 (Diana 1996). This relationship was attributed, in part, to latitudinal differences in the length of the growing season. Rapid growth continues through the second and third growing seasons but begins to slow in later years (Casselman 1996). Females grow more rapidly than males and reach a larger ultimate size. In British Columbia, males usually mature by the beginning of their third growing season, and females mature 1 or 2 years later. So far, the oldest pike aged in British Columbia (by the cleithrum) was a female in her 15th summer.

FOOD HABITS

Larval pike begin feeding at a length of about 15 mm (Morrow et al. 1997). At first, they forage on zooplankton, but as they grow their diet shifts towards larger prey—usually first to chironomid larvae, then to macrocrustaceans and, eventually, to fish, including smaller pike (Bry et al. 1995).

Although adult pike usually are characterized as voracious predators, many of the pike stomachs examined in British Columbia are empty (especially in summer). In addition, their diet is remarkably flexible (Chapman et al. 1989) and includes amphipods, aquatic insects, ducklings, and small mammals as well as fish. In British Columbia, longnose (*Catostomus catostomus*) and white (*Catostomus commersonii*) suckers are the most common fish in pike stomachs, but Arctic grayling (*Thymallus arcticus*), mountain whitefish (*Prosopium williamsoni*), small pike, small walleye (*Sander vitreus*), flathead chub (*Platygobio gracilis*), lake chub (*Couesius plumbeus*), finescale dace (*Phoxinus neogaeus*), brook sticklebacks (*Culaea inconstans*), and sculpins (*Cottus* spp.) have been recorded. Despite their propensity to take fish, insects (especially notonectids, corixids, and odonates) are often the only food items found in adult pike stomachs. In a small Alberta lake, there was a seasonal shift in diet from invertebrates early in the summer to vertebrates later in the summer (Chapman and Mackay 1990). This diet change was attributed to changes in prey abundance and distribution. Apparently, pike are active at low light levels and forage most intensively at dawn and, especially, at dusk (Casselman 1996).

Habitat Although habitat use by northern pike has not been studied in British Columbia, it has been investigated in Alberta, and there is no reason to expect major differences between B.C. and Alberta populations. Consequently, the following account is based primarily on research conducted either in Alberta or elsewhere on the northern Great Plains.

ADULTS

Northern pike occur in a variety of quiet water habitats throughout the northern and northeastern parts of the province. In lakes, adults prefer shallow (usually <4 m) weedy areas relatively close to shore (Diana et al. 1977). Adult abundance is related to cover, and apparently, 35–80% macrophyte cover is optimal (Casselman and Lewis 1996). However, the use of cover is size related: larger individuals are found along the macrophyte/open water boundary, and smaller individuals remain farther inside weed beds (Chapman and Mackay 1984). Generally, adult pike are quite sedentary and, except on spawning migrations, usually move less than 0.5 km in a day (Diana et al. 1977; Lucas 1992). If the summer temperatures in these shallow areas exceed the thermal optimum for growth (about 20 °C), pike will shift to cooler water (Casselman 1978). Interestingly, under such conditions, females tend to move into deeper water than males (Cook and Bergersen 1988).

Relative to lakes, habitat use by northern pike in rivers is not well documented. In British Columbia, pike are relatively uncommon in the mainstems of the large turbid rivers crossing the Fort Nelson Lowlands but are abundant in tributaries to these rivers. Water in most of these tributaries is either clear or stained, and since pike are visual feeders (Craig and Babaluk 1989), water transparency may influence their habitat use. In rivers, pike are associated with quiet water and cover. Typically, the cover is aquatic vegetation, but the cover often is large woody debris in the smaller streams in northeastern British Columbia. In clearwater streams with braided gravel channels, adult pike are rare but occasionally found in shallow pools downstream of riffles. In northeastern British Columbia, there appears to be an association between pike size and stream size. In the Petitot and Hay drainage systems, the largest pike (>90 cm) occurred in main river channels, and smaller pike (<40 cm) were more common in tributary streams (McPhail et al. 1998a, 1998b).

JUVENILES

In British Columbia, juvenile pike are occasionally collected in rivers, even fast-flowing, turbulent rivers. In such cases, they are associated with quiet side-channels, snyes, and embayments at the mouths of tributary streams. In the summer, such areas usually have some aquatic vegetation. In lakes, juvenile pike are associated with shallow, heavily vegetated bays.

YOUNG-OF-THE-YEAR

Young-of-the-year pike are rare in B.C. rivers. Apparently, they remain near the site (usually a shallow lake) where they were spawned for the first few weeks of life. In these areas, they stay close to shore and are associated with aquatic vegetation and shallow water. If the spawning site is a marsh or slough associated with a lake, pike fry may emigrate after about a month (Franklin and Smith 1963), or they may remain in the nursery area over the entire summer and emigrate in the fall (Derksen 1989).

Conservation Comments The northern pike populations in northeastern British Columbia appear to be healthy, and the only conservation concern with this species is keeping it within its native distribution. It is not clear why people move fish around, but they do so with surprising regularity. About a decade ago, pike were illegally introduced into the Crooked River (Peace River system above Williston Reservoir). Apparently, only a few fish were involved, and fortunately, the introduction failed. Pike were also introduced into the Pend d'Oreille River system in Idaho, where they are now established. It is probably only a matter of time before they disperse downstream into the Pend d'Oreille system in British Columbia. Although they will provide a different angling experience, their overall impact on native fishes is likely to be negative.

FAMILY OSMERIDAE — SMELTS

The osmerids are a problematic group of relatively primitive fishes, and their relationships to other early teleosts are still unclear (Fu et al. 2005; Ishiguro et al. 2003; Johnson and Patterson 1996; López et al. 2004). At present, they are included as a suborder (the Osmeroidei) within the order Salmoniformes; however, recent molecular data (Ishiguro et al. 2003; López et al. 2004) casts doubt on the closeness of the phyletic relationship between the osmeroids and the salmoniforms. The Retropinnidae (a Southern Hemisphere family) are also called smelts. In external appearance, and even smell, they are remarkably similar to some Northern Hemisphere smelts. In spite of this external similarity, their phyletic relationship with our smelts is unclear; however, recent molecular studies (Fu et al. 2005; López et al. 2004) suggest that the Southern Hemisphere smelts are the sister group to the osmeroids.

In British Columbia, there are two smelts—the longfin smelt (*Spirinchus thaleichthys*) and the eulachon (*Thaleichthys pacificus*) that spawn in fresh water. A third species, the surf smelt (*Hypomesus pretiosus*) is common in marine and estuarine waters and occasionally ascends the Fraser River as far as its confluence with the Pitt River (about 30 km upstream). Also, surf smelt have been collected, albeit rarely, about 80 km from the sea at the head end of Pitt Lake (Diewert and Henderson 1992). In spite of their presence this far inland, there is no evidence that surf smelt spawn in fresh water. In the Fraser River, there is a downstream movement of adult surf smelt in Queens Reach just below the Port Mann Bridge (Whitehouse and Levings 1989). On some tides, a saltwater wedge extends this far upstream and surf smelt may move in and out of the river with the tides. Because there is no evidence that surf smelt breed in fresh water, I have not included a detailed account for this species; however, they are included in the smelt key.

There are reports of a fourth smelt, the Arctic smelt (*Osmerus dentex*), in British Columbia. The American Fisheries Society (Nelson et al. 2004) does not recognize the Arctic smelt as a species distinct from the rainbow smelt (*Osmerus mordax*); however, there is compelling evidence—about a 7% mitochondrial sequence divergence (Taylor and Dodson 1994; Katriina Ilves, Department of Zoology, University of British Columbia, personal communication)—that the Arctic smelt is a distinct species. This anadromous species has been collected once in the marine waters of Barclay Sound on the west coast of Vancouver Island and also off Heceta Head, Oregon; however, there are no confirmed records from B.C. fresh waters. In a 1963 revision of the family Osmeridae (McAllister 1963), there is a spot on the Arctic smelt distribution map that places this species on the north coast of British Columbia (perhaps the Iskut–Stikine estuary); however, the locality is not mentioned in the text. Also, Arctic smelt are said to be present in southeastern Alaska (Mecklenburg et al. 2002; Morrow 1980), but again, no specific localities are given. Thus, it is possible that Arctic smelts occur sporadically in estuaries along the north coast of British Columbia, but so far, there is no hard evidence that this northern smelt breeds in B.C. fresh waters. For this reason, I have included the species in the smelt key but have not provided a detailed species account.

KEY TO THE SMELT SPECIES

1 (4) Lateral line short (it does not extend as far back as the origin of the dorsal fin); eye diameter about equal to snout length; upper jaw short (it does not extend back beyond the posterior margin of the eye) *(see 2)*

    2 (3) Pectoral fins long, when folded against the body, they extend back almost to the origin of the pelvic fins; lateral line almost reaches origin of dorsal fin; 15–19 anal rays   LONGFIN SMELT, *Spirinchus thaleichthys*

    3 (2) Pectoral fins short, when folded against the body, they extend less than halfway back to the origin of the pelvic fins; lateral line barely extends beyond the tips of the pectoral fins; 12–16 anal rays   SURF SMELT, *Hypomesus pretiosus*

4 (1) Lateral line complete (it extends back to the base of the caudal fin); eye diameter less than snout length; mouth large (upper jaw extends back to, or beyond, the posterior margin of the eye) *(see 5)*

    5 (6) Prominent, concentric striae on the operculum (Fig. 25); small pointed teeth on tongue and two enlarged canine teeth on vomer (teeth often lost in spawning fish); 18–23 anal rays   EULACHON, *Thaleichthys pacificus*

    6 (5) No concentric striae on the operculum; enlarged canine-like teeth on tongue and vomer; 12–16 anal rays   ARCTIC SMELT, *Osmerus dentex*

Figure 25 Diagram of a eulachon head showing the distinctive striae on the operculum

Spirinchus thaleichthys (AYRES)
LONGFIN SMELT

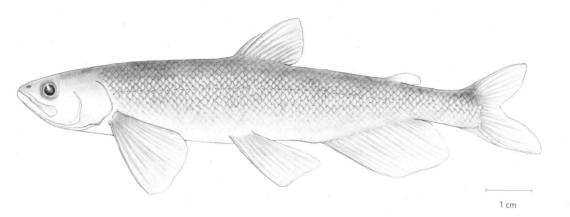

1 cm

Distinguishing Characters — This smelt has long pectoral fins that, when folded against the body, reach, or almost reach, the base of the pelvic fins; a large mouth (upper jaw extends back almost to the posterior margin of the eye); and an eye diameter that almost equals the snout length. Longfin smelts are distinguished from eulachon (*Thaleichthys pacificus*) by the absence of concentric striations on the operculum, the large number (36–44) of gill rakers on the first gill arch, and the well-developed jaw teeth on spawning fish.

Taxonomy — In British Columbia, the relationship between anadromous and lake-resident longfin smelts is unclear. In the lower Fraser system, resident longfin smelt occur in Harrison and Pitt lakes. These lake-resident smelts are small (mature adults average about 50–54 mm in length); however, in Harrison Lake, rare individuals exceed 120 mm in length. Apparently, these small adults are neotenic: they resemble anadromous juveniles in size, colour, and shape but have fully mature gonads. In addition, Harrison Lake smelt differ from anadromous longfin smelt in gill raker number and the number of teeth on the pterygoid bones. These morphological differences suggest that lake-resident smelts have diverged from their anadromous ancestors and, thus, that there may be some minor genetic differences between the lake and anadromous forms.

In Pitt Lake, the situation is more complex. A run of anadromous longfin smelt enters the lake in late summer and early fall (Diewert and Henderson 1992), but there is also a lake-resident form. Again, the lake residents appear to be neotenic. Thus, seasonally, anadromous and lake-resident smelts coexist in Pitt Lake.

A third population of lake-resident longfin smelt occurs in Lake Washington near Seattle (Dryfoos 1965; Moulton 1970, 1974). These fish are larger than the lake residents of Harrison and Pitt lakes (mature adults average about 105–130 mm FL). Although the Lake Washington population and the populations in Harrison and Pitt lakes are clearly the same species, when Schultz and Chapman (1934) described *Spirinchus dilatus* (a synonym for *S. thaleichthys*) from the Puget Sound region and Harrison Lake, they did not mention the smelts in Lake Washington. Perhaps, they were unaware of the Lake Washington smelts. Alternatively, it is possible that there were no smelts in Lake Washington at that time. A recent colonization of Lake Washington through the ship canal might explain why the Pitt and Harrison lake populations are neotenic and the Lake Washington population is not.

Sexual Dimorphism   In British Columbia, the breeding colours of both anadromous and lake-resident smelts are unrecorded. Elsewhere, however, reproductive males are much darker than females (sometimes almost black). Also, in males, the scale pockets along the midline of the body are swollen and produce a thickened midlateral ridge, there are small nuptial tubercles on the paired fins and scales, and the leading edge of the dorsal fin becomes rough. Additionally, the anal rays in mature males are longer than those in females, and thus, there is a difference in anal fin shape. In neotenic lake-resident fish, and in the anadromous form, males are larger than females at maturity.

Distribution   NATIVE DISTRIBUTION

The longfin smelt is a western North American species. The anadromous form is distributed from San Francisco Bay (the Sacramento – San Joaquin delta) northward along the coast to Prince William Sound, Alaska. Lake-resident populations are known to occur in Lake Washington, near Seattle, and in Pitt and Harrison lakes in the Fraser valley.

BRITISH COLUMBIA DISTRIBUTION

The longfin smelt is abundant in near-shore marine waters and the estuaries of large rivers along the entire coast; however, it rarely penetrates more than a few kilometres inland (Map 38). An exception is the lower Fraser River. Here, anadromous longfin smelts ascend the river at least as far upstream as Nicomen Slough and Island 22 near Chilliwack (about 130 km from the sea). The anadromous form also reaches the head end of Pitt Lake, a distance of about 80 km from the sea.

Life History   In British Columbia, the meagre life-history data available for longfin smelt come from surveys of juvenile sockeye salmon (*Oncorhynchus nerka*) in Harrison and Pitt lakes. The Lake Washington longfin smelt, however, are relatively well studied. Thus, the following account draws heavily on the Lake Washington studies (Dryfoos 1965; Moulton 1970, 1974), supplemented, where possible, with local B.C. information. It should be remembered, however, that the data on Lake Washington smelts may not be entirely applicable to the B.C. lake populations.

Spirinchus thaleichthys LONGFIN SMELT 221

Map 38 The B.C. distribution of longfin smelt, *Spirinchus thaleichthys*

REPRODUCTION

In British Columbia, the spawning season of the longfin smelt is unknown. Anadromous adults migrate into the lower Fraser River in late August and early September. Although mature, these fish are not in spawning condition. In the Nooksack River, just over the border in Washington State, the upstream migration peaks in early November, and these fish are in spawning condition. Assuming that the B.C. anadromous smelts spawn within a month or so of entry into fresh water, they probably spawn sometime between late October and early December. The lake-resident smelt in Harrison Lake are thought to spawn in late November or December (Dryfoos 1965). The situation in Lake Washington is complicated by the presence of a 2 year cycle in smelt abundance (Chigbu and Sibley 1994): the spawning peak for odd-year cohorts is in February, and the peak for even-year cohorts is in March (Chigbu 2000; Chigbu and Sibley 1994; Moulton 1970). For even-year cohorts, spawning begins when water temperature rise above 4 °C, and the peak occurs from 5.6 °C to 6.7 °C (Moulton 1970).

The spawning habitat used by longfin smelt in British Columbia is unknown; however, spawning sites in streams entering Lake Washington are usually within 2 km of the lake, and the adherent eggs are associated with gravel bars (Moulton 1970). Spawning runs enter the Cedar River after dark, and males precede females into the river (Moulton 1970). Although ripe females spawn all their eggs at one time, males may ascend and descend the spawning stream several times in a season (Moulton 1970). Egg size appears to vary among populations and cohorts; for Lake Washington smelts, average fertilized egg diameters vary from 1.07 to 1.2 mm (Dryfoos 1965; Moulton 1970). The eggs are typical smelt eggs. They have two membranes: an outer anchor membrane and an inner membrane. The anchor membrane ruptures on contact with the bottom but remains attached to the inner membrane at one point. The anchor membrane is adherent and sticks to the substrate. Once ruptured it forms a little stalk that holds the egg above the bottom. For Lake Washington females ranging in size from 108 to 126 mm, fecundity varied from 9,600 to 23,600 eggs (Dryfoos 1965). Over this limited range of body sizes, there was no relationship between size and fecundity; however, a significant reduction in fecundity was associated with reduced growth in some year classes (Chigbu and Sibley 1994). There is some fecundity data for Harrison Lake females (Dryfoos 1965; Stamford and Hume 2004): females ranged in size from 46 to 64 mm FL and fecundity varied from 337 to 2,425 eggs. Interestingly, only one ovary matures in the small-bodied smelts of Harrison Lake. Development takes 40 days at 7.0 °C and 25 days at about 10 °C (Dryfoos 1965; Moulton 1970). Size at hatching varies from 6.9 to 8.0 mm (Dryfoos 1965). Most adults die after spawning (Dryfoos 1965).

AGE, GROWTH, AND MATURITY

Although the smelt populations in Lake Washington, Harrison Lake, and Pitt Lake are dominated by young-of-the-year (0+) and adults (1+) (Dryfoos 1965; Moulton 1974; Mueller and Enzenhofer 1991; Mueller et al. 1991), 2+ females are occasionally encountered. The meagre data on anadromous populations indicate a similar age structure with, perhaps, a higher proportion of 2+ fish (Dryfoos 1965). Growth is relatively rapid but, in Lake Washington, varies from cohort to cohort, and odd-year and even-year cohorts grow at different rates (Chigbu and Sibley 1994; Moulton 1974). No comparable data on inter-cohort growth is available for Harrison and Pitt lakes or for anadromous populations. In Harrison Lake, young-of-the-year (about 20 mm FL) first appear in trawl catches in June and are about 36 mm by the end of November (Dryfoos 1965). Apparently, all populations reach maturity and spawn in their second year. Since most adults die after spawning, the occasional large individuals (>120 mm) in Harrison Lake are an enigma. Since some are as old as 8 years (Stamford and Hume 2004), presumably they have survived spawning (perhaps more than once). They may be fish that have delayed maturity, or perhaps, smelts larger than about 70 mm simply are better at eluding trawls than smaller smelts and, therefore, are under-represented in the trawl catches. Another possibility is that occasional anadromous adults still reach Harrison Lake. Distinguishing among these alternatives will require further research.

FOOD HABITS

Some diet data are available for Harrison Lake smelts (Stamford and Hume 2004), and there are seasonal and body size differences in diet. For the smallest fish, copepods are important in the spring, and cladocerans are rare; however, cladocerans are important, and copepods are less common in the fall. *Neomysis* and fish larvae are present in the diets of the larger fish. The diets of Lake Washington (Chigbu 2000; Chigbu and Sibley 1998; Dryfoos 1965) and Pitt Lake smelts (Diewert and Henderson 1992) have been examined, and in all cases, longfin smelts forage almost exclusively on zooplankton. Newly hatched larvae (about 7 mm TL) forage mainly on *Diaptomous* and *Diaphnosoma*, but as they grow, the young-of-the-year switch to *Neomysis*. In some years, *Neomysis* makes up 70% (by number) of the food taken by adults in Lake Washington (Dryfoos 1965), and there is a reciprocal relationship between the abundance of 1+ smelts and *Neomysis* that suggests that smelts regulate the abundance of *Neomysis* in Lake Washington (Chigbu and Sibley 1998). *Neomysis* also is present in Harrison and Pitt lakes, but its place in the diet of smelts in these lakes is unknown. In Pitt Lake, juveniles forage primarily on cyclopoid copepods and to a lesser extent on cladocerans (Diewert and Henderson 1992).

Habitat Again, the meagre habitat-use data for longfin smelt in British Columbia comes from surveys of juvenile sockeye salmon in Harrison and Pitt lakes. Consequently, most habitat-use data is from the Lake Washington populations (Dryfoos 1965; Moulton 1970, 1974) supplemented, where possible, with B.C information.

ADULTS

Adult longfin smelts occur in the limnetic zone of all three lakes, and within the lakes, they show diel vertical migrations and seasonal horizontal spatial movements (Chigbu et al. 1998; Dryfoos 1965; Mueller and Enzenhofer 1991). In Lake Washington at night, adults occur mostly at depths between 11 and 22 m but, during the day, are most abundant at 18–40 m (Dryfoos 1965). There is no obvious seasonal shift in the diel depth distribution of adults. In contrast, there are clear seasonal horizontal habitat shifts in Lake Washington: longfin smelts are found in both the offshore and near-shore limnetic zones in late autumn and winter but only offshore in spring and summer (Beauchamp et al. 1992). Also, in Lake Washington, the highest densities of adult smelt are found in the northern part of the lake in summer and in the southern part in the fall (Chigbu et al. 1998).

In Pitt Lake, three habitats were sampled by trawl at the southern end of the lake: the lower Pitt River, a transition zone between the river and the lake, and the limnetic zone (Diewert and Henderson 1992). Smelts reached peak abundance in the river in late summer and fall, whereas peak abundance in the transition zone was in summer. Beach seining in the lower river and in the transition zone caught no smelts.

YOUNG-OF-THE-YEAR

Because most lacustrine longfin smelt mature at 1+ in British Columbia, young-of-the-year and juveniles are the same thing. Like adults, young-of-the-year are limnetic. During the summer, they are most abundant, both day and night, at depths between 11 and 22 m. However, during the day in the winter, they occur at depths below 28 m, and there is evidence of a migration towards the surface at night (Mueller and Enzenhofer 1991; Mueller et al. 1991). In Lake Washington, the areas of maximum density of young-of-the-year shift seasonally within the lake (Chigbu et al. 1998).

Conservation Comments   Although longfin smelts are common in the estuaries of large rivers along the entire B.C. coast, the only known major run is in the Fraser River. Still, the adults are quite secretive, and there may be undiscovered runs elsewhere along the coast. Since there is no commercial, or First Nations, fishery for longfin smelts in the Fraser River, there are no data on the status of the anadromous population. There is some data on the lake-resident populations in Pitt and Harrison lakes but not enough to determine if these populations are stable. Most of these lake-resident populations appear to be neotenic (i.e., they reach sexual maturity at an unusually small size and display juvenile traits when sexually mature). Additionally, in the fall, both the anadromous and neotenic forms occur in Pitt Lake. These neotenic lake-resident forms are of considerable scientific interest and are listed by the BCCDC as critically imperiled because of extreme rarity (S1). As of this writing, they are being assessed by COSEWIC.

Thaleichthys pacificus (RICHARDSON)
EULACHON

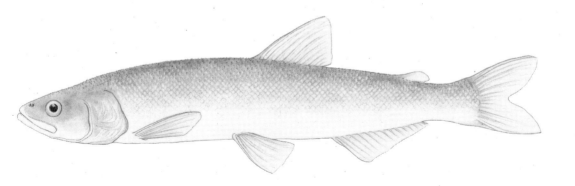

1 cm

Distinguishing Characters	This smelt, whose common name is often spelled oolichen in British Columbia, has short pectoral fins (when folded against the body, the pectoral fins reach about halfway to the base of the pelvic fins), a large mouth (reaching back to, or beyond, the hind margin of the eye), and a small eye (its diameter is about half the snout length). Eulachon differ from longfin smelt (*Spirinchus thaleichthys*) in having well-developed concentric striations on the operculum and a smaller number (17–22) of gill rakers on the first gill arch. Typically, jaw teeth are absent in breeding fish.
Sexual Dimorphism	There is no sexual dimorphism in size, but in breeding males, the nuptial tubercles are so densely packed on the head that they completely obscure the concentric striations on the operculum. In females, the tubercles are smaller and do not obscure these striae. Males also have well-developed tubercles on the dorsal surfaces of the rays in the paired fins, as well as on the dorsal, anal, and caudal fin rays, and on the scales. The scales have well-developed multiple tubercles that give the back and sides of males a velvety appearance. These tubercles are present, but poorly developed, in females. Also, there is a difference between males and females in the length of the pelvic fins: the pelvic fins typically reach the anus in males, whereas those of the female usually fall short of the anus.
Distribution	**NATIVE DISTRIBUTION** Eulachon are found only in western North American. Their marine distribution extends from southern California (Point Conception) north to Alaska (Bristol Bay and the Pribiloff Islands). Spawning occurs in the lower reaches of rivers from northern California (Mad River) to the Alaska Peninsula and southern Bering Sea (Nushagak River).

BRITISH COLUMBIA DISTRIBUTION

Eulachon regularly breed in a limited number (between 12 and 20) of suitable spawning rivers (Hay et al. 1997). These rivers rise in mountainous areas on the mainland and usually are glacial fed. Characteristically, suitable rivers have spring freshets (Hay et al. 1997). Occasionally, however, eulachon ascend rivers on islands (e.g., the Yakoun River on Graham Island), but they do not spawn regularly in small coastal rivers that are characterized by fall freshets (Hay et al. 1997). There are runs (Map 39) into most large coastal rivers (Fraser, Homathko, Klinaklini, Dean, Bella Coola, Skeena, Nass, and Stewart rivers) and a few smaller rivers in Douglas Channel and Gardner Canal (Kitimat, Kildala, Kemano, Kitlope, and Kowesas rivers). In the last 20 years, however, there has been a general decline in eulachon populations along the Pacific Coast of North America (Hay and McCarter 2000).

Life History The basics of eulachon life history were established in the middle part of the 20th century (Hart and McHugh 1944; Ricker et al. 1954; Smith and Saafield 1955). Subsequent concern over the decline in eulachon runs triggered a spate of reports on specific B.C. stocks (e.g., Langer et al. 1977). An excellent source of information on all aspects of eulachon biology is Hay and McCarter (2000).

REPRODUCTION

Eulachon spawn in the spring, but run timing varies from river to river and occurs at a wide range of river temperatures: from close to 0 °C to about 8 °C. Over the species' entire range, southern populations tend to spawn earlier than northern populations; however, in British Columbia, northern populations (e.g., in the Nass and Skeena rivers) spawn earlier (late February and early March) than southern populations (April and May in the Fraser River). Most populations spawn within a few kilometres of the sea, but most of the spawning in the Fraser system occurs between Mission and Chilliwack (about 60–120 km upstream). In this part of the river, the substrate gradually shifts from silt and sand to gravel. Most eulachon spawning rivers are turbid in the spring, and eulachon spawn at night; therefore, there are no direct observations on their spawning behaviour. The adult freshwater phase lasts about 4 weeks (or less), after which the adults die (Hay and McCarter 2000).

Fecundity varies with female size; however, the relationship between length and egg number differs among rivers. Egg numbers range from just over 3,000 in small Kitimat River females to almost 60,000 in the largest Fraser River females (Hay and McCarter 2000). Eulachon eggs are small (<1.0 mm), and like other smelts, there are two egg membranes (Parente and Snyder 1970). The outer anchor membrane is adhesive and ruptures on contact with the bottom, but it remains attached to the inner egg membrane at one point. Thus, a small stalk of anchor membrane keeps the egg above the substrate. The incubation period is temperature

Map 39 The B.C. distribution of eulachon, *Thaleichthys pacificus*

• native

dependent (DeLacy and Batts 1963): about 28–35 days at 4–5 °C and 21–25 days at 8.0 °C (Hay and McCarter 2000; Parente and Snyder 1970; Pedersen et al. 1995). The larvae are small (4–8 mm TL) and usually are flushed into estuaries immediately after hatching (Hay and McCarter 2000). Thus, the freshwater phase of eulachon life history is short: in some cases, only minutes between hatching and entry into the estuary plus the brief adult spawning period (Hay and McCarter 2000). Thus, for eulachon, the majority of their life is spent in estuaries and the sea.

AGE, GROWTH, AND MATURITY
Once in the sea, larval eulachon grow rapidly; although there are interannual differences in growth rates, they reach about 25 mm in 2 months and about 60 mm by the end of their first year (Hay and McCarter 2000). So far, age determinations in adult eulachon are unreliable (Hay and McCarter 2000; Ricker et al. 1954). Typically, otoliths are used to estimate age (DeLacy and Batts 1963; Smith and Saafield 1955); however, the ages obtained from otoliths have not been verified, and often, there are no differences in length among individuals with age differences that appear to span 3 years (Hay and McCarter 2000). Despite this uncertainty, the age at maturity of most individuals is estimated to be about 3 years (Hay and McCarter 2000). Nonetheless, there probably is some variation in the age at maturity, and a few spawners (mostly females) have been aged at 5 or 6 years (Pedersen et al. 1995).

FOOD HABITS

With the possible exception of larvae hatched 50–100 km upstream in rivers like the Columbia and Fraser, there is no exogenous feeding in fresh water. At sea, eulachon feed primarily on euphausids.

Habitat #### ADULTS

Adult eulachon are marine fishes, and before they return to spawn, the largest fish tend to occur the farthest offshore (Hay and McCarter 2000). Although most adults are caught at about 100 m, some have been reported as deep as 500 m. These deep records may involve fish captured at shallower depths as the trawl descended or when it came up (Hay and McCarter 2000). Gonadogenesis occurs during the winter prior to spawning, and the sexually maturing fish begin to migrate to spawning rivers. Apparently, they congregate for a period in estuaries before ascending to spawning sites. It is not known if eulachon home to specific spawning rivers, but genetic (McLean et al. 1999; McLean and Taylor 2001) and meristic (Hart and McHugh 1944) data, as well as the occurrence of occasional anomalous runs on Vancouver Island (Hart 1973; Hart and McHugh 1944) and elsewhere, argue against precise homing (Hay and McCarter 2000).

JUVENILES

Data on juvenile eulachon (about 30–100 mm FL) is meagre (Barraclough 1964). At this size, they avoid gear that catches larvae but they are too small for trawl nets. Based on the few observations available, it appears that juveniles disperse into inshore marine waters within their first year of life (Hay and McCarter 2000).

YOUNG-OF-THE-YEAR

Larval eulachon appear in estuaries and marine waters adjacent to spawning rivers about 6–8 weeks after the spawning period (Hay and McCarter 2000). Although they disperse towards the fjord entrances, they appear to be retained within the fjords over the next 18–20 weeks (Hay et al. 1997).

Conservation Comments Although eulachon runs have declined over several decades, their conservation status is difficult to assess for two reasons. First, there are no clear genetic differences among runs in different rivers along the B.C. coast (McLean et al. 1999; McLean and Taylor 2001), and at the population level, conservation units often are defined by genetic differences. Second, historically, eulachon runs have disappeared from some B.C. rivers for 1 or 2 years and then reappeared. This suggests rapid recolonization from adjacent rivers. Consequently, as long as the runs in some rivers are strong, it is not clear that declines in other rivers are necessarily permanent. Nevertheless, the general downward trend in spawning runs on the southern and central B.C. coasts is a cause for concern. There are a limited number of rivers (about 14) that consistently support eulachon spawning runs. These rivers probably are the source of the ephemeral populations that sporadically appear in other rivers. If we lose some of these established eulachon runs, the species may be in real trouble. Thus, the BCCDC lists the eulachon as a species of special concern. Curiously, eulachon are not listed by COSEWIC.

FAMILY SALMONIDAE — SALMONIDS

Originally native to the Northern Hemisphere, salmonids now occur in cool waters throughout the world. Although they are a predominately freshwater group, many of the 70 or so species are anadromous and spend most of their adult lives feeding in the ocean. Nonetheless, they always return to fresh water to spawn. The fossil record for salmonids stretches back at least to the early Paleocene (about 65 Ma). Anatomically, they are relatively primitive bony fish: they lack spines in their fins, their pelvic fins are positioned well back on the body (usually about halfway between the pectoral and anal fins), they have a tab-like fleshy axillary process associated with the base of pelvic fin, and they possess an adipose fin.

Curiously, for such a well-studied and charismatic group—they are large, attractive, and highly prized by anglers and gourmets—their taxonomy is still unsettled. Their closest relatives (Ishiguro et al. 2003; Johnson and Patterson 1996; López et al. 2004), number of species in the family, and the generic boundaries (Mednikov et al. 1999) within the family are all matters of constant debate. Even the limits of the family shift from time-to-time (Johnson and Patterson 1996). At present, the whitefish and graylings are included with the trout, salmon, and char in the family Salmonidae (Wilson and Williams 1992), but in the past, whitefish and graylings were placed in separate families: the Coregonidae and Thymallidae. I follow the American Fisheries Society list of common and scientific names (Nelson et al. 2004) and recognize three subfamilies within the Salmonidae—Salmoninae (trout, char, and Pacific salmon), Coregoninae (whitefish, ciscoes, and inconnu), and Thymallinae (graylings)—but the matter of the familial boundaries in these fishes is by no means settled, and two relatively recent reviews (Johnson and Patterson 1996; Sanford 1990) recommend that the whitefish be retained as a separate family, the Coregonidae.

KEY TO THE SUBFAMILIES OF SALMONIDAE

1(2) Scales small, 100 or more along lateral line; well-developed teeth in jaws (Fig. 26A)	SALMONINAE (trout, char, and salmon)
2(1) Scales large, less than 100 along lateral line; teeth in jaws small or absent (Fig. 26B) *(see 3)*	
3(4) Dorsal fin base less than head length; teeth in jaws absent	COREGONINAE (whitefishes)
4(3) Dorsal fin base equal to, or greater than, head length; teeth in jaws small	THYMALLINAE (graylings)

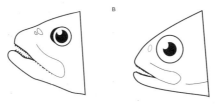

Figure 26 The heads of a trout (A) and a grayling (B) showing differences in the development of jaw teeth

Subfamily Salmoninae

The subfamily Salmoninae usually is thought to contain six genera: *Brachymystax*, *Hucho*, *Oncorhynchus*, *Salmo*, *Salmothymus*, and *Salvelinus*. However, the generic boundaries in this subfamily are still unclear. Until quite recently, our native Pacific trouts (the rainbow, *Oncorhynchus mykiss*, and cutthroat, *Oncorhynchus clarkii*, lineages) were included in the genus *Salmo*, but the weight of evidence now aligns the Pacific trouts with Pacific salmon in the genus *Oncorhynchus* (Stearley and Smith 1993). Similarly, there is some evidence that the genus *Hucho*, as now understood, is paraphyletic and should be divided into two genera: *Hucho* and *Parahucho* (Oakley and Phillips 1999). Most of the unresolved problems of generic boundaries in salmonines relate to Balkan and northeastern Eurasian groups. In contrast, the generic boundaries of the salmonines native to North America are relatively firm.

There are three native genera in North America: the Pacific trout and salmon (*Oncorhynchus*), the Atlantic salmon (*Salmo*), and the char (*Salvelinus*). They are all cool-water species that breed in fresh water; however, some of the species, and many populations, are anadromous. For most of the native species covered in this book, information on life histories and habitat use is meagre, but for the Pacific trout and salmon the opposite is true—there are mountains of information. Salmonines are the most intensively studied and managed fishes in the province. Whole books are devoted to single species, and hundreds of papers are published every year. Consequently, it is almost impossible to make short summaries of their biology and life histories from primary sources. Hence, for the trout, salmon, and char, I draw heavily on books such as *Trout and Salmon of North America* (Behnke 2002); *Trout* (Stolz and Schnell 1991); *Pacific Salmon Life Histories* (Groot and Margolis 1991), and *The Behavior and Ecology of Pacific Salmon and Trout* (Quinn 2005), as well as the salmonine status reviews produced by the U.S. National Marine Fisheries Service.

Figure 27 Diagram illustrating the presence (A, trout) or absence (B, char) of teeth on the shaft of the vomer

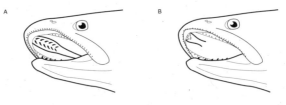

Figure 28 Diagram illustrating the presence (A, cutthroat trout) or absence (B, in rainbow trout) of basibranchial teeth

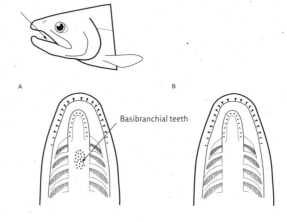

Figure 29 Head profiles of a Dolly Varden (A) and bull trout (B) showing the differences in snout shape and jaw length

 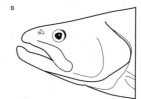

Subfamily Salmoninae

KEY TO ADULT SALMONINES

1 (10) Anal fin base longer than dorsal fin base; 13–19 major rays in the anal fin *(see 2)*

    2 (5) No distinct spots on back or tail, although sometimes there is fine speckling on the back or trailing edge of the caudal fin *(see 3)*

        3 (4) Gill rakers long, 29–44 on first gill arch; adults occur in fresh water both as silver bright residents (kokanee) and anadromous migrants (sockeye) .... SOCKEYE SALMON, *Oncorhynchus nerka*

        4 (3) Gill rakers short (relative to sockeye), 16–26 on first gill arch; adults only occur in fresh water as spawners .... CHUM SALMON, *Oncorhynchus keta*

    5 (2) Well-developed black spots on back and tail *(see 6)*

        6 (7) Large, oblong spots on back and tail, the longest spots as long as the vertical eye diameter; adults only occur in fresh water as spawners .... PINK SALMON, *Oncorhynchus gorbuscha*

        7 (6) Dark spots on back and tail small and irregular, the largest spots much smaller than the vertical eye diameter *(see 8)*

            8 (9) Spots present on upper lobe of caudal fin; gums at base of teeth in lower jaw white .... COHO SALMON, *Oncorhynchus kisutch*

            9 (8) Spots present on upper and lower lobes of caudal fin; gums at base of teeth in lower jaw black .... CHINOOK SALMON, *Oncorhynchus tshawytscha*

10 (1) Anal fin base shorter than dorsal fin base; 8–12 major rays in the anal fin *(see 11)*

    11 (20) In fresh water, spots on back and flanks are dark against a light green, blue, or brownish background, (spots may be obscure in fresh sea-run fish); some teeth on shaft of vomer (Fig. 27A) *(see 12)*

        12 (13) Some black spots on flanks surrounded by pale haloes; usually some red spots along the lateral line .... BROWN TROUT, *Salmo trutta*

        13 (12) No haloes around black spots on flanks; no red spots along the lateral line *(see 14)*

            14 (17) A distinctive red slash under each side of the lower jaw; usually some basibranchial teeth behind the tongue (Fig. 28A) *(see 15)*

                15 (16) The irregular black spots below the lateral line are evenly distributed on the front and back halves of the body .... COASTAL CUTTHROAT TROUT, *Oncorhynchus clarkii clarkii*

                16 (15) The irregular black spots below the lateral line are concentrated on the back half of the body and are almost absent below the lateral line on the front half of the body .... WESTSLOPE CUTTHROAT TROUT, *Oncorhynchus clarkii lewisi*

            17 (14) No red slashes under the lower jaw; no basibranchial teeth behind the tongue (Fig. 28B) *(see 18)*

                18 (19) Usually no spotting on the tail, if there are spots on the tail they are never arranged in radiating rows; weak teeth on shaft of vomer .... ATLANTIC SALMON, *Salmo salar*

                19 (18) Tail distinctly marked with radiating rows of black spots; strong teeth on shaft of vomer .... RAINBOW TROUT, *Oncorhynchus mykiss*

    20 (11) In fresh water, spots on back and flanks are light (cream, orange, or red) against a dark background; no teeth on shaft of vomer (Fig. 27B) *(see 21)*

        21 (22) Dorsal fin heavily marbled with dark wavy lines; spots on lower flank with blue haloes .... BROOK TROUT, *Salvelinus fontinalis*

        22 (21) Dorsal fin dusky or lightly spotted, not heavily marbled with dark wavy lines; spots on lower flank without blue haloes *(see 23)*

            23 (24) Irregularly shaped grey or whitish spots on the back, sides, dorsal fin, and caudal fin; the tail is deeply forked (the centre rays on the tail are about half the length of the longest outer rays) .... LAKE TROUT, *Salvelinus namaycush*

            24 (23) Regularly shaped light (usually coloured) spots on the back and sides, no spots on dorsal or caudal fin; the tail is almost square (the centre rays on the tail are almost the length of the longest outer rays) *(see 25)*

                25 (26) Viewed from the side, the snout is blunt (Fig. 29A); upper jaw is short (reaches to, or just beyond, hind margin of eye); 19–24 branchiostegal rays; usually 9 or 10 anal rays .... DOLLY VARDEN, *Salvelinus malma*

                26 (25) Viewed from the side, the snout is more pointed (Fig. 29B); upper jaw is long (reaches well beyond hind margin of eye); 25–29 branchiostegal rays; usually 11 or 12 anal rays .... BULL TROUT, *Salvelinus confluentus*

This key is based on colour patterns, especially the size and shape of the lateral parr marks, the stippling pattern on the adipose fin, and markings on the dorsal fin. Keep in mind that these traits change with size, age, and habitat. For example, in rainbow trout, black spots below the main row of parr marks usually do not appear until about 60–70 mm. Similarly, the light or coloured spots on the sides of brook and bull trout usually appear at about 50 mm. There are colour illustrations and photographs of most of the species in *Field Identification of Coastal Juvenile Salmonines* (Pollard et al. 1997).

KEY TO SALMONINE FRY (<50 mm TL)

1(10) Base of anal fin longer than the base of dorsal fin; horizontal eye diameter usually greater than snout length *(see 2)*

    2(3) No lateral parr marks — PINK SALMON, *Oncorhynchus gorbuscha*

    3(2) Lateral parr marks *(see 4)*

        4(7) Lateral parr marks noticeably deeper than vertical eye diameter *(see 5)*

            5(6) Anal fin sickle-shaped with a white leading edge backed with a black stripe; adipose fin with a dark edge and uniformly pigmented centre — COHO SALMON, *Oncorhynchus kisutch*

            6(5) Anal fin not sickle-shaped (the leading rays are not longer than the fin base); white leading edge of anal fin is not backed by a black stripe; adipose fin with a dark edge and clear unpigmented centre — CHINOOK SALMON, *Oncorhynchus tshawytscha*

        7(4) Lateral parr marks equal to, or smaller than, vertical eye diameter *(see 8)*

            8(9) Lateral parr marks variable in length, some roughly bisected by lateral line; lower flanks silver or white — SOCKEYE SALMON OR KOKANEE, *Oncorhynchus nerka*

            9(8) Lateral parr marks of uniform length, most of their length above the lateral line; in life, lower flanks with an iridescent green sheen — CHUM SALMON, *Oncorhynchus keta*

10(1) Base of dorsal fin longer than the base of anal fin; horizontal eye diameter equal to, or less than, snout length *(see 11)*

    11(14) Dorsal fin with dark spots or, in small (<30 mm TL) fish, only the first dorsal ray is black *(see 12)*

        12(13) Usually five or more dorsal parr marks on the back in front of the dorsal fin; black border of adipose fin with one or no breaks — RAINBOW TROUT, *Oncorhynchus mykiss*

        13(12) Usually less than five dorsal parr marks on the back in front of the dorsal fin (often none); black border of adipose fin with one or more breaks — CUTTHROAT TROUT, *Oncorhynchus clarkii*

    14(11) Dorsal fin without dark spots (sometimes a few irregular dark marks on brook trout dorsal fins); in small (<30 mm TL) fish, the first dorsal ray may be dusky but not black *(see 15)*

        15(16) Pectoral fin long, when folded against the body it extends back past the origin of the dorsal fin; adipose fin clear — ATLANTIC SALMON, *Salmo salar*

        16(15) Pectoral fin short, when folded against the body it does not extends back as far as the origin of the dorsal fin; adipose fin coloured or with a heavily stippled trailing edge *(see 17)*

            17(20) Lateral parr marks fairly regular, none wider than the eye diameter *(see 18)*

                18(19) Adipose fin orange; predorsal distance less than one-half standard length — BROWN TROUT, *Salmo trutta*

                19(18) Adipose fin not coloured; predorsal distance about equal to one-half standard length — LAKE TROUT, *Salvelinus namaycush*

            20(17) Lateral parr marks irregular, most are wider than the eye diameter *(see 21)*

                21(22) Anterior lateral parr marks usually deeper than they are wide; base of anal fin with a band of dark pigment, leading edge of anal fin white followed by some dark pigment — BROOK TROUT, *Salvelinus fontinalis*

                22(21) Anterior lateral parr marks usually as wide, or wider, than they are deep; base of anal fin without a band of dark pigment, anal fin usually immaculate although it may have a small white tip — DOLLY VARDEN OR BULL TROUT, *Salvelinus malma* or *S. confluentus*[†]

[†] In sympatry, bull trout and Dolly Varden fry can be distinguished by pigment pattern, but the differences are subtle and probably vary from site to site. Consequently, unless a series of both species is at hand, any specific identification will be suspect.

Subfamily Salmoninae 237

SALMONINE FRY

PINK

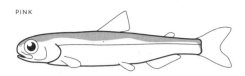

COHO

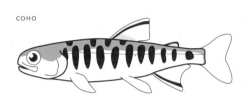

CHINOOK

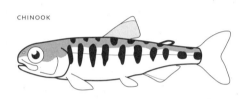

SOCKEYE

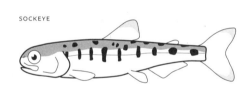

CHUM

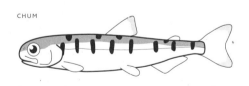

RAINBOW

CUTTHROAT

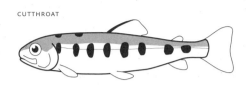

ATLANTIC

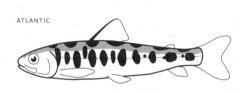

BROWN

LAKE

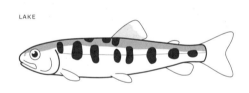

BROOK

DOLLY VARDEN/BULL

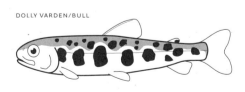

The following species are not included in this key: pink salmon, chum salmon, sockeye salmon, and lake trout. Pink and chum salmon usually migrate to the sea as fry, and sockeye salmon fry typically migrate to a lake. Consequently, stream-rearing coho and Chinook salmon usually are the only Pacific salmon parr found in fresh water; however sockeye smolts are encountered on their downstream migration to the sea. In British Columbia, juvenile lake trout (>60 mm) usually lack parr marks.

KEY TO SALMONINE PARR (≤130 mm FL)

1 (4) Anal fin base longer than dorsal fin base; 13–19 major rays in the anal fin *(see 2)*

    2 (3) Anal fin sickle-shaped with a white leading edge backed with a black stripe; adipose fin with a dark edge and uniformly pigmented centre; smolts average 7.5–12.5 cm; 13–14 branchiostegal rays (count is for one side) — COHO SALMON, *Oncorhynchus kisutch*

    3 (2) Anal fin not sickle-shaped (the leading rays are not longer than the fin base); white leading edge of anal fin is not backed by a black stripe; adipose fin with a dark edge and clear unpigmented centre; ocean-type smolts average 5.0–7.5 cm, stream-type smolts are usually over 10 cm; 16–18 branchiostegal rays (count is for one side) — CHINOOK SALMON, *Oncorhynchus tshawytscha*

4 (1) Anal fin base shorter than dorsal fin base; 8–12 major rays in the anal fin *(see 5)*

    5 (14) Dorsal fin with distinct dark spots or oblique dark marks *(see 6)*

        6 (11) A few coloured spots on lateral line; combined width of dark areas along lateral line equal to, or greater than, the width of light areas *(see 7)*

            7 (8) Pectoral fins long, when folded against the body they extend past the origin of the dorsal fin; tail forked (the centre rays about half the length of the longest rays) — ATLANTIC SALMON, *Salmo salar*

            8 (7) Pectoral fins short, when folded against the body they do not reach the origin of the dorsal fin; tail not deeply forked (the centre rays more than half the length of the longest rays) *(see 9)*

                9 (10) Adipose fin without colour except for some dark pigment at posterior base of fin; no dark spots other than extensions of lateral parr marks below lateral line — BROOK TROUT, *Salvelinus fontinalis*

                10 (9) Adipose fin with some colour (red or orange); lateral parr marks but also some small dark spots above and below lateral line — BROWN TROUT, *Salmo trutta*

        11 (6) No coloured spots along the lateral line; combined width of dark areas along lateral line less than the width of light areas *(see 12)*

            12 (13) Usually five or more dorsal parr marks in front of the dorsal fin; upper jaw does not extend back beyond the hind margin of the eye; no red slashes under the lower jaw; white tip on the dorsal fin covers the space between 3 and 5 dorsal rays — RAINBOW TROUT, *Oncorhynchus mykiss*

            13 (12) Usually less than five dorsal parr marks in front of the dorsal fin (often none); upper jaw extends back beyond the hind margin of the eye; usually red slashes under the lower jaw; white tip on the dorsal fin covers the space between 1 and 3 dorsal rays — CUTTHROAT TROUT, *Oncorhynchus clarkii*

    14 (5) Dorsal fin without distinct dark spots or oblique dark marks *(see 15)*

        15 (16) Upper jaw extends back to beyond the hind margin of the eye; usually more than 23 branchiostegal rays (total count); usually 11 or 12 anal rays — BULL TROUT, *Salvelinus confluentus*

        16 (15) Upper jaw does not extend beyond the hind margin of the eye; usually less than 23 branchiostegal rays (total count); usually 9 or 10 anal rays — DOLLY VARDEN, *Salvelinus malma*

Subfamily Salmoninae 239

SALMONINE PARR

COHO

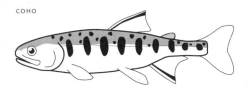

CHINOOK

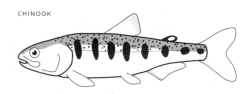

ATLANTIC

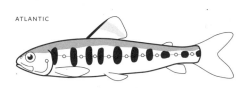

BROOK

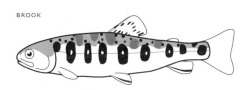

BROWN

RAINBOW

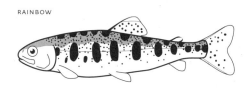

CUTTHROAT

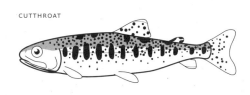

BULL

DOLLY VARDEN

Oncorhynchus clarkii (RICHARDSON)
CUTTHROAT TROUT

Distinguishing Characters — This trout has black spots on the back, flanks, and dorsal and caudal fins. It is distinguished from the trout-like chars by the black (rather than pale) spots on the flanks and the presence of teeth on the shaft of the vomer. It differs from Pacific salmon in the low number of anal rays (usually <12). Cutthroat trout are most likely to be confused with rainbow trout (*Oncorhynchus mykiss*), and the two species (especially the young) are difficult to separate.

Typically, cutthroat trout have a distinctive red slash under each side of the lower jaw; hence, the common name—cutthroat trout. On adults, these markings are a useful, but not infallible, method of distinguishing cutthroat and rainbow trout. Unfortunately, these markings usually are not obvious in fish fresh from the sea. Also, at the northern limits of their natural distribution, rainbow trout sometimes have orange slashes under their throats (McPhail and Lindsey 1970). Generally, cutthroats have longer jaws than rainbows. Thus, the upper jaw typically extends well beyond the back of the eye in cutthroat and usually ends short of the posterior margin of the eye in rainbow trout. However, like so many morphological traits in trout, jaw length is variable and in some piscivorous rainbow populations it extends back beyond the hind margin of the eye.

The most reliable morphological character for separating cutthroat and rainbow is the presence of basibranchial teeth in cutthroat trout (Leary et al. 1996; Weigel et al. 2002). Most cutthroat over 200 mm FL have basibranchial teeth, whereas rainbow trout usually lack these teeth. The fry of the two species are very difficult to distinguish.

Taxonomy — The cutthroat trout is a polytypic species—it occurs in many different morphological and life-history forms. One consequence of this variability is some uncertainty about the taxonomic relationships within the cutthroat trout lineage. In spite of this uncertainty, it is clear that the members of the cutthroat trout group are more closely related to one another than to any other trout (Behnke 1992, 2002; Shedlock et al. 1992).

Most recent reviews of the cutthroat group recognize 14 subspecies: four major (relatively old) subspecies and 10 minor (more recent) subspecies (Behnke 1992, 2002). Most of the minor subspecies are found in the Basin and Range region of the western United States and usually are allopatric (i.e., their ranges do not overlap). Although we have only two native subspecies in British Columbia, they are both major subspecies that probably diverged about a million years ago. These subspecies are the coastal cutthroat (*O. clarkii clarkii*) and the westslope cutthroat (*O. clarkii lewisi*). They differ in their morphology, genetics, chromosome number, biology, and geographic distributions. Because of the major differences between the two subspecies, I treat them separately. Originally, the natural distributions of the two subspecies did not overlap in British Columbia; however, fish culturists have introduced both subspecies into lakes and streams within the geographic range of the other subspecies.

Oncorhynchus clarkii clarkii (RICHARDSON)
COASTAL CUTTHROAT TROUT

1 cm

Distinguishing Characters In British Columbia, the main external difference between coastal and westslope cutthroat trout is the pattern of black spots on the body (Quadri 1959)—on coastal cutthroat the irregular black spots below the lateral line are evenly distributed on the front and back halves of the body. In contrast, in the westslope cutthroat, the spots are not evenly distributed below the lateral line. They are concentrated on the back half of the body and, often, are almost absent below the lateral line on the front half of the body. Although there is some overlap, the two subspecies also differ in scale counts and body proportions (Quadri 1959).

Taxonomy Although all coastal cutthroat trout belong to the same subspecies, they are not morphologically, behaviourally, or genetically uniform. There are two basic life-history types: sea-run and freshwater-resident populations. Within each of these broad categories, there are sub-types based on life-history differences, migratory patterns, and habitat use (Sumner 1953; Northcote 1997). Genetic differences among sub-types, even within the same drainage system, are common (Griswold et al. 1997; Williams et al. 1997; Zimmerman et al. 1997) and suggest restricted gene flow. Like other salmonines (and many other northwestern fishes), these divergences are mostly a response to local selection regimes and evolve independently in different places. Consequently, although sub-types may share some morphological and life-history characteristics, they are not recognized as taxonomic units.

Sexual Dimorphism   Outside the breeding season, the sexes are difficult to separate. There are, however, subtle differences between the sexes in head shape, and some experienced anglers are better at sexing non-breeding adults than would be expected by chance. When spawning, the sexes differ in colour and head shape. The males of sea-run and lacustrine populations often develop bright yellow flanks and bellies. As adults, stream-resident males often retain their parr marks and sometimes develop a reddish flush on their lower flanks. In all the life-history forms, spawning females are less colourful than males.

Hybridization   Cutthroat and rainbow trout (*Oncorhynchus mykiss*) diverged from a common ancestor around two million years ago and the coastal cutthroat came into being about one million years ago (Behnke 2002). Thus, the coastal cutthroat and the rainbow trout have coexisted along the Pacific coast of North America since about the middle Pleistocene. Nonetheless, the populations in our area are relatively recent (postglacial) immigrants, and hybrids are common. A U.S. status report on this subspecies (Johnson et al. 1999) found evidence of hybridization in about 30% of the 97 populations examined. Usually, only about 1–10% of a sample consisted of hybrids, but occasionally, 50% or more of the individuals at a site were hybrids (F_1, F_2, or backcrosses). Similarly, on Vancouver Island, 29 of 30 streams showed evidence of hybridization (Bettles 2004). In these streams, the rate of hybridization ranged from 3% to 88% and averaged about 20%. In British Columbia, hybridization between rainbow and coastal cutthroat trout is often, but not always, associated with human activities (impoundments and other environmental modifications, fish-culture practices, and selective harvesting). Nonetheless, some hybridization (e.g., the apparently introgressed trout in the upper Dean River) probably is natural.

Distribution   NATIVE DISTRIBUTION
The coastal cutthroat ranges from northern California (Eel River) to Prince William Sound, Alaska. It occurs in both large and small drainages and, on the southern coast, rarely penetrates more than about 200 km inland. The same, or a similar, subspecies may occur on Kamchatka (Behnke 1997).

BRITISH COLUMBIA DISTRIBUTION
Coastal cutthroat are found along the entire B.C. coast (including most coastal islands). In the Fraser system, coastal cutthroat occur upstream to at least Hope (about 160 km inland), and there are unverified reports of cutthroat as far upstream as the Stein River (about 240 km inland) as well as an old, unverified record from the Thompson River near Ashcroft (about 300 km inland). On the central and northern coasts, they sometimes occur well inland. For example, in the Skeena River system, they reach Morrison Lake (more than 600 km from the sea), and in the

Stikine River, they occur as far upstream as Telegraph Creek (about 200 km inland; Map 40).

The distribution of cutthroat trout on the Queen Charlotte Islands is peculiar. They are almost entirely restricted to the lowlands on Graham Island and the northeastern corner of Morseby Island (Northcote et al. 1989).

In British Columbia, introductions have expanded the distribution of coastal cutthroat outside its natural range. Coastal cutthroat have been introduced into the Similkameen, Okanagan, and Columbia river systems, but except for small populations in Kootenay and lower Arrow lakes, most of these introductions have failed.

Life History Based on patterns of migration, three general life-history forms of coastal cutthroat trout are commonly recognized (Johnson et al. 1999): a non-migratory freshwater-resident form, a migratory (often adfluvial) freshwater-resident form, and a sea-run (anadromous) form. This division into three broad life-history types is convenient but does not reflect reality. Like many anadromous salmonines, the life histories of the coastal cutthroat are surpassingly complex—the ontogeny and purpose of their migrations, the duration of marine residence, and ontogenetic niche shifts all vary within and among populations (Jones and Seifert 1997; Northcote 1997).

Reproduction In British Columbia, early sea-run populations enter small coastal streams in August or September, whereas late sea-run fish enter streams in late winter (February) or early spring (March and April). Spawning can start as early as late October (Loon Lake, U.B.C. Forest) but usually peaks in February and extends into the spring. On the central B.C. coast (Bella Coola River), the peak of spawning activity is in May (Hume 1998). Lacustrine and stream-resident populations show similar inter-population variation in stream entry and spawning times. Spawning can occur at temperatures as low as 2–4 °C (Costello and Rubidge 2003) and as high as 17 °C (Behnke and Zarn 1976). Coastal cutthroat appear to prefer small (often <1 m wide) spawning streams with relatively low gradients (Hartman and Gill 1968; Johnston 1982). In the gravel deposition reaches of the lower Fraser River, some fish spawn in side-channels.

Spawning coastal cutthroat trout are secretive and adept at using cover. In small streams, there is usually no obvious spawning aggregation. Consequently, the spawning behaviour of coastal cutthroat trout is not well known. Presumably, however, their spawning behaviour is similar to that of other salmonines (for details of nest-digging and spawning behaviours, see the rainbow trout species account). There is little information on the physical characteristics of coastal cutthroat spawning sites; however, females are known to dig nests in gravel (5–50 mm deep) at depths of 15–45 cm in the tail-outs of pools (Johnston 1982). This suggests that the physical characteristics of coastal cutthroat spawning sites are similar to those of the westslope cutthroat. In that subspecies, redds

Map 40 The B.C. distribution of coastal cutthroat trout, *Oncorhynchus clarkii clarkii*.

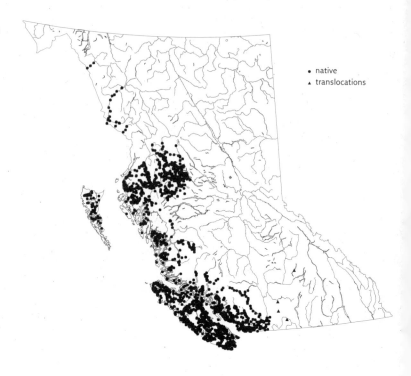

(a collective term for a sequential series of nests dug by a single female) are usually found in water 20–50 cm deep with a mean velocity of 0.3–0.4 m/s (Liknes and Graham 1988; Shepard et al. 1984). When fish in spawning condition are encountered, there is often only a single pair, or a dominant and sub-dominant male and one female. There are many sites where large sea-run, or lake-resident, trout appear to spawn in streams that also contain small, mature stream-resident trout. In such situations, the smaller resident males may act as "sneakers" and manage to fertilize some eggs.

Although coastal cutthroat have been reared in hatcheries for over 50 years, there is little published information on their fecundity. There is a positive relationship between female size and egg number in cutthroat trout (Trotter 1997); however, this relationship probably differs among populations and, perhaps, among life-history forms. In British Columbia, females in stream-resident populations isolated above barriers rarely exceed 15 cm in length and typically produce about 50–150 eggs. In contrast, females in most lake-resident and sea-run populations range in size from 25 to 50 cm and contain about 500–1,500 eggs. Fertilized eggs range in diameter from 5.0 to 6.0 mm. Fifty percent hatching occurs about 20 days after fertilization at 14 °C, 27 days at 11 °C, 41 days at 8 °C, 65–70 days at 5 °C, and 123 days at 2.0 °C (Murray 1980b). The newly hatched alevins range in length from 12 to 15 mm. They emerge from the gravel about 35–190 days after fertilization (at incubation temperatures of 14 °C and 2.0 °C for the lower and upper ends of the range, respectively).

Depending on incubation temperature and egg size, the newly emerged fry range from about 19 to 22 mm FL. Maximum survival from fertilization to emergence (72–80%) was at 8 °C (Murray 1980b). In nature, coastal cutthroat fry emerge between March and June, with peak emergence in mid-April (Giger 1972).

Cutthroat trout are iteroparous and, in the absence of a fishery, about 40% of first-time spawners live to spawn again (Trotter 1997). Some individuals live to spawn a fourth (Sumner 1953) or even fifth (Moyle 2002) time.

AGE, GROWTH, AND MATURITY

In headwaters, isolated populations of stream-resident coastal cutthroat rarely reach lengths greater than 20 cm. Initial growth is moderate (they reach about 40–60 mm in their first summer and fall) and almost double their length (85–107 mm) in their second year, but growth slows in their third year (Northcote and Hartman 1988). Some headwater males reach sexual maturity in their third summer, and most are mature by their fourth summer. Typically, females mature 1 or 2 years later than males. Stream-resident fish rarely live more than 5 years.

In lacustrine or large-river populations, growth rates accelerate after the young move from their spawning stream into a lake or large river. Depending on food availability and competitive interactions with other salmonines, adfluvial individuals can reach a large size. In big lakes on Vancouver Island (e.g., Cowichan, Sproat, and Campbell lakes), resident cutthroat occasionally reach weights of 3 kg and lengths of over 60 cm. Males in most adfluvial populations mature after three or four summers, and most females are sexually mature by age 5.

Sea-run populations spend from one to four summers in fresh water before migrating to the sea. In southern populations, most sea-run females mature at age 4, and in northern populations, at ages 5 and 6. A study off the coasts of Washington and Oregon found that 32% of their cutthroat sample (110 fish) had entered the sea after their first winter, 45% after their second winter, 19% after their third winter, and 3% after four winters (Pearcy et al. 1990). Regardless of their age at entry into the sea, however, all the fish averaged 23–30 cm. This is not surprising, since both mature and immature sea-run cutthroat return each fall to overwinter in fresh water.

Although overwintering in freshwater is the most common life-history pattern in sea-run cutthroat, some populations may be, or have been, truly anadromous. In the Bella Coola River, large (2.2–3.0 kg) cutthroat were common as late as the mid-1960s (Hume 1998). It is unlikely that they achieved such size without spending at least a full year in salt water.

FOOD HABITS

Although all coastal cutthroat trout are carnivorous, their primary prey differs among the life-history forms, and ontogenetically within life-history forms. The non-migratory stream-resident form is a drift feeder and forages mostly on insects (drifting larvae and nymphs of aquatic insects) but also takes adult insects from the surface. The adfluvial freshwater form reaches a larger size than the non-migratory form and uses a wider variety of prey. In lakes, depending on the presence of other fish species, they forage on zooplankton, surface insects, and fish (stickleback, sculpins, kokanee, *Oncorhynchus nerka*; longfin smelts, *Spirinchus thaleichthys*; and peamouth, *Mylocheilus caurinus*). In Lake Washington, near Seattle, cutthroat become increasingly piscivorous as they grow, and over 95% of the diet of individuals >400 mm is fish (Nowack et al. 2004). Even in relatively small lakes (<20 ha), their diet often includes stickleback and sculpins. In large lakes that support sockeye salmon (*Oncorhynchus nerka*), cutthroat are effective open-water predators on juvenile salmon. In Pitt and Harrison lakes, they also prey on limnetic longfin smelts. In estuaries and inshore marine areas, small sea-run cutthroat forage on amphipods, isopods, and small fishes (Trotter 1989). In early summer, cutthroats are sometimes captured in purse seines over 45 km off the coasts of Washington and Oregon. These offshore trout fed primarily on other fish (hexigrammids, scorpaenids, and anchovies), but in late summer, their diet broadens to include euphasiids, amphipods, and decapod larvae (Pearcy et al. 1990).

Habitat Although the different life-history types and complex migrations of coastal cutthroat trout make it difficult to neatly summarize their use of multiple habitats, recent experimental manipulations provide some basis for generalizations on habitat use.

ADULTS

Adult coastal cutthroat occupy a wide variety of cool-water (typically <18 °C) habitats from small, headwater streams (<2 m wide) to lowland sloughs and backwaters associated with large rivers, such as the Fraser. On Vancouver Island and the Sechelt Peninsula, they are common in large oligitrophic lakes and in small deeply stained lakes and ponds. In streams and rivers, they are associated with pools or low-gradient areas often near overhead cover (Trotter 1989). Habitat use in lacustrine cutthroat is influenced by the presence of other salmonines. When by themselves, they use all depth zones in small-to-moderate-sized lakes; however, when sympatric with rainbow trout, they tend to remain in littoral regions, usually near cover, whereas rainbow trout occupy the limnetic regions (Andrusak and Northcote 1971). In large lakes (e.g., Lake Washington near Seattle), they increase their use of the limnetic zone as they grow, and by about 250 mm, they become major predators on limnetic fish, such as longfin smelts and juvenile sockeye salmon (Nowack et al. 2004). Locally, Pitt and Harrison lakes contain the same suite of limnetic prey (juvenile sockeye salmon and longfin smelts), and cutthroat trout are common. Presumably, a similar shift into the limnetic zone occurs in these lakes. In lowland lakes on Vancouver Island and elsewhere along the B.C. coast, cutthroat are more common than freshwater-resident rainbow trout and usually grow to a larger size (Nilsson and Northcote 1981). Small, heavily stained lakes with shoreline *Sphagnum* mats are common in these coastal lowland areas, and usually, the only adult salmonines they contain are cutthroat trout and Dolly Varden (*Salvelinus malma*). These two species commonly coexist. The presence of Dolly Varden does not appear to affect habitat use by cutthroat trout; however, Dolly Varden shift into deeper water in the presence of cutthroat (Hindar et al. 1988).

JUVENILES

Analyses of habitat associations of juvenile sea-run cutthroat from 119 sites along the B.C. coast (Rosenfeld et al. 2000) indicate that bankfull channel width was the best predictor of the presence of juvenile cutthroat. They are clearly associated with small streams (<5 m channel widths). In such streams, the density of yearling, and older, cutthroat was highest in pools associated with large woody debris. Although stream size is the major determinant of cutthroat abundance, density was also highest in low- to intermediate-gradient reaches (0–5%) and gravel (rather than cobble or boulder) substrates (Rosenfeld et al. 2000). Similar associations between pools, large woody debris, and juvenile biomass have been described for stream-resident cutthroat trout (Connolly 1997).

Where they coexist with juvenile coho salmon (*Oncorhynchus kisutch*), juvenile cutthroat trout typically occur in faster water (riffles and glides) than coho (pools); however, cutthroat use all available habitat types (including pools) when coho are absent (Glova 1987). This habitat segregation in sympatry has been attributed to the social dominance of coho over cutthroat. Body size, however, is a critical factor in determining the outcome of cutthroat–coho interactions; in interactions between size-matched cutthroat and coho, cutthroat often outperform coho (Sabo and Pauley 1997).

In Oregon coastal drainages, there is a spring migration out of small, nursery streams into the mainstems of larger rivers (Giger 1972) and a movement back into nursery streams during fall freshets. A similar pattern of spring and fall movements occurs on Vancouver Island (Hartman and Brown 1987).

YOUNG-OF-THE-YEAR

Typically, newly emerged cutthroat fry inhabit either low-velocity glide areas along stream margins or shallow riffles (Glova 1984). Like the fry of many other freshwater fishes, however, it is not clear whether they prefer such habitats or whether they are forced into such areas by biotic interactions (e.g., competition and predation) with larger or more aggressive fishes. In the case of cutthroat trout fry, stream enclosure experiments support the latter explanation. Given an environment free of larger fish, cutthroat fry preferred pools, a habitat they rarely use in nature (Rosenfeld and Boss 2001). Adfluvial coastal cutthroat typically spend 1 or 2 years in stream environments before immigrating to lakes.

Conservation Comments The conservation status of most of the over 600 anadromous cutthroat trout stocks in British Columbia is unknown; however, on southern Vancouver Island and in the lower Fraser valley, at least 15 stocks have been extirpated, and another 14 populations are at high risk (Slaney et al. 1996). The main threat in these areas is habitat degradation through urbanization. Another conservation problem is the "enhancement" of coho production (often by well-meaning community stewardship groups) in cutthroat streams. The larger coho fry displace cutthroat into marginal habitats, and typically, cutthroat trout abundance declines in streams where coho have been "enhanced." Although coastal cutthroat trout are not listed by either COSEWIC or the BCCDC, many populations on southern Vancouver Island and in the lower Fraser Valley are of regional concern. Some of these populations probably merit listing as designatible units.

Oncorhynchus clarkii lewisi (GIRARD)
WESTSLOPE CUTTHROAT TROUT

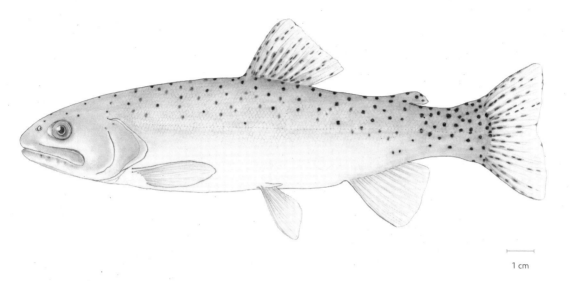

1 cm

Distinguishing Characters   In British Columbia, the main external difference between westslope and coastal cutthroat trout is the pattern of spots on the body (Quadri 1959): on westslope cutthroat, the spots below the lateral line are concentrated on the back half of the body and are almost absent on the front half of the body. In contrast, on coastal cutthroat, the irregular black spots below the lateral line are evenly distributed on the front and back halves of the body. Although there is some overlap, the two subspecies also differ in scale counts and body proportions (Quadri 1959).

Taxonomy   In the past (Carl et al. 1959; Quadri 1959), the inland subspecies of cutthroat trout in British Columbia was referred to as the Yellowstone cutthroat trout. At the time, it was thought that the cutthroat trout in the southern tributaries of the upper Missouri River system and the Snake River above Shoshone Falls were the same as the cutthroat trout of the upper Columbia River system. They are morphologically similar (Behnke 2002), but there are major genetic (Leary et al. 1987) and chromosome differences (Loudenslager and Thorgaard 1979) between the two forms. Thus, they are now treated as separate subspecies—the Yellowstone cutthroat (*O. c. bouvieri*) and the westslope cutthroat (*O. c. lewisi*). Consequently, the inland subspecies in British Columbia is properly called the westslope cutthroat trout.

Once, the small cutthroat trout in mountain lakes in the Revelstoke region were described as a distinct subspecies (Dymond 1931): the mountain cutthroat (*Salmo clarkii alpestris*). The mountain cutthroat is now considered a local form of the westslope cutthroat.

Sexual Dimorphism Outside the breeding season, the sexes are difficult to separate. Both sexes are a yellowish-green on the back and upper flanks with a blush of washed-out red on the operculum and front part of the body. In breeding males, the entire belly become a bright rosy-red, and a pale rose band extends along the midline from the head to the caudal peduncle. Also in males, the upper jaw, lower jaw, and branchiostegals darken to a dusky black. These colours are more subdued in spawning females.

Hybridization In the upper Columbia system, the evolution of the westslope cutthroat appears to have proceeded in relative isolation. Consequently, unlike their coastal cousins, they have not evolved isolating mechanisms that prevent, or reduce, hybridization with the rainbow trout (*Oncorhynchus mykiss*). In the northwestern United States, the introduction of non-native salmonines (mostly rainbows and brook trout, *Salvelinus fontinalis*) into waters that originally contained only westslope cutthroats has produced serious declines in range and numbers (Shepard et al. 2005). Hybridization with rainbows has led to the massive introgression of rainbow genes into cutthroat populations (Allendorf and Leary 1988; Leary et al. 1987; Liknes and Graham 1988). Apparently, only about 30% of the historically known populations of westslope cutthroat are now genetically pure (Shepard et al. 2005). A similar trend is evident in British Columbia: in 18 of 23 (78%) populations examined in the upper Kootenay system, there is genetic evidence of some hybridization (Rubidge et al. 2001). In these cases, hybridization can be detected; however, except in Lodgepole and lower Gold creeks, there is little genetic evidence that hybrid swarms are developing (Rubidge 2003; Rubidge and Taylor 2005).

Not all hybridization between westslope and rainbow trout is recent. In eastern Washington, there is evidence of old hybridization between steelhead and westslope cutthroat trout in a river system that now lacks cutthroat trout (Brown et al. 2004). Apparently, this pre-Columbian hybridization event occurred sometime during the Missoula Floods.

Distribution NATIVE DISTRIBUTION
A complex series of allopatric subspecies of cutthroat trout are distributed along the eastern slope of the Continental Divide from New Mexico north to southwestern Alberta, and west of the Continental Divide from the eastern slope of the Sierra Nevada, throughout the basin and range country of California, Arizona, Nevada, and Utah north into southeastern British Columbia. The subspecies at the northern end of this complex series of interior cutthroat trout is the westslope cutthroat. Its range includes most of the upper Columbia drainage system (excluding the Snake River system above Shoshone Falls) and disjunct populations in the upper John Day River (Oregon) and along the eastern slope of the Cascade Mountains in Washington (Behnke 2002; Williams 1999). Some

of these disjunct populations may be the result of early hatchery plantings. It is also native in foothills rivers and streams tributary to the South Saskatchewan River in southwestern Alberta (Nelson and Paetz 1992).

BRITISH COLUMBIA DISTRIBUTION

As the name westslope cutthroat implies, the native range of this subspecies lies along the western slope of the Rocky Mountains in the southeastern portion of the province (Map 41). Here, they occur mainly in the Kootenay and Pend d'Oreille river systems; however, they also are found in tributaries (e.g., Yard, Crazy, and Frog creeks) of the Eagle River and Mabel Lake (South Thompson drainage, Fraser River system). The presence of westslope cutthroats in Fraser tributaries is a bit of a puzzle. Most of these Fraser tributaries rise on the western sides of mountains that, on their eastern sides, drain into the Columbia system. Thus, an obvious explanation for these Fraser system populations is headwater captures from Columbia tributaries. Genetic data, however, do not support this hypothesis: two adjacent headwater streams (one a Fraser tributary and the other a Columbia tributary) are genetically distinct from one another (Taylor et al. 2003). This unexpected result supports an alternative hypothesis that postulates that westslope cutthroat were once more widely distributed in the Fraser drainage before the system was recolonized by rainbow trout (Dymond 1931). Presumably, westslope cutthroat trout were then extirpated except above barriers. The extirpation hypothesis may also explain the scattered distribution of westslope cutthroats in a few small tributaries of the Kettle River (Alder, Bitter, O'Farrell, and Sutherland creeks).

In British Columbia, westslope cutthroat have not been widely stocked outside the Columbia system, although a few introductions were made into the Fraser River system and some lakes in the lower Peace River system.

Life History In British Columbia, the westslope cutthroat typically occurs as three life-history forms: a lake-resident form, a migratory form that moves (usually for spawning) between large rivers and tributaries or between rivers and lakes, and a non-migratory stream-resident form. Liknes and Graham (1988) provide a good review of westslope cutthroat trout life history.

REPRODUCTION

Rising water temperatures (5 °C or slightly above) and a rising hydrograph appear to trigger spawning migrations in westslope cutthroat, and in southeastern British Columbia, most spawning occurs from early May to late June. Some populations in glacial streams or in streams associated with mountain lakes spawn much later (July or early August). Typically, lacustrine populations migrate to tributary streams or lake outlets to spawn, although some populations may spawn on gravel beaches (Carl and Stelfox 1989). Fluvial populations stage at the mouths of tributary streams and enter near the peak of the hydrograph (Schmetterling 2001). In the Ram River, Alberta, some fish emigrated to tributaries to spawn

Map 41 The B.C. distribution of westslope cutthroat trout, *Oncorhynchus clarkii lewisi*

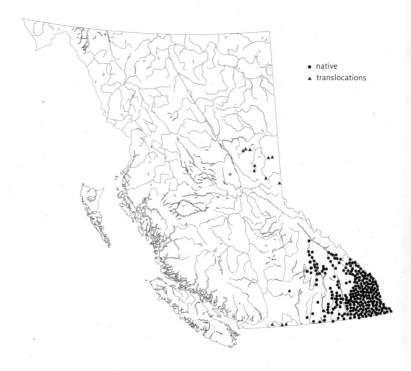

- native
- ▲ translocations

and others spawned in the main channel or in side-channels (Brown and Mackay 1995b). Like most salmonines, females choose the nest site. In the Blackfoot River, Montana, 75% of the observed redds were located upstream of the tail-outs of glides, 21% were found in riffles, and 4% in the tail-outs of scour pools (Schmetterling 2000). The average water velocity was 0.56 m/s, and the average water depth was 12.9 cm. The substrate ranged from 6 to 110 mm in diameter. These numbers are remarkably similar to those recorded in the Flathead River and its tributaries (Shepard et al. 1984). There, redds sites had water velocities of about 0.30–0.40 m/s, depths of about 15–20 cm, and fine gravel substrates (2–50 mm in diameter). The redds varied from 60 cm to 1 m in length and 30–45 cm in width. The details of courtship and spawning behaviour have not been described, but presumably, they are similar to those described for the Lahontan cutthroat trout (*O. c. henshawi*; Smith 1941), which are similar to those of the rainbow trout in most details (see the rainbow trout reproduction section for a description).

Fecundity varies with female size. Thus, fecundity in small-bodied, headwater populations is low (from about 125 to 700 eggs), whereas larger fish associated with lakes and large rivers produce about 750–2,000 eggs (Downs and White 1997). Fertilized eggs range in diameter from 5.0 to 6.0 mm. Depending on incubation temperatures, hatching occurs about 20 (14 °C) to 70 (5 °C) days after fertilization; the optimal incubation temperature appears to be about 10 °C. Newly hatched alevins range in length from 12 to 15 mm TL, and depending on temperature, emerge from the gravel 1–2 months after hatching. Newly emerged fry range from about 19 to 22 mm FL. Westslope cutthroat trout are iteroparous (they can spawn more than once); however, the proportion of repeat spawners varies among populations (<1% to about 25%).

AGE, GROWTH, AND MATURITY

In headwaters, isolated stream-resident populations of westslope cutthroat rarely exceed 20 cm TL. Initial growth is moderate (they reach 60–70 mm in their first summer and fall) and almost double their size (115–125 mm) in their second year, but growth slows in their third and fourth years (Northcote and Hartman 1988). Some headwater males reach sexual maturity in their third summer, and most are mature by their fourth summer. Typically, females mature 1 or 2 years later than males. Stream-resident fish rarely live more than 8 years (Downs and White 1997).

In lacustrine or large-river populations, growth rates accelerate after the young move from their spawning stream into a lake or large river. Depending on food availability and competitive interactions with other salmonines, migratory lacustrine populations can reach a large size (up to 60 cm). Males in most migratory populations mature after three or four summers, and most females are sexually mature by age 5.

FOOD HABITS

One of the most striking differences between the coastal and westslope subspecies of cutthroat trout is in their feeding habits. As adults, the coastal subspecies is highly piscivorous, whereas the westslope subspecies is primarily an insectivore. Small-bodied, headwater populations tend to be drift feeders and depend heavily on the drifting stages of dipterans and mayflies. Fish above 110 mm FL add caddisfly nymphs and winged insects to their diet (Shepard et al. 1984). In lakes and large rivers, winged insects are typical prey, and in some lakes and reservoirs, zooplankton is an important component of their diet.

Habitat In British Columbia, most of what is known about habitat use in westslope cutthroat trout is buried in agency files and consultant's reports. Consequently, most of this brief review is gleaned from published studies in Alberta, Montana, and northern Idaho.

ADULTS

Habitat use by adult westslope cutthroat trout varies with season, time of day, and life-history type. Typical westslope cutthroat streams are cold and nutrient poor (Liknes and Graham 1988). Adults of stream-resident populations are most often found in midwater near the upstream ends of pools (Griffith 1972). They appear to prefer low-velocity areas (<0.22 m/s) in pools with some overhead cover (e.g., large woody debris or the lee of large boulders). There are dominance hierarchies in these pools with the larger fish maintaining sites closer to the surface. In the summer, adults select pools 0.40–0.70 m deep and avoid both shallower and deeper pools. They remain close to the bottom (the average focal elevation is 0.07 m), they avoid velocities above 0.01 m/s, and their average distance to cover is 0.33 m (Spangler and Scarnecchia 2001). In the fall, they occupy deeper pools (average depth is 0.49 m), they stay closer to the bottom (average focal elevation is 0.04 m) in slower water (average focal velocity is 0.01 m/s), and their average distance to cover is 0.23 m (Spangler and Scarnecchia 2001). In the upper Fording River, B.C., adult cutthroat use reaches with large boulders or groundwater ponds as overwintering areas (Allen 1987). In the Ram River, Alberta, cutthroat trout moved from summer habitat sometime during the last half of September. Some fish moved directly to overwintering sites, but others made a two-stage habitat shift: first to staging areas and then to overwintering sites (Brown and Mackay 1995b). Apparently, the staging areas were abandoned when anchor ice formed in the staging pools. Overwintering sites were deeper pools (>0.80 m) than sites used during the summer and usually had some groundwater influence. Large numbers (up to 100) of fish aggregated in these pools (Brown and Mackay 1995b). Distances moved from spawning sites to summer feeding sites and overwintering sites are relatively short (<10 km). In contrast, populations in large rivers often make major migrations (>200 km) between spawning sites, summer foraging sites, and overwintering sites (Schmetterling 2001; Shepard et al. 1984). Lacustrine populations usually make only short migrations into tributary streams and return to the lake within days after spawning. Little is known about habitat use in lakes; however, during most of the year, westslope cutthroats are associated with near-surface waters (Shepard et al. 1984). The exception to this is in the summer, when they avoid surface areas with epilimnion temperatures above 20 °C (Shepard et al. 1984).

JUVENILES

Habitat use by juvenile westslope cutthroat trout changes seasonally and during the day (Bonneau and Scarnecchia 1998). In the summer, juveniles occupy deeper (mean depth 36 cm) water during the day than at night (mean depth 29 cm). There is also a diel shift in their focal point velocities (0.39 m/s during the day and 0.14 m/s at night). There were no diel differences in depth or focal point velocity between winter nights and days; however, winter night focal point velocities were lower (0.10 m/s) than those used during summer nights (Bonneau and Scarnecchia 1998). In fall and winter, juveniles are essentially nocturnal: during the day, they seek cover under woody debris or among large gravel or cobbles and emerge at night (Bonneau and Scarnecchia 1998; Jakober et al. 2000). This shift to nocturnal behaviour is most pronounced at temperatures below 3 °C. Lacustrine populations typically spawn in tributaries and migrate to the lake in their second or third summer. As they grow, they gradually shift from littoral areas into open water.

YOUNG-OF-THE-YEAR

Usually, newly emerged westslope cutthroat fry are found in shallow (<20 cm), quiet (0.07–0.10 m/s) water areas along stream edges. It is not clear whether they prefer such habitats or whether they are forced into these areas by biotic interactions (e.g., competition and predation) with larger or more aggressive fishes. At first, fry tend to be evenly distributed along stream margins (Shepard et al. 1984), and this suggests territorial behaviour. They use the nooks and crannies in the substrate and small woody debris as cover. As they grow, the fry move into deeper water. In lacustrine populations, some fry migrate from spawning streams to the lake in their first summer (Chisholm et al. 1989). Here, they remain in shallow littoral areas until the fall.

Conservation Comments Major declines in westslope cutthroat trout in Washington, Oregon, Idaho, and Montana probably are the result of several factors (overfishing, timber harvesting, land clearing, cattle ranching, and irrigation); however, a primary cause has been the introduction of non-native salmonines into areas that were once exclusively occupied by westslope cutthroat (Shepard et al. 2005). Brook and rainbow trout compete for resources with westslope cutthroat trout, usually depress their numbers, and sometimes cause extirpations. In British Columbia, the greatest immediate threat to westslope cutthroat is genetic swamping through hybridization with introduced rainbow trout (Rubidge and Taylor 2004, 2005; Rubidge et al. 2001). Many of the last pockets of pure westslope cutthroat in North America are in the B.C. parts the Kootenay River system. Given the well-documented effects of rainbow trout introductions on westslope cutthroats, any further introductions of rainbows into waters occupied solely by cutthroats would be ecological vandalism. COSEWIC lists the B.C. populations of westslope cutthroat as a species of special concern.

Oncorhynchus gorbuscha (WALBAUM)
PINK SALMON

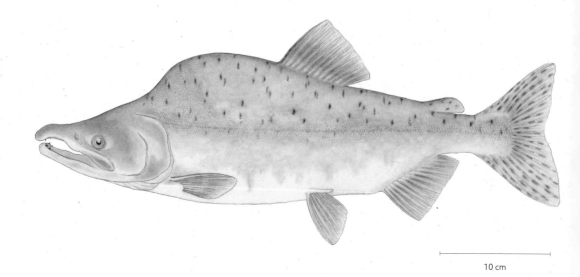

10 cm

Distinguishing Characters The distribution and shape of spots on the tail distinguish adult pink salmon from other Pacific salmon and trout. These spots are dark, elongate and occur on both lobes of the caudal fin. A less obvious morphological difference is the number of scales along the lateral line: usually 170 or more in pink salmon and less than 155 in other species of Pacific salmon. Pink salmon fry are small (usually <40 mm in fresh water) with bright silver flanks, an iridescent green back, and no lateral parr marks.

Taxonomy Although the specific status of pink salmon is not a taxonomic issue, there are unsolved problems involving the relationships and distribution of the odd-year and even-year broodlines in this species. Because virtually all native populations of pink salmon mature at 2 years, the generations spawning on odd and even years are genetically isolated from one another and often differ in life-history characteristics (Heard 1991). Usually one broodline is dominant (i.e., there is a strong run one year followed by a much smaller run the next year). At the southern end of their North American distribution (including southern British Columbia), odd-year runs are dominant; however, north of the Fraser River system, many rivers support relatively strong runs on both odd and even years. From the Queen Charlotte Islands north into Alaska, even-year runs are dominant. Presumably, both odd-year and even-year runs have evolved independently in different areas, but the reasons for the broad geographic pattern in run dominance is still a mystery. The pattern of mitochondrial variation in northern pink salmon suggests multiple Pleistocene divergences followed by a relatively recent (postglacial) expansion from different sources and, perhaps, different colonization routes for the odd-year and even-year broodlines (Churikov and Gharrett 2002).

Sexual Dimorphism   At sea, the sexes have the same body shape and bright silver colouration, but as they mature, males undergo a striking transformation. They develop a large hump on the back—hence the specific name *gorbuscha* (humpback in Russian) and the English common names humpback or humpies. Additionally, the head enlarges, large teeth grow on the jaws, and a distinct down-turned hook forms on the end of the snout and, to a lesser extent, the lower jaw hooks upward. These changes take about a month to complete and usually occur in estuaries or freshwater holding areas. Secondary sexual characteristics are not as obvious in females as in males; however, both sexes change colour in fresh water. Their heads darken, the silver colour on the flanks changes to dull white, often with purplish blotches just below the lateral line.

Hybridization   Natural hybrids between pink and chum salmon (*Oncorhynchus keta*) occur (Heard 1991; Hunter 1949) but are uncommon in British Columbia. In hatchery experiments, F_1 hybrids were fertile, and viable F_2 hybrids and backcrosses were produced (Simon and Noble 1968). Apparently, a few hybrids between introduced pink and Chinook salmon (*Oncorhynchus tshawytscha*) have been found in the St. Mary's River between Lake Huron and Lake Superior (Rosenfield et al. 2000). This hybrid combination does not occur within the natural range of these species.

Distribution   **NATIVE DISTRIBUTION**
Pink salmon are native to western North America, Siberia, and Asia. In North America, the Sacramento River contains the southernmost spawning population, and there are irregular runs along the Oregon and Washington coasts. There are no major runs in the Columbia River system; however, from Puget Sound north to Norton Sound, Alaska, there are many important runs. Additionally, there are small runs along the coasts of the Chukchi and Beaufort seas, and occasional strays reach the Mackenzie System (Dymond 1940) and the western islands of the Arctic Archipelago (Babaluk et al. 2000). On the Asian coast, pink salmon range from North Korea and Hokkaido north to the Anadyr Gulf and westward along the coast of the eastern Siberian Sea to the Lena River (Berg 1948). Pink salmon have been introduced into the Great Lakes, Hudson Bay, and the Atlantic coast. Only the Great Lakes introduction produced self-sustaining populations. Additionally, introductions from eastern Siberia to the Murmansk region have established self-sustaining runs.

BRITISH COLUMBIA DISTRIBUTION
There are spawning runs of pink salmon in most rivers and streams along the coast (Map 42). Most of these runs spawn within 100 km of the sea; however, in the Fraser system, pink salmon ascend the river into the

Map 42 The B.C. distribution of pink salmon, *Oncorhynchus gorbuscha*

• native

Thompson and Bridge river systems (about 300 km upstream), and occasionally, some individuals reach the Quesnel River (690 km upstream). In the Skeena system, some pink salmon reach the Babine counting fence (about 480 km from the sea).

Life History An excellent review of the life history of pink salmon (Heard 1991) and the U.S. pink salmon status review (Hard et al. 1996) are the major sources for the following life-history synopsis. Like all Pacific salmon (but not Pacific trout), the pink salmon is semelparous (they spawn only once in their life).

REPRODUCTION

In British Columbia, pink salmon are known to spawn in over 700 streams; however, many of the runs are relatively small, and apparently, there are only about 80 major runs (Aro and Shepard 1967). In the northern part of the province, pinks tend to enter fresh water earlier (late July and August) and spawn earlier than southern populations (August to September). Usually spawning occurs in September and October. In the Fraser and Skeena systems, the spawning run is protracted and has early and late segments. In the Fraser River, fish destined to spawn upriver or in the main river enter the Fraser 2 weeks earlier than fish that spawn in the lower tributaries (Ward 1959).

The distances that pinks ascend rivers to spawn is variable, but they are not as adept as other Pacific salmon at negotiating velocity barriers (Heard 1991). Consequently, they tend to spawn closer to the sea than other Pacific salmon, and some populations even spawn in the intertidal zone (Helle 1970).

Like other salmonines, female pink salmon choose the spawning site and apparently prefer sites with clean coarse gravel and subgravel flow. Typically, these sites are shallow riffles or channels 20–100 cm deep and with a current of 0.30–1.0 m/s (Heard 1991). Water temperature during peak spawning varies from 8 °C to 14 °C (Hunter 1959). The digging motions of the female and the actual spawning behaviour are similar to those described for other salmonines (for details, see the rainbow trout species account). Although virtually all pink salmon spawn at age 2, there is great variability in the size, shape, and colour of males. Typically, a nest-digging female is accompanied by a contingent of up to 10 or more males (Keenleyside and Dupuis 1988). The largest male in the group stays closest to the female, and the other males maintain a size-ordered hierarchy behind the pair. Frequently, however, a relatively small male with a female-like shape and colour remains close to the female on the side opposite from the dominant male. Both the dominant male and the female defend the area, each primarily attacking its own sex. When the female "crouches" in the nest—a signal that she is ready to spawn—the dominant male takes up a position alongside the female and quivers. At this time, the other males in the group crowd into the nest and release sperm. Presumably, they fertilize some of the eggs. Although spawning activity takes place throughout the day, most egg deposition occurs at night. Normally, females remain with their redd until death. In contrast, males are polygamous and move around on the spawning grounds. Since large males tend to be dominant, small males move more than large males and normally take up satellite positions relative to spawning pairs. One curious aspect of pink salmon spawning in the Fraser River near Hope, and in a few places in Alaska, is the construction of false nests. These false nests are sometimes called "exploratory nests" and may be associated with delays in entry into spawning streams (Heard 1991).

The eggs incubate in the gravel over winter; depending on incubation temperatures and egg size, they hatch in about 1.5–3 months and emerge from the gravel about 3–5 months after hatching (Murray 1980a). Some streams have temporally separated early and late spawning runs (Taylor 1980), and many pink salmon stocks appear to be adapted to different thermal regimes (Beacham and Murray 1986a). The length of newly hatched alevins varies from 16 to 18 mm, and the fry are 26–28 mm at emergence (Murray 1980a). At first, the newly hatched alevins are negatively phototactic and stay in the gravel. Emergence usually peaks about 3 hours after the onset of darkness (Godin 1980a). There is a gradual weakening of the negative phototactic response during the second week of emergence, the swimbladder is filled, and the fry begin to swim near the surface (Godin 1980b). Once they begin surface swimming, pink salmon fry quickly migrate downstream. Typically, pink salmon fry spend less time in fresh water than the other species of Pacific salmon. In the Fraser River system, the downstream fry migration begins as early as late February and lasts until late May; however, the main migration occurs in mid-April (Vernon 1966).

AGE, GROWTH, AND MATURITY

Pink salmon have the least flexible life cycle of the five North American species of Pacific salmon. The newly emerged fry migrate to sea as soon as they fill their swimbladders. Typically, they are about 30–35 mm long when they enter estuaries. Fry with longer migrations in large rivers will feed and grow in fresh water and may reach 4–5 cm before they enter the sea. Once in the ocean, they grow rapidly, and by the end of their first growing season (Heard 1991) they range from 27 to 34 cm FL. They are 45–55 cm long when, in their second year, they return to their natal streams as mature adults. Within their native range, this 2-year life cycle is virtually fixed, and 3-year-old pink salmon are extremely rare. Curiously, age at maturity in introduced Great Lakes pink salmon is highly variable.

FOOD HABITS

Although their rapid seaward migration precludes freshwater feeding in most small coastal drainages, fry migrations are more protracted in large rivers like the Skeena and Fraser, and some individuals begin feeding in fresh water (McDonald 1960). Larvae and pupae of chironomids appear to be the most common food items. Once in the sea, pink salmon fry usually remain in shallow water and forage primarily on harpacticoid and calanoid copepods. At about 5–8 cm in length, juvenile pink salmon move offshore. Here, they feed on fish, squids, euphasiids, amphipods, and copepods.

Habitat Pink salmon spend less of their lives in freshwater than any of the other species of Pacific salmon. Most populations spawn relatively close to tidewater (some spawn inter-tidally in estuaries), and upon emergence, the fry immediately migrate downstream. In the sea, they usually remain for some time feeding in estuaries or in shallow inshore water. After reaching a length of about 5–8 cm (usually by late July or August), they abandon their inshore feeding areas, move offshore, and migrate to the northwest along the North American coast. They reach the Gulf of Alaska in the fall or winter and make a long turn to the southeast. This migration carries them south to about latitude 44°N. They then turn and once again migrate to the northwest. By the end of their second summer in the sea, they have reached adult size and are once again in the Gulf of Alaska. They begin their homeward migrations in late summer or early fall. These final southeasterly migrations take them to the coast and eventually to their natal rivers (Heard 1991). The adults then enter freshwater, spawn, and die (usually after less than 2 months in fresh water). Consequently, the only significant freshwater habitat for this species is the area where they spawn and their eggs incubate.

Conservation Comments Of the five species of Pacific salmon in British Columbia, the pink salmon is least affected by human activities. There are more than 2,000 pink salmon stocks in the province, and almost 90% of the stocks are deemed to be unthreatened (Slaney et al. 1996). The same authors estimate that 17 stocks are now extirpated but caution that this is probably an underestimate: no one knows how many stocks were lost to the Hells Gate slide. The regions with the most pink salmon stocks at risk are the western coast of Vancouver Island, the Queen Charlotte Islands, and Johnstone Strait.

Oncorhynchus keta (WALBAUM)
CHUM SALMON

10 cm

Distinguishing Characters   Adult chum salmon lack the distinct black spots on their back and tail that are characteristic of all other Pacific salmon except sockeye (*Oncorhynchus nerka*). The number and length of the gill rakers on the lower limb of the first gill arch separates chum and sockeye: 11–17 stubby, smooth rakers in chum salmon, and 19–27 long, slender, serrated rakers in sockeye. The fry are small (usually <50 mm in fresh water) with a mottled iridescent green back and with regularly shaped (and spaced) lateral parr marks that are faint or absent below the lateral line.

Taxonomy   Although the specific status of chum salmon is not an issue, the species is variable, especially in life-history traits like run timing. It is not unusual to find temporally separate runs in the same river. In both Asia and North America, southern populations tend to return to the rivers of their birth at a later date than northern populations. In Asia, "summer" chum (June, July, and August) are considered different from "autumn" chum (Berg 1934), and the late-run fish have been given a formal taxonomic name (*O. k. autumnalis*). The chum salmon of the Amur River are a striking example of differences in run timing: the summer run enters the river in July and spawns within 100 km of the sea, whereas the autumn run enters in September, and some fish spawn at least 2,500 km upriver (Berg 1948).

In North America, the summer/autumn split in run timing is not as pronounced, and the different seasonal runs have no formal taxonomic status. Nonetheless, there are examples of multiple runs that enter the same river at different times and spawn at different distances from the sea (Beacham and Murray 1986b; Tallman and Healey 1994). Additionally, there appears to be two lineages of chum salmon inhabiting the coast of western North America (Seeb and Crane 1999). One lineage (the southern) is widespread and probably survived glaciation south of the Cordilleran Ice Sheet. The other lineage is northern and appears to have survived in Beringia. The contact zone between the two lineages is at the northern end of the Alaska Peninsula. Thus, most chum populations in British Columbia are of the southern lineage, but presumably, the chums in Teslin Lake (Yukon system) and the Liard River (Mackenzie system) are members of the northern lineage.

Sexual Dimorphism At sea, the sexes have the same body shape and bright silver colour, but as they mature, both sexes darken. In males, the head enlarges, large teeth develop on the jaws, and a distinct down-turned hook forms on the end of the snout and, to a lesser extent, the tip of the lower jaw develops an upturned hook. Additionally, the body changes shape—it deepens and a modest nuchal hump forms behind the head. These secondary sex characters are not as pronounced in females as in males. In most populations, these changes occur in estuaries, and chum salmon typically enter rivers ripe and in full spawning colour. The spawning colours of chum are very distinctive: the head is almost black, the back is a dark brownish green, and males develop a calico pattern (ragged purple blotches interspersed with dirty white patches on the anterior lower flanks and a jagged black line on the posterior third of the sides). Basically, female colouration is the same as in males except for the shape of the dark purple colour. In females, the purple colour forms a distinct band along the midline of the body.

Hybridization Natural hybrids between chum and pink salmon (*Oncorhynchus gorbuscha*) occur (Heard 1991; Hunter 1949) but are relatively uncommon in British Columbia. In hatchery experiments, fertile hybrids were used to make viable F_2 hybrids and backcrosses (Simon and Noble 1968).

Distribution NATIVE DISTRIBUTION
Chum salmon have the widest distribution of any Pacific salmon. They are indigenous to western North America, Siberia, and Asia. In Asia, they range from the Bering Strait south to Korea and Honshu in Japan. They also occur sporadically on Kyushu (Japan) and in northeastern China. In Siberia, they extend from Bering Strait westward along the Arctic Coast to the Lena River. In North America, chum salmon breed from Bering Strait to California (Sacramento River) with occasional individuals observed as far south as the San Lorenzo River. They also breed in rivers along the Arctic coast of Alaska, and sporadically, spawners enter the Mackenzie River system. In addition, occasional individuals have been observed as far north as Banks Island (Babaluk et al. 2000). Although other species of Pacific salmon (e.g., coho, *Oncorhynchus kisutch*; Chinook, *Oncorhynchus tshawytscha*, and sockeye) also sporadically appear as single individuals in the Mackenzie system, chum and pink salmon are the most common visitors and chum may maintain small, self-sustaining populations

in the system. Chum salmon have been introduced into Ontario and northern Europe (the Murmansk region of Russia). The Ontario introduction failed, but the Russian introduction appears to have established self-sustaining populations.

BRITISH COLUMBIA DISTRIBUTION

Chum salmon are widely distributed along the our coast but rarely penetrate more than about 200 km upstream (Map 43); however, occasional individuals ascend the Fraser River to tributaries (e.g., the Bridge and Thompson rivers) above Hell's Gate. Two exceptions to the typical coastal distribution are the Mackenzie and Yukon drainages. In the Mackenzie system, chum salmon sometimes ascend to the Liard River and beyond (at least to the Slave River). In the Liard River, they apparently spawn in British Columbia below the Grand Canyon of the Liard (Irvine and Rowland 1979); however, it is not known if the Liard chum are a self-sustaining population. In contrast, the chum population associated with Teslin Lake (3,700 km up the Yukon River system), although small, appears to be self-sustaining.

Life History Salo (1991) provides an excellent review of the biology of chum salmon. His review and the U.S. chum salmon status review (Johnson et al. 1997) are the major sources for the following life-history synopsis.

REPRODUCTION

In British Columbia, chum salmon spawn in over 300 streams. Many of the runs are relatively small, and less than 30% of the rivers support major runs (Aro and Shepard 1967). In the northern part of the province, chums enter fresh water earlier (late July and August) than those in southern areas (September and October). In the lower Fraser system, spawning occurs from September to early January in tributary streams and in the main river. Separate early and late runs are common even in relatively small streams (Beacham and Murray 1986b; Tallman and Healey 1994). The distances that chum ascend rivers to spawn are variable, and some populations spawn inter-tidally in estuaries (Bailey 1964). Most populations spawn within 100 km of tidewater; however, some chums ascend the Yukon and Mackenzie rivers all the way into British Columbia (journeys of roughly 3,700 and 2,500 km, respectively).

Like other salmonines, chum females choose the spawning site and, apparently, prefer sites with upwelling water that is often warmer than the surrounding river water (Geist et al. 2002). Females excavate the nests. The size of gravel moved is variable and depends on female size. In Columbia River tributaries, 81% of the gravel in chum redds was <15 cm, 13% was >15 cm, and 6% was sand and silt (Burner 1951). Although chum salmon will spawn in water velocities ranging from 0 to 1.67 m/s, 80% spawn at water velocities of 0.2–0.8 m/s and depths of 13–50 cm (Johnson et al. 1971). The digging motions and actual spawning behaviour are similar to those described for other salmonines (for

Map 43 The B.C. distribution of chum salmon, *Oncorhynchus keta*

details, see the rainbow trout species account). Normally, a female is accompanied by a dominant male (usually the largest male) and a variable number of satellite males. The dominant male is aggressive and uses physical force to keep satellite males at bay (Schroeder 1981, 1982). Females are also aggressive and will attack both conspecific and heterospecific females (Quinn 1999). Typically, a female deposits about 35% of her eggs in her first nest and, after a short rest period, moves a bit upstream and begins another nest. Successive nests contain fewer and fewer eggs (Schroeder 1982).

The eggs incubate in the gravel over winter; depending on incubation temperatures and egg size, they hatch in about 2–3 months and emerge from the gravel about 1.0–2.5 months after hatching (Murray 1980a). At first, the newly hatched alevins respond negatively to light; however, as they develop, they become photopositive. There is evidence that local selection on egg size influences development rate (Beacham and Murray 1986b; Tallman 1988). Both egg number and egg size vary with female size: in North America, fecundity ranges from 2,000 to 4,000 eggs (Salo 1991), and egg diameters range from 7.0 to 8.7 mm (Baakala 1970). In British Columbia, fecundities in Queen Charlotte Islands and Vancouver Island stocks are lower than in some mainland stocks (Beacham 1982). Along the coast of North America, there is a broad geographic cline in egg size (Baakala 1970), and generally, chum eggs are smaller in the northern and larger in the southern parts of their geographic range (Baakala 1970). The eggs in most B.C. populations are about 8.5 mm in diameter;

however, there are local differences in egg size, especially between early and late runs in the same river. For example, in the Vedder River, the mean egg size for the early run (mid-October) is 8.6 mm, whereas the mean egg size for the late November run is 8.2 mm (Beacham and Murray 1986b); however, in the Chehalis River the egg size relationship is reversed (8.5 mm in the early run and 7.9 mm in the late run).

Although there are often major differences in spawning times among stocks in the same river, emergence from the gravel and downstream migration usually are synchronous within a drainage system (Beacham and Murray 1986b; Tallman and Healey 1994). Evidently, this synchrony results from differences in development rates mediated through incubation temperatures and egg size. The length of newly hatched alevins varies with incubation temperature: 16.9 mm at 8 °C and 17.5 mm at 5 °C (Murray 1980a). The length of newly emerged fry also varies with temperature: 30.3 mm at 8 °C and 31.4 mm at 5 °C (Murray 1980a).

AGE, GROWTH, AND MATURITY

Growth in their first few weeks of ocean life is rapid—about 1 mm a day—and by early fall they range in size from 13 to 20 cm (Healey 1980). Apparently, growth rate is higher in the eastern Pacific than on the Asian side. Growth rate slows as they approach maturity; however, age at maturity is variable, and the number of years at sea determines their final (spawning) size. A few B.C. chums mature after one winter at sea; however, most mature after two or three winters, while some mature after four winters and a few, after five winters. In their final marine year, early maturing chums in the eastern Pacific more than double their second-year weight, but the weight gain in their final marine year is less for fish that mature at an older age.

FOOD HABITS

Newly emerged chum fry typically migrate directly to estuaries, where they begin feeding. Since most chum salmon spawn within 100 km of the sea, the downstream migration period is relatively short (<30 days) and mostly occurs at night, although some populations make diurnal migrations. Presumably, however, the downstream migration from upper Yukon and lower Liard sites takes much longer and involves some fresh water feeding. The diet of chum fry feeding in fresh water appears to consist mainly of chironomids and the larvae of aquatic insects (Salo 1991). In estuaries, the fry feed primarily on harpacticoid copepods and, in salt marshes, on amphipods, and chironomid larvae and pupae. At sea, maturing chum salmon forage primarily on invertebrates—amphipods, euphausiids, pteropods, and copepods—but also take small fishes and squid.

Habitat Because chum salmon are essentially marine fish for most of their life, there is not much information on the freshwater aspects of habitat use. The following habitat use synopsis is based on Johnson et al. (1997) and Salo (1991).

ADULTS

Adult chum salmon are found throughout the North Pacific Ocean with North American stocks concentrated in the eastern Pacific (especially the Gulf of Alaska) and Asian stocks concentrated in the western Pacific. There is some overlap between Asian and North American stocks (Seeb and Crane 1999); however, as they mature, the North American and Asian stocks migrate to the southeast and to the southwest, respectively (Salo 1991).

JUVENILES

Evidently, juvenile chum salmon in their first year migrate northward in a narrow inshore band along the North American coast. In the northern part of the Gulf of Alaska, they turn and migrate south (Hartt 1980). This southern movement is an offshore migration. In the next spring and early summer, the immature chum salmon again migrate north towards the Gulf of Alaska. This cycle of northward and southward migrations is repeated as the fish mature.

YOUNG-OF-THE-YEAR

Except for the relatively short period of downstream migration, chum fry spend their first 3–4 weeks in the upper parts of estuaries, especially in tidal creeks, sloughs, and salt marshes. At 45–60 mm, they move into shallow, near-shore marine areas and, eventually, into deeper water. Movement out of the Strait of Georgia to the ocean occurs in July (Healey 1980).

Conservation Comments   There are about 1,600 chum salmon stocks in British Columbia (Slaney et al. 1996). Most (84%) of these stocks are not at risk; however, 22 stocks have been extirpated (17 in the Vancouver area), and another 141 stocks appear to be at a high risk of extirpation. Most of the high-risk stocks are in the Nass and other northern coastal river systems, as well as the Queen Charlotte Islands and the central coast. Most of the stocks deemed at high risk are small. These stocks probably have always been small; however, this does not necessarily mean that they have little biodiversity value.

Oncorhynchus kisutch (WALBAUM)
COHO SALMON

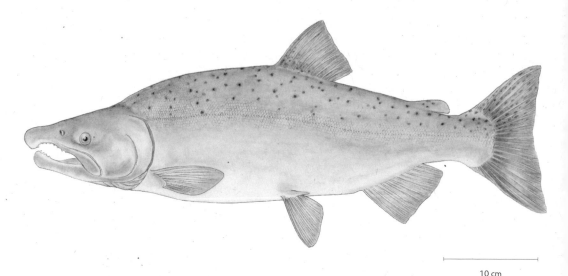

10 cm

Distinguishing Characters In the ocean, the colour patterns of adult coho and Chinook salmon (*Oncorhynchus tshawytscha*) are superficially similar—both species have silver flanks and dark spots on their backs and tails. The black spots on the tails of coho, however, are confined to the upper lobe of the caudal fin, whereas there are spots on both the upper and lower lobes of the caudal fin in Chinook. Additionally, the gums at the base of the lower jaw teeth are black in Chinooks, and the gums at the base of the lower jaw teeth in coho are white.

In British Columbia, most young coho spend 1 or 2 years in fresh water, although some individuals migrate to estuaries shortly after emergence. The fry have long narrow lateral parr marks, and the dorsal and anal fins have white leading edges that sharply contrast with a streak of dark pigment immediately behind this white leading edge. The paired fins and tail are often pale orange, and the anal fin is sickle shaped. Coho and Chinook salmon fry that migrate to estuaries can be difficult to separate. A useful, but not infallible, character for separating fry is the adipose fin. In coho fry, the centre of the adipose fin is opaque (finely stippled), whereas the centre of the adipose fin is clear in Chinook fry. Although it requires killing animals, the species differ in the number of branchiostegal rays: on the left side, there are usually 14 or fewer branchiostegal rays in coho and, usually, 16 or more in Chinook. Additionally, the pyloric caeca counts do not overlap between the two species: 45–80 pyloric caeca in coho and 135–185 in Chinook. The numbers are smaller in fry—salmonines add pyloric caeca as they grow—however, at any given size, coho usually have at least half as many cacca as Chinook.

Taxonomy   Although the taxonomic status of the coho is not an issue, the species is genetically heterogeneous, and locally adapted populations are common (Rosenau and McPhail 1987; Taylor 1991a; Weitkamp et al. 1995). Additionally, there is molecular evidence for geographically patterned genetic variation in North American coho (Gharrett et al. 2001; Smith et al. 2001). Microsatellite DNA data suggest five geographic groups: Alaska and northern British Columbia; the Queen Charlotte Islands; the B.C. mainland and northern Washington State; the Thompson River (presumably including the extirpated upper Columbia River coho); and Oregon and California (Smith et al. 2001).

Sexual Dimorphism   At sea, the sexes have the same body shape and bright silver colouration, but as they mature, both sexes darken. In males, the lower half of the head eventually becomes black, the head enlarges, the teeth on the jaws enlarge, and a strong down-turned hook forms on the end of the snout; to a lesser extent, the tip of the lower jaw develops an upturned hook. There is experimental evidence that the degree of development of these secondary sexual characteristics influences reproductive success (Fleming and Gross 1994). Although secondary sexual characteristics are not as pronounced in females as in males, they still influence reproductive success (Fleming and Gross 1994). In many populations, coho are silver-bright when they enter rivers, and these changes in shape and colour occur in fresh water. The dominant spawning colour in male coho is red. The sides of males develop a bright red colour that typically extends onto the head, while the bottom half of the head becomes black. In contrast, the nuptial colours in females are more subdued (often their sides show only a light tinge of rose).

Hybridization   Although laboratory hybrids between coho and other Pacific salmon have been produced, coho are the most difficult of the North American Pacific salmon to cross with other species. Nonetheless, occasional juvenile hybrids between coho and Chinook are encountered in fresh water. Such hybrids are rare and usually associated with streams disturbed by fish-culture activities (Bartley et al. 1990). In the mid-1980s, a few hybrid juveniles were collected in the Brunette River near Vancouver. As far as is known, no natural coho–Chinook hybrids have been found in British Columbia.

Distribution   NATIVE DISTRIBUTION

Coho are indigenous to the Pacific coasts of Asia and North America. In Asia, coho occur sporadically as far south as the eastern coast of North Korea, Peter the Great Bay, and northern Hokkaido. Apparently, they are more abundant farther north on Sakhalin, the northern coast of the Sea of Okhotsk, and both coasts of Kamchatka. The most northerly coho runs in Asia are found in the Anadyr River system. In North America, the most southerly population occurs in the San Lorenzo River near Monterey, California, although occasional individuals have been caught in the sea as far south as Baja California. The northern extent of their North American range appears to be Kotzebue Sound, Alaska; however, occasional individuals have been caught as far to the east as Great Bear Lake in the Mackenzie River system (Babaluk et al. 2000). In eastern North America, coho salmon now are well established in the upper Great Lakes.

BRITISH COLUMBIA DISTRIBUTION

Coho occur in large and small rivers along our entire coast (Map 44). Most populations spawn within 250 km of tidewater; however, in large rivers like the Fraser and Skeena, coho migrate considerable distances inland (>500 km). They reach the headwaters of the Skeena system (about 500 km), although they do not migrate as far up the Fraser system as sockeye (*Oncorhynchus nerka*) or Chinook, they reach the headwaters of the North and South Thompson rivers (Sandercock 1991), and occasional juveniles have been collected in the Nechako River near Vanderhoof.

Life History Two extensive reviews of the life history of coho salmon (Sandercock 1991; Weitkamp et al. 1995) are the major sources for the following life-history synopsis.

REPRODUCTION

In British Columbia, coho spawn in over 300 rivers and streams (Peterson 1982). The timing of river entry and spawning varies: early runs enter freshwater in September or October and spawn in late October, November, or December; in contrast, late runs enter freshwater in December or January and spawn shortly thereafter. In the lower Fraser valley, there are some exceptionally late runs. For example, in Elk Creek near Chilliwack, spawning coho have been observed in late March. The timing of river entry is influenced by water temperature and flow. In low-flow years, entry into small streams flowing into the Strait of Georgia may be delayed for a month or more. Most upstream migrations occur during the day. Typically, spawning occurs at temperatures ranging from about 1 °C to 8 °C, although in California coho apparently spawn at temperatures as high as 13.3 °C (Briggs 1953).

Like other salmonines, coho females choose the spawning site and, apparently, prefer sites with subgravel flow (e.g. in the tail-outs of pools immediately above riffles or upwelling sites). Spawning coho are the most secretive of the Pacific salmon, and most reproductive behaviour occurs at night. Often coho spawn in streams not much more than 1 m in width. Normally, a large, dominant 3-year-old male accompanies the digging female. Smaller 3-year-old males (satellites) are arrayed downstream of the redd site and, nearby, small 2-year-old jacks lurk under cover. The dominant male does not tolerate other large males near the nest site. The movements of dominant males usually are restricted to a small section of the spawning stream; however, satellite males move considerable distances (Healey and Prince 1998). Water velocities at spawning sites range

Map 44 The B.C. distribution of coho salmon, *Oncorhynchus kisutch*

from about 0.30 to 0.91 m/s, and the substrate averages 9.4 cm in diameter (range 3.9–13.7 cm) (Briggs 1953).

Fecundity in coho salmon varies with latitude and female body size. In the Karluk Lake population (Alaska), egg numbers range from 1,724 to 6,906 eggs (Drucker 1972), whereas in Minter Creek, Washington, fecundities range from 1,900 to 3,286 eggs (Salo and Bayliff 1958). Evidently, coho have smaller eggs than other Pacific salmon; however, egg size also varies with latitude (Fleming and Gross 1990). Egg diameters ranged from 4.9 to 6.9 mm in the Karluk Lake population (Drucker 1972), whereas egg diameters vary from 6.2 to 8.4 mm in the Fraser valley (Murray 1980a). Apparently, there is a trade-off between egg number and egg size, and egg number decreases as egg size increases (Fleming and Gross 1990; Quinn 2005). The eggs incubate over winter and hatch in the spring. For lower Fraser River coho eggs incubated at constant temperatures ranging from 2 °C to 14 °C, 50% hatching occurred at 159–167 days at 2 °C and 36 days at 14 °C (Murray 1980a). Generally, hatching success was high (80–100%) at temperatures ranging from 2 °C to 11 °C and almost zero at 14 °C. At the lowest temperature (2.0 °C), fry emerged from the gravel about 3 months after hatching, whereas about a month elapsed between hatching and emergence at 14 °C. Size at emergence also varied with temperature: the largest newly emerged fry (25–28 mm TL) occurred at incubation temperatures between 5 °C and 8 °C with the smallest fry (23–24 mm TL) produced at 14 °C.

AGE, GROWTH, AND MATURITY

In British Columbia, coho fry usually reach 80–90 mm FL in their first year (Sandercock 1991); however, growth rates are sensitive to temperature, food abundance, and fry density. In Chapman Creek on the Sechelt Peninsula, a manipulation of fish density and natural invertebrate drift resulted in changes in both growth rates and microhabitat use (Rosenfeld et al. 2005). Increased food increased the growth of both dominant and subdominant fish. It also appears that increased food allowed coho to exploit higher velocity habitats: average focal velocities shifted upwards to 0.65–0.84 m/s after an experimental increase in the abundance of drift. Increased fish density resulted in lower growth of subdominant but not dominant fish and shifted focal point velocities to both lower and higher levels. Although growth often ceases in the winter, feeding may continue in groundwater-fed, off-channel sites. Depending on growth conditions, yearling smolts range from about 7 to 12 cm when they migrate, whereas 2-year-old smolts range from about 10 to 15 cm. Once in the sea, growth rates increase.

FOOD HABITS

Coho fry are primarily drift-feeders. They normally take the drifting stages of aquatic insects (especially chironomids) from the water column or terrestrial insects from the surface (Mundie 1969). They rarely forage on the substrate, but in the summer, aerial prey are often as important as aquatic drift. In pools, coho fry exhibit two different foraging behaviours: they may forage as participants in a dominance hierarchy, or they may forage as nonhierarchical "floaters" (Puckett and Dill 1985). Coho in dominance hierarchies forage primarily on drift, and floaters forage opportunistically on items that drop into the water (Nielsen 1992). Yearling coho also feed on drifting insects but take a wider variety of larvae and nymphs than the fry. Here also, terrestrial insects are important in the summer. In lakes, juveniles forage in near-shore littoral areas. Again, insects are the primary prey and are usually taken off the surface. Zooplankton forms only a minor part of the diet of lake-dwelling juveniles. As they grow, juvenile coho become piscivorous. In Cultus Lake, sockeye fry were an important part of the diet of yearling coho (Ricker 1941). Coho smolts migrating to sea prey especially heavily on out-migrating pink salmon (*Oncorhynchus gorbuscha*) fry. Adult coho are highly piscivorous and in inshore waters they forage heavily on herring (*Clupea*), sand lance (*Ammodytes*), and a wide variety of small fishes and invertebrates. The general diet of offshore coho is similar, although the prey species often differ and include euphausiids and squid.

Habitat Two extensive reviews of habitat use by coho salmon (Sandercock 1991; Weitkamp et al. 1995) are the major sources for the following life-history synopsis.

ADULTS

Coho on the southern B.C. coast typically spend about 18 months at sea before returning to fresh water to spawn. Jacks, precocious males, spend about 6 months at sea. Some northern coastal coho follow the Alaska Gyre and make more extensive oceanic migrations than southern inshore coho. These northern fish spend longer at sea and return to fresh water at a larger size than southern coastal coho.

JUVENILES

In British Columbia, most smolt migration takes place from April to late June and usually peaks from early May to early June. Typically, coho migrate to sea after 1 or 2 years in freshwater. Most populations in southern British Columbia migrate after 1 year in fresh water, and only about 1–7% of the smolts are 2 years old. There are, however, exceptions. For example, in Carnation Creek on the west coast of Vancouver Island in some years, 40–50% of the coho smolts are 2 years old. These 1^+ coho use mainstem pools during the summer and also overwinter in the main channel. Here, they seek cover under cutbanks, large woody debris, and in root-wads (Shirvell 1990). They migrate to the sea in their third spring. Generally, the proportion of 2 year smolts increases towards the north, and in Alaska, most smolts migrate after 2 or 3 years and, occasionally, 4 years in fresh water. In southeastern Alaska, 1^+ or 2^+ coho usually occurred within 1.0 m of cover in water <30 cm deep at focal depths of <25 cm and focal point velocities <0.09 m/s (Dolloff and Reeves 1990).

On entry into salt water, coho tend to move north and remain inshore. In the Strait of Georgia, some coho remain within the strait and do not migrate to the open sea (Healey 1980). In their first marine summer, most B.C. coho move north along the coast. Some remain inshore, but others move directly out to sea.

YOUNG-OF-THE-YEAR

Initially, newly emerged coho fry are secretive and stay in the gravel during the day but emerge at night. In some streams (e.g., Carnation Creek), some fry move directly downstream into the upper estuary where they spend the summer. In the fall, some of these estuarine fry move into the sea, and others re-enter fresh water for the winter. Other fry show more typical movements and behaviours. After a few days, they aggregate in backwaters, side-channels, and shallow, quiet embayments along stream margins. As they grow, they move into pools; some individuals become territorial and part of a dominance hierarchy, and others become nonhierarchical floaters. Hierarchical individuals occupy territories with focal velocities >0.06 m/s, whereas floaters patrol large foraging areas with water velocities <0.06 m/s (Nielsen 1992). In southeastern Alaska, 0^+ coho usually occurred in water <30 cm deep at focal depths of <10 cm and within 50 cm of cover (Dolloff and Reeves 1990). Eventually, there is some emigration from small tributary streams to larger rivers, lakes, or estuaries (Hartman et al. 1982; Irvine and

Johnston 1992). In lakes, they may spread out along the littoral zone (Irvine and Johnston 1992) or form schools (Swain and Holtby 1989). There are often morphological and behavioural differences between stream-rearing and lake-rearing coho (Swain and Holtby 1989). In many rivers in the fall, fry move into off-channel overwintering areas (Peterson 1982), such as beaver ponds and flooded wetlands. In the winter, 0^+ coho seek cover under woody debris, undercut banks, cobbles, and penetrate deep into root-wads. They are quiescent during the day but emerge at night.

Conservation Comments   There are approximately 2,600 coho salmon stocks in British Columbia (Slaney et al. 1996). Of these, the status of about 1,200 stocks is unknown. Twenty-nine stocks have been extirpated, another 214 are at high risk, and about 20 stocks are of special concern. At one time, coho were the major recreational species in the Strait of Georgia, but in recent years, there have been precipitous declines in their numbers. Half of the extirpated stocks, and most of the stocks at high risk, are in the Strait of Georgia region or the interior of the province (Slaney et al. 1996). Most of the high-risk stocks are small and probably always have been small. Overfishing and habitat degradation, especially in urban areas, are among the causes for these declines. COSEWIC lists the Interior Fraser coho populations as endangered.

Oncorhynchus mykiss (WALBAUM)
RAINBOW TROUT

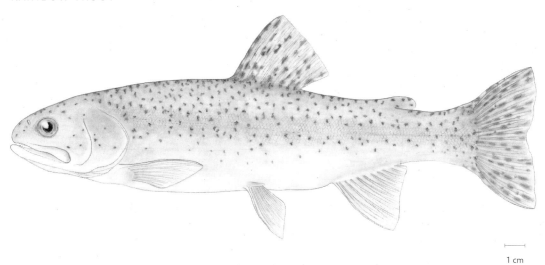

1 cm

Distinguishing Characters This trout has black spots on the back, sides, and dorsal and caudal fins; 17–22 gill rakers (11–13 on the lower limb of the first gill arch); and 12 or fewer anal rays. The anal fin base is shorter than the dorsal fin base. Fry (up to about 3 cm) have a dark stripe on leading edge of the dorsal fin; larger fish have dark spots on the dorsal fin.

A recent B.C. analysis of errors in fish identification (Haas et al. 2001) indicates that the most common mistake made by inventory personnel is confusing rainbow and cutthroat trout (*Oncorhynchus clarkii*). Understandably, most mistakes involve small specimens (<100 mm), and there are no infallible morphological markers that separate the species at this size. However, a suite of characters usually will separate adults: rainbow trout normally lack red slashes under the throat, and cutthroat trout usually have them; rainbow trout lack basibranchial teeth, and cutthroat trout usually have them; and rainbows have larger scales than cutthroat (usually <150 along the lateral line in rainbows and >150 in cutthroat). At some sites, hybridization blurs the distinctions between cutthroat and rainbow trout; however, in areas of hybridization the absence of basibranchial teeth is apparently a reliable indicator of "pure" rainbow trout (Leary et al. 1996).

Taxonomy In British Columbia, rainbow trout occur both as freshwater resident and anadromous (steelhead) populations. Behnke (2002) suggests that there are two subspecies in British Columbia—the coastal rainbow trout (*O. m. irideus*) and, in the interior, the Columbia redband trout (*O. m. gairdneri*). In British Columbia, the interior form is often called the Kamloops trout. Although the coastal and interior forms of rainbow trout generally are treated as two distinct lineages (Bagley and Gall 1998;

Nielson et al. 1994), recent molecular studies (Beacham et al. 1999; McCusker et al. 2000) indicate that, in British Columbia, there is extensive overlap in the geographic distribution of the coastal and interior clades. Thus, although many B.C. sites include both lineages, the coastal clade is dominant on the coast, and the interior clade is dominant in southern inland populations. Since the molecular markers characteristic of the two clades occur in both coastal and interior populations, the boundaries between the purported subspecies are too fuzzy to assign formal subspecific names to the two clades.

Of potential interest in British Columbia is an apparently unique population of rainbow trout in the Athabasca drainage in Alberta (Carl et al. 1994). Analyses of genetic distances (based on allozymes) indicate this population differs from both the coastal and interior clades of rainbow trout. The authors suggest that the Athabasca population has been isolated for at least 64,000 years; however, neither molecular sequence data (McCusker et al. 2000) nor a recent microsatellite study (E.B. Taylor, Department of Zoology, University of British Columbia, personal communication) supports the distinctiveness of Athabasca rainbow trout. A plausible explanation for the distinctive allozyme frequencies in these Athabasca rainbows is hybridization with introduced cutthroat trout (Nelson and Paetz 1992, McCusker et al. 2000).

Three major groups of anadromous rainbow trout (steelhead) are recognized in British Columbia (Beacham et al. 2004b; McCusker et al. 2000): a northern coastal group that includes coastal rivers from the Skeena River north; a southern coastal group that include the lower Fraser and Vancouver Island; and a southern interior group that includes rivers associated with the middle Fraser River. These groups are defined genetically; however, there are transition zones between the major groups that contain populations of uncertain relationships. Additionally, using enough loci, subgroups can be distinguished within each of the major groups (Beacham et al. 2004b). For example, within the northern coastal group the populations in the Taku and Iskut–Stikine rivers are distinct from those in the Nass and Skeena river systems.

Sexual Dimorphism Some experienced anglers and hatchery personnel can distinguish the sex of rainbow trout outside the spawning season; however, the morphological differences between males and females are subtle. As they approach spawning condition, it becomes easier to separate the sexes. The snout on large males elongates. This greatly increases their head and jaw lengths. Additionally, their lower flanks darken and the red lateral band intensifies. Similar changes occur in spawning females but are not as pronounced. In small-bodied headwater populations, juvenile characters (e.g., parr marks and bright colouration) often are retained in adults (Northcote and Hartman 1988), and even when spawning, the sexes are difficult to distinguish.

Hybridization   Rainbow and cutthroat trout (both the coastal, *O. c. clarkii*, and the westslope subspecies, *O. c. lewisi*) hybridize at many sites in northwestern North America (Hitt et al. 2003; Rubidge and Taylor 2005; Rubidge et al. 2001; Young et al. 2001). Along the B.C. coast, hybridization between coastal cutthroat and rainbow trout is often associated with environmental disturbances (e.g., reservoirs, fish-culture activities, and poor forestry practices), but there also appear to be hybrids in undisturbed sites (Parkinson et al. 1984). In eastern Washington, there is evidence of old hybridization between steelhead and westslope cutthroat trout in a river system that now lacks cutthroat trout (Brown et al. 2004). Apparently, this pre-Columbian hybridization occurred sometime during the Missoula Floods. In the B.C. interior, hybridization between westslope cutthroat and rainbow trout usually results from the introduction of rainbow trout into areas where they are not native (Rubidge et al. 2001).

Distribution   NATIVE DISTRIBUTION
Rainbow trout are native to North America and northeastern Siberia but have been introduced into cool waters around the world. In North America, except for the headwaters of a few Mackenzie tributaries (e.g., the Athabasca, upper Peace, and perhaps, the upper Liard rivers), the original distribution of rainbow trout was restricted to areas west of the Continental Divide. Here, freshwater-resident populations ranged from central Baja California to the Kuskokwim River, Alaska; however, the range of anadromous populations is more restricted. Steelhead occur from Malibu Creek, southern California, to the Alaska Peninsula. On the Asian side of the North Pacific, the original distribution included rivers flowing into the Sea of Okhotsk in the south and the Bering Sea in the north.

BRITISH COLUMBIA DISTRIBUTION
The sea-run form of rainbow trout (steelhead) occurs along the entire coast. The original B.C. distribution of the freshwater-resident form of rainbow trout has been obscured by extensive, and often unrecorded, introductions. Nonetheless, except for some native populations in the upper Peace (Parsnip and Finlay rivers) and Athabasca drainages, most of the native range probably was in west-flowing rivers (Map 45). It is not clear whether the isolated populations in the upper Liard system (Turnagain and Eagle rivers) are native or introduced. There are persistent rumours of unauthorized introductions in this region.

Life History   The rainbow trout is one of the most thoroughly studied fish in the world, and there are several succinct general descriptions of rainbow trout life history (e.g., Busby et al. 1996; Lynott et al. 1995; Smith 1991; Wydoski and Whitney 2003). The following account draws heavily on these sources.
In British Columbia, some populations are anadromous (steelhead), and other populations are freshwater residents. Embedded within each of these major life-history forms are populations that differ in size, colour,

Map 45 Approximate native distribution of rainbow trout (including steelhead), *Oncorhynchus mykiss*.

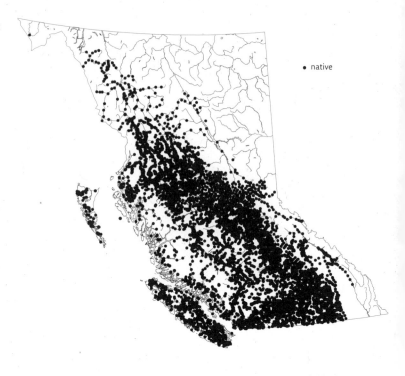

• native

migratory behaviour, run timing, and reproductive characteristics. For example, B.C. steelhead enter freshwater in the fall and winter (winter-run fish) and in the spring and summer (summer-run fish). Not only are these runs temporally separated, but the fish also differ in their state of maturation when they enter fresh water. Winter-run steelhead are almost fully mature when they enter fresh water and spawn shortly thereafter. In contrast, summer-run fish are immature when they enter the rivers and spend up to 8 months holding in fresh water before they spawn.

REPRODUCTION

In the wild, rainbow trout breed in the spring. Although they normally spawn in flowing water, a few introduced populations spawn on gravel beaches in lakes. Rainbow trout migrations into spawning streams are triggered by rising water temperatures (above 5 °C) and rising water levels (Hartman 1969). Thus, in British Columbia, the spawning migrations of freshwater-resident rainbows can begin at, or just before, ice-out (late April to May), and egg deposition usually occurs at about 8–15 °C (from late April to July depending on latitude and altitude). Often, geographically adjacent populations spawn at different times, and it is not unusual to find up to a month difference in spawning time between inlet and outlet populations in the same lake (Hartman et al. 1962).

In British Columbia, steelhead enter rivers in most months of the year: winter runs enter fresh water from November to April, and summer runs enter between May and September (Withler 1966). This difference in the timing of river entry is associated with distance to spawning grounds and seasonal differences in water levels that allow fish to surmount obstructions that become migration barriers at low water levels. Although both summer-run and winter-run populations occur in the inland and coastal rainbow trout clades (McCusker et al. 2000), most coastal populations are winter-run, and most inland populations are summer-run, fish. Typically, coastal summer-run populations are associated with migration barriers that are passable only during freshet. Usually, summer and winter runs in the same river are genetically similar; however, occasionally there are significant allele frequency differences between the seasonal runs within a river system (Beacham et al. 2004b). Steelhead usually spawn at water temperature ranging from about 4 °C to 10 °C.

A review of information on rainbow spawning habitat (Raleigh et al. 1984) indicates that water velocities of 0.30–0.90 m/s and depths of 0.15–2.5 m are normal. These numbers, however, are dependent on fish size. Thus, large-bodied populations (e.g., steelhead and the Gerrard rainbows of Kootenay Lake) spawn in faster (0.50–0.90 m/s) and deeper (1.75–2.00 m) water over larger substrates than small-bodied populations (Hartman 1969; Hartman and Galbraith 1970). Like other salmonines, female rainbow trout choose the spawning sites and, apparently, prefer sites with subgravel flow (e.g. in the tail-outs of pools immediately above riffles or upwelling sites). Females excavate a nest by turning on one side and strongly beating the caudal fin (Tautz and Groot 1975). This digging motion creates a vortex that loosens and lifts gravel that is then moved downstream by the current. The size of gravel moved is variable and depends on female size. In the later stages of nest excavation, the female begins to "probe" the nest by erecting the anal fin and sinking down into the deepest part of the nest. Eventually, the erect fin makes contact with the substrate and the female then rises out of the nest and resumes digging. The frequency of this probing behaviour increases up to about a half hour before egg release and then decreases slightly before actual spawning. Normally, the female is accompanied by a dominant male (usually the largest male) and a variable number of satellite males. In some populations, there are also small, precocious males (jacks) that may take part in spawnings. The dominant male is aggressive and attempts to drive satellite males and jacks away from the female. He courts the female by "quivering" (a series of high-frequency but low-amplitude body flexures) and "crossing-over" (a side to side movement of the male over the caudal peduncle of the female). Both quivering and crossing-over increase in frequency before spawning; however, the frequency of quivering asymptotes about 90 minutes before spawning, whereas the frequency of crossing-over asymptotes about half an hour before spawning (Tautz and Groot 1975). As actual spawning approaches, the male attempts to position himself and quiver alongside the female when she is probing the nest. At first, the female continues to lift herself out of the nest; however, she eventually "crouches" in the nest, and the male positions himself alongside. Experiments using wooden models of females (Newcome and Hartman 1980), demonstrated that a model positioned on the gravel induced male posturing and raising the model up off the substrate terminated male posturing. At gamete release, both sexes arch their backs slightly, gape, tilt their bodies to the outside, vibrate their tails rapidly, and release eggs and sperm. After spawning, the female moves slightly upstream and begins gentle digging movements. Apparently, the currents produced by these gentle tail beats do not move any gravel but force the eggs into interstices at the bottom of the nest (Tautz and Groot 1975). The digging movements gradually become more vigorous and gravel is displaced downstream over the nest. Usually a female does not deposit all her eggs in a single spawning. After a short rest, she moves to the upstream edge of the nest, digs another nest, and deposits more eggs. Each new nest is immediately upstream of the last nest. This sequence is repeated a number of times and, eventually, results in a large redd (a sequential string of nests excavated by a single female). Like many anadromous salmonines, some steelhead males do not smolt and migrate to sea. Instead, they remain in fresh water and reach sexual maturity as parr. There is indirect (genetic) evidence that, in some years, mature parr may be collectively more successful reproductively than anadromous steelhead males (Seamons et al. 2004b).

Fecundity in rainbow trout varies with female body size. Typically, lake-dwelling populations in central British Columbia produce about 1,000–3,000 eggs per female; however, fecundity is much lower in small-bodied, stream-resident populations (200 or fewer eggs per female) and much higher in large-bodied anadromous and piscivorous populations. For example, spawning Gerrard rainbows range in size from about 48 to 97 cm FL and produce about 3,000–14,000 eggs (Irvine 1978). This is comparable with the fecundity in some steelhead populations (Bulkley 1967). Egg size varies with female size and among populations. Fertilized eggs range in diameter from about 3.0 to 7.0 mm, and small-bodied rainbow trout tend to have smaller eggs than large-bodied trout and steelhead. Incubation time varies with temperature and among populations: time to 50% hatch was 60 days at 5 °C and 18 or 19 days at 14 °C (Murray 1980a). Incubation temperatures below 5 °C and above 13 °C or 14 °C produce increased egg mortality and sub-vital fry (Murray 1980a, Timoshina 1974). The optimum incubation temperature is around 8–12 °C (Murray 1980a). Newly hatched rainbow trout (alevins) are negatively phototactic and remain in the gravel until most of the yolk sac is absorbed. Depending on temperature, 50% emergence occurs from 42 (5 °C) to 32 (14 °C) days after hatching (Murray 1980a), and exogenous feeding begins shortly after emergence when the swimbladder is inflated (MacCrimmon and Twongo 1980). At hatching, rainbow trout alevins range from 11 to 13 mm TL, and the fry are about 18–21 mm when they emerge from the gravel (Murray 1980a).

AGE, GROWTH, AND MATURITY

Typically, rainbow trout and steelhead both reach a length of about 100 mm in their first summer of growth. However, temperature, food availability, habitat type, population, and interactions with other species all influence the growth rates of fry (Post and Parkinson 2001). Thus, fry in northern rivers and headwater streams in British Columbia average about 35–50 mm FL by the end of their first growing season, whereas fry in southern rivers often exceed 120 mm FL by late fall. For steelhead, body size rather than age appears to govern whether or not they will smolt in the spring. Smolting usually occurs at or above about 160 mm in length (Burgner et al. 1992; Peven et al. 1994). This threshold size is rarely reached until the end of the second or third summer of stream life.

In British Columbia, size and age at maturity varies among populations: small-bodied, stream-resident populations can mature as early as 1^+ and at <150 mm (Northcote and Hartman 1988); in contrast, large-bodied, piscivorous populations can mature as late as 4^+ and at >400 mm (Andrusak and Parkinson 1984). Regardless of population, however, males usually mature at least a year before females. Compared with char, rainbow trout are relatively short lived, and in British Columbia, fish older than 10 years (9^+) are rare. In British Columbia, most steelhead mature after 2 or 3 years at sea, but a significant proportion of some populations mature after 1 year at sea. Since most B.C. populations spend 2 or 3 years in freshwater before migrating to the ocean, their ages at maturity range from 3 to 6 years. Unlike Pacific salmon, steelhead are iteroparous: they can spawn more than once. In British Columbia, however, less than 10% of steelhead (usually females) live to spawn a second time, and very few spawn more than twice (Hootten et al. 1987). Thus, the maximum age achieved by steelhead in British Columbia is probably around 8 or 9 years.

FOOD HABITS

Like most salmonines, the diet of rainbow trout varies with size, season, time of day, and population. In streams, fry and juveniles of both rainbow trout and steelhead forage primarily on the drifting stages of aquatic insects. This food source peaks at night, and consequently, these fish consume relatively little food during the day (Tippets and Moyle 1978); however, terrestrial and emerging aquatic insects are added to the diet as they grow. Permanent stream- or river-resident adults retain this diet throughout life. At low temperatures ($<8\,°C$), the major feeding period is at night (shortly after sunset), and prey consists primarily of drifting benthic invertebrates. At temperatures above $8\,°C$, however, there is a second feeding period during daylight that targets terrestrial invertebrates and emerging aquatic insects (Elliot 1973). Interestingly, in a part of the Colorado River with a depauperate forage base, mid-size and large trout consumed more algae than small trout (McKinney and Speas 2001). Using an energy intake model, these authors argue that, in flowing water, large rainbow trout are food-limited more often than small trout. This may partly explain the widespread occurrence of "dwarf" trout and char above barriers in many headwater streams. Although persistent piscivory is rare in river-resident populations, cannibalism is common at times, and stomachs of large (>300 mm) river-resident adults often contain small trout.

In some populations, there are two foraging types: trout that forage during the day and trout that forage at night (Braennaes and Alanaerae 1997). Under laboratory conditions, these two foraging types maintained their diel foraging pattern even in the absence of the other type. Typically, the nocturnal form is smaller and grows more slowly than the diurnal form. Perhaps, nocturnal foraging allows individuals with low social status to feed at a time when dominant individuals are less aggressive (Alanaerae and Braennaes 1997). Diel patterns of habitat use by juvenile steelhead in the Bridge River were complex and varied with season, locality, and fish size (Bradford and Higgins 2001). In a reach with relatively high flows, most fish were nocturnal all year and emerged to feed at dusk. In a reach with low flow, however, some fish were day active in the summer, and others remained concealed until dusk. In summer, parr and older fish were more nocturnal than fry. All fish sizes were nocturnal during the winter.

In lakes, fry forage on both bottom organisms (amphipods, snails, and the nymphs of aquatic insects) and water-column organisms (especially chironomid pupae and cladocerans). Typically, lacustrine juveniles and adults have similar diets but take larger prey and more adult insects (terrestrial and aquatic) than fry. Potentially, diet in lacustrine populations is influenced by prey abundance and the presence of other fish species. Thus, during the summer, some populations are primarily planktivorous (Barton and Bigood 1980; Lynott et al. 1995), other populations are primarily benthivores (Buktenica and Larson 1996; Johannes and Larkin 1961), and individuals over 400 mm in large lakes (especially if kokanee, *Oncorhynchus nerka*, are present) often become piscivorous (Keeley et al. 2005).

The marine diet of steelhead is similar to that of coho (*Oncorhynchus kisutch*) and Chinook salmon (*Oncorhynchus tshawytscha*). In their first year at sea, euphausiids and fish are the main prey (Pearcy et al. 1990). Fish, squid, polychaetes, and crustaceans are important food items throughout the marine phase of their life (Burgner et al. 1992; Taylor and LeBrasseur 1957).

Habitat   Generally, rainbow trout are a cool-water species, and depending on acclimation time, the upper lethal temperature for adults is about 27 °C (Lee and Rinne 1980). They appear to prefer temperatures between 7 °C and 18 °C (Raleigh et al. 1984). Nevertheless, the habitats used by rainbow trout are as variable as their life histories. In British Columbia, five broad habitat types are recognized (Keeley et al. 2005): anadromous, lacustrine, large river, stream, and headwater habitats. There are subtle morphological differences (ecomorphs) associated with each of these broad habitat types.

ADULTS

During the summer in streams and rivers, adult rainbow trout occupy riffles, runs, glides, and pools. They tend to occur in deeper and faster water than juveniles. In the Nazko River, central British Columbia, adults occurred mostly in runs with depths of 0.5–1.0 m, cobble–boulder substrates, and average water velocities of 0.40–0.80 m/s (Porter and Rosenfeld 1999). Adult rainbows in two Kootenay River tributaries in Montana selected mid-stream depths of 0.4–0.9 m and water velocities <0.20 m/s in one stream and <0.50 m/s in the other stream (Muhlfeld et al. 2001b). In the same study, adults occurred in pools more often than expected from the availability of pools; they appeared to avoid riffles and used runs in proportion to their availability. In small streams, overhead cover (primarily riparian vegetation and large woody debris) is an important component of "good" trout habitat (Flebbe and Dolloff 1995). In small streams in the fall, adult rainbows move into primary pools (pools that span the entire channel width) to overwinter (Muhlfeld et al. 2001a). These primary pools were dominated by cobble and boulder substrates and contained large woody debris. In such habitats, adults take cover among boulders or under woody debris during the day and emerge at night (Meyer and Gregory 2000).

Lake-dwelling rainbow trout usually occur in mesotrophic or oligotrophic lakes. In these habitats, they usually stay below the 18 °C isotherm in areas where the oxygen concentration is above 3.0 mg/L (Raleigh et al. 1984). In small lakes, adult rainbows use all parts of the lake (Crossman 1959) but are often associated with cover (logs and other large woody debris) in the lower littoral zone. In a large lake (Lake Washington) during the summer, telemetered adults were relatively inactive at night and remained within 50 m of the shore (Warner and Quinn 1995). They became more active near dawn and most active at dusk. Over 90% of their time was spent within 3 m of the surface; however, they made brief dives into deeper water (an average of 6.6 m). In Lake Tahoe, adult rainbows were associated with complex boulder habitats in the littoral zone (Beauchamp et al. 1994). Usually, rainbow trout are rare in the offshore waters of large oligotrophic lakes (Luecke and Teuscher 1994). Some plankton feeding populations, however, do occur offshore and perform diel vertical migrations (Levy et al. 1991).

B.C. steelhead usually spend 1–4 years at sea before returning to fresh water on their initial spawning migration. Typically, the modal ocean age of fish returning to B.C. rivers is 2 years, but there are exceptions. For example, 68% of Nanaimo River steelhead return after only 1 year at sea (Narver and Withler 1974). Winter-run steelhead return to fresh water in the late fall or winter (November to early April) as almost fully mature adults. They ascend their natal streams, hold in pools downstream of their spawning sites for a short time, and then spawn. In contrast, summer-run fish return to fresh water in the spring or summer (May to September) with their gonads still immature. They ascend their natal streams and hold (for up to 6 months) in deep pools containing an abundance of cover (boulders, ledges, or overhanging vegetation) before spawning (Busby et al. 1996).

JUVENILES

In streams and rivers, juvenile rainbow trout also occupy riffles and runs; however, in these habitats, they are found in shallower and slower water than adults. In central British Columbia, juveniles used runs with depths of <0.25 m, cobble–boulder substrates, and average water velocities of 0.20–0.40 m/s (Porter and Rosenfeld 1999). Like adults, juveniles in flowing water appear to prefer areas that are away from vegetation and woody cover in the summer (Porter and Rosenfeld 1999). In two Idaho streams, juvenile (1^+) steelhead were territorial and associated with large substrate (>20 cm) and relatively deep (0.6–0.9 m), fast (0.60–0.90 m/s) water in the summer (Everest and Chapman 1972). The use of cover by juvenile steelhead is influenced by flow (Shirvell 1990): at high flows in the Bridge River, juveniles were nocturnal throughout the year, but in the summer at lower flows, some individuals were day-active (Bradford and Higgins 2001). In winter, juveniles shelter within the substrate during the day and emerge from cover at dusk (Bradford and Higgins 2001; Contor and Griffith 1995). In the fall in southeastern Alaska, juvenile steelhead migrated from mainstem rivers into small tributaries to overwinter (Bramblett et al. 2002). Presumably, similar fall migrations occur in rivers along the northern coast of British Columbia.

In lakes, juveniles remain inshore during winter and early spring (Beauchamp 1990b) and, by day, are associated with cover (cobble and boulder substrates or woody debris). At night, however, they leave cover and forage over sand and gravel substrates (Tabor and Wurtsbaugh 1991). In some lakes, juveniles move offshore in late spring and summer (Beauchamp 1990b).

After spending two or three summers in fresh water, juvenile steelhead transform into smolts in the spring or early summer and migrate to sea. There is, however, considerable variability in the age of smolts: some migrate after two summers in fresh water, and others migrate after four summers. Not much is known about the behaviour of smolts once they enter the ocean. They appear to move directly offshore (Hartt and Dell 1986; Welch et al. 2004), and by late summer, most North American steelhead smolts are concentrated in the western Gulf of Alaska (Burgner et al. 1992). In the fall, they apparently turn south and east towards the North American coast; however, in the winter, they are thought to again move west, and by their second marine summer, they are widely distributed in the North Pacific Ocean. Although there is broad overlap, the oceanic distributions of summer-run and winter-run steelhead are slightly different (Burgner et al. 1992). By late July, ocean age-1 steelhead are mainly distributed south of the eastern Aleutian Islands and in the western Gulf of Alaska. Apparently, in late summer, ocean age-1 summer-run fish that are returning to North American rivers begin migrating towards the coast; however, the main bulk of the oceanic population shifts south and east through the fall and winter (Burgner et al. 1992). By the spring or summer of their second year at sea, ocean age-2 fish that will spawn this year begin to migrate towards the North American coast. This migration is rapid (15–85 km/day). Once they reach coastal waters, the rate of travel appears to slow and sometimes stops at night (Ruggerone et al. 1990).

YOUNG-OF-THE-YEAR

In flowing water, newly emerged rainbow trout display agonistic behaviour as soon as they swim-up; early in their first summer, they establish territories in shallow water along stream margins. Both the level of aggression and territory size are influenced by fry density and prey abundance (Slaney and Northcote 1974). In two Montana streams, 0+ fry were associated with stream margins. They selected depths of <20 cm, small gravel substrates, and water velocities of <0.01 m/s (Muhlfeld et al. 2001b). As they grow, the largest fry move from stream margins into mid-channel areas, and most of the large fry (85–140 mm) are found in mid-channel by autumn (Mitro and Zale 2002). Cover is important to fry production (Culp et al. 1996), and fry survival during their first winter is dependent on size, cover, and temperature (Meyer and Griffith 1997; Smith and Griffith 1994). Fry less than 100 mm in length have a low probability of overwinter survival (Smith and Griffith 1994). Such size-dependent mortality is more likely in severe (cold) environments (Meyer and Griffith 1997).

Fry of most lacustrine populations migrate to their lake late in their first summer or in early fall, although fry spawned in some outlet and headwater streams overwinter in the stream and migrate the next spring (Alexander and MacCrimmon 1974; Lindsey et al. 1959). In lakes, fry remain in shallow water, typically about 2–5 m offshore (Wurtsbaugh et al. 1975). During the day, they are often associated with cover (cobble, boulders, and woody debris) and emerge to forage at night.

Steelhead fry occupy habitats similar to those used by rainbow trout fry. At first, they occur in slight currents along stream margins where they establish territories. During the summer in two Idaho streams, steelhead fry were associated with the bottom and occurred at highest densities in shallow (<50 cm), low velocity (usually <0.40 m/s) waters over rubble substrates (Everest and Chapman 1972).

Conservation Comments Rainbow trout have been cultured in British Columbia for over 100 years. Originally many, if not most, of the famous angling lakes in the Kamloops region of British Columbia that are 100 m or more above the valley floors of major rivers were fishless, and their trout populations are the result of early stocking programs. These early stocking records are not particularly reliable, and sometimes, it is not clear which populations are native and which are not. Nonetheless, there are still native populations, especially in the central part of the province, and every attempt should be made to preserve native populations from potential contamination by semi-domestic hatchery strains. Aside from protecting some native populations, freshwater resident rainbows are not a major conservation issue in British Columbia.

Steelhead, however, are another matter. In a 1996 assessment of B.C. stocks, 9 steelhead stocks (out of 867) were listed as extirpated, 8 more were listed as at a high risk of extirpation, and 143 stocks were of special concern (Slaney et al. 1996). Most of the extirpated stocks were in the urbanized areas of the Georgia Strait region or in the B.C. portion of the Columbia River system (casualties of Grand Coulee Dam). Most of the populations of special concern are found along the southern coast (especially the lower Mainland and both coasts of Vancouver Island), but some interior populations (e.g., Thompson River steelhead) are also stocks of concern. Ocean survival seems to be a special problem for the southern coastal stocks; however, over-exploitation, interception by commercial fisheries, and the loss and degradation of freshwater rearing habitat also are serious concerns. Wild steelhead are an important resource, and the chances of their survival will not be helped if Atlantic salmon (*Salmo salar*) become established on our coast.

Oncorhynchus nerka (WALBAUM)
SOCKEYE SALMON

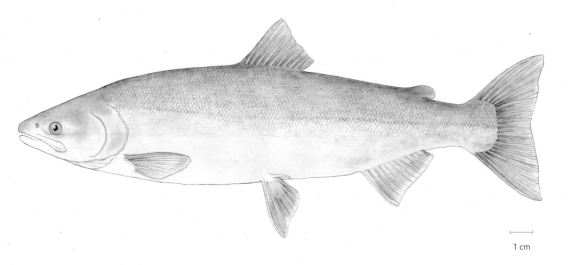

1 cm

Distinguishing Characters — These silvery, trout-like fishes have no dark spots on the back, sides, or dorsal and caudal fins (occasionally kokanee have a few speckles on the outer edge of the tail or on the dorsal surface of the body); 30–40 long, slender gill rakers (19–27 on the lower limb of the first gill arch); and 13 or more anal rays. The anal fin base is longer than the dorsal fin base. Fry (up to about 3 cm) lack a dark stripe on leading edge of the dorsal fin (a dark stripe is present in trout of the same size); larger fry and juveniles have small oval-shaped lateral parr marks, none of which are much higher than the vertical eye diameter. Some of the parr marks are divided roughly in half by the lateral line.

Taxonomy — Although the specific status of the sockeye salmon is not an issue, it is a genetically heterogeneous species and locally adapted populations are common (Gustafason et al. 1997; Taylor 1991a). Additionally, kokanee (freshwater residents) have evolved repeatedly from many different sockeye populations, and some kokanee populations spawn in the same stream and at the same time as anadromous sockeye. Nonetheless, they retain a suite of inherited morphological, physiological, and behavioural characteristics that differentiate them from sockeye (Craig 1995; Danner 1994; Foote et al. 1989; Robison 1995; Taylor and Foote 1991; Taylor et al. 1996; Wood and Foote 1990, 1996). The sockeye/kokanee dichotomy is a classic example of parallel evolution.

In some places, kokanee and anadromous sockeye act as biological species (*sensu* Mayr 1963). To avoid taxonomic chaos and to recognize that, even on different continents, the same genes are often involved in these independent divergences, members of such species pairs are referred to by the taxonomic name of the widespread species from which

Sexual Dimorphism At sea, or in the case of kokanee in lakes, the sexes have the same body shape, bright silver flanks, and a blue back. As they mature, however, sockeye and kokanee change shape and colour. In males, the head enlarges, the teeth on the jaws increase in size, and a down-turned hook develops on the end of the snout; to a lesser extent, the tip of the lower jaw develops an upturned hook. Also, mature males develop a laterally compressed body and a fleshy hump anterior to the dorsal fin. These secondary sexual characteristics are not as pronounced in females as in males. Since many sockeye populations enter fresh water months before they actually spawn, the development of their secondary sexual characteristics typically occurs in fresh water.

Spawning colours in sockeye salmon are variable. Typically, both sexes become red as they approach spawning; however, the red is more intense in males than in females, and males usually develop a green head (olive-drab in females). The undersides become dusky or black in most populations. Many kokanee populations develop secondary sex characteristics similar to those found in sockeye salmon; however, the red colour is usually less intense in kokanee. Also, some kokanee populations do not turn red but instead become a washed-out black, burgundy, or drab green above and dirty white below; in other kokanee populations, there are no obvious changes in head or body shape in either sex; and in still other populations, some males develop the full suite of sexually dimorphic traits and other males resemble (mimic?) females in colour and shape.

Distribution **NATIVE DISTRIBUTION**
Sockeye salmon are native to both the Asian and North American Pacific coasts. In Asia, sockeye range from northern Hokkaido to the Chukotsk Peninsula; however, most of the major spawning areas are in Kamchatka. In North America, anadromous sockeye are reported sporadically from the Sacramento River, California, to the Mackenzie River and other Beaufort Sea tributaries in the western Arctic; however, their region of maximum abundance is much narrower: from the Columbia River north to the Kuskokwim River, Alaska. Kokanee have a North American distribution (Columbia system to the upper Yukon system) similar to sockeye, but this life-history type is not widespread in Alaska (Rounsefell 1958) or Asia (Berg 1948).

BRITISH COLUMBIA DISTRIBUTION
Sockeye and kokanee are widely distributed (Map 46), and anadromous sockeye spawn in over 300 lakes and streams (Aro and Shepard 1967). Post-glacially, sockeye colonized most of the west-flowing rivers in British Columbia. They now ascend rivers, such as the Fraser and Skeena, to their headwaters; in the Columbia River before the construction of Grand Coulee Dam, they reached Windemere and Columbia lakes (Fulton 1970).

Map 46 The B.C. distribution of sockeye salmon (including kokanee), *Oncorhynchus nerka*

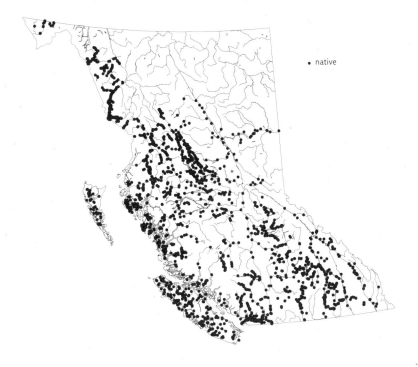

• native

The non-migratory form of sockeye (kokanee) evolved in many B.C. lakes, and some of these kokanee are now isolated from their anadromous progenitors. Still, from the Fraser system north, many kokanee populations remain in contact with anadromous sockeye. Although fish-culture activities have spread kokanee throughout the province, only two natural populations are known from the Mackenzie River system. These populations (Arctic Lake in the headwaters of the Parsnip River and Thutade Lake in the headwaters of the Finlay River) represent two, independent colonizations of the drainage: one from the upper Fraser River system and the other from the upper Skeena system.

When the reservoir behind Bennett Dam (Williston Reservoir) was filled, it created a large limnetic habitat far downstream from the original Peace River kokanee populations. Nonetheless, kokanee appeared in the reservoir before any non-native kokanee were stocked into the system. This suggests that, despite strong selection against downstream migrants, Williston Reservoir was originally colonized by native kokanee from headwater populations in the upper Peace River. Southern kokanee (mostly from Kootenay Lake stocks) were introduced, and there is now a large kokanee population in Williston Reservoir. It is not clear if the original kokanee populations still exist in the reservoir. Also, it is possible that occasional juveniles still migrate downstream out of the reservoir, and these may be the source of the kokanee reported from Great Slave Lake.

Life History The two basic life-history types in sockeye salmon are the anadromous (sockeye) life-history type and the non-anadromous (kokanee) life-history type. These life-history types differ in a suite of inherited morphological and physiological traits. Additionally, a host of variant life-history types are embedded within these basic life histories. Although the following account emphasizes the life history of kokanee and the freshwater phases of anadromous sockeye, some details of the sockeye marine phase are included. Burgner (1991) and Gustafason et al. (1997) provide excellent life-history reviews of anadromous sockeye salmon.

Although the typical sockeye life history involves a juvenile lake-rearing phase, there are some populations that do not rear in lakes (Beacham et al. 2004a; Gustafason and Winans 1999; Wood 1995; Wood et al. 1987). Some of these populations rear in streams and rivers for 1 or 2 years before entering the sea. They are referred to as river-type sockeye. There are also populations in which the fry migrate directly to estuaries and then to sea in their first summer. They are referred to as sea-type sockeye.

REPRODUCTION

Kokanee spawn in the fall, usually when water temperatures drop below 12 °C (in British Columbia, this normally occurs in September or October); however, the migration of adults to spawning sites can start as early as late July. Kokanee enter spawning streams or move onto shore-spawning sites at night (Lorz and Northcote 1965). The timing of freshwater entry by anadromous sockeye is remarkably variable. Freshwater entry for most runs peaks in July or August, and actual spawning peaks from August to November; however, some populations (e.g., Cheewat and Hobiton lakes on Vancouver Island) begin entering freshwater as early as February or late March, although spawning does not occur until the fall (September to early November). The population in Ozette Lake on the Olympic Peninsula, Washington, has a protracted period of freshwater entry (April to mid-August) and an equally protracted spawning period (November to late March).

Like other salmonines, sockeye females choose the spawning site and, apparently, prefer sites with subgravel flow: shallow riffles, the outlets of lakes, or on beaches in lakes where there is upwelling (Parsons and Hubert 1988). River-type and sea-type sockeye typically spawn in small tributaries and side-channels associated with glacial rivers. Again, they usually choose sites where there is some upwelling (Beacham et al. 2004a; Lorenz and Eiler 1989). Nest digging and courtship is typical of other salmonines (for details, see descriptions under rainbow trout). Like pink salmon (*Oncorhynchus gorbuscha*), sockeye and kokanee often form dense aggregations on spawning grounds. Depending on water velocity and female size, gravel diameters range from about 1.0 to 2.5 cm. Water velocities and depths are also variable and range from 0.15 to 0.85 m/s and from 6 to 37 cm, respectively (Acara 1977; Delisle 1962; Parsons and Hubert 1988; Smith 1973). Although males are more aggressive than females, both sexes defend the spawning site.

During the nest-digging phase, the male courts the female by nudging her side with his snout and occasionally pressing his body against the female and quivering. After spawning, the female moves slightly upstream and begins rapid digging movements. The gravel displaced by this digging covers the fertilized eggs. After depositing most of her eggs, the female remains on site and defends her redd until too weak to maintain position. Like other Pacific salmon, kokanee and sockeye die after spawning.

In lakes, kokanee and sockeye usually spawn inshore in areas where there is upwelling or some subsurface flow (Burgner 1991; Harris 1986). Again, the substrate is variable but often is too large (up to cobble size) for females to dig nests. Instead, the site is cleaned of sand and silt by the female's attempts to dig, and the fertilized eggs fall into crannies among the gravel and cobbles. The depth of lake-spawning sites usually is less than 10 m; however, these are sites that can be observed, and in Iliamna Lake, Alaska, sockeye spawn at a depth of almost 30 m (Burgner 1991). In Anderson and Seton lakes, the mysterious moribund fish that appear on the surface in late December are kokanee that probably spawn in deep water.

In rivers and streams, spawning usually involves a dominant male and a female. In anadromous populations, satellite males, jacks (small precocious males), and (where anadromous sockeye and kokanee coexist) kokanee males also take part in sockeye spawnings. When the female crouches in the nest and the dominant male comes alongside, these supernumerary males rush in and take part in the spawning. They fertilize a variable number of the eggs (Foote et al. 1997). In some small-bodied kokanee populations, non-dominant males develop female colouration and behaviour. Presumably, they gain some reproductive success by surreptitiously joining paired spawnings.

In sockeye salmon, egg size varies from about 5 to 6 mm, and fecundity from about 2,000 to 4,000 eggs (Burgner 1991). Typically, this variation is related to female size, but there is also variation within and among populations (Harris 1986). In kokanee, there can be significant differences in egg size and egg number in adjacent populations. For example, kokanee spawning in two streams entering Galena Bay, upper Arrow Lake, differed in both egg size and egg number (Murray et al. 1989). Kokanee have smaller eggs (about 4.5–5.3 mm) than sockeye salmon (about 5.3–6.6 mm). Fecundity ranges from about 200 to about 1,500 eggs in kokanee and from 1,500 to 6,000 in sockeye (Acara 1977; Burgner 1991; Gustafason et al. 1997; Lorz and Northcote 1965; Murray et al. 1989; Vernon 1957). As in most fish, development rate is a function of incubation temperature. For Fulton River sockeye incubated at constant temperatures, 50% hatching occurred in 119 days at 5 °C, 80 days at 8 °C, and 57 days at 11.0 °C (Velsen 1980). Coastal (Weaver Creek) and interior (Adams River) sockeye were incubated at 2 °C, 4 °C, 8 °C, 12 °C, and 15 °C (Beacham and Murray 1989): at all temperatures, the interior stocks developed faster, hatched earlier, and emerged earlier than the coastal stocks. For kokanee, the time from fertilization to emergence is 45 days at 14 °C and 73 days at 8 °C (Murray et al. 1989). Survival of developing eggs was highest at 6 °C and lowest at 2 °C. Kokanee and sockeye preferentially spawn in areas with upwelling ground water (Burgner 1991; Garret et al. 1998), and temperatures in groundwater-influenced redds are about 2.5 °C higher than surface flow (Garret et al. 1998). In addition, there is significantly greater survival to hatching in groundwater-influenced redds than in redds from other areas.

Upon emergence, kokanee and lake-type sockeye usually migrate to a nursery lake before starting to feed. This downstream migration occurs at night with the peak migration between dusk and midnight (Lorz and Northcote 1965). The fry are negatively phototactic and, if the migration takes more than one night, they shelter during the day under rocks and organic debris.

AGE, GROWTH, AND MATURITY

Size at hatching in *O. nerka* is primarily dependent on egg size (Murray 1980a). Consequently, kokanee produce smaller emergent fry (about 19.5 mm) than anadromous sockeye (about 25–27 mm); however, there is considerable variation in egg size and alevin size both within and among anadromous populations. Growth rate of newly emerged fry is influenced by temperature and food availability. By the end of their first summer in British Columbia, kokanee fry usually range from about 40 to 50 mm FL, whereas anadromous sockeye fry of the same age range from about 60 to 70 mm. Where they occur in the same lake with sockeye, kokanee fry usually are significantly smaller than sockeye fry (Wood et al. 1999). In Koocanusa Reservoir, 0^+ kokanee continue to grow overwinter (they add about 42 mm) and are about 180 mm by the end of their second growing season (Chisholm et al. 1989). In British Columbia, kokanee typically reach sexual maturity at the end of their third (2^+) or fourth (3^+) summer; however, even within a lake, there may be differences in the average age at maturity in different parts of the lake. For example, kokanee in the southern and western arms of Kootenay Lake matured mainly at 2^+, whereas most fish matured at 3^+ in the northern arm of the lake (Vernon 1957). Size at maturity also varies among populations and, in British Columbia, ranges from about 150 to 350 mm (Lorz and Northcote 1965; Northcote et al. 1972; Vernon 1957). When Koocanusa Reservoir was initially filled, kokanee developed a bimodal size distribution with modes at about 220 and 480 mm (Chisholm et al. 1989).

In British Columbia, lake-type sockeye typically spend 1 or 2 (occasionally 3) years in fresh water before migrating to sea. River-type sockeye also spend at least 1 year in fresh water, whereas sea-type sockeye migrate to sea in their first year of life. Smolt size varies among and within populations—about 60–90 mm for age-1 smolts and 100–120 mm for age-2 smolts. The smallest smolts (about 60 mm) in the province are produced in the turbid waters of Owikeno Lake. The length of time spent at sea also varies. In British Columbia, most sockeye spend 1–3 years at sea. Body size increases with the length of the sea phase, and most of the fish returning after 1 year at sea are precocious males (jacks).

FOOD HABITS

Migrating kokanee and lake-type sockeye fry usually begin feeding shortly after emergence. If the migration to their nursery lake is short, they may not begin feeding until the reach the lake; however, if the migration is relatively long, they begin feeding before they reach their lake. Zooplankton are the primary food of fish hatched downstream of a lake, but if zooplankton are unavailable, chironomid larvae and pupae are often important food sources. On lake entry, sockeye fry often spend a month or so in the littoral zone foraging on a variety of limnetic and benthic prey (chironomid larvae and pupae, copepods, and cladocerans). The fry of beach-spawning populations also forage in the littoral zone before moving offshore in midsummer. In some lakes, the transition from littoral to limnetic foraging is preceded by a period of diel migrations: offshore at night and inshore during the day (Burgner 1991). Once fry start limnetic feeding, their primary prey are crustacean zooplankters (Finnell and Reed 1969; Lorz and Northcote 1965); however, some kokanee populations forage on chironomids at night (Chapman et al. 1967). In Nicola Lake, adult and subadult kokanee fed on zooplankton in the fall and spring, but chironomids were important prey during the summer (Northcote and Lorz 1966).

At sea, sockeye forage primarily on macrozooplankton (e.g., euphausiids and amphipods), small fish, and squid (French et al. 1976).

Habitat Habitat use in sockeye salmon is complex. Not only do the two major life-history types (sockeye and kokanee) differ in the major environments they occupy through most of their lives, but also within life-history types (especially in the anadromous form), there are differences in the habitats used by the early life-history stages. The following account is based on the two primary reviews (Burgner 1991; Gustafason et al. 1997) that cover habitat use in this species.

ADULTS

Adult kokanee live in the offshore habitat of lakes. In many lakes, kokanee are crepuscular foragers, and in stratified lakes, they feed in the food-rich middle or upper strata at dawn and dusk and migrate down into the cool hypolimnion at night and during the day (Finnell and Reed 1969; Levy 1990). These crepuscular foraging migrations may increase the bioenergetic efficiency of kokanee and reduce predation (Levy 1987). Interestingly, the migrations stop in the fall when lakes turn over and become thermally uniform.

Sockeye salmon spend 1–4 years at sea making a counterclockwise migration to the Gulf of Alaska and then turning southwest along the Alaska Peninsula. Sometime in late autumn or early winter, they turn to the southeast, and by the summer of the next year, they have swung around and are again moving to the northwest into the Gulf of Alaska. This time, however, they are well offshore. This circular pattern of migration is repeated every year with maturing fish moving towards the coast in June and July and, eventually, returning to their natal river system in late summer.

JUVENILES

Juvenile kokanee use the same habitat as adults and show the same diel vertical migrations; they are found in the hypolimnion during the day and migrate upwards at dusk. However, after juveniles reach a threshold size in some lakes, they move inshore to forage during the day (Levy 1990; Northcote et al. 1964a). Presumably, this behaviour is associated with a size-related change in predation risk. Like kokanee, juvenile sockeye also use the littoral zone and make similar diel vertical migrations.

After one or two (rarely three) summers in fresh water, lake-type juvenile sockeye undergo a series of physiological, morphological, and behavioural changes that prepares them for their migration to, and subsequent life in, the sea (Groot 1982). In complex lake systems (e.g., Babine Lake), smolts use celestial cues to find the lake outlet. Smolt migrations usually occur at night. The timing of migrations out of nursery lakes varies with latitude and altitude (Burgner 1991) but usually occurs shortly after ice-out in the spring. In British Columbia, most smolts migrate in April, May, and June.

River-type juveniles spend 1 or 2 years in fresh water, and sea-type juveniles enter the sea in their first year of life. Once in the ocean, smolts concentrate near the river mouths. Sometime in late May, Fraser River smolts begin to move west into the Gulf Islands and north along the mainland coast. By late June or July, they exit the Strait of Georgia (usually through the northern passes between Vancouver Island and the mainland) and enter the open ocean. They migrate north along the coast towards the Gulf of Alaska and usually remain within about 50 km of the shore.

YOUNG-OF-THE-YEAR

On entering their nursery lake, the fry of some kokanee populations immediately move offshore and begin crepuscular vertical migrations. Other populations, however, remain inshore and forage in the littoral zone for variable amounts of time. These differences in fry behaviour probably are related to food availability, temperature, and predation risk.

The migrations of newly hatched lake-type sockeye fry can be complex, and there is evidence of heritable differences among populations in their responses to current and direction of movement (Brannon 1972; Quinn 1980; Raleigh 1967). Sockeye that spawn in inlet streams migrate downstream to their lake, whereas those that spawn in lake outlets migrate upstream. Weaver Creek sockeye fry first migrate downstream to the Harrison River and then turn upstream to reach Harrison Lake. Regardless of the pattern of migration, lake-rearing sockeye fry usually remain in the littoral zone for a variable period of time before making the transition to the limnetic zone. Exceptions are beach-spawning populations. Upon emergence, they sometimes move directly into the limnetic zone.

Not a lot is known about the habitat use of river-type sockeye during their first year of life, although the fry initially concentrate in quiet water along river margins. In the Taku River, a mixture of river-type and sea-type sockeye use riverine habitats, especially sloughs, backwaters, and off-channel habitats, such as tributary mouths, beaver ponds, and upland sloughs (Murphy et al. 1989). Water velocities and depths in the sloughs and backwaters ranged from 0.0 to 0.16 m/s and from 0.3 to 0.7 m, respectively. Off-channel water velocities ranged from 0.0 to 0.15 m/s, and depths were from 0.2 to 1.4 m. Sea-type sockeye fry possess physiological adaptations that allow them to enter the sea as under-yearlings (Rice et al. 1994). They usually migrate downstream to estuaries and feed in tidal creeks for 3 or 4 months before entering the sea (Levings et al. 1995). They grow rapidly and are similar in size to 1^+ lake-type smolts when they enter the sea.

Conservation Comments There are about 900 sockeye salmon stocks in British Columbia (Slaney et al. 1996) and well over 500 kokanee populations. There is information on the status of about 60% of the sockeye stocks. Twenty stocks have been extirpated, another 61 are at high risk, and at least 2 stocks are considered to be endangered (the Sakinaw and Cultus lake stocks). About 30% of the stocks at risk occur in the Johnstone Strait area, but there are stocks that are at risk along the entire coast. Dams have caused some of the extirpations. Grand Coulee Dam eliminated six stocks, and Coquitlam and Alouette dams in the Lower Mainland also eliminated sockeye runs.

An unknown number of small sockeye populations occur along the B.C. coast. These small populations are potential reservoirs of unusual life-history types. Most of these populations are not direct targets of commercial fisheries, but they are incidentally harvested in fisheries timed to intercept larger runs. Such populations pose a conundrum for managers and a test for the Species at Risk Act. Can managers maintain a viable fishing industry and, at the same time, protect small populations that are now in serious decline? The ultimate fate of the Sakinaw and Cultus Lake sockeye populations (both listed as endangered by COSEWIC) will be a test of the commitment of management agencies and politicians to the preservation of biodiversity.

Oncorhynchus tshawytscha (WALBAUM)
CHINOOK SALMON

10 cm

Distinguishing Characters   At sea, Chinook salmon are most likely to be confused with coho salmon (*Oncorhynchus kisutch*). The marine colour patterns of adult Chinook and coho salmon are superficially similar—both species have dark spots on their backs and tails. However, the black spots on the tails of Chinook are on both the upper and lower lobes of the caudal fin, whereas in coho, the spots are confined to the upper lobe of the caudal fin. Additionally, the gums at the base of the lower jaw teeth are black in Chinook and white in coho.

Also, the colour patterns of young Chinook and coho in freshwater are similar. In both species, the lateral parr marks are roughly bisected by the mid-line of the body, and most of the marks are deeper than the vertical diameter of the eye. The shape of the anal fin and the pigmentation of the leading edge of the dorsal fin distinguish the fry: the anal fin is sickle shaped (the anterior rays are much longer than the posterior rays) in coho but not in Chinook, and the leading edge of the dorsal fin is dark in Chinook and white in coho. Additionally, there usually is a white tip on the dorsal fin of Chinook fry that is absent in coho. A useful, but not infallible, character is the adipose fin: the centre of the fin is finely stippled in coho and clear in Chinook.

In coastal populations of both species, some newly emerged fry migrate directly to estuaries. These fry are difficult to separate; however, the number of branchiostegal rays (on one side) differentiates the species (usually 16 or more in Chinook and usually 14 or fewer in coho). Furthermore, the pyloric caeca counts in coho and Chinook do not overlap (135–185 in Chinook and 45–80 in coho). In small salmonines, the number of pyloric caeca increase as body size increases; however, if both species are at hand, it is not necessary to count pyloric caeca. The difference usually is obvious—at any given size, Chinook typically have about twice as many caeca as coho.

Taxonomy Although the taxonomic status of the Chinook salmon is not an issue, the species is genetically heterogeneous, and locally adapted populations are common (Healey 1991; Myers et al. 1998; Taylor 1991a). Much of the variation within this species is derived from the presence of two behavioural forms of Chinook: a stream type and an ocean type (Healey 1991). Stream-type Chinook have a relatively long period of freshwater residence (1 or more years), they make major offshore migrations at sea, and they return to their natal rivers in the spring or summer. In contrast, ocean-type Chinook usually migrate to sea within about 3 months of emergence, they spend most of their ocean life in inshore waters, and they return to their natal streams in the fall. Some of the differences between the two life-history types are genetic (Clarke et al. 1992; Rasmussen et al. 2003; Teel et al. 2000). In British Columbia, the geographic distributions of the two life-history types (see Taylor 1991b) are roughly consistent with a scenario involving post-glacial dispersal of ocean-type Chinook from a southern coastal refugium and stream-type Chinook from the unglaciated inland parts of the Columbia system (Teel et al. 2000; Waples et al. 2004). Nonetheless, the presence of stream-type fish on the southern coast of British Columbia poses a problem for this hypothesis. The close genetic relationship between coastal stream-type and ocean-type Chinook in coastal areas suggests an explanation for this apparent anomaly—the independent evolution of the stream-type life history in coastal areas (Waples et al. 2004).

Within the Fraser River system, there are seven genetically recognizable geographic groupings of Chinook salmon: an upper, middle, and lower Fraser group; a northern, southern, and lower Thompson group; and the Birkenhead River population (Beacham et al. 2003). The latter population is the most distinctive of the Fraser Chinook populations and is geographically and temporally isolated from the other populations.

Sexual Dimorphism At sea, the sexes have the same body shape and bright silver colour; however, both sexes darken as they mature. In males, the head enlarges, the teeth on the jaws increase in size. Also, the body becomes deeper in mature males, but the hooked nose is not as well developed as in the other North American species of Pacific salmon. None of these secondary sexual characters are as pronounced in females as in males. Many Chinook populations enter fresh water months before they actually spawn (hence the local B.C. name, spring salmon) and retain their bright silver colour for some time after river entry. Other populations spawn shortly after entering fresh water, and they start to change colour as soon as they enter fresh water. Spawning colours in Chinook salmon are remarkably variable. In some populations, they become dusky all over; in other populations, they develop a dull brassy sheen along the midlateral flanks; and in still other populations, the flanks become a deep rose colour.

Although nuptial colours in females are more subdued than in males, the colour differences between the sexes usually are not as striking as those in other Pacific salmon species.

Hybridization Occasional juvenile hybrids between Chinook and coho are encountered in fresh water, but they are rare and usually associated with streams disturbed by fish-culture activities. No adult hybrids have been reported in nature. Apparently, a few hybrids between Chinook and pink salmon (*Oncorhynchus gorbuscha*) are known from the St. Mary's River between Lake Huron and Lake Superior (Rosenfield et al. 2000). This hybrid combination does not occur within the natural range of these species.

Distribution NATIVE DISTRIBUTION
Chinook are indigenous to rivers on the Pacific coasts of both Asia and North America. In Asia, they occur from northern Hokkaido to the Anadyr River, Siberia. In North America, they once occurred as far south as the Ventura River in southern California. There are still runs in the Sacramento – San Joaquin system, and they extend northwards from California along the Pacific coast to Point Hope, Alaska. There are unconfirmed reports from the Arctic coast of Alaska and one confirmed record from the lower Liard River in the Mackenzie system (McLeod and O'Neil 1983).

BRITISH COLUMBIA DISTRIBUTION
Chinook occur in most medium-sized to large rivers along the B.C. coast (Map 47). They ascend the Fraser River at least as far upstream as Rear Guard Falls (680 km from the sea) and the Skeena River to its headwaters.

Life History There are two basic life-history types in Chinook salmon—ocean-type Chinook that migrate to the sea in their first year and stream-type Chinook that migrate to sea after 1 or 2 years in fresh water. These life-history types differ in a suite of inherited behavioural and physiological traits (Clarke et al. 1992; Rasmussen et al. 2003; Teel et al. 2000).
A number of variant life-history types are embedded within the two basic life histories. Healey (1991) provides an excellent review of the complexities of Chinook salmon life histories. This review, the U.S. Chinook status review (Myers et al. 1998) and Quinn (2005) are the major sources for the following life-history synopsis.

REPRODUCTION
In British Columbia, Chinook spawn in over 250 rivers and streams, but less than 20% of these runs are commercially important (Aro and Shepard 1967). Throughout their geographic range, there is considerable variation in the time mature Chinook salmon enter fresh water. In British Columbia, the peak migration in most northern populations is in June; however, fish begin entering fresh water as early as April and as late as August. South of the Skeena system, peak run timing occurs later,

Map 47 The B.C. distribution of Chinook salmon, *Oncorhynchus tshawytscha*.

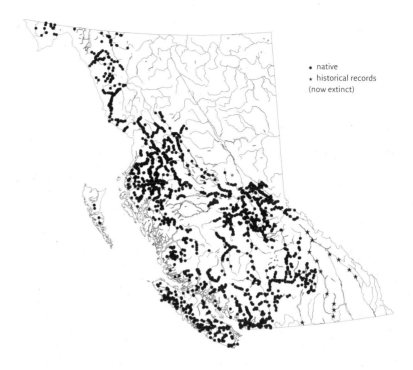

• native
∗ historical records (now extinct)

and there are two temporally separate peaks in some rivers. In the Bella Coola system, the peak run (in late June) is followed by a smaller run that peaks in August (Hume 1998). In the Fraser River, there are also two peaks: one in July and another in the fall (September–October). The Columbia River originally had three temporally separated Chinook runs: a spring run that peaked in early May, a summer run that peaked in June, and a fall run that peaked in August (Rich 1942). These separate runs not only spawned in different parts of the drainage system but also differed in their maturation state at time of river entry, the flow characteristics of their spawning sites, and the actual time of spawning (Myers et al. 1998). In rivers with multiple runs, most of the early runs are stream-type fish, whereas the later runs are predominately ocean-type fish.

Although Chinook often spawn in larger streams, in faster water, over coarser substrates than other Pacific salmon, they also spawn in small (2–3 m wide), shallow streams as well as in the mainstems and side-channels of large rivers. Chinook females choose the spawning site and appear to prefer sites with subgravel flow (e.g., in the tail-outs of pools immediately above riffles or in upwelling sites). Because of their unusually large eggs, the surface area available for oxygen diffusion is lower than in other salmonines. Consequently, subgravel flow of well-oxygenated water appears to be a primary factor governing Chinook spawning-site selection (Chapman 1943; Chapman et al. 1986; Geist 2000; Neilson and Banford 1983; Russell et al. 1983), and other factors such as water depth, water velocity, and substrate size are less critical. Water depths at redd sites range from 5 to 700 cm and average about 35 cm; water velocities range from 0.10 to 1.89 m/s and average about 0.50 m/s (Healey 1991; Groves and Chandler 1999).

The earliest description, at least in North America, of Chinook spawning is that of David Thompson. In 1807 (see Tyrrell 1916, p. 377), Thompson observed salmon spawning in the clear waters of the mainstem Columbia River just below Windemere Lake and briefly describes their nest digging and courtship behaviour. In a footnote, E.A. Preble suggested that these salmon were probably sockeye (*Oncorhynchus nerka*); however, given the distance inland and their size (Thompson indicated several were over 11 kg), they must have been Chinook salmon. Berejikian et al. (2000) provide a recent description of Chinook spawning behaviours, but the nest digging and courtship behaviours basically are similar to those seen in other salmonines (for details, see the descriptions in the rainbow trout species account). Normally, a large, dominant male accompanies the digging female. Smaller males (satellites) are arrayed downstream of the nest site, and jacks (sexually precocious males that return to freshwater usually after 1 year at sea) lurk nearby under cover. When the female crouches in the nest and the dominant male comes alongside, satellite males and jacks crowd in and join the spawning. Genetic data from other salmonines suggests that such supernumerary males successfully fertilize some proportion of the eggs (Foote et al. 1997; Seamons et al. 2004a). Mature parr (small freshwater residents with parr marks) occur in stream-type Chinook populations that spawn in headwaters hundreds of kilometres from the coast. The fertilization success of these sexually mature parr is unknown; however, their counterparts in Atlantic salmon (*Salmo salar*) do fertilize some eggs (Morán et al. 1996).

Fecundity in Chinook salmon varies within and among populations. Within populations, fecundity normally increases with female body size. Among populations, average fecundity (standardized to 740 mm post-orbital to hypural length) varied from 4,347 to 9,427 eggs (Healey and Heard 1984). These interpopulation differences are partially explained by a trend towards increased fecundity at high latitudes; however, this relationship is complicated by differences in the geographic distributions of the two major life-history types in Chinook salmon (Healey and Heard 1984). Overall, published egg counts for Chinook vary from <2,000 to >17,000. Chinook also have unusually large eggs: 6–10 mm in diameter. Egg size also varies with female size (Nicholas and Hankin 1988) and, perhaps, with latitude (Healey and Heard 1984). The fertilized eggs incubate in the gravel overwinter, and the alevins emerge in the spring. The rate of development is temperature dependent. Laboratory studies of Chinook eggs from the Kitimat, Bella Coola, and Quesnel rivers incubated at 2 °C, 4 °C, 8 °C, 12 °C, and 15 °C (Beacham and Murray 1989) showed that the time to reach 50% hatching varied among populations: 125–132 days at 4 °C, 69–71 days at 8 °C, 42–44 days at 12 °C, and 34–36 days at 15 °C. Fifty percent emergence times varied from about 211–220 days at 4 °C to 61.7–62.8 days at 15 °C. Except for zero survival at 2 °C, survival to hatching and emergence was high (60–100%) at all other temperatures. Mean alevin length at hatching varied from 19.6 to 22.7 mm TL and was related to incubation temperature, with length usually declining as temperature increased. On emergence from the gravel, mean fry length varied from 31.0 to 35.4 mm and, again, tended to decrease as temperature increased. In nature, survival to hatching and emergence probably are much lower than is indicated by these laboratory experiments.

AGE, GROWTH, AND MATURITY

Not surprisingly, ocean-type fry usually grow faster in their first year of life than stream-type fish (Healey 1991). In four B.C. estuaries, ocean-type fry ranged from 44 to 98 mm FL and averaged 53–77 mm (Healey 1991). In the Taku River, 0^+ fry ranged in fork length from 43 to 93 mm (Murphy et al. 1989); yearling smolts averaged 70–82 mm and ranged in size from 45 to 110 mm (Meehan and Siniff 1962). In the Yakima River, Washington, yearling smolts ranged from 90 to 170 mm FL (Major and Mighell 1969). Once in the sea, Chinook grow rapidly, and ocean-type fish from southern British Columbia reach a fork length of about 450 mm at age 2. In contrast, stream-type fish from southern British Columbia only reach about 110 mm FL at the same age. In the Fraser system, ocean-type jacks mature in their second year, and stream-type jacks, in their third year. Some ocean-type females mature in their third year, but most mature in their fourth or fifth year (Healey 1991).

FOOD HABITS

Depending on the salinity regime in estuaries, fry forage on chironomid larvae and pupae, cladocerans, amphipods, crab larvae, harpacticoid copepods, and possum shrimp. In fresh water, the diet of fry includes adult chironomids as well as chironomid larvae and pupae, terrestrial insects taken from the surface, and the nymphs and larvae of aquatic insects. As they grow and move away from estuaries into deeper water, the diet of juvenile Chinook shifts towards small fish (chum salmon fry, herring larvae, smelts, and other fish). The diet of juveniles in fresh water is dominated by the various life stages of terrestrial and aquatic insects. In the upper Fraser system, stream-type juveniles continued foraging throughout the winter (Levings and Lauzier 1991). Their primary prey varies among sites, but cladocerans, chironomids, and plecopterans are usually the main items in their diet. In the sea, fish (especially herring, anchovies, sand lance, and pilchards) dominate the diets of both juvenile and adult Chinook; however, invertebrates also are important in the diet of juveniles.

Habitat The ocean-type and stream-type life-history dichotomy complicates generalizations about habitat use in this species; however, Healey (1991) provides a thoughtful and detailed discussion of habitat use in Chinook salmon. The following habitat synopsis is based on this source.

ADULTS

Although knowledge of the oceanic phase of the Chinook life cycle is still meagre, there appears to be a difference in the distribution of ocean-type and stream-type Chinook salmon in the North Pacific Ocean. Relatively few adult Chinook originating from southern British Columbia and the Columbia River are caught in southeastern Alaska. This suggests that most ocean-type Chinook do not move more than about 1,000 km from their home rivers. Furthermore, most recaptures suggest that these Chinook remain relatively close to shore. In contrast, although stream-type Chinook of Columbia and southern B.C. origins are caught hundreds of kilometres offshore in both the North Pacific Ocean and the eastern Bering Sea, they probably are most abundant in the eastern North Pacific.

Maturing stream-type Chinooks return directly to the coast and to their rivers of origin. During this return migration, they are estimated to travel at more than 45 km/day (Healey and Groot 1987). In British Columbia, stream-type fish tend to enter fresh water in the spring or early summer. Typically, they are silver-bright, migrate long distances upstream, and spend several months in fresh water before they spawn. After arriving in the vicinity of their spawning area, they typically hold in deep pools. In the arid parts of the Columbia system, deep, cool, holding pools provide essential thermal refuges for early migrating Chinook (Torgersen et al. 1999). In contrast, ocean-type adults enter fresh water in late summer or fall. They are usually fully mature and beginning to colour-up. They migrate relatively short distances upstream and spawn shortly after river entry. Exceptions to this general pattern are the summer runs of ocean-type fish in the Columbia River system. They are not fully mature when they enter the river (June through mid-August), they migrate long distances upstream, and they spawn in the fall. In British Columbia before the construction of Grand Coulee Dam, some of these summer-run fish ascended the upper Columbia River almost to its source and spawned in the main river just below the outlet of Windemere Lake (Prince 1991).

JUVENILES

In British Columbia, stream-type Chinook spend a year and sometimes 2 years in fresh water before migrating to the ocean. Thus, stream-type Chinook overwinter in fresh water. In the upper Fraser system, 0^+ fry move from their natal streams into the main river channel, major tributaries, and maybe into some lakes (Levings and Lauzier 1991). In the Nicola River, Chinook appear to use off-channel ponds as overwintering sites (Swales and Levings 1989). In the Yakima River system in Washington during winter, juvenile Chinook selected slower (usually < 3.0 cm/s) focal velocities than in the summer and fall (Allen 2000). Additionally, in the winter, juveniles held focal positions close to the bottom (within 6 cm) and were only abundant close to dense cover (e.g., boulders, rip-rap, or woody debris). The migration of stream-type Chinook to the sea usually occurs in the spring and often coincides with the movement of ocean-type fry to estuaries (Healey 1991). In such cases, the ocean-type fry concentrate in the river's delta (Levy and Northcote 1982), whereas the stream-type yearlings occupy the delta front (Healey 1991).

In the Strait of Georgia, stream-type smolts occur in near-shore waters during June and July but disappear after July, whereas under-yearling ocean-type smolts are abundant in near-shore waters from June to November. A similar pattern occurs at the mouth of the Columbia River: stream-type smolts are present in May and June but are rare later in the summer. Sampling in coastal waters suggests that, in their first ocean summer, stream-type fish move northward along the outer coast. In contrast, ocean-type juveniles occurred predominately in sheltered waters. This pattern of stream-type Chinook offshore and ocean-type Chinook in coastal waters is retained as the fish grow to maturity.

YOUNG-OF-THE-YEAR

Most Chinook fry emerge from the gravel at night (Reimers 1971), and a major downstream movement of newly emerged fry is characteristic of most Chinook populations (Healey 1991). The fry of the two major life-history types (ocean-type and stream-type) differ in their habitat use. Stream-type fry spend their first year, and sometimes 2 years, in streams, rivers, and occasionally in large lakes. They are more aggressive than ocean-type fry, and their downstream migration is thought to be a dispersal mechanism that helps distribute them among suitable rearing habitats (Healey 1991). In Johnson Creek, Idaho, age-0 Chinook were initially associated with shallow (15–30 cm) edge habitats characterized by low water velocities (<0.15 m/s at the bottom) and fine substrates (Everest and Chapman 1972). In the Taku River, Chinook fry were most abundant along channel edges, in sloughs, in backwaters, and in off-channel habitats such as tributaries and tributary mouths (Murphy et al. 1989). As they grow, fry shift into deeper, faster areas, but there is some evidence that they move into quiet water areas (pools and sheltered stream margins) at night (Healey 1991). In inland populations in the fall, under-yearling stream-type Chinook often shift to overwintering habitats where they seek cover in pools or in spaces between cobbles and boulders in the substrate. This movement to overwintering sites usually involves a habitat shift from tributary streams into the mainstems of large rivers.

In short coastal rivers, the initial downstream movement of ocean-type fry often takes them into an estuary; however, not all ocean-type fry move downstream into estuaries. Ocean-type fry originating from inland populations spend 2–5 months in fresh water before migrating to tidewater. During their time in fresh water, they occupy habitats similar to those of stream-type fry. Even in short coastal rivers, not all the fry move immediately to the estuary. In the Nanaimo River, fry from three separate spawning areas differed genetically and migrated at different times and at different sizes (Carl and Healey 1984). Fry from the spawning area closest to the ocean moved directly to the estuary, many fry from the middle spawning area reared in the river for 6–8 weeks and then moved to the estuary, and fry from the uppermost spawning site remained in the river for up to a year (Carl and Healey 1984; Healey and Jordan 1982). Initially, in the Big Qualicum River, Chinook fry occupied shallow, low-velocity stream margins and were associated with bank cover (Lister and Genoe 1970). Estuaries are important rearing habitats for ocean-type fry. In the Fraser River delta, fry are scattered along the marsh edges at high tide and move down into the tidal channels that dissect the marshes as the tide ebbs (Levy and Northcote 1982). Thus, although salinities in the marshes vary with the tides, the fry are found in areas that are only slightly brackish. In smaller estuaries, fry are regularly collected in water with salinities of 20 g/kg or less. Ocean-type fry smolt as under-yearlings and are abundant in sheltered inshore waters during the summer months.

Conservation Comments   Out of the 866 Chinook stocks examined (Slaney et al. 1996) in British Columbia and the Yukon, 17 stocks were listed as extirpated, 47 were at high risk, 6 were at moderate risk, and 7 were stocks of special concern. These estimates probably are low: the status of 459 stocks is unknown. Eleven of the 17 known extirpations are the result of hydroelectric dams. Most of the populations at high risk suffer from over-exploitation and habitat degradation. COSEWIC lists the Okanagan Chinook population as endangered.

Salmo salar LINNAEUS
ATLANTIC SALMON

10 cm

Distinguishing Characters   At sea, Atlantic salmon are distinguished from Pacific salmon by the number of major rays in the anal fin: 8–12 in Atlantic salmon and 13 or more in Pacific salmon. They are most likely to be confused with steelhead (*Oncorhynchus mykiss*). The two species differ in spot distribution—Atlantic salmon often (but not always) have spots on their gill covers, and steelhead do not; Atlantic salmon usually lack spots on their tail, whereas steelhead have radiating rows of black spots on their tails.

In fresh water, Atlantic salmon fry differ from all the Pacific salmon and trout in having exceptionally long pectoral fins—they extend back to, or past, an imaginary vertical line descending from the origin of the dorsal fin. In a few places on Vancouver Island, Atlantic salmon parr can be confused with brown trout (*Salmo trutta*) parr. Both species have red dots along the lateral line; however, the long pectoral fins and clear adipose fin separates Atlantic salmon from brown trout parr (they have relatively short pectoral fins and an orange adipose fin).

Sexual Dimorphism   As they near spawning condition, the head of adult males elongates, and the lower jaw develops a strong, upturned hook (a kype) This kype is absent, or not obvious, in females and small, mature male parr. Additionally, the shape of the trailing edge of the anal fin differs in the sexes (Gruchy and Vladykov 1968): the posterior edge of the anal fin is convex in adult males and concave in females and juveniles.

Hybridization   Hybrids between Atlantic salmon and brown trout are rare where the natural ranges of both species overlap (Nyman 1970), but they are relatively common in places where brown trout have been introduced into waters that originally contained only Atlantic salmon (Verspoor 1989).

It is unlikely that Atlantic salmon would hybridize with any of the Pacific salmon or native trouts.

Distribution NATIVE DISTRIBUTION
The natural distribution of Atlantic salmon includes the Atlantic coasts of Europe and North America as well as southern Greenland and Iceland. In Europe, Atlantic salmon range from the northern Iberian Peninsula to the Arctic (Barents and White Sea drainages). In North America, their geographic range originally extended from the Delaware River to northern Quebec (Ungava Bay) and rivers in Quebec that flow into eastern Hudson Bay.

BRITISH COLUMBIA DISTRIBUTION
Adult Atlantic salmon have been reported from at least 78 streams and rivers along the B.C. coast (Morton and Volpe 2002). Most of the records are from Vancouver Island and the southern coast (Map 48), but there is at least one record from the northern coast (Stikine River). In addition, there are records of fry and juveniles in fresh water on Vancouver Island (Volpe et al. 2000).

Life History Within its native range, the Atlantic salmon displays a multitude of life-history types (reviewed by Klemetsen et al. 2003). In British Columbia, all our Atlantic salmon are derived from inbred, semi-domesticated stocks; however, if feral populations become established along our coast, some of the life-history diversity that is characteristic of this adaptable species may reappear. Although some escaped Atlantic salmon have successfully reproduced in British Columbia, it is not clear that they have established self-sustaining runs. Consequently, the following life-history account is based on anadromous Atlantic salmon within their native geographic range.

REPRODUCTION
In the Maritime Provinces, Atlantic salmon spawn in the fall or early winter (Scott and Scott 1988). Typically, spawning occurs at water temperatures below about 10 °C. Most of the feral Atlantic salmon observed in southern B.C. rivers do not show spawning colours until early winter, and farmed fish in controlled channels do not display spawning characteristics until mid-January (Volpe et al. 2001b). This suggests that, at least in Vancouver Island rivers, spawning does not occur until early winter (December–January).

Atlantic salmon spawning-site selection and spawning behaviours are similar to those of other anadromous salmonines. The female choose the spawning site, usually the tail-out of a pool just before the water shallows and become turbulent as it enters the next riffle. At spawning sites in Maine, average water velocities varied from about 0.50 to 0.65 m/s (range 0.25–0.90 m/s) and spawning depths varied from 17 to 76 cm (Beland et al. 1982). The spawning substrate is usually described as clean,

Map 48 The B.C. distribution of Atlantic salmon, *Salmo salar*

■ introduction

coarse gravel. Belding (1934) and Jones and King (1950) describe the spawning sequence. Like other salmonines, the female prepares the nest site (for a general description of salmonine reproductive behaviour, see the rainbow trout species account). The digging activity attracts males, and eventually, one male achieves dominance and keeps other males at bay. Meanwhile, the female keeps digging and occasionally probes the nest by dropping into the depression and extending her anal fin. When the nest is about 15–30 cm deep, she crouches in the nest, and the dominant male comes alongside, quivers vigorously, and eggs and sperm are released simultaneously into the nest. The female then moves upstream and rapidly digs at the upper edge of the nest. This displaces gravel downstream and covers the eggs. After a short pause, she moves farther upstream and begins digging a new nest. She repeats this process several times and lays roughly 100–200 eggs each time. Most spawnings occur at night (Belding 1934).

Unlike Pacific salmon, Atlantic salmon are iteroparous—that is, they spawn more than once (up to four or five times) in their lifetime. Not all mature individuals spawn every year, and individuals can switch from spawning in consecutive years to spawning in alternative years or vice versa (Klemetsen et al. 2003). In some populations, parr reach sexual maturity without migrating to the sea. These parr usually are males, although there are rare records of sexually mature female parr. These small males surreptitiously join spawnings involving larger fish and, thus, achieve some reproductive success (Morán et al. 1996).

Fecundity in Atlantic salmon is a function of female size and varies from 33 eggs in a stunted freshwater-resident population in Newfoundland to over 16,000 eggs in anadromous fish from the Restigouche River in New Brunswick (Gibson 1993). Egg size varies among populations (Thorpe et al. 1984) and, within populations, increases with female size. Typically, eggs range from about 5 to 7 mm in diameter. The incubation period is temperature dependent, and the time to 50% hatch varies from 63 days at 8 °C to 38 days at 12 °C (Crisp 1988). Depending on temperature, the alevins emerge from the gravel and begin exogenous feeding about 3–6 weeks after hatching. Fry emerge mainly at night and average about 26 mm at swim-up (Heggenes and Traaen 1988).

AGE, GROWTH, AND MATURITY

The growth of fry depends on fry density, water temperature, presence of competitors, and food availability. By late summer in Atlantic Canada, 0^+ fry range from about 35 to 55 mm FL, and they are about 70–90 mm long by early winter. Parr (1^+ or older) range from about 90 to 150 mm. Within a river system, parr typically migrate to sea after 1–3 (or sometimes more) years in fresh water. Some parr, however, never migrate to sea and mature in fresh water. Age at maturation is variable in anadromous Atlantic salmon. Some individuals (males and females) mature after only one winter at sea. These precocious fish are called grilse. However, most anadromous individuals mature after three to five winters at sea. Maximum age varies but rarely exceeds 13 years (Mills 1989).

FOOD HABITS

Atlantic salmon fry and juveniles forage in the water column and are primarily drift feeders (Keeley and Grant 1995). The major prey items are chironomids and the nymphs of aquatic insects; however, there is some surface feeding, and fish that hold stations close to the bottom occasionally exhibit substrate-oriented feeding (Wankowski and Thorpe 1979). In the North Atlantic, the diet of immature and adult Atlantic salmon consists of fish (especially capelin and sand lance), euphausiids, amphipods, and squid. In British Columbia, most of the adult Atlantic salmon caught in commercial fisheries are escapees from adjacent salmon farms. Most of these fish have empty stomachs, and some still contained commercial fish pellets. Nonetheless, some wild-caught salmon had wild prey (herring, sand lance, and various invertebrates) in their stomachs, and the proportion containing wild prey appeared to increase with the length of time the fish have been in the wild (Morton and Volpe 2002). Thus, escaped, farmed Atlantic salmon can learn to forage successfully on natural prey.

Habitat   The life history of anadromous Atlantic salmon usually requires the use of different habitats at different stages in their life cycle. Hence, their habitat use can be complex and shifts with size and age. The following account pertains to anadromous Atlantic salmon. In their native range, however, there are non-anadromous lacustrine populations and, occasionally, fluvial populations. Because little is known about habitat use of feral Atlantic salmon in British Columbia, the following synopsis is based on accounts from within their native range.

ADULTS

The length of the marine phase of anadromous Atlantic salmon varies from 1 to 5 years. Apparently, some salmon make long oceanic journeys, but others (e.g., those in the Baltic Sea and in the inner Bay of Fundy) stay within these geographic areas (Klemetsen et al. 2003). In the spring in the northwestern Atlantic, salmon concentrate in two areas: one along the eastern edge of the Grand Banks and the other about halfway between the southern tip of Greenland and the southern coast of Labrador (Reddin 1988). In late summer or early autumn, fish that are not going to spawn that year concentrate along the western Greenland coast, whereas those that are going to spawn start the journey back to their natal streams.

JUVENILES

In their second and later juvenile summers, parr are usually found in fast-flowing riffles over coarse substrates. Here, they defend feeding territories. The size of the territory depends on the size of the parr, the density of competitors, food availability, and current velocity (Grant et al. 1998). During the summer, 1^+ parr occur at focal point velocities ranging from 0.10 to 0.40 m/s, whereas 2^+ parr are found at focal point velocities that range from 0.30 to 0.50 m/s; their home stones (the stone most closely associated with individual positions) normally are <20 cm in diameter and usually <10 cm (Rimmer et al. 1984). At the summer–autumn transition, 2^+ parr shift to lower water velocities than smaller (1^+) parr. In winter during the day, parr hide beneath rocks (about 17–23 cm in mean diameter) in riffle–run habitats about 40 cm deep with water velocities that average around 0.40 m/s. Although they continue to feed during the winter, they are only active at night (Cunjak 1988).

Atlantic salmon parr usually remain in fresh water for 1–3 years and, in far northern regions, up to 8 years before migrating to sea as smolts. The change from parr to smolts usually occurs in the spring. The most obvious external change is that the characteristic lateral parr marks are covered with a subcutaneous layer of guanine that imparts a bright silver sheen to the sides. At the same time, there are internal physiological changes that prepare the smolt for life in salt water. There are exceptions to the typical spring smolt migration, and smolts in some rivers migrate in the fall. These fall-migrating smolts spend the winter in estuaries before moving out to sea in the spring.

YOUNG-OF-THE-YEAR

Atlantic salmon fry occupy relatively shallow (<30 cm) riffles with moderate (0.05–0.30 m/s) currents (Heggenes and Traaen 1988; Morán et al. 1996; Rimmer et al. 1984). The substrate is usually coarse gravel, and the fry defend territories and form dominance hierarchies. In the fall, fry move to deeper, faster water, and in winter during the day, they seek cover under rocks in riffle–run habitats 20–80 cm deep with water velocities that average about 0.40–0.46 m/s. During the day, they seek shelter in the substrate (Heggenes and Saltveit 1990) but continue to feed and are active at night (Cunjak 1988).

Conservation Comments Historically, sporadic attempts were made to introduce Atlantic salmon into B.C. waters. They all failed. This does not mean that Atlantic salmon could never establish viable populations in our coastal rivers. Most of the early introductions involved delicate life-history stages (eyed eggs and alevins) that suffer high mortalities even in their native waters. Moreover, our native stocks of salmon and trout, especially those with stream-rearing phases in their life histories, were strong. Consequently, the early introductions were made into highly competitive environments. This has all changed in recent years. Now, populations of our native stream-rearing salmonines (especially coho, *Oncorhynchus kisutch*; Chinook, *Oncorhynchus tshawytscha*; and steelhead) are depressed, and many stocks are either endangered or extirpated (Slaney et al. 1996).

Also, millions of Atlantic salmon are now reared in sea pens along our coast, and thousands of juveniles and adults inevitably escape into the wild (Morton and Volpe 2002). Furthermore, there is concrete evidence that some of these escaped fish have successfully spawned in B.C. rivers (Volpe et al. 2000), and there is evidence of more than one age class of young in some rivers (e.g., the Tsitika River). Still, it is not certain that any of these successful reproduction events have produced self-sustaining populations. Nonetheless, the possibility that Atlantic salmon could become established on the B.C. coast can no longer be dismissed out of hand (Volpe et al. 2001a). Consequently, it behooves government agencies responsible for the management and conservation of indigenous fishes to seriously examine the potential impacts of Atlantic salmon on native species.

Salmo trutta LINNAEUS
BROWN TROUT

1 cm

Distinguishing Characters   This trout has black, and some red, spots on the body. Usually, pale haloes surround the spots, and typically, the red spots occur only on the lower flanks. Sometimes the light haloes around the spots are absent. The red spots on the flanks, the pale haloes around both the black and red spots, and either the absence of spots on the tail or, occasionally, a few spots restricted to the upper lobe of the tail, distinguish brown trout from rainbow (*Oncorhynchus mykiss*) and cutthroat trout (*Oncorhycnus clarkii*). Juvenile brown trout have red spots along the midlateral line. This distinguishes them from juvenile rainbow and cutthroat trout (these species lack red spots), and the orange adipose fin distinguishes brown trout fry from rainbow and cutthroat fry.

    Adult brook trout (*Salvelinus fontinalis*) also have reddish spots, surrounded by haloes, on the flanks. The haloes, however, usually have a bluish tone, and there are dark worm-like markings (vermiculations) on the back and dorsal fin. Brown trout lack these vermiculations. Juveniles (parr) of both brown trout and brook trout have red spots along the midlateral line; however, the leading edge of the anal fin in brook trout has some dark pigment sandwiched between a white leading edge and the orange-red pigment on most of the fin. The anal fin in brown trout fry lack this colourful leading edge.

    Adult brown trout differ from Atlantic salmon (*Salmo salar*) in tail shape (forked in Atlantic salmon and almost square in brown trout) and in upper jaw length (the jaw extends back to about the centre of the eye in Atlantic salmon and past the eye in brown trout). The parr of both brown trout and Atlantic salmon have red spots along the midlateral line; however, the adipose fin is clear in Atlantic salmon and orange or red in brown trout. Brown trout fry are the only trout in our area with an orange adipose fin.

Taxonomy   Within its native range, the brown trout is the most variable of all the salmonines (Bachman 1991). Early taxonomists considered many of the various forms of brown trout different species and named about 50 species (Behnke 1986). Although we now use a single taxonomic name for brown trout, some of the forms appear to be biological species (i.e., they can coexist with other forms of brown trout and maintain themselves as genetically discrete units). Many of the various forms of brown trout appear to have evolved in geographic isolation during the Pleistocene (Bernatchez 2001). Fortunately, in British Columbia, we do not have to deal with the complexities of brown trout taxonomy: hatcheries probably have erased whatever genetic differences may have existed among the forms introduced into North America.

Sexual Dimorphism   As they near spawning condition, the head of adult males broadens and flattens and the lower jaw develops a strong, upturned hook (a kype) This kype is absent, or not obvious, in females and small, mature male parr. Additionally, the shape of the trailing edge of the anal fin differs in the sexes (Gruchy and Vladykov 1968): the posterior edge of the anal fin is convex in adult males and concave in females and juveniles.

Hybridization   Hybrids between brown trout and Atlantic salmon are rare where their natural ranges overlap (Nyman 1970) but are relatively common where brown trout have been introduced into waters that originally contained only Atlantic salmon (Verspoor 1989). In the past, to provide a different recreational fishing experience, tiger trout (a man-made hybrid between brown trout and brook trout) were occasionally stocked in some North American jurisdictions but not in British Columbia.

Distribution   NATIVE DISTRIBUTION
The brown trout is native to Eurasia. Originally, it was distributed throughout much of Europe (including the Iberian Peninsula), and its continuous range also included southern tributaries to the Caspian and Black seas as well as the headwaters of the Tigris and Euphrates rivers. Disjunct, but native, populations occurred farther east (Aral Sea tributaries), in North Africa (the Atlas Mountains), and in Iceland. During the 19th century, brown trout were transplanted into cool waters throughout the world. They are now established on all continents except Antarctica. Introductions into eastern North America started in 1883 and into western North America (California) in 1894.

BRITISH COLUMBIA DISTRIBUTION
Brown trout were introduced in the 1930s—first, to Vancouver Island (the Cowichan and Little Qualicum systems) and later to the Kootenays. In 1980, brown trout were introduced into Adam River, northern Vancouver Island, and now occur in both the Adam River and the lower Eve River. The Vancouver Island populations (especially in the Cowichan River) are well established and support recreational fisheries; however,

the mainland introductions appear to have failed, although occasional individuals are still caught in the Kettle River, in Kootenay Lake, and in the Columbia River downstream of Keenleyside Dam (Map 49). The brown trout caught in the Columbia mainstem probably are strays from the population in Roosevelt Reservoir, Washington. Similarly, there are brown trout in the Pend d'Oreille and Kootenay systems in Idaho, and some individuals probably wander downstream into British Columbia.

Life History Within their native range, the many life-history forms of brown trout are well studied; however, in our area, we know very little about their biology. Consequently, the following account is based on studies in Europe and other parts of North America. There is a thorough, and very readable, account of brown trout biology in the book *Trout* (Frost and Brown 1967), and there is an excellent review of the life-history diversity in this species (Klemetsen et al. 2003).

REPRODUCTION

Brown trout spawn in the fall or early winter when water temperature drops below 10 °C. Females select the nest site (Frost and Brown 1967). Typically, the nest is at the tail-out of a pool just before the water shallows and become turbulent as it enters the next riffle. Mean water velocities at spawning sites vary from about 0.30 to 0.45 m/s (Shirvell and Dungey 1983; Witzel and MacCrimmon 1983), and the substrate is usually coarse gravel. Jones and Ball (1950) give a description of the spawning sequence. Like other salmonines, the female prepares the nest site (for details of salmonine reproductive biology, see the rainbow trout species account). The digging activity attracts males. Eventually, one male achieves dominance and keeps other males at bay. Meanwhile, the female keeps digging and occasionally probes the nest by dropping into the depression and extending her anal fin. When the nest is about 7–10 cm deep, she crouches in the nest; the dominant male comes alongside and quivers vigorously, and eggs and sperm are released simultaneously into the nest. The female then moves upstream and rapidly digs at the upper edge of the nest. This displaces gravel downstream and covers the eggs. After a short pause, she moves farther upstream and begins digging a new nest. She repeats this process several times and lays roughly 100–200 eggs each time.

Fecundity is a function of female size and varies from a few hundred eggs in stream populations to several thousand in large fish. Egg size varies among populations (McFadden et al. 1965) and also increases with female size within populations. Typically, eggs are about 4–5 mm in diameter. The incubation period is temperature dependent and varies from 148 days at 2 °C to 38 days at 10.6 °C (Embody 1934). Again, depending on temperature, the alevins emerge from the gravel and begin exogenous feeding about 3–6 weeks after hatching. Newly hatched larvae are about 20 mm TL and average about 26 mm at swim-up.

Map 49 The B.C. distribution of brown trout, *Salmo trutta*

- introduction

AGE, GROWTH, AND MATURITY

Growth in brown trout is variable and influenced by the environment (e.g., temperature, food availability, habitat type, and population density) as well as the evolutionary history of specific populations (Lahti et al. 2001; Klemetsen et al. 2003). In Cowichan River tributaries, fry are about 24 mm long by the beginning of May and reach 45–55 mm by October. Yearlings (1^+) reach 64–76 mm by their second spring but grow much larger (about 142 mm) in the main river (Neave and Carl 1940). In tributaries to the Adam River, also on Vancouver Island, 0^+ brown trout averaged 60–80 mm FL by the end of their first growing season, and 1^+ fish averaged 100–135 mm by the end of their second growing season (Wilson et al. 1997). Given the shorter growing season but warmer summer temperatures, the growth rates of juveniles in the Kootenays probably are similar to those on northern Vancouver Island. Most adult brown trout in British Columbia do not exceed 400 mm in length.

In North America, male brown trout usually mature in their third (2^+) or fourth (3^+) years, whereas females normally mature at least a year later than males. In anadromous populations, some males mature as parr and remain in their nursery stream. Although brown trout sometimes live up to 20 years in Europe, they rarely exceed 10 years in North America.

FOOD HABITS

The diet of brown trout varies with locality, age, season, and genetic history. Often, they are characterized as opportunistic feeders that concentrate on whatever prey items are abundant and available. Recent evidence, however, indicates that some individuals show dietary specialization (Grey 2001). In streams, adults are bottom oriented and forage mainly on aquatic insect nymphs; however, during major insect hatches, they often feed at the surface. In the summer, most of their foraging is at dawn, dusk, and night (Bunnell et al. 1998), but they will feed during the day. During the winter, foraging activity drops, and they seek shelter in the substrate during the day but may hold position near the bottom and feed at night (Bremset 2000). Typically, large adults (>300 mm) include small fish in their diet. Fluvial juveniles feed primarily on the drifting larvae of aquatic insects, and the diet of the young-of-the-year is similar to that of juveniles but includes smaller items and, usually, more chironomids.

In lakes, the diet of adult brown trout ranges from plankton to small fish (sticklebacks, lampreys, sculpins, and juvenile salmonines). Individuals often patrol specific areas in the littoral zone and actively pursue mobile prey (large invertebrates and small fish). In lakes, young-of-the-year forage first on plankton and, as they grow, add chironomid larvae to the diet. Juveniles forage on larger and a wider variety of bottom organisms than the fry.

Habitat   Within their native range, habitat use by the many life-history forms of brown trout is well studied; however, little is known about their biology in British Columbia. Consequently, the following account is based on studies in Europe and other parts of North America.

ADULTS

In Europe, adult brown trout live in a wide variety of habitats: large lakes, small lakes, small headwater streams, and large rivers. They are more tolerant of high temperatures than other trout and can withstand temperatures above 20 °C, but below 24 °C, for a considerable time (Elliott 2000). Many European populations are anadromous, and in British Columbia, occasional anadromous individuals are rumoured to occur in the Cowichan River. Our brown trout, however, are primarily river fish. In flowing water, brown trout use pools, runs, and riffles, and there is some evidence of partial habitat segregation by sex: during the day, males use pools more extensively than females, whereas both sexes use riffles (Greenburg and Giller 2001). In New Zealand, the water velocities occupied by feeding brown trout averaged 0.67 m/s and did not differ among six rivers. In contrast, foraging depths did differ; they were deeper in rivers containing rainbow trout (Shirvell and Dungey 1983). Apparently, brown trout are quite sedentary and have a relatively small home range, at least during the day. Thus, individuals consistently occur in a single pool or a single riffle–pool combination (Burrell et al. 2000).

The same study found that brown trout were more active in fall and winter than in spring and summer.

JUVENILES

In their second spring, brown trout often move from their nursery streams into larger rivers. Here, they display seasonal microhabitat and spatial shifts (Heggenes and Saltveit 1990): focal point velocities are lower in the fall and winter (0.02–0.05 m/s) than in the summer (0.08–0.12 m/s) and are closer to the substrate. A study that monitored habitat use by marked juvenile brown trout found that, despite individual variation, coarse substrate habitats were used more often, on average, than fine substrates (Elso and Greenberg 2001). Also, juveniles changed habitat more often at night than during the day, and fish that shifted habitats survived better than those that used only one habitat.

YOUNG-OF-THE-YEAR

In streams, young-of-the-year brown trout occupy shallow (<30 cm) edge environments with moderate (0.20–0.50 m/s) currents (Roussel and Bardonett 1999). The substrate is usually coarse gravel, and the fry defend territories and form dominance hierarchies (Lahti et al. 2001). As they grow, fry move to deeper, slower parts of the stream (Heggenes 2002). In the winter during the day, they seek shelter in, or near, the substrate.

Conservation Comments The brown trout is now well established on Vancouver Island, especially in the Cowichan River system. Although it is an excellent recreational species and, in suitable habitat, is resistant to angling pressure, the time has past when any government agency could in good conscience spread any alien species into waters where it does not now occur.

Salvelinus confluentus (SUCKLEY)
BULL TROUT

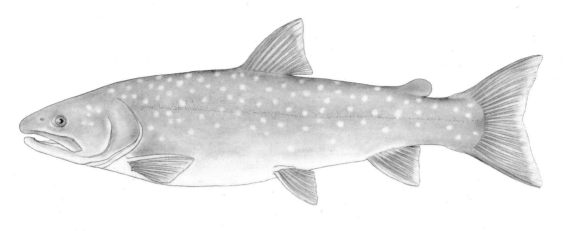

1 cm

Distinguishing Characters — The bull trout is a trout-like char with pale pink, lilac, or red spots along the flanks. In British Columbia, bull trout and Dolly Varden (*Salvelinus malma*) coexist and hybridize in most drainage systems that arise in, or cut through, the Coast Mountains. The two species (especially juveniles) are difficult to identify. Relative to Dolly Varden, the spots on the back of bull trout usually are large and well spaced—they are smaller and more crowded in Dolly Varden. The two species also differ in head shape, upper jaw length, anal fin base (noticeably smaller than upper jaw length in bull trout: average ratio is 1.7 with a range of 1.5–2.3), and total number of branchiostegal rays (usually >23 in bull trout and <23 in Dolly Varden). Where they coexist, their life histories are strikingly different (Hagen 2000; Hagen and Taylor 2001). Typically, adult bull trout are large (55–85 cm), migratory, and piscivorous, whereas adult Dolly Varden are small (10–20 cm), stream residents, and feed on drift.

Taxonomy — The relationship between bull trout and other North American chars (especially Dolly Varden and Arctic char, *Salvelinus alpinus*) has a long and tangled history. For detailed accounts of the taxonomic history of bull trout, Dolly Varden, and Arctic char and the arguments for and against considering them separate species, see Brunner et al. (2001), Cavender (1978), Elz (2003), Haas and McPhail (1991), and Phillips et al. (1999). What is clear is that the relationships among these species are complex and that they have been confounded by periodic episodes of range fragmentation through glaciation. During glaciation, fragments derived from a once-continuous distribution were isolated in different glacial refugia with one or another of the different char species, and hybridization produced limited introgression; however, in other isolated fragments of the same species, there was no introgression. The result is a set of morphological, mitochondrial, and nuclear markers that produce conflicting phylogenies. In the case of char, I have chosen sympatry as the test of species status. Thus, even though there is some hybridization in the broad contact zone between bull trout and Dolly Varden in British Columbia, I recognize them as separate species.

This does not mean that bull trout are genetically uniform in British Columbia. Molecular analyses (e.g., Taylor et al. 1999, 2001) suggest the presence of two identifiable groups of bull trout in the Pacific Northwest: a southern coastal group and an interior group. In British Columbia, the southern coastal group is clustered in the lower Fraser and Squamish systems and appears to have colonized this part of the province from the Chehalis Refugium in Washington State. In contrast, the interior group appears to have entered British Columbia from the unglaciated upper Columbia system. There is a transition zone between the two groups in the Fraser Canyon region. This division into interior and coastal forms is a repeated pattern in other B.C. fishes (e.g., Pacific salmon, trout, minnows, and sculpins).

Sexual Dimorphism In most populations, breeding males develop a hooked lower jaw (kype) that, in extreme cases, fits into an indentation on the tip of the snout. Also, breeding males usually are more vividly coloured than females, especially the markings on the paired fins. Both spawning colours and kype development vary among populations, and some stream-resident bull trout never develop a kype. Outside the breeding season, the sexes are difficult to distinguish; however, males usually have larger adipose fins than females of the same size.

Hybridization As far as is known, bull trout and Dolly Varden hybridize wherever they coexist. The hybrids are fertile, and molecular data indicate that some backcrossing occurs (Baxter et al. 1997; Redenbach 2000; Redenbach and Taylor 2002). In spite of this persistent hybridization, the two species maintain themselves as separate entities with life histories adapted to two different ecologies (Hagen 2000; Hagen and Taylor 2001). In addition, hybrids between bull trout and introduced brook trout (*Salvelinus fontinalis*) are common in the B.C. portion of the Skagit drainage system (McPhail and Taylor 1995) and probably also occur in other rivers where brook trout have been introduced.

Distribution NATIVE DISTRIBUTION
Bull trout are native to western North America. Originally, they were distributed from northern California to the extreme headwaters of the Yukon system in northwestern British Columbia; on the eastern slope of the Continental Divide, from the upper Missouri system north to the Mackenzie River system in Alberta and British Columbia; and in the Northwest Territories, at least as far north as the Bear River.

BRITISH COLUMBIA DISTRIBUTION

Bull trout are distributed in cool waters throughout the interior of the province but are absent from many of the shorter coastal rivers and from both Vancouver Island and the Queen Charlotte Archipelago (Map 50). Exceptions to this generalization are the large, west-flowing rivers that cut through the Coast Mountains (e.g., the Fraser, Homathko, Klinaklini, Skeena, Nass, Iskut–Stikine, and Taku rivers).

Life History   Like many char, bull trout display a bewildering array of different life-history patterns (McPhail and Baxter 1995). In British Columbia, three life-history patterns are common: a fluvial form that spends its entire life in flowing water but often makes extensive migrations within large river systems; an adfluvial form that migrates between rivers and lakes; and a stream-resident form that spends its entire life in small rivers and streams and, in many cases, is isolated by barriers. A fourth, anadromous, life-history form is restricted to the southwestern portion of the province. It migrates between fresh water and the sea. Anadromy in bull trout has never been investigated in British Columbia, although anglers refer to sea-run populations in the Squamish and Pitt rivers, and a fish tagged in the Squamish River was recovered in the lower Skagit River, Washington (a marine journey of about 150 km). Additionally, bull trout have been collected in the sea in Howe Sound and off Roberts Bank (Haas and McPhail 1991). Also, fish tagged above Pitt Lake are caught regularly in the estuarine bar fisheries on the Fraser River delta and only rarely recovered upstream of the confluence of the Fraser and Pitt rivers. This pattern of tag recovery implies a migration from the upper Pitt River to the Fraser Estuary. A recent radio-telemetry study (Brenkman and Corbett 2005) documented the out-migration and coastal movements of bull trout from rivers on the west side of the Olympic Peninsula, Washington. So far, all of the reports of anadromy in bull trout are from the southern edge of their coastal distribution. Bull trout life histories are reviewed in McPhail and Baxter (1995) and Rieman and McIntyre (1993).

REPRODUCTION

Like all char, bull trout spawn in the fall. Except for stream-resident populations that spawn locally, a migration typically precedes spawning. In British Columbia, fluvial populations usually begin migrating from large rivers into smaller (but still substantial rivers) in late August when temperatures are high and water levels low. These spawning migrations can involve round trips of over 400 km. Similarly, in late summer or early fall, adfluvial populations migrate from lakes or reservoirs into spawning streams. Some of these migrations are short (<10 km), but others are substantial (50–100 km). Usually, the gonads of migrating adults are not fully mature, and some populations spend a month or more in their spawning stream before breeding. At this time, their large size and voracious appetite make them particularly vulnerable to poachers.

Map 50 The B.C. distribution of bull trout, *Salvelinus confluentus*

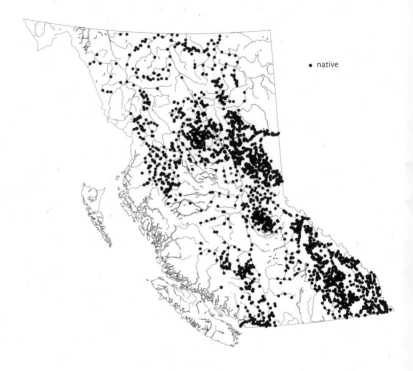

As far as is known, all bull trout spawn in flowing water. Females select the spawning site, and nest digging and spawning behaviour is similar to that of other salmonines (for details, see descriptions in the rainbow trout account). In the Arrow Lakes region, actual spawning did not start until water temperature dropped below 9 °C in mid-September (McPhail and Murray 1980). Typically, a dominant male accompanies each spawning female. Dominant males are larger and more colourful than satellite males (Baxter 1997). They vigorously defend their female from the attentions of other males. Satellite males steal fertilizations by sneaking (joining a spawning pair just before gamete release). Satellite males mimic females in colour, behaviour, and morphology (they lack a kype), and this may allow them to approach a spawning pair and sneak fertilizations (Baxter 1997). In some populations, there are also jacks (small precocious males). These dash in at the moment of egg release and often successfully fertilize some of the eggs. The persistent production of hybrids in areas where bull trout and Dolly Varden coexist may result from Dolly Varden males acting as jacks in bull trout spawnings. Typically, bull trout spawn during the day; however, in some disturbed systems (e.g., the upper Skagit in southwestern British Columbia), spawning occurs at night.

In the mainstem Skagit River, bull trout redds usually were found in runs or glides They were often associated with groundwater sources and were relatively close to cover (within 5 m). In smaller streams, spawning sites are associated with pockets of suitable gravel. Characteristically, only a small portion (relative to the total length available) of a spawning stream is used (Baxter 1997). The size and depth of redds, and the size of the spawning gravel, vary with female size: large females dig in larger gravel and at greater depth than small females. In the upper Peace system, spawning occurs in late September at water temperatures below 10 °C (Bustard 1996). Females excavate large redds (>2 m^2) usually in water 30–40 cm deep with a velocity of 0.20–0.60 m/s. Typically, 50% of the substrate is >4 cm in diameter, and most redds are located in the open (>2.0 m from cover). In the Chowade River (lower Peace system), 54 redds were examined over 2 years (Baxter 1997). Again, redds were fairly close to cover (4–5 m) in water 0.28–1.5 m deep and with coarse (an average of 3.4–3.9 mm) gravel substrates. The average nose velocity (measured 12 cm above the substrate at the upstream end of the redd) ranged from 0.39 to 0.45 m/s. Comparisons of redd sites and adjacent sites that were not selected by the fish revealed a significant difference in groundwater discharge. Generally, redd site were in areas of groundwater upwelling. Using egg capsules, alevin survival was compared between sites that were selected by wild fish and adjacent sites that were not selected (Baxter 1997). Incubation temperature was slightly higher in the selected sites, and there was a small but significant difference in alevin survival (higher in selected sites).

Like most fish, egg number in bull trout is a function of female body size and varies within and among populations. Large (typically 350–650 mm FL) fluvial and adfluvial females produce up to 9,000 eggs about 4.0–6.2 mm in diameter. Small (typically <250 mm) stream-resident populations produce fewer eggs ($<1,000$). Development rates are temperature dependent: at 2 °C, the eggs take 119–126 days to completely hatch; at 4 °C, hatching takes 92–95 days; at 6 °C, 74–78 days; at 8 °C, 70 days; and at 10 °C, approximately 50 days (McPhail and Murray 1980). Survival to hatching declines precipitously at temperatures above 8 °C. Also, the largest alevins (about 17 mm TL) and the largest fry at emergence (about 24 mm) were produced at the lowest incubation temperatures. In nature, the fertilized eggs incubate in the gravel overwinter and fry emerge near the beginning of June. In the wild, newly emerged fry are about 25 mm TL.

AGE, GROWTH, AND MATURITY

In the upper Finlay River, bull trout fry reach 30–50 mm FL by the end of their first summer (Bustard 1996). In some adfluvial populations in southern British Columbia, some fry may migrate to their lake in their first summer; however, in lakes, it is unusual to find juveniles of less than 200 mm FL. This suggests that most lakeward migration takes place in their second or third summer of growth. In Thutade Lake tributaries, the majority of juvenile bull trout remain in streams for three growing seasons (age 2$^+$), and a few individuals remain for 1 or 2 more years (Hagen 2000).

In adfluvial populations in southern British Columbia, sexual maturity is reached at about 350 mm (5^+) in males and at about 450 mm (6^+) in females. This is consistent with observations in Thutade Lake tributaries where the smallest mature bull trout was a 370 mm male, but most spawning fish were 550–850 mm long. In the Liard system, a stream-resident population reached sexual maturity at a smaller size, and several years earlier, than fluvial populations in the main river and large tributaries (Craig and Bruce 1982). Maximum age in British Columbia is unknown, but ages up to 24 years are recorded in the province (Hagen and Baxter 1992).

FOOD HABITS

Adult bull trout are piscivorous—riverine and adfluvial populations prey on trout, whitefish (especially mountain whitefish, *Prosopium williamsoni*), kokanee (*Oncorhynchus nerka*), Arctic grayling (*Thymallus arcticus*), and a variety of suckers, minnows, and sculpins. Juveniles, and adults of stream-resident populations, feed primarily on the nymphs and larvae of aquatic insects (mayflies, caddisflies, stoneflies, and chironomids). When feeding during the day, these fish remain close to the bottom, and most feeding movements are directed towards insects drifting near the bottom. Occasionally, however, bull trout rise to the surface and take terrestrial insects. Juveniles over 100 mm FL also take small fish (including young of their own species).

In Thutade Lake tributaries, the foraging behaviour and stomach contents of sympatric bull trout (juveniles) and adult Dolly Varden were compared (Hagen 2000). In both species, the proportion of feeding movements directed at the bottom increased at night, but the relative increase in bull trout was greater than in Dolly Varden. The visual acuity of bull trout under low light conditions is unknown; however, since Dolly Varden feed actively at night and can forage at low light levels (Henderson and Northcote 1985; 1988), bull trout probably also forage at low light levels. The only major difference between the stomach contents of sympatric juvenile bull trout and adult Dolly Varden was the proportion of winged insects in the diet (lower in bull trout). The fry feed primarily on chironomid larvae but gradually shift to larger prey (predominately insect nymphs) as they grow.

Habitat   Bull trout are a cold-water species and are rarely encountered in environments where the temperature exceeds 15 °C for prolonged periods. Habitat use in bull trout fry, juveniles, and adults is strongly influenced by life-history type and the presence, or absence, of other species. Like most fish, bull trout shift habitats as they grow.

ADULTS

Habitat use in adult bull trout is influenced by life-history type and the presence of other fish species. In lakes during the day, adult bull trout are rarely taken in littoral areas but, at night, appear to move into shallower water. In large rivers, they are commonly associated with the tail-outs of pools and are usually close to overhead cover. During the summer in a wilderness stream in Idaho, adult bull trout chose areas in pools that were <0.5 m deep, and on average, their focal positions were 0.38 m from cover (Spangler and Scarnecchia 2001). In the fall, they shifted to deeper water (an average depth of 0.44 m), were usually in contact with the bottom, and avoided water velocities >0.05 m/s; on average, their focal positions were 0.21 m from cover (Spangler and Scarnecchia 2001).

In British Columbia, research on habitat use in bull trout has concentrated on areas where they coexist with Dolly Varden; however, in most of the province, bull trout and Dolly Varden are allopatric. British Columbia and northwestern Washington are the only places where bull trout and Dolly Varden co-occur. In areas of sympatry, Dolly Varden are typically small stream residents, and bull trout are riverine or adfluvial. Although not tested experimentally, this pattern of habitat segregation appears to be widespread (Hagen 2000; McPhail and Taylor 1995). In areas of allopatry, both bull trout and Dolly Varden broaden their trophic and habitat niches to include resources that, in sympatry, are used by the other species. For example, in allopatry, it is common to find two bull trout life-history types in the same river: a migratory form that leaves the natal stream at 2 or 3 years of age and a non-migratory form that is resident in the natal stream and rarely exceeds 250 mm FL. Thus, in allopatry, the resident form essentially uses the niche that is filled in sympatry by Dolly Varden.

In Thutade Lake tributaries, instream habitat use was examined in juvenile bull trout and adult Dolly Varden (Hagen 2000). There was considerable overlap between the species—they used the same hydraulic habitat types (main channels pools, main channel riffles, side-channel pools, side-channel riffles) in similar proportions during both the day and night. At night, however, juvenile bull trout were found in shallower and, perhaps, slower water than adult Dolly Varden.

There is also evidence for an interaction between bull trout and lake trout (*Salvelinus namaycush*). Adults of both species are piscivores, and small northern lakes tend to contain one or the other species, but not both. Large lakes, however, often contain both species. An exception is Babine Lake in the Skeena system. This large oligitrophic lake looks like good bull trout habitat, but apparently, bull trout do not occur in the lake. Instead, the lake contains lake trout; however, the Babine River immediately below the lake is full of bull trout. In this case, bull trout apparently are competitively superior in flowing water, whereas lake trout have the edge in lakes.

JUVENILES

Regardless of life-history type, most bull trout rear in streams for 2–4 years, and there are seasonal and diel shifts in the microhabitats that they use (Bonneau and Scarnecchia 1998; Sexauer 1994). Juvenile bull trout are strongly associated with pools and deep side-channels. In an Idaho stream during summer, juvenile bull trout usually avoided water deeper than 0.3 m (0.03–0.27 m), they had focal elevations that averaged 0.02 m, they usually selected focal velocities of about 0.03 m/s (0.00–0.23 m/s), and their average distance to cover was 0.19 m (Spangler and Scarnecchia 2000). At night, they disperse and are less strongly associated with cover (Bonneau and Scarnecchia 1997; Sexauer 1994). During the fall, they stayed in shallow water (average depth about 0.19 m, they remained on the bottom in low-velocity (0.00–0.05 m/s) areas, and their average distance to cover was 0.09 m (Spangler and Scarnecchia 2001). In winter, juvenile bull trout shelter in pools under large woody debris or within the substrate (Bonneau and Scarnecchia 1997). At night, they emerge from cover and shift into shallow water away from cover (Jakober et al. 2000). Juveniles in adfluvial populations migrate to a lake in their third (2^+) or fourth (3^+) year. In lakes, juvenile bull trout are rarely taken in the littoral zone. This suggests that, when they initially enter lakes, they move into deep water.

YOUNG-OF-THE-YEAR

Newly emerged bull trout fry are denser than water and associated with shallow ($<$5 cm deep) stream edges. Here, they are found in and around gravel (20–100 mm in diameter) interspersed with boulders, especially in low water velocity ($<$0.20 m/s) areas such as side-channels and shallow bays. They are secretive during the day and, if disturbed, quickly seek shelter under rocks. In most populations, the fry remain in this habitat during their first summer but shift to deeper, faster water as they grow. In streams subject to freezing, the fry move into deeper water during winter.

Conservation Comments   In most of British Columbia, bull trout are abundant, but they have been extirpated at the southern edge of their native distribution (northern California) and are declining in most of Oregon, Washington, Idaho, and Montana. Consequently, bull trout are listed under the Endangered Species Act in the United States. In Canada, bull trout are declining in southern Alberta and in parts of southern British Columbia. Additionally, the anadromous population in the Pitt River seems to be dwindling. Whether these declines are a result of habitat loss, a general warming trend, or over-exploitation is unknown, but it is clear that bull trout are a vulnerable species. Their large size, voracious appetite, and habit of congregating below barriers before spawning make migratory bull trout particularly vulnerable to anglers (and poachers). In the past, bull trout have been "fished-out" of areas with easy road access; however, with appropriate regulations and enforcement, most B.C. bull trout populations are in no danger of extirpation.

Of particular conservation interest are sites where bull trout and Dolly Varden coexist. Apparently, these species hybridize wherever they come in contact (Baxter et al. 1997; McPhail and Taylor 1995). Areas of contact are of considerable scientific interest, and some of the sites where Dolly

Varden and bull trout coexist should be protected from disturbance. Perhaps the greatest threat to bull trout is the introduction of brook trout to augment recreational angling. Brook trout hybridize with bull trout; however, unlike Dolly Varden × bull trout hybrids, most brook trout × bull trout hybrids apparently are males (and either sterile or with reduced fertility). Hybridization with brook trout, in conjunction with competitive interactions, can induce declines in bull trout populations (Gunckel et al. 2002). Thus, introducing brook trout into areas containing bull trout may be man's final insult to one of our largest and most colourful native fishes. The BCCDC ranks the bull trout as S3 (may have lost peripheral populations).

Salvelinus fontinalis (MITCHILL)
BROOK TROUT

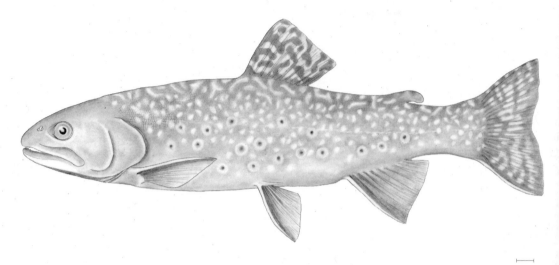

1 cm

Distinguishing Characters Like all char, brook trout have light spots along their flanks and lack teeth on the shaft of the vomer (Fig. 35). In adults, the spots are red or pink, and some on the lower flanks are surrounded by bluish halos. Brook trout differ from native char in having wavy light-green marks (vermiculations) on their backs, although occasionally both Dolly Varden (*Salvelinus malma*) and bull trout (*Salvelinus confluentus*) approach this colour pattern. In brook trout, however, the dorsal fin also is heavily marbled with dark wavy lines; in native chars, the dorsal fin is either uniformly dusky or has pale spots. Juveniles (less than about 70 mm) lack vermiculations on the back but still have dark spotted dorsal fins. Char fry 20–35 mm long are distinguished from trout and salmon fry by the large, blotchy parr marks (some of which are wider than the intervening light areas) and a triangular, dusky patch of pigment that extends about halfway out on the caudal fin. Brook trout fry are distinguished from other char in British Columbia by the band of dense pigment at the base of the anal fin and the numerous small dark spots on the dorsal surface of the body.

Taxonomy Because B.C. brook trout are derived from domesticated hatchery stocks, their taxonomic status is not a major concern. There is, however, one aspect of their morphological and ecological variation that is potentially of management interest. There are two lacustrine forms of brook trout in eastern Canada: a pelagic, plankton-feeding form, and a littoral, benthic-feeding form (Bourke et al. 1996). The two forms differ in their morphology, and there is evidence that, in some lakes, they form two partially isolated genetic stocks (Dynes et al. 1999). Interestingly, the relative proportion of the two forms in these lakes appears to be driven by competition with minnows (cyprinids) and suckers: the relative frequency of the pelagic, plankton-feeding form increases in the presence minnows and suckers, with suckers producing the most effect (Bourke et al. 1996).

Sexual Dimorphism Mature males have longer paired fins and a longer upper jaw than females. Breeding males also develop a small kype (a hooked lower jaw). At spawning, both sexes develop bright red sides. The paired fins and anal fin are red with conspicuous white leading edges followed by a thin band of black pigment. These breeding colours are more vivid in males than in females. Also, breeding males often develop black pigment under their jaw and along the lower belly.

Hybridization In British Columbia, brook trout hybridize with two of our native chars: Dolly Varden and bull trout. Hybrids between brook trout and bull trout are primarily males and are thought to be partially sterile (Markle 1992). They are not totally sterile, however, and molecular evidence clearly indicates backcrossing (Kanda et al. 2002; McPhail and Taylor 1995). In British Columbia, brook trout are stocked only into closed lakes (i.e., lakes without direct connections to other waters) or waters where they already are established. Still, lake levels fluctuate and drainage connections can be re-established. If brook trout escape these lakes, they could have a negative effect on native char and other species. The splake (a man-made hybrid between brook trout and lake trout, *Salvelinus namaycush*) has not been widely introduced in British Columbia.

Distribution NATIVE DISTRIBUTION
The brook trout is an eastern North American species. Originally, it ranged from the southern Appalachian Mountains in Georgia to Labrador and Ungava, and from the Great Lakes region to eastern Minnesota and northern Manitoba. Like many popular recreational species, brook trout have been introduced widely and now occur in cool waters in New Zealand, Africa, South America, and Eurasia. They were introduced into western North America in the late 19th century.

BRITISH COLUMBIA DISTRIBUTION
Brook trout were introduced into British Columbia in the 1920s and are now established in all parts of the province (Map 51) except for the Queen Charlotte Islands, most short coastal rivers, and most of the B.C. portions of the upper Yukon and Mackenzie drainage systems.

Life History Although brook trout were introduced into British Columbia over 80 years ago, their biology has not been studied in the province. Within their native range, brook trout exhibit considerable intra-population and inter-population life-history variation (Hutchings 1996; Power 1980). The extent of life-history variation in self-sustaining B.C. populations is unknown. Consequently, information for the following summary is derived from native populations in eastern North America and

Map 51 The B.C. distribution of brook trout, *Salvelinus fontinalis*

■ introduction

introduced populations in some western states. A major source of information on brook trout life histories is an excellent review by Power (1980).

REPRODUCTION

Like other char, brook trout spawn in the fall, usually when water temperatures drop below about 11 °C (in British Columbia, this typically occurs in late September or October); however, the migration of adults to spawning sites can begin as early as late July (O'Connor and Power 1976). Spawning occurs in both streams and lakes. In streams, spawning sites usually are associated with areas of groundwater discharge; however, this relationship is not absolute (Curry and Noakes 1995a, 1995b). Females select the actual spawning sites—often in the tail-outs of pools (Power 1980)—and the nest digging, courtship, and spawning behaviours are similar to that of other salmonines (for details, see descriptions under rainbow trout). Depending on water velocity and female size, gravel diameters range from about 0.3 to 8.0 cm (Raleigh 1982). Water velocities over spawning sites are variable and range from 0.03 to 0.9 m/s (Power 1980; Smith 1973; Witzel and MacCrimmon 1983).

Normally, a dominant male accompanies a digging female. Usually, he is positioned immediately downstream of the female. The dominant male is aggressive and attacks peripheral males that get too close to the nest. Once sexually mature, female brook trout usually spawn in several successive years; however, if food resources are limited, they may skip a year between spawnings.

In lakes and beaver ponds, brook trout usually spawn at sites where there is upwelling or subsurface flow (Blanchfield and Ridgway 1997; Carline 1980; Webster and Eriksdottir 1976; Witzel and MacCrimmon 1983). Typically, lacustrine redds are found close to shore in water about 1 m deep (Blanchfield and Ridgway 1998). In the upper Peace system (Dina #1 Lake), brook trout were observed on spawning sites in the lake. The sites were on a coarse gravel substrate within about 50 m of the inlet stream, close to shore, and in about 1 m of water. Although not measured, there was probably subsurface flow at these sites. A wide range (sand and silt up to cobbles) of spawning substrates are used in lakes and ponds. When the substrate is too large for the females to move, the site is cleaned by the female's attempts to dig, and the fertilized eggs fall into interstices between the gravel and cobbles.

As in streams, there are dominant and peripheral males involved in lacustrine spawnings. In a lake-spawning population, over half the matings involved peripheral males (Blanchfield and Ridgway 1999). The same study also observed size-assortative mating—large males preferred to mate with large females. However, large females attract more peripheral males than small females, and this results in an increase in the number of males involved in a fertilization.

Egg size in brook trout varies from 3.5 to 5.0 mm, and fecundity ranges from about 100 to 5,000 eggs depending on female size. Development rate is a function of incubation temperature, and times to hatch range from 47 days at 10 °C to 165 days at 3 °C (Flick 1991). At first, the alevins are negatively phototactic and remain in the gravel until the yolk sac is absorbed. The time from hatch until yolk absorption also is affected by temperature: 56 days at about 7.0 °C and 19 days at 19.5 °C (McCormick et al. 1972).

AGE, GROWTH, AND MATURITY

Growth rate of young-of-the-year brook trout in both streams and lakes is variable and depends on temperature (McCormick et al. 1972), productivity, and intra-specific and inter-specific interactions. Apparently, scales are not formed until near the end of the first growing season (by this time, fry in northern populations reach a length of about 40 mm) (Power 1980). Within their natural range, wild brook trout are relatively short lived, and fish over 3 or 4 years old are rare (Brash et al. 1958; Power 1980). In these short-lived populations, males typically mature at 1^+ (the end of their second growing season), and most females mature a year later (2^+). In high mountain lakes in Alberta, however, maturity is delayed, and growth and maximum age varied with temperature, food abundance, water conductivity, and altitude (Donald et al. 1980). Many of these introduced populations contained fish in their sixth growing season (5^+), and brook trout in some lakes lived for up to 10 years.

FOOD HABITS

In streams, young-of-the-year and juvenile brook trout forage primarily on amphipods and the larvae of aquatic insects (Miller 1974). Curiously, prey capture success in young-of-the-year brook trout in streams is low (<42%); however, foraging efficiency increases with size (McLaughlin et al. 2000). Additionally, there is evidence that intraspecific competition among newly emerged fry selects for divergent foraging modes: sedentary fry that feed in the lower portion of the water column and active fry that forage closer to the surface (McLaughlin et al. 1999). In lakes, underyearling brook trout consume zooplankton (e.g., ostracods, cladocerans, and copepods).

In streams, adult brook trout are generally insectivorous and feed on the nymphs and adults of aquatic insects, especially caddisflies (Flick 1991; Griffith 1972). In lakes, the diet of adults is varied, and there appear to be "pelagic" and "benthic" specialists (Bourke et al. 1996). The pelagic specialists forage in open water and feed primarily on water-column prey (plankton), whereas the benthic specialists forage in the littoral zone and feed primarily on snails, leeches, and dragonfly nymphs. The diet changes with seasons, and like many trout and char, fish can be an important winter prey.

Habitat Habitat use by introduced brook trout has not been studied in our province; however, within its native range, brook trout exhibit considerable intra-population and inter-population variation in habitat use. The following habitat summary is derived from native populations in eastern North America and introduced populations in some western states. Again, a major source of information on brook trout habitat use is the excellent review by Power (1980).

ADULTS

Like most char, brook trout prefer cool water; however, their upper thermal tolerance appears to be wider than that of our native char. Adults can tolerate temperatures ranging from near 0 °C to about 22 °C but grow and survive best at temperatures between 13 °C and 19 °C (Schofield et al. 1989). They are also more tolerant of low pH (down to about 5.0) than other salmonines (Flick 1991). Adult brook trout use a wide assortment of habitats: small streams, large rivers, beaver ponds, and lakes. In British Columbia, however, most brook trout populations are lacustrine.

In an examination of 69 lakes in Quebec (Bourke et al. 1996), the proportion of brook trout using the littoral and pelagic zones was related to the presence of other species, especially creek chub (*Semotilus atromaculatus*) and white sucker (*Catostomus commersonii*). These species forage in the littoral zone, and their presence in the same lake as brook trout was associated with major reductions in the proportion of brook trout using the littoral zone. The effect was especially strong in the presence of white suckers. Adult brook trout in two Laurentian lakes exhibited a nocturnal activity pattern during the summer: during daylight hours, they remained at inshore focal sites, and at dusk, they left their focal site and actively foraged throughout the night. At dawn, they returned to their daytime focal site (Bourke et al. 1996).

In streams and rivers, adult brook trout appear to prefer low-velocity habitats (around 0.10 m/s; Griffith 1972); however, dominant fish are found in water velocities as high as 0.34 m/s (Cunjak and Green 1984). Cover is important and includes pools, undercut banks, or areas with overhanging stream bank cover (Allan 1981; Hunt 1974). Additionally, brook trout habitat use is strongly influenced by the presence of other salmonid fishes, although not always in the same way. For example, brook trout occur closer to the bottom, in shallower water, and farther from cover than coexisting rainbow trout (*Oncorhynchus mykiss*) (McCormick et al. 1972). However, when they coexist with Arctic grayling (*Thymallus arcticus*), brook trout occupy slower water and remain closer to cover than the grayling (Byorth and Magee 1998).

During winter (water temperatures of 1–9 °C), stream-dwelling brook trout are active at night and stay concealed in the cobble–boulder substrate or under woody debris during the day (Meyer and Gregory 2000).

JUVENILES

In a small stream in the spring, water velocity was the most important variable governing habitat use by juvenile brook trout: juveniles selected areas with low velocities (Johnson and Dropkin 1996). In summer and autumn, however, cover and depth were more important than velocity. Juveniles preferred deeper water with more cover than young-of-the-year brook trout. Morphological evidence from two Laurentian lakes indicates that the ecological separation of brook trout into littoral and pelagic specialists is detectable in juveniles (Dynes et al. 1999).

YOUNG-OF-THE-YEAR

On emergence, stream-dwelling brook trout usually move into shallow edge habitats and establish territories over coarse gravel and cobble substrates; however, in slow currents, they may remain in mid-stream (Power 1980). In a small stream, the microhabitats used by brook trout fry were more uniform over both seasons and years than the microhabitats used by juveniles (0^+)—under-yearling brook trout used areas with less cover and shallower water than juveniles (Johnson and Dropkin 1996). Thus, an ontogenetic habitat shift probably occurs in brook trout between their first and second years of life (Johnson and Dropkin 1996).

In lake-spawning populations, newly emerged fry normally move into shallow edge habitats; however, lake-spawned 0^+ brook trout have been observed migrating from a lake into small inlet streams (Curry et al. 1997). These authors estimated that, by mid-summer, 81% of the fry took up residence in these streams and suggested that some young overwintered and remained in streams throughout their second summer. These inlet streams were cool and stable, and this argues that temperature maybe an important habitat variable for under-yearling brook trout.

Temperature-related seasonal microhabitat shifts are known to occur in lakes (Blanchfield and Ridgway 1997). In May, young-of-the-year brook trout foraged in the warmest available water (about 15 °C) and within 2 m of shore. By early June when the temperature was close to the upper thermal tolerance for brook trout (about 20 °C), they foraged near the bottom within 4 m of shore. By July, lake temperatures ranged from 23 °C to 27 °C, and under-yearling brook trout concentrated near the bottom in specific areas about 3–8 m from shore. The areas occupied had groundwater flow, and temperatures ranged from 18 °C to 20 °C. Apparently, during the warmest part of the summer, young-of-the-year brook trout stay on the bottom and defend cool microhabitats even at the expense of daytime foraging opportunities (Biro 1998). In contrast, young-of-the-year brook trout in two Laurentian lakes were most active during the day (Bourke et al. 1996). This suggests that there is an ontogenetic change in diel activity pattern between young-of-the-year and 1^+ brook trout.

Conservation Comments   Although introduced brook trout are a popular recreational species, they pose a threat to native salmonines. They occupy habitats similar to those used by our indigenous trout and char, compete with native species for food and space (Griffith 1972; Gunckel et al. 2002), and in some cases, appear to displace native species. Additionally, they readily hybridize with Dolly Varden and bull trout. For these native char, competitive interactions with brook trout are exacerbated by genetic contamination through hybridization. Hybrids between brook trout and bull trout are mainly males and are thought to be partially sterile (Markle 1992). Molecular analyses, however, clearly indicate backcrossing (Kanda et al. 2002; McPhail and Taylor 1995). Thus, in the upper Skagit River system, hybrids between bull trout and brook trout are more numerous than "pure" brook trout. Additionally, in small tributary streams, most of the fish that appear to be brook trout are, in fact, hybrids (including backcrosses) between Dolly Varden and brook trout. Given their potential for adverse effects on our native fauna, no further introductions should be made into waters that do not already contain brook trout.

Salvelinus malma (WALBAUM)
DOLLY VARDEN

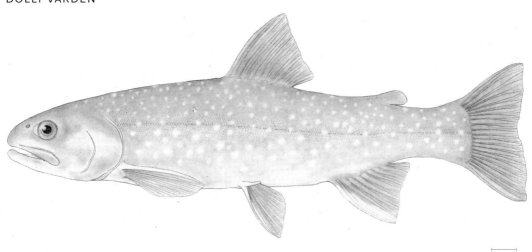

1 cm

Distinguishing Characters The Dolly Varden is a trout-like char with no teeth on the shaft of the vomer and pale pink, lilac, or red spots along the flanks. Dolly Varden and bull trout (*Salvelinus confluentus*) are difficult to separate in the field, and even experienced biologists make mistakes. They differ in head shape, upper jaw length, anal fin base (almost equal to upper jaw length in Dolly Varden), and in the total number of branchiostegal rays (usually <23 in Dolly Varden and >23 in bull trout). Relative to bull trout, the spots on the back of Dolly Varden are usually small and crowded.

Where they coexist, the ecology of adult Dolly Varden and bull trout usually are strikingly different (Hagen 2000; Hagen and Taylor 2001; McPhail and Taylor 1995): typically, adult bull trout are large (55–85 cm), migratory, and piscivorous, whereas adult Dolly Varden are small (10–25 cm) stream residents that feed on drift.

Taxonomy The relationship between the Dolly Varden and other North American chars (especially bull trout and Arctic char, *Salvelinus alpinus*) has a long and tangled history. For detailed accounts of the taxonomic history of bull trout, Dolly Varden, and Arctic char and the arguments for and against considering them separate species, see Brunner et al. (2001), Cavender (1978), Elz (2003), Haas and McPhail (1991), and Phillips et al. (1999), as well as the taxonomy section in the bull trout species account. In the case of char, I have chosen sympatry as the test of species status. Thus, even though there is hybridization in the broad contact zone between Dolly Varden and bull trout in British Columbia, I recognize them as separate species.

Some authors (e.g., Phillips et al. 1999) recognize several subspecies of Dolly Varden; however, given the confused state of specific relationships among this group of chars, I have avoided subspecific names. This does not mean that Dolly Varden in British Columbia are genetically uniform. Indeed, there is strong evidence that the Dolly Varden in southwestern British Columbia are derived from the Chehalis Refugium, whereas those on the central and northern coast are derived from a northern refugium (Taylor et al. 2001).

Sexual Dimorphism In many populations, breeding males develop a hooked lower jaw (kype) that, in extreme cases, fits into an indentation in the tip of the snout. Also, breeding males usually are more vividly coloured than females, especially the markings on the paired fins. Both spawning colours and kype development vary among populations, and many stream-resident Dolly Vardens never develop a kype; however, in the sea-run Keogh River population, even females develop a noticeable kype. In this population, the males develop dense black pigment on the underside of the head. Outside the breeding season, the sexes are difficult to distinguish, but males usually have larger adipose fins than females of the same size.

Hybridization As far as is known, Dolly Varden and bull trout hybridize wherever they coexist. The hybrids appear to be fertile, and in both the upper Skagit and upper Peace systems, molecular data indicate backcrossing (Baxter et al. 1997; McPhail and Taylor 1995; Redenbach 2000; Redenbach and Taylor 2002). In spite of this persistent hybridization, the two species maintain themselves as separate entities with life histories adapted to different ecologies (Hagen 2000; Hagen and Taylor 2001). Dolly Varden also hybridize with the introduced brook trout (*Salvelinus fontinalis*) in areas where they both occur.

Distribution **NATIVE DISTRIBUTION**
Dolly Varden are continuously distributed in coastal areas around the North Pacific Ocean. In Asia, they extend south to Korea and northern Japan, and in North America at least as far south as the Olympic Peninsula (Quinault River), Washington.

BRITISH COLUMBIA DISTRIBUTION
The Dolly Varden is primarily a coastal species (Map 52). It regularly enters the sea and occurs on most coastal islands that contain lakes or permanent streams. In the Skeena system, and perhaps in other northern coastal drainages, they penetrate much farther inland, and there have been headwater transfers of Dolly Varden from the Skeena drainage into the upper Fraser (Stuart drainage system) and upper Peace (Finlay and Ingenika rivers). There are also Dolly Varden in the B.C. headwaters of the Liard River (Mackenzie system). Presumably, they crossed into the Liard from either the Iskut–Stikine or Taku river systems.

Map 52 The B.C. distribution of Dolly Varden, *Salvelinus malma*

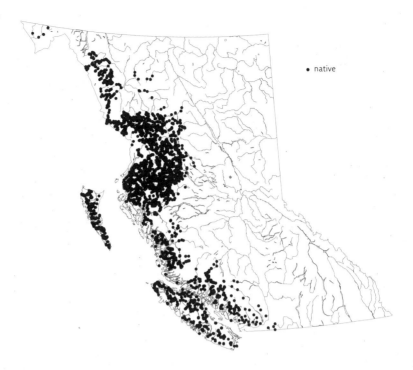

Life History In British Columbia, Dolly Varden display three general life-history forms: an anadromous form that migrates between fresh water and the sea, a stream-resident form that spends its entire life in rivers and streams, and an adfluvial form that lives most of its adult life in lakes but spawns in streams. In British Columbia, sea-run Dolly Varden tend to remain in or near estuaries. Their return migration into fresh water typically involves both immature and mature individuals. Thus, the return migration is not necessarily a breeding migration, and sea-run Dolly Varden in British Columbia probably are more amphidromous than truly anadromous (Myers 1949). In the Bering and Beaufort seas, however, the northern form of Dolly Varden is truly anadromous and migrates long distances in salt water (DeCicco 1992; Krueger et al. 1999).

REPRODUCTION

Like all char, Dolly Varden spawn in the fall, although the exact time and place of spawning varies with temperature and life-history type. With the exception of stream-resident populations that usually spawn locally, a migration typically precedes spawning. As far as is known, all Dolly Varden spawn in running water. Females select the spawning site and dig a series of nests (the redd). The size and depth of the redd, and the size of spawning gravel, vary with female size and, perhaps, with life-history type. In upper Skagit tributaries, spawning was observed throughout the day (McPhail and Taylor 1995). Spawning females were accompanied by a dominant male that vigorously defended the female from the attentions of other, usually smaller, satellite males. In some populations, there are jacks (small precocious males); however, probably because of the general small size of stream residents, jacks appear to be absent in most small stream populations. In the upper Skagit system, all observed spawnings involved a single pair; although spawning pairs were well separated (usually at least 20 m apart), spawning activity was concentrated in a small area relative to the total length of the stream. In the upper Peace system where bull trout and Dolly Varden coexist, small Dolly Varden males apparently act as jacks during bull trout spawnings. This may be the source of many of the hybrids recorded in this, and other regions, of sympatry.

In northern British Columbia (upper Peace River system), spawning for a stream-resident population peaks in late September at water temperatures of about 6 °C. In southern British Columbia (Skagit drainage system), stream residents begin to spawn in late September at 8 °C. In the main Skagit River, spawning activity peaks in early October (again at 6 °C). The Dolly Varden in the Keogh River on northern Vancouver Island is the only sea-run population that has been studied in British Columbia (Smith and Slaney 1980). This run starts in mid-July and continues into October. The early portion of the run consists mostly of parr migrating upstream from the lower river (mean FL 125 mm), subadults (immature fish) returning from their first ocean summer (mean FL 220 mm), and adults that probably had been to sea before (mean FL 340 mm). Mature adults dominated the last portion of the run. The returning adults are "green" (not ready to spawn) but quickly mature in the river. Spawning occurs in late October when water temperatures drop below 10 °C (Smith and Slaney 1980).

In upper Skagit tributaries, redd sites usually are located at the downstream ends of pools where the water is starting to break into another riffle (McPhail and Taylor 1995). In this system, the redd characteristics are similar to those described for a stream-resident populations associated with Thutade Lake, upper Peace system (Bustard 1996). In this population, females usually excavate nests in shallow (<15 cm), low-velocity (<0.40 m/s) headwaters. Typically, 50% of the substrate is <2 cm in diameter, and the redds are small (<0.5 m^2). In seepage areas, redds usually are less than 1.0 m from cover, but in the mainstems of streams, redds can be up to 4.5 m from cover. The spawning behaviour is similar to that of other salmonines (see the rainbow trout reproduction section for an account of salmonine nest digging and spawning behaviour).

In Dolly Varden, fecundity varies with female body size. Stream-resident females typically breed at 100–250 mm FL and produce about 70–500 eggs that are about 3.5 mm in diameter. In the Keogh River sea-run population, the minimum size at maturity is about 210 mm, but most spawning females range in size from 250 to 450 mm. In this population, fecundity ranges from around 100 to almost 6,000 eggs that are about 4.5 mm in diameter (Smith and Slaney 1980).

The eggs incubate in the gravel overwinter and take about 3 months to hatch. The fry emerge from the gravel in April or near the beginning of May. Females in headwater populations, such as those in the upper Finlay drainage, typically breed at a small size (<200 mm) and produce 200–500 eggs about 3.5 mm in diameter. In this region, the fry emerge near the beginning of June. Typically, Dolly Varden fry are smaller than sympatric bull trout fry and are about 20 mm TL when they emerge from the gravel.

AGE, GROWTH, AND MATURITY

Depending on stream temperature in the upper Finlay, young-of-the-year Dolly Varden reach 25–40 mm FL by the end of their first summer, about 60 mm in their second summer, and about 80 mm by the end of their third summer. Some males mature at the end of their fourth summer (>115 mm), and most fish of both sexes reach maturity in their fifth growing season (4^+). Maximum age in these populations is about 9 years (8^+). On the coast, Dolly Varden fry reach about 35–50 mm by mid-August and continue growing throughout the fall (late October). In the Keogh River, on average, they require, 3 years to reach smolt size (about 140 mm). On their first migration to the ocean, the smolts spend about 100 days at sea and almost double their size in that time (Smith and Slaney 1980). In this sea-run population, sexual maturity is reached at age 3^+ or 4^+. The maximum age reported in the Keogh River population is 8^+. .

FOOD HABITS

Stream-resident adults feed primarily on the nymphs and larvae of aquatic insects (mayflies, caddisflies, stoneflies, and chironomids). When feeding during the day, these fish remain close to the bottom, and most feeding movements are directed towards insects drifting near the bottom. Occasionally, however, they rise to the surface and take terrestrial insects that have fallen in the water. In Thutade Lake tributaries, a comparison of foraging behaviour and stomach contents in sympatric adult Dolly Varden and juvenile bull trout revealed that the proportion of feeding movements directed at the bottom by both species increased at night (Hagen 2000; Hagen and Taylor 2001). The only major difference between the stomach contents of Dolly Varden and juvenile bull trout was the relative proportion of winged insects in the diet (higher in Dolly Varden).

In coastal lakes containing both Dolly Varden and trout (*Oncorhynchus*), adult Dolly Varden forage on a wide variety of zoobenthos, especially chironomids and amphipods. In lakes without trout, zooplankton (particularly copepods) are the primary food items (Hindar et al. 1988; Hume and Northcote 1985; Schutz and Northcote 1972). In large lakes on Vancouver Island, large (>30 cm FL) Dolly Varden become piscivorous and feed on young salmonines, sticklebacks, and sculpins. Fry and juveniles feed primarily on chironomid larvae; however, in streams as they grow they shift to larger prey (predominately the drifting stages of aquatic insects). At sea, the Keogh population of Dolly Varden fed mainly on isopods and small herring (Smith and Slaney 1980).

Habitat Dolly Varden usually are associated with cool water (waters where summer temperatures rarely exceed 20 °C and normally are <15 ° C). On the eastern slope of the Coast Mountains, they are confined to small, headwater streams; however, on the western slope of the Coast Mountains and on coastal islands, they occupy more diverse habitats—small streams, rivers, small boggy lakes, large oligotrophic lakes, and near-shore marine waters.

ADULTS

Habitat use in adult Dolly Varden is influenced by life-history type and the presence of other fish species. In upper Peace drainages, all the Dolly Varden are stream residents. This may be because they are derived from stream-resident populations in the Skeena system; however, the presence of bull trout in this system probably affects their habitat use. Typically, in areas of contact between these species, Dolly Varden are strictly small stream residents, and bull trout are riverine or adfluvial. Although not tested experimentally, this pattern of habitat segregation is widespread (Hagen 2000; Hagen and Taylor 2001; McPhail and Taylor 1995). In areas where only one species occurs, however, both bull trout and Dolly Varden broaden their trophic and spatial niches to include resources that, in sympatry, are used by the other species. For example, bull trout are absent from Vancouver Island, and in the large lakes on the island, some Dolly Varden shift into the trophic niche (deep-water piscivores) occupied by bull trout on the mainland.

In Thutade Lake tributaries, in-stream habitat use was examined in adult Dolly Varden and juvenile bull trout (Hagen 2000; Hagen and Taylor 2001). There was considerable overlap between the species—they used the same hydraulic habitat types (main-channel pools, main-channel riffles, side-channel pools, and side-channel riffles) in similar proportions during both day and night. At night, however, adult Dolly Varden were found in deeper and, perhaps, faster water than juvenile bull trout.

In British Columbia, cutthroat trout (*Oncorhynchus clarkii*) and Dolly Varden coexist in many small coastal lakes. In these environments, Dolly Varden are displaced from littoral and offshore surface habitats into deeper water by the more aggressive cutthroat trout (Andrusak and Northcote 1971; Hindar et al. 1988; Hume and Northcote 1985). The visual acuity of Dolly Varden at low light levels is higher than trout (Henderson and Northcote 1985, 1988); thus, in deeper water, they probably forage more successfully than trout. The water in many of the small lowland lakes along the B.C. coast is deeply stained and the shores are boggy with floating mats of *Sphagnum*. Dolly Varden and threespine sticklebacks (*Gasterosteus aculeatus*) are often the only resident fishes in these lakes. Field observation on streams containing both Dolly Varden and cutthroat trout suggest a pattern in their distributions: cutthroat trout in the lower reaches of the streams, and Dolly Varden upstream (sometimes above barriers).

In British Columbia, the only anadromous population that has been studied is in the Keogh River (Smith and Slaney 1980). This population begins entering the river in mid-July and stops in late October. The early portion of the run consists mostly of subadults (immature fish), but mature adults dominated the last portion of the run. After spawning, the adults overwinter in lakes and return to the sea in the spring. In British Columbia, no systematic observations are available on spawning migrations of adfluvial Dolly Varden; however, casual observations in the southwestern part of the province suggest relatively short (often <1 km) migrations from lakes into spawning streams.

JUVENILES

In stream-dwelling Dolly Varden, juveniles are associated with shallow (<0.50 m deep), slow (<0.10 m/s) runs and pools. They also use side-channels. During the day, they remain close to cover, such as large rocks, woody debris, root wads, and undercut banks, but are less strongly associated with cover at night. The parr of anadromous populations sometimes migrate to the lower reaches of rivers in the spring and migrate back upstream in the late summer or fall (Smith and Slaney 1980). After 3 or 4 years in freshwater, they enter the sea but return to fresh water to overwinter in lakes. In southeastern Alaska, individuals that originated in watersheds without lakes apparently seek out watersheds with lakes as overwintering sites. Once they have chosen such a watershed, they continue to use it as an overwintering site in successive years (Bernard et al. 1995).

YOUNG-OF-THE-YEAR

Newly emerged Dolly Varden fry are denser than water, and in streams, they are primarily associated with shallows (<5 cm deep) along stream edges (Dolloff and Reeves 1990). Here, they remain close to the substrate and are found in, and around, coarse gravel and cobbles interspersed with boulders, especially in areas of low water velocities (<0.09 m/s), such as side-channels and shallow bays. They are secretive during the day and, if disturbed, quickly seek shelter under rocks, logs, and undercut banks. In most stream populations, the fry remain in this habitat during their first summer but shift to deeper water as they grow (Dolloff and Reeves 1990). In streams subject to freezing, the fry move into deeper water during winter. Although the fry of sea-run Dolly Varden in the Keogh River start life along the river margins, they move into areas of higher water velocity by mid-summer (e.g., riffles and runs).

Conservation Comments In most of coastal British Columbia, Dolly Varden are abundant and not heavily exploited. Of particular conservation interest are sites where Dolly Varden and bull trout coexist. The two species are found together in a broad contact zone associated with the Coast Mountains. Most sympatric sites are in west-flowing drainages, but both species occur in some east-flowing river systems (e.g., the Blue River in the upper Liard system and the Finlay and Ingenika rivers in the upper Peace system). Apparently, the species hybridize wherever they come in contact (Baxter et al. 1997; McPhail and Taylor 1995). The hybrids are fertile, and there is some backcrossing (Redenbach 2000; Redenbach and Taylor 2002); however, in spite of the backcrosses, the two species usually maintain their distinctness in the face of gene flow (Hagen and Taylor 2001). Nonetheless, there is some evidence (Elz 2003; Taylor et al. 2001) that the Dolly Varden of the southern coastal lineage may have originated through introgression with bull trout. This contact zone is of considerable scientific interest, and some of the areas where Dolly Varden and bull trout coexist should be protected from disturbance.

Perhaps the greatest threat to Dolly Varden is the brook trout. In the past, this alien species was introduced to augment recreational angling; however, introduced brook trout hybridize with Dolly Varden. The two species are interfertile, and it is difficult to find pure Dolly Varden in some areas (e.g., the Sumallo River). Usually, brook trout have not replaced Dolly Varden, but the two species appear to have introgressed.

Although Dolly Varden is not listed by either cosewic or BCCDC, fluvial populations receive some protection through angling regulations designed to protect bull trout.

Salvelinus namaycush (WALBAUM)
LAKE TROUT

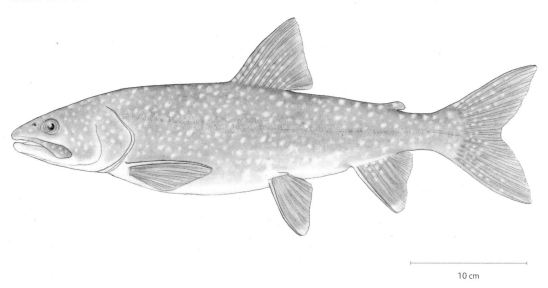

10 cm

Distinguishing Characters	This trout-like char is heavily spotted with irregularly shaped light spots on the back, sides, and dorsal and caudal fins. Unlike other char in our area, the spots on lake trout are grey or whitish but never coloured, and the tail is deeply forked: in lake trout, the centre rays on the tail are about half the length of the longest outer rays, whereas in other char, the centre rays are almost as long as the longest outer rays.
Taxonomy	At present, lake trout taxonomy is stable, but as recently as the early 1970s, some authors (e.g., Quadri 1967; Vladykov 1954, 1963) placed lake trout in a separate genus (*Cristivomer*). Still, a stable taxonomy does not imply genetic homogeneity, and there is evidence that lake trout survived (and diverged) in at least five separate refugia during the Pleistocene (Wilson and Hebert 1998). The postglacial origins of native lake trout in British Columbia are not clear, but most populations appear to be derived from two northern sources: the Bering and Nahanni refugia (Wilson and Hebert 1998).
Sexual Dimorphism	Sexual dimorphism, even in spawning fish, is not as pronounced as in other char, but the largest fish in most populations are females. There are no bright spawning colours, but the pectoral, pelvic, and anal fins usually develop white leading edges followed by a dusky pigment. Males often develop a white chin, and there is a report of "pearl" organs (Vladykov 1970) on the ventral surface of both sexes. The absence of bright colours may be associated with the nocturnal spawning behaviour in this species.

Hybridization No natural hybrids involving lake trout are known from British Columbia, and splake (the man-made hybrid between lake trout and brook trout, *Salvelinus fontinalis*) has not been widely introduced into our province (Crossman 1995).

Distribution NATIVE DISTRIBUTION
Originally, lake trout were restricted to northern North America but they have been widely introduced outside their natural range in the United States, and into Europe, South America, and New Zealand. Their natural distribution lies almost entirely within glaciated areas and includes the southern islands in the Arctic Archipelago (Baffin, Southampton, King William, Victoria, and Banks islands).

BRITISH COLUMBIA DISTRIBUTION
The lake trout's natural range in British Columbia included the upper and middle Fraser system and the upper Skeena, Nass, Iskut–Stikine, Taku, and Yukon drainage systems, as well as the Peace and Liard river systems (Map 53). Introduced populations of eastern Canadian origin have been planted outside the natural B.C. range of this species, and self-sustaining populations now occur in lakes in the Okanagan and Columbia drainage systems, as well as in Alouette Lake in the lower Fraser valley.

Life History Little is known about the life history of native lake trout in British Columbia. Consequently, most of the following information is derived from eastern and northern Great Plains populations. The following synopsis is based on an excellent review of lake trout life histories (Martin and Olver 1980) and a comprehensive bibliography (Marshall et al. 1990).

REPRODUCTION
Like other char, lake trout spawn in the fall, usually when water temperatures drop below about 10 °C (Olver 1991). For eastern North American populations, spawning begins at temperatures ranging from 8 °C to 13 °C (Sly and Evens 1996). In British Columbia, spawning dates vary with latitude and lake size: at the southern edge of their natural range and in large lakes, spawning may not start until November; however, in northern parts of the province and in small lakes, they may spawn as early as late September. So far in British Columbia, spawning has only been recorded in lakes, but elsewhere, some populations spawn in rivers. Lake trout will spawn on a variety of substrates, but broken rubble or angular rock interspersed with large boulders is a typical spawning substrate (Martin and Olver 1980). Although particles with diameters of 4–10 cm are optimal, successful spawning can occur over materials up to 30 cm in diameter (Sly and Evens 1996). Spawning depth is variable but usually ranges from about 5 to 50 m (Martin and Olver 1980); however, spawning can occur at depths as shallow as 20 cm (Dumont et al. 1982), and lake trout in small lakes sometimes spawn in water so shallow that their backs are exposed (Olver 1991).

Map 53 The B.C. distribution of lake trout, *Salvelinus namaycush*

Since lake trout spawn at night, their spawning behaviour is not as well documented as that of other salmonines. They do not construct a nest, but some prespawning activities help to clean the substrate (Royce 1951). As dusk approaches, first males and later females congregate on the spawning site. Their spawning behaviour is described in Royce (1951). The males (more than one is usually involved in a spawning) swim alongside the female, press against her, and quiver. As the eggs are released, they are fertilized and fall into cracks and crannies between the rocks. No attempt is made to cover the eggs, and sometimes more than one female is involved in a spawning (Olver 1991).

Lake trout eggs are large (4–5 mm), and fecundity is low. Depending on female body size, fecundity varies from 500 to 20,000 eggs, and females in many populations do not spawn every year (Adams and Reynolds 1997; Johnson 1973; Olver 1991). Like most fishes, the development rate of fertilized eggs is a function of incubation temperature; however, development rates differ among populations (Horns 1985), and hatching times range from about 50 days at 10 °C to about 156 days at 2 °C (Embody 1934; Garside 1959). Alevins are about 21 mm FL and negatively phototactic (Fish 1932). They remain in the substrate as the yolk sac is absorbed; however, even before the yolk sac is completely absorbed, the alevins emerge from the substrate at night (Baird and Kruger 2000).

AGE, GROWTH, AND MATURITY

Growth rates in lake trout are variable and depend on temperature, productivity, population density, and age at maturity. However, northern populations generally grow more slowly, live longer, and reach a larger size than southern populations (Martin and Olver 1980). Increased altitude, even at relatively low latitudes, also influences growth rate, and lake trout introduced into subalpine lakes in Alberta produced stunted populations (Donald and Alger 1986). Similar stunted (but natural) populations occur in lakes in British Columbia (e.g., Carbon Lake). Lake trout in northern lakes are especially long lived, and fish >20 years old are not uncommon. So far, the oldest lake trout aged by otoliths was 62 years old (Olver 1991). In Teslin Lake, juveniles (2^+) averaged 20 cm FL, and the largest age class (12^+) averaged over 750 cm (Clemens et al. 1968).

Maturity in lake trout varies among populations but usually is reached at between 5 and 13 years ((Madenjian et al. 1998a; Martin and Olver 1980). Typically, males mature 1 or 2 years before females.

FOOD HABITS

Young-of-the-year lake trout forage primarily on plankton and the larvae of aquatic insects (DeRoche 1969). As they grow, lake trout add larger benthic organisms (e.g., amphipods and mollusks) to their diet and when they reach a length of about 35–40 cm, they usually switch to fish if appropriate prey species are available (Madenjian et al. 1998b, Olver 1991). If fish are unavailable, however, lake trout remain planktivorous. Piscivorous populations grow larger, and live longer, than planktivorous populations (Martin 1955). Seasonal changes in diet are common and are associated with temperature. For example, in the spring, adults and subadults may forage in shallow water and near the surface. At this time, minnows and terrestrial insects may be important in their diet. As the lake warms and stratifies, lake trout are separated from these food sources by a thermal barrier (temperatures >15 °C). Under such conditions, they may switch to planktivory; however, as the lake cools in the fall, they may switch back to their spring diet (Olver 1991). In Teslin Lake, adult lake trout were primarily piscivores: juvenile lake whitefish (*Coregonus clupeaformis*), round whitefish (*Prosopium cylindraceum*), and least ciscoes (*Coregonus sardinella*) were the main prey (Clemens et al. 1968). In Carbon Lake near Chetwynd, an apparently stunted population contained mostly plankton.

Habitat Lake trout habitat use has not been studied in native B.C. populations. Consequently, the following account is based on studies in eastern North America, and it is possible there are minor ecological differences between our native lake trout and eastern populations.

ADULTS

Like most char, lake trout prefer cool water. Adults can tolerate temperatures ranging from near 0 °C to about 18 °C, but they prefer temperatures around 10 °C (Martin and Olver 1980). Their upper lethal temperature is 23.8 °C (Gibson and Fry 1954). Seasonally, adult lake trout shift their vertical distribution in response to temperature changes (Olver 1991). In the spring, when lakes are essentially homothermous, adult lake trout are found at all depths. As lakes stratify, however, they move to deeper, cooler water. If bottom waters become oxygen depleted (<4 mg/L), they move up in the water column. Thus, in small lakes they can be trapped between the thermocline and an oxygen-depleted bottom layer; however, at night in stratified lakes, lake trout move up through the thermocline and occupied the epilimnion (Sellers et al. 1998). Here, the water is well oxygenated (>6 mg/L) but warm (19–20 °C). This suggests that the summer behaviour of lake trout in small lakes may differ from that in large lakes. In the central British Columbia, this difference may be of some management importance.

Most studies of habitat use by lake trout have been conducted in waters in which lake trout are the only large, deep-water piscivore. In north-central British Columbia, however, bull trout (*Salvelinus confluentus*) also fill this niche. It is not known what effect bull trout have on habitat use by lake trout, but it is known that, in small lakes, there is niche overlap and that the introduction of lake trout adversely affect bull trout (Donald and Alger 1993).

JUVENILES

Juvenile lake trout occupy similar habitats to those used by adults; however, they tend to occur in deeper water than adults. This segregation of juveniles from adults may be driven by cannibalism (Olver 1991). The temperature range occupied by juveniles is about 6–13 °C. In British Columbia, the only information on juvenile habitat comes from Dease Lake. Here, in the summer, lake trout parr (1^+) are abundant in about 50 cm of water approximately 1–2 m offshore. The substrate was coarse (fist-sized) gravel, and the water temperature was 9.5 °C.

YOUNG-OF-THE-YEAR

After filling their swimbladders, lake trout alevins are neutrally buoyant and highly mobile. In some populations, they remain in shallow water in the vicinity of the spawning site for several weeks (Peck 1981), whereas in other populations, fry are thought to move directly to deep water (Olver 1991). A period of residence near the spawning site may be necessary for imprinting (Horall 1981), and there is evidence that adults in some populations home to their natal spawning sites (Olver 1991). In Lake Superior, fry occur at depths down to about 35 m (Eschmeyer 1955), and in an introduced population (Lake Tahoe, Nevada), fry are reported to occur at a depth of 100 m (Martin and Olver 1980).

Conservation Comments   Although there are management issues with some exploited populations of lake trout, so far this species is not a conservation problem in British Columbia. However, this could change if global warming continues and exploitation rates increase. For conservation purposes, it is important to distinguish between native and introduced populations. Most populations from Shuswap Lake and lakes farther north are native and mainly derived from the Nahanni refugium. Most populations south of Shuswap Lake are introduced and usually of eastern origin. Care should be taken to avoid mixing native and introduced stocks of lake trout. Unfortunately, this mistake has been made in the past, and eastern lake trout have been introduced into native populations (e.g., Morfee Lake near Mackenzie).

Subfamily Coregoninae

Whitefishes are related to the salmonines (trout, salmon, and char) and graylings. Like all salmonids, they possess an adipose fin, they have a scaly axillary process at the base of the pelvic fins, and their pelvic fins originate under the middle of the dorsal fin. They differ from trout, salmon, and chars in having large scales and weak teeth that often are only visible when stained with alizarin dye.

Whitefishes occur in cool rivers and lakes throughout much of the Northern Hemisphere. Most whitefish are restricted to fresh water, but Arctic populations of some species regularly enter the sea. Usually these anadromous populations remain close to shore, and their marine migrations are not as extensive or as regular as salmon migrations. With two exceptions (the Arctic cisco, *Coregonus autumnalis*, and the inconnu, *Stenodus leucichthys*), none of the B.C. populations are anadromous.

There are four ecological groups of whitefish in British Columbia: a large piscivorous whitefish, the inconnu (*Stenodus*); three predominately planktivorous species, the ciscoes (*Coregonus*); two predominately lacustrine benthivores, the broad and lake whitefish (*Coregonus*); and three species of round whitefishes (*Prosopium*). The phyletic relationships among, and within, these four groups are still controversial, and consequently, the taxonomic status of some genera and subgenera are unclear.

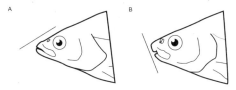

Figure 30 Head profiles of cisco (A) and lake whitefish (B) illustrating the difference in the projected slope of the upper lip.

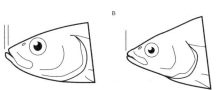

Figure 31 Head profiles of least cisco (A) and Arctic cisco (B) illustrating the difference between projecting and equal lower jaws

348 FAMILY SALMONIDAE — SALMONIDS

Figure 32 Relative position of the pelvic fins in least cisco (A) and cisco (B)

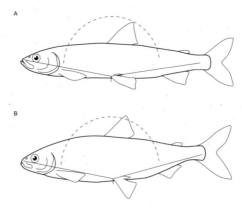

Figure 33 Body profiles of lake (A) and round (B) whitefishes illustrating the difference in the depth of the caudal peduncle relative to the dorsal fin base.

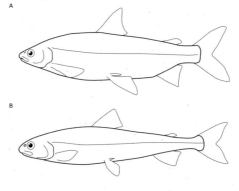

Figure 34 Body profiles of lake (A) and broad (B) whitefishes showing differences in dorsal fin shape and brow profiles.

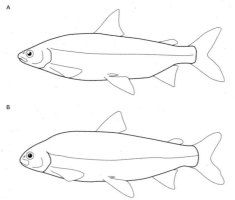

Figure 35 Dorsal views of the heads in pygmy (A) and mountain (B) whitefishes illustrating the difference in snout shape.

Figure 36 Differences in the size of anterior lateral line scales relative to adjacent body scales in pygmy (A) and mountain (B) whitefishes

 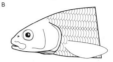

KEY TO ADULT WHITEFISH

1 (8) Profile of upper lip slopes backwards in line with the forehead (Fig. 30A); lower jaw projects beyond, or is equal to, upper jaw *(see 2)*

    2 (3) Mouth large, upper jaw extends back to below the posterior margin of the eye; small, velvet-like teeth on the anterior parts of the upper and lower jaws; 15–18 anal rays   INCONNU, *Stenodus leucichthys nelma*

    3 (2) Mouth small, upper jaw does not extend as far back as the posterior margin of the eye; 13 or 14 anal rays *(see 4)*

        4 (7) Upper and lower jaws not equal, tip of lower jaw projects slightly beyond upper jaw (Fig. 31A) *(see 5)*

            5 (6) Pelvic fins inserted forward, distance from snout to origin of pelvic fins equals the distance from origin of pelvic fins to a point on the caudal peduncle in front of the caudal flexure (Fig. 32A)   LEAST CISCO, *Coregonus sardinella*

            6 (5) Pelvic fins inserted farther back, distance from snout to origin of pelvic fins equals the distance from origin of pelvic fins to a point on the caudal fin rays behind the caudal flexure (Fig. 32B)   CISCO, *Coregonus artedi*

        7 (4) Upper and lower jaws about equal (Fig. 31B), tip of lower jaw does not project beyond upper jaw   ARCTIC CISCO, *Coregonus autumnalis*

8 (1) Profile of upper lip vertical or overhangs mouth, does not slope backward in line with the forehead (Fig. 30B); snout overhangs lower jaw *(see 9)*

    9 (12) Caudal peduncle deep, greater than or equal to the width of dorsal fin base (Fig. 33A); transparent membrane surrounding eye without a distinct notch below the pupil *(see 10)*

        10 (11) Anterior dorsal rays long, when depressed they extend well beyond the posterior dorsal rays; in profile, the brow is concave (Fig. 34A)   LAKE WHITEFISH, *Coregonus clupeaformis*

        11 (10) Anterior dorsal rays relatively short, when depressed they barely reach the tips of the posterior dorsal rays; in profile, the brow is rounded (Fig. 34B)   BROAD WHITEFISH, *Coregonus nasus*

    12 (9) Caudal peduncle narrow, less than width of dorsal fin base (Fig. 33B); transparent membrane surrounding eye has a distinct notch below the pupil *(see 13)*

        13 (14) Snout blunt (rounded) when viewed from above (Fig. 35A); dorsal rays usually 8 or 9; anterior lateral line scales about the same size as the scales immediately above and below the lateral line (Fig. 36A)   PYGMY WHITEFISH, *Prosopium coulterii*

        14 (13) Snout pointed (pinched) when viewed from above (Fig. 35B); in our area, dorsal rays usually 11–14; anterior lateral line scales about half the size of the scales immediately above and below the lateral line (Fig. 36B) *(see 15)*

            15 (16) Adipose fin large, its base more than 1.5 times eye diameter; 19–24 scales around the caudal peduncle   MOUNTAIN WHITEFISH, *Prosopium williamsoni*

            16 (15) Adipose fin small, its base less than 1.5 times eye diameter; 24–27 scales around the caudal peduncle   ROUND WHITEFISH, *Prosopium cylindraceum*

Coregonus artedi LESUEUR
CISCO

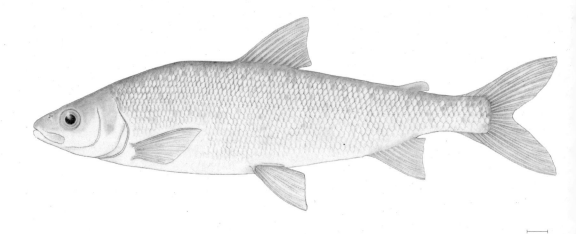

1 cm

Distinguishing Characters   This planktivorous whitefish has more than 40 gill rakers and a terminal mouth. When the mouth is closed, the tip of the lower jaw projects slightly beyond the tip of the upper jaw; the tips of the pelvic fins are dark in the B.C. population. Two other ciscoes occur in British Columbia: the Arctic cisco (*Coregonus autumnalis*) and the least cisco (*Coregonus sardinella*). The tip of the lower jaw does not project beyond the upper jaw in the Arctic cisco, and the pelvic fins in the B.C. specimens are unpigmented. The relationship between the lengths of the upper and lower jaws and pelvic fin colour are similar in the cisco and the least cisco. In British Columbia, these two species are allopatric: the least cisco is restricted to lakes in the Yukon drainage system, and the cisco is restricted to the Mackenzie drainge system. A useful character to distinguish the two species is the prepelvic distance. In the cisco, when stepped off with dividers, the distance from the snout to the origin of the pelvic fins is at least equal to the distance from the pelvic fins to the caudal flexure but falls short of the caudal flexure in the least cisco.

Taxonomy   The cisco is a morphologically and ecologically plastic species, and many subspecies have been named over its extensive geographic range. In western Canada, however, most recent authors (Nelson and Paetz 1992; Stewart and Watkinson 2004) no longer use cisco subspecific names. This does not mean that the taxonomic problems associated with the cisco are resolved—far from it! For example, it is not clear that cisco and Arctic cisco are different species. Allozyme data (Bodaly et al. 1991) indicate that the biochemical divergence between these nominal species is typical of that found among conspecific cisco populations. In contrast, other authors (e.g., Smith and Todd 1992) argue that, on the basis of a morphological analysis, the cisco and Arctic cisco are separate species.

Although the geographic distributions of the two species are mostly allopatric, they may overlap in the lower and middle Mackenzie River system. Thus, the resolution of the question of their specific distinctness probably lies in this region. If the cisco and Arctic cisco prove to be conspecific, the appropriate scientific name is *C. autumnalis*.

Another troublesome problem with the cisco is the presence of sympatric pairs (and even triplets) of ciscoes in some lakes. Typically, these sympatric pairs differ in gill raker number, growth rates, and life histories. As in many other sympatric pairs of northern freshwater fishes, morphologically similar pairs occur in geographically scattered parts of the cisco's range. Many of these pairs behave as biological species (*sensu* Mayr 1963); however, equating the pairs in different lakes is difficult and giving them scientific names based primarily on their geography would create taxonomic chaos. For now, an imperfect solution to the taxonomic problems posed by these sympatric pairs is to use a single taxonomic name but recognize the scientific and biodiversity value of these pairs of biological species (see the parallel evolution section in the introduction for a general discussion of this problem). Although there are cisco populations in the Mackenzie River system that appear to contain sympatric pairs—there are two forms in the Slave River (Tripp et al. 1981)—the only known B.C. population shows no evidence of bimodality in its morphology or life history.

Sexual Dimorphism Outside the spawning season, the sexes are difficult to distinguish. On breeding fish, the scales on the flanks, back, and undersides develop spawning tubercles. These tubercles differ from those found in lake (*Coregonus clupeaformis*) and broad (*Coregonus nasus*) whitefish: they are relatively flat and scarcely project above the scales. The tubercles are less obvious on females than on males.

Hybridization The cisco hybridizes with the lake whitefish in Alberta (Nelson and Paetz 1992) and the Great Lakes (Becker 1983). If the cisco is a species distinct from the Arctic cisco (see the taxonomy section), it probably hybridizes with the Arctic cisco in the middle and lower Mackenzie system.

Distribution NATIVE DISTRIBUTION
The cisco, as presently recognized, is restricted to North America. Its geographic range extends from the upper Mississippi River system to northern Quebec and from tributaries to the Gulf of St. Lawrence westward across the northern Great Plains to the Mackenzie River.

BRITISH COLUMBIA DISTRIBUTION
Our only cisco population is in Maxhamish Lake in the extreme northeastern portion of the province (Map 54).

Map 54 The B.C. distribution of cisco, *Coregonus artedi*

• native

Life History Little is known about the life history of cisco in British Columbia. Some age and growth data are available for the Maxhamish Lake population (DeGisi 1999), but most of the following life-history information is derived from populations in Alberta, the Northwest Territories, and eastern North America.

REPRODUCTION

The cisco spawns in the late fall when water temperatures approach 4 °C. By late August, all of the ciscoes age 3^+ or older sampled in Maxhamish Lake were mature but not yet ripe (DeGisi 1999). Thus, in this lake, spawning probably occurs near, or shortly after, freeze-up. Typical spawning sites are shallow (1–3 m deep) sandy or gravel shoals, although in the Great Lakes spawning can occur at depths exceeding 50 m, and there is even evidence of pelagic spawning (Becker 1983). The maximum depth of Maxhamish Lake is 10.6 m; however, there are cobble and gravel shoals that appear to be suitable spawning sites. Apparently, cisco do not prepare a spawning site. Their spawning behaviour has been described in a Wisconsin lake (Cahn 1927). Males arrived at the spawning site 2–5 days before the females. When the first females arrive, groups of males follow single females; however, as the number of females increased, the number of accompanying males decreased, and typically, two males followed each female. The female descends to within about 15–20 cm of the bottom. The males swim alongside with their heads near the middle of the female's body, and eggs and milt are extruded as they swim. The eggs, which are demersal and slightly adhesive, are scattered over the substrate. Egg size varies among populations. Fertilized eggs average about 2.1 mm in diameter and take about 45 days to hatch at 10 °C and 236 days at 0.5 °C (Brooke 1970). Optimal incubation temperatures in the Great Lakes region range from 2 °C to 8 °C, and hatching usually occurs shortly after ice breakup in the spring. Fecundity varies within and among populations. Within populations, egg number is related to female size and, in Lake Superior, ranges from about 4,000 to 10,000 eggs in females 269–356 mm TL. Fecundities in the two forms of cisco found in the Slave River Delta (Tripp et al. 1981) range from 506 to 2,670 in the small-bodied form (<19 cm FL) and from 700 to 10,750 in the large-bodied form (>22 cm). Newly hatched fry are 9.8–12.8 mm long (Colby and Brooke 1970) and begin feeding about 10 days after hatching (Pritchard 1930).

AGE, GROWTH, AND MATURITY
Cisco growth rates vary among populations and depend on food availability, temperature, density, and ecological interactions with other species. In Maxhamish Lake, cisco growth patterns appear to vary among years (DeGisi 1999). Some of this variability may reflect aging errors, but the data also suggest annual recruitment may vary. Extrapolating from growth curves, Maxhamish ciscoes appear to reach about 110–120 mm FL by the end of their first growing season and about 150–180 mm in their second summer (DeGisi 1999). Sexual maturity is reached in the third or fourth year of life. In British Columbia, the oldest recorded age is 8 years, but in Alberta, cisco in an introduced population lived to slightly over 30 years (Nelson and Paetz 1992).

FOOD HABITS
Cisco is primarily a planktivore. The initial diet of fry consists of rotifers, copepodites, and occasionally algae. As they grow, adult copepods and cladocerans are added to the cisco's diet. Although adults are primarily planktivores, they will consume small fish, chironomid pupae, and take adult insects from the surface.

Habitat Not much is known about the habitat use of cisco in British Columbia. Most of published information on cisco habitats pertains to large, deep lakes, and the only known B.C. population lives in a relatively shallow lake. Thus, the following summary is based mainly on what little information exists for the B.C. population.

ADULTS
Most faunal works (e.g., Becker 1983; Nelson and Paetz 1992; Scott and Crossman 1973) characterize the cisco as a fish of cool, oligotrophic lakes but note that it also occurs in shallower water than most other ciscoes. In northern Wisconsin, it is rarely caught in small lakes where the summer water temperatures exceed 17–18 °C. The only B.C. population of the

cisco is in Maxhamish Lake in the Petitot River drainage system. Most lakes in the B.C. portion of this drainage system are shallow (maximum depths of <3.0–5.0 m) and deeply stained. Maxhamish Lake, however, is of moderate size (about 5,000 ha in surface area) and, for the region, relatively deep (a mean depth of 6.0 m and a maximum depth of 10.6 m). Also, it has unstained and relatively clear water (Secchi depth of 2.8 m).

The little we know about the habitat use of adults in Maxhamish Lake fits the general habitat profile recorded from lakes farther to the east. All of the ciscoes caught in Maxhamish Lake were taken in overnight gill-net sets. Some years, the lake forms a strong thermocline, and it remains virtually isothermal in other years. The August water temperature in the first 3 or 4 m varies from 17 °C to 19 °C. If a thermocline forms, it is usually at about 5 m, and some oxygen depletion develops in the hypolimnion. Whether dissolved oxygen below the thermocline ever drops below the critical level (4.25 mg/L) for salmonids (Davis 1975) is unknown.

JUVENILES

Maxhamish Lake was sampled with variable mesh gill nets in 1982, 1987, 1990, and 1999 (DeGisi 1999). Juveniles (1^+ and most 2^+ fish) were caught in overnight gill-net sets at depths ranging from 1.3 to 4.0 m. This is the same depth range as adults and suggests that the juveniles use the same open water habitat as adults.

YOUNG-OF-THE-YEAR

Elsewhere, newly hatched cisco fry remain inshore in shallow, sheltered areas for their first month or so of life (Becker 1983) and then disappear, presumably into deeper water. There may be a similar movement of fry from shallow to deeper water in Maxhamish Lake; however, no young-of-the-year have been collected in the lake. Fine-mesh seine hauls were made in apparently suitable shallow-water habitats, but they probably were made too late in the summer to detect fry.

Conservation Comments Because the cisco is widespread and abundant in northern Canada, it is not a national conservation concern. In British Columbia, however, the cisco is ranked S1 (critically imperiled because of extreme rarity). Our only population is completely isolated from other populations in the Mackenzie drainage system. Consequently, if the Maxhamish Lake population is extirpated, there is no possibility of natural recolonization from an outside source. This makes the population especially vulnerable to disturbance. The southeastern shore of Maxhamish Lake is in a protected area (a provincial park), but the lake is adjacent to an area of active exploration for oil and gas. If exploitable quantities of hydrocarbons are discovered in the area, care must be taken not to disturb the lake.

Coregonus autumnalis (PALLAS)
ARCTIC CISCO

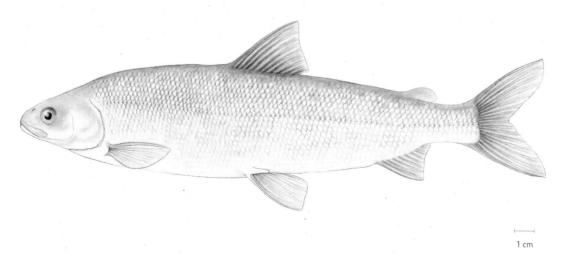

1 cm

Distinguishing Characters   This planktivorous whitefish has more than 40 gill rakers and a terminal mouth. When the mouth is closed, the tips of the lower and upper jaws are the same length, and the tips of the pelvic fins are unpigmented. Two other ciscoes occur in British Columbia: the cisco (*Coregonus artedi*) and the least cisco (*Coregonus sardinella*). In these species, when the mouth is closed, the tip of the lower jaw projects slightly beyond the upper jaw, and the pelvic fins are dark or tipped with black.

Taxonomy   In the past, the Arctic cisco has been confused with the Bering cisco (*Coregonus laurettae*); however, morphological and molecular data confirm that they are separate species (Bickham et al. 1997; McPhail 1966; Turgeon and Bernatchez 2003). The relationship between the Arctic cisco and the cisco is still unresolved. They clearly are closely related—some authors (e.g., Bodaly et al. 1991) argue that they are the same species, whereas other authors (e.g., Smith and Todd 1992) argue that they are separate species (see the taxonomy section under cisco for details).

The Arctic cisco is primarily an anadromous species with no known freshwater-resident populations in North America; however, there are landlocked populations in Lake Baikal (Russia) and in four lakes in Ireland. The Irish populations (called pollan) are glacial relicts that have been isolated from the northern populations for about 10,000 years (Ferguson et al. 1978) or, perhaps, longer (Bodaly et al. 1991). Interestingly, the Atlantic whitefish, *Coregonus canadensis*, bears a strong resemblance to the pollan. The populations in Lake Baikal are older than the last glaciation and originally were given the name *Coregonus migratorius*. Although most Russian biologists treat them as a subspecies of Arctic cisco, they may be a separate species (Politov et al. 2000).

Sexual Dimorphism Outside the breeding season, there are no obvious external differences between males and females. On average, however, females are larger than males of the same age. The scales along sides of breeding fish develop spawning tubercles. These tubercles are more obvious in males than in females.

Hybridization In the lower Mackenzie system, the Arctic cisco sometimes hybridizes with the lake whitefish (*Coregonus clupeaformis*) and the least cisco (Reist et al. 1992).

Distribution NATIVE DISTRIBUTION
As the name implies, the Arctic cisco occurs in rivers and inshore marine waters associated with the Arctic Ocean. Along the Siberian coast, it ranges from the White Sea in the west to the Chukotsk Peninsula in the east. In North America, it is found from the Point Barrow, Alaska, eastward along the Arctic coast as far as the Murchison River (Boothia Peninsula). There are disjunct populations in Ireland and in Lake Baikal, Russia (Berg 1948; Ferguson et al. 1978).

BRITISH COLUMBIA DISTRIBUTION
Arctic cisco occur in the lower Liard River and its tributaries in northeastern British Columbia (Map 55); however, it is not clear if this is a self-sustaining run. From 1978 to 1981, reproductive adults were collected from spawning runs in the B.C. portion of the lower Liard River (McLeod and O'Neill. 1983), and occasional individuals have been reported since this time.

Life History The original report documenting the presence of the Arctic cisco in British Columbia includes some life-history data (McLeod et al. 1979). However, most of the following information was gleaned from studies on Arctic cisco in the lower Mackenzie River and the Beaufort Sea, supplemented with information from populations in Siberia and Ireland.

REPRODUCTION
Arctic cisco spawn in the fall, but little is known about their spawning requirements. In the lower Liard River, ripe fish first appeared on August 12, the run peaked in late September, and it was over by October 7. At Fort Liard, about 120 km downstream from the B.C. border, ripe Arctic ciscoes were collected in mid-September (Dillinger et al. 1992). In Siberia, spawning occurs at 3–5 °C (Polyakov 1989). Actual spawning was not observed in British Columbia, but spawning in Russia apparently occurs in fast water over gravel substrates (Nikolsky 1961). There is no site preparation, and the demersal eggs are released over the substrate. Presumably, the eggs lodge in cracks and crannies in the gravel and incubate over winter. Like other whitefishes, fecundity in the Arctic cisco increases with increasing female size. In the Bratsk Reservoir (Siberia), fecundity ranges from 13,600 to 64,300 eggs in females from

Map 55 The B.C. distribution of Arctic cisco, *Coregonus autumnalis*

• native

315 to 445 mm (Polyakov 1989), whereas the species' fecundity in other parts of Russia is said to range from about 8,000 to 90,000 eggs (Nikolsky 1961). Ripe, but unfertilized, eggs sampled from the Liard River spawning run varied in size from about 2.2 to 2.4 mm in diameter. Laboratory-reared eggs of the disjunct Irish populations (pollan) incubated at about 10 °C hatched in 42 days (Dabrowski 1981). The newly hatched larvae were 10 mm long and began feeding 3 or 4 days after hatching. Presumably, since no fry or juveniles have been collected in the Liard River in British Columbia, newly hatched Arctic ciscoes are swept downstream to the Mackenzie Delta by the spring freshet.

AGE, GROWTH, AND MATURITY

In the Beaufort Sea, Arctic cisco fry (18–37 mm FL) first appeared in Wood Bay in early July (Bond and Erickson 1997). Presumably, they originated in the Anderson River. By late July and early August, the fry ranged from 50 to 69 mm in length. In the same area, age-1 fish achieved a size of 80–119 mm, and age-2 fish reached 139–159 mm by late July. In Beaufort Sea populations, sexual maturity is reached between 6 and 9 years. Some males are mature at 6, but females usually mature at 8 or 9 (Griffiths et al. 1977). The Liard River spawning run contained fish ranging from 6 to 9 years with the majority of the fish in the 6 to 8 year range (McLeod et al. 1979). Arctic cisco probably skip a year between spawnings and reaches a maximum age of about 21 years (Craig and Mann 1974).

FOOD HABITS

During their upstream spawning migrations, adults do not feed; all of the Arctic ciscoes caught in British Columbia had empty stomachs (McLeod et al. 1979). In the lagoons and inshore waters of the Beaufort Sea, the main foods of Arctic cisco are crustaceans, amphipods, mysids, and small fish (Griffiths et al. 1977). In Alaska, immature Arctic cisco captured in freshwater nursery areas close to the sea contained dipteran larvae (Bendock and Burr 1986).

Habitat The original report documenting the presence of the Arctic cisco in British Columbia includes some habitat data (McLeod et al. 1979). However, most of the following information was gleaned from studies on Arctic cisco in the lower Mackenzie River and the Beaufort Sea, supplemented with information from populations in Siberia and Ireland.

ADULTS

In North America, Arctic ciscoes are anadromous and spend most of their lives in estuaries and inshore marine environments. In the summer, adults (mostly mature fish that will not spawn that year) are common in the shallow (<4.0 m) lagoons and coastal beaches along the Beaufort Sea (Griffiths et al. 1977). In the spring (late June), the salinity in the lagoons is low (about 2.5 g/kg) but increases over the summer and reaches about 23 g/kg by late September. Arctic cisco stay in such habitats until freeze-up and then return to the lower reaches of the Colville, Mackenzie, and Anderson rivers and overwinter in fresh water (Bond and Erickson 1997; Craig and Mann 1974). Fish that will spawn in the fall leave their overwintering sites shortly after breakup and concentrate in the Mackenzie estuary. They begin moving up through the delta in May, but most upstream migration is in July and August. They ascend the river up to at least the Great Bear River (about 725 km) and have been observed in British Columbia in the lower Liard River (about 1,600 km upstream). Sometime between freeze-up and Christmas, the spawned-out adults return to the Mackenzie Delta (Wynne-Edwards 1952). They disperse along the coast in the spring on feeding migrations.

JUVENILES

Small juveniles (<150 mm), apparently of Mackenzie River origin, are abundant in inshore waters as far west as Herschel Island (about 120 km west of the Mackenzie Delta) and as far east as Wood Bay. The areas occupied by juveniles tend to be shallow and brackish with relatively warm water (Craig and Mann 1974; Griffiths et al. 1975). Older, larger juveniles are found in lagoons and beaches along the entire Beaufort Sea coast.

YOUNG-OF-THE-YEAR

Shortly after hatching, Arctic cisco fry are transported downstream into the Mackenzie Delta by the spring floods. Coastal dispersal of some of the fry from the delta is thought to be a function of the direction of the early summer winds. Thus, most, if not all, of the age-0 Arctic ciscoes found along the Beaufort Sea coast of Alaska originate from the Mackenzie River and are transported westward along the coast by inshore winds (Colonell and Gallaway 1997). Apparently, in some years, the winds also transport the fry to the east (Bond and Erickson 1997). Fry transported along the coast appear to stay inshore in shallow low-salinity areas.

Conservation Comments   In British Columbia, it is not clear that the Arctic cisco in the lower Liard River represent a self-sustaining run; however, if it is, then it migrates farther upstream (about 2,400 km) than any other North American population. Consequently, it may have unique life-history characteristics (Dillinger et al. 1992). Although the lower Liard River in British Columbia is wilderness, there was some seismic exploration in 1978 that killed Arctic ciscoes (McLeod et al. 1979). Consequently, if oil and gas exploration is renewed in this area, precautions must be taken to prevent fish kills. The Arctic cisco is not listed by either COSEWIC or the BCCDC.

Coregonus clupeaformis (MITCHILL)
LAKE WHITEFISH

1 cm

Distinguishing Characters   This silvery whitefish is flat-sided in cross section with a falcate dorsal fin (when depressed, the anterior dorsal rays extend well beyond the posterior dorsal rays). In British Columbia, lake whitefish usually have 11–13 major dorsal rays and 26–33 gill rakers. The only similar species in British Columbia is the broad whitefish (*Coregonus nasus*), although occasionally large, lake-resident mountain whitefish (*Prosopium williamsoni*) are mistaken for lake whitefish. Although the resemblance is superficial, lake-dwelling mountain whitefish are often deep-bodied and flat-sided rather than the usual round-bodied form found in rivers. The species are separated by their dorsal fins: in mountain whitefish, when depressed, the anterior dorsal rays do not extend beyond the posterior dorsal rays, whereas they extend well beyond the posterior dorsal rays in lake whitefish. In the field, broad whitefish and lake whitefish are distinguished by their head profile (convex in broad whitefish and concave in lake whitefish) and the ratio of maxillary length to interorbital width. These measurements, and those mentioned below, are described and illustrated in Lindsey (1962). The ratio of maxillary length to interorbital width changes with fish size and, in fish under 25 cm, the ratio is 0.76–0.80 in broad whitefish and 0.81–1.05 in lake whitefish; in fish over 25 cm, the ratios are 0.66–0.75 and 0.75–1.00, respectively. On preserved specimens, the ratio of the longest gill raker to interorbital width is >0.20 in lake whitefish and <0.20 in broad whitefish.

Taxonomy   The specific status of the lake whitefishes in North America is unclear. There are two widespread forms in the Northern Hemisphere: *C. clupeaformis* in North America, and *C. lavaretus* in Eurasia. Smith and Todd (1992) treat them as separate species; however, biochemical (Bodaly et al. 1991, 1994) and molecular data (Bernatchez and Dodson 1985; Bernatchez et al. 1991; Sajdak and Phillips 1997) indicate that lake whitefish in the Beringian region (including Alaska and the western Yukon) are more closely aligned to the Eurasian species (*C. lavaretus*) than to eastern North American *C. clupeaformis*. In addition, some authors (Mecklenburg et al. 2002; Page and Burr 1991) recognize two other "humpbacked" whitefish in Alaska: *Coregonus pidschian* (a Siberian species) and *Coregonus nelsoni* (named from Alaska). These little-known species may be conspecific with the Beringian form of lake whitefish.

The specific status of the Beringian form of lake whitefish is of interest in British Columbia. If it is *C. lavaretus*, then there probably are two species of lake whitefish native to the province: *C. lavaretus* in the upper Yukon and, probably, the upper Liard systems, and *C. clupeaformis* in the Mackenzie–Peace, Fraser, and Skeena systems. If, however, the two forms are treated as a single species, then *C. lavaretus* is the appropriate scientific name for our lake whitefish. At present, it is unclear which of these alternatives is correct. Until more data are available, I have retained *Coregonus clupeaformis* as the scientific name for all lake whitefish in British Columbia.

This situation is complicated further by the presence of a genetically distinctive race of lake whitefish in the Nahanni region (Foote et al. 1992b). A similar situation occurs in Arctic grayling (*Thymallus arcticus*) and lake trout (*Salvelinus namaycush*). In these species, there is evidence that Nahanni forms colonized parts of the upper Finlay system (Stamford 2001b) or the entire upper Peace system (Wilson and Hebert 1998). Thus, it is possible that the upper Peace system contains a mixture of Nahanni and eastern lake whitefish.

Like other whitefishes, the lake whitefish sometimes occurs as two sympatric forms. These forms often are reproductively isolated and use different spatial and trophic resources (i.e., they act like biological species *sensu* Mayr 1963). For now, an imperfect solution to the taxonomic problems posed by such pairs of biological species is to use a single taxonomic name but recognize their scientific and biodiversity value (see the parallel evolution section in the introduction for a general discussion of this problem). Such a species pair may have occurred in Dragon Lake near Quesnel. Unfortunately, the lake was poisoned before the whitefish could be studied; however, the range of gill raker counts for the few available museum specimens is suggestive of a species pair. Additionally, two growth forms (slow and fast growing) of lake whitefish occur in the lower Liard River (McLeod et al. 1979). These may represent either resident and migratory populations (McLeod et al. 1979) or a contact zone between the Eurasian and eastern North American forms of lake whitefish.

Sexual Dimorphism Lake whitefish are difficult to sex outside the breeding season; however, like many other fish, females are larger on average than males. Breeding whitefish develop "pearl" organs or spawning tubercles (Vladykov 1970). These tubercles form hard white lumps on scales along the midflanks but also are present on the head, back, and fins (especially the caudal fin). The tubercles are more strongly developed in males than in females, but apparently, there is some geographic variation in the development of spawning tubercles. In British Columbia, lake whitefish from Butternut Lake were examined for tubercles. These fish were not in full spawning condition, but the distribution of tubercles was similar to those described and illustrated in Vladykov (1970).

Hybridization In the northwestern portion of their range, lake whitefish hybridize with inconnu (*Stenodus leucichthys*), cisco (*Coregonus artedi*), least cisco (*Coregonus sardinella*), and Arctic cisco (*Coregonus autumnalis*) (Alt 1971; Nelson and Paetz 1992; Reist et al. 1992).

Distribution NATIVE DISTRIBUTION
The Palearctic (Eurasian) form of lake whitefish, *C. lavaretus*, ranges from western Europe eastward through Siberia to Alaska and the Yukon River system. The Nearctic form, *C. clupeaformis*, ranges from the Liard system eastward across the continent to Labrador and south to central British Columbia in the west and the Great Lakes in the east.

BRITISH COLUMBIA DISTRIBUTION
Lake whitefish are native to the Alsek, Yukon, Liard, Peace–Mackenzie, upper Fraser, and upper Skeena drainage systems but are absent from the Taku and Iskut–Stikine drainages (Map 56). In the Fraser system, their natural distribution ends in the vicinity of Williams Lake; however, in the early 20th century, lake whitefish from eastern North America were introduced into lakes in the South Thompson (Shuswap) and lower Fraser (Coquitlam, Pitt, Harrison, and Cultus lakes) systems, lakes in the Okanagan and Columbia systems, and Shawnigan Lake on Vancouver Island. Most introductions into small lakes failed, but some of those in large, deep lakes (e.g., Shuswap, Okanagan, Kootenay, and Arrow lakes) were successful.

Life History Remarkably little is known about the life history of native lake whitefish in British Columbia. Consequently, most of the following life-history information is derived from populations east of the Continental Divide. Machniak (1975a) presents an excellent literature review and bibliography for lake whitefish.

REPRODUCTION
Like other whitefishes, lake whitefish spawn in the fall, usually when water temperatures drop below about 10 °C (Machniak 1975a). Spawning occurs in lakes and rivers, and in British Columbia, spawning dates vary with latitude and lake size: at the southern edge of their natural range, or in large lakes, spawning may not start until December, but spawning in northern parts of the province or in small lakes occurs as early as late September. This pattern of northern populations spawning earlier than

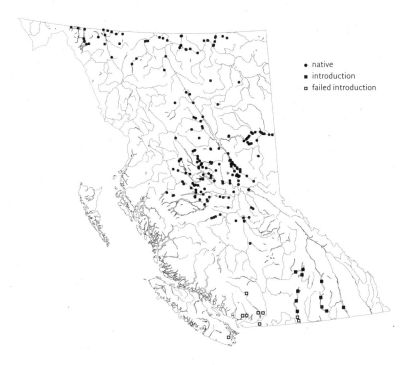

Map 56 The B.C. distribution of lake whitefish, *Coregonus clupeaformis*

southern populations is widespread (Anras et al. 1999; Kennedy 1953; Lawler 1961). In Lake Erie, spawning is delayed until water temperature drops below 8 °C (Lawler 1961). Spawning occurs at night (Quadri 1968) over a variety of substrates (sand, gravel, cobbles, and boulders) at depths ranging from 0.3 to 30 m (Machniak 1975a). In a small lake in northwestern Ontario, there were two spawning patterns in lake whitefish: offshore-oriented fish were located near the bottom and increased their depth throughout the spawning period (October–November), and inshore-oriented fish remained in the littoral zone and exhibited site fidelity during the spawning period (Anras et al. 1999). The inshore fish spawned at an average depth of 2.7 m over boulder, cobble, and detritus substrates.

Spawning in rivers usually occurs in shallow riffles or runs with gravel to cobble substrates (Machniak 1975a). A spawning run in the Parsnip River began in mid-September, and males ripened before females (McLeod et al. 1978). Actual spawning started in late October (water temperature 7 °C) and continued until mid-November (0 °C). The spawning substrate ranged from gravel (1 cm) to large cobbles (about 30 cm). A similar spawning migration occurred in the lower Liard River (McLeod et al. 1979). Here, the upstream migration started in late July (water temperature 7 °C), and catches declined after an early September peak but lasted until mid-October (3 °C). In the Athabasca River downstream of Fort McMurray, the lake whitefish spawning run started in late August (water temperature 14 °C), and fish were ripe by mid-October

(Bond 1980). Spawning occurred at 3–6 °C, and a post-spawning migration lasted from late October to the first week in November.

Since lake whitefish spawn at night, their spawning behaviour is not as well documented. Although they do not prepare a nest, they splash about at night (Quadri 1968), and some of this prespawning activity may help to clean the substrate. The well-developed spawning tubercles on their flanks suggest males make lateral contact with females during spawning. More than one male probably is involved in fertilizations. As the eggs are released, they are fertilized and then fall into cracks and crannies between rocks. No attempt is made to cover or protect the eggs. In northern populations, females often skip a year between spawnings (Johnson 1976).

Water-hardened lake whitefish eggs average 2.95 mm in diameter (Brooke 1970), and fecundity varies within and among populations (Healey and Dietz 1984; Healey and Nicol 1975). No fecundity estimates are available for B.C. populations, but Bond (1980) and Tripp et al. (1981) give counts ranging from 10,643 to 44,000 eggs in females from northern Alberta and the southern Northwest Territories. As in most fish, development rate is a function of incubation temperature. Survival is best at incubation temperatures ranging from 3 °C to 6 °C, and hatching times vary from about 40 days at 10 °C to about 180 days at 0.5 °C (Brooke 1975). Newly hatched larvae are about 10–12 mm long (Faber 1970) and positively phototactic (Shkorbatov 1966).

AGE, GROWTH, AND MATURITY

Growth rate in lake whitefish is variable and depends on temperature, productivity, population density, and age at maturity. Generally, however, northern populations grow more slowly, and live longer, than southern populations (Machniak 1975a). An uncommon, but geographically widespread, phenomenon in lake whitefish is the presence of two genetically divergent growth forms (usually "dwarfs" and "normals") in the same lake or river (Bernatchez and Dodson 1985; Fenderson 1964; Kennedy 1953; Kirkpatrick and Selander 1979; Lu and Bernatchez 1999).

Maturity in lake whitefish varies among populations but usually is reached between 4 and 10 years (Edsall 1960; Machniak 1975a). Males typically mature 1 or 2 years before females. In Teslin Lake on the British Columbia – Yukon border, the maximum age was estimated at 12 years, whereas the maximum age in northern Alberta was estimated at about 15 years (Bond 1980; Tripp et al. 1981). Most of this age data was derived from scales and probably under-estimates ages on older fish. As with other coldwater species, whitefish in northern lakes are especially long lived. So far, the oldest lake whitefish aged with scales was estimated to be 28 years old (Kennedy 1953).

FOOD HABITS

Newly hatched lake whitefish completely absorb their yolk in about 3 weeks but begin foraging on cyclopoid copepodites and small cladocerans (e.g., *Bosmina*) within a few days of hatching (Davis and Todd 1998; Reckahn 1970; Teska and Behmer 1981). During their first summer, the diet of young-of-the-year lake whitefish gradually shifts to larger water column prey: mainly *Daphnia* and adult copepods (Davis and Todd 1998). Eventually, they shift to benthic prey (Reckahn 1970). The diet of juveniles and adults consists mostly of benthic organisms (chironomid larvae, amphipods, gastropods, and occasionally, small fish). Interestingly, in lakes containing the two growth forms ("dwarfs" and "normals") of lake whitefish, there is morphological evidence for their divergence into two trophic niches (Lu and Bernatchez 1999).

Habitat Not much is known about habitat use in our native lake whitefish. Consequently, most of the following information is derived from populations east of the Continental Divide. Machniak (1975) presents an excellent literature review and bibliography on lake whitefish.

ADULTS

Like most coregonids, lake whitefish prefer cool water. Adults can tolerate temperatures ranging from near 0 °C to about 22 °C, but they prefer temperatures ranging from about 8 °C to 14 °C (Coutant 1977; Quadri 1968). In British Columbia, most lake whitefish populations are lacustrine, but there are river-resident populations in the upper Peace and Liard drainage systems. The life histories and biology of these river-resident populations need study. Adult lake whitefish are bottom oriented but seasonally shift their vertical distribution in response to temperature changes (Quadri 1968). When lakes are essentially homothermous in the spring, lake whitefish occur at all depths, but they move to deeper, cooler water as lakes warm and stratify during the summer. In the Great Lakes, they have been recorded at depths of over 100 m (Koelz 1929). In the fall, as the surface waters cool, they move back into shallow water to spawn.

JUVENILES

Juvenile lake whitefish occupy similar habitats to those used by adults; however, they are tolerant of higher temperatures and, in the summer, tend to occur in shallower water than adults. The temperature range occupied by juveniles is about 15.5–19.5 °C (Edsall 1999b). By late fall, juveniles begin moving into deep water as the adults are migrating into shallow water to spawn. In the closely related (or conspecific) lake whitefish of Eurasia, two size-related niche shifts occur in juveniles (Sandlund et al. 1992). Immature fish less than 25 cm occupied the inshore (0–30 m deep) epibenthic zone, whereas those greater than 25 cm expanded their spatial habitat to include the epibenthic zone in deeper water (30–90 m) and the pelagic zone.

YOUNG-OF-THE-YEAR

Upon filling their swimbladders, lake whitefish larvae move into shallow water, often to within 1 m of shore, and associate with emergent vegetation

(Hart 1930; Reckahn 1970). Their preferred temperature is estimated at 16.8 °C (Edsall 1999b), and in nature, they stayed close to the 17 °C isotherm (Reckahn 1970). Near the end of August, young-of-the-year lake whitefish begin to shift to deeper and colder water (Reckahn 1970). Again, in the Eurasian lake whitefish, the fry are found in shallow, littoral areas during the spring and summer (Naesje et al. 1986).

Conservation Comments Most native lake whitefish populations in British Columbia are not at risk. Exceptions are the lake whitefish that once lived in Dragon Lake near Quesnel. Little is known about the Dragon Lake whitefish, but there is some evidence that the lake contained a sympatric pair of whitefishes. Certainly, the gill raker counts made on the few available museum specimens are suggestive; however, we will never know the answer to this question. The lake was poisoned in 1956, and this population is now ranked SX (extinct) by the BCCDC.

Additionally, the taxonomic status of populations in the upper and lower Liard system needs clarification. The presence of round whitefish (*Prosopium cylindraceum*) in the upper Liard indicates a past connection with the Yukon River system, whereas the fish fauna of the lower Liard is dominated by species derived from the northern Great Plains. Consequently, the Liard River is a potential contact zone between lake whitefish derived from Beringia or the Nahanni Refugium and those derived from eastern North America. Since the taxonomic relationships of lake whitefish from these sources are contentious, a combined morphological and molecular investigation of lake whitefish in the Liard River might resolve the status of *C. clupeaformis* and *C. lavaretus*.

Coregonus nasus (PALLAS)
BROAD WHITEFISH

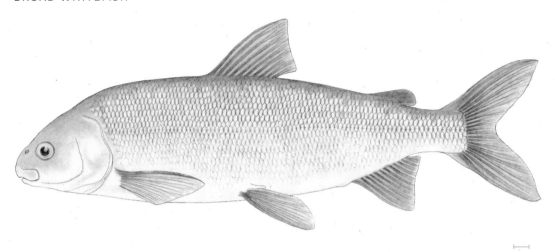

1 cm

Distinguishing Characters This large (up to 60 cm long), slab-sided whitefish has a blunt snout that projects slightly beyond the tip of the lower jaw. In profile, the head is convex and curves smoothly from the tip of the snout up to the back directly above the pectoral fins.

The only similar species in British Columbia are the lake whitefishes (either *Coregonus clupeaformis* or *Coregonus lavaretus*; see taxonomic discussion under lake whitefish). In the field, broad whitefish are distinguished from the lake whitefish species complex by their head profiles (convex in broad whitefish and concave in lake whitefish) and the ratio of maxillary length to interorbital width. These measurements, and those mentioned below, are described and illustrated in Lindsey (1962). The ratio of maxillary length to interorbital width changes with fish size: in fish under 25 cm, the ratio is 0.76–0.80 in broad whitefish and 0.81–1.05 in lake whitefish; in fish over 25 cm, the respective ratios are 0.66–0.75 and 0.75–1.00. On preserved specimens, the ratio of the longest gill raker to interorbital width provides a clear separation of broad and lake whitefish.

Taxonomy The broad whitefish has a complex taxonomic history and has often been confused with the lake whitefish (see Lindsey 1962 for details). Now, however, it is clear that the broad whitefish is a species separate from the lake whitefish species complex. It coexists with various forms of lake whitefish over its entire geographic distribution but maintains its distinctive morphological and molecular characteristics.

Sexual Dimorphism On breeding fish, hard, white tubercles develop at the centre of the scales along the flanks. Both sexes develop these spawning tubercles; however, they are less obvious in females than in males. In Teslin Lake, mature males have longer pelvic fins than females (Clemens et al. 1968).

Hybridization   Although hybridization is modestly common among whitefish species in northern environments, hybrids involving the broad whitefish are unknown.

Distribution   NATIVE DISTRIBUTION
The broad whitefish occurs in Arctic and Bering Sea drainages in Eurasia and North America. In Eurasia, the species ranges from the Pechora River in the west to the Penzhina River (Sea of Okhotsk) in the east. In North America, the broad whitefish occurs from the Kuskokwim River (a Bering Sea drainage) north to Bering Strait and then east along the Arctic coast at least as far as Bathurst Inlet. In the Mackenzie River, they range from the outer delta upstream to about Fort Simpson.

BRITISH COLUMBIA DISTRIBUTION
The only known population of broad whitefish is in Teslin Lake in the upper Yukon system (Map 57).

Life History   Along the Beaufort Sea coast of Alaska, the Yukon, and the Northwest Territories there are three life-history forms of broad whitefish: an anadromous form, a fluvial form, and a lacustrine form. Apparently, these different life histories are at least partially inherited (Reist 1997). The only population of broad whitefish in British Columbia is in Teslin Lake (Yukon River system). Its life history has never been studied. Consequently, the following summary is based on studies done in Alaska, the Yukon, and especially, the Northwest Territories. In these populations, all life-history phases make complex migrations between foraging environments (often brackish water or delta lakes), overwintering environments (usually in fresh water) and breeding sites (Tallman et al. 2002). Except for movements within the lake, the Teslin Lake population is non-migratory, and its life history probably is most similar to lake-resident broad whitefish in Travaillant Lake in the lower Mackenzie system. Nonetheless, Mackenzie and Yukon broad whitefish are genetically different (Reist 1997), and thus, it is possible that Teslin Lake broad whitefish differ in some of their life-history traits from lacustrine populations in the Mackenzie system.

REPRODUCTION
Broad whitefish spawn in the fall, probably under the ice. In the lower Mackenzie River system, there is a fall migration upstream from the delta to the Rampart Rapids, and apparently, spawning occurs in the main river from late October to early November. At this time, the water temperature is about 0 °C (Treble and Tallman 1997). In the Colville River on the northern coast of Alaska, adults migrate upstream in August (Bendock 1979). There are no data on broad whitefish spawning times or movements in Teslin Lake, but presumably, they also spawn in October or early November. The only description of spawning movements by lacustrine populations involves resident broad whitefish in Travaillant and Andrew lakes in the Northwest Territories (Harris and Howland 2005).

Map 57 The B.C. distribution of broad whitefish, *Coregonus nasus*

By September, broad whitefish fish radio-tagged during the summer in Travaillant Lake and in the Travaillant River below the lake begin congregating near the inlet of Travaillant Lake and in the Travaillant River immediately below the lake. By October, the inlet fish move up the river to three spawning sites (identified by apparently suitable gravel substrates) and a spawning site in the outlet immediately below the lake. By mid-November, most fish in the inlet had returned to Travaillant Lake, and the fish in the outlet spawning area had moved downstream into Andrew Lake. Broad whitefish overwintered in both lakes.

These observations suggest that broad whitefish usually spawn in flowing water. However, it is not known if Teslin Lake broad whitefish spawn in rivers. Presumably, there is suitable substrate in some of the inlet rivers and in the lake outlet. It is also possible that broad whitefish spawn within the lake. Teslin Lake has gravel and cobble shoals in the littoral zone that are deep enough to escape ice disturbance. Like other whitefishes, broad whitefish probably spawn at night, and the demersal eggs sink into interstices in the substrate. If Teslin broad whitefish are like those in the Mackenzie River system, they probably spawn several weeks later than sympatric lake whitefish (Stein et al. 1973).

Egg diameters reported for ripe Colville River broad whitefish varied from 2.3 to 2.6 mm (Bendock 1979). Fertilized eggs probably are larger (about 3.0 mm in diameter). Fecundity in the Colville River population ranged from 25,800 eggs in a 481 mm (FL) female to 43,860 eggs in 535 mm female. In the lower Mackenzie system, there are significant fecundity differences between anadromous and lacustrine populations (Tallman et al. 2002): fecundity ranged from about 18,000 to 70,000 eggs in the anadromous populations and from 14,000 to 51,000 in the lacustrine populations. There were also differences between these populations in the proportion of mature fish that skip spawning years.

Eggs from Mackenzie Delta fish were fertilized in late October and reared for 59 days at 4.0 °C (Ratynski and de March 1989). The temperature was then raised to 7.0 °C, and the eggs hatched. In nature, the eggs probably incubate over winter and hatch in the spring just before, or just after, breakup. Newly hatched larvae are 9.8–10.9 mm TL, and by 13.4–14.0 mm, the yolk sac is depleted (Ratynski and de March 1989). Presumably, the fry begin feeding at this size.

AGE, GROWTH, AND MATURITY

Growth rates vary among broad whitefish populations (Tallman 1997). In the Colville River, Alaska, young-of-the-year average about 71 mm FL by fall (Bendock 1979). Similarly, by late September, 0^+ fish in the Mackenzie Delta range from about 50 to 100 mm with a strong mode at about 75 mm. Growth is rapid for the first 5 or 6 years of life but begins to slow at sexual maturity. Sexual maturity is usually achieved by the eighth or ninth year of life (Bendock 1979; Bogdanov 1983); however, in the Alaskan portion of the Yukon system, broad whitefish reach maturity as early as age 5 (Alt 1976).

Broad whitefish grow to a large size—some individuals exceed 60 cm FL—but most adults are less than 50 cm long. The oldest broad whitefish reported from British Columbia was a 12^+ female. This age was based on scales; although scales give reliable ages up to about ages of 10 years, the discrepancy between scale and otolith age estimates increases after age 10. In one extreme case, scales and otoliths yielded ages of 14 and 35 years, respectively (Bond and Erickson 1985).

FOOD HABITS

There are only meagre data on the feeding habits of the Teslin Lake population, but the adults appear to be benthivores. This is consistent with stomach analysis of broad whitefish in Mackenzie Delta lakes. Here, adults consumed mostly benthic organisms (chironomids, clams, snails, caddisfly larvae, and amphipods) and, occasionally, small ninespine sticklebacks (*Pungitius pungitius*). In contrast, broad whitefish fry contained mostly zooplankton (cladocerans and copepods) and small chironomid larvae (Bond and Erickson 1985). In inland lacustrine populations, benthic organisms probably dominate the adult diet.

Habitat Nothing is known about the habitat use of broad whitefish in British Columbia. Consequently, most of the following information is based on studies from Alaska, the Yukon, and especially, the Northwest Territories. However, except for the Travaillant Lake study (Harris and

Howland 2005), most of these investigations deal with migratory fluvial and anadromous populations. Thus, much of what follows might not apply to the Teslin Lake population.

ADULTS

In the lower Mackenzie River and its delta, there are well-documented accounts of seasonal migrations and habitat shifts by broad whitefish (summarized in Reist 1997). All that is known about adult broad whitefish in Teslin Lake is that, in the summer, they are caught in gill nets set in relatively shallow water (<10 m). Given that they forage on benthic organisms, they probably also occupy much deeper water. Nothing is known about their diel activity patterns or seasonal changes in their depth distribution. A systematic sampling program is needed before any conclusions can be made about the habitat requirements of this inland population.

JUVENILES

Like the adults, nothing is known about the habitat use of juvenile broad whitefish in Teslin Lake. Yearling (1^+) juveniles are sometimes taken in beach seines. This suggests that, in the summer, they occupy littoral areas in deeper water than the fry, but again, only a systematic sampling program will supply answers to questions about their habitat use.

YOUNG-OF-THE-YEAR

In the summer, large, mixed schools of whitefish fry (including broad whitefish) occur in sheltered, shallow water (<1 m) along the shores of Teslin Lake. During windstorms, they are especially abundant on the lee sides of headlands This suggests that some movements of fry in this large lake may be wind driven. Presumably, the fry move into deeper water before freeze-up. In the Mackenzie River, in the spring, fry apparently are washed downstream into the estuary. In late July, they begin to leave the estuary and enter streams. They migrate upstream into lakes and overwinter in the deeper (>3 m) lakes.

Conservation Comments   Although the broad whitefish is not listed by COSEWIC, the BCCDC ranks it as S1 (critically imperiled) in British Columbia. The reason for the B.C. listing is that the broad whitefish is restricted to a single lake (Teslin Lake). Additionally, although there are other apparently suitable adjacent lakes (e.g., Atlin and Tagish lakes) in the Yukon drainage system, these other lakes drain into the Yukon River through a different tributary than Teslin Lake. There are barriers on the Yukon (Lewes) River that prevent the upstream movement of salmon and, presumably, whitefish into Atlin and Tagish lakes. In contrast, there is no barrier between Teslin Lake and the mainstem Yukon River. Thus, the nearest broad whitefish population to Teslin Lake is in Lake Laberge (about 200 km downstream). Although upstream dispersal from Lake Laberge into Teslin Lake is possible, gene flow between the populations is probably rare, and the Teslin broad whitefish is essentially an isolated, peripheral population. Apparently, there are genetic differences between Mackenzie and Yukon broad whitefish (Reist 1997), and their relationships are under study as of this writing.

Coregonus sardinella VALENCIENNES
LEAST CISCO

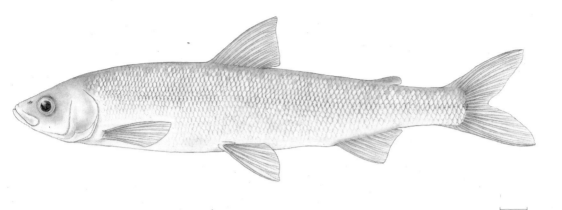

1 cm

Distinguishing Characters   This planktivorous whitefish has more than 40 gill rakers and a terminal mouth. When the mouth is closed, the tip of the lower jaw projects beyond the tip of the upper jaw, and the tips of the pelvic fins are dark. Two other ciscoes (cisco, *Coregonus artedi*, and Arctic cisco, *Coregonus autumnalis*) also occur in the province; however, the least cisco is the only cisco found in the B.C. portion of the upper Yukon River system. It differs from both the other cisco species in the relative lengths of the pre-pelvic and post-pelvic distances. In the least cisco, the post-pelvic distance is greater than the pre-pelvic distance, whereas the pre-pelvic distance in the cisco and Arctic cisco is greater than the post-pelvic distance. Thus, when stepped off with dividers, in the least cisco the distance from the tip of the lower jaw to the origin of the pelvic fin falls on the caudal peduncle well forward of the caudal flexure

Taxonomy   The taxonomic relationship between the least cisco and a Eurasian cisco (*Coregonus albula*) is contentious. They are closely related, and their Eurasian distributions are essentially complementary. On the basis of allozyme data, some authors (e.g., Bodaly et al. 1991) argue that they are the same species, and others (e.g., Perelygin 1992) argue that they are different species. On the basis of morphological analysis, other authors (e.g., Smith and Todd 1992) argue that they are separate species. This taxonomic question is still unresolved.

    In North America, the taxonomy of least ciscoes is further confounded by the presence of sympatric pairs of least ciscoes and a bewildering array of morphologically and ecologically distinctive allopatric populations. There are populations with well-developed spots and populations that lack spots; there are large-bodied populations (adults >350 mm) and small-bodied populations (adults <250 mm); and there are lake residents, river residents, and anadromous populations that make feeding migrations to estuaries and inshore coastal waters. Anadromous populations tend to have high gill-raker counts (usually around 50), and freshwater-resident populations tend to have low gill-raker counts (usually around 45).

Sometimes, anadromous and freshwater-resident populations co-occur, at least seasonally, in lakes on the Mackenzie Delta (Bond and Erickson 1985) and on the Arctic slope of Alaska and the Yukon Territory. There are two growth forms of least cisco in Ikroavik Lake, Alaska (Cohen 1954; Wolschlag 1954). One of these forms may have been anadromous. Dwarf and normal forms of freshwater-resident least ciscoes are sympatric in Trout Lake in the Babbage River system (a Beaufort Sea drainage in the Yukon Territory) and in Peter Lake, N.W.T. (McCart and Mann 1981). These sympatric pairs of least ciscoes differed in growth pattern and size at sexual maturity. Finally, there are "jumbo" spotted least ciscoes in four southern Yukon lakes, and some of these lakes also may contain sympatric populations of "dwarf" least ciscoes (Lindsey and Kratt 1982).

In the B.C. portion of the Yukon system, there are also two forms of least cisco; however, unlike the pairs reported elsewhere, our pairs do not differ in size. They are, however, strikingly different in gill raker number and colouration—one form has relatively low gill raker counts (42–49) and black tips on the pelvic and anal fins, whereas the other form has high gill raker counts (57–61) and immaculate pelvic and anal fins.

The taxonomic status of the various sympatric forms of least ciscoes is unresolved, but the populations that have been studied (e.g., McCart and Mann 1981) have the characteristics of biological species (*sensu* Mayr 1963). For now, an imperfect but practical solution to the taxonomic problems posed by pairs of biological species is to use a single taxonomic name but recognize their scientific and biodiversity value (see the parallel evolution section in the introduction for a general discussion of this problem).

Sexual Dimorphism Except when spawning, the sexes are difficult to distinguish. Unlike most other whitefishes, the least cisco does not appear to develop breeding tubercles (Alt 1971).

Hybridization In the Mackenzie Delta region, the migratory form of least cisco occasionally hybridizes with both the Arctic cisco and the lake whitefish (*Coregonus cleupeaformis*) (Reist et al. 1992).

Distribution **NATIVE DISTRIBUTION**
Least cisco occur in Arctic and Bering Sea drainages of Eurasia and North America. In Eurasia, this species ranges from the Bering Sea westward to the White Sea. The North American distribution includes Banks and Victoria islands as well as Arctic Ocean drainages east to the Melville Peninsula and Bering Sea drainages as far south as the Alaska Peninsula

(Iliamna Lake). In the Mackenzie system, the least cisco ranges from the outer delta upstream to about Fort Simpson.

BRITISH COLUMBIA DISTRIBUTION

The least cisco is restricted to the upper Yukon River system (Map 58). Here, it occurs in Atlin, Teslin, and Swan lakes, as well as a number of smaller unnamed lakes. So far, there are no records from the B.C. portion of the Mackenzie system. Since they are reported to ascend the Mackenzie River as far upstream as its confluence with the Liard River (about 250 km from the British Columbia – Yukon border), it is possible that they occasionally reach the lower Liard River in British Columbia.

Life History Little is known about the life history of least cisco in British Columbia. There are some age and growth data from Teslin Lake (Clemens et al. 1968), but most of the following information is derived from reports on least ciscoes from Alaska, the Yukon, and the Northwest Territories.

REPRODUCTION

Spawning occurs in the fall (Alt 1980; Bendock and Burr 1986; Mann 1974) just before, or after, freeze-up. Although no lake-spawning populations are recorded from North America, some B.C. populations may spawn on gravel lakeshores. The only North American descriptions of least cisco spawning are from fluvial populations in Alaska (Alt 1980). In the Chatanika River, spawning occurred mainly at night (it peaked about 19:00) and at water temperatures ranging from 0 °C to 8 °C. In the Kobuk River, most fish spawned over gravel substrates (0.6–7.6 cm in diameter) and in moderate to fast currents (0.75–1.2 m/s) at depths up to 1.0 m. In the Chatanika River, some fish spawned in slower currents, in deeper water (about 2.3 m), and over silt or sand substrates. There is no site preparation, and gamete release occurs at the surface.

Usually a male and female, but sometimes up to three males and two females, start from about 30 cm off the bottom and swim together towards the surface. As they break the surface, they roll backwards or to the side and then separate and return to the bottom. Apparently, the eggs and milt are released during this "spawning jump." Sometimes, spawning pairs swam in midwater facing upstream, turned on their sides and arched their bodies. Presumably, gamete release occurred at this time.

Fecundity varies with female size. In sympatric "dwarf" and "normal" least ciscoes, egg numbers range from 223–672 in the dwarf form to 7,886–19,261 in the normal form (McCart and Mann 1981). British Columbia least ciscoes are small-bodied, relative to the anadromous form and, probably, have fecundities similar to the normal form mentioned above. Anadromous least ciscoes are larger than most freshwater-resident populations and have correspondingly high (12,000–100,000) fecundities (Moulton et al. 1997). Egg diameters in ripe, unfertilized eggs ranges from about 1.5 to 1.9 mm (Mann 1974). The fertilized eggs are demersal

Map 58 The B.C. distribution of least cisco, *Coregonus sardinella*

• native

and incubate over winter in interstices among the gravel. The larvae begin exogenous feeding at about 7–10 mm (Shestakov 1991).

AGE, GROWTH, AND MATURITY

Growth rates differ among the various forms of least cisco and, within forms, among localities. By the end of their first growing season in Alaska, fry reach about 55–72 mm FL and are usually over 110 mm by the end of their second growing season. In the Chatanika River, most least ciscoes mature by age 4; however, in a brackish water population on the Beaufort Sea coast, only about 50% of the fish reach maturity by age 7 (Moulton et al. 1997). In the most northerly populations, mature fish probably breed every second year (Moulton et al. 1997). Maximum age varies with sex (females usually live longer than males), among forms, and among localities but is usually about 11 or 12 years. Nevertheless, some females on the Beaufort Sea coast and Victoria Island reach ages of 25 or 26 years (Moulton et al. 1997; Scott and Crossman 1973). In British Columbia, the oldest recorded age for least ciscoes is 7^+.

FOOD HABITS

Lacustrine least ciscoes are primarily planktivores. In Teslin Lake, they forage mainly on cladocerans and copepods, but in the summer, they also take some benthic prey (e.g., chironomid, amphipods, and pea clams) and occasionally terrestrial insects from the surface (Clemens et al. 1968). The diet of fluvial populations is more diverse than in lake populations and includes dipteran larvae and adults, snails, caddisfly larvae, clams, and occasionally, small fish (Bendock and Burr 1986). In brackish water, the large anadromous form forages on copepods, amphipods, and small fish.

Habitat Other than that they are associated with deep, cold lakes in the B.C. portion of the Yukon River system, little is known about the habitat use of the least cisco in our province. Throughout their geographic range, however, least cisco occur in a variety of habitats: brackish coastal waters, large rivers, and lakes. There are very little data on least cisco habitat use in lakes. Most of the following account is based on information obtained on lacustrine or fluvial populations in Alaska, Yukon Territory, and the Northwest Territories.

ADULTS

In the large, oligotrophic lakes of the upper Yukon system, adult least cisco occur in open water. Although they are pelagic in Teslin Lake, their heavy use of benthic prey (chironomid larvae) in the summer implies that, at some time during the day, they forage near the bottom (Clemens et al. 1968). Lacustrine populations appear to be non-migratory, but they probably move considerable distances within large lakes. Elsewhere, anadromous and fluvial populations are migratory and make complex migrations between spawning, feeding, and overwintering sites (Bond and Erickson 1985).

JUVENILES

Nothing is known about the habitat use of juvenile least ciscoes in British Columbia, but there is information on fluvial and anadromous juveniles on the Beaufort Sea coast and in the Mackenzie Delta (Bond and Erickson 1985). Juveniles of some populations in these areas are migratory and make complex seasonal movements between fresh and brackish waters. Juveniles in large, oligotrophic lakes like Atlin and Teslin, although non-migratory, probably move considerable distances within the lakes. It is not known whether juveniles in lakes move in schools with adults or in separate schools. The makeup of the schools may vary with predation intensity.

YOUNG-OF-THE-YEAR

In the early summer, large, mixed schools of whitefish fry (including least ciscoes) occur in sheltered, shallow water (<1 m) along the shores of Atlin and Teslin lakes. During windstorms, they are especially abundant on the lee side of bays. This suggests that some movements of fry in these large lakes may be wind driven.

Conservation Comments Although least ciscoes are abundant where they occur in British Columbia, their distribution within the province is restricted to lakes in the upper Yukon River system. This species is not ranked by COSEWIC, but the BCCDC ranks it as S2 (imperiled because of rarity) because of its restricted B.C. distribution. The common form of least cisco (with the relatively low number of gill rakers) is abundant in large, oligotrophic lakes in the upper Yukon system and is of no immediate conservation concern. So far, however, the form with more gill rakers is known only from two small, un-named lakes that also contain the normal form of least cisco. Apparently, these sympatric populations are rare and occur only under a specific, but unknown, set of ecological conditions. Such sympatric "species pairs" are of scientific interest and an important component of our aquatic biodiversity. They merit further study.

Prosopium coulterii (EIGENMANN & EIGENMANN)
PYGMY WHITEFISH

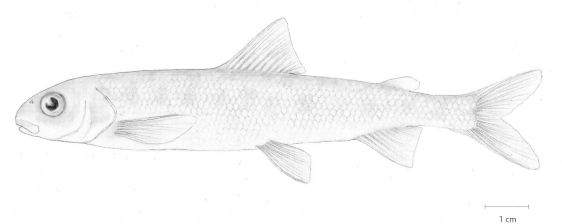

1 cm

Distinguishing Characters This large-scaled (in our area 62–67 scales along the lateral line) whitefish has a small adipose fin (its base is about equal to the eye diameter) and 9 or 10 dorsal rays. The body is round in cross section, and the snout usually is blunt and rounded when viewed from above. In the field, adult pygmy whitefish are easily confused with juvenile mountain whitefish (*Prosopium williamsoni*) and juvenile round whitefish (*Prosopium cylindraceum*). Usually, dorsal fin ray number will separate these species: 9 or 10 in pygmy whitefish and 11–15 rays in mountain and round whitefish. Under the microscope or with a hand lens, a good character is the size of the scales on the anterior portion of the lateral line: in adult pygmy whitefish, the anterior lateral line scales are about the same size as the scales immediately above and below the lateral line. In contrast, in juvenile mountain and round whitefishes the anterior lateral line scales are noticeably smaller than the scales immediately above and below the lateral line. Lateral parr marks are absent in young-of-the-year (<30 mm) pygmy whitefish; however, juveniles and most adults have well-developed parr marks.

Taxonomy There are no major taxonomic problems associated with pygmy whitefish. Like a number of other species, pygmy whitefish in British Columbia probably are derived from more than one glacial refugium. Consequently, there may be genetic differences among populations in different parts of the province, and it is possible that the upper Peace populations are a mixture of fish derived from the Bering and Pacific refugia. "Giant" pygmy whitefish are known from two lakes (McLease Lake in the Fraser system, and Tyhee Lake in the Skeena system). These populations grow faster and reach a greater size than most pygmy whitefish (McCart 1965); however, there are no mitochondrial DNA differences between them and adjacent normal populations (Rankin 1999).

The pygmy whitefish in Crescent Lake on the Olympic Peninsula, Washington, were originally described (Myers 1932) as a separate species, *Prosopium snyderi*. At that time, very few pygmy whitefish were available for comparison. Later, the locality (which was well outside the known range) was questioned (Eschmeyer and Bailey 1955), and it was suggested that the specimen probably came from Crescent Lake in northeastern Washington. Recently, however, the presence of a pygmy whitefish in Crescent Lake on the Olympic Peninsula has been confirmed (P. Mongillo, Washington Department of Fish and Game, personal communication). Since there is evidence of endemism in other freshwater fish in this region (McPhail 1967; McPhail and Taylor 1999; Schultz 1929), the status of *P. snyderi* should be re-examined.

Sexual Dimorphism Except at spawning time, the sexes are difficult to distinguish. In spawning populations, females are, on average, larger than males, and breeding fish of both sexes develop spawning tubercles on some of the scales above the lateral line, on the paired fins, and on the top of the head (Weisel and Dillon 1954). These nuptial tubercles are more pronounced on males than on females. In the upper Peace system (Dina #1 Lake), fish captured in late October had tubercles on the paired fins, the dorsal and anal fins, and the lower lobe of the caudal fin but none on the flanks.

Hybridization In the upper Peace system (e.g., Nation River), occasional individuals are encountered that are morphologically intermediate between mountain and pygmy whitefish. These may be hybrids.

Distribution NATIVE DISTRIBUTION

Until recently, the pygmy whitefish was thought to be a North American endemic; however, pygmy whitefish were discovered recently in three lakes in the Amguem River system on the Chukotsk Peninsula, Siberia (Chereshnev and Skopets 1992). In North America, pygmy whitefish are widely distributed west of the Continental Divide. Here, they occur from the Columbia River system (northern Washington, Idaho, and inter-mountain Montana) to the Bristol Bay region of Alaska. Pygmy whitefish also are found east of the Continental Divide, but here they are uncommon and noteworthy for their widely scattered distribution: Lake Superior, Lake Athabasca, Great Bear Lake, Waterton Lakes, and tributaries to the Athabasca River near Hinton, Alberta (Eschmeyer and Bailey 1955; Lindsey and Franzin 1972; Nelson and Paetz 1992).

BRITISH COLUMBIA DISTRIBUTION

Pygmy whitefish populations are scattered throughout the interior of the province but are noticeably absent from coastal lakes and rivers (Map 59). In western Washington, however, this species occurs west of the Cascade Mountains on the Olympic Peninsula (Crescent Lake) and in Chester Morse Reservoir near Seattle (Wydoski and Whitney 2003).

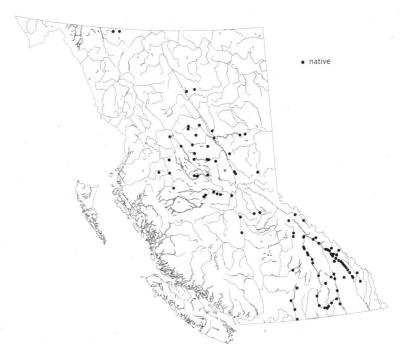

Map 59 The B.C. distribution of pygmy whitefish, *Prosopium coulterii*

Life History Throughout its geographic range, information on the life history of pygmy whitefish is meagre, and only two published papers involve B.C. populations (McCart 1965; Zemlak and McPhail 2006). Age, growth, diet, and habitat use in four populations are described in McCart (1965), and similar data are presented on one other population (McPhail and Zemlak 2001). These papers (McCart 1965; McPhail and Zemlak 2001) form the basis of the following account, supplemented with observations from western Montana (Weisel et al. 1973), western Washington (R2 Resource Consultants 1995), and the upper Peace River system (Stamford 2001a). So far, all the published information on pygmy whitefish is derived from lacustrine populations, and the life history and ecology of fluvial pygmy whitefish remains unknown.

REPRODUCTION

Pygmy whitefish spawn in late autumn or early winter. The exact time of spawning varies among populations, and fish in reproductive condition have been observed from early September to mid-January (Hallock and Mongillo 1998; McCart 1965; McPhail and Zemlak 2001; Weisel et al. 1973). Although all known spawning sites are in inlet streams, circumstantial evidence indicates that some populations spawn in lakes (Hallock and Mongillo 1998; McPhail and Zemlak 2001). Curiously, most accounts of pygmy whitefish spawning do not mention water temperature, but given the spawning times, the spawning temperature probably is usually <5 °C. There is no detailed description of either spawning sites or spawning behaviour; however, in lakes, schools of pygmy whitefish have been observed holding near the mouths of inlet streams and, in rivers, holding in pools just below riffles. If their spawning behaviour is similar to that of mountain whitefish, they probably spawn at night on riffles over coarse gravel. Presumably, there is no site preparation, and the demersal eggs lodge in interstices of the gravel and rubble on the bottom. If, as circumstantial evidence suggests, pygmy whitefish also spawn in lakes, the lacustrine spawning sites may be associated with upwelling groundwater.

Fecundity increases with body size and ranges from as low as 93 to over 1,100 eggs (Eschmeyer and Bailey 1955; Weisel et al. 1973). Pygmy whitefish eggs are smaller than those of other whitefish, and the diameter of water-hardened eggs ranges from 2.4 to 2.6 mm (Weisel et al. 1973). The incubation time is unknown; however, they presumably incubate over winter, and depending on latitude and altitude, the fry emerge in the spring or early summer (late March to early June).

AGE, GROWTH, AND MATURITY

Most pygmy whitefish are collected with gill nets. Hence, young-of-the-year are rare in samples, and descriptions of growth usually begin with fish in their second growing season (1^+). In Dease Lake, however, two collections separated by 40 years show a remarkable consistency in the growth of young-of-the-year. By late July or early August in both collections, young-of-the-year ranged from 28 to 40 mm (with both means at 34 mm). In the same collections, 1^+ pygmy whitefish ranged from 48 to 58 mm FL and averaged 53.5 mm. These fish probably had another month of growth left before winter. In Dina #1 Lake by the end of August, young-of-the-year pygmy whitefish ranged from 30 to 48 mm. By the end of the second growing season, average size varies dramatically among lakes: from about 60 mm (Heard and Hartman 1966) to about 150 mm (Rogers 1964). Although growth rate slows at maturity in all populations, interpopulation differences in size usually are maintained throughout life. The smallest recorded adults are about 65 mm FL from Brooks Lake, Alaska (Heard and Hartman 1966), and the largest known adults average about 250 mm (Tyhee Lake, B.C.). In most populations, males and females grow at the same rate until maturity; however, since females usually mature at least a year later than males and live longer, they achieve a larger adult body size than males (McCart 1965; McPhail and Zemlak 2001; Weisel et al. 1973). Typically, about 50–70% of males mature near the end of their second growing season (1^+), whereas most females do not mature until the end of their third growing season (2^+). Relatively few males survive beyond their third year, but females can reach ages of 7 or 8 years. The oldest pygmy whitefish recorded in British Columbia was 16^+ (Rankin 1999), but fish older than 9^+ are rare in most populations.

FOOD HABITS

Most studies indicate that pygmy whitefish are benthic foragers and most lake-dwelling populations feed on chironomid larvae and pupae, *Chaoborus*, cladocerans, small molluscs, ostracods, and amphipods (McCart 1965; Weisel et al. 1973). In some cases, however, they also forage on zooplankton in the pelagic zone (R2 Resource Consultants 1995) and copepods and *Daphnia* were the primary prey in Dina #1 Lake (Stamford 2001). Fluvial populations also feed heavily on chironomid larvae and pupae but include nymphs of other aquatic insects in their diet. During the breeding season, fish eggs are frequent in stomachs and what were probably salmon eggs were present in some stomachs in the fall (Heard and Hartman 1966). These same authors observed pygmy whitefish feeding during daylight hours. Most of the feeding movements were directed at the substrate, but occasionally, the fish moved up into the current and struck at drifting prey. In contrast, in Chester Morse Reservoir, peak pygmy whitefish activity occurred at night, and very few fish were active during the day (R2 Resource Consultants 1995).

Habitat   Relatively little is known about habitat use by pygmy whitefish in British Columbia. The same sources as for their life history (McCart 1965; McPhail and Zemlak 2001; R2 Resource Consultants 1995; Weisel et al. 1973; Zemlak and McPhail, 2006) provide most of the available information.

ADULTS

In British Columbia, pygmy whitefish are a cold-water species (usually found in water <10 °C) that occur in both lakes and rivers. In the southern and central parts of the province, they are restricted to relatively deep lakes, and there is probably little gene flow among populations. However, the glacial rivers of the southeastern portion of the province contain strong fluvial populations (the species was first described from the Kicking Horse River). In the northern parts of the province, both lacustrine and fluvial populations are relatively common.

Most information on habitat use by adult pygmy whitefish comes from overnight gill nets sets. Generally, they suggest pygmy whitefish are a deep-water species; however, sometimes they are taken in shallow water (Rankin 1999; Zemlak and McPhail 2006), and a number of factors may affect their intra-lacustrine distribution. For example, they have different depth distributions in lakes with, and without, other whitefish species (McCart 1965). In the summer, adult pygmy whitefish sympatric with mountain and lake whitefish (*Coregonus clupeaformis*) were restricted to depths below 10 m. In contrast, in lakes without other whitefish species, pygmy whitefish used water as shallow as 5 m. In Chester Morse Reservoir, Washington, adult pygmy whitefish were most abundant near the bottom at depths of between 30 and 40 m, but some individuals were observed in both the littoral and pelagic zones of the reservoir (R2 Resource Consultants 1995). Thus, the use of deep water in lakes is not obligatory but depends on what other fish species (predators or competitors) are present. In some northwestern lakes, there is also a seasonal component to depth use. In Priest Lake, Idaho, in the fall, there is a diel movement of adults onshore in late afternoon or early evening and back into deep water at dawn (Simpson and Wallace 1978). Since these observations were made in the autumn, the authors suggested that the onshore movements may have been associated with spawning activity. In Dina #1 Lake, however, both trapnet and gill-net sets indicate that adults aggregate close to the bottom and move into shallow water at dusk (up to, but not above, 2.5 m) and return to deep water at dawn. This onshore–offshore movement pattern was observed throughout the summer and is not confined to either adults or the spawning season.

In rivers, pygmy whitefish are found in both turbid and clear waters with moderate to swift current and, usually, over gravel or cobble substrates. In fluvial populations, there are no observations on the habitats used by either young-of-the-year or juvenile pygmy whitefish.

JUVENILES

In Dina #1 Lake, yearling (1+) pygmy whitefish were rarely captured in gill nets (bottom sets), but relatively large numbers were taken on the occasions when they were caught. This suggests that juveniles may school. Juveniles were collected more commonly in trap nets and, again, displayed the same pattern of captures as adults: they were only captured occasionally but, when they were captured, were present in large numbers. In Dease Lake (Liard system) in late summer, large schools of young-of-the-year and juvenile (1+) *Prosopium* congregate in shallow (<1 m) water along the lake margins. Some schools consist mainly of pygmy whitefish and a few round whitefish and, rarely, a mountain whitefish. Other schools contain mainly round whitefish with a few pygmy whitefish. So far, the meagre evidence suggests that, in northern British Columbia, juvenile pygmy whitefish occur in schools.

YOUNG-OF-THE-YEAR

Most habitat observations on pygmy whitefish are based on adults, and the habitat requirements of the young-of-the-year are, as yet, undocumented. However, late in their first growing season in Dina #1 Lake, 43 young-of-the-year were taken in one overnight trap-net set in shallow water (5–7 m). This suggests that, like adults, young-of-the-year are benthic oriented, school, and move inshore at night.

Conservation Comments In the southern part of their B.C. distribution, most pygmy whitefish occur as isolated populations in large, deep lakes. Consequently, there may be minor genetic differences among the southern lake populations. In addition, B.C. pygmy whitefish probably colonized our province from multiple refugia: Beringia, the unglaciated portion of the Columbia River system, and perhaps, the Great Plains. Like other species that dispersed into British Columbia from more than one refugium, there may be genetic and biological differences associated with fish from these different ice-free areas. Although we are ignorant of the status of most of our pygmy whitefish populations, some southern populations (especially in the Okanagan system) appear to be in decline. This species is not listed by COSEWIC, but the BCCDC lists the two "giant" pygmy whitefish populations—the McLeese and Tyhee (Maclure) lakes populations—as critically imperiled (S1).

Prosopium cylindraceum (PENNANT)
ROUND WHITEFISH

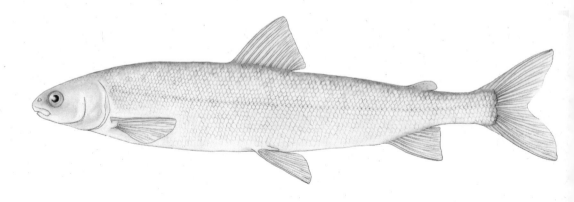

1 cm

Distinguishing Characters This round (in cross section) whitefish has small scales (over 70 lateral line scales and 24–27 scale rows around the caudal peduncle) and a small adipose fin (its base about equal to eye diameter). In British Columbia, adults often have dark spots on the top of the head.

Two other species of *Prosopium* occur in British Columbia: the mountain whitefish (*P. williamsoni*) and the pygmy whitefish (*P. coulterii*). In the field, round whitefish can be separated from mountain whitefish by the size and colour of the adipose fin. In round whitefish, the adipose is small (its length usually is less than the depth of the caudal peduncle), and it typically has a sprinkling of dark spots. In contrast, the adipose fin of mountain whitefish is large (its length usually is greater than the depth of the caudal peduncle) and is always unspotted. In addition, adult specimens of round whitefish usually have dark spots on the top of the head. In the field, adult round whitefish can be separated from adult pygmy whitefish by head shape (viewed from above, the snout is pinched and pointed in round whitefish and blunt and rounded in pygmy whitefish).

For individuals less than 70 mm FL, snout shape is unreliable; however, a combination of the shape of parr marks along the lateral line and adipose fin size will separate the three species of *Prosopium* down to about 40 mm. On small round whitefish, these parr marks—especially those along the posterior part of the midlateral line—are clearly oblong, whereas those on the sides of young pygmy and mountain whitefish are roughly round. Without a microscope, pygmy and round whitefish below a length of 40 mm are difficult to distinguish; however, for mountain whitefish, the large size of the adipose fin relative to the anal fin is a reliable identification mark even for fry.

Occasionally, young-of-the-year and 1+ round whitefish are mistaken for young Arctic grayling (*Thymallus arcticus*); however, the shape of the parr marks (oblong marks in round whitefish and thin vertical marks in Arctic grayling) and the length of the dorsal fin base easily separate these species.

Taxonomy At one time, Siberian and North American round whitefishes were considered different subspecies: *P. c. cylindraceum* in Siberia and *P. c. quadrilateralis* in North America. Although these subspecies are no longer recognized, the taxonomy of the round whitefish in North America needs study. Its North American distribution is divided into two apparently disjunct segments: populations that extend from the Bering Sea to the west side of Hudson Bay and populations in the Great Lakes region, Quebec, Labrador, and northeastern Ontario. There are slight morphological differences between round whitefish to the east and west of Hudson Bay, and this species probably dispersed from both the Bering and Mississippi refugia (McPhail and Lindsey 1970). Thus, there may be genetic and ecological differences between the B.C. populations and those in eastern Canada.

Sexual Dimorphism Except at spawning time, the sexes are difficult to distinguish (Normandeau 1969). Breeding males and females develop hard, white tubercles on the lateral scales. These tubercles are more pronounced on males than on females, and the sexes can be separated by gently rubbing the sides of the fish (Normandeau 1969).

Hybridization In the upper Liard system, occasional individuals are encountered that are intermediate in morphology between mountain and round whitefish. They may be hybrids. In this same area, however, there was no indication of hybridization in lakes where the two species were sympatric (Guinn 1982).

Distribution NATIVE DISTRIBUTION
Round whitefish occur in both Siberia and North America. In Siberia, the continuous distribution of this species is from the Chukotsk Peninsula westward along the Arctic Coast to the Yenisei River. In North America, it ranges from New England north to Labrador and west to Bering Sea drainages in Alaska (but see comments in the taxonomy section). In the Mackenzie River system, round whitefish occur from the Peace–Athabasca Delta in Alberta (Nelson and Paetz 1992) to the Mackenzie Delta.

BRITISH COLUMBIA DISTRIBUTION
Round whitefish are restricted to the northwestern corner of the province: the Alsek, Chilkat, Taku, Yukon, and Liard drainage systems (Map 60). In the Liard system, they occur mainly above the Grand Canyon and are rare, or absent, in major tributaries below the canyon (e.g., the Fort Nelson and Petitot rivers).

Map 60 The B.C. distribution of round whitefish, *Prosopium cylindraceum*

Life History   In British Columbia, information on the life history of round whitefish is meagre; however, there is published life-history information on this species in north-central and eastern North America (Armstrong et al. 1977; Jessop and Power 1972; Mackay and Power 1968; Morin et al. 1982; Normandeau 1969) and in Alaska (Bendock 1979; Craig and Wells 1975). In British Columbia, there is one unpublished thesis (Guinn 1982) that documents some aspects of the ecology of sympatric round and mountain whitefish in the upper Liard River system. In addition, there are a few scattered life-history observations in consultants' reports. These sources, supplemented with data from Alaska and eastern North America, form the core of the following account.

REPRODUCTION

Round whitefish spawn in the fall: early September in the lower Liard River (McLeod et al. 1979); late September through October in the Colville and Chandler rivers of northern Alaska (Bendock 1979; Craig and Wells 1975); November in the upper Yukon system (Bryan and Kato 1975); and October, November, or December in eastern North America (Normandeau 1969; Scott and Crossman 1973). Spawning occurs in both lakes and rivers. In Newfound Lake, New Hampshire, spawning occurred on a shallow, rocky reef with a coarse gravel and rubble substrate at depths ranging from 15 cm to 14 m (Normandeau 1969). Water temperatures during spawning ranged from 2.5 °C to 4.5 °C. In Aishihik Lake and the East Aishihik River, Yukon Territory, round whitefish spawned in both slow and fast currents over substrates ranging from silt to gravel and boulders; however, eggs deposition appeared to be densest over gravel in fast current (Bryan and Kato 1975). Most eggs were sampled from water less than 1 m deep, and water temperatures ranged from 1.4 °C to 1.6 °C.

Round whitefish aggregate on their spawning sites. Apparently, there is some pairing of individuals in eastern populations (Normandeau 1969) but no evidence of pairing in western populations (Bryan and Kato 1975). However, one lateral display was observed in the East Aishihik River (Bryan and Kato 1975), and groups of fish moved slowly in the same direction over a wide area. Although these fish generally remained close to the bottom, individuals frequently broke the surface. There was no evidence of digging or other site preparation, and the eggs appear to be broadcast over the substrate. Actual gamete release was not observed, but the authors inferred from egg sampling that spawning occurs during the day. If this is correct, it is unusual in that most whitefish spawn at night.

Fecundity in round whitefish increases with female size and in New Hampshire varied from 2,200 to over 9,000 eggs (Normandeau 1969). In the Colville River, Alaska, the average fecundity for three round whitefish was 6,646 eggs (Bendock and Burr 1986), whereas average fecundity in the Chandler River was 10,300 eggs (range 4,200–18,700) in round whitefish that ranged in size from 313 to 413 mm (Craig and Wells 1975). Unfertilized eggs vary in diameter from 2.2 to 2.9 mm. Fertilized eggs are much larger and range in diameter from 3.3 to 4.6 mm. The eggs are demersal but not adhesive and incubate over winter in interstices in the substrate. In New Hampshire, the estimated incubation time was approximately 140 days at an average temperature of 2.2 °C (Normandeau 1969). Yolk-sac fry in Newfound Lake remained quiescent on the bottom after hatching but became active when disturbed (Normandeau 1969). In the Anadyr River system, yolk-sac fry (12–15 mm TL) migrated downstream from May to July. Some of these fry had begun exogenous feeding (Shestakov 1991).

AGE, GROWTH, AND MATURITY

Apparently, growth is rapid in the first few years, and in Dease Lake, round whitefish fry reached an average fork length of 33 mm (29–38 mm) with over a month and a half still left in the growing season. Similarly, in Alaska, midsummer fry in the Chandler system averaged 35.4 mm (range 27–43 mm), and those in the Colville River achieved an average length of 63 mm (range 49–71 mm) in their first growing season (Bendock and Burr 1986; Craig and Wells 1975). In Teslin Lake, they reached 160 mm FL at age 2 and 350 mm by age 9 (Clemens et al. 1968). Like many fish, round whitefish males usually reach sexual maturity about a year earlier than females (Bailey 1963; Mraz 1964), but there are populations in which the sexes mature at about the same age (Armstrong et al. 1977) and one report of females maturing a year earlier than males (Mackay and Power 1968). The age at first maturity also varies among sites. Thus, in northern Alaska, round whitefish only begin to mature in their seventh or eighth summer (Bendock and Burr 1986; Craig and Wells 1975), whereas most individuals in eastern North America are mature at 4^+ even in Arctic environments (Armstrong et al. 1977; Jessop and Power 1972; Mackay and Power 1968; Morin et al. 1982; Normandeau 1969). Once mature, females do not necessarily spawn every year (Jessop and Power 1972).

The lifespan of round whitefish also varies with locality and generally increases at higher latitudes. Thus, in New Hampshire and the Great Lakes, 8 or 9 years is the usual maximum age. In contrast, in Ungava and Alaska, ages ranging from 12 to 22 years have been recorded (Bendock and Burr 1986; Craig and Wells 1975; Jessop and Power 1972). The oldest round whitefish recorded so far in British Columbia was in its tenth growing season (Clemens et al. 1968); however, most age estimates for round whitefish are based on scales, and ages from scales of older fish are consistently lower than ages estimated from otoliths (Jessop and Power 1972).

FOOD HABITS

The diet of round whitefish varies with environment (lake or river), habitat within environments (i.e., deep or shallow in lakes), and perhaps, with the presence or absence of other species (Guinn 1982). In Dease Lake, round whitefish fry foraging in shallow water contained mainly chironomid larvae, some plankton (*Daphnia*), and occasional flying insects. The diet of juvenile and adult round whitefish in Simmons Lake, where they coexist with mountain whitefish, overlaps substantially (Guinn 1982). Nonetheless, plankton made up a much higher proportion of the diet of immature round whitefish in shallow water (<5 m) than it did for mountain whitefish. In deep water (>5 m), adults of both species consumed mostly benthic prey organisms (chironomid larvae and pupae, trichopterans, and molluscs), whereas adults foraged extensively on cladocerans (Guinn 1982) in shallower water. At the time of sampling (2 weeks in the summer), the largest round whitefish were foraging exclusively on snails. In Atlin and Teslin lakes, important diet items for adult round whitefish were amphipods, chironomids, trichopterans, molluscs, and occasionally small fish (Clemens et al. 1968; Withler 1956). The stomachs of the few round whitefish sampled from flowing water in British Columbia contained ephemeropterans, plecopterans, and trichopterans.

Habitat ADULTS

In British Columbia, round whitefish occur in both rivers and lakes. River-dwelling round whitefish appear to avoid small, high-gradient streams and are usually associated with the mainstems of large rivers or their major tributaries. These environments are often turbid. During the summer, adults typically occur in areas of moderate current, at depths greater than 1 m, and over coarse gravel substrates. Their depth distribution in lakes where they are sympatric with both pygmy and mountain whitefish (e.g., Dease Lake) is undocumented.

JUVENILES

Habitat use by juvenile and adult round whitefish in lakes with and without mountain whitefish is described in Guinn (1982). During the summer, juvenile round whitefish occurred primarily at depths of <5 m in all the lakes examined.

YOUNG-OF-THE-YEAR

Fry typically occur close to shore in shallow water areas where the current is reduced and the substrate is sand.

Conservation Comments   Although the round whitefish has a limited geographic distribution in northwestern British Columbia, it occurs in a number of separate drainage systems and is not a major conservation concern. The only population that may be under some threat is that in the Taku River. Mineral development and road building in this river system may have a negative affect on round whitefish numbers.

Prosopium williamsoni (GIRARD)
MOUNTAIN WHITEFISH

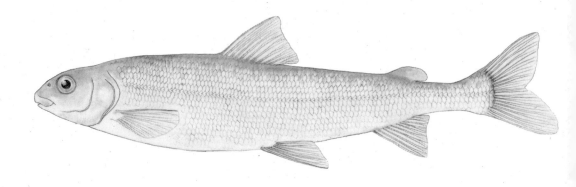

1 cm

Distinguishing Characters This small-scaled (over 70 scales along the lateral line and 19–23 scales around the caudal peduncle) whitefish has a large adipose fin (about 1.5 times the eye diameter), 11–15 dorsal rays, and 10–13 anal rays. In riverine and adfluvial populations, the body is round in cross section, and the snout usually is pinched and pointed when viewed from above. However, in strictly lacustrine populations (those that live and spawn in lakes), the body is often deep and laterally compressed. In the field, such populations are easily confused with lake whitefish (*Coregonus clupeaformis*), but the shape of the dorsal fin usually separates the species—when depressed, the anterior dorsal rays in mountain whitefish do not extend beyond the posterior dorsal rays while, in lake whitefish, they extend well beyond the posterior dorsal rays.

In the field, juvenile mountain whitefish are easily confused with adult pygmy whitefish (*Prosopium coulterii*). Usually, dorsal fin ray number separates the species: 11–15 rays in mountain whitefish and 9 or 10 in pygmy whitefish. Under a microscope or a hand lens, a good character is the size of the scales on the anterior portion of the lateral line. In mountain whitefish, the anterior lateral line scales are noticeably smaller than the scales immediately above and below the lateral line. In contrast, the anterior lateral line scales of pygmy whitefish are about the same size as the scales immediately above and below the lateral line. Juvenile mountain whitefish (up to about 200 mm) have well-developed parr marks; fry (up to about 30 mm) lack lateral parr marks but have an adipose fin that is almost as large as their anal fin.

Taxonomy Two forms of mountain whitefish coexist in many B.C. rivers. A "normal" form that in profile has the typical mountain whitefish head shape (Fig. 37A), and a "pinocchio" form that has a "turned-up" nose and an elongated snout (Fig. 37B). These differences in head shape are exaggerated in large adults; however, there are other more subtle morphological differences between the two forms, and the dimorphism is not sexual. Interestingly, in the upper Fraser system, the two forms may differ in the frequencies of some mitochondrial haplotypes (Troffe 2000). They also differ in their foraging behaviour: the "pinocchio" form spends more time digging in the substrate than the "normal" form (Troffe 2000). Although the two forms probably reflect some complex foraging dimorphism, the mitochondrial haplotype differences hint at the possibility of some genomic divergence between the two forms in the upper Fraser system.

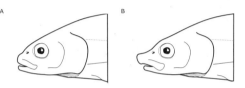

Figure 37 Head profiles of the normal (A) and pinocchio (B) forms of mountain whitefish

Sexual Dimorphism Except at spawning time, the sexes are difficult to distinguish. Breeding males and females develop lines of hard, white tubercles on the lateral scales. These tubercles are more pronounced on males than on females.

Hybridization At some sites, the distinctions between mountain whitefish and other *Prosopium* species become blurred. In the upper Peace system, occasional individuals are morphologically intermediate between mountain and pygmy whitefish. These individuals may be hybrids. Similarly, in the upper Liard system, occasional individuals are morphologically intermediate between mountain and round whitefish (*Prosopium cylindraceum*). Again, these may be hybrids.

Distribution NATIVE DISTRIBUTION
Mountain whitefish are an exclusively North American species. They are widely distributed along both slopes of the Rocky Mountains from northern Utah to the lower Mackenzie River (Norman Wells). They also occur in California, where they are native to the eastern slope of the Sierra Nevada. Although an inland species, they occur in coastal rivers in western Washington along the west side of the Olympic Peninsula and the east side of Puget Sound.

BRITISH COLUMBIA DISTRIBUTION
Mountain whitefish are primarily an interior species and only reach the coast where major rivers broach the coastal mountains. Thus, mountain whitefish are absent from coastal islands (including Vancouver Island and the Queen Charlotte Archipelago) and most short coastal rivers. The B.C. distribution of mountain whitefish extends from the U.S. border north to the Iskut–Stikine and Liard rivers (Map 61). The records from the

Map 61 The B.C. distribution of mountain whitefish, *Prosopium williamsoni*

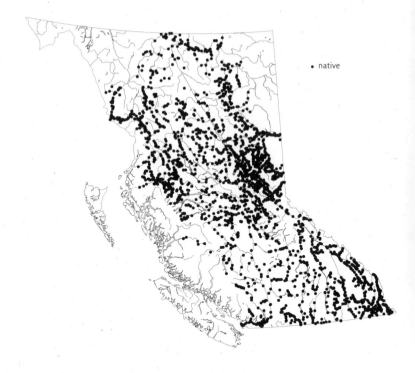

Taku River (Carl et al. 1959; Nelson and Paetz 1992) are misidentified round whitefish.

Life History Northcote and Ennis (1998) provide an excellent summary of the life history of mountain whitefish. It is the main source of the following account. In British Columbia, mountain whitefish display three basic life-history patterns: a lacustrine life history that is completed entirely in lakes, a riverine life history associated entirely with flowing water, and an adfluvial life history that involves movements between lakes and rivers.

REPRODUCTION

Mountain whitefish usually spawn in flowing water; although lake-spawning populations occur in some B.C. lakes (e.g., Kootenay Lake, Chehalis Lake, and probably Gantahaz Lake), most lake populations migrate into streams to spawn. Fluvial populations in large rivers may also migrate into smaller streams to breed, but some mainstem spawning occurs in the Peace, Columbia, and Kootenay rivers. In flowing water, spawning usually occurs at the lower end of riffles or near the head end of pools. There is no site preparation, and neither gravel size nor water velocity are critical in the choice of spawning sites. In lakes, mountain whitefish appear to preferentially spawn at sites where there is upwelling water. For example, in Chehalis Lake, they spawn at depths of over 10 m; however, the spawning site is a submerged piedmont fan deposited by a tributary stream.

Depending on latitude and altitude, mountain whitefish spawn in either the fall or early winter, usually when water temperatures drop below about 10 °C. Peak spawning activity occurs at temperatures <6 °C. Most B.C. populations spawn in October or November, but in the Columbia system, spawning can occur as late as January or early February. Apparently, spawning occurs at dusk or, perhaps, at night. Consequently, field observations on breeding fish are scarce; however, it is known that there is no site preparation and that the eggs are released over the substrate. Like other whitefish, more than one male probably is involved in a spawning.

Egg number is a function of female body size with large females producing disproportionately more eggs than small females. In British Columbia, fecundity ranges from about 1,000 to 15,000 eggs. Fertilized eggs range in size from 3 to 4 mm and average about 3.5 mm. The eggs are demersal and lodge in cracks and crannies among the gravel and rubble on the bottom. Here, they incubate over winter. The fry emerge in the spring or early summer (depending on latitude and altitude, from late March to early June).

AGE, GROWTH, AND MATURITY

Newly emerged fry are small (16–20 mm TL) and positively phototactic. In streams, immediately upon emergence, the fry drift downstream before moving into shallow, low-velocity areas along the river margins, side-channels, and backwaters. By late summer, the fry reach a size of about 60–100 mm. In lakes and reservoirs, the fry remain inshore. Growth is rapid for the first 4 years of life. In the upper Columbia and Fraser systems, most individuals are sexually mature by age 6. Males mature about a year earlier than females. The youngest mature males recorded in British Columbia were at the end of their third growing season (2^+). The maximum age recorded for this species is 23 years, but in British Columbia, relatively few individuals live beyond 12 years (McPhail and Troffe 2001).

FOOD HABITS

In lakes, newly emerged fry feed primarily on plankton, whereas in streams, mountain whitefish fry feed on the smallest life-stages of aquatic insects. Although the mouths of fry are small, the gape relative to their body size is greater than that of juveniles and adults. In rivers, adults and subadults forage on the nymphs of aquatic insects and occasional terrestrial insects, whereas the main prey in lakes are plankton, snails, surface insects, and occasionally, young fish.

Habitat Although mountain whitefish in central and northern British Columbia are often the most abundant species in the upper parts of many rivers, little is known about their habitat use. Again, the detailed life-history summary referred to earlier (Northcote and Ennis 1998) is the main source of the following account.

ADULTS

In British Columbia, adult mountain whitefish occur in both lakes and rivers. In fluvial populations, ontogenetic and seasonal habitat shifts are common. These habitat changes often involve complex, multiple migrations of over 100 km. For example, in the Sheep River, Alberta, adults perform five migrations within the drainage system (Thompson and Davies 1976): a post-spawning fall migration to overwintering sites, a spring feeding migration, a summer feeding migration, a fall pre-spawning migration, and a spawning migration. Although no equivalent study is available for B.C. populations, there are enough observations to suggest that similar migrations are a common feature of fluvial populations. For example, in the upper Fraser system, there is evidence of adult feeding migrations and the annual return of individuals to specific foraging sites (McPhail and Troffe 1998). In the Chowade River (a Peace tributary), a pre-spawning downstream migration occurs in late August and September. Also, post-spawning migrations to deep overwintering sites are known in several tributaries of the Fraser, Peace, Liard, Similkameen, and Kettle rivers.

In rivers, adults occur in loose aggregations in runs or pools (especially just below where riffles break into pools). Usually adults are close to, but not on, the bottom in water 1–2 m deep over coarse gravel or cobble substrates. In lower Kemess Creek (upper Finlay system), depth, water velocity, substrate size, and distance to cover were recorded for both juvenile (<150 mm) and adult (>200 mm) mountain whitefish (Bustard 1996). Average water depth, velocity, and substrate size all increased with body size. Adults were observed either in pools or associated with large organic debris. The average water depth for adults was 79 cm (range 60–110), their average water velocity was 0.62 m/s (range 30–80), and 50% of the substrate was >25.0 cm in diameter.

Observed underwater, mountain whitefish are remarkably active. Unlike trout or char, they rarely maintain position. Instead, they bob and weave in the current, constantly changing position within the group, and occasionally turning gravel with their snouts. This gravel turning behaviour is more common in the "pinocchio" form than in the "normal" form (Troffe 2000).

In lakes, adult mountain whitefish usually occur at depths of less than 20 m, but there are seasonal changes in adult habitat associated with temperature changes. In Koocanusa Reservoir (Kootenay system), adults are found in shallow water in the spring, but they move to deeper water as summer progresses and then return to the littoral zone in the fall. After spawning, they again move to deep water to overwinter (Chisholm et al. 1989).

JUVENILES

Juveniles in rivers appear to avoid riffles and backwaters and are associated with glides and runs (Porter and Rosenfeld 1999). They also show a preference for water about 1 m deep, large substrates (25–40 cm), and moderate currents (0.25–0.60 m/s). In lower Kemess Creek, the average depth of juveniles was about 55 cm (range 28–69), their average water velocity was about 0.45 m/s (range 23–80), and 50% of their substrate was <12 cm in diameter (Bustard 1996). In lakes, juveniles remain in shallow (<2 m), inshore habitats throughout the spring and summer. Here, they are usually found over sand and coarse gravel substrates.

YOUNG-OF-THE-YEAR

In lakes, newly emerged mountain whitefish are found in shallow water (<50 cm) over fine gravel or sand substrates. Although they are not strongly associated with cover, they often aggregate in large numbers in the lee of promontories during storms. In rivers, they are associated with shallow (<50 cm), quiet water over sand or silt substrates.

Conservation Comments The mountain whitefish is abundant in British Columbia and, at present, is not a conservation concern. Nonetheless, the nature of the relationship between the "normal" and "pinocchio" forms needs to be clarified.

Stenodus leucichthys nelma (GÜLDENSTÄDT)
INCONNU

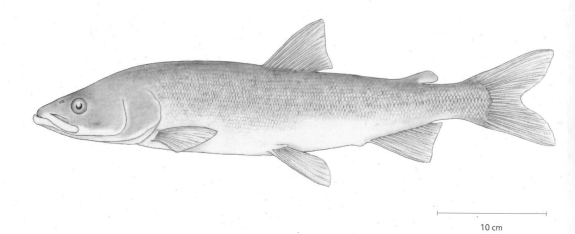

10 cm

Distinguishing Characters — This large (fork length to over 1 m) slender whitefish has a projecting lower jaw, and a large, wide mouth. The upper jaw extends back to at least the middle of the eye, and there are patches of tiny teeth on the vomer, palatines, and tongue. The unusually large mouth distinguishes the inconnu from all other whitefishes.

Taxonomy — Hamada et al. (1998) indicate that *Stenodus* and *Coregonus* share a common ancestor, and form a monophyletic group separate from *Prosopium*. This suggests that perhaps *Stenodus* may not be a valid genus. Nonetheless, morphologically and ecologically inconnu are strikingly different from *Coregonus*, and the question of its generic placement requires a more detailed analysis than is currently available. With the possible exception of the disjunct Caspian Sea form, the specific status of inconnu is not an issue; however, some authors (e.g., Shaposhnikova 1967; Walters 1955) recognized two subspecies: *S. l. leucichthys* in the Caspian region and *S. l. nelma* in Siberia and North America. This does not mean that North American inconnu are a genetically homogeneous group. There is a considerable geographic gap between Bering Sea populations and those centred around the Mackenzie River system. Thus, it is possible that the two B.C. populations (see below) belong to genetically different groups. Furthermore, there is evidence of local adaptation within drainage systems. For example, there appear to be three different life-history types (anadromous, riverine, and lacustrine) in the Mackenzie River system (Reist 1997).

Sexual Dimorphism — There are no obvious external characters that differentiate the sexes; however, females tend to live longer and grow larger than males.

Hybridization Inconnu are known to hybridize with members of the lake whitefish (*Coregonus clupeaformis*) species complex (Alt 1965; Reist et al. 1992).

Distribution NATIVE DISTRIBUTION
Inconnu are native to North America and Eurasia. In North America, there are two centres of abundance: Bering Sea drainages (Kuskokwim, Yukon, and Kobuk rivers) and the eastern Beaufort Sea drainages (Mackenzie and Anderson rivers). Curiously, inconnu are absent, or at least rare, along the Arctic coast between these two North American centres of abundance. In Eurasia, inconnu range from northern Kamchatka and the Chukchi Peninsula east along the Siberian coast to the White Sea. The subspecies, *S. l. leucichthys*, is isolated in the northern Caspian Sea and its tributaries.

BRITISH COLUMBIA DISTRIBUTION
Inconnu are restricted to the extreme northern parts of the province (Map 62). Here, they occur in two river systems (the Yukon and the Mackenzie drainages). The only population in the B.C. portion of the Yukon system is in Teslin Lake. This population is non-migratory (Clemens et al. 1968). Inconnu also occur in part of the B.C. portion of the Mackenzie system (the lower Liard River and its major tributary the Fort Nelson River).

Life History The life history of the inconnu has never been studied in British Columbia; however, life-history data are available from other jurisdictions. The major sources of information for Yukon drainages are Alt (1965, 1969, 1973, 1977, 1980) and, for the Mackenzie system, Fuller (1955), Howland et al. (2000), McLeod et al. (1979), and Tallman et al. (1996b).

REPRODUCTION
As far as is known, inconnu spawn only in flowing water. Thus, the Teslin Lake population probably spawns in one of the lake's major tributaries. The presence of ripe and spent inconnu in the B.C. portion of the lower Liard River indicates that some Mackenzie inconnu spawn in British Columbia, but the actual spawning sites are unknown. Based on descriptions of spawning sites in Alaska, the mainstem Liard above its confluence with the Toad River is a possible spawning area (McLeod et al. 1979).

In the Liard River, migratory adults begin to appear in the vicinity of supposed spawning sites shortly after ice-out in May (McLeod et al. 1979); however, upstream spawning migrations are protracted and usually do not peak until September. Before spawning, inconnu aggregate in deep water close to their spawning sites and then move from holding areas onto the actual spawning grounds. Spawning occurs from late September to early October. At this time, water temperatures in the Liard River ranged from 3.0 °C to 7.0 °C (McLeod et al. 1979) and from 5.0 °C to 10.0 °C in the Slave River (Tallman et al. 1996a). In the

Map 62 The B.C. distribution of inconnu, *Stenodus leucichthys nelma*

Chatanika River in the Alaskan portion of the Yukon system, spawning occurred at water temperatures ranging from 0.0 °C to 5.5 °C.

Inconnu spawning sites in the Kobuk River had moderate to fast currents (0.76–1.2 m/s), fine to coarse gravel substrates (0.6–7.6 cm in diameter), and moderate depths (0.3–0.9 m). There is no site preparation. In the Chatanika River, inconnu spawned in the evening (about 19:00). Here, the water was moderately deep (slightly over 2 m), and the substrate was sand and fine gravel. Presumably, the description of inconnu spawning behaviour in Alaska (Alt 1987) is typical of North American populations. Spawning starts with a pair about 30 cm from the bottom. They swim rapidly upwards with the male close behind, and below, the female. The female breaks the surface and skitters across the water (usually perpendicular to the direction of the current) expelling eggs. The male follows releasing sperm but rarely breaks the surface. Presumably, the eggs drift to the bottom and lodge in the substrate. A spawning spurt lasts about 1–3 seconds, and only a fraction of the eggs are released in each trip to the surface.

Fecundity increases with female size (Tallman et al. 1996a). In the Slave River, fecundity ranged from about 65,000 to 180,000 eggs. In the Chatanika River, fecundity ranged from 9,800 eggs in a female of 291 mm FL to 93,500 eggs in a 413 mm female; however, larger females can produce up to 450,000 eggs. Before fertilization, the eggs average about 2.5 mm in diameter and, after fertilization, about 3.0 mm (Sturm 1988). The eggs incubate in the gravel over winter and take 110 days to hatch at 4 °C (Hinrichs 1979). Newly hatched larvae range from 11 to 14 mm (Sturm 1988), and exogenous feeding begins at about 25 mm FL.

AGE, GROWTH, AND MATURITY

Inconnu growth rates vary among populations. In the Slave River, inconnu in their second year of life average about 24 cm FL and about 50 cm by age 4. In Teslin Lake, inconnu reach about 41.5 cm by age 4, whereas the Liard River population ranges from 50 to 60 cm at the same age. The maximum size and age recorded for the Teslin population is 87 cm and 17 years, respectively (Clemens et al. 1968). The largest inconnu recorded in the B.C. portion of the Liard River was 85.5 cm and 13 years old (McLeod et al. 1979). In the Slave River, the largest inconnu was over 110 cm long and 24 years old (Tallman et al. 1996a).

Typically, males mature 1 or 2 years before females. Thus, most Liard River males are mature at 5 years, and they all are mature at 6. Liard River females were mature by age 7. Similarly, males in the Slave River mature between the ages of 5 and 7, and females mature at 8 or 9 years. Once mature, most female inconnu skip 1 or 2 years between spawnings.

FOOD HABITS

Adult and juvenile inconnu are piscivores. Ten individuals from Teslin Lake contained fish remains (mostly juvenile least ciscoes, *Coregonus sardinella*). The Liard population appears to continue foraging into the late summer and early fall but probably stop feeding during the spawning period (McLeod et al. 1979). The most common food items in Liard inconnu were longnose suckers (*Catostomus catostomus*) and slimy sculpins (*Cottus cognatus*), but occasional juvenile Arctic grayling (*Thymallus arcticus*), northern pike (*Esox lucius*), and mountain whitefish (*Prosopium williamsoni*) also occurred in stomachs. In the Slave River, trout-perch (*Percopsis omiscomaycus*), northern pike, longnose suckers, flathead chub (*Platygobio gracilis*), walleye (*Sander vitreus*), and lake whitefish were all recorded from inconnu stomachs (Tallman et al. 1996a). Initially, young-of-the-year inconnu feed on small plankton, but as they grow, they switch to larger zooplankton and insect larvae (mostly chironomids). In the upper Yukon, inconnu fry begin to feed on fish late in their first growing season. In contrast, the Great Slave Lake population does not make the switch to a fish diet until they enter the lake in their third or fourth year (Fuller 1955).

Habitat Not much is known about habitat use in the B.C. populations of inconnu; however, inconnu have been studied in Alaska (Bering Sea drainages) and in the Mackenzie River system (Beaufort Sea drainages). Data from these regions are the major sources for the following habitat-use accounts.

ADULTS

Although all inconnu breed in flowing water, adults occupy three different habitats outside the breeding season: lakes, rivers, and estuaries or inshore marine waters. In North America, populations of inconnu occur in two inland lakes—Teslin Lake in the Yukon River system and Great Slave Lake in the Mackenzie River system. Nothing is known about habitat use of inconnu in Teslin Lake, but there is information on seasonal habitat shifts in Great Slave Lake. Shortly after spawning in tributary rivers, adult inconnu move downstream into the lake. Unlike the protracted upstream spawning migration, this post-spawning downstream migration is a rapid, mass movement (Fuller 1955). Once they reach the lake, the inconnu overwinter in deep water (Howland et al. 2000). Recoveries of tagged fish suggest that adults move in a counter-clockwise direction around the western basin of the lake (Tallman et al. 1996b). In the spring, they move inshore and congregate on the shallow south side of the lake near the mouths of large inlet rivers. About midsummer, they begin a protracted migration up these rivers to their spawning sites.

Anadromous populations in the lower Mackenzie system also move downstream immediately after spawning. They overwinter in the deep parts of the outer delta, coastal embayments, in mainstem channels, and perhaps, in the larger inner delta lakes (Howland et al. 2000). In the spring, they move into the productive feeding areas of the outer delta and inshore coastal waters. Apparently, there is an age, and perhaps a sex, component in the propensity to move into high-salinity areas. The strontium signature in inconnu otoliths from Arctic Red River fish indicate that they stay in freshwater for the first 12 years of life and then move annually into high-salinity areas (Howland et al. 2001). A subset of these fish (mostly males) show smaller annual strontium peaks that probably indicates estuarine rather than marine movements. Although not as well documented, Alaskan inconnu populations in the lower Yukon, Kuskokwim, Kobuk, and Selawik rivers also appear to perform post-spawning migrations downstream to deltas and estuaries to overwinter and feed (Alt 1987).

In North America, riverine populations of inconnu are confined to large, usually turbid, rivers like the Mackenzie, Yukon, and Kuskokwim rivers. Radio-tagging in the Mackenzie system indicates that the habitat use and movements of riverine populations are complex. For example, some fish tagged in the Liard River in British Columbia are anadromous and migrate over 1,700 km; others were recovered in or near Great Slave Lake; and some may be part of a riverine group that overwinters and feeds in the area near the confluence of the Liard and Mackenzie rivers (Howland et al. 2000; Stephenson et al. 2005). Riverine populations in the Yukon system are better known. The upper Yukon populations are year-round residents that perform relatively short spawning, feeding, and overwintering migrations (Alt 1977).

JUVENILES

In the lower Mackenzie River, juveniles of anadromous populations rear in fresh, or low-salinity, water for up to 12 years before moving to the outer delta and inshore coastal areas (Howland et al. 2001). In the spring, the fry of the Liard River population may be washed downstream into the upper parts of the Mackenzie Delta and then move back upstream as juveniles. Certainly, immature inconnu (4–5 years old) occur in the B.C. portion of the Liard system (McLeod et al. 1979), and at least one juvenile in its second summer has been collected in the Muskwa River near Fort Nelson. Juveniles in the Great Slave Lake population rear in the lake. Here, they forage along inshore areas during the spring and then move farther offshore during the rest of the year. There is some evidence that the juveniles move around the lake in the same counter-clockwise gyre used by the adults (Tallman et al. 1996b).

YOUNG-OF-THE-YEAR

Although inconnu from the lower reaches of the Mackenzie, Kobuk, Yukon, and Kuskokwim rivers migrate hundreds of kilometres upstream to spawn, their fry do not rear near the spawning sites. They move downstream to the lower reaches and estuaries of these rivers during the spring freshet. It is unlikely that they move directly into high-salinity water. Inconnu fry can withstand direct transfer into brackish water (10–15 g/kg) but not into seawater of 25 g/kg. Nonetheless, given a period of acclimation in brackish water, anadromous populations can osmoregulate in 25 g/kg seawater (Howland et al. 2001).

There are no records of inconnu fry from the Liard River in British Columbia or in the Mackenzie River in the vicinity of its confluence with the Liard River. Thus, Liard fry may be swept downstream by the spring freshet and rear in the upper parts of the Mackenzie Delta. As they grow, juveniles may gradually move back upstream.

Fry from the Great Slave Lake population initially move down to the lake and congregate in shallow inshore waters near river mouths. No fry have been reported from Teslin Lake, but presumably, they also concentrate in shallow productive areas near river mouths.

Conservation Comments   Neither COSEWIC nor the BCCDC list the inconnu. Nonetheless, the distribution gap along the Arctic coast of Alaska suggests that there is little or no gene flow between inconnu in Bering and Beaufort Sea drainages. Consequently, the non-migratory Teslin Lake population may be genetically different from the riverine population in the Liard River. Thus, we may have two different lineages of inconnu in British Columbia: a peripheral isolate of the Yukon lineage in Teslin Lake and the migratory inconnu in the Liard River. Furthermore, the Liard population may represent a riverine form distinct from the other Mackenzie populations (Howland et al. 2000). Both these B.C. populations are small, and both may be evolutionarily significant units. They certainly warrant study and, perhaps, a provincial listing.

Subfamily Thymallinae

The graylings are trout-like fishes with rather large scales and small, but toothed, jaws. They frequent rivers and cold lakes in Europe (including Britain), northern Asia, and North America. Although widely distributed, there are relatively few species compared with the other two subfamilies of salmonids. Some authors (e.g., Osinov and Lebedev 2002) place grayling in a separate family, but here they are included in the family Salmonidae.

The B.C. species is the Arctic grayling (*Thymallus arcticus*). It ranges throughout much of northern North America and Siberia. A second species, the European grayling (*Thymallus thymallus*) is widespread in Europe and extends into central Asia. The Amur River system contains a complex set of at least four grayling lineages (Froufe et al. 2003, 2005). Two of these lineages are sympatric (Antonov 2004) and represent two species, *Thymallus grubei* and *Thymallus burejensis*. In the arid regions of central Asia (especially Mongolia), there are a number of grayling populations isolated in separate, often internal, drainage systems. The taxonomic status of these populations is unclear and there may be four, or perhaps more, species in the region (Schöffmann 2000).

Thymallus arcticus (PALLAS)
ARCTIC GRAYLING

1 cm

Distinguishing Characters Adult Arctic grayling are unlikely to be confused with any other trout-like fish. The spectacular dorsal fin on mature males is unique, and even females and juveniles have dorsal fin bases that are longer than their heads.

Occasionally, young-of-the-year mountain whitefish (*Prosopium williamsoni*) are mistaken for young grayling; however, the shapes of their parr marks readily distinguish these species. Parr marks on fry first become apparent as a set of 7–10 dark, midlateral blotches at a size of about 25–30 mm. At this stage, although they resemble young mountain whitefish, the long dorsal fin base is obvious and, when viewed from above, there are no distinct dark pigment blotches on the dorsal surface. In contrast, dark dorsal blotches usually are well developed in similar-sized mountain whitefish. By 50 mm, the lateral parr marks in the two species are strikingly different: almost circular blotches in mountain whitefish and narrow vertical lines in grayling.

Taxonomy Fossils, genetic information, and historic distributions all indicate that Arctic grayling survived glaciation in a complex series of glacial refugia. Late Pleistocene fossils prove survival in the Yukon (Cumbaa et al. 1981), the Great Lakes region (Miller et al. 1993), and southern Alberta (Burns 1991). Additionally, genetic data suggest separate northern and southern Beringian refugia as well as Nahanni and upper Missouri refugia (Redenbach and Taylor 1999; Stamford and Taylor 2004). The presence of several populations (now extirpated) in the Great Lakes region also imply their survival in an eastern refugium, and isolated populations in the upper Missouri system in Montana indicate a Great Plains refugium.

Sexual Dimorphism In mature males, when the dorsal fin is depressed, the posterior rays of the fin extend back almost to the adipose fin, and the pelvic fins are strikingly marked with vivid orange stripes. The dorsal fin has an orange border and is almost blue with numerous orange spots highlighted with pale haloes. In mature females, the dorsal fin also is enlarged, but not to the same extent as in males, and the colours are not as vivid. Apparently, spawning fish of both sexes have small breeding tubercles (Kratt and Smith 1978), but tubercles have not been recorded in B.C. specimens.

Distribution

NATIVE DISTRIBUTION

Arctic grayling occur in North America and Siberia. As the name implies, this coldwater species ranges in a broad band across North America from the west coast of Hudson Bay to Alaska. They are absent from the Arctic Archipelago; however, historically, isolated populations occurred in upper Michigan and in the upper Missouri river system. Most of these southern populations are now extirpated; however, one native population still exists in an upper Missouri tributary (Big Hole River) in Montana (Kaya and Jeanes 1995). In Eurasia, they range from the Kara River in the west to the Chukostsk Peninsula in the east and, in Pacific drainages, to the Yalu River in North Korea.

BRITISH COLUMBIA DISTRIBUTION

Arctic grayling occur throughout the Peace and Liard rivers (Mackenzie system), the Yukon system, and in the upper Stikine, Taku, and Alsek rivers (Pacific drainages) (Map 63).

Life History

Arctic grayling display three life-history patterns. The commonest life history is riverine; however, both adfluvial (migrations between lakes and streams) and lacustrine populations occur throughout most of the species' geographic range. There is a review of grayling studies in Alaska (Armstrong 1986) and a thorough comparative review of Arctic and European grayling life histories (Northcote 1995). The following life-history summary draws heavily on the latter review supplemented, where appropriate, with more recent information.

REPRODUCTION

Arctic grayling spawn in the spring, usually shortly after ice-out when the water temperature rises to about 4 °C. In northern British Columbia, spawning occurs from early to late May. There are two accounts of grayling spawning habitat in British Columbia (Butcher et al. 1981; Stuart and Chislett 1979). Apparently, grayling usually spawn in flowing water over coarse (2–4 cm) gravel and cobble substrates. Typically, there is a modest current (0.5–1.0 m/s), and the spawning site is a shallow (10–40 cm) glide or run; however, in Alaska, spawning is reported to occur on riffles over a wide range of substrates (Tack 1973). Unlike other salmonids, the male chooses, and defends, a spawning territory (Beauchamp 1982). Consequently, the defended sites often are well separated. There is no site preparation, but the male's activity often sweeps the area clean of silt. Females remain in deeper areas near cover until ready to spawn (Tack 1971). Apparently, most spawning occurs in the late afternoon (Beauchamp 1990a; Kratt and Smith 1980; Reed 1964).

Map 63 The B.C. distribution of Arctic grayling, *Thymallus arcticus*

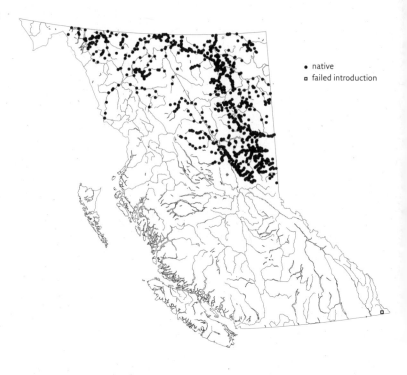

Egg number varies with female size: large females produce more eggs than small females. Little is known about the fecundity of B.C. populations, but egg number in other areas ranges from about 1,000 to 17,000 (Bishop 1971). Egg size (before fertilization) ranges from 2 to 3 mm, whereas fertilized eggs average about 4.0 mm (range 3.8–4.4 mm) in diameter. The eggs are slightly adhesive and denser than water. They are released, fertilized, and then drift to the bottom where they become lodged in cracks and crannies in the substrate. The incubation period depends on temperature, and hatching occurs after 18 days at 8 °C and 8 days at 15.5 °C (Wojcik 1955). The fry emerge from the substrate about 3 or 4 days after hatching (Kratt and Smith 1977). Yolk-sac fry are about 11 mm in length and emerge and begin feeding at 14–15 mm (Kratt and Smith 1977). They are weak swimmers and take refuge along the shallow margins of streams, especially in side-channels or quiet areas where small tributaries enter the spawning stream (Stewart et al. 1982; Stuart and Chislett 1979). Initially, the fry school; however, agonistic behaviour appears at about 3 weeks, and territories and dominance hierarchies are established by about 6 weeks (Kratt and Smith 1979).

AGE, GROWTH, AND MATURITY

Depending on stream productivity, young-of-the-year grayling can grow rapidly and occasionally reach about 120 mm (Kratt and Smith 1979) in their first year; however, in British Columbia, most populations reach 60–90 mm by the end of their first summer. At the end of their first growing season in the Table River (Parsnip system), the average fork length of grayling fry was 72.2 mm (Blackman and Hunter 2001). Growth gradually slows over the next 5 years, and by their fifth summer, they are usually over 250 mm in length and reach about 300 mm in their sixth summer. Most individuals reach sexual maturity after their fifth summer, and there appears to be little difference between males and females in age at maturity. In British Columbia, the maximum lifespan for most grayling is around 8 or 9 years, and very few individuals exceed 400 mm in length; however, in Beaufort Sea drainages, grayling reach ages of 22 years (de Bruyn and McCart 1974).

FOOD HABITS

Grayling in streams display a size-based dominance hierarchy (Hughes 1992a, 1992b). Apparently, they rank feeding positions, and larger individuals hold the most profitable sites. Adult grayling forage primarily on aquatic insects; not surprisingly, their winter diet contains more bottom-oriented insects and fewer surface insects than their summer diet. Adults and juveniles consume a similar range of prey, but adults add small fish to their diet. Lake-dwelling populations often forage on plankton. The young-of-the-year feed primarily on the smaller stages of aquatic insects, especially midge larvae, but as they grow they add drifting nymphs and terrestrial insects to their diet.

Habitat ADULTS

Arctic grayling occur in lakes, large rivers, and small streams, but in British Columbia, they are primarily a riverine species. Like the mountain whitefish, they often display complex migrations between spawning sites, feeding sites, and overwintering areas. Not only are there seasonal habitats shift but also changes in habitat as they grow. The annual movement between overwintering sites and summer feeding sites is a characteristic part of grayling life history (West et al. 1992). In the upper Peace system, adults overwinter in the lower reaches of large rivers and in embayments where rivers enter Williston Reservoir. Upstream migration occurs in the early spring (at or shortly after ice-out). In Alaska, the largest grayling migrate the farthest upstream, and this size gradient probably results from competition for preferred feeding sites (Hughes and Reynolds 1995). Like mountain whitefish, grayling show strong inter-annual fidelity to summer feeding sites (Buzby and Deegan 2000), although the probability of inter-annual migrations decreases as the fish grow (Hughes 2000). For example, during the Mesalinka River Fertilization Project, the majority of grayling tagged and recaptured (after at least 1 year) were caught in the same reach where they were originally tagged. As adults, the areas used as summer feeding sites shift from smaller streams to mainstem pools, riffles, and runs. In these habitats, adult grayling are territorial and develop dominance hierarchies, with the dominant fish occupying depressions close to where upstream riffles break into pools.

JUVENILES

Habitat use by 1+ fish is a bit of a mystery. In many studies, 1+ grayling are rare in tributary samples (although 0+ and 2+ fish are relatively common). In the upper Peace system, 1+ juveniles may use shallow glides in large, valley-floor rivers (e.g., the Parsnip River) as rearing areas. They appear to prefer relatively low velocities (<0.2 m/s) and depths between 0.4 and 1.0 m (Blackman and Hunter 2001). Because spawning typically occurs in smaller tributary streams and larger grayling usually occur farther upstream than juveniles (Hughes and Reynolds 1995), the size-based distribution pattern in the upper Peace implies substantial migrations of all size classes.

YOUNG-OF-THE-YEAR

Typically, young-of-the-year are found along the quiet margins of tributary streams or in slow (<0.05 m/s) side-channels associated with larger rivers (Butcher et al. 1981; Stewart et al. 1982; Stuart and Chislett 1979). In the Parsnip system, 0+ grayling showed a preference for water ≤ 6 cm deep with a water velocity of <0.03 m/s and a substrate of fines and gravel (Blackman and Hunter 2001). Within these areas, the fry are territorial (Kratt and Smith 1979), and the largest individuals hold the "best" territories. Interestingly, fluvial populations show an inherited response to flow that is stronger than that found in lacustrine populations, even though the latter spawn in streams (Kaya and Jeanes 1995). In British Columbia sometime during their first summer, fry move out of tributary streams and into larger rivers (McLeod et al. 1978; Stewart et al. 1982). As winter approaches they move into deep side-channels and areas associated with ground water. They overwinter in these areas and, in the spring, shift back into tributary streams.

Conservation Comments   The populations of Arctic grayling above the Peace River Canyon are of special concern. These populations have been isolated since the formation of the barrier that separates the upper Peace River from the lower Peace system. A recent microsatellite study (Stamford and Taylor 2005) found a significant ($P < 0.005$) difference in microsatellite variation in samples taken above and below this barrier. The same study concluded that current grayling population sizes are $<1\%$ of historical population sizes. Clearly, upper Peace grayling are in serious decline, and it will take a major effort to bring them back. Grayling populations in other parts of British Columbia (the Mackenzie and Yukon systems) do not appear to be in such bad shape; however, this species is especially vulnerable to angling and environmental degradation and will need constant monitoring. The BCCDC ranks grayling in the upper Peace River system as S1 (critically imperiled).

FAMILY PERCOPSIDAE — TROUT-PERCH

The Percopsiformes are a small (nine living species) order of unusual little fishes. There are three extant families in the order, all confined to North America. The living families are the trout-perch (Percopsidae), the pirate perch (Aphredoderidae), and the cavefishes (Amblyopsidae). There are two living species of trout-perch: the widely distributed trout-perch (*Percopsis omiscomaycus*) and the sand roller (*Percopsis transmontanus*). The latter species is restricted to the lower and middle Columbia River system (including the lower Snake River) but does not occur in British Columbia.

The only percopsid in British Columbia is the trout-perch. In British Columbia, it occurs only in the northeastern portion of the province; however, at one time (about 35 million years ago), an extinct genus, *Libotonius*, lived in what is now southern British Columbia and north-central Washington (Wilson 1977, 1979). The unusual name, trout-perch, alludes to their unusual morphology. They possess a curious combination of trout-like and perch-like features: an adipose fin, ctenoid scales, spines in the dorsal and anal fins (soft spines in trout-perch but strong spines in the sand roller) and pelvic fins that originate on the thorax.

Percopsis omiscomaycus (WALBAUM)
TROUT-PERCH

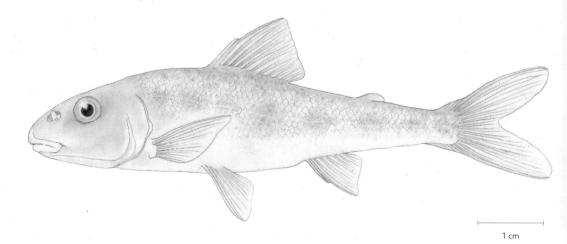

1 cm

Distinguishing Characters	This small (usually <120 mm FL) fish has a relatively large head (almost a quarter of the total body length), an adipose fin, ctenoid (rough) scales, thoracic pelvic fins, and pectoral fins that extend back past the origin of the pelvic fins. In life, the back is translucent with an iridescent purple hue, the sides are almost transparent, and the silver-white peritoneum is visible through the body wall. There are two rows of dark blotches along the flanks between the midline and the back. These pigmented areas are more conspicuous in preserved specimens than in live animals.
Sexual Dimorphism	Outside the spawning season, there is no reliable external method of determining sex in the trout-perch; however, like many fishes, mature females are on average larger than mature males. This size dimorphism results from a slightly more rapid growth rate and a longer lifespan (on average, 2 or 3 more years) in females than in males (House and Wells 1973; Magnuson and Smith 1963; Pereira and LaBar 1983).
Distribution	**NATIVE DISTRIBUTION** The trout-perch is found only in North America and is widely distributed on the east side of the Continental Divide. It occurs in Quebec and the northern portions of the Atlantic Coastal Plain and extends across the northern Great Plains to the foothills of the Rocky Mountains. In the north, the trout-perch reaches the Mackenzie Delta and has crossed the Continental Divide into the northeastern portions of the Yukon River system (the Porcupine River). In the Yukon River, it extends downstream as far as Andreafsky, Alaska.

BRITISH COLUMBIA DISTRIBUTION

The trout-perch is found only in the northeastern portion of the province (Map 64). It occurs in the lower Peace River and its tributaries downstream of Dinosaur Reservoir and in lower Liard River drainages downstream of the Liard Canyon.

Life History The life history of trout-perch has never been investigated in British Columbia. Consequently, the following account draws on studies conducted in Manitoba (Lawler 1954), Minnesota (Magnuson and Smith 1963), the Great Lakes (House and Wells 1973), and Lake Champlain (Pereira and LaBar 1983). Where possible, this information is supplemented with local observations.

REPRODUCTION

Trout-perch start to spawn in the spring (May), and in some areas, spawning continues throughout the summer and into early fall (House and Wells 1973; Lawler 1954; Magnuson and Smith 1963; Pereira and LaBar 1983). At this time, adults concentrate in shallow inshore waters. In northeastern British Columbia (Moberly Lake), gravid females were collected as early as the first week in June (water temperature 13 °C) and as late as the last week in July (17 °C). In a Minnesota stream, spawning started at dusk and continued into the night (Magnuson and Smith 1963). Egg release occurred near the stream edge and within 10–12 cm of the surface. Several males accompanied a single female. In Moberly Lake, running ripe trout-perch were captured at midnight in shallow water (<1 m) over sand and gravel beaches. Presumably, these fish were beach-spawning; however, since most B.C. populations occur in rivers, spawning in flowing water is probably more common in our area. The fertilized eggs are a little less than 2.0 mm in diameter, adhesive, and denser than water. At 20–23 °C, they take 6 or 7 days to hatch (Magnuson and Smith 1963). Fecundity increases with female size, and egg number ranges from 150 to 1,000; however, there is remarkable variation (up to a sixfold difference) in egg number among females of the same size (House and Wells 1973). This observation suggests that female trout-perch are fractional spawners (i.e., they deposit their eggs in batches, often over a period of several days).

Newly hatched larvae are about 5 mm TL and take 4 or 5 days to absorb their yolk. By this time, they are about 6.2 mm in length. Depending on when they hatched, young-of-the-year can reach about 50 mm by the end of their first growing season. By mid-August in the lower Liard system, small tributary streams often contain two size classes of trout-perch fry. The smaller group averages about 15 mm TL, and the larger group averages about 25 mm. This bimodality in the length of fry suggests the possibility that there are two spawning peaks in these populations.

414 FAMILY PERCOPSIDAE — TROUT-PERCH

Map 64 The B.C. distribution of trout-perch, *Percopsis omiscomaycus*

AGE, GROWTH, AND MATURITY

Normally, male trout-perch mature during their second summer and reproduce for the first time at the beginning of their third growing season. On average, females mature a year later than males. Apparently there is a high mortality rate in males, and they rarely live more than 4 years. British Columbia data from the Petitot River drainage also suggest a high post-spawning mortality rate in males—all the adults over 55 mm were females. In contrast, females can live for up to 8 years (House and Wells 1973). This differential mortality probably accounts for the skewed sex ratios (more females than males) reported in some studies (Magnuson and Smith 1963; Pereira and LaBar 1983).

FOOD HABITS

Adult trout-perch feed on a wide variety of benthic invertebrates (amphipods, chironomids, ostracods, and especially, the nymphs and larvae of aquatic insects). Occasionally, they also take young fish (Scott and Crossman 1973). In British Columbia, early morning samples collected in the Muskwa River had full guts, while those taken in the middle of the day were almost empty. This implies nocturnal feeding, and trout-perch have densely packed bands of papillae along the undersides of the lower jaws. These structures may be chemoreceptors that are used to locate prey at night or in turbid water. Young trout-perch (<40 mm in fork length) have a similar diet to adults but take more water column prey (cladocerans and copepods) than adults.

Habitat   Over most of its geographic range, the trout-perch is a lake-dwelling species (Scott and Crossman 1973). In British Columbia, however, this species occurs mainly in rivers and streams and, so far, is known from only three lakes (Moberly and Bearhole lakes in the lower Peace River system and Maxhamish Lake in the lower Liard system). This apparent dearth of lacustrine populations probably reflects collecting effort rather than any real ecological differences between eastern and western populations.

ADULTS

In British Columbia, adult trout-perch are strongly associated with large, turbid rivers: the mainstem Peace, the Fort Nelson drainage, and the lower Liard River. Although typically found in deposition areas (low water velocity and silt or sand bottoms), they appear to avoid seasonally flooded backwaters and side-channels. Most collections of adults were made during the day and in relatively shallow water (<1 m deep). This depth may simply reflect the limitations of fluvial collecting techniques, since in lakes they are taken in trawl nets at depths of over 60 m (Pereira and LaBar 1983). Although observations are limited, comparisons of day and night seine hauls in shallow inshore areas of Moberly Lake indicate that adult trout-perch are rare in inshore waters during the day. At night, however, they are abundant in these same areas. Similar diel onshore migrations are well documented in eastern North America (Magnuson and Smith 1963).

JUVENILES

In the streams and rivers of northeastern British Columbia, juvenile trout-perch (<50 mm TL) are most abundant in lower reaches of tributary streams associated with large turbid rivers. Usually these waters are clear, or stained, but rarely turbid. Typical sites have soft bottoms with emergent vegetation close to shore. In Maxhamish and Moberly lakes, juveniles were seined in shallow water (<1 m) over sand and gravel substrates.

YOUNG-OF-THE-YEAR

In the summer, during the day, young-of-the-year trout-perch are found mainly in tributary streams. Here, they are associated with shallow water (usually <1 m deep) and vegetation. Their nocturnal distribution is unknown. Apparently, these tributary streams are nursery areas for young trout-perch, and adults may migrate from larger rivers into these tributaries to spawn. In Maxhamish and Moberly lakes, fry were associated with shallow weedy areas.

Conservation Comments   Although the B.C. distribution of trout-perch is restricted to the northeastern part of the province, they are abundant and, at present, are not a conservation concern.

FAMILY GADIDAE — CODS

The cods (Gadidae) belong to an enigmatic order of fishes—the Gadiformes. This order contains over 500 species. Anatomically, they possess a curious mixture of primitive and advanced traits: they lack spines in their fins and have cycloid scales, but the pelvic fins (if they have any) insert in front of the pectoral fins. This order includes the true cods (Gadidae). There are about 30 living species of cods. Over half of the species lack a proper tail (the body simply tapers to a point), and many species have three dorsal and two anal fins. Most cods are marine and occur in the cooler seas of both the Northern and Southern Hemispheres; however, some species enter estuaries, and in eastern Canada, a few populations of Atlantic tomcod (*Microgadus tomcod*) live in lakes. These landlocked populations complete their entire life cycle in fresh water (Legendre and Legueux 1948; Scott and Crossman 1973). Pacific tomcod (*Microgadus proximus*) also enter estuaries, and they sporadically penetrate into fresh water in the Fraser Delta; however, there is no evidence that they breed in fresh water.

In this mainly marine family, there is one truly freshwater species, the burbot (*Lota lota*), which is locally known as ling. This circumpolar species spends its entire life in fresh water, although it also occurs in estuaries and brackish lagoons in some northern areas, such as the Gulf of Bothnia and the Mackenzie Delta (Percy 1975; Preble 1908; Pulliainen et al. 1992). Apparently, the genus has been associated with freshwater since the early Pliocene (Pietschmann 1934). In British Columbia, burbot are common in lakes throughout the interior of the province.

Lota lota (LINNAEUS)
BURBOT

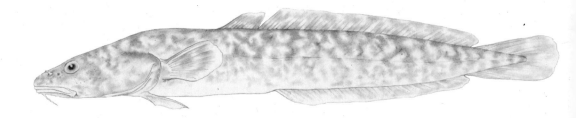

1 cm

Distinguishing Characters Burbot have two dorsal fins: the first dorsal is small, but the base of the second dorsal is long (about half the length of the body). The anal fin is also long. The pelvic fins originate in front of the pectoral fins, and the single barbel at the tip of the chin is unique among B.C. freshwater fishes.

Taxonomy In North America, burbot probably survived glaciation in multiple refugia (McPhail and Paragamian 2000; Snyder 1979; Van Houdt et al. 2005), and different morphological forms of this species occur in different regions (Hubbs and Schultz 1941). Thus, at least two subspecies have been recognized: *L. l. leptura* in Alaska and parts of the Yukon and northern British Columbia and *L. l. lacustris* (=*maculosus*) in eastern North America. Although Alaskan and Great Lakes burbot differ morphologically, the two forms appear to intergrade smoothly across the northern Great Plains (McPhail and Lindsey 1970). However, a recent broad-scale molecular analysis (Van Houdt et al. 2003) uncovered two distinct mitochondrial lineages in burbot: a Eurasian–Beringian lineage and a North American lineage. With a minor nomenclatorial change (apparently the Alaskan and Yukon subspecies, *L. l. leptura*, is a synonym for the Eurasian subspecies, *L. l. lota*), these lineages correspond to the two North American subspecies. In addition to these major lineages, the analysis also uncovered three less divergent clades (all embedded within the North American lineage) associated with the margins of the last glaciation: Mississippian, Missourian, and Pacific clades (Van Houdt et al. 2005).

Sexual Dimorphism There are no obvious colour or external morphological differences between the sexes, and except for a distended abdomen in spawning females, it is difficult to sex burbot.

Distribution NATIVE DISTRIBUTION
The burbot is one of the most widely distributed freshwater fish in the Northern Hemisphere. Their natural range extends from the British Isles (where they are now almost extirpated) eastward across Europe and Asia to the Bering Strait. In North America, it ranges from the Seward Peninsula, Alaska, to the northern parts of Washington, Idaho, and intermountain Montana. On the Great Plains, burbot occur from the Mackenzie Delta to Wyoming and, in central and eastern North America, from Labrador to Pennsylvania.

BRITISH COLUMBIA DISTRIBUTION
Burbot are widespread in the interior (Map 65) but absent from short coastal drainages and coastal islands (including Vancouver Island and the Queen Charlotte Archipelago). Burbot occasionally stray into the lower reaches of rivers like the Skeena, Fraser, and Columbia but, apparently, do not to maintain self-sustaining populations in these areas.

Life History In British Columbia, burbot exhibit three general life-history patterns: lacustrine, riverine, and adfluvial (migrating between rivers and lakes). It is not known if these life-history patterns are inherited. Riverine and adfluvial populations are normally migratory, but some form of migration usually precedes spawning even in lacustrine populations. In lakes, this migration may only involve movement from deep water into shallow areas, but migrations in riverine and adfluvial populations can involve movements of many kilometres. A recent review of burbot life history (McPhail and Paragamian 2000) is the main source of the following life-history synopsis.

REPRODUCTION
Burbot aggregate for spawning in winter or early spring (December to early March). The exact spawning period varies among populations but usually occurs at water temperatures ranging from 0 °C to 5 °C. There is disagreement about the time of day that burbot spawn. Most faunal works indicate burbot spawn at night, but some investigators have observed large "writhing balls" of burbot during the day. Similar burbot "balls," some involving 50–100 individuals, have been photographed in a spring-fed tributary of Columbia Lake. Here, the "balls" disperse at night when, apparently, actual gamete release occurs. Where burbot "balls" form there is often little overhead cover, and the "writhing balls" may result from individual burbot attempting to use other burbot as cover. Regardless of why these "balls" form, their movements excavate depressions in the substrate and clean the underlying gravel. Consequently, if gamete release occurs at these sites, this balling behaviour also may act as a form of spawning site preparation.

In lakes, spawning occurs in relatively shallow water (1.0–10.0 m) over sand or gravel bottoms. Again, in Columbia Lake, there are areas within the lake that contain depressions in the substrate that look remarkably like

Map 65 The B.C. distribution of burbot, *Lota lota*

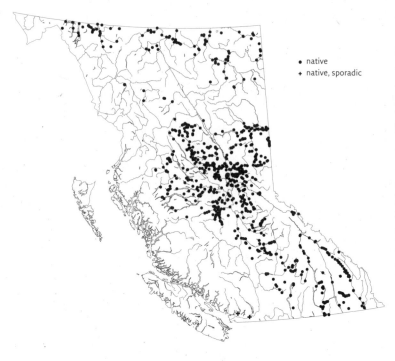

those created by burbot in the inlet stream. These areas may be spawning sites or, perhaps, simply groundwater springs. Since the sites are under ice in February (the spawning season in Columbia Lake), no direct observations are available. In rivers, burbot spawn in low-velocity areas in main channels and in side-channels behind deposition bars. The preferred substrate appears to be fine gravel, sand, or even silt. There is no evidence of site preparation in rivers and, presumably, eggs and sperm are released into the water just above the bottom.

Newly fertilized burbot eggs are only slightly denser than water, and in flowing water they probably drift a short distance before settling to the bottom. Not all mature burbot spawn every year, and the proportion of adults that skip a year varies among localities.

Egg number in burbot is a function of female size: large females produce more eggs than small females. Regardless of size, however, burbot are exceptionally fecund, and large females often produce over a million eggs. Egg size varies among populations but averages about 1 mm in diameter (0.5–1.7 mm). The eggs take from 30 to 60 days to hatch (the exact time depends on temperature), and newly hatched larvae (about 3–4 mm TL) are incompletely developed. Laboratory observations on the Columbia Lake population indicate that the larvae remain sedentary on the substrate for at least 5 days. Then, while they still have a yolk sac and well before they begin feeding, they start a peculiar behaviour where the larvae spiral upwards towards the surface and then sink back to the bottom. This behaviour resembles the movements of mosquito wrigglers. The larvae eventually fill their swim bladders and become neutrally buoyant. In the laboratory, this "wriggling" behaviour persisted for over a week before the larvae filled their swim bladders. The same behaviour occurs in newly hatched burbot in the inlet to Columbia Lake (Peter Mylechreest, personal communication). The change from "wriggler" to free-swimming larvae appears to coincide with the beginning of feeding and is accompanied by a colour change. At first the larvae are planktonic and drift passively; however, as they grow, they become more mobile.

Although larvae in lakes start their free-swimming life as part of the plankton community, their early ecology in flowing water (especially swift rivers) is a mystery. How do they avoid being flushed downstream? In Columbia Lake, burbot larvae disappear from the water column at roughly 15 mm TL and reappear inshore about a month later (at a length of 30–40 mm). By this time they are solitary, bottom-dwelling little burbot. Where they went and what they did during the intervening month is unknown.

AGE, GROWTH, AND MATURITY

Burbot grow rapidly in their first year and, depending on food resources, reach 100–120 mm by fall. They continue to grow (even in winter), and 50% of the males reached sexual maturity at age 7 in the Tanana River, Alaska; however, some males delayed maturity until 15 (Evenson 2000). Fifty percent of the females reached sexual maturity at age 8, but again, some females delayed maturity until age 14. Once mature, growth rate declines. Burbot can live for up to 20 years and reach lengths of 700 mm.

FOOD HABITS

The first food items taken by newly hatched burbot are small (phytoplankton and rotifers); however, in the laboratory, the size of food begins to increase as body size increases beyond about 15 mm. Once burbot become benthic, their diet shifts to include more bottom organisms (e.g., amphipods). As they grow, they move into deeper water and gradually increase the proportion of fish in their diet. Adult burbot are primarily piscivores and, in British Columbia, prey on a wide variety of fish species including trout, grayling, suckers, minnows, and sculpins.

Habitat Although little is known about burbot habitat use in British Columbia, a recent book on burbot biology, ecology, and management (Paragamian and Willis 2000) contains a general review of burbot habitat use and several papers that describe habitat use in specific populations. These sources form the basis for the following habitat synopsis.

ADULTS

Burbot are cool-water fish and seldom occur in either lakes or rivers where temperatures exceed 18 °C for prolonged periods. Nonetheless, in Lake Opeongo, Ontario, nocturnal activity was recorded at water temperatures as high as 20 °C (Carl 1995). These same animals moved to cooler, deeper water during the day, and generally, burbot were active at night and inactive during daylight hours. Adult burbot are benthic and rarely enter water less than 2 m deep. In Lake Superior, burbot occur at depths of up to 300 m and are associated with extensive burrows in the bottom sediments. A similar association with bottom burrows occurs in Okanagan Lake. It is not known if these burrows indicate "home" territories for individual adults; however, in Lake Opeongo, sonic-tagged burbot occupied the same local areas within seasons and between years (Carl 1995). In the fall, burbot in some lakes move into shallow areas and concentrate near the mouths of known spawning rivers (Schram 2000).

Habitat use by adult burbot in rivers is not well documented, but again, temperature probably is important: burbot are widespread in rivers where summer temperatures seldom exceed 18 °C and relatively rare in rivers where summer temperatures commonly reach 20 °C or higher. Thus, in British Columbia, riverine burbot are abundant in the north, but are restricted to cool rivers in the southern parts of the province. In northern rivers, adult burbot are associated with the deep main channels and often appear to aggregate where clear streams enter turbid rivers. In such situations, they favour the turbid water.

JUVENILES

In Columbia Lake during the summer, yearling (1+) burbot usually occupy water less than 2 m deep and are strongly associated with cover (rip-rap jetties and natural boulder areas). In Europe, however, juvenile burbot (120–160 mm TL) leave the littoral zone in the spring (Hofmann and Fischer 2002) and move to deeper water. In Columbia Lake, older (2+) juveniles (>200 mm) stay below the thermocline in the summer. In rivers, little is known about their habitat use, although they are often taken in small tributaries. In the Tanana River, Alaska, juveniles moved shorter distances than adult burbot, and their average movements did not vary among tracking periods (Evenson 2000). In contrast, the average movement of adults did vary, and movements were greatest in the spring and fall.

YOUNG-OF-THE-YEAR

In lakes, burbot start life as limnetic larvae; first, at moderate depths and then shifting toward the surface as spring progresses. By midsummer (at a length of about 15 mm), they disappear from the water column and reappear about a month later (at 30–40 mm) in near-shore areas (especially on coarse gravel along wave-swept beaches). By this time, they have shifted from a limnetic, day-active species to a benthic, night-active species. They hide during the day and remain strongly associated with the bottom for the rest of their lives.

Although young-of-the-year burbot are common in some streams, it is not known if they go through ontogenetic habitat shifts similar to those seen in lacustrine burbot. In high-gradient rivers, pelagic larvae would appear to be maladapted; however, they may concentrate behind deposition bars or be swept into backwaters. Once they shift to a benthic life, they probably use quiet water areas with coarse gravel cover similar to that found on lake beaches.

Conservation Comments Burbot are abundant in the northern half of British Columbia, and there are still some healthy populations in the southern half of the province. Nonetheless, there have been marked declines in burbot numbers in the Canadian portion of the Columbia River system. About 50 years ago, major burbot fisheries occurred in Kootenay Lake and the Columbia River. These fisheries are now gone, and burbot are a species of regional concern. Burbot are not an immediate conservation concern in most of the rest of the province, but the declining southern populations should be monitored. Although it is not known if the two distinct lineages of burbot found in British Columbia differ in their physiology, behaviour, or habitat requirements, the details of their geographic distribution and, perhaps, the details of the distribution of the three minor North American clades should be included in any province-wide burbot management plan. Neither COSEWIC nor the BCCDC list burbot.

FAMILY GASTEROSTEIDAE — STICKLEBACKS

The sticklebacks (Gasterosteidae) are small fish that together with the pipefishes, tubesnouts, seahorses, and snipefish make up the order Gasterosteiformes. Although most members of this order are marine, the sticklebacks (Gasterosteidae) occur in salt, brackish, and fresh waters. As their name implies, sticklebacks possess dorsal spines, but they also usually have an external bony pelvic girdle with modified, spinous pelvic fins. When erect, these spines are fixed in place by a simple friction lock (Hoogland 1951) and serve as a deterrent to predators (Hoogland et al. 1957). Also, the spines are used in male–male threat displays and in courtship. The earliest, dated fossil stickleback is from a Miocene (about 16 Ma) deposit in central California (Bell 1994). Other Miocene sticklebacks are known from Nevada, Sakhalin, and Kamchatka. Morphologically, these fossils are remarkably similar to the modern sticklebacks (Bell 1977, 1994). The males of all stickleback species build nests out of algae, the needles of conifers, and other bits of vegetation. The nests are glued together with a sticky mucous produced in the kidneys of the males.

Three species occur in British Columbia: the brook stickleback (*Culaea inconstans*), the threespine stickleback (*Gasterosteus aculeatus*), and the ninespine stickleback (*Pungitius pungitius*). Two of these, the brook and ninespine sticklebacks, are inland species (although the ninespine stickleback occurs in coastal areas in eastern Canada and the Arctic), and in British Columbia, they are relatively uniform in their morphology. In contrast, the threespine stickleback is a coastal species that occurs in a bewildering array of morphological, ecological, and behavioural forms. Often these different forms occur together or are in contact at ecotones (sites where two different environments abut). In spite of this close contact and some hybridization, the different forms often maintain themselves as separate genetic entities. In many B.C. localities, sympatric, or parapatric, populations behave like biological species (*sensu* Mayr 1963): they are reproductively isolated and use different trophic or spatial resources (McPhail 1994). Some of the forms appear to be on the verge of splitting into distinct species. Consequently, they provide unusual insights into the ecological and evolutionary processes involved in the formation of new species. In this regard, they are scientific treasures.

426 FAMILY GASTEROSTEIDAE—STICKLEBACKS

KEY TO THE STICKLEBACK SPECIES

1 (2) Gill membranes fused to isthmus (Fig. 38 A); dorsal spines usually 3 (sometimes 2 or 4)	THREESPINE STICKLEBACK, *Gasterosteus aculeatus*
2 (1) Gill membrane free from isthmus (Fig. 38 B); dorsal spines 4–11 *(see 3)*	
3 (4) Dorsal spines 4–6; pectoral rays branched	BROOK STICKLEBACK, *Culaea inconstans*
4 (3) Dorsal spines 7–11; pectoral rays unbranched	NINESPINE STICKLEBACK, *Pungitius pungitius*

Figure 38 Ventral views of threespine (A) and brook (B) sticklebacks heads illustrating gill membranes fused to the isthmus and gill membranes free from the isthmus

Culaea inconstans (KIRKLAND)
BROOK STICKLEBACK

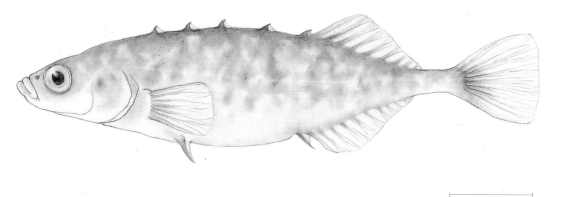

1 cm

Distinguishing Characters — A small fish with four to six short spines (often partially embedded) arranged in a straight row in front of the dorsal fin, and with the gill membranes free from the isthmus. The brook stickleback differs from other B.C. sticklebacks in dorsal spine number (typically six or fewer in brook stickleback; seven or more in ninespine stickleback, *Pungitius pungitius*; and three in threespine stickleback, *Gasterosteus aculeatus*). Additionally, it is the only B.C. stickleback with branched pectoral rays.

Taxonomy — There are no taxonomic problems associated with brook stickleback in British Columbia; however, there is evidence that this species survived glaciation in two central North American refugia (the Mississippi and Ohio river basins) and that fish from the two refugia have divergent mitochondrial DNA patterns (Gach 1996). Presumably, our populations originated from the Great Plains Refugium.

In Alberta, brook sticklebacks often have reduced pelvic spines, and sometimes, the entire pelvic girdle is missing (J.S. Nelson 1969, 1977; Reist 1981). Most B.C. populations also have reduced pelvic spines, and occasionally ($<0.05\%$), part of the girdle is missing; however, so far, no totally girdleless populations are known from within the province.

Sexual Dimorphism — Like most sticklebacks, male brook stickleback display distinctive nuptial colours—they are velvety black on the ventral surface, lower sides, and spine membranes. In contrast, in breeding females, the body has a golden hue, and the spines are uncoloured. Analyses of colour changes that occur throughout the breeding cycle in both sexes indicate that changes in the distribution and intensity of pigments on body surfaces and spines produce a complex mosaic of colour patterns (McLennan 1993a, 1993b, 1994).

These patterns correspond to different stages in both the male and female breeding cycles and provide signals on sexual motivation and aggression.

Distribution NATIVE DISTRIBUTION
The brook stickleback is endemic to North America. Here, it is almost exclusively an interior species with a geographic distribution that extends from Nova Scotia and New Brunswick to the Great Lakes and then westward across the northern Great Plains to Alberta, northeastern British Columbia, and the southwestern Northwest Territories. In the Mackenzie River Basin, it extends downstream to about Norman Wells. There is an isolated population in New Mexico that once was considered native (Koster 1957); however, a recent molecular analysis indicates that this population is probably an introduction (Gach 1996). The only known population west of the Continental Divide occurs in the Rock Creek drainage (a tributary to the Palouse River, Washington). They are a recent introduction, perhaps from an aquarium release (Wydoski and Whitney 2003).

BRITISH COLUMBIA DISTRIBUTION
The brook stickleback is restricted to the northeastern corner of the province. Here, it occurs in the lower Peace, lower Liard, Fort Nelson, Petitot, and upper Hay river systems (Map 66).

Life History Very little is known about the life history of brook sticklebacks in British Columbia. Some age and growth data are available in unpublished reports (e.g., McPhail et al. 1998a, 1998b), but most of the following life-history information is derived from populations in Manitoba (Moodie 1977, 1986).

REPRODUCTION
The details of reproductive behaviour in the brook stickleback are well known (McKenzie 1974; McLennan 1993a, 1993b; Reisman and Cade 1967; Winn 1960). The following account is a summary based on these sources supplemented with local (B.C.) field observations. Brook stickleback spawn in the spring, and in northeastern British Columbia, there is often a migration from overwintering sites in streams with deep pools into small streams and ponds. Although this migration starts shortly after ice-out, nest building and spawning usually begin in May and extend through June and into July. Spawning is triggered by a combination of increasing light and temperature. The threshold temperature for spawning is about 8 °C, and spawning typically ceases at temperatures above 19 °C.

Like other members of the family, male brook sticklebacks build nests. The nest is constructed in shallow (usually <2 m) weedy areas. Bits and pieces of vegetation and filamentous algae are glued together, often at a fork between a plant stem and a branch about 1–5 cm above the substrate. Unlike the nests of the other two sticklebacks, brook stickleback nests have no rear entrance. As nest building nears completion, the

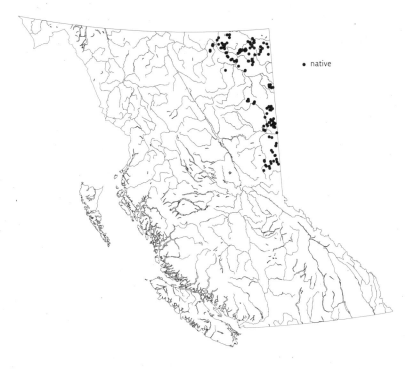

Map 66 The B.C. distribution of brook stickleback, *Culaea inconstans*

male becomes a deep velvety-black and very aggressive in his defense of the territory around the nest site. Females also change colour as they become gravid, and the variegations on their backs become more conspicuous. For both males and females, courtship involves an exchange of signals: behaviours and subtle colour changes (McLennan 1993a, 1993b) that serve to dampen aggression and synchronize the receptiveness of both sexes. Initially, a male responds aggressively to a female entering his territory. He approaches, then lunges and butts her repeatedly on the head and flanks. The female responds by sinking to the bottom and holding still. The male then slowly swims to the nest with his spines erect, his back arched, and caudal fin spread. If the female follows him to the nest, he shows the nest by fanning at the entrance. Eventually the female enters the nest, the male then prods her lower flanks and caudal peduncle, and the female vibrates and extrudes eggs into the nest. There are usually several of these bouts of vibration before the female leaves the nest (usually by forcing her way through the back of the nest). The male immediately responds aggressively and drives the female from his territory. He then enters the nest and fertilizes the eggs. The male guards the eggs and aerates them by fanning with his pectoral fins.

Egg number in brook stickleback ranges from about 50 to about 500; however, fecundity is difficult to assess. This is because brook sticklebacks have a protracted spawning season and a short (about 3 days) interclutch interval. Also, egg number varies with food availability and time within the spawning season. In a Manitoba population, it was estimated that females spawned an average of 28 times in a season and produced about 2,000 eggs per season (Moodie 1986). This estimate was based on an average fecundity of 214 eggs per clutch. Although this egg number is considerably higher than other fecundity estimates in brook stickleback (e.g., Becker 1983), it is the most reliable estimate available.

The eggs are small (1.0–1.4 mm in diameter) and yellow to pale green. They are adhesive and stick together in the nest. Incubation time varies with temperature and takes 7–9 days at 17–18 °C (McKenzie 1974; Winn 1960). At hatching, the male opens the nest. Newly hatched larvae are about 5–6 mm TL and become free-swimming 3 or 4 days after hatching.

AGE, GROWTH, AND MATURITY

The young-of-the-year grow relatively rapidly and, depending on when in the breeding season they hatched, reach 30–60 mm by late September (Becker 1983). Adult size varies among years and among populations. In British Columbia, adults average about 45 mm (35–60 mm TL). Sexual maturity is reached in their second summer (1+), and many adults die shortly after the end of their first spawning season (late summer); however, some individuals survive to a fourth summer. In northeastern British Columbia, September samples usually show three size modes (45, 65, and 75 mm), and this suggests at least three age classes (McPhail et al. 1998a, 1998b). Thus far, the largest brook stickleback recorded in British Columbia measured 81 mm TL.

FOOD HABITS

Although brook sticklebacks are primarily carnivorous (Tompkins and Gee 1983), some individuals also consume algae (Becker 1983). In one study, 24 different food categories were identified in the summer diet: 13 taxa of insects, 4 taxa of cladocerans, 6 other invertebrate taxa, and stickleback eggs (Moodie 1986). The proportion of egg cannibals in this population varied from about 9% to 47% (Salfert and Moodie 1985). Apparently, brook sticklebacks forage primarily in the afternoon, and fish in their first summer select smaller prey than adults (Tompkins and Gee 1983). An investigation of the diet of young-of-the-year brook sticklebacks penned in an enclosure with, and without, fathead minnows (*Pimephales promelas*), established that their diet becomes more diverse in the presence of fathead minnows (Abrahams 1996). When enclosed without fathead minnows, young-of-the-year brook stickleback fed mainly on copepods.

Habitat No formal habitat studies are available for brook sticklebacks in British Columbia. Consequently, this account relies on information from prairie and central North American populations supplemented, where appropriate, with casual field observations made on B.C. populations.

ADULTS

Adult brook sticklebacks are versatile and occur in a wide range of environments. In the muskeg regions of northeastern British Columbia, they are abundant in small, boggy headwater streams, and sometimes, they are the only fish species found in such habitats. Usually, however, they occur at such sites with the finescale dace (*Phoxinus neogaeus*). They are also found along the margins of small, boggy lakes (usually associated with vegetation, soft substrates, and water <1.0 m deep). Although common in heavily stained, humic waters, they also occur at clear-water sites and along the margins of large rivers. In the foothills of the northern Rocky Mountains, they even occur in moderately fast creeks (surface velocities of about 0.50 m/s) with cobble bottoms. Presumably, in these environments, brook stickleback reduces its buoyancy (Beaver and Gee 1988) and, behaviourally, they resemble young salmonines: holding position behind cobbles and darting out to feed on drift.

JUVENILES

Juveniles occur in the same array of habitats as adults; however, they are more strongly associated with vegetation, and occur closer to shore, than adults. Whereas adult males are territorial, the juveniles of both sexes aggregate. In streams with moderate gradients, they appear to avoid areas with currents >0.25 m/s.

YOUNG-OF-THE-YEAR

At the beginning of their first summer of life, newly hatched brook sticklebacks remain in vegetation, but they form schools as they grow and move out along the edges of weedy areas. Apparently, at this time, they form size-assorted schools (Reisman and Cade 1967).

Conservation Comments The brook stickleback has not been studied in British Columbia. Elsewhere, there are populations with a reduced number of dorsal spines (Edge and Coad 1983), populations in which most individuals lack pelvic girdles (Nelson and Atton 1971) or are polymorphic for girdle development (Andraso 1997), and populations that have diverged in their foraging biology in the presence of other stickleback species (Gray and Robinson 2002). Much of this variation is associated with the presence or absence of predators and competitors, and this suggests that the variation is a response to the local selection regime. Thus, in British Columbia, the brook stickleback has the potential to be a useful animal for studying the interaction between ecological factors and microevolution. For this reason, some populations may be of scientific interest. Neither COSEWIC nor the BCCDC list this species.

Gasterosteus aculeatus LINNAEUS
THREESPINE STICKLEBACK

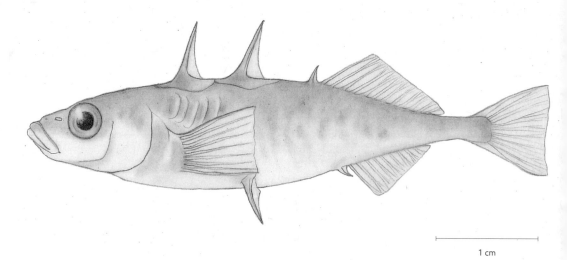

1 cm

Distinguishing Characters As its name implies, the threespine stickleback usually has three dorsal spines (the first two spines are long relative to the third spine). This species also has a pair of well-developed pelvic spines, bony lateral plates, and the gill membranes are attached to the isthmus. In British Columbia, this stickleback is remarkably variable, and some populations have reduced spines (0–3), no (or vestigial) pelvic girdles, and no lateral plates. Two other sticklebacks (the brook stickleback, *Culaea inconstans*, and the ninespine stickleback, *Pungitius pungitius*) occur in the province but do not coexist with the threespine stickleback. Usually, threespine sticklebacks have a more robust body than either brook or ninespine sticklebacks, but in some lacustrine populations, the body is long and slim. Spine number and the attachment of the gill membranes to the isthmus will distinguish this species from the other two sticklebacks that occur in the province.

Taxonomy The various forms of the threespine stickleback are classic examples of parallel evolution—under similar ecological conditions, similar phenotypes have repeatedly evolved at different localities (even on different continents). Parallel evolution also has produced sympatric and parapatric pairs of these sticklebacks (Lavin and McPhail 1986; McPhail 1994; Taylor and McPhail 1999; Thompson et al. 1998). Typically, these pairs behave as biological species (*sensu* Mayr 1963): they are reproductively isolated (albeit incompletely) and use different spatial and trophic resources. Similar species pairs also occur in other northern freshwater fishes, and they create an awkward taxonomic problem (see the discussion of parallel evolution in the introduction). For threespine sticklebacks, the imperfect, but practical, solution to this problem is to treat them as a single variable species, *Gasterosteus aculeatus*. However, the use of a single scientific name for this complex of sympatric and parapatric forms in no way lessens the scientific and biodiversity value of these remarkable examples of rapid divergence.

Sexual Dimorphism Reproductive males develop nuptial colours (typically red throats and iridescent blue eyes) that are absent or, in the case of the throat colour, weakly developed in females. In some B.C. populations, the male throat colour is black or has a bluish tinge. Many body measurements (e.g., snout length, mouth size, pelvic spine length, and gill raker length) are statistically different in males and females. The direction of these difference, however, is not consistent. For example, in some populations, the pelvic spines are longer in males than in females, whereas females in other populations have longer spines than males.

Distribution NATIVE DISTRIBUTION

Threespine sticklebacks are associated with cool seas and their adjacent coastal lowlands throughout the Northern Hemisphere. Generally, anadromous and marine populations range farther to the north than freshwater populations, and freshwater populations range farther to the south. In Europe, threespine sticklebacks extend from Novaya Zemlya in Russia, throughout Scandinavia and south to the Mediterranean region (the Iberian Peninsula, Italy, and Greece) and east to the Black Sea. Relict freshwater populations once occurred in Algeria and Syria. Both freshwater and anadromous populations are found in Iceland and the southern tip of Greenland. On the Atlantic coast of North America, threespine sticklebacks ranges from Chesapeake Bay along the coast to southern Baffin Island and Hudson and James bays. Along the Pacific coast of North America, they occur from northern Baja California (a freshwater population) to the Seward Peninsula and St. Lawrence Island, Alaska. On the Pacific Coast of the Asia, *Gasterosteus* is found in Japan (as far south as Kyushu), Manchuria, northern Korea, Kamchatka, and the Chukotsk Peninsula.

BRITISH COLUMBIA DISTRIBUTION

Threespine sticklebacks are a coastal species and, even in large rivers like the Fraser and Skeena, rarely penetrate more than 200 km inland (Map 67). There are totally marine populations, anadromous populations, and freshwater-resident populations. The freshwater-resident populations occur in most lowland waters with access to the sea and are common on islands.

Map 67 The B.C. distribution of threespine stickleback, *Gasterosteus aculeatus*

• native

Life History   For most of our small indigenous fishes, life-history information is in short supply but not for threespine stickleback. Rather, there is so much information, and such a diversity of life histories, that compressing the information into a short summary is not feasible. Thus, the following synopsis is a short, somewhat idealized version, of this species' life history. More detailed accounts are available in Baker (1994) and Wootton (1976, 1984).

REPRODUCTION

Threespine sticklebacks begin to spawn as days lengthen and water temperatures increase in the spring. The spawning period can be protracted but usually starts when water temperatures rise above 10 °C and ceases when water temperatures exceed 22 °C. Depending on the weather in British Columbia, reproduction can start in mid-March and, in large lakes, extend through the summer. Occasionally, there is a minor spawning peak in late August or early September. Usually, freshwater-resident populations start spawning earlier, and spawn over a longer period, than anadromous populations. Run timing in anadromous populations differs among even adjacent rivers. For example, in tributaries to the lower Fraser River, some runs appear in April, some in May, most in June, and a few in July. Differences in run timing and, in a few cases, marking studies all suggest that migratory threespine sticklebacks home to their natal streams. Most anadromous populations spawn a short distance inland (<5 km) above tidal influence, but in the Fraser River, anadromous threespine sticklebacks migrate over 180 km upstream.

Like other members of the family, breeding male threespine sticklebacks are territorial and build nests. The nests are built on the substrate and consist of a shallow pit covered with algae and plant material that is glued together with a sticky substance produced by the male's kidney. Freshwater stream residents prefer soft, fine substrates close to, or within, vegetation (Hagen 1967). In contrast, anadromous sticklebacks prefer to nest in the open on sand substrates. Nest sites vary among lacustrine populations: some nest on mud; some, on sand; and some, on rock outcrops and submerged logs. Their association with cover also varies: some populations nest in the open and others nest in dense cover. Most nests are located in shallow water (<1 m), but nests have been observed as deep as 20 m (Kynard 1978). In fish that nest in the open, the male often covers the nest with sand or silt. Typically, the nest has an anterior and posterior opening and is complete when the male "creeps through" the nest. This "creeping through" behaviour signals the end of the nest-building phase of reproduction and the start of the courtship phase.

As nest building nears completion, males develop their nuptial colours and become very aggressive in their defense of the territory around the nest site. As in other aspects of their reproductive biology, male nuptial colour varies among populations. Typically, males develop a red or red-orange throat and stomach and have iridescent blue eyes. In some populations, however, the males turn jet-black; in other populations, they develop dark undersides with bluish overtones. Within populations, individual males vary in the intensity of their colour: the colours are vivid in some individuals and subdued in others. Interpopulation differences in male colour are thought to be associated with differential spectral absorbance in waters of different clarity or humic content (Reimchen 1989). In contrast, individual variation is associated with male behaviour and female mate preference (Bakker 1993; McLennan and McPhail 1989, 1990).

When a gravid female enters a male's territory, he approaches using a series of zigzag movements (the zigzag dance). If the female is receptive, she raises her head and exposes her distended belly. In this position, she approaches and positions herself slightly above the male. The male then leads the female to the nest and "shows" the female the nest by poking his snout into the nest opening, turning on his side, and making jerky swimming movements. If the female enters the nest, the male rapidly quivers against her lower flanks and caudal peduncle. This stimulates the female to release her eggs. She then leaves the nest, and the male immediately drives her from his territory. The male then enters the nest and fertilizes the eggs.

Courtship is not as stereotypic as this account implies. There are interpopulation and individual differences in the behaviour of both males and females. In some populations, the males' first approach is not a zigzag dance but a head butt or a bite. Differences in the male's first approach, and the female's response to the first approach, appear to be important isolating mechanisms in sympatric pairs of threespine sticklebacks (Ridgway and McPhail 1984). Curiously, within populations in the wild, courtship is often (>50%) broken off in its final stages (nest entry). Typically, males spawn with several females before they switch over to parental behaviour. Parental behaviour involves guarding the eggs and aerating them by fanning with the pectoral fins. This fanning behaviour increases as the eggs develop and rapidly declines after hatching. Freshly fertilized threespine stickleback eggs average about 1.5–2.1 mm in diameter, and depending on female body size and population, batch fecundity can vary from less than 20 to more than 200 eggs. Generally, anadromous and fully marine females produce more, but smaller eggs, than stream-resident females. In lacustrine populations, mean egg diameter and mean batch fecundity cover the entire range found in other life-history types. The eggs are adhesive, and females usually spawn several times in a breeding season (Wootton 1976). In some populations, females produce a new batch of ripe eggs in 3–5 days. In the laboratory, a plankton-feeding female produced 29 consecutive clutches over a period of 3 months. This capacity to produce a large number of consecutive egg batches appears to be associated with populations in which most females are short lived (1 year).

Depending on incubation temperature, the eggs hatch in 4 (20 °C) to 7 (15 °C) days. The size of newly hatched young ranges from about 5 to 7 mm TL and is dependent on egg size. In some lacustrine populations, the male opens the nest immediately after the eggs hatch. The fry become neutrally buoyant 3–5 days after hatching. They stay as a group in the male's territory for variable lengths of time. The duration of their association with the male may be a function of the risk of predation on the young balanced against and the probability of the male successfully producing another batch of young. Late in the season (late August or September), in Paxton Lake on Texada Island, some limnetic males occur in schools of 20–50 fry about 20 mm in length. No data are available on the genetic relatedness of the male and fry in these schools, but late in the season, some males may remain with their offspring rather than attempting a new nest.

AGE, GROWTH, AND MATURITY

The lifespan of threespine sticklebacks is as variable as other aspects of their life history. Again, there appear to be differences among the four major lifestyles (marine, anadromous, stream-resident, and lacustrine) with lacustrine populations encompassing all of the variation found within the other groups. About 37% of the 49 populations for which there was lifespan data lived 1 year (Baker 1994). Most (65%) of the other populations lived for 2–5 years. The longest lifespan recorded for threespine stickleback is 8 years (Reimchen 1992).

Age at maturity also varies. In most populations, sexual maturity is reached in 1 year, but maturity can be delayed (especially in females) for 2, or even 3, years (Baker 1994).

FOOD HABITS

Threespine sticklebacks are carnivorous and feed on both water column and benthic prey. They are visual predators, and the habitat in which they live strongly influences their foraging behaviour (Hart and Gill 1994). Thus, marine, anadromous, and large-lake populations are primarily planktivores, whereas stream-resident and small-lake populations are primarily benthivores. Some B.C. lakes contain two foraging forms (McPhail 1994): one adapted to forage on limnetic prey (copepods and cladocerans) and the other adapted to forage on bottom organisms (amphipods, ostracods, and aquatic insect larvae). In Great Central Lake on Vancouver Island, one study (Manzer 1976) found that the diet of sticklebacks varied with fish size, sex, and reproductive condition. During the breeding season, egg cannibalism is common in most B.C. populations (Hyatt and Ringler 1989; Manzer 1976).

Habitat Threespine sticklebacks inhabit diverse aquatic environments—the sea, estuaries, ditches, streams, rivers, and standing water (bogs, swamps, ponds, and lakes). Resident populations occur in each of these environments. Typically, these resident populations display morphological and behavioural adaptations specific to their environment (Lavin and McPhail 1985); however, other populations (e.g., anadromous populations) migrate between environments. Consequently, it is difficult to generalization about habitat use in this species.

ADULTS

Except when breeding, adults are usually less strongly associated with shallow water and cover than the juveniles and fry. For example, pelagic adult sticklebacks have been collected hundreds of kilometres from shore in both the North Atlantic and North Pacific oceans (Jones and John 1978; Quinn and Light 1989). Similarly, but on a smaller scale, adults in many large B.C. lakes move offshore into open water in the late summer or fall. Even in small lakes and streams, adults often forage out beyond the edges of littoral vegetation. In the Strait of Georgia region, however, there are some small lakes that contain pairs of biological species of threespine stickleback. In these lakes, adults of one form are pelagic and forage offshore near the surface, while the other form is benthic and forages in the littoral zone.

JUVENILES

Since the lifespan in many threespine stickleback populations is only 1 year, the distinction between juveniles and young-of-the-year can be arbitrary. In these populations, loose schools of fry are associated with vegetation or some other form of cover. As they grow, they become less strongly associated with cover and, except for limnetic populations, less likely to form schools. In populations with delayed maturity, juveniles usually remain in or near cover for at least a year. In the Straight of Georgia, anadromous and marine juveniles remain inshore, and large schools of similar-sized individuals are found around docks and floats.

YOUNG-OF-THE-YEAR
Anadromous stickleback fry remain in fresh water until late July. They stay close to stream margins but usually just outside the weeds. In the Lower Mainland in late July, large schools of these anadromous fry congregate near floodgates behind dikes. Presumably, on appropriate tides, they move out through the gates and down to sea. Fully marine populations that spawn in the semi-enclosed bays and lagoons along the Sechelt Peninsula remain in these sheltered environments until late summer and then move out into the Strait of Georgia. Some individuals, however, overwinter in sheltered water and go to sea the next spring. In stream and non-pelagic lake populations, fry remain inshore in shallow water, usually close to, or in, cover. At about 25–30 mm, they move out of cover and forage in open water within the littoral zone or stream margins.

Conservation Comments   Although threespine sticklebacks are abundant in marine, brackish, and fresh waters all along the B.C. coast, all stickleback populations are not equal from a conservation perspective. Some populations are of considerable scientific interest; others are not. Some of the morphological forms of threespine sticklebacks occur around the entire Northern Hemisphere; other forms occur only in British Columbia. For example, the limnetic–benthic species pairs are known only from six lakes on three islands in the Strait of Georgia region. For evolutionary studies, these stickleback pairs are of immense scientific interest. Arguably, they are as important to evolutionary biology as the Galapagos finches or the cichlid fishes of the African Rift lakes; because these species pairs are restricted to a few small lakes, they are extremely vulnerable to human intervention. Five of the six known benthic–limnetic species pairs are listed as threatened by COSEWIC and as critically imperiled (S1) by the BCCDC. Tragically, the Hadley Lake pair on Lasqueti Island is now extirpated—the victim of an illegal introduction of the brown bullhead (*Amieurus nebulosus*) into this lake. Ironically, at the time, the loss of this unique component of the Canadian vertebrate fauna evoked no outrage, or even comment, from any conservation group (public or private). Another of the limnetic–benthic pairs (the Enos Lake pair on Vancouver Island) appears to be collapsing into a single variable population (Kraak et al. 2001). It is not clear why this is happening, but in the last decade, there have been major ecological changes in the lake. The Texada Island and Enos Lake (Vancouver Island) benthic and limnetic pairs are listed as endangered by COSEWIC and as S1 (critically imperiled) by the BCCDC. Recovery plans—required under the Species at Risk Act—are being prepared. Hopefully, the four remaining intact pairs (all on Texada

Island) will survive. Two of the many unusual populations on the Queen Charlotte Islands (the giant black stickleback and the Queen Charlotte unarmoured stickleback) also are listed by COSEWIC as species of special concern and given an S2 (imperiled) listing by the BCCDC.

Although we know a lot about threespine sticklebacks in British Columbia, our knowledge is concentrated in two regions: the Strait of Georgia and the Queen Charlotte Islands. Lakes on the islands and mainland along the intervening coast are poorly known, and there are probably other scientifically interesting populations yet to be discovered.

Pungitius pungitius (LINNAEUS)
NINESPINE STICKLEBACK

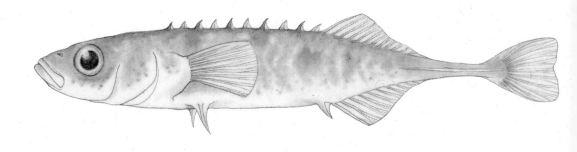

1 cm

Distinguishing Characters This small fish has 7–11 sharp dorsal spines. When viewed from above, these spines are inclined alternately to the left and right. Superficially, the ninespine stickleback resembles the brook stickleback (*Culaea inconstans*); however, spine number (typically seven or more in ninespine stickleback and six or fewer in brook stickleback), the shape of the caudal peduncle (width greater than depth in ninespine stickleback and depth greater than width in brook stickleback), and the distinctive bony caudal keel in ninespine stickleback separate the two species.

Taxonomy The taxonomy of ninespine sticklebacks is contentious. Some authors recognize up to eight species within the genus, but others (e.g., Keivany and Nelson 2000) recognize only three morphological species and five subspecies. This latter arrangement simplifies the taxonomy but obscures the underlying biological complexity in the genus. There is ample evidence (Tsuruta et al. 2002; Ziuganov and Zotin 1995) in ninespine stickleback for the existence of reproductively isolated, sympatric or parapatric, biological species (*sensu* Mayr 1963). In North America, there are at least two forms of ninespine stickleback—an interior and a coastal form that apparently survived glaciation in different glacial refugia (McPhail 1963).

Sexual Dimorphism Like other sticklebacks, the sexes in ninespine stickleback differ in colour during the breeding season; however, the nuptial colours vary among populations and, depending on the sexual state of individuals, within populations. In the coastal (Beringian) form, the underside of breeding males develops a velvety black pigment, and the membranes associated with the pelvic spines become silvery white. In contrast, breeding males of the interior (Mississippian) form develop a grayish ventral patch that becomes jet black during courtship and nest guarding (McKenzie and Keenleyside 1970). In females, colour changes associated with spawning are subtler and probably more complex than those found in males (see McLennan 1994 for an analysis of colour changes across the ovulatory cycle of female brook sticklebacks). Presumably, something similar occurs in female ninespine sticklebacks.

Hybridization   Although natural hybrids involving ninespine sticklebacks are unknown, laboratory crosses between threespine (*Gasterosteus aculeatus*) and ninespine sticklebacks are successful and produce viable offspring (Leiner 1957).

Distribution   NATIVE DISTRIBUTION
The geographic distribution of the ninespine stickleback includes western Europe, northern Siberia, coastal Asia as far south as northern China, and northern North America. In North America, the ninespine stickleback ranges from the Anchorage area in Alaska north and east along the Bering Sea and Arctic Coast to Hudson Bay and Labrador, and then south along the Atlantic Coast to the New England states. In the interior of North America, they are distributed from the Great Lakes northwestward across the northern Great Plains to central Alberta, northeastern British Columbia, and down the Mackenzie River to the Beaufort Sea. In Alaska and along the Arctic coasts of Canada and Russia, ninespine sticklebacks often occurs in inshore marine waters.

BRITISH COLUMBIA DISTRIBUTION
Ninespine sticklebacks are rare in British Columbia (Map 68)—only four specimens have ever been collected (three from the upper Petitot River and one from a site a few kilometres downstream from Fort Nelson). The Petitot River drains Bistcho Lake in Alberta, and the ninespine stickleback is listed from this lake (Nelson and Paetz 1992). So far, there is no evidence of a self-sustaining population in British Columbia, and the Petitot specimens probably are strays from the Alberta portion of the river. The specimen from near Fort Nelson is harder to explain. Again, there is no evidence of a self-sustaining population in the area, and there is no known nearby source—the Fort Nelson site is separated from the Petitot site by over 500km of swift-flowing river.

Life History   Since it is doubtful that a breeding population of ninespine stickleback exists in British Columbia, it is not surprising that we know little about its life history within the province. Consequently, sources from elsewhere in North America and Europe form the basis for the following account.

REPRODUCTION
A detailed account of the breeding behaviour of a British population of ninespine stickleback is given in Morris (1958), and the reproductive behaviour of two separate populations in the upper Great Lakes is described in McKenzie and Keenleyside (1970) and Griswold and Smith (1973). Ninespine sticklebacks usually begin to spawn in the late spring

442 FAMILY GASTEROSTEIDAE — STICKLEBACKS

Map 68 The B.C. distribution of ninespine stickleback, *Pungitius pungitius*

• native

(late May) and early summer (June and July). In Lake Superior, peak spawning occurred at temperatures between 11 °C and 12 °C (Griswold and Smith 1973). Occasionally, ninespine sticklebacks migrate between overwintering sites and spawning and nursery areas. Such a migration has been described between two Alaskan lakes that differ in depth (Harvey et al. 1997). In the spring, there was an upstream adult migration to the shallower lake followed by a mainly juvenile downstream migration to the deeper lake in the fall. These migrations were mostly nocturnal.

Like other members of the family, male ninespine sticklebacks hold territories and build nests; however, even within the Great Lakes, the substrate and depths at which nests are built vary among sites. For example, nests in Lake Superior were only found on organic–mud substrates at depths of 16–40 m (Griswold and Smith 1973). In contrast, nests in a Lake Huron population were on rocky substrates at depths of 0.25–0.8 m (McKenzie and Keenleyside 1970). In both cases, the nests were built on the substrate, whereas in Britain, nests are usually constructed on vegetation, typically at a fork between a plant stem and a branch about 10–15 cm above the substrate (Morris 1958; Wootton 1976). Like other sticklebacks, the nest is made of filamentous algae and bits of vegetation stuck together with glue produced in the male's kidney. The shape of the nests varies: some are tubular, others are ball shaped; some have an entrance and exit, others have only an entrance.

As nest building nears completion, males develop their nuptial colours and become very aggressive in their defense of the territory around the nest site. As in other aspects of their reproductive biology, male nuptial colour varies among localities: in Europe, they become entirely black and the fin membranes attached to the pelvic spines become white; in the Great Lakes, they develop a midventral grey patch and bluish spine membranes; and in coastal Alaska, they are black with white pelvic spine membranes. Some of these differences may be associated with origins from different glacial refugia (McKenzie and Keenleyside 1970).

When a gravid female enters a male's territory, he approaches her using a series of zigzag movements (the zigzag dance). If the female is receptive, she raises her head and exposes her distended belly. In this position she approaches the male and positions herself slightly below the male with her snout close to his erect pelvic spines. The male then dances towards his nest and the female follows. The male "shows" the nest by placing his snout above the nest entrance and fanning with his pectoral fins. The female enters the nest, and the male rapidly quivers against her lower flanks and caudal peduncle. This stimulates the female to release her eggs. She then leaves the nest, and the male immediately drives her from his territory. The male then enters the nest and fertilizes the eggs.

Typically, males spawn with several females before they switch entirely to parental behaviour. Parental behaviour involves guarding the eggs and aerating them by fanning with the pectoral fins. This fanning behaviour increases as the eggs develop and rapidly declines after hatching. The eggs are about 1.5 mm in diameter and fecundity varies from 20 to 120 eggs. Females usually spawn several times during the breeding season (Sokolowska and Skora 2002). Depending on temperature, the eggs hatch in 4 (19 °C) to 7 (15 °C) days. The newly hatched young are about 5–7 mm in length. Males in a British population transferred the newly hatched young to a "nursery" (Morris 1958); however, there was no evidence for a nursery in a Lake Huron population (McKenzie and Keenleyside 1970).

AGE, GROWTH, AND MATURITY

Ninespine stickleback fry fill their swim bladders about 4 days after hatching and leave the nest area shortly thereafter. They grow rapidly—a Lake Superior population reached about 50 mm TL by the end of their first growing season (Griswold and Smith 1973). In this population, males reached sexual maturity in their second summer of life (1+) and seldom lived longer than 3 years (2+). About 40% of the females matured in their second summer, and the rest, in their third summer. Females, however, survived 2 or 3 years longer than males. Fewer age classes are present in more northerly populations (Cameron et al. 1973). So far, only juvenile (49–50 mm TL) ninespine stickleback have been collected in British Columbia.

FOOD HABITS

The diet of ninespine sticklebacks is variable and changes with body size, season, locality, and the presence of other species; however, they are primarily carnivores (Coad and Power 1973; Griswold and Smith 1973; Ranta and Nuutinen 1984; Stewart and Watkinson 2004). Their diet includes both benthic prey (amphipods, ostracods, isopods, and dipteran larvae) and limnetic prey (copepods and cladocerans). A population sympatric with threespine stickleback in Matamek Lake, Quebec, foraged primarily on cladocerans and chironomids in littoral regions, whereas threespine stickleback foraged in open water (Coad and Power 1973). The diel activity patterns and diet of ninespine stickleback were examined in tidal salt-marsh pools where it coexists with two species of *Gasterosteus* (Worgan and FitzGerald 1981). The diets of all three species were similar, although ninespine stickleback had the most diverse diet. All the species were diurnal foragers. During the breeding season, males showed no diel pattern to their foraging activity and did not spend much time foraging, whereas females are voracious but feed almost exclusively in the early morning. An experimental study (Ranta and Nuutinen 1984) using a size range (0.2–4.0 mm) of plankton found that ninespine stickleback foraged preferentially on the smaller prey (0.2–2.0 mm).

Habitat Little is known about the habitat use of ninespine sticklebacks in British Columbia. Consequently, most of the following account is derived from populations in Alberta, central Canada, and Europe.

ADULTS

At first glance, information on habitat use in ninespine stickleback is confusing. The species is variously described as associated with weedy habitats in sluggish streams, open water in large rivers and lakes, deep cold lakes and their tributaries, the shallow littoral zone of lakes, and depths down to 90 m (Coad and Power 1973; Griswold and Smith 1973; Nelson and Paetz 1992; Scott and Crossman 1973; Stewart and Watkinson 2004). This wide array of habitats used by ninespine sticklebacks suggests that, like the threespine stickleback, they respond adaptively to local selection regimes. Presumably, the presence or absence of predators and competitors drives some habitat shifts and seasonal environmental changes drive others. In the spring at Apostle Island in Lake Superior, ninespine sticklebacks were fairly evenly distributed over most depths; however, in June, they moved into warm shallow water to breed, returned to a fairly even depth distribution in September, and by December most were found in deep water (Griswold and Smith 1973).

In British Columbia, the few ninespine sticklebacks that have been recorded were all associated with the margins of large, moderately swift (0.5–1.0 m/s) rivers. However, they were found at sites that were not directly exposed to current (usually shallow embayments at the mouths of tributary streams).

JUVENILES

Since ninespine sticklebacks typically reach sexual maturity at 1+, there are very few juveniles (immature 1+ fish) in most populations. Nonetheless, some small individuals (20–50 mm) are present in the early spring. Typically, these are the progeny of last year's late-season matings. They stay closer to cover than adults but grow quickly and reach maturity early in the breeding season.

YOUNG-OF-THE-YEAR

Once they become free swimming, ninespine stickleback fry (10–20 mm) are strongly associated with weedy areas and shallow water. As they grow, they become less dependent on vegetation but stay in littoral areas and form schools of similar-sized individuals (Ranta et al. 1992). In the fall, they migrate to overwintering sites (Griswold and Smith 1972; Harvey et al. 1997).

Conservation Comments Although the ninespine stickleback is native to the northeastern corner of British Columbia, it is exceedingly rare in our province. Normally, this would be a conservation concern; however, it is likely that ninespine sticklebacks are only occasional visitors to the province. This apparent failure to establish populations in the northeastern corner of British Columbia may reflect a lack of suitable habitat. The species is moderately abundant in adjacent Alberta, where one study (Nelson and Paetz 1992) lists its habitat as deep, cold lakes and their tributaries. There are no such lakes on the Fort Nelson Lowlands in British Columbia.

Although the presence of ninespine sticklebacks in British Columbia may depend on strays from Alberta, the northeastern corner of the province has not been well collected, and it is possible that a self-sustaining population of ninespine sticklebacks exists somewhere in the area. Since this region is an area of active oil and gas exploration, biologists doing environmental assessments should pay special attention to any sites where they may encounter ninespine sticklebacks. The BCCDC lists this species as S1 (critically imperiled because of extreme rarity).

FAMILY COTTIDAE — SCULPINS

The sculpins are a family in the order Scorpaeniformes. Most sculpins—usually called bullheads in British Columbia—are cool-water marine fishes; however, the family also contains some estuarine and freshwater species. Most sculpins occur in northern oceans, especially the North Pacific and its associated seas, but there are also some sculpins in the cool seas of the Southern Hemisphere. North American freshwater sculpins belong to the genus *Cottus*. Typically, *Cottus* species have large heads, preopercular spines, two partially or completely separated dorsal fins (the first fin has weak spines), and bodies that are either naked or with patches of small prickles. Their pectoral fins usually are large, and the pectoral rays are unbranched and thickened. Characteristically, their bodies taper from the head down to the base of the caudal fin. They lack a swim bladder and are bottom dwellers, although a few lacustrine species make diel vertical migrations. The males usually excavate nest cavities under rocks, and females deposit their eggs on the roof of these nests. The males guard and tend the developing eggs.

The oldest fossil freshwater sculpins known from North America occur in deposits associated with ancient (late Miocene) lakebeds in Idaho (Kimmel 1975). These deposits contain an extinct species (*Cottus calcatus*). Later (mid-Pliocene) deposits in the same area also contain sculpins (Smith 1975): extinct members of an extant northern euryhaline genus, *Myoxocephalus* and an extinct, endemic genus (*Kerocottus*). In Canada, a late Pleistocene fossil *Cottus* (probably the slimy sculpin, *C. cognatus*) is known from the Yukon Territory (Cumbaa et al. 1981). So far, no fossil sculpins are known from British Columbia.

Eight species of sculpins occur in the fresh waters of British Columbia. One species (the prickly sculpin, *Cottus asper*) has both freshwater-resident and catadromous populations. Catadromy is the opposite of anadromy. It involves a migration of reproductive adults from fresh water into the sea or, in this case, estuaries followed by a return migration of young and adults into fresh water. In some populations, the young of another B.C. species (coastrange sculpin, *Cottus aleuticus*) regularly migrate to, and grow in, estuaries but do not spawn in brackish water.

Most B.C. freshwater sculpins are small (<200 mm TL) and occur in lakes and in flowing water. With one exception, a neotenic sculpin in Cultus Lake that may migrate vertically, the adults of all B.C. species are bottom dwellers. Most of our species have an intra-gravel larval stage and by the time they emerge from the gravel, their yolk sac is absorbed, and they are miniature copies of the adults. In coastrange and prickly sculpins, there is a planktonic larval stage that drifts downstream into quiet waters (lakes, estuaries, backwaters, and side-channels) and eventually (after about a month) metamorphoses into a typical small sculpin. There is also some evidence of pelagic larvae in at least one lacustrine population of the torrent sculpin (*Cottus rhotheus*).

Despite the efforts of generations of ichthyologists, the taxonomy of northwestern *Cottus* remains tricky. Some species are relatively easy to identify, but others are genuinely difficult (especially the species in the mottled sculpin complex), and a professional ichthyologist should check questionable identifications.

All B.C. species share the same basic body plan, and at first glance, the differences between species seem subtle. Unfortunately, the morphology, behaviour, and ecology of many species are influenced by both biotic and abiotic factors. Hence, traits that reliably separate species at one site may be unreliable at another site. An example of this plasticity is the torrent sculpin. As the name implies, adults of this species are often torrent dwellers, and morphologically, their backs and sides are covered in a dense layer of fine prickles. In British Columbia, however, at sites where the torrent sculpin is the only sculpin present, adults occur in pools and other quiet water areas and appear to avoid fast water. In fact, several B.C. populations live in lakes. In addition, the dense prickling (used as a diagnostic character in most keys) is reduced or entirely lost in populations isolated above barriers. Similar complex inter-population variation in habitat and morphology is common in other northwestern sculpins (Finger 1982; Maughan and Saul 1979).

The identification of sculpins is further confounded by widespread interspecific hybridization (Godkin et al. 1982; Strauss 1986; Zimmerman and Wooten 1981). In British Columbia, hybridization appears to be especially common in the Kootenay region.

Key to the Sculpin Species None of the diagnostic traits used in the following key are infallible. Consequently, identifications should be based on a series of specimens rather than on a single individual. Additional information such as geographic locality often simplifies identification. For example, the shorthead (*Cottus confusus*) and Columbia sculpins (*Cottus hubbsi*) are known only from the Columbia River and its tributaries; the spoonhead sculpin (*Cottus ricei*) is known only from the lower Liard and lower Peace drainage systems; and the Rocky Mountain sculpin (*Cottus* sp.) is known only from the lower Flathead River. Although our knowledge of sculpin distribution is still imperfect and range extensions are possible, so far, all B.C. records of these species from outside their known ranges have proven to be errors. Consequently, if you think you have an out-of-range record, have someone familiar with sculpin taxonomy check the identification.

In addition to the eight species of freshwater sculpins in British Columbia, there are three marine species that commonly enter brackish and, in some places, fresh water. For example, the Pacific staghorn sculpin (*Leptocottus armatus*), is common in the upper parts of estuaries along the entire coast. The other two species (the sharpnose sculpin, *Clinocottus acuticeps*, and the tidepool sculpin, *Oligocottus maculosus*) are less common visitors to fresh water; however, on islands or in other coastal areas that lack a freshwater fauna, they are found occasionally in the lower

100 m or so of small streams. None of these marine sculpins breed in fresh water. Consequently, they are included in the key but I have not given them species accounts.

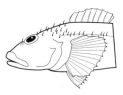

Figure 39 Diagram of the head of a sharpnose sculpin illustrating the cirri on the head and lateral line

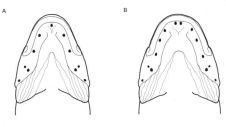

Figure 40 Single and double median chin pores: (A) prickly sculpin, and (B) shorthead sculpin

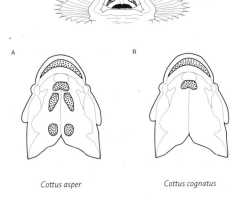

Figure 41 Diagram illustrating the presence (A) or absence (B) of palatine teeth

Cottus asper   *Cottus cognatus*

450　FAMILY COTTIDAE—SCULPINS

Figure 42 Body profiles of prickly (A) and slimy (B) sculpins showing conjoined and separate dorsal fins

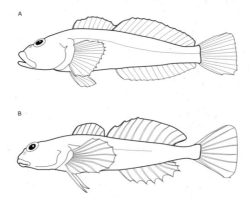

Figure 43 Diagram showing the length of the anal fin base relative to the distance from the origin of the dorsal fin to the anterior margin of the eye: (A) torrent and (B) Rocky Mountain sculpins.

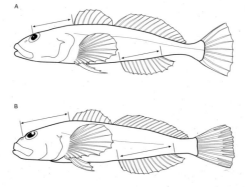

Figure 44 Single and double postmaxillary pores: (A) eastslope sculpin and (B) Rocky Mountain sculpin.

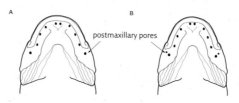

FAMILY COTTIDAE — SCULPINS

KEY TO THE SCULPIN SPECIES

1 (4) Cirri on the head (Fig. 39) and anterior part of lateral line *(see 2)*

    2 (3) Spine on the preoperculum with 2 or 3 hooks; vent directly in front of anal fin — TIDEPOOL SCULPIN, *Oligocottus maculosus*

    3 (2) Spine on the preoperculum simple; a space between the vent and the anal fin — SHARPNOSE SCULPIN, *Clinocottus acuticeps*

4 (1) No cirri on head or lateral line *(see 5)*

    5 (6) Largest spine on the preoperculum with 3 or 4 sharp hooks — PACIFIC STAGHORN SCULPIN, *Leptocottus armatus*

    6 (5) Largest spine on the preoperculum simple, no secondary hooks *(see 7)*

        7 (12) Median chin pore single (Fig. 40A) *(see 8)*

            8 (9) Palatine teeth present (Fig. 41A); anal fin base longer than head length; spinous and soft dorsal fins broadly conjoined (Fig. 42A) — PRICKLY SCULPIN, *Cottus asper*

            9 (8) Palatine teeth absent (Fig. 41B); anal fin base shorter than head length; spinous and soft dorsal fins separate, or just touching at the base of the spines (Fig. 42B) *(see 10)*

                10 (11) Caudal peduncle relatively deep, its depth greater than the horizontal eye diameter; lateral line with an abrupt downward deflection on the caudal peduncle; chin pores without raised edges — COASTRANGE SCULPIN, *Cottus aleuticus*

                11 (10) Caudal peduncle narrow, its depth about equal to horizontal eye diameter; lateral line straight, no abrupt downward deflection on the caudal peduncle; chin pores with distinctive puckered edges — SPOONHEAD SCULPIN, *Cottus ricei*

        12 (7) Median chin pore double (Fig. 40B) *(see 13)*

            13 (14) Palatine teeth absent (Fig. 41B) — SLIMY SCULPIN, *Cottus cognatus*

            14 (13) Palatine teeth present (sometimes weakly developed in Rocky Mountain and shorthead sculpins) *(see 15)*

                15 (16) When stepped off with dividers, the length of anal fin base extends from the origin of the first dorsal fin to about the middle of the eyes (Fig. 43A); caudal peduncle narrow, about 5–7 times into head length; back and sides usually heavily prickled — TORRENT SCULPIN, *Cottus rhotheus*

                16 (15) When stepped off with dividers, the length of anal fin base extends from the origin of the first dorsal fin to beyond the anterior margins of the eyes (Fig. 43B); caudal peduncle relatively deep, 3.2–4.6 times into head length; prickles absent or reduced to a patch behind the pectoral fin *(see 17)*

                    17 (18) Head moderately long, 2.7–3.2 into standard length; caudal peduncle narrow, 4.8–5.7 into head length; lateral line usually complete, but if incomplete, there is a trace line on the caudal peduncle where the lateral line would be — COLUMBIA SCULPIN, *Cottus hubbsi*

                    18 (17) Head short, 3.2–3.9 into standard length; caudal peduncle moderately deep, 3.1–4.5 into head length; lateral line usually incomplete and without a trace line on the caudal peduncle *(see 19)*

                        19 (20) Surface of head nubbly; in British Columbia, usually a single postmaxillary pore (Fig. 44A) and no prickles behind pectoral fin; in British Columbia, this species occurs only in the lower Flathead River and its tributaries — ROCKY MOUNTAIN SCULPIN, *Cottus* sp.

                        20 (19) Surface of head smooth; in British Columbia, two (rarely one) postmaxillary pores (Fig. 44B) and a dense patch of prickles behind the pectoral fin; in British Columbia, in the Kettle River below the barrier at Cascade, in the lower Columbia River and in its tributaries (below barriers), and in the lower Kootenay River (including the Slocan River) below Bonnington Falls — SHORTHEAD SCULPIN, *Cottus confusus*

Cottus aleuticus GILBERT
COASTRANGE SCULPIN

1 cm

Distinguishing Characters This smooth-skinned sculpin (there is a small patch of prickles behind the pectoral fin) has a single median chin pore and a complete lateral line. There is an abrupt downward dip in the lateral line where it traverses the caudal peduncle. The posterior nostrils in coastrange sculpin are tubular: the tubes are about the same lengths on both the anterior and posterior nostrils. The pale dorsal spot on the caudal peduncle and the unusually long pelvic fins—in adults, they almost reach the anus—are useful field characters.

Two other B.C. sculpins have single median chin pores—the spoonhead sculpin (*Cottus ricei*) and the prickly sculpin (*Cottus asper*). The geographic distributions of coastrange and spoonhead sculpins don't overlap. Thus, these species are unlikely to be confused. In contrast, the ranges of coastrange and prickly sculpin overlap extensively, and they commonly coexist in coastal rivers and streams. The coastrange sculpin is distinguished from the prickly sculpin by anal fin ray counts (usually 12–14 in the coastrange sculpin and 16–18 in prickly sculpin), the tubular posterior nostrils in coastrange sculpin, and long pelvic fins that almost reach the anus in coastrange sculpin.

Taxonomy In Cultus Lake, there are two forms of coastrange sculpin: the "normal" form that is found mainly in tributary streams and occasionally in the lake proper, and a small form that lives in the lake and migrates from near the bottom to surface waters at night (Ricker 1960). A similar, small, diurnally migrating form of the coastrange sculpin is known from Lake Washington (Ikusemiju 1975; Larson and Brown 1975). These small adults are fully mature but display some juvenile characters (e.g., enlarged head pores). Although it is not clear what developmental mechanism produces these small sculpins or what ecological factors drive this shift in body size, it is clear that some form of heterochrony is involved. Heterochrony is a change in the relative time of appearance, or rate of development, of characters present in ancestors (Gould 1977). In this case, their small size (25–50 mm vs. 60–120 mm in normal coastrange sculpin) and enlarged cephalic pores, but apparently normal age at maturity (2+ years), implicates neoteny—the retention of formerly juvenile characters by adult descendents.

In Cultus Lake, adults of this apparently neotenic form produce fewer eggs, have a more protracted spawning period (May to early September), have larger pores on the head, and have shorter pelvic fins than "normal" coastrange sculpin. The relationship between the two forms is unknown, and the problem needs investigation.

Sexual Dimorphism Breeding females are, on average, larger then males. In breeding males, the entire body, except for an orange trim on the first dorsal fin, is black. Breeding females retain their normal colour but have noticeably swollen abdomens. Outside the breeding season, adults can be sexed by examining the urogenital papillae through a hand lens. Although the urogenital papilla changes seasonally, it is an obvious structure (either triangular or bluntly rounded) immediately behind the anus in males and an inconspicuous mound or tube in non-breeding females.

At some sites in the lower Fraser River, spawning males develop curious, posteriorly directed membranous flaps on the tips of the incised anal fin rays.

Distribution NATIVE DISTRIBUTION

As the name implies, the coastrange sculpin is a coastal species. It is endemic to western North America. Here, it ranges from northern San Luis Obispo County, California (Robins and Miller 1957), northwards to the Aleutian Islands and Bristol Bay, Alaska. There is reputed to be an isolated population in the Kobuk River north of the Seward Peninsula (Evermann and Goldsborough 1907).

BRITISH COLUMBIA DISTRIBUTION

The coastrange sculpin is the common riffle sculpin in mainland and island rivers and streams along the entire coast (Map 69). Typically, this species only penetrates a short distance ($<$150 km) inland; however, in both the Fraser and Skeena drainage systems, there are isolated inland populations. In the Fraser system, coastrange sculpins are found upstream as far as the Boston Bar area (Anderson Creek and Nahatlach River tributaries), but there is also a disjunct population above the Fraser Canyon in the Bridge River system (Seton Creek). In the Skeena drainage, there are disjunct populations in streams tributary to Morice Lake.

Life History REPRODUCTION

In British Columbia, coastrange sculpin spawn in the spring (usually April to late June), but egg masses have been found in late January, February, and March in the lower Fraser River. Most spawning begins in April and

454 FAMILY COTTIDAE—SCULPINS

Map 69 The B.C. distribution of coastrange sculpin, *Cottus aleuticus*

May when water temperatures rise above 6 °C and continues into June at temperatures as high as 15 °C. Typically, spawning occurs throughout the length of the stream occupied by adults (Mason and Machidori 1976; McLarney 1968), but there is evidence of a pre-spawning downstream migration of adults in some coastal streams (Kresja 1965; Ringstad and Narver 1973; Taylor 1966). Apparently, this downstream migration stops short of the estuaries, and there is no direct evidence of coastrange sculpin spawning in brackish water (Mason and Machidori 1976; Ringstad 1974).

Males excavate a nesting site—usually under a flat rock—and defend the site against intrusions by other males. Courtship has been observed in the laboratory and, basically, is similar to that described for the mottled sculpin (*Cottus bairdii*) (for details, see the description of spawning behaviour in the slimy sculpin, *Cottus cognatus*, species account). Once in the nest, the female deposits her eggs on the rock roof of the nest. The eggs are yellow to orange in colour, adhesive, and average about 1.4 mm in diameter. Typically, a male will spawn with more than one female, and the single egg masses found in nests often contains eggs of different colours and at different developmental stages. Males aerate the eggs with their pectoral fins. They fan the eggs and guard the nest until the entire egg mass hatches. This takes about 20 days at 10–12 °C (Mason and Machidori 1976). Fecundity in the coastrange sculpin is a function of female body size. Thus, females of the neotenic sculpin in Cultus Lake that mature at a total length of about 25 mm have a relatively low fecundity

(50–150 eggs). In contrast, normal females mature at about 50 mm, and their fecundities range from about 200 to 1,000 eggs (Patten 1971).

The newly hatched larvae are transparent and about 5–6 mm TL. They are active immediately upon hatching. The larvae remain in the vicinity of the nest for several days before they are transported downstream (Mason and Machidori 1976). Apparently, this downstream movement is passive and restricted to the darkest hours of the night (Ringstad 1974). In short coastal streams, this drift carries the larvae into the upper reaches of estuaries and, in inland systems, into lakes. Where the larvae end up in large rivers like the lower Fraser is unknown, but they have not been recorded in the Fraser estuary. The larvae transform into small sculpins approximately 30 days after hatching (Mason and Machidori 1976). At this time, they are about 12 mm TL.

In coastal streams, the young-of-the-year remain in estuaries for about a year (Taylor 1966) before beginning upstream movements. In Carnation Creek, yearlings began to move upstream in late June or July, whereas fish larger than 45 mm tended to move upstream later in the summer (Ringstad 1974). In British Columbia, the movements of young in lakes and in non-coastal fluvial populations are unknown.

AGE, GROWTH, AND MATURITY

Growth in the coastrange sculpin is variable, and there are often differences in growth rates between adjacent streams (Mason and Machidori 1976). Generally, however, normal young-of-the-year average about 25 mm TL by September, and yearlings reach about 40–50 mm TL in their second growing season (Mason and Machidori 1976; Ringstad 1974). In contrast, otolith readings suggest that the neotenic sculpin in Cultus Lake takes 2 years to reach a total length of about 25 mm. In both normal and neotenic populations, sexual maturity is reached after 2 years. The maximum size recorded in British Columbia is 145 mm, and the maximum age was 7^+.

FOOD HABITS

Adult coastrange sculpin are carnivorous and forage primarily on the nymphs and larvae of aquatic insects (especially chironomids, mayflies, and stoneflies). Foraging peaks at night but continues at a low level throughout the day. In the fall, individuals over 50 mm sometimes contain salmon eggs, and they eat small numbers of salmon fry in the spring (usually pinks, *Oncorhynchus gorbuscha*, or chums, *Oncorhynchus keta*).

The pelagic larvae feed on microplankton; however, after transformation, they become bottom-dwellers, and their diet becomes similar to that of adults (except they take smaller prey).

Habitat ADULTS

Adult coastrange sculpin are primarily fluvial. During the day, they are generally found in fast water, especially riffles and glides, with surface velocities of 0.5 to >1.0 m/s and coarse gravel or cobble substrates.

Typically, the depth of capture ranges from about 20 cm to 1.0 m. At night, however, adults move into slower, shallower water along stream margins. In coastal streams, coastrange sculpin usually penetrate farther upstream than prickly sculpin. When prickly sculpin are absent, however, adult coastrange sculpin shift into the habitats normally occupied by prickly sculpins: pools and sheltered areas under log jams and cutbanks (Mason and Machidori 1976). Even when prickly sculpin are present, large individuals (>80 mm TL) sometimes occur in pools (Taylor 1966).

Although primarily a fluvial species, in coastal British Columbia and Alaska, coastrange sculpin regularly occur in large, oligotrophic lakes. Here, they are associated with coarse gravel and cobble beaches. Seasonal concentrations of this sculpin on sockeye salmon spawning beaches are known in Lake Iliamna, Alaska (Foote and Brown 1998).

In Cultus Lake, neotenic adults remain in deep water (usually >25 m) during the day; however, some, perhaps most, individuals migrate to the surface at night.

JUVENILES

Habitat use by coastrange sculpin changes with body size. Thus, after overwintering in estuaries, lakes or quiet areas in large rivers, yearling coastrange sculpin move into streams or into faster water (Mason and Machidori 1976; Ringstad and Narver 1973; Taylor 1966). Here, they occupy habitats similar to those of adults but in shallower, slower water (Ringstad 1974).

YOUNG-OF-THE-YEAR

In many coastal streams, coastrange sculpin fry overwinter in estuaries, and at low tide, they are associated with relatively faster and shallower water than sympatric prickly sculpin fry (Mason and Machidori 1976; Ringstad 1974). In lakes, the larvae remain pelagic for about 30 days and then transform and take up a benthic existence. In Cultus Lake, the neotenic form is a permanent lake resident.

Conservation Comments   COSEWIC lists the Cultus Lake pygmy sculpin as threatened, and the BCCDC ranks it as S1 (critically imperiled). Cultus Lake is the only place in Canada known to contain the neotenic form of the coastrange sculpin; however, the presence of similar small sculpins in Lake Washington near Seattle (Ikusemiju 1975; Larson and Brown 1975) suggests the parallel evolution of neotenic forms. Consequently, they may occur in other coastal lakes in British Columbia; however, even if there are other populations, the developmental processes and ecological factors that produce neotenic adults are of scientific interest. We know virtually nothing about the biology and habitat requirements of this remarkable little fish, but an effort is now underway to find out more about their habitat requirements. In the summer, Cultus Lake is a heavily used recreational site (especially by water skiers and jet skis). The impacts of this recreational use on the fish fauna are unknown.

Cottus asper RICHARDSON
PRICKLY SCULPIN

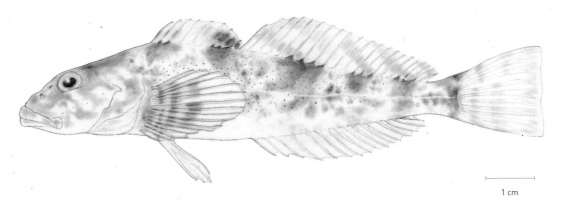

1 cm

Distinguishing Characters This large sculpin (up to 225 mm TL) has a single median chin pore, strong palatine teeth, and usually 16–18 anal rays. It is the only sculpin in British Columbia with strongly conjoined dorsal fins and an anal fin base that is noticeably longer the head.

Two other B.C. sculpins have single median chin pores—the spoonhead sculpin (*Cottus ricei*) and the coastrange sculpin (*Cottus aleuticus*). In British Columbia, the geographic distributions of prickly and spoonhead sculpins overlap in the lower Peace River and some of its tributaries. They are occasionally misidentified in this region. Although both species can be heavily prickled, they differ in caudal peduncle depth (3–4 times into head length in prickly sculpin and 5–7 times in spoonhead sculpin) and the development of the palatine teeth (present and strong in prickly sculpin and absent in spoonhead sculpin).

Prickly and coastrange sculpins coexist in the lower reaches of most coastal drainages. They are separated by the development of palatine teeth (present and strong in prickly sculpins and absent in coastrange sculpins) and the degree of connection between the spinous and soft dorsal fins (notched but clearly joined in prickly sculpins and narrowly separated in coastrange sculpins).

Taxonomy Two forms of prickly sculpin occur in British Columbia: a coastal and an inland form (Kresja 1967a). These forms differ in morphology (usually, but not always, they are lightly prickled close to the coast and heavily prickled inland), behaviour (catadromous in short coastal streams and freshwater resident inland), and physiology (inland fish have a greater capacity to retain electrolytes than coastal populations; Bohn and Hoar 1965). The two forms probably entered the province by two dispersal routes: coastal dispersal through the sea and postglacial inland dispersal through connections between the Columbia and Fraser systems.

Preliminary data suggest some minor sequence differences between the two forms, but the problem needs more study.

Sexual Dimorphism — Although there is no clear sexual size dimorphism in prickly sculpins, breeding females are, on average, larger than males. In breeding males, the entire body is black except for an orange trim on the first dorsal fin. Breeding females retain their normal colour but have noticeably swollen abdomens. Outside the breeding season, adults can be sexed by examining the urogenital papillae through a hand lens. Although the urogenital papilla changes seasonally, it is an obvious structure (either triangular or bluntly rounded) immediately behind the anus in males and an inconspicuous mound or tube in non-breeding females.

Hybridization — In Otter Creek (Similkameen system), prickly and Columbia sculpin (*Cottus hubbsi*) coexist. Some prickly sculpins in Otter Creek have mitochondrial sequences similar to those of the Columbia sculpin. Although there is no morphological indication of contemporary hybridization, these species may have hybridized sometime in the recent past.

Distribution —

NATIVE DISTRIBUTION

The prickly sculpin is endemic to western North America and ranges from the Ventura River, southern California, to the Kenai Peninsula, Alaska. Throughout most of its geographic range, the prickly sculpin is a coastal species.

BRITISH COLUMBIA DISTRIBUTION

Although primarily a coastal species, the prickly sculpin is common in lakes and streams throughout the interior portions of the Columbia, Fraser, Skeena, and Nass drainage systems. In addition, prickly sculpins have crossed the Continental Divide (from the upper Fraser system) and colonized the Peace system at least as far downstream as the Notikewin River in Alberta (Nelson and Paetz 1992). An apparent gap in their geographic distributions (Map 70) suggests that the coastal and inland forms may be allopatric in British Columbia.

Life History — Most of the available life-history information on the prickly sculpin pertains to the coastal form, and only snippets of data are available for the interior form. Preliminary mitochondrial data suggest that there may be sequence differences between the two forms. If so, there also may be life-history differences and the life history of the interior form probably warrants further study.

REPRODUCTION

In most of British Columbia, prickly sculpin spawn in the spring (April to late June) when water temperatures rise above 6 °C, and they continue to spawn until temperatures rise above 16 °C. In northern British Columbia, a gravid female was collected in Meziadin Lake (Nass system)

Map 70 The B.C. distribution of prickly sculpin, *Cottus asper*

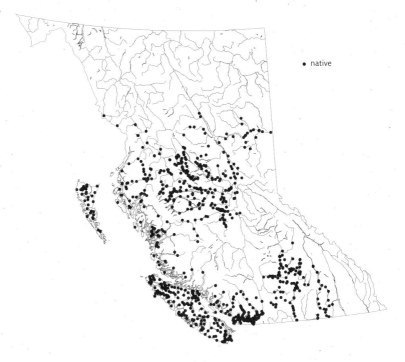

in late July (Kresja 1965). In March and April in short coastal streams, there is a pronounced downstream migration of adults into estuaries (Kresja 1965; Mason and Machidori 1976; Ringstad and Narver 1973; Taylor 1966). Apparently, inland populations do not perform extensive spawning migrations, and even in short coastal streams, some adults spawn *in situ* rather than migrating to estuaries (Brown et al. 1995; McLarney 1968; Sinclair 1968).

Regardless of where they spawn, males excavate a nesting site—usually under a flat rock or solidly embedded woody debris—and defend the site against intrusions by other males. Courtship has been observed in the laboratory and is similar to that described for the mottled sculpin (*Cottus bairdii*) (for details see the description of spawning behaviour in the slimy sculpin, *Cottus cognatus*, species account). Once in the nest, the female deposits her eggs on the roof of the nest. The yellow-orange eggs are adhesive and average between 1.6 and 1.8 mm in diameter. Normally, males spawn with more than one female, and the single egg mass commonly contain different coloured eggs at different developmental stages. One nest in the Little Campbell River was estimated (Kresja 1965) to contain eggs from 10 different females (25,000–30,000 eggs). In Carnation Creek, the number of eggs per nest ranged from 960 to 24,570, and 7 of the 16 nests examined contained over 10,000 eggs (Ringstad 1974). Fecundity in the prickly sculpin is a function of female body size and ranges from approximately 300 to over 10,000 eggs (Bond 1963; Kresja 1965; Patten 1971).

Males use their pectoral fins to aerate the eggs. They fan the eggs and guard the nest until the entire egg mass is hatched. This takes 15 or 16 days at 12 °C. The transparent larvae are about 5–6 mm TL and are active immediately upon hatching (Kresja 1965). The larvae remain planktonic for 30–35 days and then transform into small sculpins (about 12 mm TL).

AGE, GROWTH, AND MATURITY

Although growth rates vary among populations, the fry usually reach about 30–40 mm TL by September of their first summer and 50–60 mm by the end of their second growing season. Where they coexist, the prickly sculpin typically grows more rapidly than the coastrange sculpin (Brown et al. 1995; Kresja 1965; Mason and Machidori 1976; Taylor 1966). Adult size varies among populations but often exceeds 200 mm TL, and a disproportionate number of the largest individuals are females. Sexual maturity is reached in 2 or 3 years, and the maximum age recorded in British Columbia is 8+.

FOOD HABITS

Larval prickly sculpin feed on microplankton; however, when they become benthic after transformation, their diet shifts towards benthic prey. In Nicola Lake, fry <15 mm TL contained about 65% water column prey and about 35% benthic prey, but the proportion of benthic prey increased as fish size increased (Miura 1962). In the Arrow Lakes, fry fed on plankton and aquatic insect larvae (Northcote 1954). In flowing water and in lakes, juveniles and adults <70 mm forage mainly on the nymphs and larvae of aquatic insects (especially chironomids, mayflies, and caddisflies). Individuals over 70 mm begin to add fish to their diet (Northcote 1954; Patten 1962), and large adults (over 120 mm) can be significant predators on salmonine fry (Berejikian 1995; Patten 1962). Among freshwater sculpins, prickly sculpin is particularly adept at handling difficult prey. In coastal regions, large adults often contain threespine sticklebacks (*Gasterosteus aculeatus*), and in lakes where there are introduced pumpkinseeds (*Lepomis gibbosus*), prickly sculpin stomachs are often packed with young-of-the-year pumpkinseeds neatly folded head-to-tail. There is evidence of crepuscular forag-ing in yearlings; however, adults forage most actively at night (Miura 1962).

Habitat Most habitat information on prickly sculpin pertains to the coastal form, and it is possible that the interior form differs in some details. For example, in the Nazko River (upper Fraser system), adult prickly sculpin live in habitats similar to those used by adult longnose dace (*Rhinichthys cataractae*): shallow, fast water over rocky substrates (Porter and Rosenfeld 1999). In contrast, adults of the coastal form appear to prefer quiet water (Mason and Machidori 1976; Ringstad 1974; Taylor 1966).

ADULTS

In British Columbia, prickly sculpin occur in both flowing water and lakes. Also, their habitat use changes with body size. In coastal streams, adult prickly sculpin are generally found in quiet water (deep pools, under cutbanks, or associated with woody debris). In the Columbia River between Keenleyside Dam and the U.S. border, prickly sculpin were associated with boulder substrates and average water column velocities of 0.34 m/s (R. L. & L. Environmental Services Ltd, 1995a).

In lakes, adult prickly sculpin are also associated with cover, especially areas where cobbles, boulders, or woody debris are interspersed among sandy patches. During the day they remain in, or close to, cover but at night they forage in the open. Although adults are commonly observed in littoral areas, baited traps set in Harrison Lake caught large numbers of adults at depths down to 100 m.

JUVENILES

In coastal streams, juveniles occupy similar areas, but in shallower water, than adults (Mason and Machidori 1976; Ringstad 1974; Taylor 1966). In inland rivers, juveniles are found in quiet edge sites and are commonly associated with vegetation or small woody debris. In large lakes, juvenile prickly sculpin sometimes occur at, or near, the surface over deep water, and single individuals are regularly collected at night in surface trawls during juvenile sockeye salmon (*Oncorhynchus nerka*) surveys (Mueller and Enzenhofer 1991). In Nicola Lake, during the late summer, a school of yearling sculpins about 1 km long were observed slowly "streaming" along a section of steep rocky shoreline (Northcote and Hartman 1959).

YOUNG-OF-THE-YEAR

In July near Trail, newly transformed fry (about 12 mm TL) are common in seasonally flooded vegetation along the margins of the lower Columbia River. In Buttle and Campbell lakes on Vancouver Island, prickly sculpin fry were caught offshore in the limnetic zone at night (Sinclair 1968). In Nicola Lake, there was a distinct diurnal pattern of microhabitat use by young-of-the-year in inshore waters: in late summer, they were associated with weedy areas during the day and moved into more open areas (sand, gravel, and rocks) at night (Miura 1962).

The planktonic larvae in coastal populations either hatch in, or drift downstream into, estuaries. Here, they transform and, at low tide, are associated with slower water than sympatric coastrange sculpin (Ringstad 1974). The fry of some coastal populations overwinter in estuaries and, as yearlings, begin to move upstream into fresh water in July (Kresja 1967b, Taylor 1966). This upstream movement peaks in September (Ringstad 1974; Ringstad and Narver 1973) but can continue into December (Mason and Machidori 1976).

Conservation Comments At present, the prickly sculpin is not a conservation concern in British Columbia, and neither COSEWIC nor the BCCDC list this species. This may change, however, when the relationship between the coastal and inland forms is clarified.

Cottus cognatus RICHARDSON
SLIMY SCULPIN

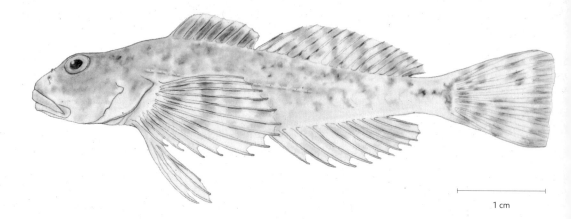

1 cm

Distinguishing Characters This slim-bodied, smooth-skinned (prickles are restricted to a small area behind the pectoral fin) sculpin has deeply incised anal fin rays. The lateral line is incomplete (it usually ends at about the middle of the second dorsal fin). There are 10–13 anal rays, 15–19 rays in the second dorsal fin, 13–15 pectoral rays, and 3 or 4 pelvic rays with the first and fourth rays usually shorter than the second and third rays. Palatine teeth are absent, and in British Columbia, there usually are two median chin pores. The dorsal surface of the head is smooth, the fins usually are barred, and the chin is lightly pigmented.

Outside the Mackenzie River system, the slimy sculpin is the only sculpin found in most northern inland waters and, therefore, is unlikely to be confused with any other species. Two other species occur in the Mackenzie system: prickly (*Cottus asper*) and spoonhead sculpin (*Cottus ricei*). Both these species have dense prickling and complete lateral lines, whereas the slimy sculpin is naked except for a small patch of prickles behind the pectoral fins and has a short, incomplete lateral line. In the Columbia River system southern British Columbia, there are three species that sometimes are confused with slimy sculpin: the Rocky Mountain sculpin (*Cottus* sp.); the shorthead sculpin (*Cottus confusus*), and the Columbia sculpin (*Cottus hubbsi*). Slimy sculpins lack palatine teeth, and the other species possess palatine teeth, although the teeth are weakly developed in Rocky Mountain sculpin.

Taxonomy In the past, different subspecific names were used for eastern and western populations of slimy sculpin; however, most recent authors (e.g., Nelson and Paetz 1992; Stewart and Watkinson 2004) do not use subspecific names for this species. Still, given what we now know about other species that are widely distributed on both sides of the Continental Divide (e.g., longnose dace, *Rhinichthys cataractae*; mountain suckers, *Catostomus platyrhynchus*; and mottled sculpins, *Cottus bairdii*), it would be surprising if eastern and western slimy sculpins did not differ substantially in their genetic makeup.

Two morphological traits (whether the last anal ray is branched or unbranched and the number of pelvic fin rays) were examined in slimy sculpins from British Columbia, Alaska, and the northern Great Plains (McAllister and Lindsey 1961). There were broad patterns of variation in these traits that suggested survival in, and subsequent postglacial dispersal from, three glacial refugia: the upper Mississippi, the ice-free regions of Alaska and the Yukon, and the unglaciated portions of the Columbia River system. They suggested two groups of slimy sculpins in British Columbia: a southern group that ranges from the Columbia drainage system north to upper Peace River drainages, and a northern group that ranges from the lower Peace River and its B.C. tributaries into Alaska and the Yukon. These groups need to be revisited using modern genetic techniques.

Sexual Dimorphism There is no obvious sexual difference in body size in slimy sculpin. In territorial males, the entire body is black except for an orange trim on the first dorsal fin. Breeding females retain their normal colouration but have noticeably swollen abdomens. Outside the breeding season, adults can be sexed by examining the urogenital papillae through a hand lens. Although the urogenital papilla changes seasonally, it is an obvious structure (either triangular or bluntly rounded) immediately behind the anus in males and an inconspicuous mound or tube in non-breeding females.

Hybridization So far, in British Columbia, the only hybrids involving slimy sculpin are found in the Flathead River system. Here, morphological and allozyme evidence indicates hybridization with the Rocky Mountain sculpin (Zimmerman and Wooten 1981).

Distribution NATIVE DISTRIBUTION
The slimy sculpin has the widest geographic distribution of any North American sculpin: it ranges from Alaska across the northern Great Plains to Labrador and, although absent from the central plains, extends south to Virginia in the east and Idaho in the west. Additionally, sometime during the late or middle Pleistocene, the slimy sculpin apparently crossed over the Bering Landbridge into the Anadyr River and Chukotsk Peninsula in eastern Siberia (Lindsey and McPhail 1986). It still occurs on St. Lawrence Island (a remnant of the Bering Landbridge).

BRITISH COLUMBIA DISTRIBUTION
The slimy sculpin is widely distributed in the inland waters of the province from the Columbia system in the south to the Yukon system in the north (Map 71). Although the slimy sculpin ranges over the entire length and most of the breadth of the province, it is an interior species and only

approaches the coast in a few northern rivers (e.g., Alsek, Taku, and Iskut–Stikine systems).

Life History Although there is no published account of the life history of the slimy sculpin in British Columbia, a number of consultants reports (e.g., Craig and Bruce 1982; Stewart et al. 1982) provide information on age distributions and diet. These reports, together with casual local observations, form the basis of the following account.

REPRODUCTION

Slimy sculpin spawn in the early spring after water temperatures rise above 4 °C. This occurs from mid-March to early April in southern British Columbia and as late as mid-May in the north. In Lake Ontario, however, slimy sculpin appear to spawn from late April to mid-October. This protracted spawning season probably reflects the seasonal cycle of water temperatures at different depths (Owens and Noguchi 1998). There is some evidence that northern slimy sculpin may spawn even before water temperature rises above 4 °C. In three small Liard River tributaries, females were gravid at temperatures ranging from 0 °C to 4.7 °C, and fry 20–29 mm TL appeared in early June (Craig and Bruce 1982).

Like most sculpins, males excavate a nest cavity under a flat rock. They are territorial and defend the nest against intruders (usually other males). Courtship behaviour has not been documented in this species; however, courtship in the mottled sculpin has been described (Savage 1963) and, apparently, spawning behaviour is basically the same in slimy and mottled sculpin (Strauss 1986). The following account is summarized from Savage (1963). When a female approaches the nest site, the male starts a series of frontal displays consisting of one or more distinctive head movements: head shaking in the horizontal plane, head nodding in the vertical plane, and flaring of the gill covers. Sometimes the body is undulated during a frontal display, and sometimes undulations occur without a head display. Some of these movements (e.g., nodding) produce sounds. Apparently, these sounds are transmitted through the substrate and, in riffle environments, this produces a clearer signal than sound transmitted through water (Whang and Janssen 1994). If the female persists in her approach, the male responds by biting—usually on the cheek, side, tail, or pectoral fin. Occasionally, he takes the female's head into his mouth. Apparently, ripe females always enter the nest after being bitten. Once in the nest, the female turns upside down against the roof-rock. The male then turns partially upside down and presses against the female. His head and all his fins become black and his body blanches. Presumably, at this time gametes are released, and the male quickly rights himself. The observation (Savage 1963) that the female remains in the nest for several days and assumes the spawning position several times, suggests that not all the eggs are released at a single spawning.

The eggs are adhesive and stick to the underside of the nest rock. The male remains with the eggs and occasionally fans them with his pectoral

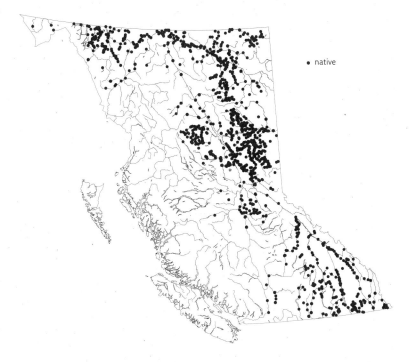

Map 71 The B.C. distribution of slimy sculpin, *Cottus cognatus*

fins. He guards and cares for the eggs until they hatch. The few slimy sculpin nests examined in British Columbia contained several clumps of eggs of different colours and at different stages of development. This suggests that the males are polygynous. In Ontario, males are primarily monogamous in some lakes but primarily polygynous in others (Mousseau et al. 1988). Although the main spawning season is spring, some populations in cool upper Columbia drainages contain reproductive males and gravid females in mid-August. Interestingly, in populations that contain gravid females in late summer, only the smallest mature females contain mature eggs. This is consistent with the observation that large females spawn before small females (Petrosky and Waters 1975).

The eggs range in colour from light yellow to amber and, when fertilized, are about 2.5 mm in diameter. Like most fish, fecundity in slimy sculpin increases with increasing female size (Foltz 1976). There is little fecundity data on slimy sculpin in British Columbia, but the available data agree with the fecundity range (from approximately 60 to 1,150 eggs) recorded in the Great Lakes region in both stream and lake populations (Foltz 1976; Owens and Noguchi 1998; Petrosky and Waters 1975). Normally, water temperatures increase during the incubation period (from about 5 °C to about 10 °C), and the eggs take about a month to hatch under these conditions (Van Vleit 1964). The newly hatched larvae are roughly 6 mm TL and have a large yolk sac. In stream-dwelling populations, larvae remain in the nest gravel for about 2 weeks and emerge as miniature adults; however, in large, oligotrophic lakes, the larvae of lacustrine populations are planktonic and don't become benthic until the fall (Wells 1976). Exogenous feeding begins when the yolk reserves are gone (about 15 days after hatching in other sculpins with similar-sized larvae).

AGE, GROWTH, AND MATURITY

Growth varies with locality. In the upper Liard system (Craig and Bruce 1982), young-of-the-year range from 30 to 32 mm TL by the end of their first summer, whereas fish 75–77 mm long were in their seventh summer (6+). In upper Columbia drainages, fry usually reach about 45 mm TL by the end of their first growing season and are over 75 mm long by the end of their fifth growing season (4+). In northern British Columbia, some males reach sexual maturity at the beginning of their fourth growing season (3+), and most females reach sexual maturity at the beginning of their fifth growing season. In southern British Columbia, sexual maturity is reached at least a year earlier than in northern populations. So far, the oldest slimy sculpin aged in British Columbia was in its eighth growing season (7+).

FOOD HABITS

Slimy sculpin are carnivorous. The newly transformed fry of fluvial slimy sculpin forage primarily on chironomid larvae, but as the fry grow, nymphs and larvae of larger aquatic insects are added to the diet. The diet of lacustrine larvae is undocumented but probably starts with microplankton and, as they grow, shifts to larger plankton. Although adults and juveniles include mayfly and stonefly nymphs and caddisfly larvae in their diet, chironomid larvae are still a major food. Large adults also eat appreciable quantities of aquatic beetles (primarily Dytiscids), snails, and amphipods. In Lake Iliamna, Alaska, slimy sculpins aggregate on beaches used by shore-spawning sockeye salmon (*Oncorhynchus nerka*) and feed extensively on salmon eggs (Foote and Brown 1998). Interestingly, the movement of sculpins to these beaches precedes the start of salmon spawning. In Lake Ontario, lacustrine slimy sculpin <35 mm long foraged mainly at night, while the adults in deeper water showed no diel feeding periodicity (Brandt 1986).

Habitat

ADULTS

Slimy sculpin are a cool-water species. Their preferred temperature is about 10 °C, and they are able to detect and avoid potentially harmful temperatures (Otto and Rice 1977). In southern British Columbia, this species is close to the southwestern edge of its geographic range and is patchily distributed. Most southwestern populations are found in cold, headwater streams or in glacier-fed rivers. In streams in eastern North America, adults are associated with moderate water depths, moderate water velocities, and cover (Johnson et al. 1992). In southern British Columbia, they are associated with riffles during the day. Here, they shelter among coarse gravel or cobbles, but they emerge at night and forage away from cover. These southern populations appear to be especially sensitive to the presence of other sculpins. Thus, they rarely co-occur with congeners and are often absent below, but abundant above, barriers that isolate them from other sculpins. When sympatric with other sculpins, southern slimy sculpin typically occupy a separate (usually cooler) part of the drainage basin. In such cases, the transition between species is often abrupt (see discussion of Rocky Mountain sculpin habitat). Southern populations also occur in large, oligotrophic lakes and reservoirs. In these environments, they are usually found in the littoral zone on gravel beaches. Although their lower depth range in British Columbia is unknown, traps set in deep water in Idaho have caught slimy sculpin down to 300 m (Wydoski and Whitney 2003).

In northern British Columbia, slimy sculpin are ubiquitous. They occur in lakes, ponds, small streams, and large rivers. In lakes, the adults are usually associated with coarse gravel or cobble substrates. Most collections from lakes in northern British Columbia are from littoral regions; however, in Lake Superior slimy sculpin occur down to 100 m (Selgeby 1988). It is not known if they reach similar depths in northern B.C. lakes.

In flowing water, northern slimy sculpin are found from small headwater streams to large lowland rivers. Usually, the adults are associated with coarse gravel and moderate to swift currents, but they are sometimes taken in quiet water over sand and silt bottoms. Apparently slimy sculpin are not sensitive to turbidity and, except in the presence of spoonhead sculpin (see discussion of spoonhead sculpin habitat), are found in waters ranging in clarity from gin-clear to very turbid. They are uncommon in muskeg regions; however, even in black-water streams, there usually are slimy sculpin if there is coarse substrate and moderate current.

JUVENILES

During the day, juveniles usually are found in shallower and slower water than adults. Here, they shelter among small rocks or coarse gravel, but they venture farther from cover at night. In Lake Ontario, slimy sculpin <35 mm in length usually inhabited water <60 m deep, whereas adults dominated catches at 75 m (Brandt 1986).

YOUNG-OF-THE-YEAR

During freshet in rivers and streams, transformed young-of-the-year are found along stream edges in quiet water (often in flooded vegetation). During the summer and fall, they usually are found in slow water along stream margins (Johnson et al. 1992). In eastern North American lakes, the larvae remain planktonic until the fall, but it is not known if this happens in any B.C. lakes.

Conservation Comments   Neither COSEWIC nor the BCCDC list the slimy sculpin; however, this species warrants further study, especially in light of an apparently relict population well south of the species' continuous distribution in Idaho (Simpson and Wallace 1978). If slimy sculpin survived glaciation west of the Continental Divide, there may be significant genetic differences between the southern B.C. populations and those in eastern North America. Additionally, McAllister and Lindsey (1961) found morphological differences between slimy sculpin of Beringian origin and those in southern British Columbia, and there is some evidence (Craig and Bruce 1982) that northern populations have evolved the capacity to breed at very low (0–4 °C) water temperatures. Thus, this widely distributed species may consist of several distinctive lineages that may deserve separate conservation assessments.

Cottus confusus BAILEY & BOND
SHORTHEAD SCULPIN

1 cm

Distinguishing Characters — This chunky, short-headed, smooth-skinned (prickles restricted to a small area behind the pectoral fin) sculpin has a lateral line that is usually incomplete (22–34 lateral line pores); although scattered remnants may extend onto the caudal peduncle, the unbroken portion usually ends before the insertion of the second dorsal fin. There are 11–13 anal rays, 15–18 rays in the second dorsal fin, 12–15 pectoral rays, and 4 (rarely 3) pelvic rays. The first and fourth pelvic rays are usually shorter than the second and third rays. There are two or three preopercular spines, usually with the third spine reduced. The palatine teeth are weak and well separated from the vomerine tooth patch. Usually, there are two median chin pores (in British Columbia, about 5% have a single median chin pore), and two postmaxillary pores. The soft dorsal, caudal, and pectoral fins are barred, but the anal fin usually is without strong bars. The dorsal surface of the head is smooth, and the chin is lightly and uniformly pigmented.

In British Columbia, there is limited overlap between shorthead and other sculpin species. In small streams, they sometimes coexist with torrent sculpin (*Cottus rhotheus*), and where tributary streams enter the Columbia River, they are occasionally taken with Columbia sculpin (*Cottus hubbsi*). In the Columbia, Kootenay, and Slocan rivers, they occasionally are collected with prickly sculpin (*Cottus asper*). In these areas, shorthead sculpin are distinguished from both torrent and prickly sculpins by the prickle pattern—the back and sides in prickly and torrent sculpins usually are densely covered in prickles. In British Columbia, the prickles in shorthead sculpin are confined to a patch behind the pectoral fin. Shorthead sculpin are distinguished from Columbia sculpins by caudal peduncle depth (4.8–5.7 into head length in the Columbia sculpin vs. 3.1–4.5 in the shorthead sculpin), and the length of the lateral line (usually complete and extending onto the caudal peduncle in Columbia sculpin and typically incomplete and ending before the end of the second dorsal fin in shorthead sculpin). A useful field character (although not infallible) is pigment on the anal fin (this fin is strongly barred in most Columbia sculpin populations and uniformly dusky in shorthead sculpins).

Taxonomy In B.C., the scientific name of the shorthead sculpin (*Cottus confusus*) aptly describes its taxonomic history—confused! Originally named from the Salmon River, Idaho, the description of the species' geographic distribution included several B.C. sites in the Flathead River (Bailey and Bond 1963). A short distance from these B.C. sites, but on the other side of the Continental Divide, a similar sculpin was identified (Henderson and Peter 1969) as the mottled sculpin (*Cottus bairdii*). This identification was consistent with the widespread presence of what were thought to be *C. bairdii* in upper Missouri tributaries in Montana (Brown 1971) and Wyoming (Baxter and Simon 1970). Later, several authors (Peden et al. 1989; Nelson and Paetz 1992) noted the similarity between the sculpins in southwestern Alberta and the species in the Flathead system in British Columbia. Consequently, the southwestern Alberta species was tentatively listed as *C. confusus*, although the authors did recognize that it could be either *C. bairdii* or an undescribed species (Nelson and Paetz 1992).

The taxonomy, however, becomes even murkier. Shorthead sculpins are present in lower Columbia tributaries in the Castlegar–Trail area and in the Kettle River below Cascade Falls (Peden et al. 1989). In addition, these *C. confusus* are morphologically different from the nominal shorthead sculpins in the Flathead River system. Thus, there were two disjunct and morphologically divergent sculpins both called *C. confusus* in British Columbia: those in the Flathead River system and those in the lower Columbia and lower Kettle drainages. Consequently, it was recommended (Peden et al. 1989) that the two forms be given separate status. Thus, in 1998 the BCCDC listed the Flathead River form as *C. bairdii* and the lower Columbia and Kettle form as *C. confusus* (Cannings and Ptolemy 1998).

A recent molecular comparison of all sculpin species in the Canadian portion of the Columbia system, as well as presumed mottled sculpins from Ontario and the eastern slope of the Rocky Mountains in Alberta and Montana sheds some light on this confusion (J.D. McPhail and E.B. Taylor, in preparation). The data unequivocally support the contention (Peden et al. 1989) that the supposed shorthead sculpin in the Flathead River and the Rocky Mountain sculpins from Alberta and Montana are the same species. Additionally, the same data indicate that the Rocky Mountain sculpins are not *C. bairdii*—molecularly they are 5–6% different from this eastern North American species. Thus, the Flathead River sculpins are not *C. bairdii*. Apparently, they are an as yet unnamed species (for details, see the discussion in the Rocky Mountain sculpin species account). Because most of the geographic range of this unnamed species lies in the Rocky Mountains, it is referred to as the Rocky Mountain sculpin. The same study indicates that the nominal shorthead sculpin in the Castlegar–Trail area have mitochondrial sequences that align with shorthead sculpin sequences from Idaho, Oregon, and Washington. Thus, the putative shorthead sculpin in the Castlegar–Trail area are assumed to be real shorthead sculpin.

Sexual Dimorphism There are no obvious body size differences between the sexes. On average, sexually mature males are larger than mature females. The entire body in breeding males is black except for an orange trim on the first dorsal fin. Breeding females retain their normal colour but have noticeably swollen abdomens. Outside the breeding season, adults can be sexed by examining the urogenital papillae through a hand lens. Although the urogenital papilla changes seasonally, it is an obvious structure (either triangular or bluntly rounded) immediately behind the anus in males and an inconspicuous mound or tube in non-breeding females.

Hybridization In the area between Keenleyside Dam and the U.S. border, occasional individuals are encountered with morphometrics (especially caudal peduncle depth) that are intermediate between shorthead and the Columbia sculpins. These fish may be hybrids. Typically, they are collected at the confluences of tributaries like Beaver, Champion, Blueberry, and Norns (Pass) creeks with the mainstem Columbia River. Ecologically, these sites are at ecotones where the two putative parental species come in contact.

Distribution NATIVE DISTRIBUTION
The shorthead sculpin is a Columbia endemic. It occurs only in the Columbia drainage system and a few rivers in western Washington that contain a derived Columbia fauna (McPhail 1967). Although disjunct populations occur in independent internal drainages on the Snake River Plain, Idaho, this species does not occur above Shoshone Falls. There are also disjunct populations on the Olympic Peninsula, Washington, and in eastern tributaries of the Willamette River in Oregon. The genetic relationships of these disjunct populations to populations directly connected to the Columbia River are unknown.

BRITISH COLUMBIA DISTRIBUTION
The shorthead sculpin is known from the Kettle River (below Cascade Falls), the Columbia River, moderate-gradient portions of Columbia tributaries below Arrow Lakes, the Kootenay River and its tributaries (e.g., the Slocan River) below Bonnington Falls, and the Pend d'Orielle River (Map 72). The species is rare in the mainstem Columbia, more common in the mainstem Kootenay, and abundant in the mainstem Slocan River. This suggests a possible association between river size and the abundance of shorthead sculpin. Alternatively, the relationship may be spurious and simply reflect the difficulty of sampling the larger rivers.

Map 72 The B.C. distribution of shorthead sculpin, *Cottus confusus*

• native

Life History   There is no published account of the life history of the shorthead sculpin in British Columbia; however, Gasser et al. (1981) and Johnson et al. (1983) provide some life-history data on an isolated population of shorthead sculpin from an internal drainage—the Big Lost River, Idaho—and there is fecundity information on this species from two streams in western Washington (Patten 1971). These accounts, together with local observations, form the basis of the following life-history summary.

REPRODUCTION

In Idaho, shorthead sculpin begin spawning in mid-April (Gasser et al. 1981), and in British Columbia, they probably start breeding in early May. The breeding period was remarkably short (about 2 weeks) in Idaho, but in British Columbia, the breeding season is protracted: nests with eyed eggs are encountered from mid-May (water temperature 8 °C) to mid-July (water temperatures 13–15 °C). Like most sculpins, males nest under rocks and are territorial. Breeding males are dark with a yellow-orange trim on the first dorsal fin. The courtship behaviour has not been documented but probably is similar to other sculpins (for details of spawning behaviour see the slimy sculpin, *Cottus cognatus*, species account). Females deposit their sticky eggs on the underside of the nest rock, where they are fertilized. The male guards the eggs until they hatch. The few nests examined in British Columbia contained several, separate clumps of eggs. This suggests that at least some shorthead sculpin males are polygynous.

Fecundity is a function of female size. No data on egg numbers are available for British Columbia, but fecundities in Washington ranged from 47 to 217 eggs in fish ranging from 61 to 86 mm in length (Patten 1971), whereas fecundities in Idaho ranged from 184 to 511 eggs in females ranging from 53 to 71 mm in length (Gasser et al. 1981). These observations suggest that there may be major inter-population difference in egg numbers. Data from nests examined in Pass Creek near Castlegar indicate a range in egg number in B.C. populations similar to that found in western Washington: one nest contained three separate egg clumps with 52 eggs in the smallest clump and 213 eggs in the largest clump. These eggs were light yellow to orange in colour and, when eyed, averaged 3.2 mm in diameter. At 10 °C, laboratory-reared eggs hatched in 29 days, and the resulting yolk-sac fry averaged 7.5 mm TL. For the first 2 weeks after hatching, the fry remained under rocks and in gravel. Exogenous feeding began 15 days after hatching. At this time, the fry were 9.5–10 mm TL.

AGE, GROWTH, AND MATURITY

Growth is moderately slow: in the Slocan River, fry reach a length of 25–35 mm TL by late August. The smallest sexually mature shorthead sculpins in Beaver Creek near Trail and in the Slocan River were a 45 mm female and a 46 mm male. All of these fish were in their third growing season (2+). So far, the oldest shorthead sculpin found in British Columbia was an 85 mm male from the Kettle River just below Cascade Falls. This fish was in its seventh growing season (6+).

FOOD HABITS

Sculpins forage mainly at night and use the lateral line system to locate prey (Janssen 1990). In Idaho, shorthead sculpin foraged on the larvae of aquatic insects, especially chironomids, caddisflies, and mayflies (Johnson et al. 1983). From the little information available in British Columbia, the diet appears to be similar. In the Slocan River, the stomachs of young-of-the-year shorthead sculpin collected in the summer contained mostly chironomid larvae. Chironomid larvae were also a major food item in juveniles and adults; however, the larger fish also contained larvae of other aquatic insects (again, mostly mayfly nymphs and caddisfly larvae). Although adults and juveniles have a more varied diet than the young-of-the-year, chironomid larvae appear to remain an important food source throughout life.

Habitat ADULTS

In the warmer parts of the Columbia Basin (Washington, Oregon, and Idaho), shorthead sculpin are considered a cold-water species, characteristic of headwater habitats (Bailey and Bond 1963). In British Columbia, however, the shorthead sculpin is a valley-floor species. In the Slocan River system, they are found in the lower reaches of tributary streams but are replaced in the cold headwaters by the slimy sculpin.

In the Castlegar–Trail region, summer stream temperatures often exceed 20 °C. In this area, shorthead sculpin appear to prefer small tributary streams (<30 m wide). Nonetheless, they are present in the lower Kootenay River, and occasionally, they are collected in the mainstem Columbia. Thus, this species is not necessarily restricted to small streams.

In summer, adults in the B.C. populations are associated with moderate to swift surface currents (0.30–0.90 m/s), depths of <50 cm, and large rock and cobble substrates. Typically, they coexist with torrent sculpin and, less commonly, Columbia sculpin. At such sites, during the day, shorthead sculpin are found in faster water than either of the other two species.

JUVENILES

During high water, juveniles are associated with shallow water (usually <20 cm deep), slow currents (<0.10 m/s at the surface), and sand substrates with scattered rocks and boulders for cover. As water levels subside, juveniles move into faster (about 0.20 m/s) and deeper water over coarse gravel or cobble substrates. By late September, some juveniles have moved into adult habitats.

YOUNG-OF-THE-YEAR

In the Slocan River in early July, newly hatched shorthead sculpin are common in flooded vegetation along the edges of the main river channel. These sites are shallow (<10 cm deep), with a slow current (<0.10 m/s), and mud–sand substrates.

Conservation Comments — The shorthead sculpin is listed as threatened by COSEWIC and as S3 (rare or uncommon) by the BCCDC. However, the shorthead sculpin's Canadian distribution is more restricted than was thought when this species was originally assessed. They are uncommon in the mainstem Columbia River but are relatively abundant in a limited number of tributary streams. Here they occur in short (1–2 km) stream reaches sandwiched between the main Columbia River and upstream velocity barriers. Thus, if these local populations are extirpated, the habitat may not be recolonized for a long time. Additionally, the narrow valley-floor regions used by shorthead sculpin are vulnerable to human encroachment (especially transportation corridors). The shorthead sculpin's conservation status probably should be re-assessed in light of recent information.

Cottus hubbsi BAILEY & DIMICK
COLUMBIA SCULPIN

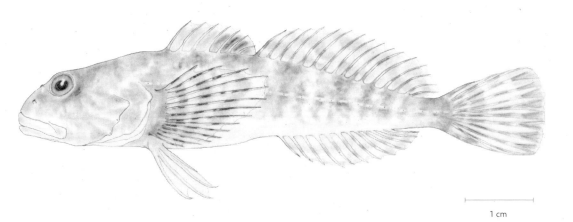

1 cm

Distinguishing Characters — The following description pertains to the Columbia sculpin found within British Columbia; however, there is an unresolved taxonomic problem with this species in the Similkameen River system (see the taxonomy section below). Usually, this sculpin is strikingly marked with broad bars on the caudal fin and oblique dark bars on the anal fin. It has a wide mouth and a relatively large head (2.8–3.3 times into standard length); the lateral line is usually complete (28–36 lateral line pores) or, if it ends on the caudal peduncle, a trace line is visible that extends back to the caudal fin. There is a large patch of prickles (usually >40, but see the taxonomy section) under the pectoral fin and a scattering of prickles (5–20) above the anterior portion of the lateral line. There are two median chin pores, well-developed palatine teeth that are slightly separated from the vomerine tooth patch, 11–14 anal rays, 16–18 (usually 17) dorsal rays, and 4 pelvic rays. The posterior nostrils are semi-tubular, and in some populations, they are almost as long as the anterior nostril tubes.

In British Columbia, the Columbia sculpin co-exists with prickly (*Cottus asper*), torrent (*Cottus rhotheus*), and shorthead sculpins (*Cottus confusus*). In the field, the length of the anal fin base distinguishes Columbia and prickly sculpins (the anal fin base less than the head length in the Columbia sculpin and longer than the head length in the prickly sculpins). The dark saddles under the soft dorsal fin are perpendicular to the long axis of the body in Columbia sculpin and slant forward in the torrent sculpin. Also, the chin in the Columbia sculpin is lightly, or evenly, pigmented vs. alternating patches of dark and light pigment on the chin of torrent sculpin. Columbia sculpin are distinguished from the shorthead sculpin by caudal peduncle depth (4.8–5.7 into head length in the Columbia sculpin vs. 3.1–4.5 in the shorthead sculpin) and the length of the lateral line (usually complete and extending onto the caudal peduncle in Columbia sculpin and typically incomplete and ending before the end of the second dorsal fin in shorthead sculpin). A useful field character (although not infallible) is pigment on the anal fin (this fin is usually strongly barred in Columbia sculpin and uniformly dusky in shorthead sculpin).

Taxonomy The Columbia sculpin has a checkered taxonomic history. Originally, it was confused with the torrent sculpin but, eventually, recognized as a distinct species and described as *Cottus hubbsi* (Bailey and Dimick 1949). Later, in a review of western North American sculpins, Bailey and Bond (1963) synonymized the Columbia sculpin with the mottled sculpin (*Cottus bairdii*). Consequently, since 1963, the mottled sculpin in the Columbia River system has been called *C. bairdii*; however, *C. bairdii* was named from Ohio (Girard 1850), and there is a major range disjunction between eastern and western North American mottled sculpins (Bailey and Bond 1963; Lee 1980).

Recently, mitochondrial sequences of mottled sculpins from Ontario (Great Lakes drainages) and the Columbia River system were compared (J.D. McPhail and E.B Taylor, in preparation), and there is a substantial divergence (about 5–6%) between the two areas. The depth of this divergence suggests that eastern and western mottled sculpins have been isolated since the Pliocene (3–5 Ma). Consequently, I recognize the Columbia sculpin as a species distinct from *C. bairdii*.

An additional taxonomic complication is the presence in the Harney Basin, Oregon, of both the Columbia sculpin and a related species, the Malheur mottled sculpin (*Cottus bendirei*). Apparently, there is a complex geographic pattern of hybridization between Columbia sculpins and the Malheur mottled sculpin (Markle and Hill 2000). These authors also contend that the latter species occurs in the Columbia system outside the Harney Basin and suggest (on the basis of the morphological data of Peden et al. 1989) that this species may occur in British Columbia. Although *C. bendirei* has never been recorded from British Columbia or Washington, specimens of *C. hubbsi* in Otter Creek (and other streams in the upper Similkameen system) approach *C. bendirei* in some aspects of their morphology—they have relatively few prickles (0–10) behind the pectoral fins, lack prickles above the lateral line, lack bold markings on the fins and flanks, and have fewer pectoral rays (13–15, usually 14) than most *C. hubbsi* populations (14–16, usually 15). Additionally, some Otter Creek fish differ from *C. hubbsi* in the mainstem Columbia in their mitochondrial sequences. Although these sequences align with *C. hubbsi* in phylogenies, the differences are substantial (3–4%). Until *C. bendirei* sequences from Oregon become available, the taxonomic status of these unusual sculpins will remain unclear; however, it is possible that *C. bendirei*, or hybrids between *C. bendirei* and *C. hubbsi*, occur in British Columbia. Until this problem is resolved, I have retained *C. hubbsi* as the scientific name for all the B.C. populations of Columbia sculpins; however, in the upper Similkameen system and, perhaps, elsewhere in the Canadian portion of the Columbia system, this should be viewed as a provisional identification.

Sexual Dimorphism On average, sexually mature male Columbia sculpin are larger than mature females. The entire body in breeding males is black except for an orange trim on the first dorsal fin. Breeding females retain their normal colour but have noticeably swollen abdomens. Outside the breeding season, adults can be sexed by examining the urogenital papillae through a hand lens. Although the urogenital papilla changes seasonally, it is an obvious structure (either triangular or bluntly rounded) immediately behind the anus in males and an inconspicuous mound or tube in non-breeding females.

Hybridization The Columbia sculpin appears to hybridize occasionally with torrent sculpin and, more commonly, with shorthead sculpin (see the hybridization sections under these species). This suggestion is based on the presence of morphologically intermediate individuals in areas where the species co-occur; however, given our ignorance about the limits of variation in these species, such unusual individuals simply may be extreme examples of one of the species. In Oregon, Columbia sculpin appear to hybridize extensively with the Malheur mottled sculpin (Markle and Hill 2000).

Distribution **NATIVE DISTRIBUTION**
The Columbia sculpin is endemic to the Columbia River system below Shoshone Falls. It occurs only in Washington, Oregon, Idaho, and British Columbia.

BRITISH COLUMBIA DISTRIBUTION
The Columbia sculpin is known from the Similkameen, Tulameen, and Kettle (below the barrier at Cascade) rivers, the Columbia River and its tributaries below Arrow Lakes, and the Kootenay River and its tributaries below Bonnington Falls (Map 73).

Life History No life-history study is available for the Columbia sculpin. Consequently, little is known about its life history in British Columbia. The following account is derived mainly from observations made in the Tulameen drainage (Otter Creek) and may pertain to another species (see the taxonomic section above).

REPRODUCTION
The Columbia sculpin spawns in the spring. In Otter Creek, nests with eggs were found from late May (water temperature 7 °C) to mid-June (water temperature 12 °C). Only prickly and Columbia sculpins occur in Otter Creek (but see the taxonomic section above), and they differ strikingly in egg size; thus, the nests of the two species are easily distinguished. Like most freshwater sculpins, Columbia sculpin males nest under rocks and are territorial. Breeding males are dark with a yellow-orange trim

Map 73 The B.C. distribution of Columbia sculpin, *Cottus hubbsi*

on the first dorsal fin. Courtship behaviour has not been documented in this species. Presumably, however, it is similar to that described for mottled sculpins (for details see the description of spawning behaviour in the slimy sculpin, *Cottus cognatus*, species account). Females deposit their sticky eggs on the underside of the nest rock, where the male fertilizes them. The male guards the eggs until they hatch. In Otter Creek, all the egg masses collected were close to hatching. Most nests contained two separate egg clumps. There were no nests with more than two clumps; however, two nests contained a single egg clump. Presumably, each egg clump represents the clutch of a single female and, when there were two, the egg clumps were separated by at least 1 cm. The egg number in each clump was estimated to be between 50 and 100. This is consistent with fecundity estimates made in the Yakima River, Washington. Here, egg numbers ranged from 46 in a female 46 mm TL to 275 in a 91 mm female (Patten 1971). Egg size, as measured from Otter Creek nests, averaged 2.8 mm. Presumably, the incubation period is temperature dependent. Newly hatched larvae are unpigmented and range from 7.5 to 8.2 mm TL. They remain hidden in the gravel until the yolk sac is absorbed (about 2 weeks at 12 °C). After transformation, they emerge from the gravel at about 9.5–10.5 mm. At this stage, they are pigmented and begin exogenous feeding. In Lawless Creek, Similkameen River system, newly transformed fry (about 12 mm) were observed in early July.

AGE, GROWTH, AND MATURITY

In British Columbia, the young-of-the-year range from 13 to 18 mm TL by mid-July, and they are 25–35 mm long by late September. Males begin to mature at the end of their second summer, and most spawn for the first time at the beginning of their third summer. Most females mature a year later than males, but some also mature in their third summer. So far, the oldest Columbia sculpin recorded in British Columbia was a 106 mm TL male in his sixth summer (5+).

FOOD HABITS

Like other sculpins, Columbia sculpin are carnivores. Newly transformed individuals feed on small chironomid larvae but quickly add larger larvae and nymphs of other aquatic insects to the diet. Adults and juveniles have a more varied diet than the young-of-the-year; however, aquatic insects (especially mayflies, caddisflies, and stoneflies) are still the main food. Large adults (>70 mm long) have relatively large mouths and include larger more agile prey (e.g., small fish) in their diet.

Habitat

ADULTS

Most B.C. records of Columbia sculpin are from rivers and streams that range in size from the mainstem Columbia to creeks less than 5 m wide; however, they once occurred in lakes in the Similkameen system. Unfortunately, most of these lakes were "rehabilitated" in the 1950s before the biology of lacustrine Columbia sculpin could be studied. In the Castlegar–Trail area, adult Columbia sculpin typically coexist with torrent, shorthead, and prickly sculpins. Here, they appear to favour larger rivers than the shorthead sculpin and are rarely collected in small tributaries. In contrast, in the Similkameen system (where there are no shorthead sculpin), Columbia sculpins are abundant in small streams. They share these small stream environments with prickly sculpin, and the two species appear to use different microhabitats—prickly sculpins in pools over fine substrates and Columbia sculpins in riffles over cobble substrates. Like most sculpins, Columbia sculpin are night active and difficult to observe during the day. In the summer, in the Columbia River, adults are associated with moderate currents (0.3–0.6 m/s), cobble to boulder substrates, and depths of about 40–100 cm (R. L. & L Envronmental Services Ltd. 1995a).

JUVENILES

During high water, juveniles typically occur in shallower and quieter water (such as seasonally flooded embayments) than adults. As water levels subside, juveniles move into riffle habitats similar to, but less extreme, than those used by adults (water <40 cm deep, currents <0.2 m/s, and coarse gravel substrates). In a series of exclusion experiments in cages, juveniles of the related mottled sculpins preferred the same microhabitats as adults (Freeman and Stouder 1989). When adults were added to cages, however, juveniles shifted to shallower microhabitats.

Similar interactions among age classes, rather than habitat preferences, may influence the distribution of juvenile Columbia sculpins.

YOUNG-OF-THE-YEAR
In July and August, Columbia sculpin fry occur in quiet, shallow water (typically <0.1 m/s and <20 cm deep). In the mainstem Columbia, they are associated with submerged vegetation adjacent to patches of open sand. During spates, fry shelter in flooded vegetation as close as possible to the stream margins.

Conservation Comments COSEWIC listed the Columbia sculpin as a species of special concern (as a subspecies of mottled sculpin) and the BCCDC listed it as S3 (rare or uncommon). We now know these listings were based on two allopatric species: the Columbia sculpin and the Rocky Mountain sculpin. Consequently, both of their geographic distributions are more restricted than the combined distribution on which the original assessments were based. Presumably, these species will be reassessed in light of the change in their taxonomic status. Additionally, the taxonomic problems with the B.C. populations of Columbia sculpins (see the taxonomy section) suggest that this species needs a detailed review.

Cottus rhotheus (SMITH)
TORRENT SCULPIN

1 cm

Distinguishing Characters — This large-headed sculpin has a narrow caudal peduncle (about 5–7 times into head length), densely prickled skin, a complete lateral line (28–34 lateral line pores), and well-developed palatine teeth that are in contact with the vomerine tooth patch. The dorsal surface of the head is densely covered with raised papillae that often extend onto the tops of the eyes. The fins usually are conspicuously barred, and the chin and lower jaw are mottled (i.e., with alternating areas of light and dark pigment). Typically, there are two or three dark, forward-slanting, saddle-like patches below the second dorsal fin. There are usually 3 or 4 preopercular spines, 11–13 anal rays, and 15–17 rays in the second dorsal fin. There are 14–16 pectoral rays and 4 pelvic rays, with the first and fourth ray usually shorter than the two middle rays.

In British Columbia, the only other heavily prickled sculpin south of the Mackenzie River system is the prickly sculpin (*Cottus asper*). Prickly and torrent sculpins coexist in most of the Columbia River system and in part (the North Thompson River) of the Fraser River system. Median chin pores (two in the torrent sculpin and one in the prickly sculpin) and anal fin ray number (11–13 in torrent sculpin and 16–19 in prickly sculpins) distinguish the species.

Taxonomy — At present, the torrent sculpin pose no taxonomic problems; however, the species is in need of study. The DNA of a few individuals has been sequenced in British Columbia, Washington, and Oregon, and the sequence differences among Willamette, Snake, upper Columbia, and Chehalis torrent sculpins appear to be greater than normally found among geographically scattered populations within a river system. Additionally, there are morphological differences in prickle development and body shape that show geographic patterns. For example, in western Washington (the Olympic Peninsula and southeastern Puget Sound drainages) and in some lower Columbia streams, there is a form of torrent sculpin that lacks prickles and has a deep caudal peduncle. Elsewhere in the Columbia Basin, torrent sculpins are heavily prickled and have narrow caudal peduncles. This geographically patterned variation suggests some divergence in allopatry.

In British Columbia, there is also some variation in colour pattern and in the degree of prickling on the body. Populations isolated above barriers typically lose their prickles and have more sharply contrasting markings than populations from below barriers.

Sexual Dimorphism On average, adult males and females are about the same size, although most of the largest specimens are females. Breeding males are dark with a yellow-orange trim on the first dorsal fin. Breeding females retain their normal colour but have noticeably swollen abdomens. Outside the breeding season, adults can be sexed by examining the urogenital papillae through a hand lens. Although the urogenital papilla changes seasonally, it is an obvious structure (either triangular or bluntly rounded) immediately behind the anus in males, and an inconspicuous mound or tube in non-breeding females.

Hybridization In British Columbia, hybrids involving torrent sculpin are rare. Occasional unusual specimens of the Columbia sculpin (*Cottus hubbsi*) are collected in the mainstem Columbia River in the Trail–Castlegar area. These rare individuals have narrower caudal peduncles and more prickles than typical Columbia sculpins from the same area. In general body shape, they look intermediate between torrent and Columbia sculpins, but there is no hard evidence that they are hybrids.

Distribution NATIVE DISTRIBUTION

The torrent sculpin is a Columbia endemic. As such, it occurs only within the Columbia River system (from the estuary to the source) plus a few rivers in western Washington (McPhail 1967), western Oregon (Finger 1982), and the North Thompson River in British Columbia. It does not occur above Shoshone Falls on the Snake River.

BRITISH COLUMBIA DISTRIBUTION

The torrent sculpin occurs in the Similkameen and Kettle rivers (below barriers), the Pend d'Oreille River, the Kootenay River upstream to the Cranbrook area, and the Columbia River and its tributaries from the U.S. border to Columbia Lake. Postglacially, torrent sculpins dispersed from some upper Columbia tributary (perhaps the Canoe River) into the North Thompson River (Fraser River system). In the North Thompson River and in the lower reaches of tributary streams above Heffley, torrent sculpins are moderately common (Map 74).

Map 74 The B.C. distribution of torrent sculpin, *Cottus rhotheus*

Life History   Curiously, no life-history study is available for the torrent sculpin, and little is known about its life history in British Columbia. The following account is derived mainly from original observations made in the Similkameen and Slocan rivers and Columbia Lake. This information was augmented with data collected by R. L. & L. Environmental Services Ltd. (1995a) in the mainstem Columbia and its tributaries between Keenleyside Dam and the U.S. border.

REPRODUCTION

In Columbia Lake, torrent sculpin spawn shortly after ice-out (usually mid-April); however, in flowing water, eyed torrent sculpin eggs have been found as early as April (water temperature 7 °C) and as late as early June (water temperature 15 °C). Apparently, this species sometimes performs pre-spawning migrations. An early spring (mid-March to mid-April) migration of torrent sculpins was observed in western Washington (Thomas 1973). Spawning migrations into tributary streams might explain the observation that, in the mainstem Columbia River, torrent sculpin are less abundant in the spring than in the fall (R. L. & L. Environmental Services Ltd. 1995a).

Like most freshwater sculpins, males nest under rocks and are territorial. Courtship behaviour has not been documented in this species, but presumably, it is similar to that described for other species (for a description of spawning behaviour, see the slimy sculpin, *Cottus cognatus*, species account). Females deposit their sticky eggs on the underside of a nest-rock,

where the male fertilizes them. The male guards the eggs until they hatch. Apparently, most males are polygamous, and each nest usually contains eggs of several shades of yellow-orange and at different development stages. Presumably, these represent the clutches of several females.
In the Yakima River, Washington, egg numbers ranged from 100 in a female 69 mm TL to 412 in a 108 mm female (Patten 1971). Egg size, as measured by eggs taken from nests, averages about 2.8 mm. The incubation period is temperature dependent, and newly hatched young range from about 7.5 to 8 mm TL. They remain in gravel near the nest site for about 2 weeks. Newly transformed fry (about 10–12 mm) are found from early June to the beginning of July. Exogenous feeding begins at about 10 mm (apparently before they emerge from the gravel). By about 12 mm, the young-of-the-year exhibit the adult colour pattern and are found in shallow, edge habitats. In lakes, newly hatched larvae (still with yolk sacs) are sometimes taken at night in plankton tows.

AGE, GROWTH, AND MATURITY

In British Columbia, young-of-the-year range from 20 to 30 mm TL by mid-July and, by late September, are 30–40 mm long. Males begin to mature at the end of their second summer, and most spawn for the first time at the beginning of their third summer. Most females mature a year later than males, but some also breed in their third summer. So far, the oldest torrent sculpin recorded in B.C. was a female 135 mm TL. She was entering her eighth summer (7+).

FOOD HABITS

In the summer, adult and juvenile torrent sculpin feed primarily at night; however, large adults sometimes forage during the day and have been observed to burrow into sand substrates and from there ambush cyprinid and sucker fry. In flowing water, the main prey of adults and juveniles are the nymphs and larvae of aquatic insects (especially caddisflies and mayflies). Adults become modestly piscivorous—small minnows, suckers, and sculpins are major food items—nonetheless, insect nymphs and larvae remain their primary food. In lakes, the planktonic larvae feed on microplankton, but chironomid larvae are their primary prey once they transform. In lacustrine populations, torrent sculpin fry include copepods, ostracods, and amphipods in their diet (Northcote 1954). Information on the winter diet of the torrent sculpin is scarce; however, in the Columbia and Kootenay rivers, aquatic insects were the primary winter prey (R. L. & L. Environmental Services Ltd. 1995a).

Habitat ADULTS

In British Columbia, torrent sculpin are found in lakes, streams, and rivers. As their name implies, adults appear to prefer fast water (surface velocities ranging from 0.30 to >1.0 m/s) with boulder and cobble substrates; however, the microhabitats (interstices among rocks and rubble) in which they shelter during the day have much lower water velocities of 0.0–0.4 m/s (R. L. & L. Environmental Services Ltd. 1995a).

In flowing water, the habitats used by adult and juvenile torrent sculpin shift seasonally and in response to the presence of other sculpin species (Finger 1982). In Columbia and Kootenay River tributaries, they often coexist with shorthead sculpin (*Cottus confusus*). At such sites, torrent sculpin shift habitat to quiet water (pools with surface velocities of 0.1–0.3 m/s and fine gravel or sand substrates). Here, they often use woody debris as cover and sometimes burrow into soft substrates. In the Arrow Lakes, adults occurred in shallow water (often <1 m deep) on cobble and rubble substrates (Northcote 1954). Torrent sculpin also occur in relatively small lakes like Windemere and Columbia lakes. In Columbia Lake, adult torrent sculpin occupy shallow water (<1.0 m deep) on gravel beaches.

JUVENILES

In lakes and streams, juvenile torrent sculpin occupy habitats similar to, but less extreme than, those used by adults. Thus, they occur in shallower (<30 cm deep) and slower (<0.20 m/s) water than adults. An exception, however, is the spring freshet. At this time, fluvial juveniles shelter among seasonally flooded vegetation in still-water areas.

YOUNG-OF-THE-YEAR

During periods of high water, newly emerged (about 12 mm TL) torrent sculpin occur in quiet water, usually in or near vegetation, and over sand–mud bottoms. By late summer, they move into deeper water with more current and gravel substrates. In Columbia Lake in early May, torrent sculpin larvae (8–9 mm TL and still with yolk sacs) were collected in inshore tow-net samples at night over coarse gravel substrates. By mid-May, larvae taken in tow-net samples at night were about 10 mm long, and their yolk sacs were almost completely absorbed. They were captured at least 1.5 m off the bottom. Curiously, in Columbia Lake, electro-fishing along gravel and rock shores throughout the summer produced juveniles and adults but no young-of-the-year. Then, in late August, young-of-the-year suddenly recruited to the shallow littoral zone. Where they were in the intervening 2 months is unknown.

Conservation Comments   The torrent sculpin has not been assessed by either the BCCDC or COSEWIC; however, most populations appear to be healthy. In the west Kootenay region, there are at least two populations isolated above barriers. Both these isolated populations have lost their prickles and have unusually bold markings. Additionally, they both appear to prefer quiet water and pools to the typical riffle habitat. They may be unusual enough to qualify as designated units.

Cottus ricei (NELSON)
SPOONHEAD SCULPIN

1 cm

Distinguishing Characters   This slim, heavily prickled sculpin has a broad, flat head; elongate and strongly curved upper preopercular spines, a deeply wrinkled chin, and a single median chin pore. The only other B.C. sculpins with a single median chin pore are the coastrange sculpin (*Cottus aleuticus*) and the prickly sculpin (*Cottus asper*). The geographic ranges of spoonhead and coastrange sculpins do not overlap in British Columbia. Consequently, the only places where they might be confused are classrooms. In contrast, the ranges of spoonhead and prickly sculpins overlap in the lower Peace River and some of its tributaries. Because both species usually are heavily prickled and have a single median chin pore; they are occasionally misidentified in this region. The depth of the caudal peduncle and the development of palatine teeth separate the species—spoonhead sculpin have an exceptionally narrow caudal peduncle (five to seven times into head length), whereas the caudal peduncle in the prickly sculpin is moderately deep (three to four times into head length); prickly sculpin have strong palatine teeth, and spoonhead sculpin lack palatine teeth. For adults, the peculiarly puckered chin pores on spoonhead sculpin are a useful field character.

Taxonomy   The spoonhead sculpin is the most distinctive of the eight *Cottus* species found in British Columbia. In their 1961 review of the freshwater sculpins of British Columbia, McAllister and Lindsey argued that the spoonhead sculpin was more closely related to some Eurasian sculpins (e.g., *C. sibiricus*, *C. spinulosus*, and *C. gobio*) than to any of the North American species of *Cottus*. A recent molecular review (Kinziger et al. 2005) supports the close relationship of the spoonhead sculpin to some Eurasian species.

Sexual Dimorphism   On average, mature males and females are about the same size. In breeding males, the entire body is dark except for a yellowish trim on the first dorsal fin. Breeding females retain their normal colour but have noticeably swollen abdomens. Outside the breeding season, adults can be sexed by examining the urogenital papillae through a hand lens. Although the urogenital papilla changes seasonally, it is an obvious structure (either triangular or bluntly rounded) immediately behind the anus in males and an inconspicuous mound or tube in non-breeding females.

In British Columbia, the sides and backs of adult females (and juveniles of both sexes) are heavily prickled, whereas adult males appear to be smooth. Usually, however, the bases of the prickles are embedded in the skin of adult males. This suggests that the protruding portions of the prickles on males have eroded, and since adult males guard nests under rocks, they may be more subject to abrasion than females.

Distribution   NATIVE DISTRIBUTION

The spoonhead sculpin is native to North America where it is distributed from the lower St. Lawrence River in Quebec, throughout the Great Lakes, and westward across the northern Great Plains to the eastern slope of the Rocky Mountains. In the west, it ranges from the Oldman and, perhaps, the Milk (see comments in Nelson and Paetz 1992) rivers in Alberta north to the Mackenzie Delta.

BRITISH COLUMBIA DISTRIBUTION

In British Columbia, the spoonhead sculpin is restricted to the large rivers east of the Continental Divide: the lower Peace River and its major tributaries and, the lower Liard River and its major tributaries (Map 75).

Life History   Little is known about the life history and ecology of the spoonhead sculpin in British Columbia. Data on eastern populations are summarized in Scott and Crossman (1973); however, these are mainly lacustrine populations, and most B.C. populations are fluvial. There is some life-history information on spoonhead sculpins in Alberta (Bond 1980; Roberts 1988). The life-history summary in Roberts (1988) is the most complete for western populations and is the primary source of the following account supplemented, where appropriate, with original B.C. observations.

REPRODUCTION

Spoonhead sculpin spawn shortly after ice-out. Apparently, spawning starts when water temperatures reach 4 °C, and depending on year and latitude, spawning may start as early as April or as late as May (Roberts 1988). Like most freshwater sculpins, males nest under rocks and are territorial. Breeding males are dark with a yellow-orange trim on the first dorsal fin. Courtship behaviour has not been documented in this species. Presumably, however, it is similar to that described for other freshwater sculpins (for details of spawning behaviour, see the slimy sculpin, *Cottus cognatus*, species account). Females deposit their sticky eggs on the

Map 75 The B.C. distribution of spoonhead sculpin, *Cottus ricei*

underside of the nest-rock, where the male fertilizes them. The male guards the eggs until they hatch. Although some males are polygamous and may spawn with up to three females, slightly over 60% of the nests examined contained only a single egg mass (Roberts 1988).

Fecundity is a function of female size and, in Alberta, varies from 280 to 1,200 eggs. No data are available on the size of fertilized eggs; however, a gravid female taken from Lake Superior in late April had a mean egg size of 1.4 mm (Becker 1983). At 8 °C, the eggs hatch in about 21 days. The newly hatched larvae are roughly 7 mm TL and have a large yolk sac. Apparently, at this stage, they are day active and begin exogenous feeding when the yolk reserves are gone (about 2 weeks after hatch in other sculpins with similar-size larvae). The switch to nocturnal foraging occurs at about 10 mm. By this size, the fry are miniature copies of the adults.

AGE, GROWTH, AND MATURITY

In the Muskwa River, young-of-the-year spoonhead sculpin reach an average size of around 25 mm by late July and are 35–45 mm long by mid-September. By the end of their second growing season, they are about 60–70 mm in length and, apparently, reach sexual maturity at this time. In Alberta, they probably first spawn at the beginning of their third summer (Roberts 1988).

FOOD HABITS

Spoonhead sculpin are carnivorous. The fry forage primarily on chironomids, but as they grow, larger nymphs and larvae of other aquatic insects are added to the diet. Adults and juveniles have a more varied diet than the young-of-the-year, but the larvae and nymphs of aquatic insects (especially mayflies, stoneflies, and caddisflies) are still their primary food.

Habitat So far, in British Columbia, the spoonhead sculpin has only been collected in flowing water; however, in eastern North America this species is strongly associated with lakes (Dadswell 1972; Houston 1990).

ADULTS

In British Columbia, adult spoonhead sculpin are rarely collected, and most records of this species are based on young-of-the-year or juveniles. Typically, when adults are collected, one or two individuals are taken along the edges of large, turbid rivers (e.g., the lower Liard, Fort Nelson, Muskwa, and lower Peace rivers) over coarse gravel substrates in areas of moderate (<0.75 m/s) surface currents. The only large catches of adults (up to 10 individuals at a site) were made in tributaries to larger rivers. Here, they were found in fast (>0.75 m/s), shallow (<25 cm) riffles with a substrate of loose, fist-sized rocks.

JUVENILES

In the summer, juvenile spoonhead sculpin are common along the margins of large rivers. Typically, the water at such sites is shallow (<50 cm) and slow (<0.25 m/s), and the substrate is fine gravel or sand. They often congregate in embayments associated with the confluence of small tributary streams and large rivers where clear tributary water and turbid river water meet. At such sites, juvenile spoonhead sculpins usually are found in the turbid water, and juvenile slimy sculpins in the clear water.

YOUNG-OF-THE-YEAR

In the summer, most spoonhead fry are found in shallow backwaters and isolated pools left by receding water levels. The substrate is usually soft silt or sand, and there is little or no current. When disturbed, the fry burrow into the bottom.

Conservation Comments The apparent rarity of adult spoonhead sculpins in British Columbia may be a reflection of the difficulty of sampling turbulent water in large rivers rather than actual rarity, and occasionally, adults are encountered in large numbers. For example, in Buick Creek near Prestapou, over 50 adults were collected with a kick-net at night by disturbing rocks in about 10 m of swift riffle. Because of its extensive distribution in the rest of Canada, COSEWIC does not list this species, and the BCCDC has not assessed the spoonhead sculpin in British Columbia.

Cottus sp.
ROCKY MOUNTAIN SCULPIN

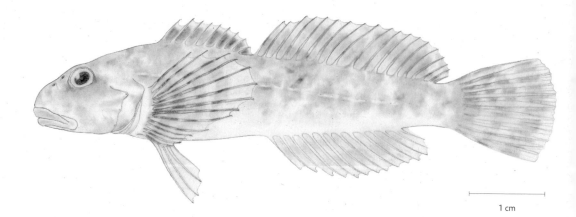

1 cm

Distinguishing Characters   This slender sculpin has a relatively short head (about 3.0–3.8 times into standard length) covered in small papillae. The palatine teeth are weak, there are no prickles (at least in British Columbia) on the body, the lateral line is incomplete (21–26 lateral line pores), and there are 13–15 (usually 14) pectoral fin rays. In British Columbia, this sculpin usually has a single postmaxillary pore.

In British Columbia, the Rocky Mountain sculpin only occurs in the lower Flathead River system. This greatly simplifies identification, because the only other sculpin in this drainage is the slimy sculpin (*Cottus cognatus*). The two species are separated by their palatine teeth (absent in slimy sculpin and weak, but present, in Rocky Mountain sculpin), anal fin ray number (usually more than 12 in Rocky Mountain sculpin and less than 12 in slimy sculpin), the width of the isthmus (usually wider than the eye in Rocky Mountain sculpin and about equal to the eye in slimy sculpin), and the presence of papillae (nubbles) on the head of Rocky Mountain sculpin (absent in slimy sculpin).

Taxonomy   There are two sculpins in the Flathead River system: the well-known slimy sculpin and another sculpin whose identity has been debated for decades. The first B.C. mention of this species is in McAllister and Lindsey (1961). These authors recognized that it was not *C. cognatus* and listed it as *Cottus* sp. Later, a new species of sculpin (the shorthead sculpin, *Cottus confusus*) was named from the Salmon River, Idaho (Bailey and Bond 1963). The geographic distribution of *C. confusus* given in the original description included B.C. records from the Flathead River (Bailey and Bond 1963). The ecology and taxonomy of the two sculpins in the Flathead River were re-examined in 1984 (Hughes and Peden 1984), and the two species were referred to as *C. cognatus* and *C. confusus*.

Later, shorthead sculpin were collected in lower Columbia River tributaries in the Castlegar–Trail region and in the Kettle River below Cascade Falls (Peden et al. 1989). These lower Columbia *C. confusus* were shown to be morphologically different from the nominal shorthead sculpin in the Flathead River system (Peden et al. 1989). Thus, there appeared to be two disjunct, and morphologically divergent, forms of *C. confusus* in British Columbia: those in the Flathead River system and those in the lower Columbia and lower Kettle drainages. In 1998, the Flathead River form was listed as the mottled sculpin (*Cottus bairdii*) and the lower Columbia and Kettle form as *C. confusus* (Cannings and Ptolemy 1998).

The identification of the Flathead form as *C. bairdii* was consistent with reports of the widespread presence of *C. bairdii* in upper Missouri tributaries in Alberta (Henderson and Peter 1969), Montana (Brown 1971), and Wyoming (Baxter and Simon 1970), especially since the Flathead sculpin was shown to be morphologically similar to the sculpin in the Milk (an upper Missouri tributary) and South Saskatchewan rivers in southwestern Alberta (Peden et al. 1989).

However, the story does not end here. *Cottus bairdii* is an eastern species named from Ohio (Girard 1850), and there is a substantial range disjunction between eastern and western North American mottled sculpins (Lee 1980). Recently, J.D. McPhail and E.B. Taylor (in preparation) compared mitochondrial sequences of mottled sculpins from Ontario (Great Lakes drainages) with those of the Flathead River system. The comparison revealed a divergence of about 5.5%. The depth of this divergence suggests that eastern and western mottled sculpins have been separated since sometime in the Pliocene (about 3–5 Ma) and argues that the Flathead sculpin is not *Cottus bairdii*. What, then, is the correct scientific name for the Flathead sculpin? A recent Ph.D. dissertation examined relationships within the *C. bairdii* species complex and concluded that the Flathead sculpin (and the sculpins in the South Saskatchewan and upper Missouri River system) are an unnamed species (Dr. D.A. Neely, Saint Louis University, personal communication). Dr. Neely gave this species the common name, Rocky Mountain sculpin; however, until a formal description is published, the scientific name of the Rocky Mountain sculpin in British Columbia is back where it began, *Cottus* sp.

Sexual Dimorphism Breeding males are dark with a yellow-orange trim on the first dorsal fin. Breeding females retain their normal colour but have noticeably swollen abdomens. Outside the breeding season, adults can be sexed by examining the urogenital papillae through a hand lens. Although the urogenital papilla changes seasonally, it is an obvious structure (either triangular or bluntly rounded) immediately behind the anus in males, and an inconspicuous mound or tube in non-breeding females.

Hybridization Slimy and Rocky Mountain sculpins are known to hybridize in the Flathead River (Zimmerman and Wooten 1981).

Distribution NATIVE DISTRIBUTION
The Rocky Mountain sculpin occurs along the eastern slope of the Rocky Mountains from the South Saskatchewan system to the upper Missouri system in Montana and Wyoming. The only populations west of the Continental Divide are in the Flathead River system in British Columbia and Montana.

BRITISH COLUMBIA DISTRIBUTION
The Rocky Mountain sculpin occurs only in the lower 24 km of the B.C. portion of the Flathead River (Map 76) and the lower reaches of its tributary streams (e.g., Middlepass, Commerce, Howell, Cabin, Burnham, Couldrey, Sage, and Kishìnena creeks). In the cooler parts of the upper Flathead River and in headwaters of tributaries, the Rocky Mountain sculpin is replaced by the slimy sculpin (Hughes and Peden 1984).

Life History Little is known about the life history of the Rocky Mountain sculpin in our area. Some life-history data were published (Hughes and Peden 1984) for this species from the Flathead River (under the name *C. confusus*), but the most complete description of the species' life history is a study of a population in southwestern Montana published under the name *C. bairdii* (Bailey 1952). In addition, there is some information on the reproductive biology of Rocky Mountain sculpin in southwestern Alberta (Roberts 1988). The following life-history summary is derived from these sources supplemented, where appropriate, with field observations made in the Flathead River.

REPRODUCTION
In Montana, ripe males were found in late April, and nests with eggs were recorded in early June (Bailey 1952). Water temperatures recorded over the breeding season ranged from 3.8 °C to 11.0 °C (Bailey 1952). In southwestern Alberta, males were found guarding eggs in a tributary of the St. Mary River in mid-May (water temperature 15 °C), and gravid females were present in the main river (water temperature 7.5 °C). Like most sculpins, males nest under rocks and are territorial. Breeding males are dark with a yellow-orange trim on the first dorsal fin.

Courtship behaviour has not been documented in this species. Presumably, however, it is similar to that described for other freshwater sculpins (for details see the description of spawning behaviour in the slimy sculpin species account). Females deposit their sticky eggs on the underside of the nest-rock, where they are fertilized. The male guards and cares for the eggs until they hatch. The number of eggs per nest varied from 54 to 1,884 (Bailey 1952). The number of eggs per female ranged from 69 in a 57 mm female to 406 in a 94 mm female. Consequently, nests with more than 400 eggs probably contain eggs from several females, and this suggests that some Rocky Mountain sculpin males are polygynous.

Map 76 The B.C. distribution of Rocky Mountain sculpin, *Cottus* sp.

Egg size, as measured in nests, averaged 2.5 mm and ranged from 2.5 to 2.9 mm. The incubation period is temperature dependent; at temperatures fluctuating between 5.3 °C and 7.0 °C, hatching began 30 days after fertilization but was not complete until 10 days later (Bailey 1952). Newly hatched young ranged from 5.8 to 8.1 mm, and exogenous feeding began at about 9 mm.

AGE, GROWTH, AND MATURITY

Growth is relatively slow, and young-of-the-year in southwestern Montana averaged 19.5 mm in length by the end of August (Bailey 1952). This is consistent with the observation that, by mid-September, young-of-the-year in the Flathead River averaged 19.2 mm TL. Reports (Hughes and Peden 1984; Roberts 1988) of 0+ fish attaining lengths of 30–40 mm by late autumn probably result from mistaking 1+ fish for young-of-the-year. By the end of their second summer, Rocky Mountain sculpins in the Flathead River range from 36 to 43 mm TL. Most males reach sexual maturity by their third summer (2+), and most females, by their fourth summer (3+). So far, the oldest Rocky Mountain sculpin found in British Columbia was a 95 mm male collected in Howell Creek. This fish was in its seventh growing season (6+).

FOOD HABITS

Young-of-the-year Rocky Mountain sculpin forage mainly on chironomid larvae, but as the fry grow, larvae of other aquatic insects are added to the diet. Although adults and juveniles have a more varied diet than young-of-the-year, chironomid larvae remain an important food source. In the Flathead River, adults consume the larvae and nymphs of larger aquatic insects: mayflies, stoneflies, caddisflies, and coleopterans (Hughes and Peden 1984). The largest sculpins sometimes contain fish remains (Bailey 1952).

Habitat   Little is known about the habitat requirements of the Rocky Mountain sculpin. There is a brief account of the sites where this sculpin was collected (under the name *C. confusus*) in the Flathead River (Hughes and Peden 1984). In addition, there is some information on the ecology of Rocky Mountain sculpin in Montana and southwestern Alberta (Bailey 1952; Roberts 1988). The following habitat-use summary is derived from these sources supplemented, where appropriate, with field observations made in the Flathead River.

ADULTS

In late summer in the lower Flathead River, adult Rocky Mountain sculpin are abundant in shallow (<50 cm) areas with coarse rubble substrates and a moderate surface current (0.2–0.5 m/s). During the day, they shelter in the substrate, but at night, adults are out foraging along the river edges in water <30 cm deep with surface velocities of <0.10 m/s. Interestingly, in September, large males shift into areas with faster surface velocities (>0.6 m/s) and are associated with large rocks and boulders. In such areas, there appeared to be one male per boulder, and these males looked to be in spawning colouration—they were black with conspicuous orange-trimmed first dorsal fins. Thus, although spawning occurs in the spring, adult males may establish breeding territories in the autumn. Apparently, adult Rocky Mountain sculpin are relatively sedentary. A mark and recapture study conducted in Montana found that, after 8 months, 80% of all recaptures were within 50 m of the original marking site (McCleave 1964).

JUVENILES

Although juvenile Rocky Mountain sculpin occupy habitats similar to those used by adults, they usually are found closer to shore in shallower and quieter water than adults. As found in a study of mottled sculpins, this may be a response to the presence of adults in deeper, faster areas rather than a preference for the supposed juvenile habitat (Freeman and Stouder 1989).

YOUNG-OF-THE-YEAR

In autumn, young-of-the-year in the Flathead River were associated with sand and detritus substrates in quiet water areas (pools, root-wads, back-channels, and shallow embayments).

Conservation Comments   COSEWIC lists the Alberta populations, but not the B.C. population, of Rocky Mountain sculpin as threatened. Nonetheless, its restricted distribution in British Columbia makes this species vulnerable to habitat changes. Although the Flathead River valley is isolated, it is not exempt from major development projects. In the past, the area has seen wildcat oil exploration, a proposal for a major open pit coalmine and town site, and extensive clear-cut logging. The increasing international demand for coal will almost certainly generate new interest in the Flathead coal deposits. If modern coal mining comes to the Flathead valley, it will take a major effort by government agencies and conservation groups to ensure the survival of the Rocky Mountain sculpin in British Columbia.

FAMILY CENTRARCHIDAE — SUNFISHES

Although the centrarchids are a relatively small family (about 30 living species) of perch-like fishes, the family belongs to the largest and most diverse order of living vertebrates: the Perciformes. This order contains about 160 families and almost 10,000 species. Most perciform fishes are marine, but the centrarchids live in fresh water. Relative to most B.C. freshwater fish, they are anatomically advanced fishes—they have spines in their dorsal, anal, and pelvic fins; they have two united dorsal fins and lack an adipose fin; their pelvic fins are placed well forward on their bodies; and they have ctenoid (rough) scales. The males of most species dig nest pits in the substrate and guard their eggs and fry.

Initially, the centrarchids were restricted to North America, but several species of sunfish and bass have been planted into temperate waters throughout the world. With one exception, the natural ranges of all the living species of centrarchids originally were restricted to eastern North America. Nevertheless, preglacial fossil centrarchids are known from Alaska, Idaho, Oregon, and Washington. Additionally, there is a living remnant of this ancient western North American centrarchid fauna, Sacramento perch (*Archoplites interruptus*), that is native to the Sacramento – San Joachim drainage system and the adjacent Pajaro and Salinas rivers in central California. Also, species assignable to the same genus occur in Miocene and Pliocene deposits associated with the Columbia and Snake rivers (Kimmel 1975; Smith 1975; Smith et al. 2000; Van Tassell 2001). Apparently, tectonic activity and the onset of glaciation eliminated these ancient bass from northwestern North America. Consequently, all of the centrarchids that now occur in British Columbia are introduced.

FAMILY CENTRARCHIDAE — SUNFISHES

KEY TO SUNFISHES, CRAPPIE, AND BASS

1 (8) Three spines in the anal fin; length of dorsal and anal fin bases unequal, anal fin base at least 1.5 times into dorsal fin base *(see 2)*

    2 (5) Body deep, greatest depth 2.0–2.5 times into standard length (snout to caudal fin base); opercular flap present (Fig. 45A) *(see 3)*

        3 (4) Opercular flap with black centre, light margins, and a prominent red spot at its posterior edge; no dusky spot at posterior base of soft dorsal fin — PUMPKINSEED, *Lepomis gibbosus*

        4 (3) Opercular flap black, no light margins, and no red spot at its posterior edge; dusky spot present at posterior base of soft dorsal fin — BLUEGILL, *Lepomis macrochirus*

    5 (2) Body depth moderate, greatest depth 3.0–3.3 times into standard length; opercular flap absent (Fig. 45B) *(see 6)*

        6 (7) Upper jaw extends past eye; deep notch between spiny and soft dorsal fins (shortest posterior dorsal spine less than half as long as the spine preceding the soft dorsal fin) — LARGEMOUTH BASS, *Micropterus salmoides*

        7 (6) Upper jaw does not extend past eye; notch between spiny and soft dorsal fins not as deep as above (shortest posterior dorsal spine more than half as long as the spine preceding the soft dorsal fin) — SMALLMOUTH BASS, *Micropterus dolomieu*

8 (1) Five spines in the anal fin; dorsal and anal fin bases about equal in length — BLACK CRAPPIE, *Pomoxis nigromaculatus*

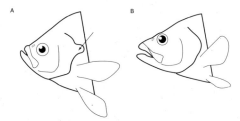

Figure 45 Opercular flap present A) pumpkinseed, and absent B) smallmouth bass.

Lepomis gibbosus (LINNAEUS)
PUMPKINSEED

1 cm

Distinguishing Characters   This small (in British Columbia, rarely longer than 200 mm), colourful, deep-bodied fish has a laterally compressed body, 35–45 scales along the lateral line, and 10 or 11 strong spines in the dorsal fin. Breeding males are particularly beautiful—they have conspicuous wavy blue lines on the sides of the head, an orange or red spot at the tip of a black opercular flap, bronze flanks dappled with an irregular (but interconnected) pattern of lighter wavy lines, and an orange-red belly.

     Aside from the pumpkinseed, the only other *Lepomis* in British Columbia is the bluegill (*L. macrochirus*). Both species occur in Osoyoos Lake and, perhaps, elsewhere in the Okanagan drainage system. Adults of the two species differ in the colour of the opercular flap (tipped in red in the pumpkinseed and blue-black in the bluegill). Additionally, bluegills usually have a diffuse dark spot at the posterior base of the dorsal fin. This spot is faint in juvenile bluegills and absent in pumpkinseeds. Morphologically, the two species can be separated by their gill-raker morphology: the gill rakers on the first gill arch of pumpkinseeds are short and stubby, whereas the gill rakers in bluegills are long and pointed. Apparently, the young-of-the-year also differ in the shape and colour of markings on the last few interradial membranes near the end of the soft dorsal fin (Becker 1983). These marks form well-defined, dark ovals in young bluegills and diffuse, irregular brown markings in young pumpkinseeds.

In British Columbia, pumpkinseeds also co-occur with the black crappie (*Pomoxis nigromaculatus*) in the lower Pend d'Oreille drainage system and in the lower Fraser valley. Although superficially similar, they are easily separated by the lengths of the anal and dorsal fin bases—these are about equal in the black crappie, whereas the anal fin base in the pumpkinseed is less than half as long as the dorsal fin base.

Sexual Dimorphism Although adult pumpkinseeds of both sexes are highly coloured, the colour pattern in breeding females is noticeably more subdued than in males. Additionally, when viewed through water, the light blue trailing edges of the soft dorsal and caudal fins are conspicuous in breeding males and inconspicuous in females, while the dark vertical bands are more prominent in breeding females than in breeding males.

Hybridization Within their native range, pumpkinseeds often hybridize with bluegills (Konkle and Philipp 1992). So far, hybrids pose no identification problem in British Columbia. Still, if bluegills become common in the Osoyoos–Oliver region, hybridization between these species could occur.

Distribution NATIVE DISTRIBUTION
Although the pumpkinseed is native to eastern North America, it has been widely introduced into western North America and elsewhere in the world. The original distribution included lowland areas east and west of the Appalachian Mountains. Along the Atlantic Coastal Plain, it ranged from New Brunswick south to Georgia, and west of the Appalachians from the Great Lakes south to Ohio, and from western Ontario and eastern Manitoba south to Missouri.

BRITISH COLUMBIA DISTRIBUTION
Pumpkinseeds (Map 77) occur in the lower Fraser valley, southeastern Vancouver Island, and in the Columbia system (Columbia and Windemere lakes, the upper Kootenay River, and the Pend d'Oreille and Okanagan river systems). Originally, pumpkinseeds in the Canadian portion of the Columbia system probably dispersed into British Columbia from Washington and Idaho; however, their present scattered distribution suggests human intervention. The original Vancouver Island populations may have been introduced along with smallmouth bass (*Micropterus dolomieu*) (Carl et al. 1959); however, they now occur in many lakes that lack bass.

Life History The only study of the biology of pumpkinseeds in our area is a life-history account included in study of introduced fishes in the Creston Valley Wildlife Management Area (Forbes 1989). Consequently, the following account draws heavily on eastern North American studies.

Map 77 The B.C. distribution of pumpkinseed, *Lepomis gibbosus*

■ introduction

REPRODUCTION

Pumpkinseeds start spawning in the spring when water temperatures reach, or exceed, 15 °C and continue to spawn until surface temperatures reach about 25 °C. In the Creston Valley Wildlife Management Area, pumpkinseeds spawn from mid-May until early August at water temperatures ranging from 16 °C to 23 °C (Forbes 1989). Males excavate a spawning pit by cleaning a shallow depression in the substrate (typically sand or gravel). The pit usually is located in, or near, aquatic vegetation and normally several nests are clumped together in shallow water (<1 m). In Duck Lake near Creston, 94 nests were observed in 100 m^2 (Forbes 1989). Courtship involves circle swimming and lateral displays. Spawning takes place with the male swimming upright and in circles, and the female swimming with the male but tilted at an angle (Breder and Rosen 1966). Typically, several females spawn in the nest of a single male. Depending on body size, females produce from 600 to 12,000 eggs, but only a few eggs are released during each spawning bout. The eggs are about 1 mm in diameter, demersal, and sticky. The male defends his nest diligently, and the vigor of his defense increases with the developmental stage of his clutch (Colgan and Brown 1988). At 18–20 °C, hatching takes 2 or 3 days; the fry are 2.5–3.0 mm TL at hatch (Konkle and Philipp 1992), and about 5 mm TL at swim-up. The male guards the fry for about a week, catching them in his mouth if they stray and returning them to the nest pit. Eventually, however, the fry disperse from the nest.

AGE, GROWTH, AND MATURITY

Depending on food and temperature, pumpkinseed fry grow quite quickly, and in British Columbia, they reach lengths ranging from 17 to 34 mm by the end of August. The minimum age at maturity is 1 year; however, depending on environmental conditions, size and age at maturity vary (Deacon and Keast 1987; Danylchuk and Fox 1994; Fox 1994). Generally, early maturation is associated with high population density and fluctuating environments (Fox and Keast 1990, 1991). Consequently, small pond populations often are stunted. Lifespan varies among populations: it can be as short as 2 years in stunted populations and over 8 years in populations with low adult mortality (Fox and Keast 1990). In British Columbia, pumpkinseeds rarely exceed 15 cm TL and 5 years of age.

FOOD HABITS

Apparently, pumpkinseeds are morphologically specialized for foraging on hard-bodied prey (Hardy 1978a); however, their diet varies with age (Robinson et al. 1993), prey availability (Fox and Keast 1990; Mittelbach et al. 1992), habitat (Kieffer and Colgan 1993), season (Collins and Hinch 1993), and the presence, or absence, of other fish species (Robinson et al. 1993). In some lakes that lack bluegills, there are two behavioural and morphological forms of pumpkinseeds—an open water (limnetic) form that feeds on plankton, and an inshore (littoral) form that forages on snails and the nymphs of aquatic insects (Robinson et al. 1993). Although the peak foraging periods for pumpkinseeds typically are dawn and dusk, they also forage during the day and to a lesser extent at night (Collins and Hinch 1993). Interestingly, this nocturnal foraging is directed mainly at plankton. In spite of this variability, snails make up a large part of the adult diet in most populations. In summer, adults also consume chironomids, the nymphs of aquatic insects, and in some populations, plankton. In the absence of other sunfish, aquatic insects often dominate the diet. Juvenile pumpkinseeds forage on softer bodied prey than adults (Osenberg et al. 1992), and the larvae feed on plankton. Foraging stops when the water temperature drops below 6.5 °C.

Habitat In British Columbia, not much is known about how pumpkinseeds use their habitat, but casual observation suggests that habitat use is similar to that described in eastern North America. Consequently, most of the following account is derived from populations outside our area.

ADULTS

In British Columbia, pumpkinseeds usually are associated with clear, quiet water (ponds, small lakes, sluggish streams, and sloughs). Although, pumpkinseeds usually favour warm littoral areas with dense vegetation and large populations of snails or aquatic insects, under some circumstances (e.g., the absence of bluegills), they diverge into two forms: a littoral form and an open-water form (Robinson et al. 1993). Since most pumpkinseed populations in British Columbia exist in environments that should favour divergence (see Robinson et al. 2000), the two forms may exist in British Columbia.

JUVENILES

Juvenile pumpkinseeds occupy the same habitat as adults.

YOUNG-OF-THE-YEAR

After newly hatched larvae fill their swimbladders, they leave the nest pit and move into open water (Faber 1967). At first, they are transparent but develop pigment as they grow and eventually move back into littoral areas.

Conservation Comments This attractive little fish is a pest. It competes with native species and has extirpated some scientifically valuable stickleback populations on Vancouver Island. It is still spreading on Vancouver Island and the B.C. mainland. Apparently, people keep small pumpkinseeds as aquarium fish but release them when they get too big or becomes a nuisance. Consequently, in British Columbia, it is almost always spread by humans.

Lepomis macrochirus RAFINESQUE
BLUEGILL

1 cm

Distinguishing Characters This small (rarely longer than 200 mm in British Columbia), deep-bodied fish has a laterally compressed body, 39–45 scales along the lateral line, and 10 strong spines in the dorsal fin. Breeding males have a rusty-coloured breast and a bluish sheen over the flanks. Adults have a black opercular flap and a prominent dusky spot on the posterior end of the dorsal fin.

Bluegills and pumpkinseeds (*Lepomis gibbosus*) co-exist in Osoyoos Lake and perhaps elsewhere in the Okanagan system. Adults of the two species differ in the colour of the opercular flap (blue-black in the bluegill and tipped with red in the pumpkinseed). Additionally, bluegills usually have a diffuse dark spot at the posterior base of the dorsal fin. This spot is faint in juvenile bluegills and absent in pumpkinseeds. Morphologically, the two species can be separated by their gill raker morphology: the gill rakers on the first gill arch of bluegills are long and pointed but are short and stubby in pumpkinseeds. Apparently, the young-of-the-year also differ in the shape and colour of markings on the last few interradial membranes near the end of the soft dorsal fin (Becker 1983): these marks form well-defined, dark ovals in young bluegills and diffuse, irregular brown markings in young pumpkinseeds.

Sexual Dimorphism   The breast and ventral parts of the body in front of the pectoral and pelvic fins develop a copper-orange colour in breeding males. This region is yellowish in breeding females and non-breeding males. Outside the breeding season, the shape of the urogenitial opening differs in males and females (McComish 1968).

Hybridization   Within their native range, bluegills and pumpkinseeds often hybridize (Konkle and Philipp 1992). So far, hybrids pose no identification problem in British Columbia. Nonetheless, bluegills are still relatively uncommon in the Osoyoos–Oliver region, whereas pumpkinseeds are abundant. Such discrepancies in abundance often lead to hybridization.

Distribution   NATIVE DISTRIBUTION
Bluegills are native to central and eastern North America. Here, they range from the St. Lawrence River in Quebec; through southern Ontario to eastern Minnesota; and south to eastern New Mexico, Texas, and northeastern Mexico. East of the Appalachian Mountains, native populations occur from Virginia to Florida. They have been widely introduced outside their native range.

BRITISH COLUMBIA DISTRIBUTION
The bluegill was first recorded from Osoyoos Lake in 2001 (Map 78). Although bluegills are not listed from the Okanagan system in Washington (Wydoski and Whitney 2003), they probably entered British Columbia from the Washington portion of Osoyoos Lake.

Life History   Nothing is known about the biology of the bluegill in British Columbia. Consequently, the following is based on accounts of the species' life history within its native range. The age and growth data are primarily from adjacent parts of eastern Washington.

REPRODUCTION
Bluegills begin spawning in the spring when water temperatures reach, or exceed, 20 °C and continue to spawn into August or until surface temperatures reach about 25 °C. Males move into littoral areas and select spawning sites (usually in water 0.3–1.5 m deep). They excavate a spawning pit by cleaning a shallow depression in the fine gravel or sand substrate (Miller 1963). Males usually nest in colonies of 40–50 males, and the edges of their pits sometimes almost touch. Each male defends his pit against neighbouring males and other intruders. With the approach of other fish or females, the male increases the speed and frequency of his circling. Bluegill males use three alternative mating strategies (Gross 1982)—some males are parental, some are sneakers, and others are female mimics (satellite males). Parental males delay maturation, excavate nest sites, court females, and guard their nest and eggs. Sneakers mature early and steal fertilizations by lurking near nests and rushing in and ejaculating

Map 78 The B.C. distribution of bluegill, *Lepomis macrochirus*

■ introduction

during normal spawnings. Satellite males also mature early. They enter the nest area by mimicking female behaviour and colour pattern and join normal spawnings. These sneaker and satellite males take no part in the rearing of any offspring they may have sired. This burden is left to the parental males that they have cuckolded. Female bluegills are fractional spawners and only release a few eggs at each spawning. Typically, they do not release all their eggs in one nest, and both males and females spawn with several mates.

Fecundity is function of female size and varies from about 2,000 to 40,000 eggs. The eggs are small (1.1–1.4 mm after fertilization), adhesive, and demersal (Meyer 1970). They sink to the bottom of the spawning pit, and the male aerates the eggs by fanning with his pectoral fins. Hatching occurs in 71 hours at 22.6 °C, 34 hours at 26.9 °C, and 32.5 hours at 27.3 °C (Carlander 1977). Apparently, in nature, larval survival is highest at temperatures above 23.5 °C (Garvey et al. 2002). The newly hatched larvae are 2–3 mm long and become free-swimming about 3 days after hatching. At this time, they are about 5 mm TL (Meyer 1970). A single nest may contain up to 70,000 larvae (Wydoski and Whitney 2003). The male abandons the fry shortly after they swim-up.

AGE, GROWTH, AND MATURITY

Typically, young-of-the-year bluegills grow rapidly and, at the northern edge of their range, reach about 30–50 mm by September. Growth slows at maturity (usually in their second or third summer); however, some populations delay maturity in response to predation pressure (Belk 1998). Usually, bluegills do not grow very large, although some individuals in eastern Washington reach 30 cm TL (Wydoski and Whitney 2003). Bluegills are prone to stunting, especially in waters without major predators. For small fish, they live a relatively long time—up to 11 years in Washington (Wydoski and Whitney 2003).

FOOD HABITS

Bluegills begin feeding at about 5–6 mm, and their first food consists of rotifers and copepodites; however, as they grow, cladocerans become the dominant food item (Siefert 1972). Juveniles and adults feed primarily on the larvae, pupae, and adults of insects (mainly chironomids), macro-invertebrates (e.g., amphipods), and zooplankton (especially *Daphnia*). In South Dakota, the diets of three size groups of bluegills (80–149, 150–199, and >200 mm) were followed throughout the growing season, April to October (Harris et al. 1999). Cladocerans (especially *Daphnia*), chironomids, and corixids were the dominant prey. The smallest size group fed primarily on zooplankton. Early in the season (April), the two larger size groups fed mainly on *Daphnia* but switched to corixids and chironomids in the summer. In Nebraska, bluegill diets in two lakes were examined (Paukert and Willis 2002). In one of the lakes, but not the other, they found that larger fish switched prey to chironomids during the summer.

Habitat Again, little is known about bluegill habitat use in British Columbia. Consequently, this account is based on accounts of the species' habitat use in Washington and in its native range.

ADULTS

Within their native range, bluegills generally are found in warm, shallow lakes and are associated with weed-beds in littoral zones. Their distribution in Osoyoos Lake (the only site in British Columbia where they are known to occur) fits this general habitat description. It is the warmest lake in the province, and within the lake, bluegills appear to concentrate in warm peripheral areas (e.g., shallow weedy bays, ponds, and lagoons) rather than in the main lake. In Nebraska, radio-tagged adults (>200 mm) were followed from April to September (Olson et al. 2003). Male bluegills selected emergent vegetation during April, June, and July, whereas females showed no preference for emergent vegetation and used all available habitats. They found no evidence of diel habitat shifts in females. In contrast, adult male bluegills spent significantly less time in macrophytes at night than during the day (Shoup et al. 2003).

JUVENILES

The role of body size, the presence of a caged predator, and the role of food on diel habitat use in bluegills were examined experimentally (Shoup et al. 2003). Juveniles spent significantly more time in cover (artificial macrophytes) in the presence of the predator and with food in open water than in either the control treatment (no predator and food) or the food-only treatment. In addition, regardless of the treatment, juveniles spent more time in cover during the day than at night. This suggests that juvenile bluegills have the potential to make diel littoral–pelagic habitat shifts.

YOUNG-OF-THE-YEAR
Shortly after swim-up, bluegill fry move from the nest into nearby vegetation. At about 10–12 mm TL, they move from vegetation offshore into open water (Werner 1967). Here, they feed on planktonic crustaceans for 1 or 2 months and then return to the littoral zone (Werner 1969). In September, at Goodman's Pond, Osoyoos Lake, groups of small (about 25 mm) bluegills were observed foraging amongst the vegetation in shallow (<1 m) water.

Conservation Comments The bluegill is a relatively new introduction into B.C. waters, and it is too early to assess their impact on native species. So far, they are confined to the south Okanagan, and the only conservation concern is keeping them from spreading to other drainage systems. Still, except for competitive interactions with native planktivorous species, they probably are not a serious threat to our aquatic biodiversity.

Micropterus dolomieu LACEPÈDE
SMALLMOUTH BASS

1 cm

Distinguishing Characters — This large (up to 350 mm FL in British Columbia), deep-bodied fish has rough scales and well-developed spines in the dorsal and anal fins. There are two introduced species of *Micropterus* in British Columbia: the smallmouth bass and the largemouth bass (*M. salmoides*). As their common names imply, the two species are distinguished by mouth size: the upper jaw extends backwards to a point below about the middle of the eye in smallmouth bass, whereas the upper jaw extends backwards to a point below, and beyond, the hind margin of the eye in largemouth bass. Also, in smallmouth bass, there is a shallow notch between the spinous and soft dorsal fins, and the shortest spine in the notch is greater than half the length of the longest spine. This notch is deeper in largemouth bass, and the shortest spine in the notch is less than half the length of the longest spine. Additionally, in smallmouth bass, there is a membrane that connects the pelvic fins. This membrane is absent in largemouth bass. Although colour is variable, there are usually two or three dark lines on the head radiating backwards from the eye in smallmouth bass. Juvenile smallmouths have thin, vertical bars along the midline, a conspicuous orange mark at the base of the tail and radiating cheek stripes. Juvenile largemouths have a conspicuous, dark midlateral stripe that extends from the snout through the eye and backwards to the base of the caudal fin. Also, juvenile largemouth lack the orange mark at the base of the tail and the radiating cheek stripes.

Sexual Dimorphism The overall colour darkens in spawning males, and the colour pattern intensifies in females. For largemouth bass, Benz and Jacobs (1986) indicate that the sexes usually can be distinguished (even in the fall) by inserting a probe into the urogenital opening and noting the depth and angle of the probe. Presumably, this technique also works for smallmouth bass.

Distribution NATIVE DISTRIBUTION
The smallmouth bass is native to eastern North America. The original geographic range was from southern Minnesota through the Great Lakes (except the north shore of Lake Superior) to southwestern Quebec and New York, and south to northern Alabama and eastern Oklahoma. Apparently, it is not native to the Atlantic Coastal Plain (Hubbs and Bailey 1938). The smallmouth bass has been widely introduced outside its natural range and, in western North America, is now found in British Columbia, California, Oregon, Idaho, and Washington. It has also been introduced into cool waters on other continents.

BRITISH COLUMBIA DISTRIBUTION
The Dominion Fisheries Department introduced smallmouth bass into the province (Christina Lake in the Kootenays and Florence and Langford lakes on Vancouver Island) in 1901. The original stock came from Ontario. In 1908, they were transplanted from Christina Lake into Moyie Lake (Kootenay system) and, in 1920, from Langford Lake to St. Mary Lake, Saltspring Island. In the 1960s, smallmouth bass appeared in Osoyoos Lake (presumably by dispersal from Washington), and another transplant was made from Christina Lake in 1987—this time into Vaseux and Skaha lakes (southern Okanagan system). Smallmouth bass now occur in most lowland lakes in the Okanagan system and many lakes on southern Vancouver Island (Map 79). Recently, someone illegally introduced this species into the middle Fraser system (Quesnel area).

Life History Little is known about the biology of smallmouth bass in British Columbia. Consequently, the following account is based on a review (Coble 1975) of the species' life history within its native range, supplemented, where possible, with information from British Columbia and Washington.

REPRODUCTION
Smallmouth bass spawn in the late spring, and in waters where both bass species occur, they usually spawn a little later than largemouth bass. Males move onto the spawning grounds and select a spawning site when the water temperature reaches about 15 °C; however, a substantial portion of the adult population (both sexes) does not breed every year (Raffetto et al. 1990). Typically, experienced males return to the same spawning area on successive years (Ridgway et al. 1991a), and older, larger males tend to nest earlier than smaller males (Ridgway et al. 1991b). The nest is a saucer-shaped depression (5–10 cm deep and about 50 cm in diameter) and is swept out of the sand or gravel bottom with the caudal fin.

Map 78 The B.C. distribution of smallmouth bass, *Micropterus dolomieu*

■ introduction

Typically, nests are located in quiet water at a depth of less than 1 m and near cover (stumps, logs, or large rocks). The male assiduously guards the nest. Smallmouth are solitary nesters, and the nests usually are spread out at densities of one nest every 15–50 m.

Courtship involves an initial phase away from the nest site and a final phase at the nest site (Ridgway et al. 1989). Spawning occurs after this final courtship phase, usually either at dawn or at dusk. The eggs are deposited in a series of spawning bouts. Fertilized eggs are about 1.2–2.5 mm in diameter and adhesive (Hardy 1978b). Fecundity is a function of female size, and egg numbers range from about 2,000 to over 20,000. The incubation period depends on temperature (usually about 10 days at 12 °C and 3 days at 25 °C). The male guards and fans the incubating eggs; although he does not eat at this time, he will aggressively attack intruders in his territory. The newly hatched fry are 4–5 mm TL and transparent. They remain in crevices in the substrate while the yolk is absorbed (about 10–12 days after hatching). At this stage, the fry are about 6 mm long and black. They become neutrally buoyant and rise above the nest in a dense swarm. They stay with the male for up to a month, after which they gradually turn green and disperse (Ridgway 1988).

AGE, GROWTH, AND MATURITY

Early growth is rapid, and young-of-the-year range in size from 50 to 110 mm by the end of their first growing season. In Christina Lake near Grand Forks, young-of-the-year smallmouth average about 110 mm by the end of August. Growth is steady until they reach sexual maturity (2–4 years in males and 3–5 years in females), after which the rate of growth slows but never completely stops. The maximum age of smallmouth bass in Canada is around 15 years (Scott and Crossman 1973). No maximum age is available for British Columbia; however, in Washington they can live for 10 years (Wydoski and Whitney 2003).

FOOD HABITS

Smallmouth bass are noted for their voracious appetites, and their diet is strongly influenced by the range of available prey. In British Columbia, adults in interior waters feed primarily on fish (especially redside shiners, *Richardsonius balteatus*; peamouth, *Mylocheilus caurinus*, and chiselmouth, *Acrocheilus alutaceus*), and macroinvertebrates, principally crayfish. On Vancouver Island, adults mainly consume crayfish and odonate nymphs. At one time, most small lowland lakes on Vancouver Island contained sticklebacks, and those in the Nanaimo area also contained peamouth. These native species are now gone from most of the lakes with introduced smallmouth bass.

In the summer, adult smallmouths are crepuscular and show activity peaks in the morning and evening (Emery 1973). They are less active during the day than largemouth bass (Reynolds and Casterlin 1976). The young start life foraging on zooplankton and gradually add insects to their diet. They first start taking small fish at a size of about 50 mm, and the proportion of fish in their diet increases as they grow.

Habitat Again, little is known about smallmouth bass habitat use in our province. Consequently, this account draws heavily on a review (Coble 1975) of habitat use within the species' native range, supplemented, where possible, with local information.

ADULTS

Within their natural range, smallmouth bass occur in cool (17–21 °C), clear lakes and rivers. In rivers, adults prefer areas of moderate current over sand or rock substrates and, in lakes, they are usually are associated with inshore rocky areas. Typically, by day, adult smallmouth are strongly associated with shelter (submerged logs or jumbles of rocks), and less strongly associated with dense weed beds than largemouth bass. During the summer, smallmouth may have spatially separate nocturnal and diurnal home ranges (Savitz et al. 1993). As water temperature rises in the summer, they shift into deeper, cooler water. In the fall, when water temperatures drop below 15 °C, they move into deep water, and they virtually cease feeding at about 4 °C. Apparently, they survive the winter on stored energy (Keast 1968). In the spring, there is a return to shallow water where they resume feeding before spawning.

Since the smallmouth bass in British Columbia were introduced into lakes, they are rarely encountered in rivers. Most of our lakes that contain smallmouth bass are small to medium sized, and relatively shallow.

Where they coexist with largemouth bass in British Columbia, the habitat segregation between the species is similar to that found in their natural ranges: largemouth in warm, soft-bottomed, weedy bays and smallmouth in cooler, rock-bottomed or sand-bottomed areas associated with submerged cover.

JUVENILES

Apparently, habitat use by juveniles is similar to that of adults, except that they occupy shallower water.

YOUNG-OF-THE-YEAR

After young-of-the-year leave the nest, they disperse into the shallow, calm, vegetated, margins of lakes and streams.

Conservation Comments   Although the smallmouth bass is a popular recreational species, it is not a native species, and none of our native fishes have co-evolved with this efficient predator. Consequently, it has a devastating impact on small species. On Vancouver Island, it has eliminated the native fish species and macroinvertebrates in most of the lakes where it has been introduced. Deliberately introducing this species to waters where it does not occur is an inexcusable act of ecological vandalism. The recent illicit introduction of this species into the Fraser system (in the Quesnel area) is cause for concern. If the smallmouth bass spreads in the Fraser system, it could have a serious impact on Pacific salmon populations.

Micropterus salmoides (LACEPÈDE)
LARGEMOUTH BASS

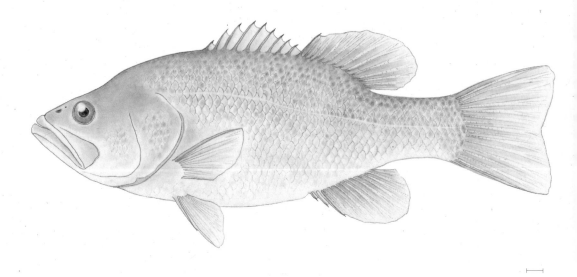

1 cm

Distinguishing Characters   This large (up to 500 mm FL in British Columbia), deep-bodied fish has coarse, rough scales and well-developed spines in the dorsal and anal fins. There are two introduced species of *Micropterus* in British Columbia: the largemouth bass and the smallmouth bass (*M. dolomieu*). As their common names imply, the two species are distinguished by mouth size: the upper jaw extends backwards to a point below, and beyond, the hind margin of the eye in largemouth bass, whereas the upper jaw extends backwards to a point below about the middle of the eye in smallmouth bass. Also, in largemouth bass there is a deep notch between the spinous and soft dorsal fins, and the shortest spine in the notch is less than half the length of the longest spine. This notch is shallower in smallmouth bass, and the shortest spine in the notch is greater than half the length of the longest spine. Additionally, in smallmouth bass, there is a membrane that connects the pelvic fins. This membrane is absent in largemouth bass. Juvenile largemouth bass have a conspicuous, dark midlateral stripe that extends from the snout through the eye and backwards to the base of the caudal fin, whereas juvenile smallmouth bass have thin, vertical bars along the midline. Also, juvenile smallmouth bass have a conspicuous orange mark at the base of the tail and radiating cheek stripes. These features are absent in juvenile largemouth bass.

Sexual Dimorphism The sexes can be reliably distinguished (even in the fall) by inserting a probe into the urogenital opening and noting the depth and angle of the probe (Benz and Jacobs 1986).

Distribution NATIVE DISTRIBUTION
The largemouth bass is native to eastern North America but has been introduced into cool waters on other continents. The original range extended from the Great Lakes to Florida and east of the Appalachian Mountains from Florida to Virginia. West of the Appalachian Mountains, its range extended from Minnesota to the Gulf Coast and northeastern Mexico. In the 19th century, largemouth bass were widely introduced into western North America. They are now found in British Columbia, California, Oregon, Idaho and Washington.

BRITISH COLUMBIA DISTRIBUTION
In British Columbia, largemouth bass occur in lakes and ponds in the upper Columbia and Kootenay drainage systems, the Okanagan River system, and the lower Fraser valley (Map 80). Most B.C. populations probably originated through natural dispersal from Idaho and Washington. Introductions into the northwestern states started in 1890 (Wydoski and Whitney 2003), and the first B.C. specimen was recorded from Vaseux Lake (Okanagan system) in 1909 (Field and Dickie 1987). This species was recorded in the Kootenay River system in 1921 (Dymond 1936). The upper Columbia populations (Columbia and Windemere lakes) appear to be the result of an unsanctioned introduction made sometime in the 1950s or 1960s (Griffith 1994). Subsequently, largemouth bass have been transplanted within limited areas in the Okanagan and Kootenay regions. Dispersal into the lower Fraser valley was through the Sumas River system; however, this species did not start spreading in the lower Fraser system until sometime in the 1970s. It is now well established in the Sumas River, Hatzic and Silvermere lakes, and sloughs associated with the lower Fraser River.

Life History Little is known about the biology of largemouth bass in British Columbia. Consequently, this account is based on a review (Heidinger 1975) of the species' life history within its native range supplemented, where appropriate, with local information. Most of this local information is from a study of introduced fishes in the Creston Valley Wildlife Management Area (Forbes 1989).

REPRODUCTION
Largemouth bass spawn in the spring when water temperatures rise above 15 °C. In Duck Lake, Creston valley, spawning occurred between May 20 and June 24 at water temperatures ranging from 20 °C to 23 °C. The spawning peak is in early June. Within its natural range, largemouth bass sometimes spawn twice in a summer (Heidinger 1975). There is no direct evidence for two spawning periods in British Columbia; however, by late

Map 80 The B.C. distribution of largemouth bass, *Micropterus salmoides*

■ introduction

August in some years, two barely overlapping size classes of largemouth fry are present in some lakes. This suggests either two reproductive peaks or a prolonged spawning period with different growth rates in early and late fry. Certainly, within their natural range, the largemouth spawning season can be protracted, with larger adults spawning earlier than smaller adults (Goodgame and Miranda 1993) and with fry produced for up to 70 days (Phillips et al. 1995). Because fry that hatch early in the season grow faster than late-hatched fry, the length–frequency distribution of young-of-the-year can be bimodal by late summer (Maceina and Isely 1986).

In largemouth bass, the male selects the spawning site, and there is some evidence that males return to the same general area on successive years (Heidinger 1975). The nest is a saucer-shaped depression in the bottom, usually at a water depth of 1 m or less. The average water depth over nests in Duck Lake is 73.5 cm (Forbes 1989). Typically, nests are located near cover (often stumps or logs), and males diligently guard the area. The nests of adjacent males usually are more than 2 m apart. When a gravid female enters a male's territory, spawning occurs after a short courtship. Most spawning occurs at dawn or at dusk. The eggs are deposited during a series of spawning bouts, and females do not necessarily deposit all their eggs in a single nest. Similarly, males may spawn with more than one female. Fertilized eggs are about 1.5–2 mm in diameter and adhesive.

Fecundity is a function of female body size, and egg number ranges from 2,000 to over 100,000 eggs. The incubation period depends on temperature (about 13 days at 10 °C and 3 days at 28 °C). The male guards and fans the incubating eggs; although he does not eat at this time, he will aggressively attack intruders in his territory. The newly hatched fry are 3–4 mm TL and transparent (Hardy 1978c). About 10 days after hatching, the yolk is absorbed, and the fry are about 6 mm long. At this stage, they are pale green in colour and become neutrally buoyant. They then rise above the nest in a dense swarm. During the day, schools of fry are loosely aggregated, but the schools become more compact at night and are associated with submerged cover. The fry are inactive until morning (Elliott 1976). They remain in schools and are guarded by the male for up to a month (or a size of about 25–30 mm), after which they disperse (Kramer and Smith 1960).

AGE, GROWTH, AND MATURITY
Early growth is rapid, and young-of-the-year range in size from 50 to 130 mm by the end of their first growing season. In Duck Lake (Forbes 1989), the fry grew 20 mm in a month (July 20 to August 24). Growth is steady until they reach sexual maturity (3 or 4 years in males and 4 or 5 years in females), after which the rate of growth slows but never completely stops. So far, the oldest known largemouth reached an age of 23 years (Green and Heidinger 1994); however, the maximum recorded age in British Columbia is 12 years (Forbes 1989), and fish over 4 kg are uncommon.

FOOD HABITS
Largemouth bass are noted for their voracious appetites, and adults feed primarily on fish; however, they also take crayfish and other macroinvertebrates. In the summer, they show peak feeding periods in the morning and evening but are more active during the day than smallmouth bass (Reynolds and Casterlin 1976). The young start life foraging on zooplankton and, as they grow, gradually add insects to their diet. The shift to fish starts at about 40–80 mm (Phillips et al. 1995), and the proportion of fish in the diet increases with size.

Habitat Not much is known about habitat use by largemouth bass in British Columbia. Consequently, this account draws heavily on a review (Heidinger 1975) of habitat use within the species' native range, supplemented, where possible, with local information.

ADULTS
Within their natural range, largemouth bass normally occur in warm (25–28 °C), shallow lakes and, less commonly, in slow-moving rivers. In the summer, adults are associated with soft substrates in areas with dense beds of emergent and submerged vegetation. As water temperature rises in late summer, largemouth bass remain in shallow water, but they become inactive during the day and shift to nocturnal activity at temperatures above 27 °C. In the fall when water temperatures drop below 10 °C,

foraging activity decreases, and they move into deeper water; however, unlike smallmouth bass, they continue feeding and remain moderately active throughout the winter (Brandt and Flickinger 1987). In the spring, they return to shallow water where they resume feeding before spawning.

In British Columbia, most of the lakes containing largemouth bass are shallow, warm-water lakes, and this species does not do well in the large, oligotrophic lakes. Where they coexist with smallmouth bass, the habitat segregation between the species is similar to that found in their natural ranges (i.e., largemouth in warm, soft-bottomed, weedy bays and smallmouth in cooler, rock or sand bottomed areas associated with submerged cover).

JUVENILES
Apparently, habitat use by juveniles is similar to that of adults except that, in the summer, they form small schools and cruise closer to shore in shallower water than the solitary adults.

YOUNG-OF-THE-YEAR
After the fry leave the nest, they disperse into the shallow, calm, often vegetated, margins of lakes.

Conservation Comments Largemouth bass, like smallmouth bass, provide angling opportunities; however, they wreak havoc with native fishes, and little can be done to control their spread once they are introduced into an open system like the lower Fraser River. Only time will reveal the full extent of the impact of largemouth bass on aquatic biodiversity in the Fraser valley; however, illicit introductions of this voracious predator into drainage systems where they do not occur constitute an inexcusable act of ecological vandalism.

Pomoxis nigromaculatus (LESUEUR)
BLACK CRAPPIE

1 cm

Distinguishing Characters This moderate-sized (rarely longer than 300 mm in British Columbia) sunfish has a deep, laterally compressed body, 36–40 scales along lateral line, and usually 7 or 8 (rarely 6–9) well-developed spines in the dorsal fin. No other spiny-rayed fish in British Columbia has an anal fin base that is equal to, or greater than, the dorsal fin base. The adult colour pattern on the dorsal, anal, and caudal fins is striking—irregular pale (yellow to light green) spots against a black background.

Although the black crappie is the only *Pomoxis* recorded from British Columbia, another species, the white crappie (*P. annularis*), occurs in the Columbia system in Washington State (Wydoski and Whitney 2003). The two species have similar body shapes; however, as the name implies, the background colour of white crappies is a much paler than that of black crappies. Additionally, the dorsal spine count in the white crappie usually is 5 or 6 as opposed to 7 or 8 in the black crappie.

Sexual Dimorphism The dark pigment on breeding males (especially on the head) becomes a deep velvety black, and the colour pattern is more vivid than the colours in breeding females.

Hybridization In disturbed habitats (e.g., reservoirs), black crappies often hybridize with white crappies (Guy and Willis 1995; Epifano et al. 1999). Since the white crappie does not occur in British Columbia, there is no identification problem; however, this could change if white crappies from the central Columbia system in Washington spread north into British Columbia.

Distribution NATIVE DISTRIBUTION
Black crappie are native to eastern North America. The eastern boundary of the natural range was from southern Quebec and the Great Lakes south along the west side of the Appalachian Mountains to the Gulf of Mexico; the western boundary extended from southern Manitoba and eastern Montana to central Texas.

BRITISH COLUMBIA DISTRIBUTION
Black crappie occur in reservoirs in the lower Pend d'Oreille system near Trail, in Osoyoos Lake, and in shallow lakes, ponds, and some sluggish streams in the lower Fraser valley (Map 81).

Life History Black crappie life history has not been studied in British Columbia. Consequently, the following account draws heavily on eastern North American sources and especially on papers from a symposium on crappie biology and management published in the *North American Journal of Fisheries Management* (e.g., Hooe 1991). The age and growth data are from Washington (Wydoski and Whitney 2003).

REPRODUCTION
Black crappie spawn in the spring when water temperatures rise above 15 °C. The male excavates a circular nest, usually at a depth of <3 m. Apparently, substrate type is not critical for nest construction, and nests occur in a wide variety of substrates (mud to gravel). The nests are usually adjacent to cover and are separated from their nearest neighbour by 1.5–3 m. Curiously, the spawning behaviour of black crappie is not well described. Most females spawn once a season, but some may spawn twice (Pope et al. 1996). Nest defense increases with the developmental stage of the eggs (Colgan and Brown 1988). Fecundity is strongly correlated with female body size and ranges from 34,000 to 348,000 (Baker and Heidinger 1994). The eggs are small (about 1 mm), demersal, and adhesive. Development rate is temperature dependent but usually varies from 3 to 5 days (Hardy 1978d). Newly hatched fry are 2–3 mm in length and about 8 mm at swim-up. The male guards the fry for 3 or 4 days (Faber 1967) before they disperse into open water.

AGE, GROWTH, AND MATURITY
In British Columbia, black crappies reach about 40–60 mm in their first growing season and about 12–15 cm by the end of their second year. Growth rate is positively correlated with the size of the pond or lake (Guy and Willis 1995). Males usually mature at the end of their second growing season (1^+), and females mature a year later. Although there are no data

Map 81 The B.C. distribution of black crappie, *Pomoxis nigromaculatus*

■ introduction

on the age or size of black crappies in British Columbia, they usually live 5 or 6 years (maximum of 10) and, in Washington, they reach a length of about 25 cm.

FOOD HABITS

Adult black crappie forage mainly in mid-water. Their diet consists of aquatic insects, small fish, and some plankton. The diet of juveniles is similar to that of adults except they consume more plankton than the adults. Copepods dominate the diet of the pelagic larvae (Schael et al. 1991).

Habitat ADULTS

Black crappie typically occur in clear, warm lakes and sluggish streams and sloughs. Often, they are associated with dense weed beds or submerged logs. Guy et al. (1992) studied the movement patterns of ultrasonic tagged black crappies in a South Dakota Lake. Rate of movement ranged from 0 to 584 m/h and differed seasonally and within diel periods. Activity peaked in April and July. Diel movements increased from evening to morning and were least during the day (Guy et al. 1992). Generally, black crappie were found in shallow water during evening and at night in the spring and during the day and evening in the summer (Guy et al. 1992).

JUVENILES

Juveniles occupy habitats similar to those used by adults but usually remain in water <3 m deep.

YOUNG-OF-THE-YEAR

Free-swimming fry are pelagic, and early in the season, they are distributed across open water near the surface. They become relatively more abundant near shore later in their pelagic phase and, in fall, disappear from the pelagic zone into the littoral. In the pelagic phase, the fry tend to have a clumped distribution (Post et al. 1994).

Conservation Comments Like the pumpkinseed, the black crappie is not a benign addition to our fish fauna. Most of the small native fishes have disappeared in lakes containing black crappie (e.g., Whonnock Lake near Ruskin). It is not clear if these extirpations result from competitive interactions, predation, or both. What is clear is that this species should never be introduced into waters where it does not now occur.

FAMILY PERCIDAE — PERCHES

The perches (Percidae) belong to the largest and most diverse order of living fishes—the Perciformes. This order contains 160 families and almost 10,000 species. Most of these species inhabit tropical or subtropical seas, but a few families occur in the temperate fresh waters of Eurasia and North America. One such family is the perches. There are about 370 species in eastern North America (about half are small bottom dwellers called darters), but only two species in British Columbia: the yellow perch (*Perca flavescens*) and the walleye (*Sander vitreus*). The walleye is indigenous to the northeastern corner of the province, but most, if not all, of our yellow perch populations are introduced.

The B.C. species have two separate dorsal fins. The first fin is made up entirely of spines, while the second has 1–3 spines at the front of the fin and the rest of the fin is made up of soft rays. In addition, the anal fin has two spines and the body is covered in ctenoid (rough) scales.

KEY TO PERCH AND WALLEYE

1 (2) Adults have 6 or 7 dark vertical bars on their sides; no white tip on the lower lobe of the caudal fin; 12–15 soft rays in the second dorsal fin	YELLOW PERCH, *Perca flavescens*
2 (1) Adults lack dark vertical bars on their sides; white tip on the lower lobe of the caudal fin; 18–22 soft rays in the second dorsal fin	WALLEYE, *Sander vitreus*

Perca flavescens (MITCHILL)
YELLOW PERCH

1 cm

Distinguishing Characters   This relatively small (adults usually <300 mm) perch is laterally compressed in cross section. Adults have 6–8 dark vertical bars on the sides, 2 spines, and 6–8 soft rays in the anal fin. The teeth are small and arranged in brush-like patches on the jaws and roof of the mouth. Occasionally, small yellow perch are confused with small walleye (*Sander vitreus*); however, the resemblance is superficial, and canine teeth are clearly visible even in small walleye and absent in yellow perch.

Taxonomy   Although the yellow perch is endemic to North America, a closely related species, *Perca fluviatilis*, is widely distributed in Eurasia. The morphological similarity and geographic distribution of these species led to questions about the validity of the North American species (e.g., Svetovidov and Dorofeeva 1963). However, mitochondrial DNA sequences indicates an ancient (late Miocene) divergence between North American and Eurasian yellow perch (Song 1995). Consequently, the weight of evidence supports the view that North American and Eurasian yellow perch are separate species.

Sexual Dimorphism   Adult female yellow perch are on average larger than adult males, and in the breeding season the pectoral and pelvic fins are less colourful than those of males. Adult yellow perch can be sexed by examining the urogenital opening. This structure is immediately behind the anus. It is round and surrounded by a raised ring of pale tissue in females, and oval and lacking the ring of pale tissue in males (Packham 1988).

Perca flavescens YELLOW PERCH

Distribution   NATIVE DISTRIBUTION

The native distribution of yellow perch ranges from Nova Scotia along the Atlantic Coastal Plain to South Carolina and, west of the Appalachian Mountains, from Quebec to Missouri. On the Great Plains, they occur naturally from Great Slave Lake south to Kansas. Although originally restricted to areas east of the Continental Divide, yellow perch have been introduced into British Columbia, California, Oregon, Idaho, and Washington.

BRITISH COLUMBIA DISTRIBUTION

It is not clear if there is a native population of yellow perch in British Columbia. Certainly, most B.C. populations are introduced; however, one population in the Peace system (Swan Lake near Tupper) may be an exception. In *The Fishes of Alberta*, Nelson and Paetz (1992) list yellow perch as native to the Peace drainage. Swan Lake is in the Peace system and lies on the British Columbia–Alberta border. It contains yellow perch. Apparently, yellow perch were introduced into Swan Lake in 1981 (Crossman 1991); however, yellow perch were present in the lake at least 25 years before this introduction. Consequently, the original Swan Lake population may have been native. In 1986, yellow perch were transplanted from Swan Lake to Charlie Lake near Fort St. John and, later, from Charlie Lake into Stony and Bearhole lakes near Tumbler Ridge.

Yellow perch were introduced into eastern Washington in the early 1890s (Wydoski and Whitney 2003), and most of the populations in the Kettle, Kootenay, Pend d'Oreille, and Okanagan river systems probably dispersed into British Columbia from Washington. More recently, clandestine introductions have moved yellow perch to southern Vancouver Island (Elk Lake) and several small lakes in the Shuswap region (Map 82).

Life History   Little is known about the biology of yellow perch in British Columbia; however, there is a review of the life history of the yellow perch and its relatives (Craig 1987). In addition, there have been a number of studies of yellow perch in Alberta (e.g., Jacobson 1989; Mance 1987; Norris 1984; Packham 1988) and one in Washington (R.E. Nelson 1977). These articles and theses are the major sources of the following life-history synopsis.

REPRODUCTION

Yellow perch begin to spawn when water temperatures rise above 7 °C. This temperature usually is reached in April on southern Vancouver Island, May or June (depending on altitude) in the Interior, and late June in northeastern British Columbia. Spawning is preceded by the movement of adults into littoral areas, and gamete release usually is associated with vegetation or submerged woody debris but can occur over almost any substrate (Thorpe 1977). Since spawning occurs at night or at dawn, accounts of yellow perch spawning behaviour are limited. Nonetheless, the spawning behaviour of both wild and aquarium populations have been observed (Harrington 1947; Hergenrader 1969). Harrington (1947)

Map 82 The B.C. distribution of yellow perch, *Perca flavescens*

■ introduction

describes an obviously gravid female accompanied by a long (about 1 m) queue of males (15–20). The first two males in the queue prodded the female with their snouts, and the other males formed a double row with the individuals so close together that they moved as a single entity. This whole unit snaked through branches and vegetation at depths ranging from 0.5 to 1.0 m. At the time of egg release, milt was shed close to the female's vent. Fecundity is high. In Lake Washington, estimated egg numbers for fish 15–33 cm TL were 18,000–139,000 (R.E. Nelson 1977), and fecundity is said to range from 950 to 210,000 eggs (Thorpe 1977). The fertile eggs are about 1.5–2.0 mm in diameter and are deposited in long (up to 2 m) rope-like strands encased in a gelatinous sheath. These egg masses are folded like accordion pleats and are semi-buoyant and adhesive. Typically, they are draped over vegetation or submerged tree branches. These apparently vulnerable ribbons of eggs are rarely subject to predation, and the jelly-like sheath may contain protective chemicals (Newsome and Tompkins 1985). Investigations involving the spatial distribution and manipulation (removal) of egg masses over several years suggest that female yellow perch return to breed near the location where they themselves were spawned (Aalto and Newsome 1990). During the breeding season, females mate with more than one male. The eggs take about a month to hatch at low (<10 °C) temperatures; however, development is faster at higher temperatures and hatching usually occurs in little over a week. Newly hatched larvae are transparent and about 5 mm long.

AGE, GROWTH, AND MATURITY

Although, female yellow perch generally grow faster than males (Danehy et al. 1991), growth rates are notoriously variable (Craig 1987). In enclosure experiments, first-year growth varied by greater than an order of magnitude both among populations and among cohorts within a population (Post and McQueen 1994). Although the type of food available (benthos or plankton) affected growth rate, the dominant mechanism appeared to be density-dependent intra-age-class competition. This variability in first-year growth is reflected in adult size differences among populations and among year classes within populations. Given their high fecundity, yellow perch can be prolific, and stunted populations are common in western North America. Age at maturity is also variable and influenced by diet, adult mortality, and the energetic costs of reproduction (Hayes and Taylor 1990). In Lake Washington, male yellow perch matured between 1 and 2 years of age, whereas females matured at 2 or 3 (R.E. Nelson 1977). In Lake Superior, yellow perch became sexually mature at 4 years (Bronte et al. 1993). The only B.C. data on the growth of young-of-the-year yellow perch is for the Koocanusa Reservoir on the Kootenay River. Here, young-of-the-year reached 25–59 mm by September (Chisholm et al. 1989). Yellow perch are not long lived (6 or 7 years is the maximum age in most populations); however, occasionally older individuals (>10+) are encountered (Craig 1987).

FOOD HABITS

Adult yellow perch consume a wide variety of prey: plankton, chironomid pupae, aquatic insect nymphs, amphipods, clams, and small fishes (Chisholm et al. 1989; Danehy et al. 1991; Parrish and Margraf 1994; Pazkowski and Tonn 1994). In one Alberta Lake during the summer, feeding intensity increased throughout the day, peaked in late evening, and ceased after sunset (Jansen and Mckay 1992). Juveniles have a diet similar to adults but consume more *Daphnia* and less fish than adults. The diet of young-of-the-year perch contains both plankton (cladocerans and calanoid and cyclopoid copepods) and benthos (Faurot and White 1994; Post and McQueen 1994).

Habitat Little is known about habitat use by yellow perch in British Columbia; however, habitat data on the yellow perch and its relatives has been reviewed (Craig 1987). In addition, there are a number of yellow perch studies in Alberta (Jacobson 1989; Mance 1987; Norris 1984; Packham 1988) and one in Washington (R.E. Nelson 1977). These articles are the source of the following life-history synopsis.

ADULTS

In British Columbia, yellow perch usually are lacustrine, although they are the dominant fish species in some adjacent rivers (e.g., the Box Canyon portion of the Pend d'Oreille River; Barber et al. 1990). In lakes in the spring, adults move inshore to spawn, and schools usually remain in water about 3–15 m deep (Beauchamp et al. 1995; Danehy et al. 1991; Lyons 1987b) until water temperatures rise above 20 °C (Ferguson 1958). At this time, they move into offshore bottom areas. During the summer, yellow perch in a Wyoming Reservoir were significantly more numerous in littoral and offshore bottom areas than in offshore surface areas (Hubert and O'Shea 1992), and yellow perch were most abundant in diverse, structurally complex weed beds (Weaver et al. 1997). Growth is significantly higher in individuals caught in shallow water over cobble/rubble shoals than for individuals collected in shallow water over open sandy areas (Danehy et al. 1991). There is a strong tendency for displaced yellow perch to return to the site of initial capture (Hodgson et al. 1998). This suggests that, in the summer, yellow perch maintain a home range.

JUVENILES

In summer, juveniles remain in the littoral zone and usually are associated with vegetation (Bryan and Scarnecchia 1992).

YOUNG-OF-THE-YEAR

As larvae, yellow perch inhabit the limnetic zone of lakes (Bryan and Scarnecchia 1992). Initially, they remain in the limnetic zone day and night; however, at about 20–30 mm TL larvae begin a diurnal migration. They move inshore at dawn and offshore at dusk (Post and McQueen 1988; Wahl et al. 1993). As they grow, the strength of this migration diminishes and, eventually, the fry remain in the littoral zone and are associated with vegetation.

Conservation Comments In British Columbia, with the possible exception of Swan Lake near Tupper, all our yellow perch are introduced. For a long time, they were confined to the Okanagan and Kootenay systems, but they have been spreading in the last two decades. New populations have appeared on Vancouver Island and in a number of lakes in the Kamloops and Shuswap regions. These new populations are a result of illicit introductions, but it is not clear why people are moving yellow perch. They do provide some angling opportunities, but they tend to stunt in most small lakes and usually have negative impacts on native species.

Sander vitreus (MITCHILL)
WALLEYE

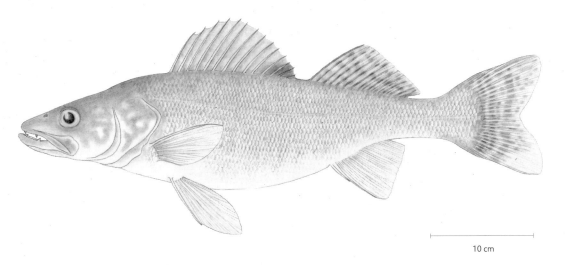

10 cm

Distinguishing Characters	This relatively large (adults often >300 mm) perch is cylindrical in cross section. Adults lack dark, vertical bars on the sides and have two spines and 11–13 soft rays in the anal fin. There are conspicuous canine teeth on the jaws and on the roof of the mouth. In air, the pupils glow with a faint ethereal light. This is light reflected from a mirror-like structure (the *tapetum lucidum*) associated with the retina. Like a similar structure in the eyes of cats, it increases visual acuity under low-light conditions. Occasionally, small yellow perch (*Perca flavescens*) are confused with small walleye; however, the resemblance is superficial, and the canine teeth are obvious even in small walleye.
Taxonomy	Although walleye are restricted to North America, a closely related species, *Sander lucioperca*, is widely distributed in Eurasia. Until recently the generic name for both these species, as well as for the sauger (*Sander canadensis*) was *Stizostedion*.
Sexual Dimorphism	Outside the breeding season, the sexes are difficult to distinguish; however, on average, adult females are larger than adult males.
Distribution	**NATIVE DISTRIBUTION** The native distribution of walleye extends from western Quebec along the west side of the Appalachian Mountains to Alabama. Most of the populations on the Atlantic Coastal Plain probably are introduced, although they may be native in some drainages (Barila 1980). Walleye are native throughout the Mississippi valley and across the Great Plains to Alberta, northeastern British Columbia, and the Mackenzie Delta. Originally, walleye were restricted to regions east of the Continental

Divide, but they have been widely introduced in the west (Arizona, British Columbia, California, Idaho, Oregon, Nevada, and Utah).

BRITISH COLUMBIA DISTRIBUTION

Walleye are native to the northeastern corner of the province (the lower Peace, lower Liard, and Hay drainage systems); however, even within this region, at least one population (Charlie Lake near Fort St. John) is an introduction. Also, in the late 1960s and early 1970s, walleye were introduced into Washington, Oregon, and Idaho. They increased rapidly in Roosevelt Reservoir (above Grand Coulee Dam) and, from there, spread into south-central British Columbia (Map 83). In British Columbia, walleye now are seasonally abundant in the Columbia mainstem from Keenleyside Dam south to the U.S. border. They also occur in Waneta and Seven Mile reservoirs (Pend d'Oreille system), and recent reports from the "Narrows" between Upper and Lower Arrow lakes suggest that they are still spreading in the upper Columbia system.

Life History   Little is known about the life history of native walleyes in British Columbia, and most of the following data are derived from Alberta or eastern North American populations.

REPRODUCTION

Typically, walleye spawn in the spring when water temperatures rise above 5 °C (Bodaly 1980); however, pre-spawning behaviours (e.g., aggregations and migrations) often begin at lower temperatures. In northeastern British Columbia, this threshold temperature is reached sometime between late March and early June. Males arrive at spawning sites before females, usually outnumber females on the spawning sites, and remain on site for a few days after the females leave (Bodaly 1980; Bond 1980; Machniak and Bond 1979). It is not clear whether walleyes home to natal breeding sites. Some authors (e.g., Bodaly 1980) report a low incidence of homing to natal streams; however, genetic evidence implies some degree of natal homing (Stepien and Faber 1998; Todd 1990). In lakes, walleye prefer to spawn in shallow water, usually over rocky wave-washed shoals. In contrast, in rivers, the preferred substrate appears to be gravel and cobbles. There is evidence that the preference for these alternative spawning habitats (rocky shoals in lakes and gravelly areas in rivers) is at least partially inherited (Jennings et al. 1996). On suboptimal substrates (sand and mud), egg survival is low (Auer and Auer 1990; Johnson 1961). There is a review (Pitlo 1989) of the physical characteristics of walleye spawning habitat in the upper Mississippi River. Here, they spawn at depths ranging from 0.6 to 6.1 m and water velocities of 0.4–1.16 m/s. Elsewhere, slower (0.15–0.25 m/s) water appears to be preferred (DiStefano and Hiebert 1999).

Spawning occurs mostly at dusk, but observations on spawning behaviour are rare; however, some aspects of walleye courtship and spawning have been described (Ellis and Giles 1965). There is no site

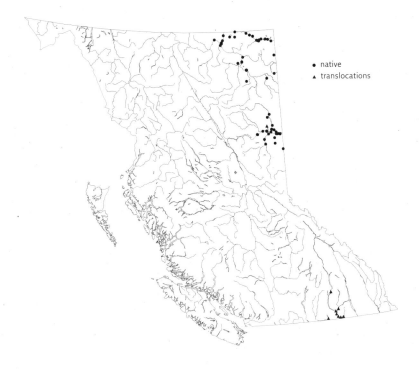

Map 83 The B.C. distribution of walleye, *Sander vitreus*

preparation, but apparently, a form of courtship precedes spawning and can be initiated by either sex. Courtship consists of an approach (from behind, from the side, or by drifting backwards) and contact (described as pushing) with another fish. The first dorsal fin is alternately raised and lowered during these approaches. As courtship activity increases in frequency and intensity, individual fish begin to dart forward and upwards to the surface. Finally, one or more males and one or more females form a compact group and rush upwards to the surface. Here, they swim vigorously until swimming abruptly stops, the female(s) are pushed violently onto their sides, and gametes are released near the surface. The fertilized eggs (about 2 mm in diameter) are adhesive and denser than water. They sink and drop into interstices in the substrate. Fecundity is high and, depending on female size, ranges from about 40,000 to 600,000 eggs (Bond 1980; Machniak 1975b). The eggs take 26 days to hatch at 4.5 °C and 7 days at 14 °C (Niemuth et al. 1959); however, at optimal incubation temperatures (9–15 °C), the eggs hatch in about 12–18 days (Koenst and Smith 1976). Newly hatched fry are about 8.0 mm TL and begin exogenous feeding at about 12 mm (i.e., about 5 or 6 days after hatching).

AGE, GROWTH, AND MATURITY

Growth rate is variable and dependent on temperature and food availability, although females usually grow faster and attain a larger size than males. In northeastern British Columbia and northern Alberta, young-of-the-year walleye typically reach about 110 mm in their first growing season (Bond 1980; McPhail et al. 1998a). Age at maturity is also variable. In northeastern British Columbia (Petitot River), males mature at about 8+, and females, a year later (McPhail et al. 1998a). In contrast, males in northern Alberta mature at 3+ or 4+, and females, at 4+ or 5+ (Bond 1980). In these northern populations, walleye are modestly long lived: 18+ in British Columbia (McPhail et al. 1998a), 18+ in the Slave River (Tallman et al. 1996b), and 14+ in the Athabasca system (Bond and Berry 1980). In the Slave River population, about 30% of the mature fish were 12 or older (Tallman et al. 1996b). So far, the largest walleye angled from a population native to British Columbia was a 69 cm FL female.

FOOD HABITS
Adult and juvenile walleyes are primarily piscivores (Bryan et al. 1995; Niemuth et al. 1959; Stephenson and Momet 1991), although some populations include a high proportion of aquatic insects in their diet. In northeastern British Columbia and northern Alberta, walleye consume a wide variety of fish: suckers (*Catostomus*), mountain whitefish (*Prosopium williamsoni*), burbot (*Lota lota*), lake chub (*Couesius plumbeus*), flathead chub (*Platygobio gracilis*), yellow perch, Arctic grayling (*Thymallus arcticus*), northern pike (*Esox lucius*), and other walleye (Bond 1980; McPhail et al. 1998a). They are most active at night, dawn, and dusk (Prophet et al. 1989), and both larvae and adults are effective predators at low light levels (Rieger and Summerfelt 1997; Ryder 1977). This does not mean that walleye feed exclusively at night or in low light. There are seasonal shifts in foraging time (Swenson and Smith 1973): apparently, walleye feed night and day in June but switch to nocturnal feeding in July, August, and September.

Larval walleye start life as planktivores, but in many populations, they quickly shift to fish. For example, in Lake Roosevelt, calanoid copepods were the most abundant food item in larval walleye stomachs, but they started preying on larval suckers when they reached 9–24 mm in length (Faurot and White 1994). Similarly, young-of-the-year walleye (>25 mm) were almost entirely piscivorous (Leis and Fox 1996), and in northern Alberta, 85% of the 0+ walleye examined contained fish remains (Bond 1980). Apparently, an early switch to fish in the diet of young-of-the-year walleye depends on the presence of suitable prey. In Lake Oneida, walleye in the first month of life were planktivores; however, as they grew they shifted to larger and larger prey (Graham and Sprules 1992). Thus, larvae 10–13 mm TL consumed mostly cyclopoid copepods; larvae 13–19 mm consumed mostly calanoid copepods; and fish 29–38 mm foraged primarily on cladocerans (*Daphnia*). Like adults, juvenile walleye are mainly piscivorous (Lyons 1987a).

Habitat
ADULTS
Although walleye occur in rivers as well as in lakes and reservoirs, they appear to prefer slow water. A study of their swimming performance found relatively low sustained swimming speeds (0.3–0.73 m/s) but good

(1.6–2.6 m/s) fast start performance (Peake et al. 2000). Despite their preference for slow water, some walleye populations undertake lengthy migrations. For example, tagging studies suggest that walleye in the Canadian portion of the Columbia system migrate over 250 km to spawn in, or near, Roosevelt Reservoir, Washington (Hildebrand 1991).

In northern Alberta, one tagged fish moved 600 km (Bond 1980). There are two seasonal migrations in the Slave River Delta (Tallman et al. 1996): one in the spring (which may be a spawning migration) and one in the fall (which may be an overwintering migration to Great Slave Lake).

In flowing water, walleye apparently prefer rivers with reduced light penetration (turbid or deeply stained waters) or deep pools (Paragamian 1989). In the Cedar River, Iowa, walleye occurred in low-velocity shallow areas with current breaks (boulders or submerged logs). In winter, however, they moved into pools and occupied a wider range of depths.

In lakes, adults avoid well-lit areas and, typically, are found in moderately shallow water (1–15 m) close to cover (Colby et al. 1979; Ryder 1977); however, they may perform diurnal movements into shallow water as day fades and into deeper water at sunrise (Ryder 1977).

JUVENILES

In rivers during the summer, juvenile walleye occupy shallow edge habitats close to vegetation or other forms of cover and are usually in areas with a slight current (Paragamian 1989; Ryder 1977). In lakes, the summer habitat of juveniles is inshore and, typically, in water less than 2 m deep (Johnson 1977).

YOUNG-OF-THE-YEAR

In lakes, newly hatched walleye are pelagic and feed offshore in the limnetic zone, but they move inshore by early summer, become demersal, and are strongly associated with thick weed-beds at depths of 2–5 m (Pratt and Fox 2001). In late summer, they remain inshore in shallow water but shift to areas with less cover (Pratt and Fox 2001) and, in winter, they move into deeper water.

Conservation Comments For conservation purposes, it is important to distinguish between native and introduced walleye populations. There may be valid management reasons for husbanding introduced walleye populations—they are popular with anglers and delicious—however, they should not be transplanted outside their native range. They are efficient predators and are capable of eliminating native species. For example, Charlie Lake near Fort St. John once supported a diverse community of native minnows. Now, they are all gone, and spottail shiners (*Notropis hudsonius*) had to be introduced to provide forage for pike and walleye. Introduced walleye may not be the only factor involved in this local extirpation, but they certainly didn't help. The walleye in the Columbia system are not native.

FAMILY PLEURONECTIDAE — RIGHTEYE FLOUNDERS

The righteye flounders (Pleuronectidae) belong to a curious order of fishes—the Pleuronectiformes—that undergo a remarkable metamorphosis when they shift from their pelagic larval stage to benthic life. The entire skull is twisted until both eyes are on one side of the head. The animal then settles onto the bottom on its left side and develops pigment on its right side. You can tell if a flounder is right-eyed or left-eyed by holding it on its side with the head pointing away from you. If the pigmented (eyed) side is on your right, the flounder is right-eyed. If the pigmented side is on your left, the fish is left-eyed. Most pleuronectids are strictly right-eyed; however, for some reason, the starry flounder (*Platichthys stellatus*) can be either right-eyed or left-eyed. Along the B.C. coast, just less than half of the starry flounders are left-eyed. Farther south on the North American coast, they are predominately right-eyed, and they are predominately left-eyed farther north near Bering Strait.

As a group, the right-eyed flounders are of considerable economic importance, and many of the species (e.g., the halibuts, *Hippoglossoides*) support large commercial fisheries. Most members of the family are strictly marine, but the starry flounder regularly enters fresh water and is common in estuaries and often ascends rivers to above tidewater. It is the only flatfish on our coast with alternating pale and black stripes on the dorsal and anal fins. In the Fraser River, it penetrates about 120 km upstream from the sea and has been recorded from Pitt Lake and the lower Sumas River near Chilliwack. Only juveniles are found above tidewater. They spend 1 or 2 years in fresh water but return to the sea before they reach sexual maturity. Because they do not breed in fresh water, I have not included an account of this species' life history or habitat use.

BIBLIOGRAPHY

Aalto, K.R., and G.E, Newsome. 1990. Additional evidence supporting demic behaviour of a yellow perch (*Perca flavescens*) population. Can. J. Fish. Aquat. Sci. 47: 1959–1962.

Aalto, K.R., W.D. Sharp, and P.R. Renne. 1998. $^{40}Ar/^{39}Ar$ dating of detrital micas from Oligocene-Pleistocene sandstones of the Olympic Peninsula, Klamath Mountains, and northern California Coast Ranges: provenance and paleodrainage patterns. Can. J. Earth Sci. 35: 735–745.

Abell, R., D.M. Olson, E. Dinerstein, P. Hurley, J.T. Diggs, W. Eichbaum, S. Walters, P. Hurley, W. Wettengal, T. Allnutt, C. Loucks, and P. Hedao. 2000. A conservation assessment of the freshwater ecoregions of North America. Island Press, Washington, DC.

Ableson, D.R. 1973. Contributions to the life history of the brassy minnow (*Hybognathus hankinsoni*). M.Sc. thesis, University of Michigan, Ann Arbor.

Abrahams, M.V. 1996. Interaction between young-of-the-year fathead minnows and brook sticklebacks: effects on growth and diet selection. Trans. Am. Fish. Soc. 125: 480–485.

Acara, A.H. 1977. The Meadow Creek spawning channel. Can. Fish. Mar. Serv. Tech. Rep. No. 744. 90 pp.

Adams, F.J., and J.B. Reynolds. 1997. Relative abundance and maturity of lake trout from Walker Lake, Gates of the Arctic National Park and Preserve, Alaska, 1987–1988. pp. 104–108. *In* J. Reynolds (ed.). Fish ecology in arctic North America. Am. Fish. Soc. Symp. No. 19.

Adams, P.B., C.B. Grimes, S.T. Lindley, and M.L. Moser. 2002. Status review for North American green sturgeon, *Acipenser medirostris*. US Departement of Commerce, National Marine Fisheries Service, Washington, DC. NOAA Tech. Memo. No. NMFS-SWFSC-630.

Ahlgren, M.O. 1990. Diet selection and contribution of detritus to the diet of the juvenile white sucker (*Catostomus commersoni*). Can. J. Fish. Aquat. Sci. 47: 41–48.

Ahsan, S.N. 1966. Effects of temperature and light on the cyclical changes in the spermatogenetic activity of the lake chub, *Couesius plumbeus* (Agassiz). Can. J. Zool. 44: 149–159.

Alanaerae, A, and E. Braennaes. 1997. Diurnal and nocturnal feeding activity in Arctic charr (*Salvelinus alpinus*) and rainbow trout (*Oncorhynchus mykiss*). Can. J. Fish. Aquat. Sci. 54: 2894–2900.

Alexander, D.R., and H.R. MacCrimmon. 1974. Production and movement of juvenile rainbow trout (*Salmo gairdneri*) in a headwater of Bothwell's Creek, Georgian Bay, Canada. J. Fish. Res. Board Can. 31: 117–121.

Allan, J.D. 1981. Determinants of diet of brook trout (*Salvelinus fontinalis*) in a mountain stream. Can. J. Fish. Aquat. Sci. 38: 184–192.

Allen, J.E., M. Burns, and S.C. Sargent. 1986. Cataclysms on the Columbia. Timberlands Press, Portland, OR.

Allen, J.H. 1987. Fisheries investigations in Line Creek—1987. Report prepared for Line Creek Resources Ltd., Sparwood, BC.

Allen, M.A. 2000. Seasonal microhabitat use by juvenile spring salmon in the Yakima River Basin, Washington. Rivers, 7: 314–332.

Allendorf, F.W., and R.F. Leary. 1988. Conservation and distribution of genetic variation in a polytypic species, the cutthroat trout. Conserv. Biol. 2: 170–184.

Alt, K.T. 1965. Food habits of inconnu in Alaska. Trans. Am. Fish. Soc. 94: 272–274.

Alt, K.T. 1969. Taxonomy and ecology of the inconnu, *Stenodus leucichthys nelma*, in Alaska. University of Alaska, Fairbanks. Biol. Pap. Univ. Alaska, 12: 1–61.

Alt, K.T. 1971. Occurrence of hybrids between inconnu, *Stenodus leucichthys nelma* (Pallas), and humpback whitefish, *Coregonus pidschian* (Linnaeus), in the Chatinika River, Alaska. Trans. Am. Fish. Soc. 171: 362–365.

Alt, K.T. 1973. Age and growth of the inconnu, *Stenodus leucichthys*, in Alaska. J. Fish. Res. Board Can. 30: 457–459.

Alt, K.T. 1976. Age and growth of Alaskan broad whitefish, *Coregonus nasus*. Trans. Am. Fish. Soc. 105: 526–528.

Alt, K.T. 1977. Inconnu, *Stenodus leucichthys*, migration studies in Alaska, 1961–74. J. Fish. Res. Board Can. 34: 129–133.

Alt, K.T. 1980. A life history study of sheefish and whitefish in Alaska. Alaska Department of Fish and Game, Juneau. Fed. Aid Fish Restor. Stud. No. R11. 31 pp.

Alt, K.T. 1987. Review of sheefish (*Stenodus leucichthys*) studies in Alaska. Alaska Department of Fish and Game, Juneau. Fish. Manuscr. No. 3. 69 pp.

Andraso, G.M. 1997. A comparison of startle response in two morphs of the brook stickleback (*Culaea inconstans*): further evidence for a trade-off between defensive morphology and swimming ability. Evol. Ecol. 11: 83–90.

Andrusak, H., and T.G. Northcote. 1971. Segregation between adult cutthroat trout (*Salmo clarki*) and Dolly Varden (*Salvelinus malma*) in small, coastal British Columbia lakes. J. Fish. Res. Board Can. 28: 1259–1268.

Andrusak, H., and E. Parkinson. 1984. Food habits of Gerrard stock rainbow in Kootenay Lake, British Columbia. BC Ministry of the Environment, Victoria. Fish. Tech. Circ. No. 60. 14 pp.

Anras, M.L.B., P.M. Cooley, R.A. Bodaly, L. Anras, and R.J.P. Fudge. 1999. Movement and habitat use by lake whitefish during spawning in a boreal lake: integrating acoustic telemetry and geographic information systems. Trans. Am. Fish. Soc. 128: 939–952.

Antonov, A.L. 2004. A new species of grayling, *Thymallus burejensis*, from the Amur Basin. J. Ichthyol. 44: 401.

Apperson, K.A., and P.J. Anders. 1991. Kootenai River white sturgeon investigations and experimental culture. Report to Bonneville Power Administration. Fisheries Research Section, Idaho Department of Fish and Game, Boise.

Armstrong, J.E. 1981. Post-Vashon Wisconsin glaciation, Fraser Lowland, British Columbia. Bull. Geol. Surv. Can. 322: 1–24.

Armstrong, J.W., C.R. Liston, P.I. Tack, and R.C. Anderson. 1977. Age, growth, maturity, and seasonal food habits of round whitefish, *Prosopium cylindraceum*, in Lake Michigan near Ludington, Michigan. Trans. Am. Fish. Soc. 106: 151–155.

Armstrong, R.H. 1986. A review of Arctic grayling studies in Alaska, 1952–1982. University of Alaska, Fairbanks. Biol. Pap. Univ. Alaska No. 23: 3–17.

Arnold, M.L. 1997. Natural hybridization and evolution. Oxford Series in Ecology and Evolution. Oxford University Press, London.

Aro, K.V., and M.P. Shepard. 1967. Pacific salmon in Canada. pp. 225–327. *In* Salmon of the North Pacific Ocean. Part lV. Spawning populations of North Pacific salmon. Int. N. Pac. Fish. Comm. Bull. No. 23.

Aspinwall, N., and J.D. McPhail. 1995. Reproductive isolating mechanisms between the peamouth, *Mylocheilus caurinus*, and the redside shiner, *Richardsonius balteatus*, at Stave Lake, British Columbia, Canada. Can. J. Zool. 73: 330–338.

Aspinwall, N, D. Carpenter, and J. Bramble. 1993. The ecology of hybrids between the peamouth, *Mylocheilus caurinus*, and the redside shiner, *Richardsonius balteatus*, at Stave Lake, British Columbia, Canada. Can. J. Zool. 71: 83–90.

Auer, M.T., and N.A. Auer. 1990. Chemical suitability of substrates for walleye egg development in the lower Fox River, Wisconsin. Trans. Am. Fish. Soc. 119: 871–876.

Baakala, R.G. 1970. Synopsis of biological data on the chum salmon (*Oncorhynchus keta* (Walbaum 1792)). FAO Fish. Synopses No. 41. 89 pp.

Babaluk, J.A., J.D. Reist, J.D. Johnson, and L. Johnson. 2000. First records of sockeye (*Oncorhynchus nerka*) and pink salmon (*O. gorbuscha*) from Banks Island and other records of Pacific salmon in Northwest Territories, Canada. Arctic, 53: 161–164.

Bachman, R.A. 1991. Brown trout, *Salmo trutta*. pp. 208–228. *In* J. Stolz and J. Snell (eds.). Trout. Stackpole, Harrisburg, PA.

Bagley, M.J., and G.A.E. Gall. 1998. Mitochondrial and nuclear DNA sequence variability among populations of rainbow trout (*Oncorhynchus mykiss*). Mol. Ecol. 7: 945–961.

Bailey, J.E. 1952. Life history and ecology of the sculpin *Cottus bairdi punctulatus* in southwestern Montana. Copeia, 1952: 243–255.

Bailey, J.E. 1964. Intertidal spawning of pink and chum salmon in Olsen Bay, Prince William Sound, Alaska. US Fish and Wildlife Laboratory, Auke Bay, MS. Rep. No. MR 64-6. 23 pp.

Bailey, M.M. 1963. Age, growth, and maturity of round whitefish of the Apostles Islands and Isle Royal regions, Lake Superior. Fish. Bull 63: 63–75.

Bailey, R.M. 1954. Distribution of the American cyprinid fish, *Hybognathus hankinsoni*, with comments on its original description. Copeia, 1954: 289–291.

Bailey, R.M. 1980. Comments on the classification and nomenclature of lampreys—an alternative view. Can. J. Fish. Aquat. Sci. 37: 1626–1629.

Bailey, R.M., and M.O. Allum. 1962. Fishes of South Dakota. University of Michigan, Ann Arbor. Misc. Publ. Univ. Mich. Mus. Zool. No. 119. 132 pp.

Bailey, R.M., and C.E. Bond. 1963. Four new species of freshwater sculpins, genus *Cottus*, from western North America. University of Michigan, Ann Arbor. Occas. Pap. Mus. Zool. Univ. Mich. No. 634. 27 pp.

Bailey, R.M., and M.F. Dimick. 1949. *Cottus hubbsi*, a new cottid fish from the Columbia River System in Washington and Idaho. University of Michigan, Ann Arbor. Occas. Pap. Mus. Zool. Univ. Mich. No. 513. 18 pp.

Bailey, R.M., W.C. Latta, and G.R. Smith. 2004. An atlas of Michigan fishes with keys and illustrations for their identification. University of Michigan, Ann Arbor. Univ. Mich. Mus. Zool. Spec. Publ. No. 2.

Baird, O.E., and C.C. Kruger. 2000. Behavior of lake trout sac fry: vertical movement at different developmental stages. J. Great Lakes Res. 25: 141–151.

Baker, J.A. 1994. Life history variation in female threespine stickleback. pp. 144–187. *In* M.A. Bell and S.A. Foster (eds.). The evolutionary biology of the threespine stickleback. Oxford University Press, Oxford.

Baker, S.C., and R.C. Heidinger. 1994. Individual and relative fecundity of black crappie (*Pomoxis nigromaculatus*) in Baldwin Cooling Pond. Trans. Ill. State Acad. Sci. 87: 145–150.

Bakker, T.C.M. 1993. Positive genetic correlation between female preference and preferred male ornament in sticklebacks. Nature (London), 363: 255–257.

Balon, E.K. 1995. The common carp, *Cyprinus carpio*: its wild origin, domestication in aquaculture, and selection as colored nishikigoi. Guelph Ichthyol. Rev. 3: 3–54.

Barber, M.R., B.L. Renberg, J.J. Vella, K.L. Scholz, and K.L. Woodward. 1990. Assessment of fishery improvement opportunities on the Pend Oreille River. Annual Report to the Bonneville Power Administration. Oregon Division of Fish and Wildlife, Portland.

Bardach, J.E., J.H. Todd, and R. Crickmer. 1967. Orientation by taste in fish of the genus *Ictalurus*. Science (Washington, DC), 155: 1276–1278.

Barendregt, R.W. and A. Duk-Rodkin. 2004. Chronology and extent of late Cenozoic ice sheets in North America: a magnetostratigraphic assessment. pp. 1–7. *In* J. Ehlers and P.L. Gibbard (eds.). Quaternary glaciations, extent and chronology. Part II: North America. Elsevier, Amsterdam.

Barendregt, R.W., and E. Irving. 1998. Changes in the extent of the North American ice sheets during the late Cenozoic. Can. J. Earth Sci. 35: 504–509.

Barila, T.Y. 1980. *Stizostedion vitreum* (Mitchill), walleye. pp. 747–748. *In* D.S Lee, C.R. Gilbert, C.H. Hocutt, R.E. Jenkins, D.E. McAllister, and J.R. Stauffer (eds.). Atlas of North

American freshwater fishes. North Carolina State Museum of Natural History, Raleigh.

Barraclough, W.E. 1964. Contributions to the marine life history of the eulachon (*Thaleichthys pacificus*). J. Fish. Res. Board Can. 21: 1333–1337.

Bartley, D.M., G.A. Gall, and B. Bentley. 1985. Preliminary description of the genetic structure of white sturgeon, *Acipenser transmontanus*, in the Pacific Northwest. pp. 105–109. *In* F.P. Binkowski and S.I. Doroshov (eds.). North American sturgeons: biology and aquaculture potential. Dr. W. Junk Publishers, Dordrecht, the Netherlands.

Bartley, D.M., G.A.E. Gall, and B. Bentley. 1990. Biochemical genetic detection of natural and artificially hybridization of chinook and coho salmon in northern California. Trans. Am. Fish. Soc. 119: 431–437.

Bartnik, V.G. 1972. Comparison of the breeding habits of two subspecies of longnose dace, *Rhinichthys cataractae*. Can. J. Zool. 50: 83–86.

Barton, B.A. 1980. Spawning migrations, age and growth, and summer feeding of white and longnose suckers in an irrigation reservoir. Can. Field-Nat. 94: 300–304.

Barton, B.A., and B.F. Bigood. 1980. Competitive feeding habits of rainbow trout, white sucker, and longnose sucker in Paine Lake, Alberta. Alberta Fish and Wildlife Division, Edmonton. Fish. Res. Rep. No. 16. 27 pp.

Battle, H.I. 1940. The embryology and larval development of the goldfish (*Carassius auratus* L.) from Lake Erie. Ohio J. Sci. 40: 82–93.

Battle, H.I., and W.M. Sprules. 1960. A description of the semi-buoyant eggs and early developmental stages of the goldeye, *Hiodon alosoides* (Rafinesque). J. Fish. Res. Board Can. 17: 245–266.

Baxter, G.T., and J.R. Simon. 1970. Wyoming fishes. Wyoming Game and Fish Department, Cheyenne.

Baxter, J.S. 1997. Aspects of the reproductive ecology of bull trout (*Salvelinus confluentus*) in the Chowade River, British Columbia. M.Sc. thesis, Department of Zoology, University of British Columbia, Vancouver.

Baxter, J.S., E. Taylor, R. Devlin, J. Hagen, and J.D. McPhail. 1997. Evidence for natural hybridization between Dolly Varden (*Salvelinus malma*) and bull trout (*Salvelinus confluentus*) in a north central British Columbia watershed. Can. J. Fish. Aquat. Sci. 54: 421–429.

Baxter, J.S., G.J. Birch, and W.R. Olmsted. 2003. Assessment of a constructed fish migration barrier using radiotelemetry and Floy tagging. N. Am. J. Fish. Manage. 23: 1030–1035.

Beacham, T.D. 1982. Fecundity of coho salmon (*Oncorhynchus kisutch*) and chum salmon (*O. keta*) in the northeast Pacific Ocean. Can. J. Zool. 60: 1463–1469.

Beacham, T.D., and C.B. Murray. 1986a. Comparative developmental biology of pink salmon, *Oncorhynchus gorbuscha*, in southern British Columbia. J. Fish Biol. 28: 233–246.

Beacham, T.D., and C.B. Murray. 1986b. Comparative developmental biology of chum salmon (*Oncorhynchus keta*) from the Fraser River, British Columbia. Can. J. Fish. Aquat. Sci. 43: 252–262.

Beacham, T.D. and C.B. Murray. 1989. Variation in the developmental biology of sockeye salmon (*Oncorhynchus nerka*) and chinook salmon (*Oncorhynchus tshawytscha*) in British Columbia. Can. J. Zool. 67: 2081–2089.

Beacham, T.D., S. Pollard, and D.L. Khai. 1999. Population structure and stock identification of steelhead in southern British Columbia, Washington, and the Columbia River based on microsatellite DNA variation. Trans. Am. Fish. Soc. 128: 1068–1084.

Beacham, T.D., K.J. Supernault, M. Wetklo, B. Deagle, K. Larabee, J.R. Irvine, J.R. Candy, K.M. Miller, R.J. Nelson, and R.E. Withler. 2003. The geographic basis for population structure in Fraser River chinook salmon (*Oncorhynchus tshawytscha*)). Fish. Bull. 101: 229–242.

Beacham, T.D., B. McIntosh, and C. MacConnachie. 2004a. Population structure of lake-type and river-type sockeye salmon in transboundary rivers of northern British Columbia. J. Fish Biol. 65: 389–402.

Beacham, T.D., K.D. Le, and J.R. Candy. 2004b. Population structure and stock identification of steelhead trout (*Oncorhynchus mykiss*) in British Columbia and the Columbia River based on microsatellite variation. Environ. Biol. Fishes, 69: 95–109.

Beamesderfer, R.C. 1992. Reproduction and early life history of northern squawfish, *Ptychocheilus oregonensis*, in Idaho's St. Joe River. Environ. Biol. Fishes, 35: 231–241.

Beamesderfer, R.C., J.C. Elliot, and C.A. Foster. 1989. Report A. pp. 5–52. In A.A. Nigro (ed.). Status and habitat requirements of white sturgeon populations in the Columbia River downstream of McNary Dam. Bonneville Power Administration, Portland, OR. 1989 Progress Report, Project 86-50.

Beamesderfer, R.C., D.L. Ward, and A.A. Nigro. 1996. Evaluation of the biological basis for a predator control program on northern squawfish (*Ptychocheilus oregonensis*) in the Columbia and Snake rivers. Can. J. Fish. Aquat. Sci. 53: 2898–2908.

Beamish, R.J. 1980. Adult biology of the river lamprey (*Lampetra ayresii*) and the Pacific lamprey (*Lampetra tridentata*) from the Pacific coast of Canada. Can. J. Fish. Aquat. Sci. 37: 1906–1923.

Beamish, R.J. 1982. *Lampetra macrostoma*, a new species of freshwater parasitic lamprey from the west coast of Canada. Can. J. Fish. Aquat. Sci. 39: 736–747.

Beamish, R.J. 1985. Freshwater parasitic lamprey on Vancouver Island and a theory of the evolution of the freshwater parasitic and nonparasitic life history types. pp. 123–140. In R.E. Foreman, A. Gorbman, J.M. Dodd, and R. Olsson (eds.). Evolutionary biology of primitive fishes. Plenum Press, New York. NATO Adv. Sci. Inst. Ser. A Life Sci. No. 103.

Beamish, R.J. 1987. Evidence that parasitic and nonparasitic life history types are produced by one population of lamprey. Can. J. Fish. Aquat. Sci. 44: 1779–1782.

Beamish, R.J., and C.D. Levings. 1991. Abundance and freshwater migrations of the anadromous parasitic lamprey, *Lampetra tridentata*, in a tributary of the Fraser River, British Columbia. Can. J. Fish. Aquat. Sci. 48: 1250–1263.

Beamish, R.J., and T.G. Northcote. 1989. Extinction of a population of anadromous parasitic lamprey, *Lampetra tridentata*, upstream of an impassable dam. Can. J. Fish. Aquat. Sci. 44: 525–537.

Beamish, R.J., and R.E. Withler. 1986. A polymorphic population of lampreys that may produce parasitic and nonparasitic varieties. pp. 31–45. In T. Ueno, R. Arai, T. Taniuchi, and K. Matsuura (eds.). Indo-Pacific Fish Biology. Proceedings of the Second International Conference on Indo-Pacific Fishes. Ichthyological Society of Japan, Tokyo.

Beamish, R.J., and J.H. Youson. 1987. Life history and abundance of young adult *Lampetra ayresii* in the Fraser River and their possible impact on salmon and herring stocks in the Strait of Georgia. Can. J. Fish. Aquat. Sci. 46: 420–425.

Beauchamp, D.A. 1982. The life history, spawning behaviour, and intraspecific interactions of Arctic grayling (*Thymallus arcticus*) in upper Granite Lake. M.Sc. thesis, School of Fisheries, University of Washington, Seattle.

Beauchamp, D.A. 1990a. Movements, habitat use, and spawning strategies of Arctic grayling in a subalpine lake tributary. Northwest Sci. 64: 195–207.

Beauchamp, D.A. 1990b. Seasonal and diel food habits of rainbow trout stocked as juveniles in Lake Washington. Trans. Am. Fish. Soc. 119: 475–482.

Beauchamp, D.A., S.A. Vecht, and G.L. Thomas. 1992. Temporal, spatial, and size-related foraging of wild cutthroat trout in Lake Washington. Northwest Sci. 66: 149–159.

Beauchamp, D.A., E.R. Byron, and W.A. Wurtsbaugh. 1994. Summer habitat use by littoral-zone fishes in Lake Tahoe and the effects of shoreline structures. N. Am. J. Fish. Manage. 14: 385–394.

Beauchamp, D.A., M.G. LaRiviere, and G.L. Thomas. 1995. Evaluation of competition and predation as limits to juvenile kokanee and sockeye salmon production in Lake Ozette,

Washington. N. Am. J. Fish. Manage. 15: 193–207.

Beaver, B.J., and J.H. Gee. 1988. Role of water current and related variables in determining buoyancy in the sticklebacks *Culaea inconstans* and *Pungitius pungitius*. Can. J. Zool. 66: 2006–2014.

Becker, G.C. 1983. Fishes of Wisconsin. University of Wisconsin Press, Madison.

Beers, C.E., and J.M. Culp. 1990. Plasticity in the foraging behaviour of a lotic minnow (*Rhinichthys cataractae*) in response to different light intensities. Can. J. Zool. 68: 101–105.

Behnke, R.J. 1986. Brown trout. Trout, 27: 42–47.

Behnke, R.J. 1992. Native trout of western North America. Am. Fish. Soc. Monogr. No. 6.

Behnke, R.J. 1997. Evolution, systematics, and structure of *Oncorhynchus clarki clarki*. pp. 3–6. *In* J.D. Hall, P.A. Bisson, and R.E. Gresswell (eds.). Sea-run cutthroat trout: biology, management, and future conservation. Oregon Chapter, American Fisheries Society, Corvallis.

Behnke, R.J. 2002. Trout and salmon of North America. The Free Press, Toronto, ON.

Behnke, R.J., and M. Zarn. 1976. Biology and management of threatened and endangered western trout. USDA For. Serv. Gen. Tech. Rep. RM-28. 45 pp.

Beland, K.F., R.M. Jordan, and A.L. Meister. 1982. Water depth and velocity preferences of spawning Atlantic salmon in Maine rivers. N. Am. J. Fish. Manage. 2: 11–13.

Belding, D.L. 1934. The spawning habits of Atlantic salmon. Trans. Am. Fish. Soc. 64: 211–216.

Belk, M.C. 1998. Predator-induced maturity in bluegill sunfish (*Lepomis macrochirus*): variation among populations. Oecologia 113: 203–209.

Bell, M.A. 1977. A late Miocene marine threespine stickleback, *Gasterosteus aculeatus aculeatus*, and its zoogeographic and evolutionary significance. Copeia, 1977: 277–282.

Bell, M.A. 1994. Palaeobiology and evolution of threespine stickleback. pp. 438–471. *In* M.A. Bell and S.A. Foster (eds.). The evolutionary biology of the threespine stickleback. Oxford University Press, Oxford, UK.

Bendell, B.E., and D.K. McNichol. 1987. Cyprinid assemblages and physical and chemical characteristics of small northern Ontario lakes. Environ. Biol. Fishes, 19: 229–234.

Bendock, T.N. 1979. Inventory and cataloging of Arctic area waters. Alaska Department of Fish and Game, Juneau. Fed. Aid Fish Restor. Anad. Fish Stud. No. G-I-1. 64 pp.

Bendock, T.N., and J.M. Burr. 1986. Arctic area trout studies. Alaska Department of Fish and Game, Juneau. Fed. Aid Fish Restor. Anad. Fish Stud. No. T-7-1. 75 pp.

Benz, G.W., and R.P. Jacobs. 1986. Practical field methods of sexing largemouth bass. Prog. Fish-Cult. 48: 221–225.

Berejikian, B.A. 1995. The effects of hatchery and wild ancestry and experience on the relative ability of steelhead trout fry (*Oncorhynchus mykiss*) to avoid a benthic predator. Can. J. Fish. Aquat. Sci. 52: 2476–2482.

Berejikian, B.A., E.P. Tezal, and A.L. LaRae. 2000. Female mate choice and spawning behaviour of chinook salmon under experimental conditions. J. Fish Biol. 57: 647–661.

Berg, L.S. 1934. Vernal and hiemal races among anadromous fishes. [English translation published 1959 in J. Fish. Res. Board Can. 16: 515–537.]

Berg, L.S. 1948. Freshwater Fishes of the USSR and adjacent countries. Akad. Nauk SSSR Zool. Inst. No. 1. 493 pp. [Translated from Russian by the Israel Program for Scientific Translation, Jerusalem. 1962.]

Bernard, D.R, K.R. Hepler, J.D. Jones, M.E. Whalen, and D.N. McBride. 1995. Trans. Am. Fish. Soc. 124: 297–307.

Bernatchez, L. 2001. The evolutionary history of brown trout (*Salmo trutta* L.) inferred from phylogeographic. Nested clade, and mismatch analyses of mitochondrial DNA variation. Evolution, 55: 351–379.

Bernatchez, L., and J.J. Dodson. 1985. Influence of temperature and current speed on the swimming capacity of the lake whitefish

(*Coregonus clupeaformis*) and cisco (*Coregonus artedii*). Can. J. Fish. Aquat. Sci. 42: 1522–1529.

Bernatchez, L., F. Colombani, and J.J. Dodson. 1991. Phylogenetic relationships among the subfamily Coregoninae as revealed by mitochondrial DNA restriction analysis. J. Fish Biol. 39(Suppl. A): 283–290.

Bettles, C.M. 2004. Interspecific hybridization between sympatric coastal cutthroat and coastal rainbow/steelhead trout on Vancouver Island, British Columbia: a conservation and evolutionary examination. M.Sc. thesis, University of Windsor, Windsor, ON.

Bickham, J., J. Patten, S. Minzenmayer, L.L. Moylton, J. Benny, and B.J. Gallaway. 1997. Identification of Arctic and Bering ciscoes in the Colville River Delta, Beaufort Sea. pp. 224–228. *In* J.B. Reynolds (ed.). Fish ecology in arctic North America. Am. Fish. Soc. Symp. No. 19.

Billard, R. 1996. Reproduction of pike: gametogenesis, gamete biology and early development. pp 13–43. *In* J.F. Craig (ed.). Pike: biology and exploitation. Chapman & Hall, London. Fish Fish. Ser. No. 19.

Biro, P. 1998. Staying cool: behavioral thermoregulation during summer by young-of-the-year brook trout in a lake. Trans. Am. Fish. Soc 127: 212–222.

Birstein, V.J., and R. DeSalle. 1998. Molecular phylogeny of Acipenserinae. Mol. Phylogenet. Evol. 9: 141–155.

Bishop, F.G. 1971. Observations on spawning habits and fecundity of Arctic grayling. Prog. Fish-Cult. 33: 12–19.

Bishop, F.G. 1975. Observations on the fish fauna of the Peace River in Alberta. Can. Field-Nat. 89: 423–430.

Bisson, P.A., and P.E. Reimers. 1977. Geographic variation among Pacific Northwestern populations of longnose dace, *Rhinichthys cataractae*. Copeia, 1977: 518–522.

Bisson, P.A., and C.E. Bond. 1971. Origin and distribution of the fishes of the Harney Basin, Oregon. Copeia, 1971: 268–281.

Blackman, B.G., and M.J. Hunter. 2001. 1998 Arctic grayling surveys in the Table, Anzac, and Parsnip Rivers. B.C. Hydro, Northern Region, Prince George. Peace/Williston Fish Wildl. Compens. Prog. Rep. No. 237. 39 pp.

Blanchfield, P.J., and M.S. Ridgway. 1997. Reproductive timing and use of redd sites by lake-spawning brook trout (*Salvelinus fontinalis*). Can. J. Fish. Aquat. Sci. 54: 747–756.

Blanchfield, P.J., and M.S. Ridgway. 1998. Brook trout spawning areas in lakes. Ecol. Freshw. Fish, 7: 140–145.

Blanchfield, P.J., and M.S. Ridgway. 1999. The cost of peripheral males in a brook trout mating system. Anim. Behav. 57: 537–544.

Bodaly, R.A. 1980. Pre-spawning and post-spawning movements of walleye, *Stizostedion vitreum*, in southern Indian Lake, Manitoba. Can. Tech. Rep. Fish. Aquat. Sci. No. 931. 30 pp.

Bodaly, R.A., J. Vuorinen, R.D. Ward, M. Lucyzynski, and J.D. Reist. 1991. Genetic comparisons of New and Old World coregonid fishes. J. Fish Biol. 38: 37–51.

Bodaly, R.A., J.A. Vuorinen, Y.S. Reshetnikov, and J.D. Reist. 1994. Genetic relationships of five species of coregonid fishes from Siberia. [In Russian.] Vopr. Ictiol. 34: 195–203. [English translation published in J. Icthyol. 34: 117–130.]

Bogdanov, V.D. 1983. Species-specific differences of larvae of some whitefishes (Coregoninae) at hatching. J. Ichthyol. 23: 86–96.

Bohn, A., and W.S. Hoar. 1965. The effect of salinity on the iodine metabolism of coastal and inland prickly sculpins, *Cottus asper* Richardson. Can. J. Zool. 43: 977–985.

Bond, C.E. 1963. Distribution and ecology of freshwater sculpins, genus *Cottus*, in Oregon, Ph.D. thesis, University of Michigan, Ann Arbor.

Bond, C.E. 1973. Keys to Oregon freshwater fishes. Oreg. State Univ. Agric. Exp. Stn. Tech. Bull. No. 58. 42 pp.

Bond, W.A. 1980. Fishery resources of the Athabasca River downstream of Fort McMurray: Vol. 1. Prepared for the Alberta Oil Sands Environmental Research Program. Department of Fisheries Oceans, Freshwater

Institute, Edmonton, AB. AOSERP Rep. No. 89.

Bond, W.A., and D.K. Berry. 1980. Fisheries resources of the Athabasca River downstream of Fort McMurray. Vol. II. Prepared for Alberta Oil Sands Environmental Research Program. Department of Fisheries Oceans, Freshwater Institute. Edmonton, AB. AOSERP Project AF 4.3.2, 158 pp.

Bond, W.A. and R.N. Erickson. 1985. Life history studies of anadromous coregonid fishes in two freshwater lakes on the Tuktoyaktuk Peninsula, Northwest Territories. Can. Tech. Rep. Fish. Aquat. Sci. No. 1336. 61 pp.

Bond, W.A., and R.N. Erickson. 1997. Coastal migrations of Arctic ciscoes in the eastern Beaufort Sea. pp. 155–164. In J.B. Reynolds (ed.). Fish ecology in arctic North America. Am. Fish. Soc. Symp. No. 19.

Bonneau, J.L., and D.L. Scarnecchia. 1998. Seasonal and diel changes in habitat use by juvenile bull trout (*Salvelinus confluentus*) and cutthroat trout (*Oncorhynchus clarki*) in a mountain stream. Can. J. Zool. 76: 783–790.

Bourke, P., P. Magnan, and M.A. Rodriguez. 1996. Diel locomotor activity of brook charr, as determined by radiotelemetry. J. Fish Biol. 49: 1174–1185.

Bourke, P., P. Magnan, and M.A. Rodriguez. 1999. Phenotypic responses of lacustrine brook charr in relation to the intensity of competition. Evol. Ecol. 13: 19–31.

Bradford, M.J., and P.S. Higgins. 2001. Habitat-, season-, and size-specific variation in diel activity patterns of juvenile chinook salmon (*Oncorhynchus tshawytscha*) and steelhead trout (*Oncorhynchus mykiss*). Can. J. Fish. Aquat. Sci. 58: 365–374.

Braennaes, E., and A. Alanaerae. 1997. Is diel dualism in feeding activity influenced by competition between individuals? Can. J. Zool. 75: 661–669.

Bramblett, R.G., M.D. Bryant, B.E. Wright, and R.G. White. 2002. Seasonal use of small tributary and main-stem habitats by juvenile steelhead, coho salmon, and Dolly Varden in a southeastern Alaska drainage basin. Trans. Am. Fish. Soc. 131: 498–506.

Brandt, S.B. 1986. Ontogenetic shifts in habitat, diet, and diel-feeding periodicity of slimy sculpin in Lake Ontario. Trans. Am. Fish. Soc. 115: 711–715.

Brandt, T.M., and S.A. Flickinger. 1987. Feeding largemouth bass during cool and cold weather. Prog. Fish-Cult. 49: 286–290.

Brannon, E.L. 1972. Mechanisms controlling migration of sockeye salmon fry. Int. Pac. Salmon Comm. Bull. 21: 1–86.

Brannon, E.L., and A. Setter. 1992. Movements of white sturgeon in Roosevelt Lake. Bonneville Power Administration, Portland, OR. Final Report, Project No. 89-44. 35 pp.

Brannon, E.L., S.D. Brewer, A. Setter, M. Miller, F. Utter, and W. Hershberger. 1985. Columbia River white sturgeon (*Acipenser transmontanus*) early life history and genetics study. Bonneville Power Administration, Portland, OR. Final Report, Project No. 83-316. 68 pp.

Brash, J., J. McFadden, and S. Kimiotek. 1958. Brook trout: life history, ecology, and management. Wisconsin Department of Natural Resources, Madison. Publ. No. 226.

Braum, E., N. Peters, and M. Stolz. 1996. The adhesive organ of larval pike, *Esox lucius* L. (Pisces). Int. Rev. Gesamten Hydrobiol. 81: 101–108.

Brazo, D.C., C.R. Liston, and R.C. Anderson. 1978. Life history of the longnose dace, *Rhinichthys cataractae*, in the surge zone of eastern Lake Michigan near Ludington, Michigan. Trans. Am. Fish. Soc. 107: 550–556.

Breder, C.M., and D.E. Rosen. 1966. Modes of reproduction in fishes. Natural History Press, New York.

Bremset, G. 2000. Seasonal and diel changes in behaviour, microhabitat use and preferences by young pool-dwelling Atlantic salmon, *Salmo salar*, and brown trout, *Salmo trutta*. Environ. Biol. Fishes, 59: 163–179.

Brenkman, S.J., and S.C. Corbett. 2005. Extent of anadromy in bull trout and implications for conservation of a threatened species. N. Am. J. Fish. Manage. 25: 1073–1081.

Bretz, J.H. 1913. Glaciation of the Puget Sound region. Wash. Geol. Surv. Bull. 8: 1–244.

Briggs, J.C. 1953. The behaviour and reproduction of salmonid fishes in a small coastal stream. Calif. Dep. Fish Game Fish Bull. No. 94. 62 pp.

Bronte, C.R., J.H. Selgeby, and D.V. Swedberg. 1993. Dynamics of a yellow perch population in western Lake Superior. N. Am. J. Fish. Manage. 13: 511–523.

Brooke, H.E. 1970. Speciation parameters in coregonine fishes: I. Egg size, II. Karyotype. pp. 61–66. In C.C. Lindsey and C.S. Woods (eds.). Biology of coregonid fishes. University of Manitoba Press, Winnipeg.

Brooke, L.T. 1975. Effect of different constant temperatures on egg survival and embryonic development in lake whitefish (Coregonus clupeaformis). Trans. Am. Fish. Soc. 104: 555–559.

Brown, C.J.D. 1971. Fishes of Montana. Big Sky Books, Montana State University, Bozeman.

Brown, J.H., U.T. Hammer, and G.D. Koshinsky. 1970. Breeding biology of the lake chub, Couesius plumbeus, at Lac la Ronge, Saskatchewan. J. Fish. Res. Board Can. 27: 1005–1015.

Brown, J.R., A.T. Beckenbach, and M.J. Smith. 1992. Influence of Pleistocene glaciations and human intervention upon mitochondrial DNA diversity in white sturgeon (Acipenser transmontanus) populations. Can. J. Fish. Aquat. Sci. 49: 358–367.

Brown, J.R., A.T. Beckenbach, and M.J. Smith. 1993. Intraspecific DNA sequence variation of the mitochondrial control region of white sturgeon (Acipenser transmontanus). Mol. Biol. Evol. 10: 326–341.

Brown, K.H., S.J. Patton, K.E. Martin, K.M. Nichols, R. Armstrong, and G. Thorgaard. 2004. Genetic analysis of interior Pacific Northwest Oncorhynchus mykiss reveals apparent ancient hybridization with westslope cutthroat trout. Trans. Am. Fish. Soc. 133: 1078–1088.

Brown, L.R., S.A. Matern, and P.B. Moyle. 1995. Comparative ecology of prickly sculpin, Cottus asper, and coastrange sculpin, Cottus aleuticus, in the Eel River, California. Environ. Biol. Fishes, 42: 329–343.

Brown, M.L., D.W. Willis, and B.G. Blackwell. 1999. Physiochemical and biological influences on black bullhead populations in eastern South Dakota glacial lakes. J. Freshwater Ecol. 14: 47–60.

Brown, R.S., and W.C. Mackay. 1995a. Spawning ecology of cutthroat trout (Oncorhynchus clarki) in the Ram River, Alberta. Can. J. Fish. Aquat. Sci. 52: 983–992.

Brown, R.S., and W.C. Mackay. 1995b. Fall and winter movements and habitat use by cutthroat trout in the Ram River, Alberta. Trans. Am. Fish. Soc. 124: 873–885.

Bruce, W.J., and R.F. Parsons. 1976. Age, growth, and maturity of lake chub (Couesius plumbeus) in mile 66 brook, Tenmile Lake, western Labrador. Tech. Rep. Fish. Mar. Serv. Can. Res. Dev. No. 683.

Brunner, D.M., M.R. Douglas, A. Osinov, C.C. Wilson, and L. Bernatchez. 2001. Holarctic phylogeography of Arctic charr (Salvelinus alpinus L.) inferred from mitochondrial DNA sequences. Evolution, 55: 573–586.

Bry, C. 1996. Role of vegetation in the life cycle of pike. pp. 45–67. In J.F. Craig (ed.). Pike: biology and exploitation. Chapman & Hall, London. Fish Fish. Ser. No. 19.

Bry, C., F. Bonamy, J. Manelphe, and B. Duranthon. 1995. Early life characteristics of pike, Esox lucius, in rearing ponds: temporal survival patterns and ontogenetic diet shifts. J. Fish Biol. 46: 99–113.

Bryan, J.E., and D.A. Kato. 1975. Spawning of lake whitefish, Coregonus clupeaformis, and round whitefish, Prosopium cylindraceum, in Aishihik Lake and East Aishihik River, Yukon Territory. J. Fish. Res. Board Can. 32: 283–288.

Bryan, M.D., and D.L. Scarnecchia. 1992. Species richness, composition, and abundance of fish larvae and juveniles inhabiting natural and developed shorelines of a glacial Iowa lake. Environ. Biol. Fishes, 35: 329–341.

Bryan, S.D., T.D. Hill, S.T. Lynott, and W.G. Duffy. 1995. On the food habits of walleye in Lake Oahe, South Dakota. J. Freshw. Ecol. 10: 1–10.

Buddington, R.K., and J.P. Christofferson. 1985. Digestive and feeding characteristics of the chondrosteans. pp. 31–42. In F.P. Binkowski and S.I. Doroshov (eds.). North American sturgeons: biology and aquaculture potential. Dr. W. Junk Publishers, Dordrecht, the Netherlands.

Buktenica, M.W., and G.L. Larson. 1996. Ecology of kokanee salmon and rainbow trout in Crater Lake, Oregon. Lake Reservoir Manage. 12: 298–310.

Bulkley, R.V. 1967. Fecundity of steelhead trout, *Salmo gairdneri*, from the Alsea River, Oregon. J. Fish. Res. Board Can. 24: 917–926.

Bulmer, L.S. 1985a. Reproductive natural history of the brown bullhead, *Ictalurus nebulosus*. Am. Midl. Nat. 114: 318–330.

Bulmer, L.S. 1985b. The significance of biparental care in the brown bullhead, *Ictalurus nebulosus*. Environ. Biol. Fishes, 12: 231–236.

Bulmer, L.S. 1986. The function of parental care in the brown bullhead, *Ictalurus nebulosus*. Am. Midl. Nat. 115: 234–238.

Bunnell, D.B., J.J. Isely, K.H. Burrell, and D.H. van Lear. 1998. Diel movement of brown trout in a southern Appalachian river. Trans. Am. Fish. Soc.127: 630–636.

Burgner, R.L. 1991. Life history of sockeye salmon. pp. 3–117. In C. Groot and L. Margolis (eds.). Pacific salmon life histories. University of British Columbia Press, Vancouver.

Burgner, R.T., J.T. Light, L. Margolis, T. Okazaki, A. Tautz, and S. Ito. 1992. Distribution and origins of steelhead trout (*Oncorhynchus mykiss*) in offshore waters of the North Pacific Ocean. Int. North Pac. Fish. Comm. Bull. No. 51. 92 pp.

Burner, C.J. 1951. Characteristics of spawning nests of Columbia River Salmon. Fish. Bull. 61: 97–110.

Burns, J.A. 1991. Mid-Wisconsinan vertebrates and their environment from January Caves, Alberta, Canada. Quat. Res. 35: 130–143.

Burrell, K.H., J.J. Isely, D.B. Bunnell, D.H. van Lear, and C.A. Dolloff. 2000. Seasonal movement of brown trout in a southern Appalachian river. Trans. Am. Fish. Soc. 129: 1373–1379.

Busby, P.J., T.C. Wainwright, G.J. Bryant, L.J. Lierheimer, R.S. Waples, F.W. Waknitz, and I.V. Lagomarsino. 1996. Status review of west coast steelhead from Washington, Oregon, and California. US Department of Commerce, National Marine Fisheries Service, Washington, DC. NOAA Tech. Memo. No. NMFS-NWFSC-27. 261 pp.

Bustard, D. 1996. Kemess South Project, 1995 fisheries studies. Final report. Prepared for El Condor Resources Ltd. Dave Bustard and Associates, Smithers, BC. 118 pp. + 4 appendices.

Butcher, G.A., J.R. Ellis, and R.B. Davidson. 1981. Aspects of the life history of Arctic grayling (*Thymallus arcticus*) in the Liard River drainage, British Columbia. Working Report. BC Aquatic Studies Branch, Victoria.

Buzby, K.M., and L.A. Deegan. 2000. Inter-annual fidelity to summer feeding sites in Arctic grayling. Environ. Biol. Fishes, 59: 319–327.

Byorth, P.A., and J.P. Magee. 1998. Competitive interactions between Arctic grayling and brook trout in the Big Hole River Drainage, Montana. Trans. Am. Fish. Soc 127: 921–931.

Cahn, A.R. 1927. An ecological study of southern Wisconsin fishes. Ill. Biol. Monogr. 11: 1–151.

Callan, W.T., and S.L. Sanderson. 2003. Feeding mechanisms in carp: crossflow filtration, palatal protrusions, and flow reversals. J. Exp. Biol. 206: 883–892.

Cameron, J.N., J. Kostoris, and P.A. Penhale. 1973. Preliminary energy budget for the nine-spine stickleback (*Pungitius pungitius*) in an Arctic lake. J. Fish. Res. Board Can. 30: 1179–1189.

Campbell, J.S., and H.R. MacCrimmon. 1970. Biology of the emerald shiner, *Notropis atherinoides* Rafinesque, in Lake Simcoe, Canada. J. Fish Biol. 2: 259–273.

Cannings, S.G., and J. Ptolemy. 1998. Rare freshwater fish of British Columbia. BC Environment, Victoria.

Carl, G.C. 1945. Three apparently unrecorded fresh-water fishes in British Columbia. Can. Field-Nat. 59: 25.

Carl, G.C. 1953. Limnobiology of Cowichan Lake, British Columbia. J. Fish. Res. Board Can. 9: 417–449.

Carl, G C., W.A. Clemens, and C.C. Lindsey. 1959. The fresh-water fishes of British Columbia. Queen's Printer, Victoria, BC. Prov. Mus. Handb. No. 5.

Carl, L.M. 1995. Sonic tracking of burbot in Lake Opeongo, Ontario. Trans. Am. Fish. Soc. 124: 77–83.

Carl, L.M., and M.C. Healey. 1984. Differences in enzyme frequency and body morphology among three juvenile life history types of chinook salmon (*Oncorhynchus tshawytscha*) in the Nanaimo River, British Columbia. Can. J. Fish. Aquat. Sci. 41: 1070–1077.

Carl, L.M., and J.D. Stelfox. 1989. A meristic, morphometric, and electrophoretic analysis of cutthroat trout, *Salmo clarki*, from two mountain lakes in Alberta. Can. Field-Nat. 103: 80–84.

Carl, L.M., C. Hunt, and P.E. Ihssen. 1994. Rainbow trout of the Athabasca River, Alberta: a unique population. Trans. Am. Fish. Soc. 123: 129–140.

Carlander, K.D. 1977. Handbook of freshwater fishery biology. Vol. 2. Iowa State University Press, Ames.

Carline, R.F. 1980. Features of successful spawning site development for brook trout in Wisconsin ponds. Trans. Am. Fish. Soc 109: 453–457.

Carlson, C.C., and K. Klein. 1996. Late Pleistocene salmon of Kamloops Lake. pp. 274–280. *In* R. Ludvigsen (ed.). Life in stone: a natural history of British Columbia's fossils. University of British Columbia Press, Vancouver.

Carr, M.G., J.E. Carr, and S.R. Johnson. 1987. Seasonal changes in territoriality and frequency of agonistic behavior in two densities of juvenile brown bullhead, *Ictalurus nebulosus*. Environ. Biol. Fishes, 19: 175–181.

Carra, P.E., E.P. Kiver, and D.F. Strading. 1996. The southern limit of Cordilleran ice in the Colville and Pend Oreille valleys of northeastern Washington during the late Wisconsinan glaciation. Can. J. Earth Sci. 33: 769–778.

Cartwright, J.W. 1956. Contributions to the life history of the northern squawfish, *Ptychocheilus oregonensis* (Richardson). Honours B.A. thesis, Department of Zoology, University of British Columbia, Vancouver.

Casselman, J.M. 1974. External sex determination of northern pike, *Esox lucius* Linnaeus. Trans. Am. Fish. Soc. 103: 343–347.

Casselman, J.M. 1978. Effects of environmental factors on growth, survival, activity and exploitation of northern pike. Am. Fish. Soc. Spec. Publ. 11: 114–128.

Casselman, J.M. 1996. Age, growth and environmental requirements of pike. pp. 69–101. *In* J.F. Craig (ed.). Pike: biology and exploitation. Chapman & Hall, London. Fish Fish. Ser. No. 19.

Casselman, J.M., and C.A. Lewis. 1996. Habitat requirements of northern pike (*Esox lucius*). Can. J. Fish. Aquat. Sci. 53(Suppl. 1): 161–174.

Cavender, T.M. 1978. Taxonomy and distribution of the bull trout, *Salvelinus confluentus* (Suckley) from the American Northwest. Calif. Fish Game, 64: 139–174.

Cavender, T.M. 1986. Review of the fossil history of North American freshwater fishes. pp. 699–724. *In* C.H. Hocutt and E.O. Wiley (eds.). The zoogeography of North American freshwater fishes. John Wiley & Sons, New York.

Cavender, T.M. 1992. The fossil record of the Cyprinidae. pp. 34–54. *In* I.J. Winfield and J.S. Nelson (eds.). Cyprinid fishes: systematics, biology, and exploitation. Chapman & Hall, London.

Chadwick, H.K. 1959. California sturgeon tagging studies. Calif. Fish Game, 45: 297–301.

Chapman, D.W., H.J. Campbell, and J.D. Fortune. 1967. Summer distribution and food of kokanee salmon in Elk Lake, Oregon. Trans. Am. Fish. Soc. 96: 308–312.

Chapman, D.W., D.E. Weitcamp, T.L. Welsh M.B. Dwell, and T.H. Schadt. 1986. Effects of river flow on the distribution of chinook salmon redds. Trans. Am. Fish. Soc. 115: 537–547.

Chapman, L.J., and W.C. Mackay. 1984. Direct observation of habitat utilization by northern pike. Copeia, 1984: 255–258.

Chapman, L.J., and W.C. Mackay. 1990. Ecological correlates of feeding flexibility in northern pike (*Esox lucius*). J. Freshw. Ecol. 5: 313–322.

Chapman, L.J., W.C. Mackay, and C.W. Wilkinson. 1989. Feeding flexibility in northern pike (*Esox lucius*): fish versus invertebrate prey. Can. J. Fish. Aquat. Sci. 46: 666–669.

Chapman, W.M. 1943. The spawning of chinook salmon in the main Columbia River. Copeia, 1943: 168–170.

Chen, Y., and H.H. Harvey. 1994a. Growth, abundance, and food supply of white sucker. Trans. Am. Fish Soc. 124: 262–270.

Chen, Y., and H.H. Harvey. 1994b. Maturation of white sucker, *Catostomus commersoni*, populations in Ontario. Can. J. Fish. Aquat. Sci. 51: 2066–2076.

Chereshnev, I.A., and M.B. Skopets. 1992. A new record of the pygmy whitefish, *Prosopium coulteri*, from the Amguem River Basin (Chukotski Peninsula). J. Ichthyol. 32: 46–55.

Chigbu, P. 2000. Population biology of longfin smelt and aspects of the ecology of other major planktivorous fishes in Lake Washington. J. Freshw. Ecol. 15: 543–557.

Chigbu, P., and T.H. Sibley. 1994. Relationship between abundance, growth, egg size and fecundity in a landlocked population of longfin smelt, *Spirinchus thaleichthys*. J. Fish Biol. 45: 1–15.

Chigbu, P., and T.H. Sibley. 1998. Predation by the longfin smelt (*Spirinchus thaleichthys*) on the mysid *Neomysis mercedis* in Lake Washington. Freshw. Biol. 2: 295–304.

Chigbu, P., T.H. Sibley, and D.A. Beauchamp. 1998. Abundance and distribution of *Neomysis mercedis* and a major predator, longfin smelt (*Spirinchus thaleichthys*) in Lake Washington. Hydrobiologia, 386: 167–182.

Chisholm, B.A. 1975. An experimental study of nocturnal feeding in the northern squawfish (*Ptychocheilus oregonensis*). Honours B.Sc. thesis, Department of Zoology, University of British Columbia, Vancouver.

Chisholm, I., M.E. Hensler, B. Hansen, and D. Skaar. 1989. Quantification of Libby Reservoir levels needed to maintain or enhance reservoir fisheries. Summary Report 1981–1987. Bonneville Power Administration, Division of Fish and Wildlife, Portland, OR.

Churikov, D., and A.J. Gharrett. 2002. Comparative phylogeography of the two pink salmon broodlines: an analysis based on a mitochondrial genealogy. Mol. Ecol. 11: 1077–1101.

Clague, J.J. 1981. Late Quaternary geology and geochronology of British Columbia, Part 2: Summary and discussion of radio-carbon-dated Quaternary history. Geol. Surv. Can. Pap. No. 80-35. 40 pp.

Clague J.J., J.R. Harper, R.J. Hebda, and D.E. Howes. 1982. Late Quaternary sea levels and crustal movements, coastal British Columbia. Can. J. Earth Sci. 19: 597–618.

Clark, C.F. 1950. Observations on the spawning habits of the northern pike, *Esox lucius*, in northwestern Ohio. Copeia, 1950: 285–287.

Clark, W.D., and J.E. McInerney. 1974. Emigration of the peamouth chub, *Mylocheilus caurinus*, across a dilute seawater bridge: an experimental biogeographic study. Can. J. Zool. 52: 457–469.

Clarke, W.C., and Beamish, R.J., 1988. Response of recently metamorphosed anadromous parasitic lamprey (*Lampetra tridentata*) to confinement in fresh water. Can. J. Fish. Aquat. Sci. 45: 42–47.

Clarke, W.C., R.E. Withler, and J. Shelbourn. 1992. Genetic control of juvenile life history pattern in chinook salmon (*Oncorhynchus tshawytscha*). Can. J. Fish. Aquat. Sci. 49: 2300–2306.

Clemens, W.A., D.S. Rawson, and J. McHugh. 1939. A biological survey of Okanagan Lake,

British Columbia. Bull. Fish. Res. Board Can. No. 56. 70 pp.

Clemens, W.A., R.V. Boughton, and J.A. Rattenbury. 1968. A limnological study of Teslin Lake, Canada. BC Fish and Wildlife, Victoria.

Close, D.A., M.S. Fiztpatrick, and H.W. Li. 2002. The ecological and cultural importance of a species at risk of extinction, Pacific lamprey. Fisheries, 27: 19–25.

Coad, B.W., and G. Power. 1973. Observations on the ecology and meristic variation of the ninespine stickleback, *Pungitius pungitius* (L. 1758) of the Matamek River system, Quebec. Am. Midl. Nat. 90: 498–503.

Coble, D.W. 1975. Smallmouth bass. pp. 21–33. *In* H. Clepper (ed.). Black bass biology and management. Sport Fishing Institute, Washington, DC.

Cochran, P.A., D.M. Lodge, J.R. Hodgson, and P.G. Knapik. 1988. Diets of syntopic finescale dace, *Phoxinus neogaeus*, and northern redbelly dace, *Phoxinus eos*: a reflection of trophic morphology. Environ. Biol. Fishes, 22: 235–240.

Cohen, D.M. 1954. Age and growth studies on two species of whitefishes from Point Barrow, Alaska. Stanford Ichthyol. Bull. 4: 168–187.

Colby, P.J., and L.T. Brooke. 1970. Survival and development of lake herring (*Coregonus artedii*) eggs at various incubation temperatures. pp. 417–428. *In* C.C. Lindsey and C.S. Woods (eds.). Biology of coregonid fishes. University of Manitoba Press, Winnipeg.

Colby, P.J., R.E. McNichol, and R.A. Ryder. 1979. Synopsis of biological data on the walleye, *Stizostedion vitreum* (Mitchill 1818). FAO, Rome. FAO Fish. Synopsis No. 119. 139 pp.

Colgan, P.W., and J.A. Brown. 1988. Dynamics of nest defense by male centrarchid fish. Behav. Processes, 17: 17–26.

Collins, N.C., and S.G. Hinch. 1993. Diel and seasonal variation in foraging activities of pumpkinseeds in an Ontario pond. Trans. Am. Fish. Soc. 122: 357–365.

Colonell, J.M., and B.J. Gallaway. 1997. Wind-driven transport and dispersion of age-0 Arctic ciscoes along the Alaska Beaufort coast. pp. 90–103. *In* J.B. Reynolds (ed.). Fish ecology in arctic North America. Am. Fish. Soc. Symp. No. 19.

Colosimo, P.F., C.L. Peichel, K.S. Nereng, B.K. Blackman, M.D. Shapiro, D. Schluter, and D.M. Kingsley. 2004. The genetic architecture of parallel armor plate reduction in threespine sticklebacks. PloS Biol. 2: 0635–0641.

Colosimo, P.F., K.E. Hosemann, S. Balabhadra, G. Villarreal, M. Dickson, J. Grimwood, J. Schmutz, R. M. Myers, D. Schluter, and D.M. Kingsley. 2005. Widespread parallel evolution in sticklebacks by repeated fixation of ectodysplasin alleles. Science (Washington, D.C.), 307: 1928–1933.

Connolly, P.J. 1997. Influence of stream characteristics and age-class interactions on populations of coastal cutthroat trout. pp. 173–174. *In* J.D. Hall, P.A. Bisson, and R.E. Gresswell (eds.). Sea-run cutthroat trout: biology, management, and future conservation. Oregon Chapter, American Fisheries Society, Corvallis.

Contor, C.R., and J.S. Griffith. 1995. Nocturnal emergence of juvenile rainbow trout from winter concealment relative to light intensity. Hydrobiologia, 299: 179–183.

Cook, M.F., and E.P. Bergersen. 1988. Movement, habitat selection, and activity periods of northern pike in Eleven Mile Reservoir, Colorado. Trans. Am. Fish. Soc. 117: 495–502.

Cope, E.D. 1874. On the Plagopterinae and the ichthyology of Utah. Proc. Am. Philos. Soc. 14: 129–139.

Copes, E. 1975. Ecology of the brassy minnow, *Hybognathus hankinsoni* Hubbs. University of Wisconsin, Stevens Point. Mus. Nat. Hist. Fauna Flora Wis. Part III Rep. No. 10: 46–72.

Corbett, B., and P.M. Powles. 1983. Spawning and early-life ecological phases of the white sucker in Jack Lake, Ontario. Trans. Am. Fish Soc. 112: 308–313.

Costello, A.B., and E. Rubidge. 2003. Status report on cutthroat trout (*Oncorhynchus clarki spp.*). Prepared for the Committee on the

Status of Endangered Wildlife in Canada (COSEWIC), Ottawa, ON.

Coutant, C.C. 1977. Compilation of temperature preference data. J. Fish. Res. Board Can. 34: 739–745.

Craig, J.F. 1987. The biology of perch and related fish. Croom Helm Ltd., Kent, UK.

Craig, J.F. (ed.). 1996. Pike: biology and exploitation. Chapman & Hall, London. Fish Fish. Ser. No. 19.

Craig, J.F., and J.A. Babaluk. 1989. Relationship of condition of walleye (*Stizostedion vitreum*) and northern pike (*Esox lucius*) to water clarity with special reference to Dauphin Lake, Manitoba. Can. J. Fish. Aquat. Sci. 46: 1581–1586.

Craig, J.K. 1995. Genetic divergence in spawning color of sympatric forms of sockeye salmon (*Oncorhynchus nerka*): investigation of an unknown form. M.Sc. thesis, University of Washington, Seattle.

Craig, J.K., and C.J. Foote. 2001. Countergradient variation and secondary sexual color: phenotypic convergence promotes genetic divergence in carotenoid use between sympatric anadromous and non-anadromous morphs of sockeye salmon (*Oncorhynchus nerka*). Evolution, 55: 380–391.

Craig, P.C., and K.A. Bruce. 1982. Fish resources in the upper Liard River drainage. Report prepared by LGL Ltd. for B.C. Hydro and Power Authority, Vancouver. 184 pp.

Craig, P., and G. Mann. 1974. Life history and distribution of the Arctic cisco (*Coregonus autumnalis*) along the Beaufort Sea coastline in Alaska and the Yukon Territory. pp. 1–33. *In* P. McCart (ed.). Life histories of anadromous and freshwater fishes in the western Arctic. Canadian Arctic Gas Study Ltd., Calgary, AB. Arct. Gas Biol. Rep. Ser. No. 20(4).

Craig, P., and J. Wells. 1975. Fisheries investigations in the Chandalar River Region, Northeast Alaska. pp. 1–114. *In* P. Craig (ed.). Fisheries investigations in a coastal region of the Beaufort Sea. Canadian Arctic Gas Study Ltd., Calgary, AB. Arct. Gas Biol. Rep. Ser. No. 34(1).

Crass, D.W, and R.H. Gray. 1982. Snout dimorphism in the white sturgeon, *Acipenser transmontanus*, in the Columbia River at Hanford, Washington. Fish. Bull. 80: 158–160.

Crisp, D.T. 1988. Prediction from water temperature of eyeing, hatching, and swim-up time for salmonid embryos. Freshw. Biol. 19: 41–48.

Cross, F.B., R.L. Mayden, and J.D. Stewart. 1986. Fishes in the western Mississippi Basin (Missouri, Arkansas and Red rivers). pp. 363–412. *In* C.H. Hocutt and E.O. Wiley (eds.). The zoogeography of North American freshwater fishes. John Wiley & Sons, New York.

Crossman, E.J. 1959. Distribution and movement of a predator, the rainbow trout, and its prey, the redside shiner, in Paul Lake, British Columbia. J. Fish. Res. Board Can. 16: 247–267.

Crossman, E.J. 1991. Introduced freshwater fishes: a review of the North American perspective with emphasis on Canada. Can. J. Fish. Aquat. Sci. 41(Suppl. 1): 46–57.

Crossman, E.J. 1995. Introductions of the lake trout (*Salvelinus namaycush*) in areas outside its native distribution: a review. J. Great Lakes Res. 21(Suppl. 1): 17–29.

Culp, J.M. 1989. Nocturnally constrained foraging of a lotic minnow (*Rhinichthys cataractae*). Can. J. Zool. 67: 2008–2012.

Culp, J.M., G.J. Scrimgeour, and J.D. Townsend. 1996. Simulated fine woody debris accumulations in a stream increase rainbow trout fry abundance. Trans. Am. Fish. Soc. 125: 472–479.

Cumbaa, S.L., D.E. McAllister, and R.E. Morlan. 1981. Late Pleistocene fish fossils of *Coregonus, Stenodus, Thymallus, Lota*, and *Cottus* from the Old Crow Basin, northern Yukon, Canada. Can. J. Earth Sci. 18: 1740–1754.

Cunjak, R.A. 1988. Behaviour and microhabitat of young Atlantic salmon (*Salmo salar*) during winter. Can. J. Fish. Aquat. Sci. 45: 2156–2160.

Cunjak, R.A., and J.M. Green. 1984. Species dominance by brook trout and rainbow trout in a simulated stream environment. Trans. Am. Fish. Soc 113: 737–743.

Curry, R.A., and D.L.G. Noakes. 1995a. Groundwater and the selection of spawning sites by brook trout (*Salvelinus fontinalis*). Can. J. Fish. Aquat. Sci. 52: 1733–1740.

Curry, R.A., and D.L.G. Noakes. 1995b. Groundwater and the incubation and emergence of brook trout (*Salvelinus fontinalis*). Can. J. Fish. Aquat. Sci. 52: 1741–1749.

Curry, R.A., C. Brady, D.L.G. Noakes, and R.G. Danzmann. 1997. Use of small streams by young brook trout spawned in a lake. Trans. Am. Fish. Soc 126: 77–83.

Dabrowski, K.R. 1981. The spawning and early life history of the pollan (*Coregonus pollan* Thompson) in Lough Neagh, northern Ireland. Int. Rev. Gesamten Hydrobiol. 66: 299–326.

Dadswell, M.J. 1972. Postglacial dispersal of four deepwater fishes on the basis of new distribution records in eastern Ontario and western Quebec. J. Fish. Res. Board Can. 29: 545–553.

Danehy, R.J., N.H. Ringler, and J.E. Gannon. 1991. Influence of nearshore structure on growth and diets of yellow perch (*Perca flavescens*) and white perch (*Morone americana*) in Mexico Bay, Lake Ontario. J. Great Lakes Res. 17: 183–193.

Danner, G.R. 1994. Behavioral, physiological and genetic differences among anadromous and resident *Oncorhynchus nerka* populations. Master's thesis, University of Idaho, Boise.

Danylchuk, A.J., and M.G. Fox. 1994. Age and size dependent variation in the seasonal timing and probability of reproduction among mature female pumpkinseed, *Lepomis gibbosus*. Environ. Biol. Fishes, 39: 119–127.

Darnell, R.M., and R.R. Meieretto. 1965. Diurnal periodicity in the black bullhead. Trans. Am. Fish. Soc. 94: 1–8.

Das, M., and J.S. Nelson. 1990. Spawning time and fecundity of northern redbelly dace, *Phoxinus eos*, finescale dace, *Phoxinus neogaeus*, and their hybrids in upper Pierre Grey Lake, Alberta. Can. Field-Nat. 104: 409–413.

Dauble, D.D. 1980. Life history and ecology of the bridgelip sucker (*Catostomus columbianus*) in the central Columbia River. Trans. Am. Fish. Soc. 109: 92–98.

Dauble, D.D. 1986. Life history and ecology of the largescale sucker (*Catostomus macrocheilus*) in the Columbia River. Am. Midl. Nat. 116: 356–357.

Dauble, D.D., and R.L. Buschbom. 1981. Estimates of hybridization between two species of catostomids in the Columbia River. Copeia, 1981: 802–810.

Davis, B.M., and T.N. Todd. 1998. Competition between larval lake herring (*Coregonus artedi*) and lake whitefish (*Coregonus clupeaformis*) for zooplankton. Can. J. Fish. Aquat. Sci. 55: 1140–1148.

Davis, J.C. 1975. Minimum dissolved oxygen requirements of aquatic life with emphasis on Canadian species. J. Fish. Res. Board Can. 32: 2295–2332.

de Bruyn, M., and P.J. McCart. 1974. Life history of the arctic grayling (*Thymallus arcticus*) in Beaufort Sea drainages in the Yukon Territory. pp. 1–41. *In* P.J. McCart (ed.). Fisheries research associated with proposed gas pipeline routes in Alaska, Yukon and Northwest Territories. Canadian Arctic Gas Study Ltd., Calgary, AB. Arct. Gas Biol. Rep. No. 15.

Deacon, L.I., and J.A. Keast. 1987. Patterns of reproduction in two populations of pumpkinseed sunfish, *Lepomis gibbosus*, with differing food resources. Environ. Biol. Fishes, 19: 281–296.

DeCicco, A.L. 1992. Long distance movements of anadromous Dolly Varden between Alaska and the U.S.S.R. Arctic, 45: 120–123.

DeGisi, J.S. 1999. Reconnaissance (1:20,000) fish and fish habitat inventory of Maxhamish Lake. Report prepared for BC Parks, Peace–Liard District, Fort St. John.

DeLacy, A.C., and B.S. Batts. 1963. Possible population heterogeneity in the Columbia River smelt. Fisheries Research Institute, College of Fisheries, University of Washington, Seattle. Circ. No. 198.

Delisle, G.E. 1962. Water velocities tolerated by spawning kokanee salmon. Calif. Fish Game, 48: 77–78.

DeMarais, B.D., T.E. Dowling, M.E. Douglas, W.L. Minckley, and P.C. Marsh. 1992. Origin of *Gila seminuda* (Teleostei; Cyprinidae) through introgressive hybridization: implications for evolution and conservation. Proc. Natl. Acad. Sci. U.S.A. 89: 2747–2752.

Deng, X., J.P. Van Eenennaam, and S.I. Doroshov. 2002. Comparison of early life stages and growth of green and white sturgeon. pp. 237–248. *In* W. Van Winkle, P.J. Anders, D.H. Secor, and D.D. Dixon (eds.). Biology, management, and protection of North American sturgeon. Am. Fish. Soc. Symp. No. 28.

Dennison, S.G., and R.V. Bulkely. 1972. Reproductive potential of the black bullhead, *Ictalurus melas*, in Clear Lake, Iowa. Trans. Am. Fish. Soc. 101: 483–487.

Derksen, A.J. 1989. Autumn movements of underyearling northern pike, *Esox lucius*, from a large Manitoba marsh. Can. Field-Nat. 103: 429–431.

DeRoche, S.E. 1969. Observations on the spawning habits and early life history of lake trout. Prog. Fish-Cult. 31: 109–113.

Diana, J.S. 1996. Energetics. pp. 103–124. *In* J.F. Craig (ed.). Pike: biology and exploitation. Chapman & Hall, London. Fish Fish. Ser. No. 19.

Diana, J.S., W.C. Mackay, and M. Ehrman. 1977. Daily movements and habitat preference of northern pike (*Esox lucius*) in Lac Ste. Anne, Alberta. Trans. Am. Fish. Soc. 106: 560–565.

Diewert, R.E. and M.A. Henderson. 1992. The effect of competition and predation on production of juvenile sockeye salmon (*Oncorhynchus nerka*) in Pitt Lake. Can. Tech. Rep. Fish. Aquat. Sci. No. 1853. 51 pp.

Dillinger, R.E., T.P. Birt, and J.M. Green. 1992. Arctic cisco, *Coregonus autumnalis*, distribution, migration and spawning in the Mackenzie River. Can. Field-Nat. 106: 175–180.

DiStefano, R.J., and J.F. Hiebert. 1999. Distribution and movement of walleye in reservoir tailwaters during spawning season. J. Freshw. Ecol. 15: 145–155.

Docker, M.F., Youson, J.H., R.J. Beamish, and R.H. Devlin. 1999. Phylogeny of the lamprey genus *Lampetra* inferred from mitochondrial cytochrome b and ND_3 sequences. Can. J. Fish. Aquat. Sci. 56: 2340–2349.

Dolloff, C.A., and G.H. Reeves. 1990. Microhabitat partioning among stream-dwelling juvenile coho salmon, *Oncorhynchus kisutch*, and Dolly Varden, *Salvelinus malma*. Can. J. Fish. Aquat. Sci. 47: 2297–2306.

Donald, D.B. 1997. Relationship between year-class strength for goldeyes and selected environmental variables during their first year of life. Trans. Am. Fish. Soc. 126: 361–368.

Donald, D.B., and D.J. Alger. 1986. Stunted lake trout (*Salvelinus namaycush*) from the Rocky Mountains. Can. J. Fish. Aquat. Sci. 43: 608–612.

Donald, D.B., and D.J. Alger. 1993. Geographic distribution, species displacement, and niche overlap for lake trout and bull trout in mountain lakes. Can. J. Zool. 71: 238–247.

Donald, D.B., and A.H. Kooyman. 1977a. Food, feeding habits, and growth of the goldeye, *Hiodon alosoides* (Rafinesque), in waters of the Peace-Athabasca Delta. Can. J. Zool. 55: 1038–1047.

Donald, D.B., and A.H. Kooyman. 1977b. Migration and population dynamics of the Peace-Athabasca Delta goldeye population. Canadian Wildlife Service, Ottawa, ON. Occ. Pap. No. 31. 19 pp.

Donald, D.B., R.S. Anderson, and D.W. Mayhood. 1980. Correlations between brook trout growth and environmental variables for mountain lakes in Alberta. Trans. Am. Fish. Soc 109: 603–610.

Donald, D.B., J.A. Babaluk, J.F. Craig, and W.A. Musker. 1992. Evaluation of the scale and operculum methods to determine age of adult goldeyes with special reference to a dominant year-class. Trans. Am. Fish. Soc. 121: 792–796.

Dowling, T.E., and B.D. DeMarais. 1993. Evolutionary significance of introgressive hybridization in cyprinid fishes. Nature (London), 362: 444–446.

Dowling, T.E., and C.L. Secor. 1997. The role of hybridization and introgression in the diversification of animals. Annu. Rev. Ecol. Syst. 28: 593–619.

Downs, C.C., and R.G. White. 1997. Age at sexual maturity, sex ratios, fecundity, and longevity of isolated headwater populations of westslope cutthroat trout. N. Am. J. Fish. Manage. 17: 85–92.

Drucker, B. 1972. Some life history characteristics of coho salmon of the Karluk River system, Kodiak Island, Alaska. Fish. Bull. 70: 79–94.

Dryfoos, R.L. 1965. The life history and ecology of the longfin smelt in Lake Washington. Ph.D. thesis, University of Washington, Seattle.

Duk-Rodkin, A., and O.L. Hughes. 1994. Tertiary–Quaternary drainage of the pre-glacial Mackenzie basin. Quat. Int. 22–23: 221–241.

Duk-Rodkin, A., R.W. Barendregt, D.G. Frose, F. Weber, R. Enkin, I.R. Smith, G.D. Zazula, P. Waters, and R. Klassen. 2004. Timing and extent of Plio-Pleistocene glaciations in northwestern Canada and east-central Alaska. pp. 313–345 . In J. Ehlers and P. L. Gibbard (eds.). Quaternary glaciations, extent and chronology. Part 11: North America. Elsevier, Amsterdam.

Dumont, P., R. Pariseau, and J. Archambault. 1982. Spawning of lake trout (*Salvelinus namaycush*) in shallow depths. Can. Field-Nat. 96: 353–354.

Dyke, A.S. 2004. An outline of North American deglaciation with emphasis on central and northern Canada. pp. 373–424. In J. Ehlers and P.L. Gibbard (eds.). Quaternary glaciations, extent and chronology. Part 11: North America. Elsevier, Amsterdam.

Dymond, J.R. 1926. The fishes of Lake Nipigon. Univ. Toronto Stud. Biol. Ser. Publ. Ont. Fish. Res. Lab. 27: 1–108.

Dymond, J.R. 1931. Description of two new forms of British Columbia trout. Contrib. Can. Biol. Fish. 6: 391–395.

Dymond, J.R. 1936. Some fresh-water fishes of British Columbia. Rep. Comm. Fish. 1935: 60–73.

Dymond, J.R. 1940. Pacific salmon in the Arctic Ocean. p. 435. In Proceedings of the Sixth Pacific Science Congress, 1939. University of California Press, Berkeley.

Dynes, J., P. Magnan, L. Bernatchez, and M. A. Rodriguez. 1999. Genetic and morphological variation between two forms of lacustrine brook charr. J. Fish Biol. 54: 955–972.

Echols, J.C. 1995. Review of Fraser River white sturgeon (*Acipenser transmontanus*). Report prepared for the Fraser River Action Plan. Department of Fisheries and Oceans, Vancouver, BC. 33 pp.

Edge, T.A., and B.W. Coad. 1983. Reduced dorsal spine numbers in two isolated populations of the brook stickleback (*Culaea inconstans*) from eastern Canada. Nat. Can. 110: 99–101.

Edsall, T.A. 1960. Age and growth of whitefish, *Coregonus clupeaformis*, in Munising Bay, Lake Superior. Trans. Am. Fish. Soc. 89: 323–332.

Edsall, T.A. 1999a. The growth-temperature relation of juvenile lake whitefish. Trans. Am. Fish. Soc. 128: 962–964.

Edsall, T.A. 1999b. Preferred temperatures of juvenile lake whitefish. J. Great Lakes Res. 25: 583–588.

Elliot, J.M. 1973. The food of brown and rainbow trout (*Salmo trutta* and *S. gairdneri*) in relation to the abundance of drifting invertebrates in a mountain stream. Oecologia, 12: 329–348.

Elliott, G.V. 1976. Diel activity and feeding of schooled largemouth bass fry. Trans. Am. Fish. Soc. 105: 624–627.

Elliott, J.M. 2000. Pools as refugia for brown trout during two summer droughts: trout responses to thermal and oxygen stress. J. Fish Biol. 56: 938–948.

Ellis, D.V., and M.A. Giles. 1965. The spawning behaviour of the walleye, *Stizostedion vitreum* (Mitchill). Trans. Am. Fish. Soc. 94: 358–362.

Elso, J.I., and L.A. Greenberg. 2001. Habitat use and survival of individual 0^+ brown trout (*Salmo trutta*) during winter. Arch. Hydrobiol. 152: 279–295.

Elz, A.E. 2003. A hierarchical analysis of historic processes and phylogeographic patterns in *Salvelinus* (Pisces: Salmonidae). M.Sc. thesis,

Department of Zoology, University of British Columbia, Vancouver.

Embody, G.C. 1934. Relation of temperature to the incubation periods of the eggs of four species of trout. Trans. Am. Fish. Soc. 64: 281–291.

Emery, A.R. 1973. Preliminary comparisons of night and day habits of freshwater fish in Ontario lakes. J. Fish. Res. Board Can. 30: 761–774.

Epifano, J.M., M. Hooe, D.H. Buck, and D.P. Philipp. 1999. Reproductive success and assortative mating among *Pomoxis* species and their hybrids. Trans. Am. Fish Soc. 128: 104–120.

Eschmeyer, P.H. 1955. The reproduction of lake trout in southern Lake Superior. Trans. Am. Fish. Soc 4: 47–74.

Eschmeyer, P.H., and R.M. Bailey. 1955. The pygmy whitefish, *Coregonus coulteri*, in Lake Superior. Trans. Am. Fish. Soc. 84: 161–199.

Evenson, M.J. 2000. Reproductive traits of burbot in the Tanana River, Alaska. pp. 61–70. *In* V.L. Paragamian and D.W. Willis (eds.). Burbot biology, ecology, and management. American Fisheries Society, Spokane, WA. Fish. Manage. Sect. Publ. No. 1.

Everest, F.H., and D.W. Chapman. 1972. Habitat selection and spatial interactions by juvenile chinook salmon and steelhead trout in two Idaho streams. J. Fish. Res. Board Can. 29: 91–100.

Evermann, B.W. 1893. Description of a new sucker, *Pantosteus jordani*, from the upper Missouri basin. Bull. U. S. Fish. Comm. 1891: 3–60.

Evermann, B.W., and E.L. Goldsborough. 1907. Fishes of Alaska. Bull. U. S. Bur. Fish. 26: 219–360.

Faber, D.J. 1967. Limnetic larval fish in northern Wisconsin lakes. J. Fish. Res. Board Can. 24: 927–937.

Faber, D.J. 1970. Ecological observations on newly hatched lake whitefish in South Bay, Lake Huron. pp. 481–500. *In* C.C. Lindsey and C.S. Woods (eds.). Biology of coregonid fishes. University of Manitoba Press, Winnipeg.

Fabricius, E., and K.-J. Gustafason. 1958. Some new observations on the spawning behaviour of the pike, *Esox lucius* L. Rep. Inst. Freshw. Res. Drottningholm, 39: 23–54.

Farlinger, S.P., and R.J. Beamish. 1984. Recent colonization of a major salmon-producing lake in British Columbia by the Pacific lamprey (*Lampetra tridentata*). Can. J. Fish. Aquat. Sci. 41: 278–285.

Faurot, M.W., and R.G. White. 1994. Feeding ecology of larval fishes in Lake Roosevelt, Washington. Northwest Sci. 68: 189–196.

Fava, J.A., and C.F. Tsai. 1974. The life history of the pearl dace, *Semotilus margarita*, in Maryland. Chesapeake Sci. 15: 159–162.

Fenderson, O.C. 1964. Evidence of subpopulations of lake whitefish, *Coregonus clupeaformis*, involving a dwarf form. Trans. Am. Fish. Soc. 93: 77–94.

Ferguson, A., and G. Osborn. 1981. Minimum age of deglaciation of upper Elk Valley, British Columbia. Can. J. Earth Sci. 18: 1635–1636.

Ferguson, A., K.-J.M. Himberg, and G. Svärdson. 1978. Systematics of the Irish pollan (*Coregonus pollan* Thompson): an electrophoretic comparison with other Holarctic Coregoninae. J. Fish. Biol. 12: 221–233.

Ferguson, R.G. 1958. The preferred temperature of fish and their midsummer distribution in temperate lakes and streams. J. Fish. Res. Board Can. 15: 607–624.

Fernet, D.A., and R.J.F. Smith. 1976. Agonistic behavior of captive goldeye (*Hiodon alosoides*). J. Fish. Res. Board Can. 33: 695–702.

Field, P., and T. Dickie. 1987. Investigations of black bass in the Okanagan sub-unit. Report prepared for the BC Ministry of the Environment. BC Conservation Foundation, Surrey. 90 pp.

Finger, T.E. 1982. Interactive segregation among three species of sculpins (*Cottus*). Copeia, 1982: 680–694.

Finnell, L.M., and E.B. Reed. 1969. The diel vertical migration of kokanee salmon (*Oncorhynchus nerka*) in Granby Reservoir, Colorado. Trans. Am. Fish. Soc. 98: 245–252.

Firth, H.R., and R.W. Blake. 1991. Mechanics of the startle response in the northern pike, *Esox lucius*. Can. J. Zool. 69: 2831–2839.

Firth, H.R., and R.W. Blake. 1995. The mechanical power output and hydrodynamic efficiency of northern pike (*Esox lucius*) fast-starts. J. Exp. Biol. 198: 1863–1873.

Fish, M.P. 1932. Contributions to the early life histories of sixty-two species of fishes from Lake Erie and its tributary waters. US Bur. Comm. Bur. Fish. Bull. No. 10: 293–398.

Fisher, S.J., D.W. Willis, M.M. Olson, and S.C. Krentz. 2002. Flathead chubs, *Platygobio gracilis*, in the upper Missouri River: the biology of a species at risk in an endangered habitat. Can. Field-Nat. 116: 26–41.

Flebbe, P.A., and C.A. Dolloff. 1995. Trout use of woody debris and habitat in Appalachian wilderness streams of North Carolina. N. Am. J. Fish. Manage. 15: 579–590.

Fleming, I.A., and M.R. Gross. 1990. Latitudinal clines: a trade-off between egg number and size in Pacific salmon. Ecology, 71: 1–11.

Fleming, I.A., and M.R. Gross. 1994. Breeding competition in Pacific salmon (coho: *Oncorhynchus kisutch*): measures of natural and sexual selection. Evolution, 48: 637–657.

Flick, W.A. 1991. Brook trout (*Salvelinus fontinalis*). pp. 196–207. *In* J. Stolz and J. Schnell (eds.). Trout. Stackpole Books, Harrisburg, PA.

Flittner, G.A. 1964. Morphology and life history of the emerald shiner *N. atherinoides*. Ph.D. thesis, University of Michigan, Ann Arbor. (The original was not seen; the citation is based on excerpts published in Becker 1983.)

Foltz, J.W. 1976. Fecundity of the slimy sculpin, *Cottus cognatus*, in Lake Michigan. Copeia, 1976: 802–804.

Foote, C.J. 1988. Male mate choice dependent on male size in salmon. Behaviour, 106: 63–80.

Foote, C.J., and G.S. Brown. 1998. Ecological relationship between freshwater sculpins (genus *Cottus*) and beach-spawning sockeye salmon (*Oncorhynchus nerka*) in Iliamna Lake, Alaska. Can. J. Fish. Aquat. Sci. 55: 1524–1533.

Foote, C.J., and P.A. Larkin. 1988. The role of mate choice in the assortative mating of anadromous and non-anadromous sockeye salmon (*Oncorhynchus nerka*). Behaviour, 106: 43–62.

Foote, C.J., C.C. Wood, and R.E. Withler. 1989. Biochemical genetic comparison of sockeye salmon and kokanee, the anadromous and nonanadromous forms of *Oncorhynchus nerka*. Can. J. Fish. Aquat. Sci. 46: 149–158.

Foote, C.J., I. Mayer, C.C. Wood, W.C. Clarke, and J. Blackburn. 1992a. Circannual cycle of seawater adaptability in *Oncorhynchus nerka*: genetic differences between sympatric sockeye salmon and kokanee. Can. J. Fish. Aquat. Sci. 49: 99–109.

Foote, C.J., J.W. Clayton, C.C. Lindsey, and R.A. Bodaly. 1992b. Evolution of lake whitefish (*Coregonus clupeaformis*) in North America during the Pleistocene: evidence for a Nahanni glacial refuge race in the northern Cordillera region. Can. J. Fish. Aquat. Sci. 49: 760–768.

Foote, C.J., G.S. Brown, and C.C. Wood. 1997. Spawning success of males using alternative mating tactics in sockeye salmon, *Oncorhynchus nerka*. Can. J. Fish. Aquat. Sci. 54: 1785–1795.

Foote, C.J., G.S. Brown, and C.W. Hawryshyn. 2004. Female colour and mate choice in sockeye salmon: implications for the phenotypic convergence of anadromous and non-anadromous morphs. Anim. Behav. 67: 69–83.

Forbes, L.S. 1989. Spawning, growth, and mortality of three introduced fishes at Creston, British Columbia. Can. Field-Nat. 103: 520–523.

Forbes, L.S., and D.R. Flook. 1985. Notes on the occurrence and ecology of the black bullhead, *Ictalurus melas*, near Creston. Can. Field-Nat. 99: 110–111.

Fox, M.G. 1994. Growth, density, and interspecific influences on pumpkinseed life histories. Ecology 75: 1157–1171.

Fox, M.G., and A. Keast. 1990. Effects of winterkill on population structure, body size, and prey consumption patterns of pumpkinseed in isolated ponds. Can. J. Zool. 68: 2489–2498.

Fox, M.G., and A. Keast. 1991. Effect of overwinter mortality on reproductive life history characteristics of pumpkinseed (*Lepomis gibbosus*). Can. J. Fish. Aquat. Sci. 48: 1792–1799.

Franklin, D.R., and L.L. Smith. 1963. Early life history of the northern pike, *Esox lucius* L., with special reference to the factors influencing the numerical strength of year classes. Trans. Am. Fish. Soc. 92: 91–110.

Freeman, M.C., and D.J. Stouder. 1989. Intraspecific interactions influence size specific depth distribution in *Cottus bairdi*. Environ. Biol. Fishes, 24: 231–236.

French, R., H. Bilton, M. Osako, and A. Hartt. 1976. Distribution and origin of sockeye salmon (*Oncorhynchus nerka*) in offshore waters of the North Pacific Ocean. Int. North Pacific Fish. Comm. Bull. 34. 113 pp.

Frost, W.E., and M.E. Brown. 1967. The trout. Collins, London.

Frost, W.E., and C. Kipling. 1967. A study of reproduction, early life, weight–length relationship, and growth of pike (*Esox lucius*) in Windemere. J. Anim. Ecol. 36: 651–693.

Froufe, E., I. Knizhin, M.T. Koskinen, C.R. Primmer, and S. Weiss. 2003. Identification of reproductively isolated lineages of Amur grayling (*Thymallus grubei* Dybowski 1869): concordance between phenotypic and genetic variation. Mol. Ecol. 12: 2345–2355.

Froufe, E., I. Knizhin, and S. Weiss. 2005. Phylogenetic analysis of the genus *Thymallus* (grayling) based on mtDNA control region and ATPase 6 genes, with inferences on the control region constraints and broad-scale Eurasian phylogeography. Mol. Phylogenet. Evol. 34: 106–117.

Fu, C., J. Luo, J. Wu, J. A. López, Y. Zhong, G. Lei, and J. Chen. 2005. Phylogenetic relationships of salangids fishes (Osmeridae, Salanginae) with comments on phylogenetic placement of the salangids based on mitochondrial DNA sequences. Mol. Phylogenet. Evol. 35: 76–84.

Fuchs, E.H. 1967. Life history of the emerald shiner, *Notropis atherinoides*, in Lewis and Clark Lake, South Dakota. Trans. Am. Fish. Soc. 96: 247–256.

Fuiman, LA., and J.P. Baker. 1981. Larval stages of the lake chub, *Couesius plumbeus*. Can. J. Zool. 59: 218–224.

Fukutomi, N., T. Nakamura, T. Doi, K. Takeda, and N. Oda. 2001. Records of *Entosphenus tridentatus* from the Naka River system, central Japan; physical characteristics of possible spawning redds and spawning behaviour in the aquarium. Jpn. J. Ichthyol. 49: 53–58.

Fuller, W.A. 1955. The inconnu, *Stenodus leucichthys mackenzii*, in Great Slave Lake and adjoining waters. J. Fish. Res. Board Can. 12: 768–780.

Fulton, L.A. 1970. Spawning areas and abundance of steelhead trout and coho, sockeye, and chum salmon in the Columbia River Basin—past and present. US Natl. Mar. Fish. Serv. Spec. Sci. Rep. Fish. No. 618. 37 pp.

Fulton, R.J. 1969. Glacial lake history, southern interior Plateau, British Columbia. Geol. Surv. Can. Pap. No. 69-37. 14 pp.

Fulton, R.J., and R.A. Archard. 1985. Quaternary sediments, Columbia River Valley, Revelstoke to the Rocky Mountain Trench, British Columbia. Geol. Surv. Can. Pap. 84-13. 14 pp.

Fulton, R.J., J.M. Ryder, and S. Tang. 2004. The Quaternary glacial record of British Columbia, Canada. pp. 39–50. In J. Ehlers and P.L. Gibbard (eds.). Quaternary glaciations, extent and chronology. Part 11: North America. Elsevier, Amsterdam.

Gach, M.H. 1996. Geographic variation in mitochondrial DNA and biogeography of *Culaea inconstans* (Gasterosteidae). Copeia, 1996: 563–575.

Gadomski, D.M., and C.A. Barfoot. 1998. Diel and distributional abundance patterns of fish embryos and larvae in the lower Columbia and Deschutes rivers. Environ. Biol. Fishes, 51: 353–368.

Galbreath, J.L. 1985. Status, life history, and management of Columbia River white sturgeon, *Acipenser transmontanus*. pp. 119–125. In F.P. Binkowski and S.I. Doroshov (eds.).

North American sturgeons: biology and aquaculture potential. Dr. W. Junk Publishers, Dordrecht, the Netherlands.

Garrett, J.W., D.H. Bennett, F.O. Frost, and R.F. Thurow. 1998. Enhanced incubation success for kokanee spawning in groundwater upwelling sites in a small Idaho stream. N. Am. J. Fish. Manage. 18: 925–930.

Garside, E.T. 1959. Some effects of oxygen in relation to temperature on the development of lake trout embryos. Can. J. Zool. 37: 689–698.

Garvey, J.E., T.P. Herra, and W.C. Leggett. 2002. Protracted reproduction in sunfish: the temporal dimension in fish recruitment revisited. Ecol. Appl. 12: 194–205.

Gasser, K.W., D.A. Cannamela, and D.W. Johnson. 1981. Contributions to the life history of the shorthead sculpin, *Cottus confusus*, in the Big Lost River, Idaho: age, growth, and fecundity. Northwest Sci. 55: 174–181.

Gee, J.H., and K. Machniak. 1972. Ecological notes on a lake-dwelling population of longnose dace (*Rhinichthys cataractae*). J. Fish. Res. Board Can. 29: 330–332.

Gee, J.H., and T.G. Northcote. 1963. Comparative ecology of two sympatric species of dace (*Rhinichthys*) in the Fraser River system, British Columbia. J. Fish. Res. Board Can. 20: 105–118.

Geen, G.H. 1955. Some features of the life history of the lake chub (*Couesius plumbeus greeni* Jordan) in British Columbia. Honours B.A. thesis, Department of Zoology, University of British Columbia, Vancouver.

Geen, G.H. 1958. Reproduction of three species of suckers (Catostomidae) in British Columbia. M.A. thesis, Department of Zoology, University of British Columbia, Vancouver.

Geen, G.H., T.G. Northcote, G.F. Hartman, and C.C. Lindsey. 1966. Life histories of two species of catostomid fishes in Sixteenmile Lake, British Columbia, with particular reference to inlet stream spawning. J. Fish. Res. Board Can. 23: 1761–1788.

Geist, D.R. 2000. Hyporheic discharge of river water into fall chinook (*Oncorhynchus tshawytscha*) spawning areas in the Hanford Reach, Columbia River. Can. J. Fish. Aquat. Sci. 57: 1647–1656.

Geist, D.R., T.P. Hanrahan, E.V. Arntez, G.A. McMichael, C.J. Murray, and Y.-J. Chien. 2002. Physiochemical characteristics of the hyporheic zone affect redd site selection by chum salmon and fall chinook salmon in the Columbia River. N. Am. J. Fish. Manage. 22: 1077–1085.

Gharrett, A.J., A.K. Gray, and V. Brykov. 2001. Phylogeographic analysis of mitochondrial DNA variation in Alaskan coho salmon, *Oncorhynchus kisutch*. Fish. Bull. 99: 528–544.

Gibbons, J.R.H., and J.H. Gee. 1972. Ecological segregation between longnose and blacknose dace (genus *Rhinichthys*) in the Mink River, Manitoba. J. Fish. Res. Board Can. 20: 1245–1252.

Gibson, E.S., and F.E.J. Fry. 1954. The performance of lake trout, *Salvelinus namaycush*, at various levels of temperature and oxygen pressure. Can. J. Zool. 32: 252–260.

Gibson, R.J. 1993. The Atlantic salmon in freshwater: spawning, rearing, and production. Rev. Fish Biol. Fish. 3: 39–73.

Giger, R.D. 1972. Ecology and management of coastal cutthroat trout in Oregon. Oregon State Game Commission, Corvallis. Fish. Res. Rep. No. 6. 61 pp.

Gilbert, C.H., and B.A. Evermann. 1895. A report upon investigations in the Columbia River basin, with descriptions of four new species of fish. Bull. U.S. Fish Comm. (1894) XIV: 169–207.

Gill, H.S., C.B. Renaud, F. Chapleau, R.L. Mayden, and I.C. Potter. 2003. Phylogeny of living parasitic lampreys (Petromyzontiformes) based on morphological data. Copeia, 2003: 687–703.

Girard, C. 1850. A monograph of the freshwater *Cottus* of North America. Proc. Am. Assoc. Adv. Sci., 1849: 409–411.

Glova, G.J. 1984. Management implications of the distribution and diet of sympatric populations of juvenile coho salmon and coastal cutthroat trout in small streams in British Columbia. Progr. Fish-Cult. 46: 269–277.

Glova, G.J. 1987. Comparison of allopatric cutthroat trout stocks with those sympatric with coho salmon and sculpins in small streams. Environ. Biol. Fishes, 20: 274–284.

Glozier, N.E., J.M. Culp, and G J. Scrimgeour. 1997. Transferability of habitat suitability curves for a benthic minnow, *Rhinichthys cataractae*. J. Freshw. Ecol. 12: 379–393.

Goddard, K.A., and R.M. Dawley. 1990. Clonal inheritance of a diploid nuclear genome by a hybrid freshwater minnow (*Phoxinus eos-neogaeus*, Pisces, Cyprinidae). Evolution, 44: 1052–1065.

Goddard, K.A., O. Megwinoff, L.L. Wessner, and F. Giaimo. 1998. Confirmation of gynogenesis in *Phoxinus eos-neogaeus* (Pisces: Cyprinidae). J. Hered. 89: 151–157.

Godin, J.G.J. 1980a. Temporal aspects of juvenile pink salmon (*Oncorhynchus gorbuscha* Walbaum) emergence from a simulated gravel redd. Can. J. Zool. 58: 735–734.

Godin, J.G.J. 1980b. Ontogenetic changes in the daily rhythms of swimming activity and vertical distribution in juvenile pink salmon (*Oncorhynchus gorbuscha* Walbaum). Can. J. Zool. 58: 745–753.

Godkin, C.M., Christie, W.J., and D.E. McAllister. 1982. Problems of species identity in the Lake Ontario sculpins *Cottus bairdii* and *Cottus cognatus*. Can. J. Fish. Aquat. Sci. 39: 1373–1382.

Golder Associates Ltd. 2004. A synthesis of white sturgeon investigations in Arrow Lakes Reservoir, B.C. 1995–2003. Draft Report prepared for BC Hydro, Castlegar.

Goodgame, L.S., and L.E. Miranda. 1993. Early growth and survival of age-0 largemouth bass in relation to parental size and swim-up time. Trans. Am. Fish. Soc. 122: 131–138.

Gould, S.J. 1977. Ontogeny and phylogeny. Harvard University Press, Cambridge, MA.

Graham, D.M., and W.G. Sprules. 1992. Size and species selection of zooplankton by larval and juvenile walleye (*Stizostedion vitreum vitreum*) in Oneida Lake, New York. Can. J. Zool. 70: 2059–2067.

Grant, J.W.A., S.O. Steingrimsson, E.R. Keeley, and R.A. Cunjak. 1998. Implications of territory size for the measurement and prediction of salmonoid abundance in streams. Can. J. Fish. Aquat. Sci. 55(Suppl. 1): 181–190.

Gray, S.M., and B.W. Robinson. 2002. Experimental evidence that competition between stickleback species favors adaptive character divergence. Ecol. Lett. 5: 264–272.

Green, D.M. 2005. Designatable units for status assessment of endangered species. Conserv. Biol. 19: 1813–1820.

Green, D.M., and R.C. Heidinger. 1994. Longevity record for largemouth bass. N. Am. J. Fish. Manage. 14: 464–465.

Greenburg, L.A., and P.S. Giller. 2001. Individual variation in habitat use and growth of male and female brown trout. Ecography, 24: 212–224.

Grey, J. 2001. Ontogeny and dietary specialization in brown trout (*Salmo trutta* L.) from Loch Ness, Scotland, examined using stable isotopes of carbon and nitrogen. Ecol. Freshw. Fish, 10: 168–176.

Griffith, J.S. 1972. Comparative behaviour and habitat utilization of brook trout (*Salvelinus fontinalis*) and cutthroat (*Salmo clarki*) in small streams in northern Idaho. J. Fish. Res. Board Can. 29: 265–273.

Griffith, R.P. 1994. Inventory and assessment of bass in the upper Columbia River drainage. Report prepared for the Mica Fisheries Compensation Program. BC Hydro and BC Environment, Nelson. 86 pp.

Griffiths, W., P. Craig, G. Walder, and G. Mann. 1975. Fisheries investigations in a coastal region of the Beaufort Sea (Nunaluk Lagoon, Yukon Territory). pp. 1–219. *In* P. Craig (ed.). Fisheries investigations in a coastal region of the Beaufort Sea. Canadian Arctic Gas Study Ltd., Calgary, AB. Arct. Gas Biol. Rep. Ser. No. 34 (2).

Griffiths, W.B., J.K. Den Beste, and P.C. Craig. 1977. Fisheries investigations in a coastal region of the Beaufort Sea (Kaktovik Lagoon, Alaska). pp. 70–86. *In* P.J. McCart (ed.). Fisheries investigations along the north slope from Prudhoe Bay, Alaska, to the Mackenzie

Delta, NWT. Canadian Arctic Gas Study Ltd., Calgary, AB. Arct. Gas Biol. Rep. Ser. No. 40 (2).

Griswald, K.E., K.P. Currans, and G.H. Reeves. 1997. Genetic and meristic divergence of coastal cutthroat trout residing above and below barriers in two coastal basins. pp. 167–169. *In* J.D. Hall, P.A. Bisson, and R.E. Gresswell (eds.). Sea-run cutthroat trout: biology, management, and future conservation. Oregon Chapter, American Fisheries Society, Corvallis.

Griswold, B.L., and L.L. Smith. 1972. Early survival and growth of the ninespine stickleback, *Pungitius pungitius*. Trans. Am. Fish. Soc. 101: 350–352.

Griswold, B.L., and L.L. Smith. 1973. The life history and trophic relationship of the ninespine stickleback, *Pungitius pungitius*, in the Apostle Islands area of Lake Superior. Fish. Bull. 71: 1039–1060.

Groot, C. 1982. Modifications on a theme—a perspective on migratory behavior of Pacific salmon. pp. 1–21. *In* E.L. Brannon and E.O. Salo (eds.). Proceedings of the Salmon and Trout Migratory Behavior Symposium. School of Fisheries, University of Washington, Seattle.

Groot, C., and L. Margolis (eds.). 1991. Pacific salmon life histories. University of British Columbia Press, Vancouver.

Gross, M.R. 1982. Sneakers, satellites, and parentals: polymorphic mating strategies in North American sunfishes. Z. Tierpsychol. 60: 1–26.

Groves, P.A., and J.A. Chandler. 1999. Spawning habitat used by fall Chinook salmon in the Snake River. N. Am. J. Fish. Manage. 19: 912–922.

Gruchy, C.G., and V.D. Vladykov. 1968. Sexual dimorphism in the anal fin of brown trout, *Salmo trutta*, and close relatives. J. Fish. Res. Board Can. 25: 813–815.

Guinn, B.R. 1982. Ecology and morphology of sympatric and allopatric populations of mountain whitefish, *Prosopium williamsoni*, and round whitefish, *Prosopium cylindraceum*, in lakes in western Canada. M.Sc. thesis, Department of Zoology, University of Manitoba, Winnipeg.

Gunckel, S.L., A.R. Hemmingsen, and J.L. Li. 2002. Effect of bull trout and brook trout interactions on foraging habitat, feeding behavior, and growth. Trans. Am. Fish. Soc. 131: 1119–1130.

Gustafason, R.G., and G.A. Winans. 1999. Distribution and population genetic structure of river-type and sea-type sockeye salmon in western North America. Ecol. Freshw. Fish, 8: 181–193.

Gustafason, R.G., T.C. Wainwright, G.A. Winans, F.W. Waknitz, L.T. Parker, and R.S. Waples. 1997. Status review of sockeye salmon from Washington and Oregon. US Department of Commerce, National Marine Fisheries Service, Wasington, DC. NOAA Tech. Memo. NMFS-NWFSC-33. 282 pp.

Guy, C.S., and D.W. Willis. 1995. Growth of crappies in South Dakota waters. J. Freshw. Ecol. 10: 151–161.

Guy, C.S., R.M. Neumann, and D.W. Willis. 1992. Movement patterns of adult black crappie, *Pomoxis nigromaculatus*, in Brant Lake, South Dakota. J. Freshw. Ecol. 7: 137–147.

Haas, G.R. 2001. The evolution through natural hybridization of the Umatilla dace (Pisces: *Rhinichthys umatilla*), and their associated ecology and systematics. Ph.D. thesis, Department of Zoology, University of British Columbia, Vancouver.

Haas, G.R., and J.D. McPhail. 1991. Systematics and distributions of Dolly Varden (*Salvelinus malma*) and bull trout (*Salvelinus confluentus*) in North America. Can. J. Fish. Aquat. Sci. 48: 2191–2211.

Haas, G.R., J.D. McPhail, and L.M. Ritchie. 2001. British Columbia fish inventory fish species voucher program: identification errors, problems, and their consequences with recommendations for management and conservation. pp. 71–74. *In* M.K. Brewin, A.J. Paul, and M. Monita (eds.). Ecology and management of northwest salmonids. Trout Unlimited, Calgary, AB.

Hagen, D.W. 1967. Isolating mechanisms in threespine sticklebacks (*Gasterosteus aculeatus*). J. Fish. Res. Board Can. 24: 1637–1692.

Hagen, J. 2000. Reproductive isolation between Dolly Varden (*Salvelinus malma*) and bull trout (*Salvelinus confluentus*) in sympatry: the role of ecological factors. M.Sc. thesis, Department of Zoology, University of British Columbia, Vancouver.

Hagen, J., and J. Baxter. 1992. Bull trout populations of the North Thompson River Basin, British Columbia: initial assessment of a biological wilderness. BC Ministry of Environment, Lands, and Parks, Fisheries Branch, Kamloops. 37 pp.

Hagen, J., and E.B. Taylor. 2001. Resource partitioning as a factor limiting gene flow in hybridizing populations of Dolly Varden char (*Salvelinus malma*) and bull trout (*Salvelinus confluentus*). Can. J. Fish. Aquat. Sci. 58: 2037–2047.

Hallock, M., and P.E. Mongillo. 1998. Washington State status report for the pygmy whitefish. Washington Department Fish and Wildlife, Olympia. 20 pp.

Hamada, M., M. Himberg, R.A. Bodaly, J. D. Reist, and N. Okada. 1998. Monophyletic origin of the genera *Stenodus* and *Coregonus* inferred from an analysis of the insertion of SINEs (short interspersed repetitive elements). Arch. Hydrobiol. Spec. Issues Adv. Limnol. 50: 383–389.

Hanson, A.J, and H.D. Smith. 1967. Mate selection in a population of sockeye salmon (*Oncorhynchus nerka*) of mixed age groups. J. Fish. Res. Board Can. 24: 1955–1977.

Hard, J.J., R G. Kope, W.S. Grant, F.W. Waknitz, L.T. Parker, and R.S. Waples. 1996. Status review of pink salmon from Washington, Oregon, and California. US Department of Commerce, National Marine Fisheries Service, Wasington, DC. NOAA Tech. Memo. NMFS-NWFSC-25. 131 pp.

Hardman, M., and L.M. Page. 2003. Phylogenetic relationships among bullhead catfishes of the genus *Ameiurus* (Siluriformes: Ictaluridae). Copeia, 2003: 20–33.

Hardy, J.D. 1978a. *Lepomis gibbosus*. pp. 199–208. *In* Development of fishes of the mid-Atlantic Bight. Vol. 3. US Department of the Interior, Fish and Wildlife Service, Washington, DC.

Hardy, J.D. 1978b. *Micropterus dolomieui*. pp. 236–246. *In* Development of fishes of the mid-Atlantic Bight. Vol. 3. US Department of the Interior, Fish and Wildlife Service, Washington, DC.

Hardy, J.D. 1978c. *Micropterus salmoides*. pp. 247–262. *In* Development of fishes of the mid-Atlantic Bight. Vol. 3. US Department of the Interior, Fish and Wildlife Service, Washington, DC.

Hardy, J.D. 1978d. *Pomoxis nigromaculatus*. pp. 273–277. *In* Development of fishes of the mid-Atlantic Bight. Vol. 3. US Department of the Interior, Fish and Wildlife Service, Washington, DC.

Harington, C.R. 1978. Quaternary vertebrate faunas of Canada and Alaska and their suggested chronological sequence. Can. Nat. Hist. Mus. Syllogeus No. 15. 105 pp.

Harington, C.R. 1996. Quaternary animals: vertebrates of the Ice Age. pp. 259–273. *In* R. Ludvigsen (ed.). Life in stone: a natural history of British Columbia's fossils. University of British Columbia Press, Vancouver.

Harlan, J.R., and E.B. Speaker. 1956. Iowa fish and fishing. Iowa State Conservation Commission, Des Moines.

Harrington, R.W. 1947. Observations on the breeding habits of yellow perch, *Perca flavescens* (Mitchill). Copeia, 1947: 199–200.

Harris, B.S. 1986. Enhancement of kokanee shore spawning sites in Okanagan Lake, British Columbia. pp. 46–48. *In* J.H. Patterson (ed.). Proceedings of the Workshop on Habitat Improvement. Can. Tech. Rep. Fish. Aquat. Sci. No. 1483.

Harris, L., and K. Howland. 2005. Tracking the movements of a lacustrine population of broad whitefish (*Coregonus nasus*) in the Travaillant Lake system, Northwest Territories. Gwich'in Renewable Resources Board, Inuvik, NWT. Rep. No. 05-03. 40 pp.

Harris, N.J., G.F. Galinat, and D.W. Willis. 1999. Seasonal food habits of bluegills in Richmond Lake. Proc. S.D. Acad. Sci. 78: 79–86.

Harris, P.M., and R.L. Mayden. 2001. Phyletic relationships of major clades of Catostomidae (Teleostei: Cypriniformes) as inferred from mitochondrial SSU and LSU rDNA sequences. Mol. Phylogenet. Evol. 20: 225–237.

Hart, J.L. 1930. The spawning and early life history of the whitefish, *Coregonus clupeaformis* (Mitchill), in the Bay of Quinte, Ontario. Contrib. Can. Biol. Fish. 6: 165–214.

Hart, J.L. 1973. Pacific fishes of Canada. Bull. Fish. Res. Board Can. No. 180.

Hart, J.L., and J.L. McHugh. 1944. The smelts (Osmeridae) of British Columbia. Bull. Fish. Res. Board Can. No. 64.

Hart, P.J.B., and A.B. Gill. 1994. Evolution of foraging behaviour in the threespine stickleback. pp. 205–239. *In* M.A. Bell and S.A. Foster (eds.). The evolutionary biology of the threespine stickleback. Oxford University Press, Oxford, UK.

Hartman, G.F. 1969. Reproductive biology of Gerrard stock rainbow trout. pp. 53–67. *In* T.G. Northcote (ed.). Symposium on salmon and trout in streams. H.R. MacMillan Lectures in Fisheries, University of British Columbia, Vancouver.

Hartman, G.F., and T.G. Brown. 1987. Use of small, temporary floodplain tributaries by juvenile salmonids in a west coast rain-forest drainage basin, Carnation Creek, British Columbia. Can. J. Fish. Aquat. Sci. 44: 262–270.

Hartman, G.F., and D.M. Galbraith. 1970. The reproductive environment of the Gerrard stock rainbow trout. B.C. Fish and Wildlife Branch, Victoria. Fish Manage. Publ. No. 15. 51 pp.

Hartman, G.F., and C.A. Gill. 1968. Distribution of juvenile steelhead and cutthroat trout (*Salmo gairdneri* and *S. clarki clarki*) within streams in southwestern British Columbia. J. Fish. Res. Board Can. 25: 33–48.

Hartman, G.F., T.G. Northcote, and C.C. Lindsey. 1962. Comparison of inlet and outlet spawning runs of rainbow trout in Loon Lake, British Columbia. J. Fish. Res. Board Can. 19: 173–200.

Hartman, G.F., B.C. Andersen, and J.C. Scrivener. 1982. Seaward movement of coho salmon (*Oncorhynchus kisutch*) fry in Carnation Creek, an unstable coastal stream in British Columbia. Can. J. Fish. Aquat. Sci. 39: 588–597.

Hartt, A.C. 1980. Juvenile salmonids in the oceanic ecosystem—the critical first summer. pp. 25–57. *In* W.J. McNeil and D.C. Himsworth (eds.). Salmonid ecosystems of the North Pacific. Oregon State University Press, Corvallis.

Hartt, A.C., and M.B. Dell. 1986. Early oceanic migrations and growth of juvenile Pacific salmon and steelhead trout. North Pac. Fish. Comm. Bull. No. 46. 105 pp.

Harvey, C.J., G.T. Ruggerone, and D.E. Rogers. 1997. Migrations of three-spined stickleback, nine-spined stickleback, and pond smelt in the Chignik catchment, Alaska. J. Fish Biol. 50: 1133–1137.

Hauser, W.J. 1969. Life history of the mountain sucker, *Catostomus platyrhynchus*, in Montana. Trans. Am. Fish. Soc. 98: 209–215.

Hay, D.E., and P.B. McCarter. 2000. Status of the eulachon, *Thaleichthys pacificus*, in Canada. Fisheries and Oceans Canada, Canadian Stock Assessment Secretariat, Ottawa, ON. Res. Doc. No. 2000/145. 92 pp.

Hay, D.E., J. Boutillier, M. Joyce, and G. Langford. 1997. The eulachon (*Thaleichthys pacificus*) as an indicator species in the North Pacific. pp. 509–530. *In* Forage fishes in marine ecosystems. American Fisheries Society, Bethesda, MD. Lowell Wakefield Fish. Symp. Ser. No. 18.

Hayes, D.B., and W.W. Taylor. 1990. Reproductive strategy in yellow perch: effects of diet ontogeny, mortality, and survival costs. Can. J. Fish. Aquat. Sci. 47: 921–927.

Haynes, J.M., R.H. Gray, and J.C. Montgomery. 1978. Seasonal movements of white sturgeon (*Acipenser transmontanus*) in the mid-Columbia River. Trans. Am. Fish. Soc. 107: 275–280.

Healey, M.C. 1980. The ecology of juvenile salmon in Georgia Straight, British Columbia.

pp. 203–229. *In* W.J. McNeil and D.C. Himsworth (eds.). Salmonid ecosystems of the North Pacific. Oregon State University Press, Corvallis.

Healey, M.C. 1991. Life history of chinook salmon (*Oncorhynchus tshawytscha*). pp. 314–393. *In* C. Groot and L. Margolis (eds.). Pacific salmon life histories. University of British Columbia Press, Vancouver.

Healey, M.C., and K. Dietz. 1984. Variation in fecundity of lake whitefish (*Coregonus clupeaformis*) from Lesser Slave and Utikuma lakes in northern Alberta. Copeia 1984: 238–242.

Healey, M.C., and C. Groot. 1987. Marine migration and orientation of ocean-type chinook and sockeye salmon. pp. 298–312. *In* M.J. Daswell, R.J. Klanda, C.M. Moffitt, R.L. Saunders, R.A. Rulifson, and J.E. Cooper (eds.). Common strategies of anadromous and catadromous fishes. Am. Fish. Soc. Symp. No. 1.

Healey, M.C., and W.R. Heard. 1984. Interpopulation and intra-population variation in the fecundity of chinook salmon (*Oncorhynchus tshawytscha*) and its relevance to life history theory. Can. J. Fish. Aquat. Sci. 41: 476–483.

Healey, M.C., and F.P. Jordan. 1982. Observations on juvenile chum and chinook and spawning chinook in the Nanaimo River, British Columbia. Can. MS Rep. Fish. Aquat. Sci. 1659. 31 pp.

Healey, M.C., and C.W. Nicol. 1975. Fecundity comparisons for various stocks of lake whitefish, *Coregonus clupeaformis*. J. Fish. Res. Board Can. 32: 404–407.

Healey, M.C., and A. Prince. 1998. Alternative tactics in the breeding of male coho salmon. Behaviour, 135: 1099–1124.

Heard, W.R. 1991. Life history of pink salmon. pp. 119–230. *In* C. Groot and L. Margolis (eds.). Pacific salmon life histories. University of British Columbia Press, Vancouver.

Heard, W.R., and W.L. Hartman. 1966. Pygmy whitefish, *Prosopium coulteri*, in Naknek River system of southwest Alaska. Fish. Bull. 65: 555–579.

Hebert, C.E., and G.D. Haffner. 1991. Habitat partitioning and contaminant exposure in cyprinids. Can. J. Fish. Aquat. Sci. 48: 261–266.

Heggenes, J. 2002. Flexible summer habitat selection by wild, allopatric brown trout in lotic environments. Trans. Am. Fish. Soc. 131: 287–298.

Heggenes, J., and S.J. Saltveit. 1990. Seasonal and spatial microhabitat selection and segregation in young Atlantic salmon, *Salmo salar* L., and brown trout, *Salmo trutta* L., in a Norwegian river. J. Fish Biol. 36: 707–720.

Heggenes, J., and T. Traaen. 1988. Downstream migration and critical water velocities in stream channels for fry of four salmonid species. J. Fish Biol. 32: 717–727.

Heidinger, R.C. 1975. Life history and biology of the largemouth bass. pp. 11–20. *In* H. Clepper (ed.). Black bass biology and management. Sport Fishing Institute, Washington, DC.

Held, J.W., and J.J. Peterka. 1974. Age, growth and food habits of the fathead minnow, *Pimephales promelas*, in North Dakota saline lakes. Trans. Am. Fish. Soc. 103: 743–756.

Helle, J.H. 1970. Biological characteristics of intertidal and fresh-water spawning pink salmon at Olsen Creek, Prince William Sound, Alaska. US Fish and Wildlife Service, Washington, DC. Spec. Sci. Rep. Fish. No. 602. 19 pp.

Henderson, M.A., and T.G. Northcote. 1985. Visual prey detection and foraging in sympatric cutthroat trout (*Salmo clarki clarki*) and Dolly Varden (*Salvelinus malma*). Can. J. Fish. Aquat. Sci. 42: 785–790.

Henderson, M.A., and T.G. Northcote. 1988. Retinal structures of sympatric and allopatric populations of cutthroat trout (*Salmo clarki clarki*) and Dolly Varden (*Salvelinus malma*) in relation to their spatial distribution. Can. J. Fish. Aquat. Sci. 45: 1321–1336.

Henderson, N.E., and R.E. Peter. 1969. Distribution of fishes of southern Alberta. J. Fish. Res. Board Can. 26: 325–338.

Hergenrader, G.L. 1969. Spawning behaviour of *Perca flavescens* in aquaria. Copeia, 1969: 839–841.

Hetherington, R., J.V. Barrie, R.G.B. Reid, R. MacLeod, D.J. Smith, T.S. James, and R. Kung. 2003. Late Pleistocene coastal paleogeography of the Queen Charlotte Islands, British Columbia, Canada, and its implications for terrestrial biogeography and early postglacial human occupation. Can. J. Earth Sci. 40: 1755–1766.

Hey, J., R.S. Waples, M.L. Arnold, R.K. Butlin, and R.G. Harrison. 2003. Understanding and confronting species uncertainty in biology and conservation. Trends Ecol. Evol. 18: 597–603.

Hildebrand, L. 1991. Lower Columbia River fisheries inventory. 1990 studies. Vol. I. Main Report. Prepared for B.C. Hydro, Environmental Resources by R. L. & L. Environmental Services Ltd., Edmonton, AB. 166 pp.

Hildebrand, L., T. Clayton, and S. McKenzie. 1995. Lower Columbia River fisheries inventory program: 1990–1994. R. L. & L. Environmental Services Ltd., Castlegar, B.C. 156 pp.

Hill, C.W. 1962. Observations on the life histories of the peamouth (*Mylocheilus caurinus*) and the northern squawfish (*Ptychocheilus oregonensis*) in Montana. Proc. Mont. Acad. Sci. 22: 27–44.

Hindar, K., B. Jonsson, J.H. Andrew, and T.G. Northcote. 1988. Resource utilization of sympatric and experimentally allopatric cutthroat trout and Dolly Varden charr. Oecologia, 74: 481–491.

Hinrichs, M.A. 1979. A description and key to the eggs and larvae of five species of fish in the subfamily Coregoninae. M.Sc. thesis, University of Wisconsin, Stevens Point.

Hitt, N.P., C.A. Frissell, C.C. Muhlfeld, and F. Allendorf. 2003. Spread of hybridization between native westslope cutthroat trout, *Oncorhynchus clarki lewisi*, and nonnative rainbow trout, *Oncorhynchus mykiss*. Can. J. Fish. Aquat. Sci. 60: 1440–1451.

Hodgson, J.R., D.E. Schindler, and Xi He. 1998. Homing tendency of three piscivorous fishes in a north temperate lake. Trans. Am. Fish Soc 127: 1078–1081.

Hofmann, N., and P. Fischer. 2002. Temperature preferences and critical thermal limits of burbot: implications for habitat selection and ontogenetic habitat shift. Trans. Am. Fish. Soc. 131: 1164–1172.

Holland, S.S. 1964. Landforms of British Columbia, a physiographic outline. BC Department of Mines and Petroleum Resources, Victoria. Bull. No. 48. 138 pp.

Hooe, M. L. 1991. Crappie biology and management. N. Am. J. Fish. Manage. 11: 483–489.

Hoogland, R.D. 1951. On the fixing-mechanism in the spines of *Gasterosteus aculeatus* L. K. Ned. Akad. Weten. Proc. Ser. C, 54: 171–180.

Hoogland, R.D., D. Morris, and N. Tinbergen. 1957. The spines of sticklebacks (*Gasterosteus* and *Pygosteus*) as a means of defense against predators (*Perca* and *Esox*). Behaviour, 10: 205–237.

Hooten, R.S., B.R. Ward, V.A. Lewynsky, M.G. Lirette, and A.R. Facchin. 1987. Age and growth of steelhead in Vancouver Island populations. BC Ministry of Environment and Parks, Victoria. Fish. Tech. Circ. No. 77. 39 pp.

Horns, W.H. 1985. Differences in early development among lake trout (*Salvelinus namaycush*) populations. Can. J. Fish. Aquat. Sci. 42: 737–743.

Horrall, R.M. 1981. Behavioural stock isolating mechanisms in Great Lakes fishes with special reference to homing and site imprinting. Can. J. Fish. Aquat. Sci. 38: 1481–1496.

House, R., and L. Wells. 1973. Age, growth, spawning season and fecundity of trout-perch (*Percopsis omiscomaycus*) In southeastern Lake Michigan. J. Fish. Res. Board Can. 30: 1221–1225.

Houston, J. 1990. Status of the spoonhead sculpin, *Cottus ricei*, in Canada. Can. Field-Nat. 104: 14–19.

Howland, K.L., R.F. Tallman, and W.M. Tonn. 2000. Migration patterns of freshwater and anadromous inconnu in the Mackenzie River system. Trans. Am. Fish. Soc. 129: 41–59.

Howland, K.L., W.M. Tonn, and G. Goss. 2001. Contrasts in the hypo-osmoregulatory abilities of a freshwater and an anadromous population of inconnu. J. Fish Biol. 59: 916–927.

Hubbs, C.L. 1955. Hybridization between fish species in nature. Syst. Zool. 4: 1–20.

Hubbs, C.L., and R.M. Bailey. 1938. The smallmouthed bass. Cranbrook Inst. Sci. Bull. No. 10. 92 pp.

Hubbs, C.L., and G.P. Cooper. 1936. Minnows of Michigan. Cranbrook Inst. Sci. Bull. No. 8. 95 pp.

Hubbs, C.L., and L.P. Schultz. 1941. Contributions to the ichthyology of Alaska with descriptions of two new fishes. University of Michigan, Ann Arbor. Univ. Mich. Mus. Zool. Occas. Pap. No. 431.

Hubbs, C.L., L.C. Hubbs, and R.E. Johnson. 1943. Hybridization in nature between species of catostomid fishes. Contrib. Lab. Vert. Biol. Univ. Mich. No. 22. 76 pp.

Hubbs, C.L., R.R. Miller, and L.C. Hubbs. 1974. Hydrographic history and relict fishes of the north-central Great Basin. Memoirs Calif. Acad. Sci. VII.

Hubert, W.A., and D.T. O'Shea. 1992. Use of spatial resources by fishes in Grayrocks Reservoir, Wyoming. J. Freshw. Ecol. 7: 219–225.

Huesser, C.J. 1971. North Pacific coastal refugia. Ecology, 52: 727–728.

Hughes, G.W., and A.E. Peden. 1984. Life history and status of the shorthead sculpin (*Cottus confusus*: Pisces, Cottidae) in Canada and the sympatric relationship to the slimy sculpin (*Cottus cognatus*). Can. J. Zool. 62: 306–311.

Hughes, N.F. 1992a. Ranking of feeding positions by drift-feeding Arctic grayling (*Thymallus arcticus*) in dominance hierarchies. Can. J. Fish. Aquat. Sci. 491: 1994–1998.

Hughes, N.F. 1992b. Selection of positions by drift-feeding salmonids in dominance hierarchies: model and test for Arctic grayling (*Thymallus arcticus*) in subarctic mountain streams, interior Alaska. Can. J. Fish. Aquat. Sci. 491: 1999–2008.

Hughes, N.F. 2000. Testing the ability of habitat selection theory to predict interannual movement patterns of a drift-feeding salmonid. Ecol. Freshw. Fish, 9: 4–8.

Hughes, N.F., and J.B. Reynolds. 1995. Why do Arctic grayling (*Thymallus arcticus*) get bigger as you go upstream? Can. J. Fish. Aquat. Sci. 51: 2154–2163.

Hughes, R.M., E. Rexstad, and C.E. Bond. 1987. The relationship of aquatic ecoregions river basins, and physiographic provinces to the ichthyogeographic regions of Oregon. Copeia, 1987: 423–432.

Hughes, T. 1960. Pike. *In* Lupercal. Harper, New York.

Hume, J.M.B., and T.G. Northcote. 1985. Initial changes in use of space and food by experimentally segregated populations of Dolly Varden (*Salvelinus malma*) and cutthroat trout (*Salmo clarki clarki*). Can. J. Fish. Aquat. Sci. 42: 101–109.

Hume, M. (with H. Thommasen). 1998. River of the angry moon. GreyStone Books, Vancouver, B.C.

Hunt, R.L. 1974. Overwinter survival of wild fingerling brook trout in Lawrence Creek during eleven years. Wis. Dep. Nat. Resour. Tech. Bull. No. 82.

Hunter, J.G. 1949. Occurrence of hybrid salmon in the British Columbia commercial fishery. Prog. Rep. Pac. Biol. Stn. Fish. Res. Board Can. 79: 33–34.

Hunter, J.G. 1959. Survival and production of pink and chum salmon in a coastal stream. J. Fish. Res. Board Can. 16: 835–886.

Hutchings, J.A. 1996. Adaptive phenotypic plasticity in brook trout, *Salvelinus fontinalis*, life histories. Ecoscience, 3: 25–32.

Hyatt, K.D., and N.H. Ringler. 1989. Egg cannibalism and the reproductive strategies of threespine stickleback (*Gasterosteus aculeatus*) in a coastal British Columbia lake. Can. J. Zool. 67: 2036–2046.

Ikusemiju, K. 1975. Aspects of the ecology and life history of the sculpin, *Cottus aleuticus* (Gilbert), in Lake Washington. J. Fish Biol. 7: 235–245.

Imamura, K.K. 1975. Life history of the brown bullhead (*Ictalurus nebulosus* Lesueur) in Lake Washington. M.Sc. thesis, University of Washington, Seattle.

Irvine, J.R. 1978. The Gerrard rainbow trout of Kootenay Lake, British Columbia— a discussion of their life history with management, research and enhancement recommendations. BC Ministry of the Environment, Victoria. Fish. Manage. Rep. No. 72. 58 pp.

Irvine, J.R., and N.T. Johnston. 1992. Coho (*Oncorhynchus kisutch*) use of lakes and streams in the Keogh River drainage, British Columbia. Northwest Sci. 66: 15–25.

Irvine, J., and D. Rowland. 1979. Aquatic resources of the Liard River and tributaries relative to proposed hydroelectric development in British Columbia. Environment and Socio-economic Services, BC Hydro and Power Authority, Vancouver. Rep. No. SE 7905. 179 pp.

Ishiguro, N.B., M. Miya, and M. Nishida 2003. Basal euteleostean relationships: a mitogenomic perspective on the phylogenetic reality of the "Protacanthopterygii." Mol. Phylogenet. Evol. 27: 476–488.

Israel, J., J.F. Cordes, M.A. Blumberg, and B. May. 2004. Geographic patterns of genetic differentiation among collections of green sturgeon. N. Am. J. Fish. Manage. 24: 922–931.

Jacobson, T.-L. 1989. Ecology of a population of stunted yellow perch (*Perca flavescens* (Mitchill)). M.Sc. thesis, University of Alberta, Edmonton.

Jakober, M.J., T.E. McMahon, and R.F. Thurow. 2000. Diel habitat partitioning by bull charr and cutthroat trout during fall and winter in Rocky Mountain streams. Environ. Biol. Fishes, 59: 79–89.

Jansen, W.A., and W.C. Mckay. 1992. Foraging in yellow perch, *Perca flavescens*: biological and physical factors affecting diel periodicity in feeding, consumption, and movement. Environ. Biol. Fishes, 34: 287–303.

Janssen, J 1990. Localization of substrate vibrations by the mottled sculpin, *Cottus bairdi*. Copeia, 1990: 349–355.

Jennings, M.J., J.E. Claussen, and D.P. Philipp. 1996. Evidence for heritable preferences for spawning habitat between two walleye populations. Trans. Am. Fish. Soc. 125: 978–972.

Jeppson, P.W., and W.S. Platts. 1959. Ecology and control of the Columbia squawfish in northern Idaho lakes. Trans. Am. Fish. Soc. 88: 197–202.

Jessop, B.M., and G. Power. 1972. Age, growth, and maturity, of round whitefish (*Prosopium cylindraceum*) from the Leaf River, Ungava, Quebec. J. Fish. Res. Board Can. 30: 299–304.

Johannes, R.E. 1958. The feeding relationship of *Rhinichthys cataractae* and *Rhinichthys falcatus* in British Columbia. Honours B.Sc. thesis, Department of Zoology, University of British Columbia, Vancouver.

Johannes, R.E., and P.A. Larkin. 1961. Competition for food between redside shiners (*Richardsonius balteatus*) and rainbow trout (*Salmo gairdneri*) in two British Columbia lakes. J. Fish. Res. Board Can. 18: 203–220.

John, K.R. 1963. The effects of torrential rains on the reproductive cycle of *Rhinichthys osculus* in the Chiracahua Mountains, Arizona. Copeia, 1963: 286–291.

Johnson, D.W., D.A. Cannamela, and K.W. Gasser. 1983. Food habits of the shorthead sculpin (*Cottus confusus*) in the Big Lost River, Idaho. Northwest Sci. 57: 229–239.

Johnson, F.H. 1961. Walleye egg survival during incubation on several types of bottom in Lake Winnibigoshish, Minnesota, and connecting waters. Trans. Am. Fish. Soc. 90: 312–322.

Johnson, F.H. 1977. Responses of walleye (*Stizostedion vitreum vitreum*) and yellow perch (*Perca flavescens*) populations to removal of white suckers (*Catostomus commersoni*) from a Minnesota Lake, 1966. J. Fish. Res. Board Can. 43: 1633–1642.

Johnson, G.D., and C. Patterson. 1996. Relationships of lower Euteleostean fishes. pp. 251–332. *In* M.L.J. Stiassny, L.R. Parenti, and G.D. Johnson (eds.). Interrelationships of fishes. Academic Press, London.

Johnson, J.H., and D.S. Dropkin. 1996. Seasonal habitat use by brook trout, *Salvelinus fontinalis*

(Mitchill), in a second order stream. Fish. Manage. Ecol. 3: 1–11.

Johnson, J.H., D.S. Dropkin, and P.G. Schaffer. 1992. Habitat use by a headwater stream fish community in north-central Pennsylvania. Rivers, 3: 69–79.

Johnson, L. 1973. Stock and recruitment in some unexploited Canadian Arctic lakes. Rapp. Rens. Cons. Int. Explor. Mer, 164: 219–227.

Johnson, L.E. 1976. Ecology of Arctic populations of lake trout, *Salvelinus namaycush*, lake whitefish, *Coregonus clupeaformis*, Arctic char, *S. alpinus*, and associated species in unexploited lakes of the Canadian Northwest Territories. J. Fish. Res. Board Can. 33: 2459–2488.

Johnson, O.W., W.S. Grant, R.G. Kope, K. Neely, F.W. Waknitz, and R. Waples. 1997. Status review of chum salmon from Washington, Oregon, and California. US Department of Commerce, Washington, DC. NOAA Tech. Memo. No. NMFS-NWFSC-32. 280 pp.

Johnson, O.W., M.A. Ruckelshaus, W.S. Grant, F.W. Waknitz, A.M. Garrett, G.J. Bryant, K. Neely, and J. Hard. 1999. Status review of coastal cutthroat trout from Washington, Oregon, and California. US Department of Commerce, National Marine Fisheries Service, Washington, DC. NOAA Tech. Memo. NMFS-NWFSC-37. 292 pp.

Johnson, R.C., R.J. Gerke, D.W. Heiser, R.F. Orrell, S.B. Mathews, and J.G. Olds. 1971. Pink and chum salmon investigations, 1969: supplementary progress report. Research Division, Washington Department of Fisheries, Olympia. 66 pp.

Johnston, J. 1982. Life history of anadromous cutthroat with emphasis on migratory behavior. pp.123–127. *In* E. Brannon and E. Salo (eds.). Proceedings of the salmon and trout migratory behavior symposium. School of Fisheries, University of Washington, Seattle.

Jones, D.H., and A.W.G. John. 1978. The three-spined stickleback, *Gasterosteus aculeatus* L., from the north Atlantic. J. Fish Biol. 13: 231–236.

Jones, J.D., and C.L. Seifert. 1997. Distribution of mature sea-run cutthroat trout overwintering in Auke Lake and Lake Eva in southeastern Alaska. pp. 27–28. *In* J.D. Hall, P.A. Bisson, and R.E. Gresswell (eds.). Sea-run cutthroat trout: biology, management, and future conservation. Oregon Chapter, American Fisheries Society, Corvallis.

Jones, J.W., and J.N. Ball. 1950. The spawning behaviour of brown trout and salmon. Br. J. Anim. Behav. 2,3: 103–114.

Jones, J.W., and G.M. King. 1950. Further experimental observations on the spawning behaviour of the Atlantic salmon (*Salmo salar* Linn.). Proc. Zool. Soc. London, 120: 317–323.

Jones, P.W., F.D. Martin, and J.D. Hardy. 1978a. Development of fishes of the mid-Atlantic Bight. Vol. 1. Acipenseridae through Ictaluridae. Biological Services Program, US Fish and Wildlife Service, Washington, DC. pp. 265–268.

Jones, P.W., F.D. Martin, and J.D. Hardy. 1978b. *Ictalurus natalis*. pp. 317–318. *In* Development of fishes of the mid-Atlantic Bight. Vol. 1. US Department of the Interior, Fish and Wildlife Service., Washington, DC.

Jones, P.W., F.D. Martin, and J.D. Hardy. 1978c. *Ictalurus nebulosus*. pp. 319–322. *In* Development of Fishes of the mid-Atlantic Bight. Vol. 1. US Department of the Interior, Fish and Wildlife Service, Washington, DC.

Kanda, N., R.F. Leary, and F.W. Allendorf. 2002. Evidence of introgressive hybridization between bull trout and brook trout. Trans. Am. Fish. Soc 131: 772–782.

Karrow, P.F., A. Ceska, R.J. Hebda, B.B. Miller, K.L. Seymour, and A.J. Smith. 1995. Diverse non-marine biota from the Whidbey Formation (Sangamonian) at Point Wilson, Washington. Quat. Res. 44: 433–437.

Kaya, C.M. 1991. Laboratory spawning and rearing of speckled dace. Prog. Fish-Cult. 53: 259–260.

Kaya, C.M., and E.D. Jeanes. 1995. Retention of adaptive rheotactic behaviour by F_1 fluvial Arctic grayling. Trans. Am. Fish. Soc. 124: 453–457.

Keast, A. 1968. Feeding of some Great Lakes fishes at low temperatures. J. Fish. Res. Board Can. 25: 1199–1218.

Keast, A. 1985. Implications of chemosensory feeding in catfishes: an analysis of the diets of *Ictalurus nebulosus* and *I. natalis*. Can. J. Zool. 63: 590–602.

Keast, A., and L. Welsh. 1968. Daily feeding periodicities, food uptake rates, and some dietary changes with hour of day in some lake fishes. J. Fish. Res. Board Can. 23: 1845–1867.

Keeley, E.R., and J.W.A. Grant. 1995. Allometric and environmental correlates of territory size in juvenile Atlantic salmon (*Salmo salar*). Can. J. Fish. Aquat. Sci. 52: 186–196.

Keeley, E.R., E.A. Parkinson, and E.B. Taylor. 2005. Ecotypic differentiation of native rainbow trout (*Oncorhynchus mykiss*) populations from British Columbia. Can. J. Fish. Aquat. Sci. 62: 1523–1539.

Keenleyside, M.H.A. 1954. First record of the brassy minnow, *Hybognathus hankinsoni* Hubbs, from British Columbia. Can. Field-Nat. 68: 43.

Keenleyside, M.H.A., and H. Dupuis. 1988. Courtship and spawning competition in pink salmon (*Oncorhynchus gorbuscha*). Can. J. Zool. 66: 262–265.

Keivany, Y., and J.S. Nelson. 2000. Taxonomic review of the genus *Pungitius*, ninespine sticklebacks (Gasterosteidae). Cybium, 24: 107–122.

Kennedy, W.A. 1953. Growth, maturity, fecundity, and mortality in the relatively unexploited whitefish, *Coregonus clupeaformis*, in Great Slave Lake. J. Fish. Res. Board Can. 10: 413–441.

Kennedy, W.A., and W.M. Sprules. 1967. Goldeye in Canada. Bull. Fish. Res. Board Can. No. 161. 45 pp.

Kershaw, A.C. 1978. The Quaternary history of the Okanagan. pp. 27–42. *In* The forty-second annual report of the Okanagan Historical Society. Wayside Press Ltd., Vernon, BC.

Kieffer, J.D. and P.W. Colgan. 1993. Foraging flexibility in pumpkinseed (*Lepomis gibbosus*): influence of habitat structure and prey type. Can. J. Fish. Aquat. Sci. 50: 1699–1705.

Kimmel, P.G. 1975. Fishes of the Miocene–Pliocene Deer Butte Formation, southeastern Oregon. Museum of Paleontology, University of Michigan, Ann Arbor. Pap. Paleontol. No. 14: 69–85.

Kinziger, A.P., R.W. Wood, and D.A. Neely. 2005. Molecular systematics of the genus *Cottus* (Scorpaeniformes: Cottidae). Copeia, 2005: 303–311.

Kirkpatrick, M., and R. Selander. 1979. Genetics of speciation in lake whitefishes in the Allegash Basin. Evolution, 33: 478–485.

Klemetsen, A., P.-A. Amundsen, J.B. Dempson, B. Jonsson, N. Jonsson, M.F. O'Connell, and E. Mortensen. 2003. Atlantic salmon, *Salmo salar* L., brown trout, *Salmo trutta* L., and Arctic charr, *Salvelinus alpinus* (L.): a review of aspects of their life histories. Ecol. Freshw. Fish, 12: 1–59.

Kline, J.L., and B.M. Wood. 1996. Food habits and diet selectivity of the brown bullhead, *Ameiurus nebulosus*. J. Freshw. Ecol. 11: 145–151.

Koelz, W. 1929. Coregonid fishes of the Great Lakes. Bull. US Bur. Fish. 43: 297–643.

Koenst, W.M., and L.L. Smith. 1976. Thermal requirements of the early life history stages of walleye, *Stizostedion vitreum*, and sauger, *Stizostedion canadense*. J. Fish. Res. Board Can. 35: 1130–1138.

Konkle, B.R., and D.P. Philipp. 1992. Asymmetric hybridization between two species of sunfishes (*Lepomis*: Centrarchidae). Mol. Ecol. 1: 215–222.

Koster, W.J. 1957. Guide to the fishes of New Mexico. University of New Mexico Press, Albuquerque.

Kraak, S.B., B. Mundwiler, and P.J. Hart. 2001. Increased number of hybrids between benthic and limnetic threespine sticklebacks in Enos Lake, Canada: the collapse of a species pair? J. Fish. Biol. 58: 1458–1464.

Kramer, R.H., and L.L. Smith. 1960. First year growth of the largemouth bass, *Micropterus salmoides* (Lacépède), and some related ecological factors. Trans. Am. Fish Soc. 89: 222–233.

Kratt, L.F., and R.J.F. Smith. 1977. A post-hatching sub-gravel phase in the life history of the Arctic grayling, *Thymallus arcticus*. Trans. Am. Fish. Soc. 106: 241–243.

Kratt, L.F., and R.J.F. Smith. 1978. Breeding tubercles on male and female Arctic grayling (*Thymallus arcticus*). Copeia, 1978: 185–188.

Kratt, L.F., and R.J.F. Smith. 1979. Agonistic behaviour of age 0, age 1 and non-breeding adult Arctic grayling *Thymallus arcticus* (Pallas). J. Fish Biol. 15: 389–404.

Kratt, L.F., and R.J.F. Smith. 1980. An analysis of the behaviour of the Arctic grayling, *Thymallus arcticus* (Pallas) with observations on mating success. J. Fish Biol. 17: 661–666.

Krejsa, R.J. 1965. The systematics of the prickly sculpin, *Cottus asper*: an investigation of genetic and non-genetic variation within a polytypic species. Ph.D. thesis, University of British Columbia, Vancouver.

Krejsa, R.J. 1967a. The systematics of the prickly sculpin, *Cottus asper* Richardson, a polytypic species: Part I. Synonymy, nomenclatural history, and distribution. Pac. Sci. XXI: 241–251.

Krejsa, R.J. 1967b. The systematics of the prickly sculpin, *Cottus asper* Richardson, a polytypic species: Part II. Studies on life history, with special reference to migration. Pac. Sci. XXI: 414–422.

Kristensen, J. 1980. Large flathead chub (*Platygobio gracilis*) from the Peace-Athabasca Delta, Alberta, including a Canadian record. Can. Field-Nat. 94: 342.

Krueger, C.C., R.L. Wilmer, and R.J. Everett. 1999. Stock origins of Dolly Varden collected from the Beaufort Sea coastal sites of arctic Alaska and Canada. Trans. Am. Fish. Soc. 128: 49–57.

Kynard, B.E. 1978. Breeding behaviour of a lacustrine population of threespine sticklebacks (*Gasterosteus aculeatus* L.). Behaviour, 67: 178–207.

Kynard, B., and E. Parker. 2005. Ontogenetic behavior and dispersal of Sacramento River white sturgeon, *Acipenser transmontanus*, with a note on body color. Environ. Biol. Fishes, 72: 19–30.

Kynard, B., E. Parker, and T. Parker. 2005. Behavior of early life history intervals of Klamath River green sturgeon, *Acipenser medirostris*, with a note on body colour. Environ. Biol. Fishes, 72: 85–97.

Lahti, K., A. Laurila, K. Enberg, and J. Piironen. 2001. Variation in aggressive behaviour and growth rate between populations and migratory forms in the brown trout, *Salmo trutta*. Anim. Behav. 62: 935–944.

Lampman, B.H. 1946. The coming of the pond fishes. Binfords & Mort, Portland, OR.

Lane, E.D., and M.L. Rosenau. 1995. The conservation of sturgeon stocks in the lower Fraser watershed. A baseline investigation of habitat, distribution, and age and population of juvenile white sturgeon (*Acipenser transmontanus*) in the lower Fraser River downstream of Hope, B.C. Final Report. Habitat Conservation Tust Fund Project, Surrey, BC.

Langer, O.E., B.G. Shepherd, and P.R. Vroom. 1977. Biology of the Nass River eulachon. BC Ministry of the Environment, Victoria. Dep. Fish. Environ. Tech. Rep. Ser. No. PAC/T-77-10. 56 pp.

Langlois, T.H. 1929. Breeding habits of the northern pearl dace. Ecology, 10: 161–163.

Larson, K.W., and G.W. Brown. 1975. Systematic status of a midwater population of freshwater sculpin (*Cottus*) from Lake Washington, Seattle, Washington. J. Fish. Res. Board Can. 32: 21–28.

Lassuy, D.R. 1990. Herbivory by a north temperate fish, *Acrocheilus alutaceus* (Agassiz and Pickering). Ph.D. thesis, Oregon State University, Corvallis.

Latham, S. 2002. Historical and anthropogenic influences on genetic variation in bull trout (*Salvelinus confluentus*) in the Arrow Lakes, British Columbia. M.Sc. thesis, Department of Zoology, University of British Columbia, Vancouver.

Lavin, P.A., and J.D. McPhail. 1985. The evolution of freshwater diversity in threespine stickleback (*Gasterosteus aculeatus*): site-specific differentiation of trophic morphology. Can. J. Zool. 63: 2632–2638.

Lavin, P.A., and J.D. McPhail. 1986. Adaptive divergence of trophic phenotype among freshwater populations of threespine stickleback (*Gasterosteus aculeatus*). Can. J. Fish. Aquat. Sci. 44: 1820–1829.

Lawler, G.H. 1954. Observations on the trout-perch *Percopsis omiscomaycus* (Walbaum) at Heming Lake, Manitoba. J. Fish. Res. Board Can. 11: 1–4.

Lawler, G.H. 1961. Egg counts of Lake Erie whitefish. J. Fish. Res. Board Can. 18: 293–294.

Leary, R.F., F.W. Allendorf, S.R. Phelps, and K.L. Knutsen. 1987. Genetic divergence and identification of seven cutthroat trout subspecies and rainbow trout. Trans. Am. Fish. Soc. 116; 580–587.

Leary, R.F., W.R. Gould, and G.K. Sage. 1996. Success of basibranchial teeth in indicating pure populations of rainbow trout and failure to indicate pure populations of westslope cutthroat trout. N. Am. J. Fish. Manage. 16: 210–213.

Lecointre, G. and G. Nelson. 1996. Clupeomorpha, sister-group of Ostariophysi. pp. 193–207. *In* M.L.J. Stiassny, L.R. Parenti, and G.D. Johnson (eds.). Interrelationships of fishes. Academic Press, London.

Lee, D.S. 1980. *Cottus bairdi* Girard, mottled sculpin. p. 805. *In* D.S. Lee, C.R. Gilbert, C.H. Hocutt, R.E. Jenkins, D.E. McAllister, and J.R. Stauffer (eds.). Atlas of North American freshwater fishes. North Carolina State Museum of Natural History, Raleigh.

Lee, R.M., and J.N. Rinne. 1980. Critical thermal maxima of five trout species in the southwestern United States. Trans. Am. Fish. Soc. 109: 632–635.

Legendre, P. 1970. The bearing of *Phoxinus* hybridity on the classification of its North American species. Can. J. Zool. 48: 1167–1179.

Legendre, V., and R. Legueux. 1948. The tomcod (*Microgadus tomcod*) as a permanent resident of Lake St. John, Province of Quebec. Can. Field-Nat. 65: 157.

Leggett, W.C., and J.E. Carscadden. 1978. Latitudinal variation in reproductive characteristics of American shad (*Alosa sapidissima*): evidence for population specific life history strategies in fish. J. Fish. Res. Board Can. 35: 1469–1478

Leidy, R.A., and P.B. Moyle. 1997. Conservation status of the world's fish fauna: an overview. pp. 187–227. *In* P.A. Fiedler and P.M. Karieva (eds.). Conservation biology for the coming decade. Chapman & Hall, New York.

Leim, A.H. 1924. The life history of the shad (*Alosa sapidissima* (Wilson), with special reference to the factors limiting its abundance. Contrib. Can. Biol. 2: 161–284.

Leiner, M. 1957. Stichlingsbastarde. Naturwiss. Volk, 87: 299–300.

Leis, A.L., and M.G. Fox. 1996. Feeding, growth and habitat associations of young-of-the-year walleye (*Stizostedion vitreum*) in a river affected by a mine-tailing spill. Can. J. Fish. Aquat. Sci. 53: 2408–2417.

Lemmen, D.S., A. Duc-Rodkin, and J.M. Bednarski. 1994. Late glacial drainage systems along the northwestern margin of the Laurentide Ice-sheet. Quat. Sci. Rev. 13: 805–828.

Levings, C.D., and R.B. Lauzier. 1991. Extensive use of the Fraser River basin as winter habitat by juvenile chinook salmon (*Oncorhynchus tshawytscha*). Can. J. Zool. 69: 1759–1767.

Levings, C.D., D.E. Boyle, and T.R. Whitehouse. 1995. Distribution and feeding of juvenile Pacific salmon in freshwater tidal creeks of the lower Fraser River, British Columbia. Fish. Manage. Ecol. 2: 299–308.

Levy, D.A. 1987. Review of the ecological significance of diel vertical migration by juvenile sockeye salmon (*Oncorhynchus nerka*). pp. 44–52. *In* H.A. Smith, L. Margolis, and C.C. Wood (eds.). Sockeye salmon (*Oncorhynchus nerka*) population biology and future management. Can. Spec. Publ. Fish. Aquat. Sci. No. 96.

Levy, D.A. 1990. Acoustic analysis of diel migration behaviour of *Mysis relicta* and kokanee, *Oncorhynchus nerka*, within Okanagan

Lake, British Columbia. Can. J. Fish. Aquat. Sci. 48: 67–72.

Levy, D.A., and T.G. Northcote. 1982. Juvenile salmon residency in a marsh area of the Fraser River estuary. Can. J. Fish. Aquat. Sci. 39: 270–276.

Levy, D.A., R.L. Johnson, and J.M. Hume. 1991. Shifts in fish vertical distribution in response to an internal seiche in a stratified lake. Limnol. Oceanogr. 36: 187–192.

Li, G.-Q., M.V.H. Williams, and L. Grande. 1997. Review of *Eohiodon* (Teleostei: Osteoglossomorpha) from western North America, with a phylogenetic reassessment of Hiodontidae. J. Paleontol. 71: 1109–1124.

Liknes, G., and P. Graham. 1988. Westslope cutthroat trout in Montana: life history, status, and management. pp. 53–60. In R.E. Gresswell (ed.). Status and management of interior stocks of cutthroat trout. Am. Fish. Soc. Symp. No. 4.

Lindsey, C.C. 1962. Distinctions between the broad whitefish, *Coregonus nasus*, and other North American whitefishes. J. Fish. Res. Board Can. 19: 687–714.

Lindsey, C.C., and W.G. Franzin. 1972. New complexities in zoogeography and taxonomy of the pygmy whitefish (*Prosopium coulteri*). J. Fish. Res. Board Can. 29: 1772–1775.

Lindsey, C.C., and L.F. Kratt. 1982. Jumbo spotted form of least cisco *Coregonus sardinella* in lakes of southern Yukon Territory. Can. J. Zool. 60: 2783–2786.

Lindsey, C.C., and J.D. McPhail. 1986. Zoogeography of the fishes of the Mackenzie and Yukon systems. pp. 639–673. In C.H. Hocutt and E.O. Wiley (eds.). The zoogeography of North American freshwater fishes. John Wiley & Sons, New York.

Lindsey, C.C., and T.G. Northcote. 1963. Life history of redside shiners, *Richardsonius balteatus*, with particular reference to movements in and out of Sixteenmile Lake streams. J. Fish. Res. Board Can. 20: 1001–1030.

Lindsey, C.C., T.G. Northcote, and G.F. Hartman. 1959. Homing of rainbow trout to inlet and outlet spawning streams at Loon Lake, British Columbia. J. Fish. Res. Board Can. 16: 695–719.

Lister, D.B., and H.S. Genoe. 1970. Stream habitat utilization by cohabiting underyearlings of chinook (*Oncorhynchus tshawytscha*) and coho (*Oncorhynchus kisutch*) salmon in the Big Qualicum River, British Columbia. J. Fish. Res. Board Can. 27: 1215–1224.

Litvak. M.K., and R.I.C. Hansell. 1990. Investigations of food habit and niche relationships in a cyprinid community. Can. J. Zool. 68: 1873–1879.

Loeb, H.A. 1954. Submergence of brown bullheads in bottom sediments. N.Y. Fish Game J. 11: 119–124.

Loew, E.R., and A.J. Sillman. 1998. An action spectrum for the light-dependent inhibition of swimming behavior in newly hatched white sturgeon, *Acipenser transmontanus*. Vision Res. 38: 111–114.

Logan, C., E.A. Trippel, and F.W.H. Beamish. 1991. Thermal stratification and benthic foraging patterns of white sucker. Hydrobioligia, 213: 125–132.

López, J.A., W-J. Chen, and G. Orti. 2004. Esociform phylogeny. Copeia, 2004: 449–464.

Lorenz, J.M., and J.H. Eiler. 1989. Spawning habitat and redd characteristics of sockeye salmon in the glacial Taku River, British Columbia and Alaska. Trans. Am. Fish. Soc. 118: 495–502.

Lorz, H.W., and T.G. Northcote. 1965. Factors affecting stream location, timing and intensity of entry by spawning kokanee (*Oncorhynchus nerka*) into an inlet of Nicola Lake, British Columbia. J. Fish. Res. Board Can. 22: 665–687.

Loudenslager, E.J., and G.H. Thorgaard. 1979. Karyotypic and evolutionary relationships of the Yellowstone (*Salmo clarki bouvieri*) and westslope (*S. c. lewisi*) cutthroat trout. J. Fish. Res. Board Can. 6: 630–635.

Lu, G., and L. Bernatchez. 1999. Correlated trophic specialization and genetic divergence in sympatric whitefish ecotypes (*Coregonus clupeaformis*): support for the ecological speciation hypothesis. Evolution, 53: 1491–1505.

Lucas, M.C. 1992. Spawning activity of male and female pike, *Esox lucius* L., determined by acoustic tracking. Can. J. Zool. 70: 191–196.

Ludwig, A., N.M. Belfiore, C. Pitra, V. Svirsky, and I. Jenneckens. 2001. Gene duplication events and functional reduction of ploidy levels in sturgeon (*Acipenser*, *Huso*, and *Scaphyrhynchus*). Genetics, 158: 1203–1215.

Luecke, C., and D. Teuscher. 1994. Habitat selection by lacustrine rainbow trout within gradients of temperature, oxygen, and food availability. pp. 133–150. *In* D.J. Stouder, K.L. Fresh, and R.J. Feller (eds.). Theory and applications of fish feeding ecology. University of South Carolina Press, Columbia.

Lundberg, J.G. 1992. The phylogeny of ictalurid catfishes: a synthesis of recent work. pp. 392–420. *In* R.L. Mayden (ed.). Systematics and historical ecology of North American freshwater fishes. Stanford University Press, Stanford, CA.

Lynott, S.T., S.D. Bryan, T.D. Hill, and W.G. Duffy. 1995. Monthly and size-related changes in the diet of rainbow trout in Lake Oahe, South Dakota. J. Freshw. Ecol. 10: 399–407.

Lyons, J. 1987a. Prey choice among piscivorous juvenile walleyes (*Stizostedion vitreum*). Can. J. Fish. Aquat. Sci. 44: 758–764.

Lyons, J. 1987b. Distribution, abundance, and mortality of small littoral zone fish in Sparkling Lake, Wisconsin. Environ. Biol. Fishes, 18: 93–107.

MacCrimmon, H.R. 1968. The carp in Canada. Fish. Res. Board Can. Bull. No. 165. 93 pp.

MacCrimmon, H.R., and T.K. Twongo. 1980. Ontogeny of feeding behaviour in hatchery-reared rainbow trout, *Salmo gairdneri* Richardson. Can. J. Zool. 58: 20–26.

Maceina, M.J., and J.J. Isely. 1986. Factors affecting growth of an initial largemouth bass year class in a new Texas reservoir. J. Freshw. Ecol. 3: 485–492.

Machniak, K. 1975a. The effects of hydroelectric development on the biology of northern fishes (reproduction and population dynamics) 1. Lake whitefish *Coregonus clupeaformis* (Mitchill). A literature review and bibliography. Environment Canada, Ottawa, ON. Fish. Mar. Serv. Dev. Tech. Rep. No. 527. 67 pp.

Machniak, K. 1975b. The effects of hydroelectric development on the biology of northern fishes. (reproduction and population dynamics). III. Yellow walleye (*Stizostedion vitreum*). Environment Canada, Ottawa, ON. Fish. Mar. Serv. Tech. Rep. No. 529. 68 pp.

Machniak, K., and W.A. Bond. 1979. An intensive study of the fish fauna of the Steepbank River watershed, northeastern Alberta. Prepared for the Alberta Oil Sands Environmental Research Program by Department of Fisheries Oceans, Freshwater Institute, Winnipeg, MB. AOSERP Rep. No. 89. 81 pp.

Mackay, I., and G. Power. 1968. Age and growth of round whitefish (*Prosopium cylindraceum*) from Ungava. J. Fish. Res. Board Can. 25: 657–666.

MacLeod, J.C. 1960. The diurnal migration of peamouth chub, *Mylocheilus caurinus* (Richardson), in Nicola Lake, British Columbia. M.Sc. thesis, Department of Zoology, University of British Columbia, Vancouver.

Madenjian, C.P., C.P. DeScorcie, and R.M. Stedman. 1998a. Ontogenetic and spatial patterns in diet and growth of lake trout in Lake Michigan. Trans. Am. Fish. Soc 127: 236–252.

Madenjian, C.P., C.P. DeScorcie, and R.M. Stedman. 1998b. Maturity schedules of lake trout in Lake Michigan. J. Great Lakes Res. 24: 404–410.

Magnuson, J.L., and L.L. Smith. 1963. Some phases of the life history of the trout-perch. Ecology, 44: 83–95.

Major, R.L., and J.L. Mighell. 1969. Egg-to-migrant survival of spring chinook salmon (*Oncorhynchus tshawytscha*) in the Yakima River, Washington. Fish. Bull. 67: 347–359.

Mance, C.H. 1987. The fecundity and histology of ovarian recrudescence in the yellow perch (*Perca flavescens*) from selected lakes in Alberta. M.Sc. thesis, University of Alberta, Edmonton.

Mann, G.J. 1974. Life history types of the least cisco (*Coregonus sardinella*, Valenciennes), in the Yukon Territory, north slope and eastern Mackenzie River Delta drainages. M.Sc. thesis, Department of Zoology, University of Alberta, Edmonton.

Manzer, J.I. 1976. Distribution, food, and feeding of the threespine stickleback, *Gasterosteus aculeatus*, in Great Central Lake, Vancouver Island, with comments on competition for food with juvenile sockeye salmon, *Oncorhynchus nerka*. Fish. Bull. 74: 647–668.

Markle, D.F. 1992. Evidence of bull trout × brook trout hybrids in Oregon. pp. 58–67. *In* P.J. Howell and D.V. Buchanan (eds.). Proceedings of the Gearhart Mountain Bull Trout Workshop. Oregon Chapter, American Fisheries Society, Corvallis.

Markle, D.F., and D.L. Hill. 2000. Taxonomy and distribution of the Malheur mottled sculpin, *Cottus bendirei*. Northwest Sci. 74: 202–211.

Markus, H.C. 1934. Life history of the blackhead minnow (*Pimephales promelas*). Copeia, 1934:116–122.

Marshall, K.E., M. Layton, and C. Stobbe. 1990. A bibliography of the lake trout, *Salvelinus namaycush* (Walbaum), 1984 through 1990. Can. Tech. Rep. Fish. Aquat. Sci. No. 1749. 25 pp.

Martin, N.V. 1955. The effect of drawdowns on lake trout reproduction and use of artificial spawning beds. Trans. N. Am. Wildl. Conf. 20: 263–271.

Martin, N.V., and C.H. Olver. 1980. The lake charr, *Salvelinus namaycush*. pp. 205–272. *In* E.K. Balon (ed.). Charrs, salmonid fishes of the genus *Salvelinus*. W. Junk, The Hague.

Mason, J.C., and S. Machidori. 1976. Populations of sympatric *Cottus aleuticus* and *Cottus asper* in four adjacent salmon-producing coastal streams on Vancouver Island. Fish. Bull. 74: 131–141.

Mathews, W.H. 1944. Glacial lakes and ice-retreat in south-central British Columbia. Trans. R. Soc. Can. Ser. 2, No. 38: 39–57.

Mathews, W.H. 1980. Retreat of the last ice sheets in northeastern British Columbia and adjacent Alberta. Geol. Surv. Can. Bull. No. 331. 22 pp.

Maughan, O.E., and G.E. Saul. 1979. Distribution of sculpins in the Clearwater River basin, Idaho. Great Basin Nat. 39: 59–62.

Maxwell, H. 1904. British fresh-water fishes. Hutchinson & Co., London.

Mayden, R.L., and E.O. Wiley. 1992. The fundamentals of phylogenetic systematics. pp. 114–185. *In* R.L. Mayden (ed.). Systematics and historical ecology of North American freshwater fishes. Stanford University of Press, Stanford, CA.

Mayr, E. 1942. Systematics and the origin of species. Columbia University Press, New York.

Mayr, E. 1963. Animal species and evolution. Harvard University Press, Cambridge, MA.

McAdam, S.O., C.J. Walters, and C. Nistor. 2005. Linkages between white sturgeon (*Acipenser transmontanus*) recruitment and altered substrates in the Nechako River, Canada. Trans. Am. Fish. Soc. 134: 1448–1456.

McAllister, D.E. 1963. A revision of the smelt family, Osmeridae. Bull. Natl. Mus. Can. No. 191. Biol. Ser. No. 71. 53 pp.

McAllister, D.E., and C.C. Lindsey. 1961. Systematics of the freshwater sculpins (*Cottus*) of British Columbia. Bull. Natl. Mus. Can. Contrib. Zool. No. 172 (1959): 66–89.

McCabe, G., and C.A. Tracy. 1994. Spawning and early life history of white sturgeon, *A. transmontanus*, in the lower Columbia River. Fish. Bull. 92: 760–772.

McCabe, G., S.A. Hinton, and R.J. McConnell. 1989. Report D. pp. 167–207. *In* A.A. Nigro (ed.). Status and habitat requirements of white sturgeon populations in the Columbia River downstream of McNary Dam. 1989. Bonneville Power Administration, Portland, OR. Annual Progress Report, Project 86-50.

McCarraher, D.B., and R.E. Thomas. 1972. Ecological significance of vegetation to northern pike, *Esox lucius*, spawning. Trans. Am. Fish. Soc. 101: 560–563.

McCart, P.J. 1965. Growth and morphology of four British Columbia populations of pygmy whitefish (*Prosopium coulteri*). J. Fish. Res. Board Can. 22: 1229–1259.

McCart, P.J., and G.J. Mann. 1981. Comparisons of sympatric dwarf and normal populations of least cisco (*Coregonus sardinella*) inhabiting Trout Lake, Yukon Territory. Can. J. Fish. Aquat. Sci. 38: 240–244.

McCart, P.W., and N. Aspinwall. 1970. Spawning habits of the largescale sucker, *Catostomus macrocheilus*, in Stave Lake, British Columbia. J. Fish. Res. Board Can. 27: 1154–1158.

McCleave, J.D. 1964. Movement and population of the mottled sculpin (*Cottus bairdi* Girard) in a small Montana stream. Copeia, 1964: 506–513.

McComish, T.S. 1968. Sexual differentiation of bluegills by the urogenital opening. Prog. Fish-Cult. 30: 28.

McCormick, J.H., K.E.F. Hokanson, and B.R. Jones. 1972. Effects of temperature on growth and survival of young brook trout, *Salvelinus fontinalis*. J. Fish. Res. Board Can. 29: 1107–1112.

McCusker, M.R., E. Parkinson, and E.B. Taylor. 2000. Mitochondrial DNA variation in rainbow trout (*Oncorhynchus mykiss*) across its native range: testing biogeographical hypotheses and their relevance to conservation. Mol. Ecol. 9: 2089–2108.

McDonald, J. 1960. The behaviour of Pacific salmon fry during their downstream migration to freshwater and saltwater nursery areas. J. Fish. Res. Board Can. 17: 655–676.

McFadden, J.T., E.L. Cooper, and J.K. Andersen. 1965. Some effects of environment on egg production in brown trout (*Salmo trutta*). Limnol. Oceanogr. 10: 88–95.

McIntyre, J.D. 1969. Spawning behaviour of the brook lamprey, *Lampetra planeri*. J. Fish. Res. Board Can. 26: 3252–3254.

McKenzie, J.A. 1974. The parental behaviour of the male brook stickleback, *Culaea inconstans* (Kirtland). Can. J. Zool. 52: 649–652.

McKenzie, J.A., and M.H.A. Keenleyside. 1970. Reproductive behaviour of ninespine sticklebacks (*Pungitius pungitius* (L.)) in South Bay, Manitoulin. Ontario. Can. J. Zool. 55–61.

McKinney, T., and D.W. Speas. 2001. Observations of size-related asymmetries in diet and energy intake of rainbow trout in a regulated river. Environ. Biol. Fishes, 61: 435–444.

McLarney, W.O. 1968. Spawning habits and morphological variation in the coastrange sculpin, *Cottus aleuticus*, and the prickly sculpin, *Cottus asper*. Trans. Am. Fish. Soc. 97: 46–48.

McLarney, W.O., D.G. Engstrom, and J.H. Todd 1974. Effects of increasing temperature on social behavior in a group of yellow bullhead (*Ictalurus natalis*). Environ. Pollut. 7: 111–119.

McLaughlin, R.L., M.M. Ferguson, and D.L.G. Noakes. 1999. Adaptive peaks and alternative foraging tactics in brook charr: evidence of short-term divergent selection for sitting-and-waiting and actively feeding. Behav. Ecol. Sociobiol. 45: 386–395.

McLaughlin, R.L., J.W.A. Grant, and D.L.G. Noakes. 2000. Living with failure: the prey capture success of young brook charr in streams. Ecol. Freshw. Fish, 9: 81–91.

McLean, J.E., and E.B. Taylor. 2001. Resolution of population structure in a species with high gene flow: microsatellite variation in the eulachon (Osmeridae: *Thaleichthys pacificus*). Mar. Biol. 139: 411–420.

McLean, J.E., Hay, D.E., and E.B. Taylor. 1999. Marine population structure in an anadromous fish: life-history influences patterns of mitochondrial DNA variation in the eulachon, *Thaleichthys pacificus*. Mol. Ecol. 8(Suppl. 1): 143–158.

McLennan, D.A. 1993a. Changes in female breeding behaviour across the ovulatory cycle in the brook stickleback, *Culaea inconstans* (Kirtland). Behaviour, 126: 191–218.

McLennan, D.A. 1993b. Temporal changes in the structure of the male nuptial signal in the brook stickleback, *Culaea inconstans* (Kirtland). Can. J. Zool. 71: 1111–1119.

McLennan, D.A. 1994. Changes in female colour across the ovulatory cycle in the brook stickleback, *Culaea inconstans* (Kirtland). Can. J. Zool. 72: 144–153.

McLennan, D.A., and J.D. McPhail. 1989. Experimental investigations of the evolutionary significance of sexually dimorphic nuptial colouration in *Gasterosteus aculeatus* (L.): the relationship between male colour and male behaviour. Can. J. Zool. 67: 1767–1777.

McLennan, D.A., and J.D. McPhail. 1990. Experimental investigations of the evolutionary significance of sexually dimorphic nuptial colouration in *Gasterosteus aculeatus* (L.): the relationship between male colour and female behaviour. Can. J. Zool. 68: 482–492.

McLeod, C.L., and J.P. O'Neill. 1983. Major range extensions of anadromous salmonids and first record of chinook salmon in the Mackenzie River drainage. Can. J. Zool. 61: 2183–2184.

McLeod, C., J. O'Neil, and M. Psuka. 1978. McGregor River Diversion Project. Fisheries and benthic fauna. Vol. II. Inventory of aquatic resources. Report prepared by Renewable Resources Consulting Services Ltd. for BC Hydro and Power Authority, Vancouver.

McLeod, C., J. O'Neil, L. Hildebrand, and T. Clayton. 1979. An examination of fish migrations in the Liard River, British Columbia, relative to proposed hydroelectric development at Site A. R. L. & L. Environmental Services Ltd., Edmonton, AB. 199 pp.

McMahon, T.E., and D.H. Bennett. 1996. Walleye and northern pike: boost or bane to northwest fisheries? Fisheries, 21: 6–11.

McMillan, V. 1972. Mating of the fathead minnow. Nat. Hist. 81: 73–78.

McPhail, J.D. 1963. Geographic variation in North American ninespine sticklebacks, *Pungitius pungitius*. J. Fish. Res. Board Can. 20: 27–44.

McPhail, J.D. 1966. The *Coregonus autumnalis* complex in Alaska and northwestern Canada. J. Fish. Res. Board Can. 23: 141–148.

McPhail, J.D. 1967. Distribution of freshwater fishes in western Washington. Northwest Sci. 41: 1–11.

McPhail, J.D. 1994. Speciation and the evolution of reproductive isolation in the sticklebacks (*Gasterosteus*) of south-western British Columbia. pp. 399–437. *In* M.A. Bell and S.A. Foster (eds.). The evolutionary biology of the threespine stickleback. Oxford University Press, Oxford, UK.

McPhail, J.D. 2001. Report on the biology and taxonomic status of lake chub, *Couesius plumbeus*, populations inhabiting the Liard Hotsprings thermal complex. Report submitted to BC Ministry of the Environment, Lands, and Parks, Fort St. John.

McPhail, J.D., and J.S. Baxter. 1995. A review of bull trout (*Salvelinus confluentus*) life-history and habitat use in relation to compensation and improvement opportunities. Report prepared for the Habitat Management Division, Department of Fisheries and Oceans, Vancouver.

McPhail, J.D., and C.C. Lindsey. 1970. Freshwater fishes of northwestern Canada and Alaska. Bull. Fish. Res. Board Can. No. 173.

McPhail, J.D., and C.C. Lindsey. 1986. Zoogeography of the freshwater fishes of Cascadia (the Columbia system and rivers north to the Stikine). pp. 615–637. *In* C.H. Hocutt and E.O. Wiley (eds.). The zoogeography of North American freshwater fishes. John Wiley & Sons, New York.

McPhail, J.D., and C.B. Murray. 1980. The early life-history and ecology of Dolly Varden (*Salvelinus malma*) in the upper Arrow Lakes. Report to BC Hydro an Power Authority and the Kootenay Region Fish and Wildlife Branch, Nelson.

McPhail, J.D., and V.L. Paragamian. 2000. Burbot biology and life history. pp. 11–23. *In* V.L. Paragamian and D.W. Willis (eds.). Burbot biology, ecology, and management. Fisheries Management Section, American Fisheries Society, Spokane, WA. Publ. No. 1.

McPhail, J.D., and E.B. Taylor. 1995. Skagit char. Final report to the Skagit Environmental Endowment Commission, North Vancouver, BC. Project No. 94-1. 39 pp.

McPhail, J.D., and E.B. Taylor. 1999. Morphological and genetic variation in northwestern longnose suckers, *Catostomus catostomus*: the Salish sucker problem. Copeia, 1999: 884–893.

McPhail, J.D., and P.M. Troffe. 1998. The mountain whitefish (*Prosopium williamsoni*): a potential indicator species for the Fraser system. Technical Report prepared for the Fraser River Action Plan. Environment Canada, Vancouver, BC.

McPhail, J.D., and P.M. Troffe. 2001. The mountain whitefish (*Prosopium williamsoni*): a brief review of the distribution, biology and life history of a neglected species. pp. 17–21. *In* Proceedings of the Bull Trout II Conference. Trout Unlimited, Calgary, AB.

McPhail, J.D., and R J. Zemlak. 2001. Pygmy whitefish studies on Dina Lake #1, 2000. Peace/Williston Fish and Wildlife Comp. Prog. Rep. No. 245. 36 pp.

McPhail, J.D., D. O'Brien, and J. DeGisi. 1998a. Overview of the distribution and biology of the fishes in the Petitot River system, northeastern British Columbia. Report submitted to BC Environment, Fisheries Branch, Peace Subregion, Fort St. John. 73 pp.

McPhail, J.D., D. O'Brien, and J. DeGisi. 1998b. Overview of the distribution and biology of the fishes in the Hay River system, northeastern British Columbia. Report submitted to BC Environment, Fisheries Branch, Peace Subregion, Fort St. John. 73 pp.

Mecklenburg, C.W., T.A. Mecklenburg, and L.K. Thorsteinson. 2002. Fishes of Alaska. American Fisheries Society, Bethesda, MD.

Mednikov, B.M., E.A. Shubina, M.H. Melinikova, and K.A. Savvaitova. 1999. The genus status problem in Pacific salmons and trouts: a genetic systematics investigation. J. Ichthyol. 39: 10–17.

Meehan, W.R., and D.B. Siniff. 1962. A study of the downstream migrations of anadromous fishes in the Taku River, Alaska. Trans. Am. Fish. Soc. 91: 399–407.

Meeuwig, M.H., and J.M. Bayer. 2005. Morphology and aging precision of statoliths from larvae of Columbia River Basin lampreys. N. Am. J. Fish. Manage. 25: 38–48.

Meeuwig, M.H., J.M. Bayer, and J.G. Seelye. 2005. Effects of temperature on survival and development of early life stage Pacific and western brook lampreys. Trans. Am. Fish. Soc. 134: 19–27.

Mesa, M.G., and T.M. Olson. 1993. Prolonged swimming performance of the northern squawfish. Trans. Am. Fish. Soc. 122: 1104–1110.

Metcalf, A.L. 1966. Fishes of the Kansas River system in relation to the zoogeography of the Great Plains. University of Kansas, Lawrence. Publ. Mus. Nat. Hist. Univ. Kans. No. 17: 23–189.

Meyer, F.A. 1970. Development of some larval centrarchids. Prog. Fish-Cult. 32: 130–136.

Meyer, K.A. and J.S. Griffith. 1997. First winter survival of rainbow trout and brook trout in the Henrys Fork of the Snake River, Idaho. Can. J. Zool. 75: 59–63.

Meyer, K.A., and J.S. Gregory. 2000. Evidence of concealment behaviour in adult rainbow and brook trout in winter. Ecol. Freshw. Fish, 9: 138–144.

Michael, J.H. 1980. Repeat spawning of Pacific lamprey. Calif. Fish Game, 66: 186–187.

Michael, J.H. 1984. Additional notes on the repeat spawning by Pacific lamprey. Calif. Fish Game, 70: 186–187.

Miller, B.B., D.F. Palmer, W.D. McCoy, A.J. Smith, and M.L. Colburn. 1993. A pre-Illinoisan fossil assemblage from near Connersville, southeastern Indiana. Quat. Res. 40: 254–261.

Miller, H.C. 1963. The behavior of the pumpkinseed sunfish, *Lepomis gibbosus* (Linnaeus), with notes on the behavior of other *Lepomis* and pygmy sunfish, *Elassoma evergladei*. Behaviour, 22: 88–151.

Miller, J.M. 1974. The food of brook trout, *Salvelinus fontinalis* (Mitchill), fry from different subsections of Lawrence Creek, Wisconsin. Trans. Am. Fish. Soc 103: 130–134.

Miller, R.R. 1965. Quaternary freshwater fishes of western North America. pp. 569–581. *In* H.E. Wright and D.G. Frey (eds.). The Quaternary of the United States. Princeton University Press, Princeton, NJ.

Miller, R.R. 1984. *Rhinichthys deaconi*, a new species of dace (Pisces: Cyprinidae) from southern Nevada. University of Michigan, Ann Arbor. Occas. Pap. Mus. Zool. Univ. Mich. No. 707: 1–21.

Mills, D. 1989. Ecology and management of Atlantic salmon. Chapman & Hall, London.

Mitro, M.G., and A.V. Zale. 2002. Seasonal survival, movement, and habitat use of age-0 rainbow trout in the Henrys Fork of the Snake River, Idaho. Trans. Am. Fish. Soc. 131: 271–286.

Miura, T. 1962. Early life history and possible interaction of five inshore species of fish in Nicola Lake, British Columbia. Ph.D. thesis, Department of Zoology, University of British Columbia, Vancouver.

Moen, T. 1959. Sexing of channel catfish. Trans. Am. Fish. Soc. 88: 149.

Moodie, G.E.E. 1966. Some factors affecting the distribution and abundance of the chiselmouth (*Acrocheilus alutaceus*). M.Sc. thesis, Department of Zoology, University of British Columbia, Vancouver.

Moodie, G.E.E. 1977. Meristic variation, asymmetry, and aspects of habitat of *Culaea inconstans* (Kirtland), the brook stickleback, in Manitoba. Can. J. Zool. 52: 398–404.

Moodie, G.E.E. 1986. The population biology of *Culaea inconstans*, the brook stickleback, in a small prairie lake. Can. J. Zool. 64: 1709–1717.

Moodie, G.E.E., and C.C. Lindsey. 1972. Life-history of a unique cyprinid fish, the chiselmouth (*Acrocheilus alutaceus*), in British Columbia. Syesis, 5: 55–61.

Moon, D.N., S J. Fisher, and D.W. Willis. 1998. Goldeye recruitment and growth in two Missouri River backwaters. Proc. S.D. Acad. Sci. 77: 139–144.

Morán, P. A.M. Pendás, E. Beall, and E. Garcia-Vázquez. 1996. Genetic assessment of the reproductive success of Atlantic salmon precocious parr by means of VNTR loci. Heredity, 77: 655–660.

Morin, R., J.J. Dodson, and G. Power. 1982. Life history variations of anadromous cisco (*Coregonus artedii*), lake whitefish (*Coregonus clupeaformis*) and round whitefish (*Prosopium cylindraceum*) populations of eastern James–Hudson Bay. Can. J. Fish. Aquat. Sci. 39: 958–967.

Morris, D. 1958. The reproductive behaviour of the ten-spined stickleback (*Pygosteus pungitius* L.). Behav. Suppl. 6: 1–154.

Morrow, J.E. 1980. Fishes of Alaska. Alaska Northwest Publishing Co., Anchorage.

Morrow, J.V., G.L. Miller, and K.J. Killgore. 1997. Density, size, and foods of larval northern pike in natural and artificial wetlands. N. Am. J. Fish. Manage. 17: 210–214.

Morton, A., and J. Volpe. 2002. A description of escaped farmed Atlantic salmon *Salmo salar* captures and their characteristics in one Pacific salmon fishery area in British Columbia, Canada in 2000. Alaska Fish. Res. Bull. 9: 102–110.

Moulton, L.L. 1970. The 1970 longfin smelt spawning run in Lake Washington with notes on egg development and changes in the population since 1964. M.Sc. thesis, University of Washington, Seattle.

Moulton, L.L. 1974. Abundance, growth, and spawning of the longfin smelt in Lake Washington. Trans. Am. Fish. Soc. 103: 46–52.

Moulton, L.L., L.M. Philo, and J.C. George. 1997. Some reproductive characteristics of least ciscoes and humpback whitefish in Dease Inlet, Alaska. pp. 119–126. *In* J.B. Reynolds (ed.). Fish ecology in arctic North America. Am. Fish. Soc. Symp. No. 19.

Mousseau, T.A., N.C. Collins, and G. Cabana. 1988. A comparative study of sexual selection and reproductive investment in the slimy sculpin, *Cottus cognatus*. Oikos, 51: 156–162.

Moyle, P.B. 2002. Inland fishes of California. University of California Press, Berkeley.

Mraz, D. 1964. Age and growth of the round whitefish in Lake Michigan. Trans. Am. Fish. Soc. 93: 46–52.

Mueller, C.W. and H.J. Enzenhofer. 1991. Trawl catch statistics from seven sockeye rearing lakes of the Fraser River drainage basin: 1975–1985. Can. Data Rep. Fish. Aquat. Sci., No. 825. 131 pp.

Mueller, C.W., H.J. Enzenhofer, and J.M.B. Hume. 1991. Trawl catch statistics from seven sockeye salmon rearing lakes of the Fraser River drainage basin: 1986–1991. Can. Data Rep. Fish. Aquat. Sci., No. 864. 87 pp.

Mueller, G.A. 1984. Spawning by *Rhinichthys osculus* (Cyprinidae), in the San Francisco River, New Mexico. Southwest. Nat. 29: 354–356.

Muhlfeld, C.C., D.H. Bennett, and B. Marotzs. 2001a. Fall and winter habitat use and movement by Columbia River redband trout in a small stream in Montana. N. Am. J Fish. Manage. 21: 170–177.

Muhlfeld, C.C., D.H. Bennett, and B. Marotzs. 2001b. Summer and winter habitat use by Columbia River redband trout in the Kootenai River drainage in Montana. N. Am. J. Fish. Manage. 21: 223–235.

Muir, W.D., G.T. McCabe, M.J. Parsley, and S.A. Hinton. 2000. Diet of first-feeding larvae and young-of-the-year white sturgeon in the lower Columbia River. Northwest Sci. 74: 25–33.

Mullen, D.M., and T.M. Burton. 1998. Experimental tests of intraspecific competition in stream riffles between juvenile and adult longnose dace (*Rhinichthys cataractae*). Can. J. Zool. 76: 855–862.

Mundie, J.H. 1969. Ecological implications of the diet of juvenile coho in streams. pp. 135–152. *In* T.G. Northcote (ed.). Symposium on Salmon and Trout in Streams. H. R. MacMillan Lectures in Fisheries. Institute of Fisheries, University of British Columbia, Vancouver.

Murphy, M.L., J. Heifetz, J.F. Thedinga, S.W. Johnson, and K.V. Koski. 1989. Habitat utilization by juvenile Pacific salmon (*Oncorhynchus*) in the glacial Taku River, southeastern Alaska. Can. J. Fish. Aquat. Sci. 46: 1677–1685.

Murray, A.M., and M.V.H. Wilson. 1996. A new Palaeocene genus and species of percopsiform (Teleostei: Paracanthopterygii) from the Paskapoo formation, Smoky Tower, Alberta. Can. J. Earth Sci. 33: 429–438.

Murray, C.B. 1980a. Some effects of temperature on zygote and alevin survival, rate of development and size at hatching and emergence of Pacific salmon and rainbow trout. M.Sc. thesis, University of British Columbia, Vancouver.

Murray, C.B. 1980b. Temperature, development rates, and zygote survival in coastal cutthroat trout. Unpublished manuscript, Native Fish Research Group, University of British Columbia, Vancouver.

Murray, C.B., J.D. McPhail, and M.L. Rosenau. 1989. Reproductive and developmental biology of kokanee salmon from Upper Arrow Lake, British Columbia. Trans. Am. Fish Soc. 118: 503–509.

Muus, B.J., and P. Dahlstrom. 1971. Collins guide to the freshwater fishes of Britain and Europe. Collins, London.

Myers, G.S. 1932. A new whitefish, *Prosopium snyderi*, from Crescent Lake, Washington. Copeia, 1932: 62–64.

Myers, G.S. 1949. Usage of anadromous, catadromous and allied terms for migratory fishes. Copeia, 1949: 89–97.

Myers, J.M., R.G. Kope, G.J. Bryant, D. Teel, L.J. Lierheimer, T.C. Wainwright, W.S. Grant, F.W. Waknitz, K. Neely, S.T. Lindley, and R.S. Waples. 1998. Status review of chinook salmon from Washington, Idaho, Oregon, and California. US Department of Commerce, National Marine Fisheries Service, Washington, DC. NOAA Tech. Memo. NMFS-NWFSC-35. 443 pp.

Naesje, T., O.T. Sandlund, and B. Jonnson. 1986. Habitat use and growth of age-0 *Coregonus lavaretus* and *Coregonus albula*. Environ. Biol. Fishes, 15: 309–314.

Narver, D.W. 1967. Primary production in two small lakes of the northern Interior Plateau, British Columbia. J. Fish. Res. Board Can. 24: 2189–2193.

Narver, D.W., and F.C. Withler. 1974. Steelhead of the Nanaimo River—aspects of their biology and fishery from 3 years of angler's catches. Fish. Res. Board Can. Nanaimo Biol. Stn. Circ. No. 99. 25 pp.

Naud, M., and P. Magnan. 1988. Diel offshore migrations in northern redbelly dace, *Phoxinus eos* (Cope), in relation to prey distribution in a small oligotrophic lake. Can. J. Zool. 66: 1249–1253.

Neave, F., and G.C. Carl. 1940. The brown trout on Vancouver Island. pp. 341–343. *In* Proceedings of the Sixth Pacific Science Congress. Vol. 3. Pacific Science Association, Honolulu, HI.

Neilson, J.D., and C.E. Banford. 1983. Chinook salmon (*Oncorhynchus tshawytscha*) spawner characteristics in relation to redd physical features. Can. J. Zool. 61: 1524–1531.

Nelson, J.S. 1968. Hybridization and isolating mechanisms between *Catostomus commersonii* and *C. macrocheilus* (Pisces: Catostomidae). J. Fish. Res. Board Can. 25: 101–150.

Nelson, J.S. 1969. Geographical variation in the brook stickleback, *Culaea inconstans*, and notes on nomenclature and distribution. J. Fish. Res. Board Can. 26: 2431–2447.

Nelson, J.S. 1974. Hybridization between *Catostomus commersoni* (white sucker) and *Catostomus macrocheilus* (largescale sucker) in Williston Reservoir, British Columbia. Syesis, 7: 187–194.

Nelson, J.S. 1977. Evidence of a genetic basis for absence of the pelvic skeleton in brook stickleback, *Culaea inconstans*, and notes on the geographic distribution and origin of the loss. J. Fish. Res. Board Can. 34: 1314–1320.

Nelson, J.S., and F.M. Atton. 1971. Geographic and morphological variation in the presence and absence of the pelvic skeleton in the brook stickleback, *Culaea inconstans* (Kirtland), in Alberta and Saskatchewan. Can. J. Zool. 49: 343–352.

Nelson, J.S., and M.J. Paetz. 1992. The fishes of Alberta. University of Alberta Press, Edmonton.

Nelson, J.S., E.J. Crossman, H. Espinosa-Pérez, L.T. Findley, C.R. Gilbert, R.N. Lea, and J.D. Williams. 2004. Common and scientific names of fishes from the United States, Canada, and Mexico. 6th ed. Am. Fish. Soc. Spec. Publ. No. 29.

Nelson, R.E. 1977. Life history of the yellow perch (*Perca flavescens* (Mitchill)) in Lake Washington. M.Sc. thesis, University of Washington, Seattle.

Newcome, C.P., and G.F. Hartman. 1980. Visual signals in the spawning behaviour of rainbow trout. Can. J. Zool. 58: 1751–1757.

Newsome, G.E., and J. Tompkins. 1985. Yellow perch egg masses deter predators. Can. J. Zool. 63: 2882–2884.

Nicholas, J.W., and D.G. Hankin. 1988. Chinook salmon populations in Oregon coastal river basins: descriptions of life histories and assessment of recent trends in run strength. Oregon Department of Fish and Wildlife, Corvallis. Inf. Rep. No. 88-1. 359 pp.

Nielsen, J.L. 1992. Microhabitat-specific foraging behaviour, diet, and growth of juvenile coho salmon. Trans. Am. Fish. Soc. 121: 617–634.

Nielson, J.L., C.A. Gan, J.M. Wright, D.B. Morris, and W.K. Thomas. 1994. Biogeographic distributions of mitochondrial and nuclear markers for southern steelhead. Mol. Mar. Biol. Biotechnol. 3: 281–293.

Niemuth, W., W. Churchill, and T. Wirth. 1959. The walleye, its life history, ecology, and management. Wisconsin Conservation Department, Madison. Publ. No. 227. 14 pp.

Nikolski, G.V. 1961. Special ichthyology. [Translated from Russian by the Israel Program for Scientific Translation, Jerusalem.]

Nilsson, N.-A., and T.G. Northcote. 1981. Rainbow trout (*Salmo gairdneri*) and cutthroat trout (*S. clarki*) interactions in coastal British Columbia lakes. Can. J. Fish. Aquat. Sci. 28: 1228–1246.

Normandeau, D.A. 1969. Life history and ecology of the round whitefish, *Prosopium cylindraceum* (Pallas), of Newfound Lake, Bristol, New Hampshire. Trans. Am. Fish. Soc. 98: 7–13.

Norris, H.J. 1984. A comparison of aging techniques and growth of yellow perch (*Perca flavescens*, Mitchill) from selected Alberta lakes. M.Sc. thesis, University of Alberta, Edmonton.

North, J.A., R.A. Farr, and P. Vescei. 2002. A comparison of the meristic and morphometric characteristics of green sturgeon, *Acipenser medirostris*. J. Appl. Ichthyol. 18: 234–239.

Northcote, T.G. 1954. Observations on the comparative ecology of two species of fish,

Cottus asper and *Cottus rhotheus*, in British Columbia. Copeia, 1954: 25–28.

Northcote, T.G. 1964. Occurrence and distribution of seawater in Sakinaw Lake, British Columbia. J. Fish. Res. Board Can. 21: 1321–1324.

Northcote, T.G. 1995. Comparative biology and management of Arctic and European grayling (Salmonidae, *Thymallus*). Rev. Fish Biol. Fish. 5: 141–194.

Northcote, T.G. 1997. Why sea-run? An exploration into the migratory/residency spectrum of coastal cutthroat trout. pp 20–26. *In* J.D. Hall, P.A. Bisson, and R.E. Gresswell (eds.). Sea-run cutthroat trout: biology, management, and future conservation. Oregon Chapter, American Fisheries Society, Corvallis.

Northcote, T.G., and G.L. Ennis. 1998. Mountain whitefish biology and habitat use in relation to compensation and improvement possibilities. Rev. Fish. Sci. 2: 347–371.

Northcote, T.G., and G.F. Hartman. 1959. A case of "schooling" behaviour in the prickly sculpin, *Cottus asper* Richardson. Copeia, 1959: 156–158.

Northcote, T.G., and G F. Hartman. 1988. The biology and significance of stream trout populations (*Salmo* spp.) living above and below waterfalls. Pol. Arch. Hydrobiol. 35: 409–442.

Northcote, T.G., and H.W. Lorz. 1966. Seasonal and diel changes in the food of adult kokanee (*Oncorhynchus nerka*) in Nicola Lake. British Columbia. J. Fish. Res. Board Can. 23: 1259–1263.

Northcote, T.G., H.W. Lorz, and J.C. MacLeod. 1964a. Studies of diel vertical movement of fishes in a British Columbia lake. Verh. Int. Verein. Theor. Agnew. Limnol. 15: 940–966.

Northcote, T.G., M,S. Wilson, and D.R. Hurn. 1964b. Some characteristics of Nitinat Lake, an inlet on Vancouver Island, British Columbia. J. Fish. Res. Board Can. 21: 1069–1081.

Northcote, T.G., T.G. Halsey, and S.J. MacDonald. 1972. Fish as indicators of water quality in the Okanagan basin lakes, British Columbia. Canada–B.C. Okanagan Basin Agreement. Prelim. Rep. No. 22. 78 pp.

Northcote, T.G., A.E. Peden, and T.E. Reimchen. 1989. Fishes of the coastal marine, riverine and lacustrine waters of the Queen Charlotte Islands. pp. 147–174. *In* G.G.E. Scudder and N. Gressit (eds.). The outer shores. Queen Charlotte Islands Museum Press, Skidegate, BC.

Nowak, G.M., R.A. Tabor, E.J. Warner, K.L. Fresh, and T.P. Quinn. 2004. Ontogenetic shifts in habitat and diet of cutthroat trout in Lake Washington, Washington. N. Am. J. Fish. Manage. 24: 624–635.

Nyman, O.L. 1970. Electrophoretic analysis of hybrids between salmon (*Salmo salar* L.) and brown trout (*Salmo trutta* L.). Trans. Am. Fish. Soc. 99: 229–236.

O'Connor, J.F., and G. Power. 1976. Production of brook trout (*Salvelinus fontinalis*) in four streams in the Matamek watershed, Quebec. J. Fish. Res. Board Can. 33: 6–18.

Oakey, D.D., M.E. Douglas, and M.R. Douglas. 2004. Small fish in a large landscape: diversification of *Rhinichthys osculus* (Cyprinidae) in western North America. Copeia, 2004. 207–221.

Oakley, T.H., and R.B. Phillips. 1999. Phylogeny of Salmonine fishes based on growth hormone introns: Atlantic salmon (*Salmo*) and Pacific (*Oncorhynchus*) salmon are not sister taxa. Mol. Phylogenet. Evol. 11: 381–393.

Ogilvie, R.T. 1989. Disjunct vascular flora of Northwestern Vancouver Island in relation to Queen Charlotte Islands' endemism and Pacific Coast refugia. pp. 127–130. *In* G.G.E. Scudder and N. Gressler (eds.). The outer shores. Queen Charlotte Islands Museum Press, Skidegate, BC.

Olney, F.E. 1975. Life history and ecology of the northern squawfish, *Ptychocheilus oregonensis* (Richardson) in Lake Washington. M.Sc. thesis, School of Fisheries, University of Washington, Seattle.

Olsen, D.E., and W.J. Scidmore. 1963. Homing tendency of spawning white suckers in Many Point Lake, Minnesota. Trans. Am. Fish Soc. 92: 13–16.

Olson, N.W., C.P. Paukert, D.W. Willis, and J.A. Klammer. 2003. Prey selection and diets of bluegill, *Lepomis macrochirus*, with differing population characteristics in two Nebraska natural lakes. Fish. Manage. Ecol. 10: 31–40.

Olund, L.J., and F.B. Cross. 1961. Geographic variation in the North American cyprinid fish, *Hybopsis gracilis*. University of Kansas, Lawrence. Univ. Kans. Publ. Mus. Nat. Hist. 13: 323–348.

Olver, C.H. 1991. Lake trout, *Salvelinus namaycush*. pp. 286–299. *In* J. Stolz and J. Schnell (eds.). Trout. Stackpole Books, Harrisburg, PA.

Orti, G., M.A. Bell, T.E. Reimchen, and A. Meyer. 1994. Global survey of mitochondrial DNA sequences in the threespine stickleback: evidence of recent migrations. Evolution, 48: 608–622.

Osenberg, C.W., G.H. Mittlebach, and P.C. Wainwright. 1992. Two-stage life histories in fish: the interaction between juvenile competition and adult performance. Ecology, 73: 255–267.

Osinov, A.G., and V.S. Lebedev. 2002. Genetic divergence and the phylogeny of the Salmoninae based on allozyme data. J. Fish Biol. 57: 354–381.

Otto, R.G., and J.O. Rice. 1977. Responses of freshwater sculpin (*Cottus cognatus gracilis*) to temperature. Trans. Am. Fish. Soc. 106: 89–94.

Owens, R., and G.E. Noguchi. 1998. Intra-lake variation in maturity, fecundity, and spawning of slimy sculpins (*Cottus cognatus*) in southern Lake Ontario. J. Great Lakes Res. 24: 383–391.

Packham, M.J. 1988. Attributes of male and female yellow perch in Mink Lake, Alberta: external differences, maturation, survival, abundance, and productivity. M.Sc. thesis, University of Alberta, Edmonton.

Page, L.M., and B.M. Burr. 1991. A field guide to freshwater fishes of North America north of Mexico. The Peterson Field Guide Series, Houghton Mifflin Co., Boston.

Paragamian, V.L. 1989. Seasonal habitat use by walleye in a warm water river system, as determined by radiotelemetry. N. Am. J. Fish. Manage. 9: 392–401.

Paragamian, V.L., and G. Krause. 2001. Kootenai River white sturgeon spawning and migration behavior and a predictive model. N. Am. J. Fish. Manage. 21: 10–21.

Paragamian, V.L., and D.W. Willis. 2000. Burbot biology, ecology, and management. Fisheries Management Section, American Fisheries Society, Spokane, WA. Publ. No. 1.

Paragamian, V.L., G. Krause, and V. Wakkinen. 2001. Spawning habitat of Kootenai River white sturgeon, post-Libby Dam. N. Am. J. Fish. Manage. 21: 22–33.

Parente, W.D., and G.R. Snyder. 1970. A pictorial record of the hatching and early development of the eulachon (*Thaleichthys pacificus*). Northwest Sci. 44: 50–57.

Parker, R.M., M.P. Zimmerman, and D.L. Ward. 1995. Variability in biological characteristics of northern squawfish in the lower Columbia and Snake rivers. Trans. Am. Fish. Soc. 124: 335–346.

Parkham, H.J. 1937. A nature lover in British Columbia. H.F. & G. Witherby, London.

Parkinson, E.A., R.J. Behnke, and W. Pollard. 1984. A morphological and electrophoretic comparison of rainbow trout (*Salmo gairdneri*) above and below barriers on five streams on Vancouver Island, British Columbia. Fish. Manage. Rep. No. 83. 16 pp.

Parrish, D.L., and F.J. Margraf. 1994. Spatial and temporal patterns of food use by white perch and yellow perch in Lake Erie. J. Freshw. Ecol. 9: 29–35.

Parsley, M.J., and K.M. Kappenman. 2000. White sturgeon spawning areas in the lower Snake River. Northwest Sci. 74:192–201.

Parsons, B.G.M., and W.A. Hubert. 1988. Influence of habitat availability on spawning site selection by kokanee in streams. N. Am. J. Fish. Manage. 8: 426–431.

Patten, B.G. 1962. Cottid predation upon salmon fry in a Washington stream. Trans. Am. Fish. Soc. 91: 427–429.

Patten, B.G. 1971. Spawning and fecundity of seven species of northwest American species of *Cottus*. Am. Midl. Nat. 85: 493–506.

Paukert, C.P., and D.W. Willis. 2002. Seasonal and diel habitat selection by bluegills in a shallow natural lake. Trans. Am. Fish Soc. 131: 1131–1139. .

Pazkowski, C.A., and W.M. Tonn. 1994. Effects of prey size, abundance, and population structure on piscivory by yellow perch. Trans. Am. Fish. Soc. 123: 855–865.

Peake, S., R.S, McKinley, and D.A. Scruton. 2000. Swimming performance of walleye (*Stizostedion vitreum*). Can. J. Zool. 78: 1686–1690.

Pearcy, W.G., R.D. Brodeur, and J.P. Fisher. 1990. Distribution and biology of juvenile cutthroat trout, *Oncorhynchus clarki clarki* and steelhead, *O. mykiss*, in coastal waters off Oregon and Washington. Fish. Bull. 88: 697–711.

Pearson, M.P., and M.C. Healey. 2003. Life-history characteristics of the endangered Salish sucker (*Catostomus* sp.) and their implications for management. Copeia, 2003: 758–768.

Peck, J.W. 1981. Dispersal of lake trout fry from an artificial spawning reef in Lake Superior. Minnesota Department of Natural Resources, St. Paul. Fish. Res. Rep. No. 1892. 13 pp.

Peden, A.E. 1991. Status of the leopard dace, *Rhinichthys falcatus*, in Canada. Can. Field-Nat. 105: 179–188.

Peden, A.E. 1994. Updated status report on Canadian populations of speckled dace, *Rhinichthys osculus*. Unpublished report submitted to COSEWIC Subcommittee on Fish and Marine Mammals. Committee on the Status of Endangered Wildlife in Canada (COSEWIC), Ottawa, ON.

Peden, A.E., and G.W. Hughes. 1981. Life history notes relevant to the Canadian status of the speckled dace (*Rhinichthys osculus*). Syesis, 14: 21–31.

Peden, A.E., and G.W. Hughes. 1984. Status of the speckled dace, *Rhinichthys osculus*, in Canada. Can. Field-Nat. 98: 98–103.

Peden, A.E., and G.W. Hughes. 1988. Sympatry in four species of *Rhinichthys* (Pisces), including the first documented occurrences of *R. umatilla* in the Canadian drainages of the Columbia River. Can. J. Zool. 66: 1846–1856.

Peden A.E., and S. Orchard. 1993. Vulnerable dace populations of the Similkameen River. Interim report submitted to the British Columbia Habitat Conservation Trust Fund, Victoria. 27 pp.

Peden, A.E., G.W. Hughes, and W.E. Roberts. 1989. Morphologically distinct populations of the shorthead sculpin, *Cottus confusus*, and mottled sculpin, *Cottus bairdii* (Pisces, Cottidae), near the western border of Canada and the United States. Can. J. Zool. 67: 2711–2720.

Pedersen, R.V.K., U.N. Orr, and D.E. Hay. 1995. Distribution and preliminary stock assessment (1993) of the eulachon, *Thaleichthys pacificus*, in the lower Kitimat River, British Columbia. Can. MS. Rep. Fish. Aquat. Sci. No. 2330. 23 pp.

Percy, R. 1975. Fishes of the outer Mackenzie Delta. Environment Canada, Beaufort Sea Project, Winnipeg. Tech. Rep. No. 8.

Pereira, D.L., and G.W. LaBar. 1983. Age and growth of trout-perch in Lake Champlain. N.Y. Fish Game J. 30: 201–209.

Perelygin, A.A. 1992. Genetic variability of proteins in the populations of vendace (*Coregonus albula*) and least cisco (*Coregonus sardinella*). Nord. J. Freshwater Res. 67: 99–100.

Perrin, C.J., A. Heaton, and M.A. Laynes. 1999. White sturgeon (*Acipenser transmontanus*) spawning habitat in the lower Fraser River. Report prepared by Limnotek Research and Development Inc. for the BC Ministry of Fisheries, Victoria. 53 pp.

Perrin, C.J., L.L. Rempel, and M.L. Rosenau. 2003. White sturgeon spawning habitat in an unregulated river: Fraser River, Canada. Trans. Am. Fish. Soc. 132: 154–165.

Peters, E.J., and R.S. Holland. 1994. Biological and economic analyses of the fish communities in the Platte River: modifications and tests of habitat suitability criteria for fishes in the Platte River. Completion Report. University of Nebraska, Lincoln. Fed. Aid Fish Restor. Proj. No. F-78-R, Study III: Job III-2.

Peterson, N.P. 1982. Immigration of juvenile coho salmon (*Oncorhynchus kisutch*) into riverine ponds. Can. J. Fish. Aquat. Sci. 39: 1308–1310.

Petrosky, C.E., and T.F. Waters. 1975. Annual production by the slimy sculpin population in a small Minnesota trout stream. Trans. Am. Fish. Soc. 104: 237–244.

Peven, C.M., R.R. Whitney, and K.R. Williams. 1994. Age and length of steelhead smolts from the mid-Columbia Basin, Washington. N. Am. J. Fish. Manage. 14: 77–86.

Phillips, J.M., J.R. Jackson, and R.L. Noble. 1995. Hatching date influence on age-specific diet and growth of age-0 largemouth bass. Trans. Am. Fish Soc. 124: 370–379.

Phillips, R.B., L.I. Gudex, K.M. Westrich, and A.L. DeCicco. 1999. Combined phylogenetic analysis of ribosomal ITS1 sequences and new chromosome data supports three subgroups of Dolly Varden char (*Salvelinus malma*). Can. J. Fish. Aquat. Sci. 56: 1504–1511.

Pietschmann, V. 1934. *Lota hulai*, eine neue fischart aus dem Wiener Becken. Paleontol. Z. 16: 48–52.

Pitlo, J. 1989. Walleye spawning habitat in Pool 13 of the upper Mississippi River. N. Am. J. Fish. Manage. 9: 303–308.

Pletcher, F.T. 1963. The life history and distribution of lampreys in the Salmon and certain other rivers in British Columbia, Canada. M.Sc. thesis, Department of Zoology, University of British Columbia, Vancouver.

Politov, D.V., N.Y. Gordon, K.I. Atanasiev, Y.P. Altukov, and J.W. Bickham. 2000. Identification of Palearctic coregonid fish using mtDNA and allozyme genetic markers. J. Fish Biol. 57: 51–71.

Pollard W.R., G.F. Hartman, C. Groot, and P. Edgell. 1997. Field identification of coastal juvenile salmonids. Harbour Publishing, Maderia Park, BC.

Polyakov, O.A. 1989. Biology of Baikal omul, *Coregonus autumnalis migratorius*, in Bratsk Reservoir. J. Ichthyol. 29: 40–46.

Pope, K.L., W.T. Geraets, and D.W. Willis. 1996. Egg development in a high-density black crappie (*Pomoxis nigromaculatus*) population. J. Freshw. Ecol. 11: 451–458.

Porter, M., and J. Rosenfeld. 1999. Microhabitat selection and partitioning by an assemblage of fish in the Nazko River. BC Ministry of Fisheries, Victoria. Fish. Proj. Rep. No. RD77. 28 pp.

Post, J.R., and D.J. McQueen. 1988. Ontogenetic changes in the distribution of larval and juvenile yellow perch (*Perca flavescens*): a response to prey or predators. Can. J. Fish. Aquat. Sci. 45: 1820–1826.

Post, J.R., and D.J. McQueen. 1994. Variability in first year growth of yellow perch (*Perca flavescens*): predictions from a simple model, observations, and an experiment. Can. J. Fish. Aquat. Sci. 51: 2501–2512.

Post, J.R., and E.A. Parkinson. 2001. Energy allocation strategy in young fish: allometry and survival. Ecology, 82: 1040–1051.

Post, J.R., L.G. Rudstam, and D.M. Schael. 1994. Temporal and spatial distribution of pelagic age-0 fish in Lake Mendota, Wisconsin. Trans. Am. Fish Soc. 124: 84–93.

Potter, I.C. 1980. The Petromyzontiformes with particular reference to paired species. Can. J. Fish. Aquat. Sci. 37: 1595–1615.

Power, G. 1980. The brook charr, *Salvelinus fontinalis*. pp. 141–203. *In* E.K. Balon (ed.). Charrs, salmonid fishes of the genus *Salvelinus*. W. Junk, The Hague.

Powles, P.M., H.R. MacCrimmon, and D.A. Macrae. 1983. Seasonal feeding of carp, *Cyprinus carpio*, in the Bay of Quinte watershed, Ontario. Can. Field-Nat. 97: 293–298.

Powles, P.M., S. Finucan, M. Haaften, and R.A. Curry. 1992. Preliminary evidence for fractional spawning by the northern redbelly dace, *Phoxinus eos*. Can. Field-Nat. 106: 237–240.

Pratt, T.C., and M.G. Fox. 2001. Biotic influences on habitat selection by young-of-year walleye

(*Stizostedion vitreum*) in the demersal stage. Can. J. Fish. Aquat. Sci. 58: 1058–1069.

Preble, E.A. 1908. Fishes of the Athabasca–Mackenzie Region. U.S. Biological Survey, Washington, DC. N. Am. Fauna, 27: 502–515.

Price, C.J., W.M. Tonn, and C.A. Paszkowski. 1991. Intraspecific patterns of resource use by fathead minnows in a small boreal lake. Can. J. Zool. 69: 2109–2115.

Prince, A. 1991. A natural history of Columbia River fisheries in British Columbia. Report. to the Columbia–Kootenay Fishery Renewal Partnership. Westslope Fisheries, Cranbrook, BC. 50 pp.

Pritchard, A.L. 1930. Spawning habits and fry of the cisco (*Leucichthys artedi*) in Lake Ontario. Contrib. Can. Biol. Fish. Stud. Biol. Stn. Can. 6: 227–240.

Prophet, C.W., Brungardt, T.B., and N.K. Prophet. 1989. Diel behavior and seasonal distribution of walleye, *Stizostedion vitreum* (Mitchill), in Marion Reservoir, based on ultrasonic telemetry. J. Freshw. Ecol. 5: 177–185.

Puckett, K.J. and L.M. Dill. 1985. The energetics of feeding territoriality in juvenile coho salmon (*Oncorhynchus kisutch*). Behaviour, 42: 97–111.

Pulliainen, E., K. Korhonen, L. Kankaaranta, and K. Maeki. 1992. Non-spawning burbot on the northern coast of the Bothnia Bay. Ambio, 21: 170–175.

Quadri, S.U. 1959. Some morphological differences between the subspecies of cutthroat trout, *Salmo clarkii clarkii* and *Salmo clarkii lewisi*, in British Columbia. J. Fish. Res. Board Can. 16: 903–922.

Quadri, S.U. 1967. Morphological comparison of three populations of the lake char, *Cristivomer namaycush*, from Ontario and Manitoba. J. Fish. Res. Board Can. 24: 1407–1411.

Quadri, S.U. 1968. Growth and reproduction of the lake whitefish, *Coregonus clupeaformis*, in Lac la Ronge, Saskatchewan. J. Fish. Res. Board Can. 25: 2091–2100.

Quinn, T.P. 1980. Evidence for celestial and magnetic compass orientation in lake migrating sockeye salmon fry. J. Comp. Physiol. 137: 243–248.

Quinn, T.P. 1999. Variation in Pacific salmon reproductive behaviour associated with species, sex and levels of competition. Behaviour, 136: 179–204.

Quinn, T.P. 2005. The behavior and ecology of Pacific salmon and trout. University of Washington Press, Seattle.

Quinn, T.P., and D.J. Adams. 1996. Environmental changes affecting the migratory timing of American shad and sockeye salmon. Ecology, 77: 1151–1162.

Quinn, T.P., and J.T. Light. 1989. Occurrence of threespine stickleback (*Gasterosteus aculeatus*) in the open North Pacific Ocean: migration or drift. Can. J. Zool. 67: 2850–2852.

R. L. & L. Environmental Services Ltd. 1993. Columbia River development—Lower Columbia fisheries inventory. 1993 Study. R. L. & L. Environmental Services Ltd., Edmonton, AB. Data Rep. No.381D.

R. L. & L. Environmental Services Ltd. 1995a. Shallow-water habitat use by dace spp. and sculpin spp. in the lower Columbia River Basin Development Area. Report for BC Hydro, Environmental Resources Division, Vancouver. R. L. & L. Environmental Services Ltd., Edmonton, AB.

R. L. & L. Environmental Services Ltd. 1995b. White sturgeon in the Columbia River, B.C., 1994 study results. Report prepared for BC Hydro Environmental Affairs, Vancouver. R. L. & L. Environmental Services Ltd., Edmonton, AB. R. L. & L. Rep. No. 377D. 77 pp.

R. L. & L. Environmental Services Ltd. 1998. Columbia River white sturgeon spawning studies, 1998 investigations. Report prepared for Cominco Ltd., Trail Operations. R. L. & L. Environmental Services Ltd., Edmonton, AB. R. L. & L. Rep. No. 614F. 18 pp.

R. L. & L. Environmental Services Ltd. 2001. Columbia River white sturgeon spawning studies, 2000 investigations. Data Report prepared for BC Ministry of Environment, Lands, and Parks. Nelson, B.C. R. L. &

Environmental Services Ltd., Edmonton, AB. R. L. & L. Rep. No. 853. 24 pp.

R2 Resource Consultants. 1995. Pygmy whitefish status report. Chapter 6. pp. 1–37. In Draft report: upper Cedar River watershed fisheries study. Seattle Water Department, Seattle, WA.

Radtke, L. 1966. Distribution of smelt, juvenile sturgeon and starry flounder in the Sacramento – San Joaquin Delta. pp. 115–119. In S.L. Turner and D.W. Kelly (eds.). Ecological studies or the Sacramento – San Joaquin Delta, Part II. Calif. Dep. Fish Game Fish Bull. No. 136.

Raffetto, N.S., J.R. Baylis, and S.L. Serns. 1990. Complete estimates of reproductive success in a closed population of smallmouth bass (*Micropterus dolomieui*). Ecology, 71: 1523–1535.

Raleigh, R.F. 1967. Genetic control in lakeward migrations of sockeye salmon (*Oncorhynchus nerka*)) fry. J. Fish. Res. Board Can. 24: 2613–2622.

Raleigh, R.F. 1982. Habitat suitability index models: brook trout. US Department of the Interior, Fish and Wildlife Service, Washington, DC. FWS/OBS-82/10.24. 42 pp.

Raleigh, R.F, T. Hickman, R.C. Solomon, and P.C. Nelson. 1984. Habitat suitability information: rainbow trout. US Department of the Interior, Fish and Wildlife Service, Washington, DC. FWS/OBS-82/10.60, 64 pp.

Rankin, L. 1999. Phylogenetic and ecological relationships between "giant" pygmy whitefish (*Prosopium* spp.) and pygmy whitefish (*Prosopium coulteri*) in north-central British Columbia. M.Sc. thesis, Department of Biology, University of Northern British Columbia, Prince George.

Ranta, E.S., and V. Nuutinen. 1984. Zooplankton predation by rock-pool fish (*Tinca tinca* L. and *Pungitius pungitius* L.): an experimental study. Ann. Zool. Fenn. 21: 441–449.

Ranta, E., S-K. Juvonen, and N. Peuhkuri. 1992. Further evidence for size-assortative schooling in sticklebacks. J. Fish Biol. 41: 627–630.

Rasmussen, C., C.O. Osterberg, D.R. Clifton, and R.J. Rodriguez. 2003. Identification of a genetic marker that discriminates ocean-type and stream-type chinook salmon in the Columbia River basin. Trans. Am. Fish. Soc. 132: 131–142.

Ratynski, R.A., and B.G.E. deMarch. 1989. *Coregonus nasus* larvae. Can. Tech. Rep. Fish. Aquat. Sci. No. 1670.

Read, R.B. 1993. Geology of northeast Taseko Lakes map area, southwestern British Columbia. Geol. Surv. Can. Rep. 93-1A: 159–166.

Reckahn, J.A. 1970. Ecology of young lake whitefish in South Bay, Manitoulin Island, Lake Huron. pp. 437–460. In C.C. Lindsey and C.S. Woods (eds.). Biology of coregonid fishes. University of Manitoba Press, Winnipeg.

Reddin, D.G. 1988. Ocean life of Atlantic salmon (*Salmo salar* L.) in the northwest Atlantic. pp. 483–511. In D. Mills and D. Piggins (eds.). Atlantic salmon: planning for the future. Croom Helm, London.

Redenbach, Z.R. 2000. Patterns of hybridization between Dolly Varden (*Salvelinus malma*) and bull trout (*S. confluentus*) in nature. M.Sc. thesis, Department of Zoology, University of British Columbia, Vancouver.

Redenbach, Z., and E.B. Taylor. 1999. Zoogeographical implications of variation in mitochondrial DNA of Arctic grayling (*Thymallus arcticus*). Mol. Ecol. 8: 23–35.

Redenbach, Z.R., and E.B. Taylor. 2002. Evidence for historical introgression along a contact zone between two species of char (Pisces: Salmonidae) in northwestern North America. Evolution, 56: 1021–1035.

Reebs, S.G., L. Boudreau, P. Hardie, and R.A. Cunjak. 1995. Diel activity patterns of lake chub and other fishes in a temperate stream. Can. J. Zool. 73: 1221–1227.

Reed, R.J. 1964. Life history and migration patterns of Arctic grayling. Alaska Department of Fish and Game, Juneau. Res. Rep. No. 2. 30 pp.

Reeves, B.O.K. 1973. The nature and the age of the contact between the Laurentide and Cordilleran ice sheets in the western interior of North America. Arct. Alp. Res. 5: 1–16.

Reimchen, T.E. 1989. Loss of nuptial colour in threespine sticklebacks (*Gasterosteus aculeatus*). Evolution, 43: 450–460.

Reimchen, T.E. 1992. Extended longevity in a large-bodied *Gasterosteus* population. Can. Field-Nat. 106: 122–125.

Reimers, P.E. 1971. The length of residence of juvenile fall chinook in Sixes River, Oregon. Ph.D. thesis, Oregon State University, Corvallis.

Reisman, H.M., and T.J. Cade. 1967. Physiological and behavioral aspects of reproduction in the brook stickleback, *Culaea inconstans*. Am. Midl. Nat. 77: 257–295.

Reist, J.D. 1981. Variation in frequencies of pelvic phenotypes of the brook stickleback, *Culaea inconstans*, in Redwater drainage, Alberta. Can. Field-Nat. 95: 178–182.

Reist, J.D. 1997. Stock structure and life history types of broad whitefish in the lower Mackenzie River Delta—a summary of research. pp. 85–93. *In* R.F. Tallman and J.D. Reist (eds.). The proceedings of the broad whitefish workshop: the biology, traditional knowledge and scientific management of broad whitefish (*Coregonus nasus* (Pallas)) in the lower Mackenzie River. Can. Tech. Rep. Fish. Aquat. Sci. No. 2193.

Reist, J.D., J. Vuorinen, and R.A. Bodaly. 1992. Genetic and morphological identification of coregonids hybrid fishes from the Arctic Canada. Pol. Arch. Hydrobiol. 39: 551–561.

Remple, L.L., and D.G. Smith. 1998. Postglacial fish dispersal from the Mississippi refuge to the Mackenzie River Basin. Can. J. Fish. Aquat. Sci. 55: 893–899.

Reynolds, W.W., and M.E. Casterlin. 1976. Activity rhythms and light intensity preferences of *Micropterus salmoides* and *M. dolomieui*. Trans. Am. Fish Soc. 105: 400–403.

Reynolds, W.W., and M.E. Casterlin. 1978. Ontogenetic changes in preferred temperature and diel activity of the yellow bullhead, *Ictalurus natalis*. Comp. Biochem. Physiol. 59A: 409–411.

Ricciardi, A., and J.B. Rasmussen. 1999. Extinction rates of North American freshwater fauna. Conserv. Biol. 13: 1220–1222.

Rice, S.D., R.E. Thomas, and A. Moles. 1994. Physiological and growth differences in three stocks of underyearling sockeye salmon (*Oncorhynchus nerka*) on early entry into the sea. Can. J. Fish. Aquat. Sci. 51: 974–980.

Rich, W.H. 1942. The salmon runs of the Columbia River in 1938. Fish. Bull. 50: 101–147.

Richards, J.E., R.J. Beamish, and F.W.H. Beamish. 1982. Descriptions and keys for ammocoetes of lampreys of British Columbia, Canada. Can. J. Fish. Aquat. Sci. 39: 1484–1495.

Ricker, W.E. 1941. The consumption of young sockeye salmon by predaceous fish. J. Fish. Res. Board Can. 5: 293–313.

Ricker, W.E. 1960. A population of dwarf coastrange sculpins, *Cottus aleuticus*. J. Fish. Res. Board Can. 17: 929–939.

Ricker, W.E., D.F. Manzer, and E.A. Neave. 1954. The Fraser River eulachon fishery, 1941–1953. Fish. Res. Board Can. MS. Rep. Pac. Biol. Stn. No. 583. 35 pp.

Ridgway, M.S. 1988. Developmental stage of offspring and brood defense in smallmouth bass (*Micropterus dolomieui*). Can. J. Zool. 66: 1722–1728.

Ridgway, M.S., and J D. McPhail. 1984. Ecology and evolution of sympatric sticklebacks (*Gasterosteus*): mate choice and reproductive isolation in the Enos Lake species pair. Can. J. Zool. 62: 1813–1818.

Ridgway, M.S., G.P. Goff, and M.H.A. Keenleyside. 1989. Courtship and spawning behaviour in smallmouth bass (*Micropterus dolomieui*). Am. Midl. Nat. 122: 209–213.

Ridgway, M.S., J.A. MacLean, and J.C. MacLeod. 1991a. Nest-site fidelity in a centrarchid fish, the smallmouth bass (*Micropterus dolomieui*). Can. J. Zool. 69: 3103–3105.

Ridgway, M.S., B.J. Schuter, and E.E. Post. 1991b. The relative influence of body size and territorial behaviour on nesting synchrony in male smallmouth bass, *Micropterus dolomieui*

(Pisces: Centrarchidae). J. Anim. Ecol. 60: 665–681.

Rieger, P.W., and R.C. Summerfelt. 1997. The influence of turbidity on larval walleye, *Stizostedion vitreum*, behaviour and development in tank culture. Aquaculture, 159: 19–32.

Rieman, B.E., and J.D. McIntyre. 1993. Demographic and habitat requirements for conservation of bull trout. USDA For. Serv. Gen. Tech. Rep. INT-302. 38 pp.

Rimmer, D.M., U. Paim, and R.L. Saunders. 1984. Changes in the selection of microhabitat by juvenile Atlantic salmon (*Salmo salar*) at the summer–autumn transition in a small river. Can. J. Fish. Aquat. Sci. 41: 469–475.

Ringstad, N.R. 1974. Food competition between freshwater sculpins (genus *Cottus*) and juvenile coho salmon (*Oncorhynchus kisutch*): an experimental and ecological study in a British Columbia coastal stream. Environment Canada, Fisheries Marine Service, Ottawa, ON. Tech. Rep. No. 457. 88 pp.

Ringstad, N.R., and D.W. Narver. 1973. Some aspects of the ecology of two species of sculpin (*Cottus*) in a west coast Vancouver Island stream. Environment Canada, Fisheries Marine Service, Ottawa, ON. MS Rep. No. 1267. 69 pp.

Robb, T., and M.V. Abrahams. 2003. Variation in tolerance to hypoxia in a predator and prey species: an ecological advantage of being small. J. Fish Biol. 62: 1067–1081.

Roberts, J.H., and G.D. Grossman. 2001. Reproductive characteristics of female longnose dace in the Coweeta Creek drainage, North Carolina. Ecol. Freshw. Fish, 10: 184–190.

Roberts, W.E. 1988. The sculpins of Alberta. Alberta Nat. 18: 121–127.

Robins, C.R., and R.R. Miller. 1957. Classification, variation, and distribution of the sculpins, genus *Cottus*, inhabiting Pacific slope waters in California and southern Oregon, with a key to the species. Calif. Fish Game, 43: 213–233.

Robins, C.R., R.M. Bailey, C.E. Bond, J.R. Brooker, E.A. Lachner, R.N. Lea, and W.B. Scott. 1991. List of common and scientific names of fishes from the united States and Canada. 5th ed. Am. Fish. Soc. Spec. Publ. No. 20.

Robinson, B.W., D.S. Wilson, A.S. Margosian, and P.T. Lotito. 1993. Ecological and morphological differentiation of pumpkinseed sunfish in lakes without bluegill sunfish. Evol. Ecol. 7: 451–464.

Robinson, B.W., D.S. Wilson, and A.S. Margosian. 2000. A pluralistic analysis of character release in pumpkinseed sunfish (*Lepomis gibbosus*). Ecology, 81: 2799–2812.

Robison, B.D. 1995. Genetic relationships between maternal lineages of sympatric anadromous and resident forms of *Oncorhynchus nerka* determined through PCR RFLP analysis of mtDNA. Master's thesis, University of Idaho, Boise.

Rogers, D.E. 1964. Some morphological and life history data on pygmy whitefish from Wood River. Res. Fish. 1963. College of Fisheries, University of Washington, Seattle. Contrib. No. 166: 24–25.

Rosenau, M.L., and J.D. McPhail. 1987. Inherited differences in agonistic behaviour between two populations of coho salmon. Trans. Am. Fish. Soc. 116: 646–654.

Rosenfeld, J. 1996. Fish distribution and habitat use in the Similkameen watershed. BC Ministry of Environment, Lands, and Parks, Victoria. Fish. Proj. Rep. No. 52. 40 pp.

Rosenfeld, J.R., B. Wicks, M. Porter, and P. van Dishoeck. 1998. Habitat use by chiselmouth (*Acrocheilus alutaceus*) in the Blackwater River. BC Ministry of Environment, Lands, and Parks, Victoria. Fish. Proj. Rep. No. RD75. 22 pp.

Rosenfeld, J.R., M. Porter, M. Pearson, B. Wicks, P. van Dishoeck, T. Patton, E. Parkinson, G. Haas, and D. McPhail. 2001. The influence of temperature and habitat on the distribution of chiselmouth, *Acrocheilus alutaceus*, in British Columbia. Environ. Biol. Fishes, 62: 403–413.

Rosenfeld, J.S., and S. Boss. 2001. Fitness consequences of habitat use for juvenile cutthroat trout: energetic costs and benefits in pools and riffles. Can. J. Fish. Aquat. Sci. 58: 585–593.

Rosenfeld, J.S., M. Porter, and E. Parkinson. 2000. Habitat factors affecting the abundance and distribution of juvenile cutthroat trout (*Oncorhynchus clarki*) and coho salmon (*Oncorhynchus kisutch*). Can. J. Fish. Aquat. Sci. 57: 766–774.

Rosenfeld, J.S., T. Leiter, G. Linder, and L. Rothman. 2005. Food abundance and fish density alters habitat selection, growth, and habitat suitability curves for juvenile coho salmon (*Oncorhynchus kisutch*). Can. J. Fish. Aquat. Sci. 62: 1691–1701.

Rosenfield, J.A., T. Todd, and R. Greil. 2000. Asymmetrical hybridization and introgression between pink salmon and chinook salmon in the Laurentian Great Lakes. Trans. Am. Fish. Soc. 129: 670–679.

Rounsefell, G.A. 1958. Anadromy in North American Salmonidae. Fish. Bull. 131: 171–185.

Roussel, J.M., and A. Bardonnet. 1999. Ontogeny of diel pattern of stream-margin habitat use by emerging brown trout, *Salmo trutta*, in experimental stream channels: influence of food and predator presence. Environ. Biol. Fishes, 56: 253–262.

Royce, W.F. 1951. Breeding habits of lake trout in New York. Fish. Bull. 52: 59–76.

Rubidge, E.M. 2003. Molecular analysis of hybridization between native westslope cutthroat trout (*Oncorhynchus clarki lewisi*) and introduced rainbow trout (*O. mykiss*) in southeastern British Columbia. M.Sc. thesis, Department of Zoology, University of British Columbia, Vancouver.

Rubidge, E.M., and E.B. Taylor. 2004. Hybrid zone structure and the potential role of selection in hybridizing populations of native westslope cutthroat trout (*Oncorhynchus clarki lewisi*) and introduced rainbow trout (*O. mykiss*). Mol. Ecol. 13: 3735–3749.

Rubidge, E.M., and E.B. Taylor. 2005. An analysis of spatial and environmental factors influencing hybridization between native westslope cutthroat trout (*Oncorhynchus clarki lewisi*) and introduced rainbow trout (*O. mykiss*) in the upper Kootenay River drainage, British Columbia. Conserv. Genet. 6: 369–384.

Rubidge, E.M., P. Corbett, and E.B. Taylor. 2001. A molecular analysis of hybridization between native westslope cutthroat trout and introduced rainbow trout in southeastern British Columbia, Canada. J. Fish Biol. 59 (Suppl. A): 42–54.

Ruggerone, G.T., T.P. Quinn, I.A. McGregor, and T.D. Wilkinson. 1990. Horizontal and vertical movements of adult steelhead trout. *Oncorhynchus mykiss*, in the Dean and Fisher channels, British Columbia. Can. J. Fish. Aquat. Sci. 47: 1963–1969.

Russell, J.E., F.W.H. Beamish, and R.J. Beamish. 1987. Lentic spawning by the Pacific lamprey, *Lampetra tridentata*. Can. J. Fish. Aquat. Sci. 44: 476–478.

Russell, L.R., K.R. Conlin, O.K. Johansen, and U. Orr. 1983. Chinook salmon studies in the Nechako River: 1980, 1981, 1982. Can. MS Rep. Fish. Aquat. Sci. No. 1728. 185 pp.

Ryder, R.A. 1977. Effects of ambient light variations on behaviour of yearling, subadult, and adult walleyes (*Stizostedion vitreum vitreum*). J. Fish. Res. Board Can. 34: 1481–1491.

Sabo, J.L, and G.B. Pauley. 1997. Competition between stream dwelling cutthroat trout (*Oncorhynchus clarkii*) and coho salmon (*Oncorhynchus kisutch*): effects of relative size and population origin. Can. J. Fish. Aquat. Sci. 54: 2609–2617.

Sajdak, S.L, and R.B. Phillips. 1997. Phylogenetic relationships among *Coregonus* species inferred from the DNA sequence of the first internal transcribed spacer (ITS1) of ribosomal DNA. Can. J. Fish. Aquat. Sci. 54: 1494–1503.

Salfert, I.G., and G.E.E. Moodie. 1985. Filial egg-cannibalism in the brook stickleback, *Culaea inconstans* (Kirtland). Behaviour, 93: 82–100.

Salo, E.O. 1991. Life history of chum salmon. pp. 231–309. *In* C. Groot and L. Margolis (eds.). Pacific salmon life histories. University of British Columbia Press, Vancouver.

Salo, E.O., and W.H. Bayliff. 1958. Artificial and natural production of silver salmon, *Oncorhynchus kisutch*, of Minter Creek,

Washington. Wash. Dep. Fish. Res. Bull. No. 4. 76 pp.

Sandercock, F.K. 1991. Life history of coho salmon (*Oncorhynchus kisutch*). pp. 396–445. *In* C. Groot and L. Margolis (eds.). Pacific salmon life histories. University of British Columbia Press, Vancouver.

Sandlund, O.T., T. Naesje, and B. Jonnson. 1992. Ontogenetic changes in habitat use by whitefish, *Coregonus lavaretus*. Environ. Biol. Fishes, 33: 341–349.

Sanford, C.P.J. 1990. the phylogenetic relationships of salmonoid fishes. Bull. Br. Mus. (Nat. Hist.) Zool. 56: 145–153.

Savage, T. 1963. Reproductive behaviour of the mottled sculpin, *Cottus bairdi* Girard. Copeia, 1963: 317–325.

Savitz, J., L.G. Bardygula, T. Harder, and K. Stuecheli. 1993. Diel and seasonal utilization of home ranges in a small lake by smallmouth bass (*Micropterus dolomieui*). Ecol. Freshw. Fish, 2: 31–39.

Sawicki, O., and D.G. Smith. 1992. Glacial Lake Invermere, upper Columbia River valley, British Columbia: a paleogeographic reconstruction. Can. J. Earth Sci. 29: 687–692.

Sayigh, L., and R. Morin. 1986. Summer diet and daily consumption of periphyton of the longnose sucker, *Catostomus catostomus*, in the lower Matamek River, Quebec. Nat. Can. 113: 361–368.

Scarsbrook, J.R., and J. McDonald. 1973. Purse seine catches of sockeye salmon (*Oncorhynchus nerka*) and other species of fish at Babine Lake, British Columbia. Fish. Res. Board Can. Tech. Rep. No. 390. 46 pp.

Scarsbrook, J.R., and J. McDonald. 1975. Purse seine catches of sockeye salmon (*Oncorhynchus nerka*) and other species of fish at Babine Lake, British Columbia. Fish. Res. Board Can. Tech. Rep. No. 515. 43 pp.

Schael, D.M., L.G. Rudstam, and J.R. Post. 1991. Gape limitation and prey selection in larval yellow perch (*Perca flavescens*), freshwater drum (*Aplodinotus grunniens*), and black crappie (*Pomoxis nigromaculatus*). Can. J. Fish. Aquat. Sci. 48: 1919–1925.

Scheurer, J.A., K.R. Bestgen, and K.D. Fausch. 2003. Resolving taxonomy and historic distribution for conservation of rare Great Plains fishes: *Hybognathus* (Teleostei: Cyprinidae) in eastern Colorado Basins. Copeia, 2003: 1–12.

Schlosser, I.J., M.R. Doeringsfeld, J.F. Elder, and L.F. Arzayus. 1998. Niche relationships of clonal and sexual fish in a heterogeneous landscape. Ecology, 79: 953–968.

Schmetterling, D.A. 2000. Redd characteristics of fluvial westslope cutthroat trout in four tributaries to the Blackfoot River, Montana. N. Am. J. Fish. Manage. 20: 776–783.

Schmetterling, D.A. 2001. Seasonal movements of fluvial westslope cutthroat trout in the Blackfoot River Drainage, Montana. N. Am. J. Fish. Manage. 21: 507–520.

Schmidt, R.E. 1983. Variation in barbels of *Rhinichthys cataractae* (Pisces: Cyprinidae) in southwestern New York with comments on phylogeny and functional morphology. J. Freshw. Ecol. 2: 239–246.

Schöffmann, J. 2000. The grayling species (Thymallinae) of three different catchment areas of Mongolia. J. Grayl. Soc. 2000 (Spring): 41–44.

Schofield, C.L., S.P. Gloss, B. Plonski, and R. Spateholts. 1989. Production and growth efficiency of brook trout (*Salvelinus fontinalis*) in two Adirondack mountain (New York) lakes following liming. Can. J. Fish. Aquat. Sci. 46: 333–341.

Schram, S.T. 2000. Seasonal movements and mortality estimates of burbot in Wisconsin waters of western Lake Superior. pp. 90–95. *In* V.L. Paragamian and D.W. Willis (eds.). Burbot biology, ecology, and management. Fisheries Management Section, American Fisheries Society, Spokane, WA. Publ. No. 1.

Schroeder, S.L. 1981. The role of sexual selection in determining overall mating patterns and mate choice in chum salmon. Ph.D. thesis, School of Fisheries, University of Washington, Seattle.

Schroeder, S.L. 1982. The influence of intrasexual competition on the distribution of chum salmon in an experimental stream. pp. 275–285. *In* E.L. Brannon and E.O. Salo (eds.). Proceedings of the salmon and trout migratory behavior symposium. School of Fisheries, University of Washington, Seattle.

Schultz, L.P. 1929. Description of a new type of mudminnow from western Washington with notes on related species. University of Washington, Seattle. Univ. Wash. Publ. Fish. No. 1: 73–82.

Schultz, L.P. 1930. The life history of *L. planeri* (Bloch) with a statistical analysis of the rate of growth in western Washington. University of Michigan, Ann Arbor. Occas. Pap. Mus. Zool. Univ. Mich. No. 221: 1–35.

Schultz, L.P. 1935. The spawning habits of the chub, *Mylocheilus caurinus*—a forage fish of some value. Trans. Am. Fish. Soc. 65: 143–147.

Schultz, L.P. 1936. Keys to the fishes of Washington, Oregon and closely adjoining regions. University of Washington, Seattle. Univ. Wash. Publ. Zool. No. 2: 103–228.

Schultz, L.P. 1941. Fishes of Glacier National Park Montana. U.S. Department of the Interior, National Park Service, Washington, DC. Conserv. Bull. No. 22. 42 pp.

Schultz, L.P., and W.M. Chapman. 1934. A new osmerid fish, *Spirinchus dilatus*, from Puget Sound. Annu. Mag. Nat. Hist. Ser. 10, 13: 67–78.

Schultz, L.P., and R.J. Thompson. 1936. *Catostomus syncheilus palouseanus*, a new subspecies of catostomid fish from the Palouse River (Columbia system). Proc. Biol. Soc. Wash. 49: 71–76.

Schutz, D.C., and T.G. Northcote. 1972. An experimental study of feeding behaviour and interaction of coastal cutthroat trout (*Salmo clarki clarki*) and Dolly Varden (*Salvelinus malma*). J. Fish. Res. Board Can. 29: 555–565.

Scott, W.B., and E.J. Crossman 1973. The freshwater fishes of Canada. Bull. Fish. Res. Board Can. No. 184. 966 pp.

Scott, W.B., and M.G. Scott. 1988. Atlantic fishes of Canada. Can. Bull. Fish. Aquat. Sci. No. 219. 731 pp.

Scudder, G.G.E. 1989. The adaptive significance of marginal populations: a general perspective. Can. Spec. Publ. Fish Aquat. Sci. No. 105: 180–185.

Seamons, T.R., P. Bentzen, and T.P. Quinn. 2004a. The mating system of steelhead, *Oncorhynchus mykiss*, inferred by molecular analysis of parents and progeny. Environ. Biol. Fishes, 69: 333–344.

Seamons, T.R., P. Bentzen, and T.P. Quinn. 2004b. The effects of adult length and arrival date on individual reproductive success in wild steelhead trout, *Oncorhynchus mykiss*. Can. J. Fish. Aquat. Sci. 61: 193–204.

Seeb, J.E., L.W. Seeb, D.W. Oates, and F.M. Utter. 1987. Genetic variation and postglacial dispersal of populations of northern pike (*Esox lucius*) in North America. Can. J. Fish. Aquat. Sci. 44: 556–561.

Seeb, L.W., and P.A. Crane. 1999. Allozyme and mitochondrial DNA discriminate Asian and North American populations of chum salmon in mixed stock fisheries along the south coast of the Alaska Peninsula. Trans. Am. Fish. Soc. 128: 88–103.

Selgeby, J.H. 1988. Comparative biology of the sculpins of Lake Superior. J. Great Lakes Res. 14: 44–51.

Sellers, T.J., B.R. Parker, D.W. Schindler, and W.M. Tonn. 1998. Pelagic distribution of lake trout (*Salvelinus namaycush*) in small Canadian Shield lakes with respect to temperature, dissolved oxygen, and light. Can. J. Fish. Aquat. Sci. 55: 170–179.

Semakula, S.N, and P.A. Larkin. 1968. Age, growth, food, and yield of white sturgeon (*Acipenser transmontanus*) of the Fraser River, British Columbia. J. Fish. Res. Board Can. 25: 2589–2602.

Sexauer, H.S. 1994. Life history aspects of bull trout, *Salvelinus confluentus*, in the eastern Cascades, Washington. M.Sc. thesis, Central Washington University, Ellensburg.

Shapiro, M.D., M.E. Marks, C.L. Peichel, B.K. Blackman, K.S. Nereng, B. Jonsson, D. Schluter, and D.M. Kingsley. 2004. Genetic and developmental basis of evolutionary pelvic reduction in threespine sticklebacks. Nature (London), 428: 717–723.

Shaposhnikova, G.K. 1967. Comparative characteristics of *Stenodus leucichthys nelma* (Pallas) and *Stenodus leucichthys leucichthys*, (Güldenstatd). Vopr. Ikhtiol. 7: 227–239.

Shedlock, A., J. Parker, D. Crispin, T. Pietsch, and C. Burmer. 1992. Evolution of the salmonoid mitochondrial control region. Mol. Phylogenet. Evol. 1: 179–192.

Shepard, B., B. May, and W. Urie. 2005. Status and conservation of westslope cutthroat trout within the western United States. N. Am. J. Fish. Manage. 25: 1426–1440.

Shepard, B., K. Pratt, and P. Graham. 1984. Life histories of westslope cutthroat trout and bull trout in the upper Flathead River Basin, Montana. Montana Department of Fisheries, Wildlife, and Parks, Kalispell.

Shestakov, A.V. 1991. Features of morphology of larvae of the Siberian cisco, round whitefish, and nelma of the Anadyr River Basin. J. Ichthyol. 31: 867–871.

Shirvell, C.S. 1990. Role of instream rootwads as juvenile coho (*Oncorhynchus kisutch*) and steelhead trout (*O. mykiss*) cover habitat under varying streamflows. Can. J. Fish. Aquat. Sci. 47: 852–861.

Shirvell, C.S., and R.G. Dungey. 1983. Microhabitats chosen by brown trout for feeding and spawning in rivers. Trans. Am. Fish. Soc. 112: 355–367.

Shkorbatov, G.L. 1966. The selected temperature and phototaxis of *Coregonus* larvae. Zool. Zhurn. 14: 1515–1525.

Shoup, D.E., R.E. Carso, and R.T. Heath. 2003. Effects of predation risk and foraging return on the diel use of vegetated habitat by two size classes of bluegills. Trans. Am. Fish Soc. 132: 590–597.

Siefert, R.E. 1972. First food of larval yellow perch, white sucker, bluegill, emerald shiner, and rainbow smelt. Trans. Am. Fish Soc. 101: 219–225.

Simon, R.C., and R.E. Noble. 1968. Hybridization in *Oncorhynchus* (Salmonidae). I. Viability and inheritance in artificial crosses of chum and pink salmon. Trans. Am. Fish. Soc. 97: 109–118.

Simpson, G.G. 1944. Tempo and mode in evolution. Columbia University Press, New York.

Simpson, G.G. 1953. The major features of evolution. Columbia University Press, New York.

Simpson, J.C., and R L. Wallace. 1978. Fishes of Idaho. University of Idaho Press, Moscow.

Sinclair, D.C. 1968. Diel limnetic occurrence of young *Cottus asper* in two British Columbia lakes. J. Fish. Res. Board Can. 25: 1997–2000.

Sivak, J.G. 1973. Interrelationship of feeding behaviour and accommodative lens movements in some species of North American fishes. J. Fish. Res. Board Can. 30: 1141–1146.

Slack, T. and D. Stace-Smith. 1996. Distribution of the green sturgeon rarely seen in B.C. waters. Cordillera, 3: 39–43.

Slaney, P.A., and T.G. Northcote. 1974. Effects of prey abundance on density and territorial behaviour of young rainbow trout (*Salmo gairdneri*) in laboratory stream channels. J. Fish. Res. Board Can. 31: 1201–1209.

Slaney, T.L., K.D. Hyatt, T.G. Northcote, and R. Fielden. 1996. Status of anadromous salmon and trout in British Columbia and Yukon. Fisheries, 21: 20–35.

Sly, P.G., and D.O. Evans. 1996. Suitability of habitat for spawning lake trout. J. Aquat. Ecosyst. Health, 5: 153–157.

Smith, A.K. 1973. Development and application of spawning velocity and depth criteria for Oregon salmonids. Trans. Am. Fish. Soc. 102: 312–316.

Smith, C.T., R.J. Nelson, C.C. Wood, and B.F. Koop. 2001. Glacial biogeography of North American coho salmon (*Oncorhynchus kisutch*). Mol. Ecol. 10: 2775–2785.

Smith, C.T., R.J. Nelson, S. Pollard, E. Rubidge, S.J. McKay, J. Rodzen, B. May, and B. Koop. 2002. Population genetic analysis of white sturgeon (*Acipenser transmontanus*) in the Fraser River. J. Appl. Ichtyol. 18: 307–312.

Smith, G.R. 1966. Distribution and evolution of the North American catostomid fishes of

Smith, G.R. 1966. Distribution and evolution of the subgenus *Pantosteus*, genus *Catostomus*. University of Michigan, Ann Arbor. Misc. Publ. Mus. Zool. Univ. Mich. No. 19. 132 pp.

Smith, G.R. 1975. Fishes of the Pliocene Glenns Ferry Formation, southwest Idaho. University of Michigan, Ann Arbor. Mus. Paleontol. Univ. Mich. Pap. Paleontol. No. 14: 1–68.

Smith, G.R. 1981. Late Cenozoic freshwater fishes of North America. Annu. Rev. Ecol. Syst. 12: 163–193.

Smith, G.R. 1992. Phylogeny and biogeography of the Catostomidae, freshwater fishes of North America and Asia. pp. 778–826. *In* R.L. Mayden (ed.). Systematics and historical ecology of North American freshwater fishes. Stanford University Press, Stanford, CA.

Smith, G.R., and T.N. Todd. 1992. Morphological cladistic study of coregonine fishes. Pol. Arch. Hydrobiol. 39: 479–490.

Smith, G.R., N. Morgan, and E.A. Gustafson. 2000. Fishes of the Mio-Pliocene Ringold Formation, Washington: Pliocene capture of the Snake River by the Columbia River. University of Michigan, Ann Arbor. Univ. Mich. Mus. Paleontol. Pap. Paleontol. No. 32. 47 pp.

Smith, H.A., and P.A. Slaney. 1980. Age, growth, survival and habitat of anadromous Dolly Varden (*Salvelinus malma*) in the Keogh River, British Columbia. BC Ministry of the Environment, Victoria. Fish. Manage. Rep. No. 76. 50 pp.

Smith, L.L., and R.H. Kramer. 1964. The spottail shiner in Lower Red Lake, Minnesota. Trans. Am. Fish. Soc. 93: 35–45.

Smith, O.R. 1941. The spawning habits of cutthroat and eastern brook trouts. J. Wildl. Manage. 5: 461–471.

Smith, R.H. 1991. Rainbow trout, *Oncorhynchus mykiss*. pp. 304–322. *In* J. Stolz and J. Schnell (eds.). Trout. Stackpole Books, Harrisburg, PA.

Smith, R.J.F., and A. Lamb. 1976. Fathead minnows (*Pimephales promelas* Rafinesque) in northeastern British Columbia. Can. Field-Nat. 90: 188.

Smith, R.J.F., and B.D. Murphy. 1974. Functional morphology of the dorsal pad in fathead minnows (*Pimephales promelas* Rafinesque). Trans. Am. Fish. Soc. 103: 65–72.

Smith, R.W., and J.S. Griffith. 1994. Survival of rainbow trout in their first winter in the Henrys Fork of the Snake River, Idaho. Trans. Am. Fish. Soc. 123: 747–756.

Smith, W.E., and R.W. Saafield. 1955. Studies on Columbia River smelt, *Thaleichthys pacificus*. Fish. Res. Pap. Wash. Dep. Fish. 1: 3–26.

Snyder, D.E. 1979. Burbot—larval evidence for more than one North American species. pp. 204–219. *In* Proceedings of the third symposium on larval fish. Western Kentucky University, Bowling Green.

Snyder, D.E. 1983. Identification of catostomid larvae in Pyramid Lake and the Truckee River, Nevada. Trans. Am. Fish. Soc. 112: 333–348.

Sokolowska, E., and K.E. Skora. 2002. Reproductive cycle and related spatial and temporal distribution on the ninespine stickleback (*Pungitius pungitius*, L.) in Puck Bay (Baltic Sea). Oceanologia, 44: 475–490.

Song, C.B. 1995. Genetic divergence and speciation of Eurasian and American yellow perch based on the nucleotide sequence of cytochrome b gene. Bull. Kor. Fish. Soc. 28: 699–707.

Spangler, R.E., and D.L. Scarnecchia. 2001. Summer and fall microhabitat utilization of juvenile bull trout and cutthroat trout in a wilderness stream, Idaho. Hydrobiologia, 452: 145–154.

Stamford, M. 2001a. Dina Lake #1 pygmy whitefish (*Prosopium coulteri*) study, stomach contents analysis. Peace/Williston Fish and Wildlife Compensation Program. Progress Report. 9 pp. + appendices.

Stamford, M.D. 2001b. Mitochondrial and microsatellite DNA diversity throughout the range of a cold adapted freshwater salmonid: phylogeography, local population structure, and conservation genetics of Arctic grayling (*Thymallus arcticus*) in North America. M.Sc. thesis, Department of Zoology, University of British Columbia, Vancouver.

Stamford, M.D., and J.M.B. Hume. 2004. Status report on pygmy longfin smelt, *Spirinchus* sp. COSEWIC Interim Report. Committee on the Status of Endangered Wildlife in Canada (COSEWIC), Ottawa, ON. 39 pp.

Stamford, M.D., and E.B. Taylor. 2004. Evidence for multiple phylogeographic lineages of Arctic grayling (*Thymallus arcticus*) in North America. Mol. Ecol. 13: 1533–1549.

Stamford, M.D., and E.B. Taylor. 2005. Population subdivision and genetic signatures of demographic changes in Arctic grayling (*Thymallus arcticus*) from an impounded watershed. Can. J. Fish. Aquat. Sci. 11: 2548–2559.

Stanley, D.R. 1988. Sexual dimorphism of pelvic fin shape in four species of Catostomidae. Trans. Am. Fish. Soc. 117: 600–602.

Stasiak, R.H. 1978a. Food, age and growth of the pearl dace, *Semotilus margarita*, in Nebraska. Am Midl. Nat. 100: 463–466.

Stasiak, R.H. 1978b Reproduction, age and growth of the finescale dace, *Chrosomus neogaeus*, in Minnesota. Trans. Am. Fish. Soc. 107: 720–723.

Stearley, R.F., and G.R. Smith. 1993. Phylogeny of the Pacific trouts and salmons (*Oncorhynchus*) and genera of the family Salmonidae. Trans. Am. Fish. Soc. 122: 1–33.

Steigenberger, L.W., and P.A. Larkin. 1974. Feeding activity and rates of digestion of northern squawfish (*Ptychocheilus oregonensis*). J. Fish. Res. Board Can. 31: 411–420.

Stein, J.N., C.S. Jessop, J.R. Porter, and K.T.J. Chang-Kue. 1973. An evaluation of the fish resources of the Mackenzie River valley as related to pipeline development. Vol. 1. Department of the Environment, Fisheries Service, Winnipeg, MB.

Stephenson, S.A., and W.T. Momet. 1991. Food habits and growth of walleye, *Stizostedion vitreum*, smallmouth bass, *Micropterus dolomieui*, and northern pike, *Esox lucius*, in the Kaministiquia River, Ontario. Can. Field-Nat. 105: 517–521.

Stephenson, S.A., J.A. Burrows, and J.A. Babaluk. 2005. Long-distance migrations by inconnu (*Stenodus leucichthys*) in the Mackenzie River system. Arctic, 58: 21–25.

Stepien, C.A., and J.E. Faber. 1998. Population genetic structure, phylogeography and spawning phylopatry in walleye (*Stizostedion vitreum*) from mitochondrial DNA control region sequences. Mol. Ecol. 7: 1757–1769.

Stewart, K.W. 1966. A study of hybridization between two species of cyprinid fishes, *Acrocheilus aleutaceus* and *Ptychocheilus oregonensis*. Ph.D. thesis, Department of Zoology, University of British Columbia, Vancouver.

Stewart, K.W., and D.A. Watkinson. 2004. The freshwater fishes of Manitoba. University of Manitoba Press, Winnipeg.

Stewart, N.H. 1926. Development, growth, and food habits of the white sucker, *Catostomus commersoni* LeSueur. Bull. U.S. Bur. Fish. 42, Doc. 1007: 147–184.

Stewart, R.J., R.E. McLenehan, J.D. Morgan, and W.R. Olmsted. 1982. Ecological studies of Arctic grayling (*Thymallus arcticus*), Dolly Varden char (*Salvelinus malma*) and mountain whitefish (*Prosopium williamsoni*) in the Liard River drainage, B.C. Report prepared by E.V.S. Consultants Ltd. for Westcoast Transmission Company Ltd., Vancouver, and Foothills Pipe Lines (North B.C.) Ltd., Calgary, AB. 98 pp.

Stolz, J., and J. Schnell (eds.). 1991. Trout. Stackpole Books, Harrisburg, PA.

Stone, G. 2000. Phylogeography, hybridization and speciation. Trends Ecol. Evol. 15: 354–355.

Stone, J., and S. Barndt. 2005. Spatial distribution and habitat use of Pacific lamprey (*Lampetra tridentata*) ammocoetes in a western Washington stream. J. Freshw. Ecol. 20: 171–185.

Strauss, R.E. 1986. Natural hybrids of the freshwater sculpins *Cottus bairdi* and *Cottus cognatus* (Pisces: Cottidae): electrophoretic and morphometric evidence. Am. Midl. Nat. 115: 87–105.

Stuart, K.M., and G.R. Chislett. 1979. Aspects of the life history of Arctic grayling in the Sukunka drainage. Internal Report. BC Fisheries Branch, Prince George. 111 pp.

Sturm, E.A. 1988. Description and identification of larval fishes in Alaskan freshwaters. M.Sc. thesis, University of Alaska, Fairbanks.

Sumner, F.H. 1953. Migration of salmonids in Sand Creek, Oregon. Trans. Am. Fish. Soc. 82: 139–150.

Suzuki, A. 1992. Karyotype and DNA content of the Chinese catostomid fish, *Myxocyprinus asiaticus*. Chromosome Inf. Serv. 53: 25–27.

Svärdson, G. 1949. Note on spawning habits of *Leuciscus erythropthalmus* (L.), *Abramis brama* (L.), and *Esox lucius* L. Rep. Inst. Freshw. Res. Drottningholm, 29: 102–107.

Svetovidov, A.N., and E.A. Dorofeeva. 1963. Systematics, origin and history of the distribution of the Eurasian and North American perches and pike-perches (*Perca*, *Lucioperca*, and *Stizostedion*). English translation from Vopr. Ichthyol. 3: 625–651.

Swain, D.P., and L.B. Holtby. 1989. Differences in morphology and behaviour between juvenile coho salmon (*Oncorhynchus kisutch*) rearing in a lake or its tributary stream. Can. J. Fish. Aquat. Sci. 46: 1406–1414.

Swales, S., and C.D. Levings. 1989. Role of off-channel ponds in the life cycle of coho salmon (*Oncorhynchus kisutch*) and other juvenile salmonids in the Coldwater River, British Columbia. Can. J. Fish. Aquat. Sci. 46: 232–242.

Swee, U.B., and H.R. MacCrimmon. 1966. Reproductive biology of the carp, *Cyprinus carpio* L., in Lake St. Lawrence, Ontario. Trans. Am. Fish. Soc. 95: 372–380.

Swenson, W.A., and L.L. Smith. 1973. Gastric digestion, food consumption, feeding periodicity and food conversion efficiency in walleye (*Stizostedion vitreum vitreum*). J. Fish. Res. Board Can. 30: 1327–1336.

Swift, D.F. 1965. Effect of temperature on mortality rate and development of the eggs of the pike (*Esox lucius*) and the perch (*Perca fluviatilis*). Nature (London), 206: 528.

Tabor, R.A., and W.A. Wurtsbaugh. 1991. Predation risk and the importance of cover for juvenile rainbow trout in lentic systems. Trans. Am. Fish. Soc. 120: 728–738.

Tack, S.L. 1971. Distribution, abundance, and natural history of Arctic grayling in the Tanana River drainage. Annual Progress Report 1970–1971, Alaska Department of Fish and Game, Juneau. Project No. F-9-3, 12 (R-I). 35 pp.

Tack, S.L. 1973. Distribution, abundance, and natural history of Arctic grayling in the Tanana River drainage. Annual Progress Report 1972–1973. Alaska Department of Fish and Game, Juneau. Project No. F-9-5, 14 (R-I). 52 pp.

Tallman, R.F. 1988. Seasonal isolation and adaptation among chum salmon, *Oncorhynchus keta* (Walbaum), populations. Ph.D. thesis, Department of Zoology, University of British Columbia, Vancouver.

Tallman, R.F. 1997. Interpopulation variation in growth rates of broad whitefish. pp. 184–193. *In* J. Reynolds (ed.). Fish ecology in arctic North America. Am. Fish. Soc. Symp. 19.

Tallman, R.F., and J.H. Gee. 1982. Intraspecific resource partitioning in a headwaters stream fish, the pearl dace, *Semotilus margarita* (Cyprinidae). Environ. Biol. Fishes, 7: 243–249.

Tallman, R.F., and M.C. Healey. 1994. Homing, straying, and gene flow among seasonally separated populations of chum salmon (*Oncorhynchus keta*). Can. J. Fish. Aquat. Sci. 51: 577–588.

Tallman, R.F., K.H. Mills, and R.G. Rotter. 1984. The comparative ecology of the pearl dace (*Semotilus margarita*) and fathead minnow (*Pimephales promelas*) in Lake 114, the experimental lakes area, northwestern Ontario, with an appended key to the cyprinids of the experimental lakes area. Can. MS Rep. Fish. Aquat. Sci. No. 1756. 31 pp.

Tallman, R.F., W. Tonn, and K.J. Howland. 1996a. Life history variation of inconnu (*Stenodus leucichthys*) and burbot (*Lota lota*). Lower Slave River, June to December 1994. Northern River Basins Study, Edmonton, AB. 49 pp.

Tallman, R.F., W. Tonn, K.J. Howland, and A. Little. 1996b. Synthesis of fish distribution, movements, critical habitat and food web for the lower Slave River north of the 60th parallel: a food chain perspective. Northern River Basins Study, Edmonton, AB. Synth. Rep. No. 13. 152 pp.

Tallman, R.F., M.V. Abraham, and D.N. Chudobiak. 2002. Migrations and life history alternatives in a high latitude species, the broad whitefish, *Coregonus nasus* Pallas. Ecol. Freshw. Fish, 11: 101–111.

Tautz, A.F., and C. Groot. 1975. Spawning behaviour of chum salmon (*Oncorhynchus keta*) and rainbow trout (*Salmo gairdneri*). J. Fish. Res. Board Can. 32: 633–642.

Taylor, E.B. 1991a. A review of local adaptation in Salmonidae, with particular reference to Pacific and Atlantic salmon. Aquaculture, 98: 185–207.

Taylor, E.B. 1991b. Environmental correlates of life-history variation in juvenile chinook salmon, *Oncorhynchus tshawytscha* (Walbaum). J. Fish Biol. 37: 1–17.

Taylor, E.B, and J.J. Dodson 1994. A molecular analysis of relationships and biogeography within a species complex of holarctic fish (genus, *Osmerus*). Mol. Ecol. 3: 235–248.

Taylor, E.B., and C.J. Foote. 1991. Critical swimming velocities of juvenile sockeye salmon and kokanee, the anadromous and nonanadromous forms of *Oncorhynchus nerka* (Walbaum). J. Fish Biol. 38: 407–419.

Taylor, E.B., and J.D. McPhail. 1999. Evolutionary history of an adaptive radiation in species pairs of threespine sticklebacks (*Gasterosteus*): insights from mitochondrial DNA. Biol. J. Linn. Soc. 66: 271–291.

Taylor. E.B., and J.D. McPhail. 2000. Historical contingency and ecological determinism interact to prime speciation in sticklebacks, *Gasterosteus*. Proc. R. Soc. London Ser. B: Biol. Sci. 267: 2375–2384.

Taylor, E.B., C.J. Foote, and C.C. Wood. 1996. Molecular genetic evidence for parallel life history evolution within a Pacific salmon (sockeye and kokanee, *Oncorhynchus nerka*). Evolution, 50: 401–416.

Taylor, E.B., S. Pollard, and D. Louie. 1999. Mitochondrial DNA variation in bull trout (*Salvelinus confluentus*) from northwestern North America: implications for zoogeography and conservation. Mol. Ecol. 8: 1155–1170.

Taylor, E.B., Z.R. Redenbach, A.B. Costello, S.J. Pollard, and C.J. Pacas. 2001. Nested analysis of genetic diversity in northwestern North American char, Dolly Varden (*Salvelinus malma*) and bull trout (*Salvelinus confluentus*). Can. J. Fish. Aquat. Sci. 58: 406–420.

Taylor, E.B., M.B. Stamford, and J.S. Baxter. 2003. Population subdivision in westslope cutthroat trout (*Oncorhynchus clarki lewisi*) at the northern periphery of its range: evolutionary inferences and conservation implications. Mol. Ecol. 12: 2609–2622.

Taylor, G.D. 1966. Distribution and habitat response of the coastrange sculpin, *Cottus aleuticus*, and prickly sculpin, *Cottus asper*, in the Little Campbell River, BC. M.Sc. thesis, Department of Zoology, University of British Columbia, Vancouver.

Taylor, G.T., and R.J. LeBrasseur. 1957. Distribution, age and food of steelhead trout, *Salmo gairdneri*, caught in the northeast Pacific Ocean. Fish. Res. Board. Can. Pac. Biol. Stn. Prog. Rep. No. 109: 9–11.

Taylor, S.G. 1980. Marine survival of pink salmon fry from early and late spawners. Trans. Am. Fish. Soc. 109: 79–82.

Teel, D.J., G.B. Milner, G.A. Winans, and W.S. Grant. 2000. Genetic population structure and origin of life history types in chinook salmon in British Columbia, Canada. Trans. Am. Fish. Soc. 129: 194–209.

Teller, J.T., and L. Clayton (eds.). 1983. Glacial Lake Agassiz. Geol. Assoc. Can. Spec. Pap. No. 26. 451 pp.

Templeman-Kluit, D. 1980. Evolution of physiography and drainage in southern Yukon. Can. J. Earth Sci. 17: 1189–1203.

Teska, J.D., and D.J. Behmer. 1981. Zooplankton preference of larval lake whitefish. Trans. Am. Fish. Soc. 110: 459–461.

Thomas, A. 1973. Spawning migration and intragravel movement of the torrent sculpin, *Cottus rhotheus*. Trans. Am. Fish. Soc. 102: 620–622.

Thompson, C.E., E.B. Taylor, and J.D. McPhail. 1998. Parallel evolution of lake-stream pairs of threespine stickleback (*Gasterosteus aculeatus*) inferred from mitochondrial DNA variation. Evolution, 51: 1955–1965.

Thompson, G.E., and R.W. Davies. 1976. Observations on the age, growth, reproduction, and feeding of mountain whitefish (*Prosopium williamsoni*) in the Sheep River, Alberta. Trans. Am. Fish. Soc. 105: 208–219.

Thompson, K.R., and D.W. Beckman. 1995. Validation of age estimates from white sucker otoliths. Trans. Am. Fish. Soc. 124: 637–639.

Thorpe, J.E. 1977. Morphology, physiology, behaviour, and ecology of *Perca fluviatilus* L. and *Perca flavescens* (Mitchill). J. Fish. Res. Board Can. 34: 1504–1514.

Thorpe, J.E., M.S. Miles, and D.S. Keay. 1984. Development rate, fecundity, and egg size in Atlantic salmon, *Salmo salar* L. Aquaculture, 43: 289–305.

Thorson, R.M. 1980. Ice-sheet glaciation of the Puget Lowland, Washington, during the Vashon Stade (late Pleistocene). Quat. Res. 13: 303–321.

Timoshina, L.A. 1974. Embryonic development of the rainbow trout (*Salmo gairdneri irideus* (Gibbons)) at different temperatures. J. Ichthyol. 12: 425–432.

Tipper, H.W. 1971. Glacial geomorphology and Pleistocene history of central British Columbia. Bull. Geol. Surv. Can. No. 196: 1–89.

Tippets, W.E., and P.B. Moyle. 1978. Epibenthic feeding by rainbow trout (*Salmo gairdneri*) in the McCloud River, California. J. Anim. Ecol. 47: 549–559.

Titus, K. 1996. Food habits and habitats of juvenile chiselmouth, *Acrocheilus alutaceus*, in the Yakima River, Washington. M.Sc. thesis, Central Washington University, Ellensberg.

Todd, J.H., J. Atema, and J.E. Bardach. 1967. Chemical communications in social behavior of a fish, the yellow bullhead. Science (Washington, DC), 158: 672–673.

Todd, T.N. 1990. Genetic differentiation of walleye stocks in Lake St. Clair and western Lake Erie. U.S. Fish and Wildl. Serv. Tech. Rep. No. 28. 26 pp.

Toline, C.A., and A.J. Baker. 1993. Foraging tactic as a potential selection pressure influencing geographic differences in body shape among populations of dace (*Phoxinus eos*). Can. J. Zool. 71: 2178–2184.

Toline, C.A., and A.J. Baker. 1994. Genetic differentiation among populations of the northern redbelly dace (*Phoxinus eos*) in Ontario. Can. J. Fish. Aquat. Sci. 51: 1218–1228.

Tompkins, A.M., and J.S. Gee. 1983. Foraging behaviour of brook stickleback, *Culaea inconstans* (Kirtland): optimization of time, space, and diet. Can. J. Zool. 61: 2482–2490.

Toregson, C.E., and T.A. Close. 2004. Influence of habitat heterogeneity on the distribution of larval Pacific lamprey (*Lampetra tridentata*) at two spatial scales. Freshw. Biol. 49: 614–630.

Torgersen, C.E., D.M. Price, H.W. Li, and B.A. McIntosh. 1999. Multiscale thermal refugia and stream habitat associations of chinook salmon in northeastern Oregon. Ecol. Appl. 9: 301–319.

Trautman, M.B. 1957. The fishes of Ohio. Ohio State University Press, Columbus.

Treble, M.A., and R.F. Tallman. 1997. An assessment of the exploratory fishery and investigation of the population structure of broad whitefish from the Mackenzie Delta, 1989–1993. Can. Tech. Rep. Fish. Aquat. Sci. 2180. 65 pp.

Tripp, D.B., P.J. McCart, R.D. Saunders, and G.W. Hughes. 1981. Fisheries studies in the Slave River delta, NWT. Final Report, Mackenzie River Basin Study. Aquatic Environments Ltd., Calgary, AB. 262 pp.

Troffe, P.M. 2000. Fluvial mountain whitefish (*Prosopium williamsoni*) in the upper Fraser River: morphological, behavioural, and genetic comparison of foraging forms. M.Sc. thesis, Department of Zoology, University of British Columbia, Vancouver.

Trotter, P.C. 1987. Cutthroat: native trout of the west. Colorado Associated University Press, Boulder.

Trotter, P.C. 1989. Coastal cutthroat trout: a life history compendium. Trans. Am. Fish. Soc. 118: 463–473.

Trotter, P.C. 1997. Sea-run cutthroat trout: life history profile. pp. 7–15. In J.D. Hall, P.A. Bisson, and R.E. Gresswell (eds.). Sea-run cutthroat trout: biology, management, and future conservation. Oregon Chapter, American Fisheries Society, Corvallis.

Tsuruta, T., H. Takahashi, and A. Goto. 2002. Evidence for type assortative mating between the freshwater and Omono types of nine-spined sticklebacks in natural freshwater. J. Fish Biol. 61: 230–241.

Turgeon, J., and L. Bernatchez. 2003. Reticulate evolution and phenotypic diversity in North American ciscoes, *Coregonus* ssp. (Teleostei: Salmonidae): implications for the conservation of an evolutionary legacy. Conserv. Genet. 4: 67–81.

Tyrrell, J.B. (ed.) 1916. David Thompson's narrative of his explorations in western North America, 1784–1812. The Champlain Society, Toronto, ON. 740 pp.

Van Eenennaam, J.P., M.A.H. Webb, X. Deng, and S.I. Doroshov. 2001. Artificial spawning and larval rearing of Klamath River green sturgeon. Trans. Am. Fish. Soc. 130: 159–165.

Van Houdt, J.K.J., L. De Cleyn, A. Perretti, and F.A.M. Volckaert. 2005. A mitogenic view on the evolutionary history of the holarctic freshwater gadoid, burbot (*Lota lota*). Mol. Ecol. 14: 2445–2457.

Van Houdt, J.K., B. Hellemans, and F.A.M. Volckaert. 2003. Phylogenetic relationships among Palearctic and Nearctic burbot (*Lota lota*): Pleistocene extinctions and recolonization. Mol. Phylogenet. Evol. 29: 599–612.

Van Tassell, J., M. Ferns, V. McConnell, and G.R. Smith. 2001. The mid-Pliocene Imbler fish fossils, Grande Ronde Valley, Union County, Oregon, and the connection between Lake Idaho and the Columbia River. Oreg. Geol. 63: 77–84, 89–96.

Van Vleit, W.H. 1964. An ecological study of *Cottus cognatus* Richardson in northern Saskatchewan. M.Sc. thesis, University of Saskatchewan, Saskatoon.

Varely, M.E. 1967. British freshwater fishes. Fishing News Ltd., London.

Veinott, G., T.G. Northcote, M.L. Rosenau, and R.D. Evans. 1999. Concentrations of strontium in the pectoral fin rays of the white sturgeon (*Acipenser transmontanus*) by laser ablation sampling—inductively coupled plasma—mass spectrometry as an indicator of marine migration. Can. J. Fish. Aquat. Sci. 56: 1981–1990.

Velsen, F.P. 1980. Embryonic development in the eggs of sockeye salmon, *Oncorhynchus nerka*. Can. Spec. Publ. Fish. Aquat. Sci. No. 49. 19 pp.

Vernon, E.H. 1957. Morphometric comparison of three races of kokanee (*Oncorhynchus nerka*) within a large British Columbia lake. J. Fish. Res. Board Can. 14: 573–598.

Vernon, E.H. 1966. Enumeration of migrant pink salmon (*Oncorhynchus gorbuscha*) fry in the Fraser River estuary. Int. Pac. Salmon Fish. Comm. Bull. No. 19. 83 pp.

Verspoor, E. 1989. Widespread hybridization between native Atlantic salmon, *Salmo salar*, and introduced brown trout, *Salmo trutta*, in eastern Newfoundland. J. Fish. Biol. 34: 41–46.

Vescei, P., M.L. Litvak, D.L.G. Noakes, T. Rien, and M. Hochleitner. 2003. A non-invasive technique for determining sex of live North American sturgeons. Environ. Biol. Fishes, 68: 333–338.

Vladykov, V.D. 1954. Taxonomic characters of the eastern North American chars (*Salvelinus*) and (*Cristivomer*). J. Fish. Res. Board Can. 13: 904–932.

Vladykov, V.D. 1963. A review of salmonid genera and their broad geographic distribution. Trans. R. Soc. Can. (Series IV) 1: 459–504.

Vladykov, V.D. 1970. Pearl tubercles and certain cranial peculiarities useful in the taxonomy of coregonid genera. pp. 167–193. In C.C. Lindsey and C.S. Woods (eds.). Biology of coregonid fishes. University of Manitoba Press, Winnipeg.

Vladykov, V.D., and W.I. Follett. 1958. Redescription of *Lampetra ayresii* (Günther) of western North America, a species of lamprey (Petromyzontidae) distinct from *Lampetra fluviatilis* (Linnaeus) of Europe. J. Fish. Res. Board Can. 15: 47–77.

Vladykov, V.D., and E. Kott. 1979. Satellite species among the holarctic lampreys. Can. J. Zool. 57: 808–823.

Volpe, J.P., B.R. Anholt, and B.W. Glickman. 2001a. Competition among juvenile Atlantic salmon (*Salmo salar*) and steelhead (*Oncorhynchus mykiss*): relevance to invasion potential in British Columbia. Can. J. Fish. Aquat. Sci. 58: 197–207.

Volpe, J.P., B.R. Anholt, and B.W. Glickman. 2001b. Reproduction of Atlantic salmon (*Salmo salar*) in a controlled stream channel on Vancouver Island, British Columbia. Trans. Am. Fish. Soc. 130: 489–494.

Volpe, J.P., E.B. Taylor, D.W. Rimmer, and B.W. Glickman. 2000. Natural reproduction of aquaculture escaped Atlantic salmon (*Salmo salar*) in a coastal British Columbia river. Conserv. Biol. 14: 899–903.

Wahl, C.M., E.L. Mills, W.N. McFarland, and J.S. DiGisi. 1993. Ontogenetic changes in prey selection and visual acuity of the yellow perch, *Perca flavescens*. Can. J. Fish. Aquat Sci. 50: 743–749.

Wallace, C.R. 1967. Observations on the reproductive behavior of black bullhead (*Ictalurus melas*). Copeia, 1967: 852–853.

Wallace, C.R. 1972. Spawning of *Ictalurus natalis*. Tex. J. Sci. 24: 307–310.

Walters, V. 1955. Fishes of western Arctic America and eastern Arctic Siberia. Bull. Am. Mus. Nat. Hist. No. 106: 255–368.

Wang, Y.L., F.P. Binkowski, and S.I. Doroshov. 1985. Effect of temperature on early development of white and lake sturgeon, *Acipenser transmontanus* and *A. fulvescens*. Environ. Biol. Fishes, 14: 43–50.

Wankowski, J.W.J., and J.E. Thorpe. 1979. Spatial distribution and feeding in Atlantic salmon, *Salmo salar* L., juveniles. J. Fish. Biol. 14: 239–247.

Waples, R.A., D.J. Teel, J.M. Myers, and A.R. Marshall. 2004. Life history divergence in chinook salmon: historical contingency and parallel evolution. Evolution, 58: 386–403.

Waples, R.S. 1995. Evolutionary significant units and the conservation of biodiversity under the Endangered Species Act. pp. 8–27. *In* J.L. Neilson and D.A. Powers (eds.). Evolution and the aquatic ecosystem: defining unique units in population conservation. Am. Fish. Soc. Symp. No. 17.

Ward, F.J. 1959. Character of the migration of pink salmon to Fraser River spawning grounds in 1957. Int. Pac. Salmon Fish. Comm. Bull. No. 10. 70 pp.

Warner, B.G., R.W. Mathewes, and J.J. Clague. 1982. Ice-free conditions on the Queen Charlotte Islands, British Columbia, at the height of late Wisconsin glaciation. Science (Washington, DC), 218: 675–677.

Warner, E.J., and T.P. Quinn. 1995. Horizontal and vertical movements of telemetered rainbow trout (*Oncorhynchus mykiss*) in Lake Washington. Can. J. Zool. 73: 146–153.

Weatherley, A.H. 1959. Some features of the biology of the tench (*Tinca tinca* Linnaeus) in Tasmania. J. Anim. Ecol. 28: 73–87.

Weaver, M.J., J.J. Magnuson, and M.K. Clayton. 1997. Distribution of littoral fishes in structurally complex macrophytes. Can. J. Fish. Aquat. Sci. 54: 2277–2289.

Webb, M.A.H., G.W. Feist, E.P. Foster, C.B. Schreck, and M.S. Fitzpatrick. 2002. Potential classification of sex and stage of gonadal maturity of wild white sturgeon using blood plasma indicators. Trans. Am. Fish. Soc. 131: 132–142.

Webster, D.A., and G. Eriksdottir. 1976. Upwelling water as a factor influencing choice of spawning sites by brook trout (*Salvelinus fontinalis*). Trans. Am. Fish. Soc. 105: 416–421.

Weigel, J.K., J. Peterson, and P. Spruell. 2002. A model using phenotypic characteristics to detect introgressive hybridization in wild westslope cutthroat trout and rainbow trout. Trans. Am. Fish. Soc. 131: 389–403.

Weisel, G.F., and J.B. Dillon. 1954. Observations on the pygmy whitefish, *Prosopium coulteri*, in western Montana. Copeia, 1954: 124–127.

Weisel, G.F., and H.W. Newman. 1951. Breeding habits, development and early life history of *Richardsonius balteatus*, a northwestern minnow. Copeia, 1951: 187–194.

Weisel, G.F., D.A. Hanzel, and R.L. Newell. 1973. The pygmy whitefish, *Prosopium coulteri*, in western Montana. U.S. Fish Wildl. Serv. Fish. Bull. No. 71: 587–596.

Weitkamp, L.A., T.C. Wainwright, G.J. Bryant, G.B. Milner, D.J. Teel, R.G. Kope, and R.S. Waples. 1995. Status review of coho salmon from Washington, Oregon, and California. U.S. Department of Commerce, National Marine Fisheries Service, Washington, DC. NOAA Tech. Memo. NMFS-NWFSC-24. 258 pp.

Welch, D.W., B.R. Ward, and S.D. Batten. 2004. Early ocean survival and marine movements of hatchery and wild rainbow trout (*Oncorhynchus mykiss*) determined by an acoustical array: Queen Charlotte Strait, British Columbia. Deep Sea Res. 51: 897–909.

Welker, T.L., and D.L. Scarnecchia. 2003. Differences in species composition and feeding ecology of catostomid fishes in two distinct segments of the Missouri River, North Dakota, U.S.A. Environ. Biol. Fishes, 68: 129–141.

Wells, A.W. 1978. Systematics, variation and zoogeography of two North American cyprinid fishes. Ph.D. thesis, Department of Zoology, University of Alberta, Edmonton.

Wells, L. 1968. Seasonal depth distribution of fish in southeastern Lake Michigan. Fish. Bull. 67: 1–15.

Wells, L., and R. House. 1974. Life history of the spottail shiner (*Notropis hudsonius*) in southeastern Lake Michigan, Kalamazoo River and western Lake Erie. U.S. Bureau of Sport Fisheries, Washington, DC. Wildl. Res. Rep. No. 78. 10 pp.

Werner, R.G. 1967. Intralacustrine movements of bluegill fry in Crane Lake, Indiana. Trans. Am. Fish Soc. 96: 416–420.

Werner, R.G. 1969. Ecology of limnetic bluegill (*Lepomis macrochirus*) in Crane Lake, Indiana. Am. Midl. Nat. 81: 164–181.

Werner, R.G. 1979. Homing mechanism of spawning white suckers in Wolfe Lake, New York. N.Y. Fish Game, 26: 48–58.

West, R.L., M.W. Smith, W.E. Barber, J.B. Reynolds, and H. Hop. 1992. Autumnal migration and over-wintering of Arctic grayling in coastal streams of the Arctic National Wildlife Refuge, Alaska. Trans. Am. Fish. Soc. 121: 709–715.

Whang, A., and J. Janssen. 1994. Sound production through the substrate during reproduction in the mottled sculpin, *Cottus bairdi* (Cottidae). Environ. Biol. Fishes, 40: 141–148.

Wheeler, A. 1969. The fishes of the British Isles and north-western Europe. Macmillan Co., Toronto, ON.

White, J.M., R.W. Matthews, and W.J. Mathews. 1979. Radiocarbon dates from Boone Lake and their relation to the "ice free corridor" in the Peace River district of Alberta. Can. J. Earth Sci. 16: 1870–174.

White, J.M., R.W. Matthews, and W.J. Mathews. 1985. Late Pleistocene chronology and environment of the "ice free corridor" of northwestern Alberta. Quat. Res. 24: 173–186.

Whitehouse, T.R., and C.D. Levings. 1989. Surface trawl catch data from the lower Fraser River at Queens Reach during 1987 and 1988. Can. Data Rep. Fish. Aquat. Sci. No. 768. 53 pp.

Williams, I.V., and P. Gilhousen. 1968. Lamprey parasitism on Fraser River sockeye and pink salmon during 1967. Int. Pac. Salmon Comm. Prog. Rep. No. 18. 22 pp.

Williams, J.E., J.E. Johnson, D.A. Hendrikson, S. Contras-Balderas, J.D. Williams, M. Navarro-Mendoza, D.E. McAllister, and J.E. Deacon. 1989. Fishes of North America, endangered, threatened, or of special concern, 1989. Fisheries, 14: 2–20.

Williams, K.R. 1999. Washington westslope cutthroat status report. Washington Department of Fish Wildlife, Olympia.

Williams, P.M., W.H. Mathews, and G.L. Pickard. 1961. A lake in British Columbia containing old seawater. Nature (London), 191: 830–832.

Williams, T.H., K.P. Currans, N.E. Ward, and G.H. Reeves. 1997. Genetic population structure of coastal cutthroat trout. pp. 16–17. *In* J.D. Hall, P.A. Bisson, and R.E. Gresswell (eds.). Sea-run cutthroat trout: biology, management, and future conservation. Oregon Chapter, American Fisheries Society, Corvallis.

Wilson, C.C. and P.D.N. Hebert. 1998. Phylogeography and postglacial dispersal of lake trout (*Salvelinus namaycush*) in North America. Can. J. Fish. Aquat. Sci. 55: 1010–1024.

Wilson, G., K. Ashley, S. Ewing, P. Slaney, S. Jennings, and R. Land. 1997. Development of a resident trout fishery on the Adam River through increased habitat productivity: year three (1996) of low-level inorganic nutrient addition. BC Ministry of Environment, Lands, and Parks, Fisheries Branch, Victoria. Fish Project Rep. No. RD 63. 62 pp.

Wilson, G.M., W.K. Thomas, and A.T. Beckenbach. 1987. Mitochondrial DNA analysis of Pacific Northwestern populations of *Oncorhynchus tshawytscha*. Can. J. Fish. Aquat. Sci. 44: 1301–1305.

Wilson, M.K.A., and R.R.G. Williams. 1992. Phylogenetic, biogeographic, and ecological significance of early fossil records of North American freshwater teleostean fishes. pp. 224–244. *In* R.L. Mayden (ed.). Systematics and historical ecology of North American freshwater fishes. Stanford University Press, Stanford, CA.

Wilson, M.V.H. 1977. Middle Eocene freshwater fishes from British Columbia. Royal Ontario Museum, Toronto. Life Sci. Contrib. No. 113: 1–61.

Wilson, M.V.H. 1979. A second species of *Libotonius* (Pisces: Percopsidae) from the Eocene of Washington State. Copeia, 1979: 400–405.

Wilson, M.V.H. 1980. Oldest known *Esox* (Pisces: Esocidae) part of a new Paleocene teleost fauna from western Canada. Can. J. Earth Sci. 17: 307–312.

Wilson, M.V.H. 1982. A new species of the fish *Amia* from the middle Eocene of British Columbia. Paleontology, 25: 413–424.

Wilson, M.V.H. 1996. Fishes from the Eocene lakes of the interior. pp. 212–224. *In* R. Ludvigsen (ed.). Life in stone: a natural history of British Columbia's fossils. University of British Columbia Press, Vancouver.

Wilson, M.V.H., and R.R.G. Williams. 1991. New Paleocene genus and species of smelt (Teleostei: Osmeridae) from freshwater deposits of the Paskapoo Formation Alberta, Canada, and comments on osmerid phylogeny. J. Vert. Paleontol. 11: 434–451.

Wilson, M.V.H., D.B. Brinkman, and A.G. Neuman. 1992. Cretaceous Esocoidei (Teleostei): early radiation of the pikes in North American fresh waters. J. Paleontol. 66: 839–846.

Winn, H.E. 1960. Biology of the brook stickleback, *Eucalia inconstans* (Kirkland). Am. Midl. Nat. 63: 424–438.

Withler, F.C. 1955. Coho salmon fingerling attacked by young lamprey. Fish. Res. Board Can. Pac. Prog. Rep. No. 104. 15 pp.

Withler, I.L. 1956. A limnological survey of Atlin and Southern Tagish lakes. BC Game Commission, Victoria. Manage. Publ. No. 5. 36 pp.

Withler, I.L. 1966. Variability in life history characteristics of steelhead trout (*Salmo gairdneri*) along the Pacific Coast of North America. J. Fish. Res. Board Can. 23: 365–393.

Witzel, L.D., and H.R. MacCrimmon. 1983. Redd-site selection by brook trout and brown trout in southwestern Ontario streams. Trans. Am. Fish. Soc. 112: 760–771.

Wohlschlag, D.E. 1954. Growth peculiarities of the cisco, *Coregonus sardinella* (Valenciennes), in the vicinity of Point Barrow, Alaska. Stanford Ichthyol. Bull. 4: 189–209.

Wojcik, F.J. 1955. Life history and management of the grayling in interior Alaska. M.Sc. thesis, University of Alaska, Fairbanks.

Wood, C.C., and C.J. Foote. 1990. Genetic differences in the early development and growth of sympatric sockeye salmon and kokanee (*Oncorhynchus nerka*), and their hybrids. Can. J. Fish. Aquat. Sci. 47: 2250–2260.

Wood, C.C., and C.J. Foote. 1996. Evidence for sympatric genetic divergence of anadromous and non-anadromous morphs of sockeye salmon (*Oncorhynchus nerka*). Evolution, 50: 1265–1279.

Wood, C.C., B.E. Riddell, and D.T. Rutherford. 1987. Alternative life histories of sockeye salmon (*Oncorhynchus nerka*) and their contribution to production in the Stikine River, northern British Columbia. pp. 12–24. *In* H.D. Smith, L. Margolis, and C.C. Wood (eds.). Sockeye salmon (*Oncorhynchus nerka*) population biology and future management. Can. Spec. Publ. Fish. Aquat. Sci. 96.

Wood, C.C., C.J. Foote, and D.T. Rutherford. 1999. Ecological interactions between juveniles of reproductively isolated anadromous and non-anadromous morphs of sockeye salmon, *Oncorhynchus nerka*, sharing the same nursery lake. Environ. Biol. Fishes, 54: 161–173.

Wood, P.M., and L. Flahr. Taking endangered species seriously? British Columbia's Species-at-Risk policies. Can. Public Policy, 30: 381–399.

Woodman, D.A. 1992. Systematic relationships within the cyprinid genus *Rhinichthys*. pp. 374–391. *In* R.L. Mayden (ed.). Systematics and historical ecology of North American freshwater fishes. Stanford University Press, Stanford, CA.

Wootton, R.J. 1976. The biology of sticklebacks. Academic Press, New York.

Wootton, R.J. 1984. A functional biology of sticklebacks. Croom Helm, London.

Worgan, J.P., and G.J. FitzGerald. 1981. Diel activity and diet of three sympatric sticklebacks in tidal salt marsh pools. Can. J. Zool. 59: 2375–2379.

Wurtsbaugh, W.A., R.W. Brockensen, and C.R. Goldman. 1975. Food and distribution of underyearling rainbow trout in Castle Lake, California. Trans. Am. Fish. Soc. 104: 88–95.

Wydoski, R.G., and R.S. Wydoski. 2002. Age growth and reproduction of mountain suckers in Lost Creek Reservoir, Utah. Trans. Am. Fish. Soc. 131: 320–328.

Wydoski, R.S., and R.R. Whitney. 1979. Inland fishes of Washington. 1st ed. University of Washington Press, Seattle.

Wydoski, R.S., and R.R. Whitney. 2003. Inland fishes of Washington. 2nd ed. University of Washington Press, Seattle.

Wynne-Edwards, V.C. 1952. Freshwater vertebrates of the arctic and subarctic. Fish. Res. Board Can. Bull. No. 94. 24 pp.

Young, B.A., T.L. Welker, M.L. Wildhaber, C.R. Berry, and D. Scarnecchia. 1997. Population structure and habitat use of benthic fishes along the Missouri and lower Yellowstone River. US Army Corps of Engineers and US Bureau of Reclamation, Washington, DC. 1997 Annu. Rep. Mo. River Benth. Fish Stud. PD-95-5832.

Young, W.P., C.O. Ostberg, P. Keim, and G.H. Thorgaard. 2001. Genetic characterization of hybridization and introgression between anadromous rainbow trout (*Oncorhynchus mykiss irideus*) and coastal cutthroat trout (*O. clarki clarki*). Mol. Ecol. 10: 921–930.

Zemlak, R.J., and J.D. McPhail. 2006. The biology of pygmy whitefish, *P. coulterii*, in a closed sub-boreal lake: spatial distribution and diel movements. Environ. Biol. Fishes, 76: 317–327.

Zhang, S., Q. Wu, and Y. Zhang. 2001. On the taxonomic status of the Yangtze sturgeon, Asian and American green sturgeon based on mitochondrial control region sequences. Acta Zool. Sin. 47: 632–639.

Zhou, J.F., Q.J. Wu, Y.Z. Ye, and J.G. Tong. 2003. Genetic divergence between *Cyprinus carpio carpio* and *Cyprinus carpio haematopterus* as assessed by mitochondrial DNA analysis, with emphasis on the origin of European domesticated carp. Genetica, 119: 93–97.

Zimmerman, C.E., K.P. Currans, and G.H. Reeves. 1997. Genetic population structure of coastal cutthroat trout in the Muck Creek Basin, Washington. pp. 170–172. *In* J.D. Hall, P.A. Bisson, and R.E. Gresswell (eds.). Sea-run cutthroat trout: biology, management, and future conservation. Oregon Chapter, American Fisheries Society, Corvallis.

Zimmerman, E.G., and M.C. Wooten. 1981. Allozyme variation and natural hybridization in sculpins, *Cottus confusus* and *Cottus cognatus*. Biochem. Syst. Ecol. 9: 341–346.

Ziuganov, V.V., and A.A. Zotin. 1995. Pelvic girdle polymorphism and reproductive barriers in the ninespine stickleback *Pungitius pungitius* (L.) from northwest Russia. Behaviour, 132: 1095–1098

APPENDIX I

TABLE 1 A checklist of British Columbia freshwater fishes.

Common name	Scientific name	Region of origin
RIVER LAMPREY	*Lampetra ayresii*	Native
VANCOUVER LAMPREY	*Lampetra macrostoma*	Native
WESTERN BROOK LAMPREY	*Lampetra richardsoni*	Native
PACIFIC LAMPREY	*Lampetra tridentata*	Native
GREEN STURGEON	*Acipenser medirostris*	Native
WHITE STURGEON	*Acipenser transmontanus*	Native
GOLDEYE	*Hiodon alosoides*	Native
CHISELMOUTH	*Acrocheilus alutaceus*	Native
GOLDFISH	*Carassius auratus*	Introduced (Eurasia)
LAKE CHUB	*Couesius plumbeus*	Native
COMMON CARP	*Cyprinus carpio*	Introduced (Europe)
BRASSY MINNOW	*Hybognathus hankinsoni*	Native
NORTHERN PEARL DACE	*Margariscus margarita*	Native
PEAMOUTH	*Mylocheilus caurinus*	Native
EMERALD SHINER	*Notropis atherinoides*	Native
SPOTTAIL SHINER	*Notropis hudsonius*	Native, introduced (Alberta)
NORTHERN REDBELLY DACE	*Phoxinus eos*	Native
FINESCALE DACE	*Phoxinus neogaeus*	Native
FATHEAD MINNOW	*Pimephales promelas*	Introduced (Alberta?)
NORTHERN PIKEMINNOW	*Ptychocheilus oregonensis*	Native
LONGNOSE DACE	*Rhinichthys cataractae*	Native
LEOPARD DACE	*Rhinichthys falcatus*	Native
SPECKLED DACE	*Rhinichthys osculus*	Native
UMATILLA DACE	*Rhinichthys umatilla*	Native
REDSIDE SHINER	*Richardsonius balteatus*	Native
TENCH	*Tinca tinca*	Introduced (Eurasia)
LONGNOSE SUCKER	*Catostomus catostomus*	Native
BRIDGELIP SUCKER	*Catostomus columbianus*	Native
WHITE SUCKER	*Catostomus commersonii*	Native
LARGESCALE SUCKER	*Catostomus macrocheilus*	Native
MOUNTAIN SUCKER	*Catostomus platyrhynchus*	Native
BLACK BULLHEAD	*Ameiurus melas*	Introduced (eastern North America)
YELLOW BULLHEAD	*Ameiurus natalis*	Introduced (eastern North America)
BROWN BULLHEAD	*Ameiurus nebulosus*	Introduced (eastern North America)
NORTHERN PIKE	*Esox lucius*	Native
LONGFIN SMELT	*Spirinchus thaleichthys*	Native
EULACHON	*Thaleichthys pacificus*	Native
CUTTHROAT TROUT	*Oncorhynchus clarkii*	Native
PINK SALMON	*Oncorhynchus gorbuscha*	Native
CHUM SALMON	*Oncorhynchus keta*	Native
COHO SALMON	*Oncorhynchus kisutch*	Native
RAINBOW TROUT	*Oncorhynchus mykiss*	Native
SOCKEYE SALMON	*Oncorhynchus nerka*	Native
CHINOOK SALMON	*Oncorhynchus tshawytscha*	Native
ATLANTIC SALMON	*Salmo salar*	Introduced (Europe)
BROWN TROUT	*Salmo trutta*	Introduced (Europe)
BULL TROUT	*Salvelinus confluentus*	Native
BROOK TROUT	*Salvelinus fontinalis*	Introduced (eastern North America)
DOLLY VARDEN	*Salvelinus malma*	Native
LAKE TROUT	*Salvelinus namaycush*	Native, introduced (eastern North America)
CISCO	*Coregonus artedi*	Native
ARCTIC CISCO	*Coregonus autumnalis*	Native
LAKE WHITEFISH	*Coregonus clupeaformis*	Native, introduced (eastern North America)
BROAD WHITEFISH	*Coregonus nasus*	Native
LEAST CISCO	*Coregonus sardinella*	Native
PYGMY WHITEFISH	*Prosopium coulterii*	Native
ROUND WHITEFISH	*Prosopium cylindraceum*	Native
MOUNTAIN WHITEFISH	*Prosopium williamsoni*	Native

TABLE 2 *(continued)*

INCONNU	*Stenodus leucichthys*	Native
ARCTIC GRAYLING	*Thymallus arcticus*	Native
TROUT-PERCH	*Percopsis omiscomaycus*	Native
BURBOT	*Lota lota*	Native
BROOK STICKLEBACK	*Culaea inconstans*	Native
THREESPINE STICKLEBACK	*Gasterosteus aculeatus*	Native
NINESPINE STICKLEBACK	*Pungitius pungitius*	Native
COASTRANGE SCULPIN	*Cottus aleuticus*	Native
PRICKLY SCULPIN	*Cottus asper*	Native
SLIMY SCULPIN	*Cottus cognatus*	Native
SHORTHEAD SCULPIN	*Cottus confusus*	Native
COLUMBIA SCULPIN	*Cottus hubbsi*	Native
TORRENT SCULPIN	*Cottus rhotheus*	Native
SPOONHEAD SCULPIN	*Cottus ricei*	Native
EASTSLOPE SCULPIN	*Cottus* sp.	Native
PUMPKINSEED	*Lepomis gibbosus*	Native, introduced (eastern North America)
BLUEGILL	*Lepomis macrocheirus*	Native, introduced (eastern North America)
SMALLMOUTH BASS	*Micropterus dolomieu*	Native, introduced (eastern North America)
LARGEMOUTH BASS	*Micropterus salmoides*	Native, introduced (eastern North America)
BLACK CRAPPIE	*Pomoxis nigromaculatus*	Native, introduced (eastern North America)
YELLOW PERCH	*Perca flavescens*	Native(?), introduced (eastern North America)
WALLEYE	*Sander vitreus*	Native, introduced (eastern North America)

GLOSSARY

Adfluvial	Life histories that include migrations between rivers, or streams, and lakes.
Alevins	Newly hatched salmonids before the yolk sac is absorbed.
Alleles	Two or more alternate forms of a gene.
Allopatric	Geographically separated, usually refers to populations or species.
Ammocoetes	The larval stage of lampreys; in British Columbia, they are commonly called ditch eels.
Amphidromous	Fishes that, at stages in their life cycle, migrate between fresh and salt water, but the migration is not a breeding migration (e.g., juvenile Dolly Varden).
Anadromous	Fishes in which the major growing phase of their life cycle is in the sea but return to fresh water to breed.
Axillary process	Fleshy tab in the axillae of the paired fins; the pelvic axillary process is especially obvious in large salmonids.
Axillae	The area immediately behind the paired fins.
Barbel	Fleshy processes located on the chin, snout, or in the corners of the upper jaw. Typically, they are covered in chemoreceptors. Usually the barbels are long and thin in sturgeons and catfishes, and relatively small in minnows.
Basibranchial teeth	The teeth on one (usually the most anterior) of the three median bones immediately behind the tongue. They are present in cutthroat trout and absent in rainbow trout.
BCCDC	An acronym for the British Columbia Conservation Data Centre.
Benthic	Pertaining to the bottom.
Biological species pairs	Sympatric, reproductively isolated forms of the same taxonomic species (e.g., the benthic and limnetic forms of threespine stickleback).
Branchiostegal rays	Bony rays that support the pleated membranes that cover the lower part of the gill cavities.
Carotenoids	Refers to fat-soluble yellow, orange, or red pigments.
Catadromous	Fishes that spend most of their adult lives in fresh water but migrate to marine or brackish water to breed.
Caudal flexure	The point where the caudal rays hinge on the hypural plates. This flexure becomes apparent when the tail is bent from side-to-side.
Caudal peduncle	The narrow, wrist-like area between the insertion of the anal fin and the origin of the caudal fin.

Chromosome	Microscopic thread-like bodies within cells on which genes are linearly arranged.
Cirri	Slender filaments found on the head and lateral line; in British Columbia found on some marine sculpins that sporadically enter fresh water.
Cline	(*adjective:* clinal) A graded series of structures or features that follow a geographic gradient (e.g., egg size in many salmonid species gradually changes with latitude).
Cloaca	The shared external opening of the excretory and genital tracts.
Congeneric	Species in the same genus.
Conspecific	Individuals of the same species.
Coregonines	Fishes in the subfamily Coregoninae (whitefishes).
COSEWIC	An acronym for the Canadian Committee on the Status of Endangered Wildlife in Canada.
Crepuscular	Refers to organisms that are most active at dawn and dusk.
Ctenoid scales	Scales that are rough to the touch (e.g., scales on bass and perch)
Cycloid scales	Scales that are almost round and smooth to the touch (e.g., the scales on salmonids).
Demes	A term in population biology for genetically recognizable populations in which individuals are more likely to breed within their local population than with individuals from adjacent populations (see stocks).
Demersal	On the bottom; demersal eggs are denser than water and sink to the bottom.
Dentary	(*plural:* dentaries) A bony element(s) of the lower jaw, often with teeth.
Diel	A 24-hour period that includes both day and night.
Diploid	An organism or cell with two sets of chromosomes. In sexually reproducing organisms, diploids are formed by the fusion of the nuclei of two haploid cells (gametes) to produce a diploid zygote.
DNA	Deoxyribonucleic acid, the self-replicating material in chromosomes.
Designated units	(DUS) Infraspecific units that are distinguishable from, and have different extinction probabilities than, a species as a whole.
Ecomorphs	Morphological forms of a species that are associated with specific ecological conditions (e.g., body shape differences between river-dwelling and lake-dwelling populations of the same species).
Ecoregion	A major drainage system (e.g., the Columbia or Mackenzie river systems) or a series of physiographically and faunistically similar rivers (e.g., the North Pacific Coastal Ecoregion). In British Columbia, ecoregions are defined by the primary source of their fish fauna.

Ecotone	A transition zone between two different environments (e.g., the abrupt shift from flowing to standing water where an inlet stream enters a lake).
Ecological drainage unit	(EDU) A river system within an ecoregion or subregion that differs from adjacent drainages in their fauna and physiography.
Endemic	Found only in one geographic area (e.g., the Umatilla dace is endemic to the Columbia River system).
Epibenthic zone	In lakes, the epibenthic zone is the area immediately above the bottom.
Epilimnion	The upper layer in a thermally stratified lake. It is warmer and better oxygenated than the hypolimnion (bottom layer).
Evolutionarily significant unit	(ESU) A population or group of populations constitute an ESU if some unique part of a species evolutionary history would be lost if they became extinct (e.g., the Salish sucker).
Euryhaline	Refers to the tolerance of a wide range of osmotic environments (e.g., fishes that can move between salt and fresh waters).
Eutrophic	A lake, usually shallow, that is rich in nutrients.
Extirpation	The loss of a local population as opposed to the global loss of a species (extinction).
Extinction	The loss of a species.
Falcate	Sickle-shaped, in fish it usually refers to the shape of the fins.
Feral	A domesticated animal that has escaped to the wild (e.g., feral goldfish).
Fluvial	Pertaining to rivers and organisms that live in rivers.
Fork length	(FL) The distance from the tip of the snout to the tip of the middle caudal ray.
Fontanelle	An opening in the top of the skull that is only visible when the skin is removed.
Frenum	A fleshy bridge connecting the snout and upper jaw.
Gametogenesis	The process by which cells produce haploid gametes (eggs and sperm) by meiosis.
Genes	The units of heredity that, in conjunction with environmental factors, determine the characteristics of individuals; usually a specific nucleotide sequence on a chromosome.
Genetic distance	Measures of the depth of genetic differentiation between populations or species.
Genotype	The genetic constitution of an individual.

Gill rakers	The hard, often slender projections on the anterior edge of the gill arches.
Glide	A shallow, low-gradient stream or river habitat without surface turbulence.
Grilse	In the strictest sense an anadromous Atlantic salmon on its first return to freshwater. In British Columbia, anglers use the word as a general term for small, immature Pacific salmon caught at sea.
Gynogenesis	A mode of reproduction where sperm activates an egg, but the normal fusion of male and female nuclei (syngamy) never occurs. Consequently, only the female genome participates in further development.
Haploid	An organism or cell (e.g., eggs and sperm) having one set of chromosomes.
Heterocercal tail	A tail where the fleshy upper lobe of the caudal fin is much longer than the lower lobe (e.g., the tail seen in sturgeons).
Heterochrony	A change from the normal state in the relative time of appearance, or disappearance, of characters (see neoteny).
Holarctic	A biogeographic term used for the polar and temperate regions of North America and Eurasia.
Hydrograph	A graphical depiction of stream flow over a season or over a year.
Hypolimnion	The bottom layer in a thermally stratified lake. It is colder than the epilimnion. In eutrophic and mesotrophic lakes, the deepest part of the hypolimnion is often oxygen deficient by late summer.
Hypural plate	The flat projections off of the ventral side of the last few vertebrae on which the caudal rays articulate.
Infraspecific	Units (e.g., subspecies, metapopulations, and populations) below the level of species.
Insertion	Pertaining to dorsal and anal fins, the most posterior point of attachment to the body.
Interorbital width	The width of the space between the eyes.
Introgression	The movement of genes between species through hybridization.
Isthmus	The narrow bridge of tissue on the underside of the head that connects the thorax to the gill covers (see Fig. 38A).
Iteroparous	Fishes that spawn more than once in their lifetime.
Jacks	Sexually precocious males that attempt to take part in spawnings (usually by sneaking).
ka	An abbreviation for thousands of years ago, used mainly in Pleistocene geology.

Keratin	A hard fibrous protein that forms the "teeth" in lampreys and sheathes the lower jaw in the chiselmouth (*Acrocheilus*).
Kype	The upturned hook at the tip of the lower jaw on the breeding males of some salmonids. In British Columbia, the kype is especially well developed in large bull trout. According to Day (1887), the term is derived from the word kip, a Scots-English term for things that are turned-up (e.g., a kip-nosed girl has a turned-up nose).
Lacustrine	Refers to lakes.
Lateral line	A line of pores along the sides that lead to a canal containing sensory structures. In B.C. fishes, the length of the lateral line varies within, and among, species and is a useful taxonomic character.
Lateral plates	The bony plates along the sides of some fishes. In threespine sticklebacks, the lateral plates can completely cover the sides or be completely absent.
Lentic	Pertaining to standing water (e.g., swamps, ponds, lakes, and some reservoirs).
Limnetic zone	The offshore surface zone in lakes; usually defined by the limit of light penetration.
Littoral zone	The inshore zone in lakes; usually defined by the depth limits of macrophytes.
Locus (loci)	In its strictest sense, the location of a gene, or genes, on a chromosome. The term is often used as a synonym for a gene.
Lotic	Pertaining to running water (e.g., creeks, streams, and rivers).
Ma	An abbreviation for millions of years ago, usually used in geology.
Macrophytes	Large aquatic plants.
Major genes	Genes with large effects, often involved in the expression of major anatomical traits (e.g., the formation of the pelvic girdle in sticklebacks).
Maxillary length	The length of the maxillary bone in the upper jaw.
Mesotrophic	A lake with nutrient levels intermediate between eutrophic and oligotrophic lakes.
Modifiers	Genes that produce minor effects on the phenotypic expression of major genes.
Neoteny	(adjective: neotenic, neotenous) The retention of juvenile characters into adult life stages (see heterochrony).
Nubbles	Small, soft bumps on the heads of some sculpins.
Nuchal hump	In fishes, a hump on the nape (the area immediately behind the head). It usually occurs in breeding males (e.g., some species of Pacific salmon).

Nuptial colours	Colours associated with breeding (e.g., spawning colours).
Oligotrophic	Refers to a lake that is poor in nutrients. Oligotrophic lakes in B.C. typically are large, deep, and cold.
Ontogenetic	(*adverb:* ontogenetically) Refers to the development of an individual (e.g., the changes in resource use, behaviour, or morphology as a fish grows).
Operculum	(opercle) The thin bones on the sides of the head that cover the gill cavities. Typically, there are four bones in the operculum: the preopercle, opercle, subopercle, and interopercle.
Opercular flap	Refers to the elongate fleshy flap at the posterior end of the gill covers. The structure is especially conspicuous in sunfish (*Lepomis*).
Origin	Pertaining to fins, the anterior-most point of attachment to the body.
Osmeroids	Fishes in the suborder Osmeroidei (e.g., the smelts and eulachon).
Palatines	A pair of bones on the roof of the mouth, one on each side of the centrally located vomer. They often bear teeth.
Palatine teeth	The tooth patches on the palatine bones. The presence or absence of palatine teeth is used as a major character in sculpin identification.
Papilla	A small fleshy projection.
Parallel evolution	The independent evolution, from the same ancestral form, of similar phenotypes (morphological, behavioural, or ecological) in similar environments.
Parapatric	Refers to a situation (often at an ecotone) where the distributions of two species, or forms, abut. Parapatric forms are not strictly sympatric but the implication is that, without some form of reproductive isolation, there could be gene flow at the contact site.
Parasphenoid	The median bone that forms the base of the fish skull. It extends from the posterior end of the vomer back to the basioccipital bone. In most fishes, the parasphenoid does not bear teeth, but the goldeye has teeth on the parasphenoid bone.
Parr	A juvenile, freshwater stage of salmonids. It usually refers to individuals in their second, or later, growing season.
Parr marks	The vertical dark marks on the sides of many juvenile salmonids.
Peritoneum	The membrane lining the inside of the body cavity.
Phenotype	The physical expression of the interaction of an individual's genotype and the environment.
Phototaxis	(phototactic) Movement in response to light.
Phylogeny	(phylogenetic) Refers to the evolutionary history and relationships of a group of organisms.

Plicate	Has plicae (wrinkle-like folds of skin).
Polygynous	In males, this refers to having more than one female mate.
Polyploidy	Increases in numbers of chromosomes, usually by multiples of the normal diploid number. Polyploidy can occur spontaneously within a species (autoploidy) or through hybridization (alloploidy). Polyploidy has been implicated in the evolution of several groups of fishes (e.g., the sturgeons, salmonids, and suckers).
Pool	A habitat unit in a river or stream that is deeper than the habitats immediately above or below it.
Postdorsal	The area behind the insertion of the dorsal fin. The postdorsal distance is the distance from the insertion of the dorsal fin to the caudal flexure.
Postmaxillary pores	In sculpins, the pores on the ventral surface of the chin immediately behind the corner of the mouth (Fig. 44). Most B.C. species have two postmaxillary pores, but the Rocky Mountain sculpin and some slimy sculpin populations usually have only a single postmaxillary pore.
Postpelvic distance	The distance from the origin of the pelvic fins to the base of the caudal peduncle.
Predorsal	The area in front of the origin of the dorsal fin. The predorsal distance is the distance from the tip of the snout to the origin of the dorsal fin.
Premaxillae	The paired bones at the front of the upper jaw. In some fishes (e.g., salmonids), the premaxillae form the front part of the upper jaw, whereas in other fishes (e.g., walleye), they form the entire upper jaw.
Prepelvic distance	The distance from the tip of the snout to the origin of the pelvic fins.
Prickles	Modified scales that make the skin feel rough. In sculpins (*Cottus*), the presence, absence, and distribution of prickles on the body are useful in species identification.
Protractile	Capable of being protruded. In fishes without a frenum, the word is used to describe an upper jaw that can be extended.
Pterygoid	A pair of bones on the roof of the mouth situated behind the palatine bones. The pterygoids bear teeth in the smelts (Osmeridae).
Pyloric caecum	(caeca) The finger-like projections connected to the upper end of the gut immediately behind the stomach.
Redd	A collective term for the sequential series of nests dug by a single female salmonid. Maxwell (1904) indicates the word "redd" in lowland Scottish means to prepare (in the sense of make ready).
Reproductive isolating mechanisms	Behavioural, anatomical, or physiological mechanisms that prevent, or reduce, the probability of successful matings between populations or species.

Resource partitioning	The differential use of biotic or physical resources (e.g., space, food, diel period), often by closely related species or forms.
Riffle	A shallow, moderate-gradient stream or river habitat with a turbulent surface.
Riverine	Pertaining to rivers.
Run	A deep (relative to a riffle), moderate-gradient stream or river habitat with little surface turbulence.
Salmonids	Refers to members of the family Salmonidae (in the broad sense).
Salmonines	Refers to members of the subfamily Salmoninae (in British Columbia, Pacific salmon, trout, and chars).
Scute	A bony plate (e.g., the plates on sturgeons and sticklebacks).
Semelparous	Fishes that spawn only once in their lifetime.
Serrated	Notched like a saw.
Size-assortative mating	The tendency in many fishes to mate with fish of a specific size.
Snye	A blind (dead end) narrow meandering side channel.
Speciation	The process by which species are formed.
Standard length	(SL) The distance from the tip of the snout to the posterior end of the hypural plate.
Stays	Fleshy or membranous structures that connect the inner rays of the pelvic fins to the body. They are seen best by viewing from above and pulling the pelvic fin forward (see Fig. 22).
Stock	A term used by fisheries managers for a local interbreeding population. Stocks are similar to demes but they are not necessarily genetically differentiated from adjacent stocks. The separation between adjacent stocks may be spatial or temporal (e.g., in a lake, there may be inlet and outlet spawning stocks of trout).
Striae	Fine lines or marks, usually on the opercle (e.g., the marks on the gill covers of eulachon and American shad).
Sympatric	In the same place: the ability of species to coexist and retain their genetic integrity is the essence of the biological species concept.
Syngamy	The fusion of two haploid nuclei to produce a diploid cell.
Taxonomic species	Formally described species recognized by most taxonomists (in contrast see biological species pairs).
Terminal mouth	A mouth situated at the tip of the snout.
Thermocline	The layer of water between the epilimnion and hypolimnion. This layer is characterized by a rapid change in temperature with depth (at least 1.0 °C/m).

Total length (TL) Refers to the distance from the tip of the snout to the posterior end of the caudal fin.

Triploids Organisms with one extra set of chromosomes above the diploid number. The condition sometimes occurs in interspecific hybrids (e.g., some hybrids between northern redbelly dace and finescale dace). In British Columbia, artificially induced triploid rainbow trout have been used to create trophy fisheries. Triploids are usually sterile.

Tubercles Hard projections on scales, skin, and fin rays; they are usually temporary and associated with the reproductive season.

Vent A synonym for anus.

Vomer A median bone on the roof of the mouth.

Vomerine teeth Teeth on the vomer. The presence or absence of vomerine teeth is a major trait used in the identification of sculpin species.

INDEX TO SCIENTIFIC AND COMMON NAMES

Species marked with asterisks do not have species accounts but are mentioned in relation to the freshwater fishes of British Columbia. *Italic* page numbers are illustrations of the species.

Acipenser medirostris (green sturgeon) 25–29
Acipenser transmontanus (white sturgeon) 30–38
Acrocheilus alutaceus (chiselmouth) 56–60
Agosia chrysogaster (longfin dace)* 134
Alaskan lake whitefish (*Coregonus nelsoni*)* 361
Alosa sapidissima (American shad) 46–49
Ameiurus melas (black bullhead) 196–199
Ameiurus natalis (yellow bullhead) 200–203
Ameiurus nebulosus (brown bullhead) 204–207
American shad (*Alosa sapidissima*) 46–49
Amur grayling (*Thymallus grubei*)* 405
Archoplites interruptus (Sacramento perch)* 495
Arctic cisco (*Coregonus autumnalis*) 355–359
Arctic grayling (*Thymallus arcticus*) 406–410
Arctic char (*Salvelinus alpinus*)* 317, 333
Arctic smelt (*Osmerus dentex*)* 217, 218
Atlantic salmon (*Salmo salar*) 305–310
Atlantic tomcod (*Microgadus tomcod*)* 417
Atlantic whitefish (*Coregonus canadensis*)* 355
Bering cisco (*Coregonus laurettae*)* 355
black bullhead (*Ameiurus melas*) 196–199
black crappie (*Pomoxis nigromaculatus*) 519–522
blacknose dace (*Rhinichthys atratulus*)* 134
bluegill (*Lepomis macrochirus*) 504–508
brassy minnow (*Hybognathus hankinsoni*) 76–80
bridgelip sucker (*Catostomus columbianus*) 169–174
broad whitefish (*Coregonus nasus*) 367–372
brook stickleback (*Culaea inconstans*) 427–431
brook trout (*Salvelinus fontinalis*) 326–332
brown bullhead (*Ameiurus nebulosus*) 204–207
brown trout (*Salmo trutta*) 311–316
bull trout (*Salvelinus confluentus*) 317–325
burbot (*Lota lota*) 418–423
Bureya grayling (*Thymallus burejensis*)* 405
Carassius auratus (goldfish) 61–64
Catostomus catostomus (longnose sucker) 162–168
Catostomus columbianus (bridgelip sucker) 169–174
Catostomus commersonii (white sucker) 175–180
Catostomus macrocheilus (largescale sucker) 181–186
Catostomus platyrhynchus (mountain sucker) 187–193

Chinese sucker (*Myxocyprinus asiaticus*)* 159
Chinook salmon (*Oncorhynchus tshawytscha*) 296–304
chiselmouth (*Acrocheilus alutaceus*) 56–60
chum salmon (*Oncorhynchus keta*) 262–267
cisco (*Coregonus artedi*) 350–354
Clinocottus acuticeps (sharpnose sculpin)* 446, 447, 449
Clupea pallasii (Pacific herring)* 45, 46
coastal cutthroat trout (*Oncorhynchus clarkii clarkii*) 241–248
coastrange sculpin (*Cottus aleuticus*) 452–456
coho salmon (*Oncorhynchus kisutch*) 268–274
Columbia sculpin (*Cottus hubbsi*) 475–480
common carp (*Cyprinus carpio*) 72–75
Coregonus albula (Eurasian cisco)* 373
Coregonus artedi (cisco) 350–354
Coregonus autumnalis (Arctic cisco) 355–359
Coregonus canadensis (Atlantic whitefish)* 355
Coregonus clupeaformis (lake whitefish) 360–366
Coregonus laurettae (Bering cisco)* 355
Coregonus lavaretus (Eurasian lake whitefish)* 360, 362, 366, 367
Coregonus nasus (broad whitefish) 367–372
Coregonus nelsoni (Alaskan lake whitefish)* 361
Coregonus pidschian (Siberian lake whitefish)* 361
Coregonus sardinella (least cisco) 373–378
Cottus aleuticus (coastrange sculpin) 452–456
Cottus asper (prickly sculpin) 457–461
Cottus bairdii (mottled sculpin)* 452, 457, 461, 462, 468, 474, 477, 478, 489, 490, 492
Cottus bendirei (Malheur mottled sculpin)* 474, 475
Cottus cognatus (slimy sculpin) 462–468
Cottus confusus (shorthead sculpin) 469–474
Cottus gobio (bullhead)* 484
Cottus hubbsi (Columbia sculpin) 475–480
Cottus rhotheus (torrent sculpin) 481–485
Cottus ricei (spoonhead sculpin) 486–489
Cottus sibricus (sculpin) 484*
Cottus sp. (Rocky Mountrain sculpin) 490–495
Cottus spinulosus (sculpin) 484*
Couesius plumbeus (lake chub) 65–71
creek chub (*Semotilus atromaculatus*) 330*
Culaea inconstans (brook stickleback) 427–431
cutthroat trout (*Oncorhynchus clarkii*) 240
Cymatogaster aggregata (shiner perch) xx*
Cyprinus carpio (common carp) 72–75
Dolly Varden (*Salvelinus malma*) 333–340
emerald shiner (*Notropis atherinoides*) 92–95
Esox lucius (northern pike) 210–215
eulachon (*Thaleichthys pacificus*) 225–229

Eurasian cisco (*Coregonus albula*) 373*
Eurasian lake whitefish (*Coregonus lavaretus*) 360, 362, 366, 367*
Eurasian perch (*Perca fluviatilis*)* 522
European grayling (*Thymallus thymallus*)* 405
fathead minnow (*Pimephales promelas*) 111–115
finescale dace (*Phoxinus neogaeus*) 106–110
flathead chub (*Platygobio gracilis*) 116–119
Gasterosteus aculeatus (threespine stickleback) 432–439
goldeye (*Hiodon alosoides*) 40–44
goldfish (*Carassius auratus*) 61–64
green sturgeon (*Acipenser medirostris*) 25–29
Hiodon alosoides (goldeye) 40–44
Hiodon tergisus (mooneye)* 39, 40
Hybognathus hankinsoni (brassy minnow) 76–80
Hypomesus pretiosus (surf smelt)* 217, 218
inconnu (*Stenodus leucichthys nelma*) 398–403
kokanee (*Oncorhynchus nerka*) 287–295
lake chub (*Couesius plumbeus*) 65–71
lake trout (*Salvelinus namaycush*) 341–346
lake whitefish (*Coregonus clupeaformis*) 360–366
Lampetra ayresii (river lamprey) 3–7
Lampetra macrostoma (Vancouver lamprey) 8–11
Lampetra richardsoni (western brook lamprey) 12–15
Lampetra tridentata (Pacific lamprey) 16–21
largemouth bass (*Micropterus salmoides*) 514–518
largescale sucker (*Catostomus macrocheilus*) 181–186
Las Vegas dace (*Rhinichthys deaconi*)* 134
least cisco (*Coregonus sardinella*) 373–378
leopard dace (*Rhinichthys falcatus*) 133–138
Lepomis gibbosus (pumpkinseed) 499–503
Lepomis macrochirus (bluegill) 504–508
Leptocottus armatus (staghorn sculpin)* 446, 449
loach minnow (*Tiaroga cobitis*)* 134
longfin dace (*Agosia chrysogaster*)* 134
longfin smelt (*Spirinchus thaleichthys*) 219–224
longnose dace (*Rhinichthys cataractae*) 126–132
longnose sucker (*Catostomus catostomus*) 162–168
Lota lota (burbot) 418–423
Margariscus margarita nachtriebi (northern pearl dace) 81–85
Malheur mottled sculpin (*Cottus bendirei*)* 474, 475
Microgadus proximus (Pacific tomcod)* 417
Microgadus tomcod (Atlantic tomcod)* 417
Micropterus dolomieu (smallmouth bass) 509–513
Micropterus salmoides (largemouth bass) 514–518
mooneye (*Hiodon tergisus*)* 39, 40
mottled sculpin (*Cottus bairdii*)* 452, 457, 461, 462, 468, 474, 477, 478, 489, 490, 492

mountain sucker (*Catostomus platyrhynchus*) 187–193
mountain whitefish (*Prosopium williamsoni*) 392–397
Mylocheilus caurinus (peamouth) 86–91
Myxocyprinus asiaticus (Chinese sucker)* 159
ninespine stickleback (*Pungitius pungitius*) 440–445
northern pearl dace (*Margariscus margarita nachtriebi*) 81–85
northern pike (*Esox lucius*) 210–215
northern pikeminnow (*Ptychocheilus oregonensis*) 120–125
northern redbelly dace (*Phoxinus eos*) 101–105
Notropis atherinoides (emerald shiner) 92–95
Notropis hudsonius (spottail shiner) 96–100
Oligocottus maculosus (tidepool sculpin)* 446, 449
Oncorhynchus clarkii (cutthroat trout) 240
Oncorhynchus clarkii clarkii (coastal cutthroat trout) 241–248
Oncorhynchus clarkii lewisi (westslope cutthroat trout) 249–255
Oncorhynchus gorbuscha (pink salmon) 256–261
Oncorhynchus keta (chum salmon) 262–267
Oncorhynchus kisutch (coho salmon) 268–274
Oncorhynchus mykiss (rainbow trout, steelhead) 275–286
Oncorhynchus nerka (sockeye salmon, kokanee) 287–295
Oncorhynchus tshawytscha (Chinook salmon) 296–304
Osmerus dentex (Arctic smelt)* 217, 218
Osmerus mordax (rainbow smelt)* 217
Pacific herring (*Clupea pallasii*)* 45, 46
Pacific lamprey (*Lampetra tridentata*) 16–21
Pacific tomcod (*Microgadus proximus*)* 417
peamouth (*Mylocheilus caurinus*) 86–91
Perca flavescens (yellow perch) 524–528
Perca fluviatilis (Eurasian perch)* 522
Percopsis omiscomaycus (trout-perch) 412–415
Percopsis transmontanus (sand roller)* 411
Phoxinus eos (northern redbelly dace) 101–105
Phoxinus neogaeus (finescale dace) 106–110
Pimephales promelas (fathead minnow) 111–115
pink salmon (*Oncorhynchus gorbuscha*) 256–261
Platichthys stellatus (starry flounder)* 533
Platygobio gracilis (flathead chub) 116–119
Pomoxis annularis (white crappie)* 517
Pomoxis nigromaculatus (black crappie) 519–522
prickly sculpin (*Cottus asper*) 457–461
Prosopium coulterii (pygmy whitefish) 379–385
Prosopium cylindraceum (round whitefish) 386–391
Prosopium williamsoni (mountain whitefish) 392–397
Ptychocheilus oregonensis (northern pikeminnow) 120–125
pumpkinseed (*Lepomis gibbosus*) 499–503
Pungitius pungitius (ninespine stickleback) 440–445
pygmy whitefish (*Prosopium coulterii*) 379–385

rainbow trout (*Oncorhynchus mykiss*) 275–286
redside shiner (*Richardsonius balteatus*) 150–154
Rhinichthys atratulus (blacknose dace)* 134, 458, 461
Rhinichthys cataractae (longnose dace) 126–132
Rhinichthys deaconi (Las Vegas dace)* 134
Rhinichthys evermanni (Umpqua dace)* 134
Rhinichthys falcatus (leopard dace) 133–138
Rhinichthys obtusus (western blacknose dace)* 134
Rhinichthys osculus (speckled dace) 139–143
Rhinichthys umatilla (Umatilla dace) 144–149
Richardsonius balteatus (redside shiner) 150–154
river lamprey (*Lampetra ayresii*) 3–7
Rocky Mountrain sculpin (*Cottus* sp.) 490–495
round whitefish (*Prosopium cylindraceum*) 386–391
Salmo salar (Atlantic salmon) 305–310
Salmo trutta (brown trout) 311–316
Salvelinus alpinuis (Arctic char)* 317, 333
Salvelinus confluentus (bull trout) 317–325
Salvelinus fontinalis (brook trout) 326–332
Salvelinus malma (Dolly Varden) 333–340
Salvelinus namaycush (lake trout) 341–346
sand roller (*Percopsis transmontanus*)* 411
Sander vitreus (walleye) 529–533
Semotilus atromaculatus (creek chub)* 330
sharpnose sculpin (*Clinocottus acuticeps*)* 446, 447, 449
shiner perch (*Cymatogaster aggregata*)* xx
shorthead sculpin (*Cottus confusus*) 469–474
Siberian lake whitefish (*Coregonus pidschian*)* 361
slimy sculpin (*Cottus cognatus*) 462–468
smallmouth bass (*Micropterus dolomieu*) 509–513
sockeye salmon (*Oncorhynchus nerka*) 287–295
speckled dace (*Rhinichthys osculus*) 139–143
splake* 327, 342
Spirinchus thaleichthys (longfin smelt) 219–224
spoonhead sculpin (*Cottus ricei*) 486–489
spottail shiner (*Notropis hudsonius*) 96–100
staghorn sculpin (*Leptocottus armatus*)* 446, 449
starry flounder (*Platichthys stellatus*)* 533
steelhead (*Oncorhynchus mykiss*) 275–286
Stenodus leucichthys nelma (inconnu) 398–403
surf smelt (*Hypomesus pretiosus*)* 217, 218
tench (*Tinca tinca*) 155–157
Thaleichthys pacificus (eulachon) 225–229
threespine stickleback (*Gasterosteus aculeatus*) 432–439
Thymallus arcticus (Arctic grayling) 406–410
Thymallus burejensis (Bureya grayling)* 405
Thymallus grubei (Amur grayling)* 405

Thymallus thymallus (European grayling)* 405
tidepool sculpin (*Oligocottus maculosus*)* 446, 449
Tiaroga cobitis (loach minnow)* 134
Tinca tinca (tench) 155–157
torrent sculpin (*Cottus rhotheus*) 481–485
trout-perch (*Percopsis omiscomaycus*) 412–415
Umatilla dace (*Rhinichthys umatilla*) 144–149
Umpqua dace (*Rhinichthys evermanni*)* 134
Vancouver lamprey (*Lampetra macrostoma*) 8–11
walleye (*Sander vitreus*) 529–533
western blacknose dace (*Rhinichthys obtusus*)* 134
western brook lamprey (*Lampetra richardsoni*) 12–15
westslope cutthroat trout (*Oncorhynchus clarkii lewisi*) 249–255
white crappie (*Pomoxis annularis*)* 517
white sturgeon (*Acipenser transmontanus*) 30–38
white sucker (*Catostomus commersonii*) 175–180
yellow bullhead (*Ameiurus natalis*) 200–203
yellow perch (*Perca flavescens*) 524–528